D1715625

Aquatic Redox Chemistry

ACS SYMPOSIUM SERIES **1071**

Aquatic Redox Chemistry

Paul G. Tratnyek, Editor
Oregon Health & Science University

Timothy J. Grundl, Editor
University of Wisconsin–Milwaukee

Stefan B. Haderlein, Editor
Eberhard-Karls Universität Tübingen

Sponsored by the
ACS Division of Environmental Chemistry
ACS Division of Geochemistry

American Chemical Society, Washington, DC

Distributed in print by Oxford University Press, Inc.

Library of Congress Cataloging-in-Publication Data

Aquatic redox chemistry / Paul G. Tratnyek, Timothy J. Grundl, Stefan B. Haderlein, editor[s] ; sponsored by the ACS Division of Environmental Chemistry and ACS Division of Geochemistry.

p. cm. -- (ACS symposium series ; 1071)

Includes bibliographical references and index.

ISBN 978-0-8412-2652-4

1. Groundwater recharge--Congresses. 2. Oxidation-reduction reaction--Congresses. 3. Groundwater--Carbon content--Congresses. I. Tratnyek, Paul G. II. Grundl, Timothy J., 1953- III. Haderlein, Stefan B. IV. American Chemical Society. Division of Environmental Chemistry. V. American Chemical Society. Division of Geochemistry.

GB1197.77.A67 2011

551.49--dc23

2011031438

The paper used in this publication meets the minimum requirements of American National Standard for Information Sciences—Permanence of Paper for Printed Library Materials, ANSI Z39.48n1984.

Distributed in print by Oxford University Press, Inc.

PRINTED IN THE UNITED STATES OF AMERICA

Foreword

The ACS Symposium Series was first published in 1974 to provide a mechanism for publishing symposia quickly in book form. The purpose of the series is to publish timely, comprehensive books developed from the ACS sponsored symposia based on current scientific research. Occasionally, books are developed from symposia sponsored by other organizations when the topic is of keen interest to the chemistry audience.

Before agreeing to publish a book, the proposed table of contents is reviewed for appropriate and comprehensive coverage and for interest to the audience. Some papers may be excluded to better focus the book; others may be added to provide comprehensiveness. When appropriate, overview or introductory chapters are added. Drafts of chapters are peer-reviewed prior to final acceptance or rejection, and manuscripts are prepared in camera-ready format.

As a rule, only original research papers and original review papers are included in the volumes. Verbatim reproductions of previous published papers are not accepted.

ACS Books Department

Contents

Indexes

Preface

Life and cycling of inorganic and organic matter on earth is driven to a large extent by electron transfer (i.e. redox) reactions. This makes understanding aquatic redox processes essential to all aspects of biogeochemistry, from remediation of legacy contamination problems, to sustaining environmental health, to managing ecosystem services. Aquatic redox processes exert their influence by driving metabolic processes, mobilization and sequestration of metals, and transformation of organic and inorganic contaminants. Thus, aquatic redox processes control the chemical speciation, bioavailability, toxicity, and mobility of both natural and anthropogenic compounds.

Despite the breadth and centrality of aquatic redox chemistry in the environmental sciences, there have been few attempts to provide a comprehensive perspective on this topic. A unique opportunity to bring together a wide range of the community of aquatic redox chemists arose from a symposium at the 239th ACS National Meeting (21-25 March 2010 in San Francisco, CA) in honor of the contributions of Donald L. Macalady. Throughout his career, Prof. Macalady made influential contributions to many aspects of aquatic redox chemistry, including inorganic, organic, and biogeochemical electron transfer processes in natural waters and sediments.

The symposium—which was co-sponsored by the ACS Divisions of Environmental Chemistry and Geochemistry—attracted a large number of high-quality contributions from a diverse group of leading scientists (representing environmental and aquatic chemistry, surface chemistry, electrochemistry, photochemistry, theoretical chemistry, soil chemistry, geochemistry, geology, microbiology, hydrology, limnology, and oceanography) and engineers (environmental, civil, and chemical). At the symposium, the synergy between these researchers was palpable, which led to the idea that a volume based on the symposium would be timely and constructive.

The scope of this volume was planned to provide a comprehensive overview of the state of the art in aquatic redox chemistry. Major areas of interest include interactions between iron, natural organic matter (NOM), and contaminants including metals, metalloids, and organic pollutants. The contributed chapters were selected and edited to highlight recent developments in the field, but also to introduce fundamental aspects and approaches of aquatic redox chemistry in a systematic and didactic way. To this end, this volume should be effective as teaching material for upper level students in environmental science or engineering, as well as being a valuable resource for scientists and practitioners.

In the future, the frontiers in aquatic redox chemistry will be transformed by increasingly interdisciplinary research efforts and emerging analytical methods. These developments should greatly improve our ability to characterize the interrelated spatial and temporal dynamics that complicate the speciation of reactants and mechanisms of electron transfer in natural and engineered aqueous systems. Current knowledge gaps are particularly large with regard to the mechanistic understanding of heterogeneous electron transfer processes at interfaces, especially those between water and minerals or bacteria. Thus, for the forseeable future, aquatic redox chemistry will continue to be a dynamic and challenging field of research.

The editors gratefully acknowledge all those who contributed to the planning and implementation of this volume and the symposium on which it was based. We are particularly indebted to the authors and reviewers of each chapter, all of whom fulfilled their roles with very high standards. We also thank the attendees of the symposium for their numerous comments and thought provoking suggestions, which helped to shape the final outcome. Finally, we thank the staff of the ACS Division of Environmental Chemistry, Division of Geochemistry, and Books Department who contributed the symposium and book.

Stefan B. Haderlein

Center for Applied Geosciences
Eberhard-Karls Universität Tübingen
D-72076, Tübingen
+49 7071 2973148 (telephone)
+49 7071 5059 (fax)
haderlein@uni-tuebingen.de (e-mail)

Timothy J. Grundl

Geosciences Department
University of Wisconsin–Milwaukee
Milwaukee, WI 53201
(414) 229-4765 (telephone)
(414) 229-5452 (fax)
grundl@uwm.edu (e-mail)

Paul G. Tratnyek

Division of Environmental and Biomolecular Systems
Oregon Health & Science University
20000 NW Walker Road
Beaverton, OR 97006
(503) 748-1023 (telephone)
(503) 748-1464 (fax)
tratnyek@ebs.ogi.edu (e-mail)

Chapter 1

Introduction to Aquatic Redox Chemistry

Timothy J. Grundl,[1,*] Stefan Haderlein,[2] James T. Nurmi,[3] and Paul G. Tratnyek[3]

[1]Geosciences Department and School of Freshwater Sciences, University of Wisconsin-Milwaukee, Milwaukee, WI 53201
[2]Center for Applied Geosciences, Eberhard-Karls Universität Tübingen, D-72076, Tübingen
[3]Division of Environmental and Biomolecular Systems, Oregon Health & Science University, Beaverton, OR 97006
*grundl@uwm.edu

Oxidation-reduction (redox) reactions are among the most important and interesting chemical reactions that occur in aquatic environmental systems, including soils, sediments, aquifers, rivers, lakes, and water treatment systems. Redox reactions are central to major element cycling, to many sorption processes, to trace element mobility and toxicity, to most remediation schemes, and to life itself. Over the past 20 years, a great deal of research has been done in pursuit of process-level understanding aquatic redox chemistry, but the field is only beginning to converge around a unified body of knowledge. This chapter provides a very broad overview of the state of this convergence, including clarification of key terminology, some relatively novel examples of core thermodynamic concepts (involving redox ladders and Eh-pH diagrams), and some historical perspective on the persistent challenges of how to characterize redox intensity and capacity of real, complex, environmental materials. Finally, the chapter attempts to encourage further convergence among the many facets of aquatic redox chemistry by briefly reviewing major themes in this volume and several past volumes that overlap partially with this scope.

Definitions and Scope

Historically, the terms oxidation and reduction arose from experimental observations: oxidation reactions consumed O_2 by incorporating O into products and reduction reactions reduced the mass or volume of products by expelling O (*1*). Chlorine substitution is equivalent to oxygen in this context, so chlorination is oxidation and dechlorination is reduction. A similarly empirical definition of reduction is that it usually involves incorporation of hydrogen, and, therefore, oxidation can be regarded as dehydrogenation (e.g., dehydrogenase enzymes catalyze oxidation).

More rigorously, oxidation-reduction (redox) reactions are commonly understood to occur by the exchange of electrons between reacting chemical species. Electrons are lost (or donated) in oxidation, and gained (or accepted) in reduction. Oxidation of a species is caused by an oxidizing agent (or oxidant), which accepts electrons (and is thereby reduced). Similarly, reduction results from reaction with a reducing agent (or reductant), which donates electrons (and is oxidized).

These definitions are adequate for most purposes, but not all. Just as acid-base concepts have proton-specific definitions (the Brönsted model) and more general definitions (e.g., the Lewis model), redox concepts can be extended from electron transfer specific definitions to more general definitions that are based on electron density of chemical species ((*2*), *and references cited therein*). The latter allows for redox reactions that occur by atom-transfer as well as electron transfer mechanisms. While often ignored, the role of atom-transfer mechanisms can be important, particularly in redox reactions involving organic compounds.

Redox reactions, defined inclusiv, are central to many priority and emerging areas of research in the aquatic sciences. This scope includes all aspects of the aquatic sciences: not just those involving the hydrosphere, but also aquatic (i.e., aqueous) aspects of environmental processes in the atmosphere, lithosphere, biosphere, etc. (*3*). As a field of study, *aquatic redox chemistry* also has multidisciplinary roots (spanning mineralogy to microbiology) and interdisciplinary applications (e.g., in removal of contaminants from water, sediment, or soil). Despite its cross-cutting appeal, however, very little prior work has used aquatic redox chemistry as a niche-defining theme. The main exception to this appears to be several publications by Donald Macalady (e.g, (*4*)), which is convenient and appropriate—and not entirely coincidental—given the origins of this volume (see Preface).

Core Concepts

Any redox reaction can be formulated as the sum of half-reactions for oxidation of the reductant and reduction of the oxidant. The overall free energy of a redox reaction is determined by the contributing half-reactions, and the free energy of each half-reaction depends on the reactants, products, and solution conditions. At a common set of standard conditions, the free energies—or corresponding redox potentials—can be used to compare the relative strength of oxidants and reductants and thereby determine the thermodynamic favorability of the overall reaction between any particular combination of half-reactions. This type of analysis is well suited for a variety of graphical representations, the two most common of which are redox ladders and Eh-pH (or Pourbaix) diagrams. The fundamentals of constructing these diagrams are presented in numerous texts on aquatic chemistry (*3*, *5*, *6*), geochemistry (*7*, *8*), and other fields (*9*). Some new data that could be used in constructing such diagrams are given in Chapters 2, 3 and 4 of this volume.

Figure 1 is a redox ladder that summarizes a diverse range of redox couples that are significant in aquatic redox chemistry. The top of the figure is bounded by several strong oxidants (e.g., hypochlorite, monochloramine, and ozone) that are capable of oxidizing essentially any compound found in aquatic environments. Similarly, the bottom of the figure is bounded by strong reductants (zerovalent metals) that are capable of reducing essentially any compound found in aquatic environments. These oxidants and reductants fall outside the stability field of water, so they are not persistent natural species, but they often form the basis of engineered water treatment systems. Hypochlorite and other strong chlorine-based oxidants are discussed in Chapters 2 and 11 of this volume, and zerovalent metals such as iron and zinc are discussed in Chapter 18

The first column of Fig. 1 is devoted to the redox couples that form the major terminal electron accepting processes (TEAPs) of microbial metabolism. The overall redox conditions of most aquatic system are ultimately determined by these TEAPs. The TEAP that provides the most energy recovery (those at the top of the redox ladder) favors the types of microorganisms that utilize that process. As the most favorable electron acceptor is depleted, the next TEAP on the redox ladder becomes most favorable. This process can result in sequential progression (in space or time) from TEAPs higher on the redox ladder to those below. This basic understanding of TEAPs and their effect on aquatic redox chemistry is well established, but a detailed understanding of the fundamental controls on these processes is still emerging, as discussed in Chapter 4 of this volume.

Once environmental conditions are established, however, many important redox reactions proceed without further mediation by organisms. These reactions are considered to be abiotic when it is no longer practical (or possible) to link them to any particular biological activity (*4*, *10*). Thus, many of the half-reactions represented in the 2nd-6th columns in Fig. 1 can be more or less a/biotic—depending on conditions—and the overall favorability of these processes is not necessarily affected by microbiological mediation (i.e., the redox ladder applies either way). However, systems where biotic and abiotic controls on contaminant fate are closely coupled currently are frontier areas of research (e.g., Chapters 19-24).

The 2nd-6th columns in Fig. 1 arranged so they represent families of major redox active species in order from relatively oxidized (and oxidizing) to relatively reduced (and reducing). Thus, the second column includes the reactive oxygen species that arise mostly by photochemical processes in natural waters. The chemistry of some of these processes is described in Chapters 8 and 9. Other oxidants that arise mainly in water treatment processes are not shown because they plot above the scale used in the figure, but two are discussed in later:

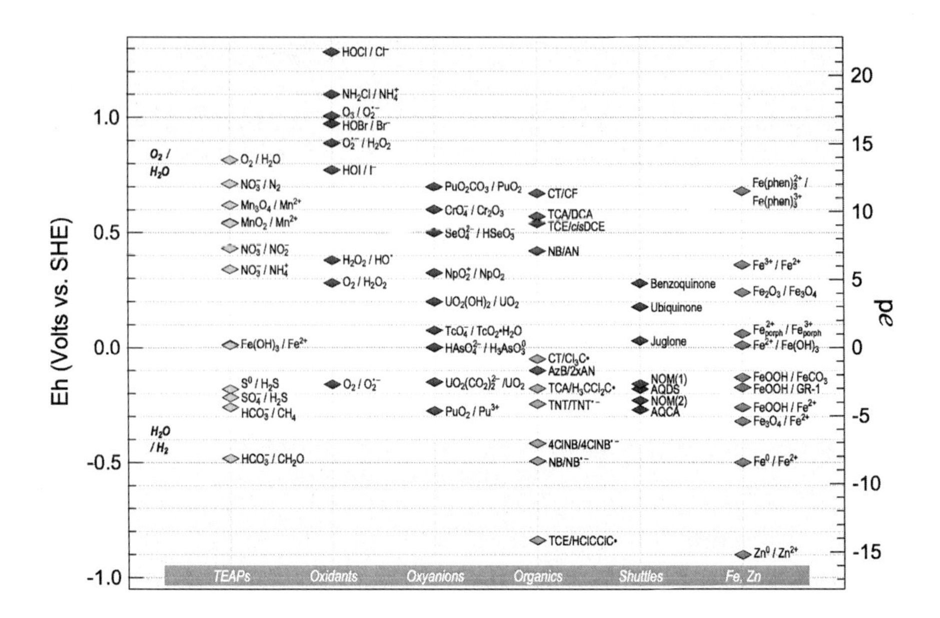

Figure 1. Redox ladder summarizing representative redox couples for six categories of species that are important in aquatic chemistry. The potentials shown are for environmental aquatic conditions (pH 7, most other solutes at 1 mM). TEAPs are terminal electron accepting processes that define regimes of microbial metabolism. For the organic contaminant category, the upper group of potentials (red symbols) shows multi-electron couples to stable species and the lower set (green symbols) are reduction potentials for the first electron transfer. The chlorinated aliphatic organic contaminants are given by their usual abbreviations; nitro aromatics include nitrobenzene (NB), 4-chloro-nitrobenzene (4ClNB), and 2,4,6-trinitrotoluene (TNT). AzB is azobenzene and AN is aniline. The electron shuttle category includes model quinones (only oxidized forms listed), anthraquinone disulfonate (AQDS) and anthraquinone carboxylic acid (AQCA). The two values for natural organic matter (NOM) are described in Chapter 7. In the last column, GR-1 refers to carbonate substituted green rust. Data for the this figure were obtained from Chapters 2, 3, and 22 in this volume and a variety of other sources, especially (7, 9, 11, 12).

hydroxyl radical (from photoactivation with TiO_2, Chapter 10) and sulfate radical (from persulfate, Chapter 12).

The 3rd and 4th columns of Fig. 1 are major classes of redox-active contaminants: the metal oxyanions, chlorinated aliphatic hydrocarbons (CACs), and nitro aromatic compounds (NACs). For the metal oxyanions, the oxidized form is the most mobile and toxic, and reduction results in sequestration of insoluble products and lower risk (see Chapters 21, 22); for CACs and NACs, reduction of these compounds may produce more or less toxic end products depending on the latter steps of reaction (see Chapters 19, 20, 23, 24). In the column for organic contaminants, another important distinction is illustrated between the overall reduction potentials (upper, red symbols) and first electron potentials (lower, green symbols). The overall reduction potentials are always highly favorable, but the first electron transfer is much less favorable and this step is usually rate determining.

The last two columns in Fig. 1 show organic electron transfer mediators (shuttles), especially those that might be associated with natural organic matter (NOM), and some of the many redox couples involving species of iron. These two groups can be regarded as bulk electron donors, or as mediators of electron transfer from other bulk donors. The role of electron shuttles is discussed further below, and in several other chapters in this volume, especially Chapter 6.

A major limitation of redox ladders such as the one shown in Fig. 1 is that all the potentials are shown for a common set of conditions, usually "standard" environmental conditions. The most important of these conditions is pH, which strongly effects the redox reactions of some species. These effects are represented by Eh-pH (or p*e*-pH, or Pourbaix) diagrams, such as the example shown in Fig. 2. On top of the familiar stability fields for iron, Fig. 2 shows the—perhaps not so familiar—stability fields for an organic compound: e.g., model quinone, juglone. This combination shows that the oxidized forms of juglone (QH and Q^-) will be reduced by any Fe^{II} species above pH 5, with the most common product being the partially deprotonated hydroquinone (QH_2^-) over the range of most relevant pH's; whereas juglone is stable in the presence of Fe^{II} at pH below 5.

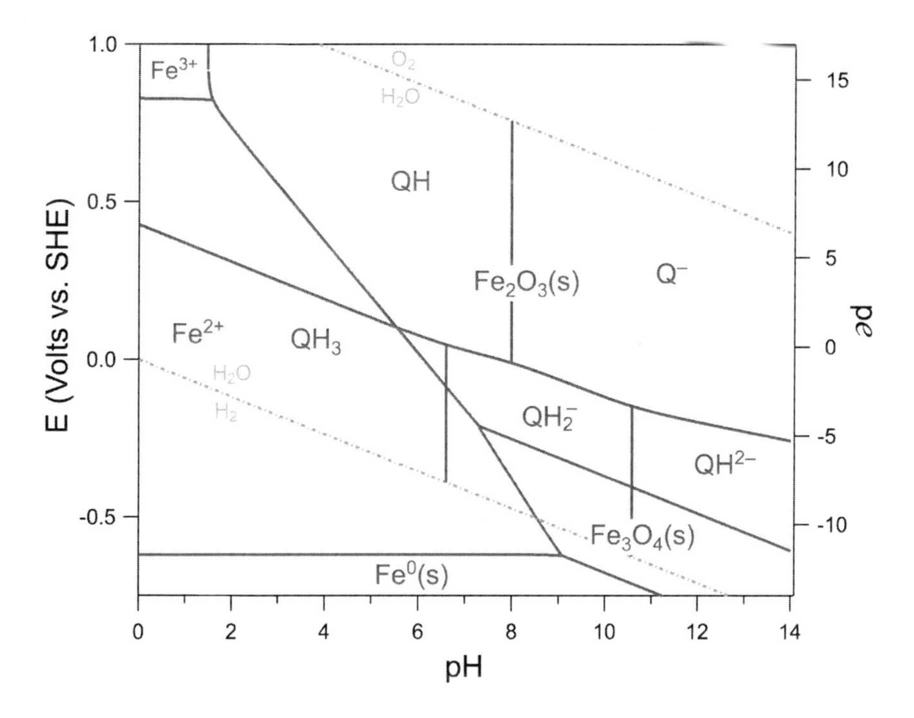

Figure 2. Comparison of the Eh-pH stability diagrams for juglone (a model quinone and organic electron shuttle, QH) vs. iron (as a bulk electron donor). Thermodynamic data for juglone obtained from (1). Total aqueous iron concentration assumed to be 10^{-6} M.

Applied Concepts

The core concepts highlighted in the preceding section are powerful heuristics for understanding, and teaching, aquatic redox chemistry. But, as is often the case with such heuristics, they are simplifications that can obscure complicating factors that sometimes dominate real-world behavior. In particular, the concept of Eh (or p*e*) as a "master" variable, which—along with pH—defines the "reaction space" of aquatic systems, has proven to be so attractive that the many limitations to this concept tend to be neglected. This issue must be addressed when attempting to relate the relatively-unambiguous thermodynamic analysis of well-defined half-reactions (illustrated in the section above) to experimentally observable indicators of redox conditions, such as electrode potential measurements (this section).

A fundamental reason for complexity in assessing the redox conditions of aquatic systems is that most aqueous redox reactions in such systems are kinetically limited and therefore not in equilibrium with each other. Under these circumstances, potential measurements made with an inert electrode (e.g., Pt) are mixed potentials in which each redox couple in contact with the electrode exchanges electrons independently and the electrode response is the sum of anodic (reductive) and cathodic (oxidative) currents, each weighted by the corresponding exchange current density (a measure of the sensitivity of the electrode to particular species). This mixed potential does not necessarily reflect equilibrium among the couples or between the electrode and any particular couple, so the relationship between mixed potentials measured on complex environmental samples and specific redox active species in the sample is not well defined. The theoretical and practical difficulties with interpretation of direct potentiometry as a means to define the redox intensity of aquatic systems was a major issue in the early literature on aquatic redox chemistry (*13–18*).

Despite these complications, there is a general correspondence between Eh (both the theoretical thermodynamic quantity and the measured mixed potential) and qualitative characterizations of redox conditions, such as the concentration of key indicator species like oxygen, sulfide, or carbon (*19*), and hydrogen (*20*). Taken together, these criteria can be used to locate various types of environmental waters on an Eh-pH diagram, as shown in Fig. 3. The labels for types of environmental waters are positioned in Fig. 3 based on entirely qualitative considerations (and "ideal" behavior), but the light gray points show the distribution of measured data on water samples ranging from highly aerobic to transitional systems to highly anaerobic conditions.

Like redox intensity (potential) measurement, characterization of redox capacity (e.g., "poising", the redox capacity property analogous to buffer capacity with respect to pH (*25*, *26*)) is also ambiguous for natural systems. In principle, oxidative capacity should be the stoichiometric sum of all oxidants present minus the sum or reductants present (*27*, *28*), but the lack of equilibrium between the key redox active species means that redox capacity is, at best, a conditional property strongly dependent on the operational conditions used in its determination. For example, it has been proposed that redox capacity be determined by titrating (reducing) samples with dissolved oxygen, and the changes in Eh measured with a Pt electrode be used to quantify its capacity with respect to S(II), Fe(II), etc.

(29). Of course, using stronger oxidants (than O_2, e.g., permanganate) or longer (or shorter) contact times will result in different amounts of oxidant consumption and therefore different indications of oxidant capacity.

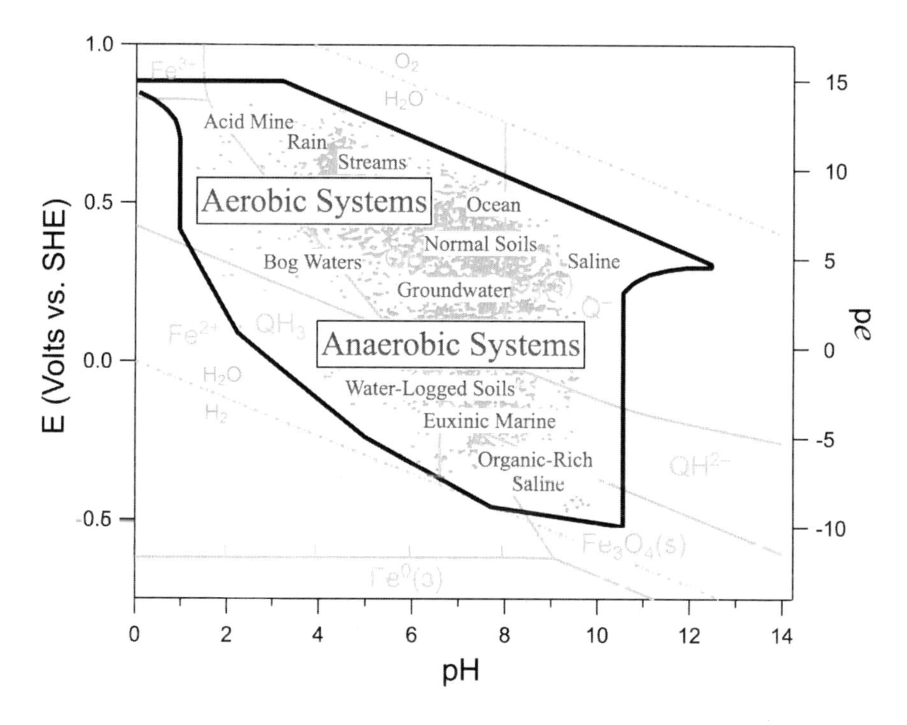

Figure 3. Eh-pH diagram showing approximate regions of typical environmental systems superimposed on the stability fields for iron and juglone from Fig. 2. Labels for representative types of environmental waters are adapted from the figures in various sources (7, 21–23). Light gray data points in the background are measured Eh data, adapted from the classic figure by Baas-Becking (24).

Diverse Perspectives

The concepts of aquatic redox chemistry (such as those summarized above) play such a central role in many aspects of environmental science that they are prominent in several previous monographs on other aspects of aquatic science. An early example of this is an ACS Advances in Chemistry Series volume on interfacial and interspecies aspects of aquatic chemistry (*30*), which contains chapters dealing with the development of redox zones, specific redox reactions within large water bodies, photochemically driven reductive dissolution of iron oxides, and reduction of halocarbon compounds is discussed especially when mediated by NOM. Another, similarly-relevant monograph (*31*) covers numerous topics that are pertinent to aquatic redox chemistry including reactivity of NOM and microbially mediated contaminant degradation. NOM mediation is an important mechanism and is a prominent theme in the current volume.

The theme of heterogeneous electron transfer is explored in some detail in an ACS Symposium Series volume focused on interfacial reactions (*32*). The scope of this work includes mineral oxidation, mineral reduction, and the effects of surface layers on the reacting minerals; the ability of ferric oxides to not only exchange structural electrons, but to also mediate electron exchange from surface bound Fe^{II}; and NOM mediation of redox reactions by algal exudates that have been photochemically activated.

The usefulness of electrochemical techniques, including amperometric, potentiometric and voltammetric techniques, for the direct determination of redox speciation and elemental cycling in general is highlighted in (*33*). This topic is extended in the current volume (Chapters 7 and 13 in this volume). These authors show that when applied carefully, voltammetric methods are particularly useful in deciphering redox mechanisms.

A recent special issue of the journal Environmental Science and Technology (Vol. 44, No. 1, 2010) on biogeochemical redox processes is a particularly germane compendium of environmental redox processes as they apply to the fate of contaminants, primarily involving inorganics. Both abiotic and biotically driven electron transfer between solution and a variety of mineral phases and the resultant mobilization of contaminants is explored in great detail ((*34*), *and references cited therein*). The role of real world complexities and the relative importance of physical limitations (diffusion limited mass transport, sediment heterogeneity, seasonal variability) versus chemical limitations (reaction kinetics, dynamics of microbial consortia) are long recognized questions that are beginning to be addressed by contributions to this compendium. New advances in both electrode based (*35–37*) and spectroscopic techniques for the measurement of redox rates and processes are also described.

Signs of Convergence

Although the literature on the diverse range of aspects of aquatic redox chemistry has grown greatly in quantity and sophistication over the last 20 years, there is no single volume focused on aquatic redox chemistry. The centrality of redox to much of environmental chemistry means that contributions to this field have come from a wide range of disciplines, but there has been little convergence between the contributions of these disciplines. The goal of this volume is to provide a compilation of papers that, together, define the scope and fundamental concepts of the field of aquatic redox chemistry together with a selection of the most significant new research developments.

One common theme that is shared by many chapters in this volume is that redox reactions can be facilitated by sequences of coupled electron transfer reactions—herein referred to electron transfer systems (ETS)—where the intermediate species serve as mediators or shuttles for the process. Broadly defined, the general ETS model includes electron exchange between solids and aqueous phases (*38*) via external electron mediators (*39*) via atom exchange in solid phases (*40*), or outer membrane enzymatic "nanowires" (*41*, *42*).

The ETS model is illustrated in Fig. 4. Starting in the upper left of the diagram, electrons are transferred from donor species to acceptor species through a mediator species, with the overall path represented by the curved red arrow. The narrow black arrows indicate the step-wise electron exchanges as they are commonly shown in electron shuttle/mediator schemes (e.g., Fig. 1 in Chapter 6). As reactions proceed, the net flow of electrons is from upper left to lower right (i.e., "downhill") along the curved red arrow. (Note that the Y-axis of Fig. 4 presents negative potentials at the top and positive potentials at the bottom, which is opposite the usual arrangement in redox ladders and Eh-pH diagrams). Individual electron transfers occur at interfaces (represented by the diagonal dashed lines), which can be physical boundaries (such as between two distinct layers of sediment, a mineral-water interface, or the outer wall of a cell membrane), or can be conceptual boundaries (where the donor, mediator and acceptor species are in the same phase).

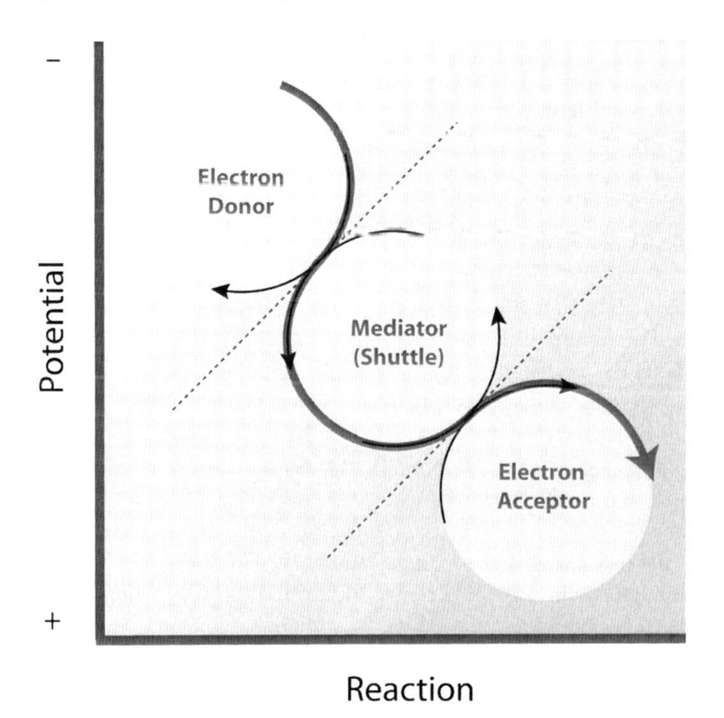

Figure 4. General representation of a 3-cell electron transfer system (ETS). The reaction coordinate can also represent time or space. Contacts between the cells (dashed lines) can be pairs of reactants in the same medium or physical interfaces. Physical separation of the cells may be necessary to avoid short-circuiting, especially for ETS with more than 3 cells.

Mediation is not constrained to the direct transfer from donor through mediator to acceptor but is free to make use of a variety of physiochemical processes including the intervalence band electron delocalization or semiconducting properties of mineral lattices, coupling among redox active moieties in NOM, or the rapid photolytic regeneration of reactive oxidants. Interfaces are viewed broadly and vary widely from one ETS to another. Interfaces range from the surface of a molecule in the case of a dissolved mediator to an entire mineral grain in the case of semiconducting minerals. Depending on the ETS, interfaces remain active by continuous replenishment of donor species; by successful removal of reaction products (acceptor species); avoidance of surface passivation; activation/inhibition by adsorbed ligands or maintenance of an appropriate microbial consortia. Interfaces of larger scale are also applicable in this context: e.g. aquitard/aquifer boundaries where aquitards act as a source of electrons to adjacent aquifers or to sediments with sequential TEAP zones allow the ETS paradigm to be extended to field scale processes.

Volume Organization

The volume is organized into four sections. Chapters 2 through 7 deal with the core aspects of aquatic redox chemistry including recent advances in theoretical understanding of these inherently disequilibrium processes and an in-depth treatment of the redox behavior of NOM. The second section considers the formation of a variety of reactive oxygen species (Chapters 8 through 10) and finishes with a look at reactions driven by two specific oxidants, chlorine and sulfate radicals (Chapters 11 and 12). An equivalent treatment of environmental reductants (Chapters 13 through 18) followed by an elucidation of specific reduction reactions (Chapters 19 through 23) constitutes the third section. The final section (Chapters 24 through 26) contains in-the-field studies that describe the primary redox processes at play in natural systems.

Several themes are common to multiple chapters in this volume, the most apparent of which is the complicated, subtle manner in which electrons move within the environment. This supports the need to develop a fundamental, process-level understanding of the reactions in question. Highly specific and complicated redox chemistry is found within any system of related redox reactions and the redox activity within the iron system is one clear example.

The reductive activity of Fe^{II} in solution depends strongly on Fe-ligand complexing. The ligand involved either stabilizes the original Fe^{II} reactant or the resulting Fe^{III} product with the resultant inhibition or enhancement of reaction rate (Chapter 14 this volume). Similarly, Fe^{II}-ligand enhancement of reaction rates allows the Fenton reaction to operate at environmentally relevant pHs in natural systems that contain organic ligands (Chapter 9 this volume). NOM coatings can either inhibit or enhance the reactivity of nano-sized ZVI (Chapter 18 this volume). In this case the inhibition is due to the formation of a semi-permeable passivating layer and enhancement is due to coatings containing reactive moieties that act as electron shuttles. In iron-rich clays the reversibility and extent of reactivity of is a function of many variables including the location of the iron center within the mineral lattice, the ability of the lattice to accommodate the charge imbalance induced by either the oxidation or reduction of iron, electron delocalization through Fe^{III}-O-Fe^{II} intervalence electron transfer and the nature of the reductant (Chapter 17 this volume). Oxidation of dissolved Fe^{II} at the surface of solid ferric oxides is extremely complex because of the semi-conducting nature of these mineral phases. Electrons released by Fe^{II} oxidation either become delocalized within the oxide lattice or trapped in a lattice defect (Chapter 15 this volume). Delocalized electrons are free to move within the lattice and become involved in ancillary reduction reactions with other oxidants in solution or to leave the lattice entirely as desorbed Fe^{II} (Chapter 15 this volume).

It is important to recognize that the need for a process-level understanding is not limited to the specifics of a given electron transfer reaction but also extends to external constraints affecting the progress of the reaction. External constraints include field-scale effects such as mass transfer limitations and heterogeneity in both the flow regime and the reactivity that cause competition between the residence time and the characteristic reaction time of solutes within a moving parcel of water (Chapters 21 and 25 this volume) or between aquifer and aquitard (Chapter 26 this volume). Whether the reaction is driven biotically or abiotically will affect the rates of reaction directly (Chapter 23 this volume) or in the case of smectite reduction, completely change the mechanism of electron transfer (Chapter 17 this volume). Obviously the microbial consortia that drive biotic processes are subject to electron donor/acceptor and nutrient availability as well as to the buildup of metabolic waste products. Chapter 4 (this volume) presents the concept of measured energy thresholds that are related to the energy ideally conserved by microbial ATP synthesis to determine if the microbes are limited by thermodynamic constraints or by mass transfer constraints.

A second theme that becomes apparent is that much of this process-level understanding of electron flow is due to advances in analytical techniques. Some are clever variants of established techniques and some are truly new. Variants of older techniques include the use of voltammetry at mercury electrodes for the detection of FeS nanoparticles (Chapter 13 this volume) or complex waveforms with microelectrdes for study of NOM (Chapter 7 this volume); and the use of probe compounds (Chapters 17, 19 and 24 this volume). New analytical techniques that are proving useful include the use of redox driven isotopic fractionation via compound specific isotope analysis (CSIA) (Chapters 16 and 17 this volume). The isotope specificity of Mossbauer spectroscopy has been a powerful tool in the understanding of redox behavior of ferric oxides (Chapter 15 this volume)

Closing

This volume summarizes the maturing understanding of redox processes in the environment on the part of researchers in the field. Long standing gaps in our knowledge are falling in the face of advances in analytical techniques and the attendant process-level understanding of electron flow. We hope that as the reader progresses through this volume these new conceptualizations of electron flow and redox processes in general will stimulate new avenues of research in this fascinating and important field.

References

1. Clark, W. M. *Oxidation-Reduction Potentials of Organic Systems*; Williams & Wilkins: Baltimore, 1960.
2. Tratnyek, P. G.; Macalady, D. L. Oxidation-reduction reactions in the aquatic environment. In *Handbook of Property Estimation Methods for Chemicals: Environmental and Health Sciences*; Mackay, D., Boethling, R. S., Eds.; Lewis: Boca Raton, FL, 2000; pp 383−415.
3. Stumm, W.; Morgan, J. J. *Aquatic Chemistry: Chemical Equilibria and Rates in Natural Waters*, 3rd ed.; Wiley: New York, 1996.
4. Wolfe, N. L.; Macalady, D. L. New perspectives in aquatic redox chemistry: Abiotic transformations of pollutants in groundwater and sediments. *J. Contam. Hydrol.* **1992**, *9*, 17–34.
5. Morel, F. M. M.; Hering, J. G. *Principles and Applications of Aquatic Chemistry*; Wiley: New York, 1993.
6. Pankow, J. F. Aquatic Chemistry Concepts. 1991.
7. Langmuir, D. *Aqueous Environmental Geochemistry*; Prentice-Hall, Inc.: Upper Saddle River, NJ, 1997.
8. Drever, J. I. *The Geochemistry of Natural Waters: Surface and Groundwater Environments*, 3rd ed.; Prentice-Hall: New York, 1997.
9. Pourbaix, M. *Atlas of Electrochemical Equilibria in Aqueous Solutions*; National Association of Corrosion Engineers: Houston, TX, 1974.
10. Macalady, D. L.; Tratnyek, P. G.; Grundl, T. J. Abiotic reduction reactions of anthropogenic organic chemicals in anaerobic systems. *J. Contam. Hydrol.* **1986**, *1*, 1–28.
11. Amonette, J. E. Iron redox chemistry of clays and oxides: Environmental applications. In *Electrochemistry of Clays*; Fitch, A., Ed.; Clay Minerals Society: Aurora, CO, 2002, Vol. 10; pp 89−147.
12. Sawyer, D. T. *Oxygen Chemistry*; Oxford: New York, 1991.
13. Morris, J. C.; Stumm, W. Redox equilibria and measurements of potentials in the aquatic environment. In *Equilibrium Concepts in Natural Water Systems*; ACS Symposium Series No. 67; American Chemical Society: Washington, DC, 1967; pp 270−285.
14. Whitfield, M. Thermodynamic limitations on the use of the platinum electrode in Eh measurements. *Limnol. Oceanogr.* **1974**, *19*, 857–865.
15. Hostettler, J. D. Electrode electrons, aqueous electrons, and redox potentials in natural waters. *Am. J. Sci.* **1984**, *284*, 734–759.
16. Thorstenson, D. C. The concept of electron activity and its relation to redox potentials in aqueous geochemical systems. U.S. Geological Survey Open-File Report 84-072; 1984.
17. Peiffer, S.; Klemm, O.; Pecher, K.; Hollerung, R. Redox measurements in aqueous solutions—A theoretical approach to data interpretation based on electrode kinetics. *J. Contam. Hydrol.* **1992**, *10*, 1–18.
18. Lindberg, R. D.; Runnells, D. D. Ground water redox reactions: An analysis of equilibrium state applied to Eh measurements and geochemical modeling. *Science* **1984**, *225*, 925–927.

19. Berner, R. A. A new geochemical classification of sedimentary environments. *J. Sediment. Petrol.* **1981**, *51*, 359–365.
20. Chapelle, F. H.; Vroblesky, D. A.; Woodward, J. C.; Lovley, D. R. Practical considerations for measuring hydrogen concentrations in groundwater. *Environ. Sci. Technol.* **1997**, *31*, 2873–2877.
21. Krumbein, W. C.; Garrells, R. M. Origin and classification of chemical sediments in terms of pH and oxidation-reduction potentials. *J. Geol.* **1952**, *60*, 1–33.
22. Garrells, R. M.; Christ, C. L. *Solutions, Minerals, and Equilibria*; Harper & Row: New York, 1965.
23. Arbestain, M. C.; Macías, F.; Chesworth, W. Near-Neutral Soils. In *Encyclopedia of Soil Science*; Chesworth, W., Ed.; Springer, 2008; pp 487−488.
24. Baas-Becking, L. G. M.; Kaplan, I. R.; Moore, D. Limits of the natural environment in terms of pH and oxidation-reduction potentials. *J. Geol.* **1960**, *68*, 243–283.
25. Nightingale, E. R., Jr. Poised oxidation-reduction systems. *Anal. Chem.* **1958**, *30*, 267–272.
26. Grundl, T. A review of the current understanding of redox capacity in natural, disequilibrium systems. *Chemosphere* **1994**, *28*, 613–626.
27. Scott, M. J.; Morgan, J. J. Energetics and conservative properties of redox systems. In *Chemical Modeling of Aqueous Systems II*; ACS Symposium Series 416; Henry, S. M., Ed.; American Chemical Society: Washington, DC, 1990; pp 368−378.
28. Barcelona, M. J.; Holm, T. R. Oxidation-reduction capacities of aquifer solids. *Environ. Sci. Technol.* **1991**, *25*, 1565–1572.
29. Frevert, T. Can the redox conditions in natural waters be predicted by a single parameter? *Aquat. Sci. (Schweiz. Z. Hydrol.)* **1984**, *46*, 269–296.
30. Huang, C. P., O'Melia, C. R., Morgan, J. J., Eds.; *Aquatic Chemistry: Interfacial and Interspecies Processes*; Advances in Chemistry Series 244; American Chemical Society: Washington, DC, 1995.
31. Macalady, D. L. Perspectives in Environmental Chemistry. 1998.
32. Sparks, D. L., Grundl, T. J., Eds. *Mineral-Water Interfacial Reactions: Kinetics and Mechanisms*; ACS Symposium Series No. 715; American Chemical Society: Washington, DC, 1998.
33. Taillefert, M., Rozan, T. F., Eds.; *Environmental Electrochemistry: Analyses of Trace Element Bigeochemistry*; ACS Symposium Series No. 811; American Chemical Society: Washington, DC, 2002.
34. Borch, T.; Kretzschmar, R.; Kappler, A.; Cappellen, P. V.; Ginder-Vogel, M.; Voegelin, A.; Campbell, K. Biogeochemical redox processes and their impact on contaminant dynamics. *Environ. Sci. Technol.* **2010**, *44*, 15–23.
35. Farnsworth, C. E.; Hering, J. G. Hydrous manganese oxide doped gel probe sampler for measuring in situ reductive dissolution rates. 1. Laboratory development. *Environ. Sci. Technol.* **2010**, *44*, 34–40.
36. Farnsworth, C. E.; Griffis, S. D.; Wildman, R. A., Jr.; Hering, J. G. Hydrous manganese oxide doped gel probe sampler for measuring in situ reductive

dissolution rates. 2. Field deployment. *Environ. Sci. Technol.* **2010**, *44*, 41–46.

37. Aeschbacher, M.; Sander, M.; Schwarzenbach, R. P. Novel electrochemical approach to assess the redox properties of humic substances. *Environ. Sci. Technol.* **2010**, *44*, 87–93.
38. Klausen, J.; Troeber, S. P.; Haderlein, S. B.; Schwarzenbach, R. P. Reduction of substituted nitrobenzenes by Fe(II) in aqueous mineral suspensions. *Environ. Sci. Technol.* **1995**, *29*, 2396–2404.
39. Dunnivant, F. M.; Schwarzenbach, R. P.; Macalady, D. L. Reduction of substituted nitrobenzenes in aqueous solutions containing natural organic matter. *Environ. Sci. Technol.* **1992**, *26*, 2133–2141.
40. Handler, R. M.; Beard, B. L.; Johnson, C. M.; Scherer, M. M. Atom exchange between aqueous Fe(II) and goethite: An Fe isotope tracer study. *Environ. Sci. Technol.* **2009**, *43*, 1102–1107.
41. Gorby, Y. A.; Yanina, S.; McLean, J. S.; Rosso, K. M.; Moyles, D.; Dohnalkova, A.; Beveridge, T. J.; Chang In, S.; Kim, B. H.; Kim, K. S.; et al. Electrically conductive bacterial nanowires produced by *Shewanella oneidensis* strain MR-1 and other microorganisms. *Proc. Natl. Acad. Sci. U.S.A.* **2006**, *103*, 11358–11363.
42. Nielsen, L. P.; Risgaard-Petersen, N.; Fossing, H.; Christensen, P. B.; Sayama, M. Electric currents couple spatially separated biogeochemical processes in marine sediment. *Nature* **2010**, *463*, 1071–1074.

Chapter 2

Thermodynamic Redox Calculations for One and Two Electron Transfer Steps: Implications for Halide Oxidation and Halogen Environmental Cycling

George W. Luther, III*

School of Marine Science and Policy, University of Delaware, Lewes, DE 19958, U.S.A.
***luther@udel.edu**

In oxygenated waters, chloride and bromide are the thermodynamically stable halogen species that exist whereas iodate, the thermodynamically stable form of iodine, and iodide can co-exist. The stability and oxidation of halides in the environment is related to the unfavorable thermodynamics for the first electron transfer with oxygen to form X· atoms. However, reactive oxygen species (ROS) such as $^1O_2$, H_2O_2 and O_3 can oxidize the halides to X_2 and perhaps HOX in two electron transfer processes; these reactions become less favorable with increasing pH. Fe(III) and Mn(III,IV) solid phases can oxidize halides with similar patterns as ROS. The ease of oxidation increases from $Cl^- < Br^- < I^-$. X_2 can also form HOX in water, and both halogen species can react with natural organic matter with formation of organo-halogen (R-X) compounds. During the treatment of drinking water, unwanted R-X disinfectant byproducts can form when the oxidant is not capable of quantitatively converting iodide to iodate. Natural and anthropogenic volatile R-X compounds are photochemically active and lead to X· atoms in the atmosphere which undergo reaction with O_3 via an O atom transfer step (two electron transfer step) resulting in O_3 destruction. In the case of iodine, iodine oxide species form aerosol nanoparticles leading to cloud condensation nuclei.

Introduction

To understand redox transformations in the environment, thermodynamic calculations from experimental aqueous redox potential or free energy data are used to describe reactions in which one or more of the reactants undergo an electron transfer of many electrons (n>2) from most reduced to most oxidized chemical species (e.g., (*1*)) as in the oxidation of hydrogen sulfide to sulfate. However, reactions proceed in several discrete one or two electron transfer steps along the entire reaction coordinate or pathway of 8 electrons as in the oxidation of sulfide to sulfate. In a previous paper (*2*), the thermodynamics of several multi-step reactions were shown to have a thermodynamic barrier when a sequence of one and/or two electron transfer reactions along the entire reaction pathway were considered. Thermodynamic calculations were performed over all pH for reactions between the elemental cycles of C, N, O, S, Fe, Mn and Cu using common environmental oxidants and reductants. The transformations included (i) the oxidation of Fe^{2+} and Mn^{2+} by O_2, reactive oxygen species (ROS) and NO_x species to Fe(III) and Mn(III,IV) solid phases, (ii) the oxidation of NH_4^+ to N_2 and N_2O by oxygen species as well as Fe(III) and Mn(III,IV) solid phases and (iii) the reduction of Fe(III) and Mn(III,IV) solid phases by H_2S to Fe^{2+} and Mn^{2+}. The calculations showed that the one-electron reaction of H_2S or HS^- with O_2 was unfavorable over all pH, but the two electron transfer was favorable. The reason is due to the unlikely formation of HS· and $O_2^{\cdot -}$ (superoxide) versus the formation of S(0) and H_2O_2 as products for the one versus two electron transfer, respectively. The pH dependence on the kinetics of Mn(II) and Fe(II) oxidation was also shown to be related to the first electron transfer step and $O_2^{\cdot -}$ formation.

In this paper, the same thermodynamic principles are used to show the stability or reactivity of Cl^-, Br^- and I^- to oxidation in the environment by oxygen and ROS species, (oxy)hydroxides of Fe(III) and Mn(III,IV) and NOX species. Possible initial products of halide oxidation are the halide radical (X·, an one electron transfer) or dihalogens (X_2; a two electron transfer), which can react with water or hydroxide to form hypohalous species (e.g., HOX or OX^-). HOX species can also form directly on halide oxidation in two electron transfer processes involving O atom transfer. Soluble HOX can react with organic matter to form R-X compounds (*3*, *4*) as HOX is a source of positive halogen ($HO^- X^+$) whereas gaseous HOX leads to radicals (HO· + ·X). The formation of volatile and non-volatile R-X compounds can occur in waters and sediments. In addition, R-X compounds (more recently R-I) form in drinking water that is treated with chloro-amine (*5*) and manganese dioxide (*6*) rather than bleach alone which oxidizes iodide directly to iodate. Because halides are good leaving groups in many carbon compounds, they can be displaced by each other (e.g., normally Cl^- displaces Br^- which displaces I^-) or other nucleophiles such as HS^-. This leads to toxic organic compounds including toxic disinfectant byproducts in drinking water.

For aqueous I^-, the continued inorganic oxidation reaction of HOI or its disproportionation leads to IO_3^- (*7*, *8*), the thermodynamically favored form in the environment, whereas Cl^- and Br^- are the only thermodynamically favored forms in solution (*9*). The organic and inorganic HOI reactions permit a buildup of I^- even in fully oxygenated natural waters (*10, 11*).

In the atmosphere, X· forms from the photochemical decomposition of C-X bonds in volatile organic compounds (*12*). The resulting X· undergoes many reactions with ROS and in the case of iodine leads to IO· and eventual aerosol or nanoparticle formation (*13*).

In this chapter, experimentally derived aqueous thermodynamic data are used for all calculations. In chapter 2 of this volume, Bylaska et al (*14*) demonstrate how one-electron potentials can be obtained from theoretical calculations when experimental data are not readily available.

Methods

Calculating Aqueous Redox Potentials from Half-Reactions

Aqueous thermodynamic data used (at 25 ^{0}C and 1 atm) are from Stumm and Morgan (*15*), Bard et al (*16*) and Stanbury (*17*), who tabulated reduction potentials for aqueous inorganic free radicals. The value used for the Gibbs free energy for Fe^{2+} (-90.53 kJ/mole) is that discussed in Rickard and Luther (*18*). Values of the free energy for IO_2H and IO_2^- are from Schmitz (*19*). The basic mathematical approach has been fully developed in standard textbooks and used in previous publications (*2, 20*). First, a reduction half-reaction for each redox couple is written as for the case of aqueous oxygen reduction to water in eqn. 1a. From the known Gibbs free energies or standard redox potentials, a $p\varepsilon^0$ (=log K) is calculated at the standard state conditions for each half-reaction (eqns. 1b and 1c), which is normalized to a one electron reaction. For example, eqn. 1a is normalized to become eqn. 1d.

$$O_{2(aq)} + 4\ H^+ + 4\ e^- \rightarrow 2\ H_2O \qquad (1a)$$

$$16.32 \qquad 0 \qquad 0 \qquad 2(-237.18) \qquad \Delta G^o_f\ (kJ/mol)$$

The standard state ΔG^o for the reaction = - 490.68 kJ / 2 mole H_2O or 4 mole of electrons. The equilibrium constant (K^0_4) is given in eqn. 1b where {} indicates activity for each chemical species and the activity of H_2O is defined as 1. Log $K^0_4 = \Delta G^0 / [(-2.303)RT\ ]$ = 85.96 for 4 mole of electrons or 21.49 for 1 mole of electrons where

$$K^0_4 = \{H_2O\}^2 / [\{O_2\}\ \{H^+\}^4\ \{e^-\}^4] \qquad (1b)$$

$$\text{or } \log K_4 = -\log \{O_{2(aq)}\} - \log \{H^+\}^4 - \log \{e^-\}^4 \qquad (1c)$$

For a one-electron reaction, we have

$$\tfrac{1}{4}\, O_{2(aq)} + \quad H^+ + e^- \rightarrow \tfrac{1}{2}\, H_2O \tag{1d}$$

and equation 1c becomes eqn. 1e

$$\tfrac{1}{4} \log K^0{}_4 = - \tfrac{1}{4} \log \{O_{2(aq)}\} - \log \{H^+\} - \log \{e^-\}$$
$$\tfrac{1}{4} \log K^0{}_4 = - \tfrac{1}{4} \log \{O_{2(aq)}\} + pH + p\varepsilon \tag{1e}$$

which is rearranged to eqn. 1f

$$p\varepsilon = \tfrac{1}{4} \log K^0{}_4 + \tfrac{1}{4} \log \{O_{2(aq)}\} - pH \tag{1f}$$

From the Nernst Equation, $p\varepsilon^o = \tfrac{1}{4} \log K^0{}_4 = 21.49$ (the standard state value). Thus,

$$p\varepsilon = \quad p\varepsilon^o + \tfrac{1}{4} \log \{O_{2(aq)}\} - pH$$
$$p\varepsilon = 21.49 + \tfrac{1}{4} \log \{O_{2(aq)}\} - pH \tag{1g}$$

At 250 µM O_2 (250 x 10^{-6} M; 100% saturation at ~ 25 °C; the activity coefficient for O_2 ~ 1), this expression becomes

$$p\varepsilon = 21.49 + \tfrac{1}{4} \log \{250 \times 10^{-6}\, M\} - pH = 20.59 - pH \tag{1h}$$

and at 1µM O_2 (10^{-6} M), this expression becomes

$$p\varepsilon = 21.49 + \tfrac{1}{4} \log \{10^{-6}\, M\} - pH = 19.99 - pH \tag{1i}$$

At unit activity for all reagents $p\varepsilon = p\varepsilon^o$. At unit activity of all reagents other than the H^+, we can calculate the approximate pH dependence for all reactions. In the O_2 example, eqn. 1f becomes eqn. 1j, which is used for all calculations in this paper.

$$p\varepsilon = p\varepsilon^o - pH = 21.49 - pH \tag{1j}$$

A pε(pH) for a half-reaction at a given pH can then be calculated and the reactions for this paper are presented in Appendix 1. The calculated value for each half-reaction is given as a function of pH as in the examples in Appendix 1, and these can easily be entered into spreadsheets for quick calculations of full reactions from two half-reactions (see next section). [When H^+ or OH^- is not in a balanced equation for a half-reaction, there is no pH dependence on the half-reaction.] The pε calculated is termed pε(pH) which provides a log K for each half-reaction at a given pH. Concentration dependence for the other reactants are not considered in the calculation; thus, these are considered standard state calculations (note that eqns. 1i and 1j show a 1.50 log change for an oxygen concentration range from 1 µM to unity activity so the calculations could vary an order of magnitude or more in either direction when concentration dependence is included). However, comparisons can be more easily made when combining different half-reactions at a given pH. This permits an assessment of which combined half-reactions are thermodynamically favorable and thus more likely to react in a given environmental setting.

Coupling Half-Reactions

As an example of the coupling of two half reactions to determine whether a reaction is favorable, I use the data in Appendix 1 for the reaction of O_2 (eqn. O1) and Mn^{2+} (eqn. Mn1) to form H_2O and MnO_2 at pH = 7. In this case, log K [for the complete reaction] = pε (O_2) - pε(Mn^{2+}) = 14.49 - (6.8) = 7.69. The negative sign in front of pε(Mn^{2+}) indicates that the half-reaction is now an oxidation half-reaction. This calculation is for the full reduction of O_2 to water.

For this work, Appendix 1 lists the pε(pH) values for Mn, Fe, oxygen and halide species for the relevant one and two-electron transfer reactions considered. Dissolved Fe(II) and Mn(II) are primarily hexaaquo species until the pH is > 7 so the concentration and activity are equal, where hydroxo complexes start to become important. The latter are not considered in this analysis, but the pH dependence for the known oxidation kinetics of Fe(II) and Mn(II) with oxygen are predicted by the thermodynamics, and a linear free energy relationship exists for Fe(II) oxidation by O_2 (*2*).

Results and Discussion

Oxidation of Halides to X· by O_2 and ROS: One Electron Transfer

Figure 1 shows thermodynamic calculations for the reaction of halides to halide radicals (X·) with triplet O_2 ($^3O_2$) and ROS for the full reduction of O_2 to H_2O in four one-electron steps. For this and all subsequent figures, a positive Δlog K on the y-axis indicates a favorable complete reaction and a negative Δlog K indicates an unfavorable reaction as $\Delta G^o = -RT \ln K = -2.303RT \log K$.

Figure 1a indicates that the first electron transfer reaction of halides with O_2 is unfavorable as the free radical products, X· and $O_2^{\cdot-}$, are formed. This reaction requires that halides must unpair electrons in an outer sphere electron process so that one electron from an halide can be donated to the singly occupied molecular orbital of $^3O_2$ (*11*). Figure 1b shows the reaction of $O_2^{\cdot-}$ with halides is also unfavorable except for iodide at low pH. Thus, iodide may react with $O_2^{\cdot-}$ in acid rain. The reaction of peroxide with halides (Figure 1c) is also unfavorable, but the reaction of OH· radical is favorable for iodide and bromide.

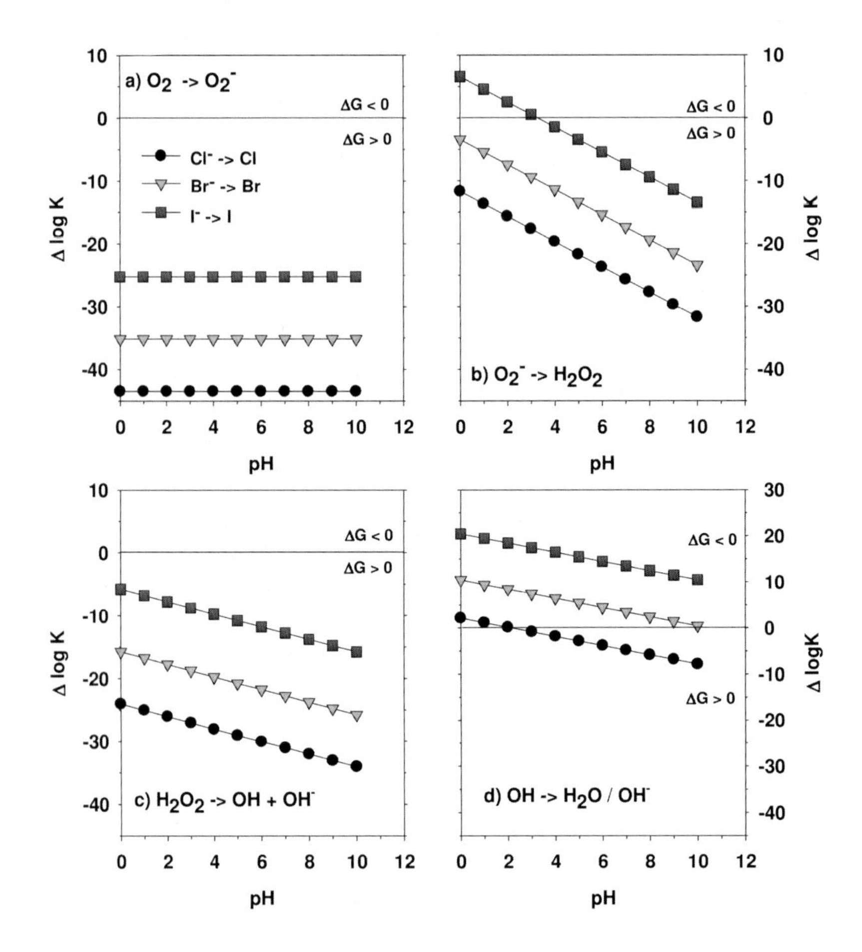

Figure 1. One-electron transfer reactions of X^- with oxygen species from Appendix 1 to form $X\cdot$; (a) reaction O6 coupled with reactions CL3, Br3, Io3. (b) reaction O7 coupled with reactions CL3, Br3, Io3. (c) reaction O8 coupled with reactions CL3, Br3, Io3. (d) reaction O9 coupled with reactions CL3, Br3, Io3.

Oxidation of Halides to X_2 by O_2 and ROS

Figure 2 shows the thermodynamic calculations for the possible reaction of halides to dihalogens, X_2, with the same oxidants.

Figure 2a again shows that O_2 reactions are unfavorable whereas $O_2^{\cdot -}$ (figure 2b) and peroxide (figure 2c) are predicted to be more favorable, especially for iodide and bromide. There is no apparent thermodynamic barrier to OH· radical reaction (figure 2d).

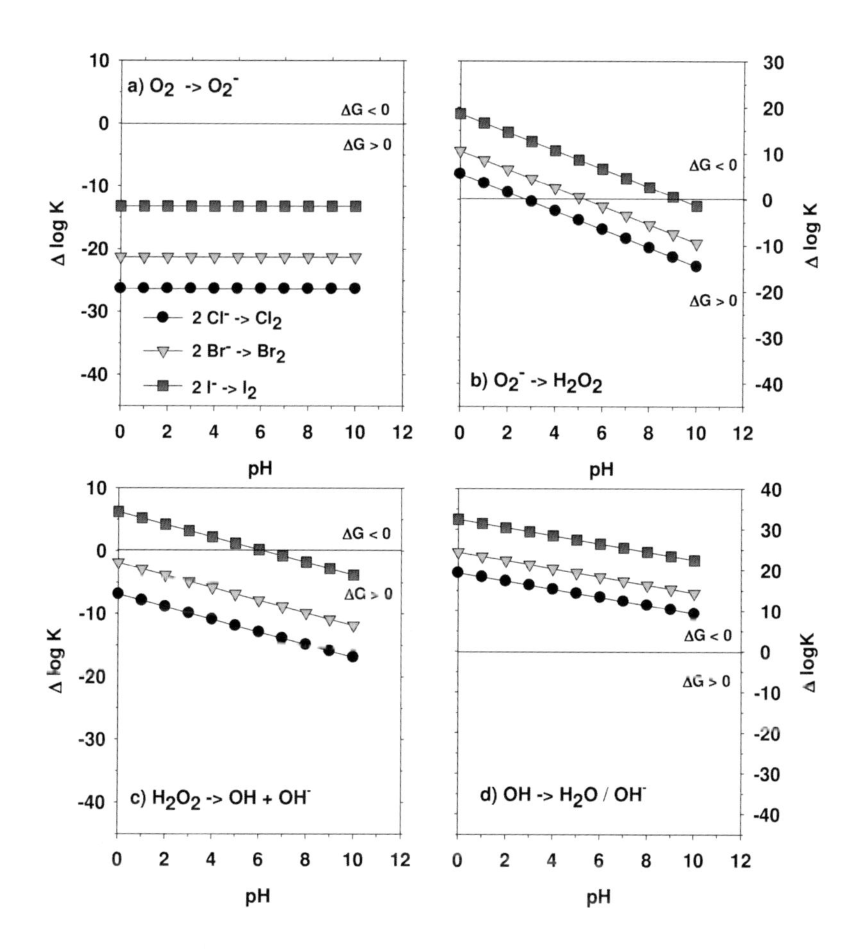

Figure 2. One-electron transfer reactions of X^- with oxygen species from Appendix 1 to form X_2; (a) reaction O6 coupled with reactions CL1, Br1, Io1. (b) reaction O7 coupled with reactions CL1, Br1, Io1. (c) reaction O8 coupled with reactions CL1, Br1, Io1. (d) reaction O9 coupled with reactions CL1, Br1, Io1.

Oxidation of Halides by O_3, $^1O_2$, NO_x: One Electron Transfer

Figures 3a and 3b show the thermodynamics for one electron transfer of halides to halide radicals (X·) with singlet O_2 ($^1O_2$) and ozone, respectively, and these reactions are independent of pH.

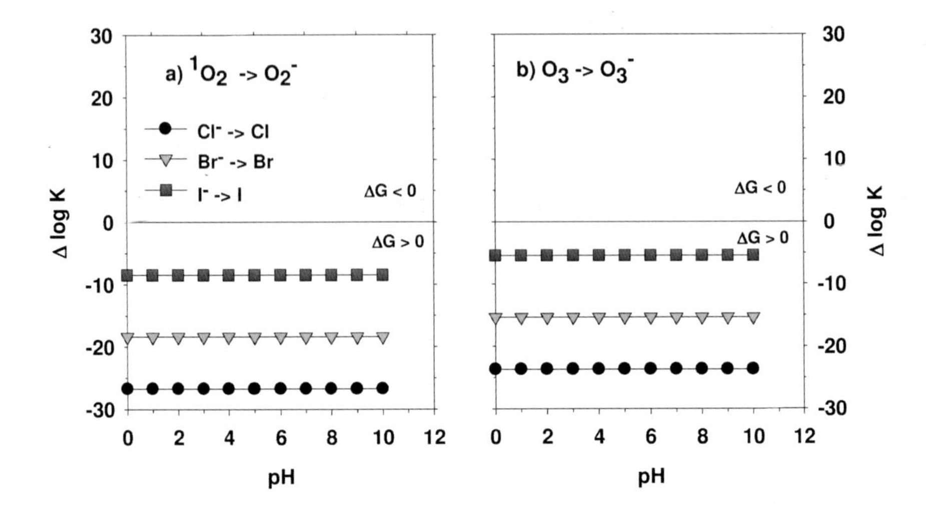

Figure 3. One-electron transfer reactions of X^- with $^1O_2$ and O_3 to form $X\cdot$; (a) reaction O6 coupled with reactions CL3, Br3, Io3. (b) reaction O7 coupled with reactions CL3, Br3, Io3.

The unpairing of electrons in the highest occupied molecular orbital of an halide in an outer sphere electron process is unfavorable despite the fact that $^1O_2$ and O_3 have lowest unoccupied molecular orbitals that can readily accept an electron (*11*). Both of these reactions are also unfavorable over all pH. The first electron transfer step in the oxidation of halides by the free radical NO_2 to form NO_2^- and the free radical NO to form NO_2 are also thermodynamically unfavorable over all pH (data not shown).

Oxidation of Halides by $^3O_2$, $^1O_2$, H_2O_2, O_3 To Form X_2 or HOX: Two Electron Transfer

Figure 4 shows the thermodynamic calculations for the reaction of halides to form dihalogens, X_2, in two-electron transfer reactions. $^3O_2$ (Figure 4a) again has a significant thermodynamic barrier to halide oxidation and H_2O_2 formation whereas $^1O_2$ (Figure 4b) can oxidize iodide. H_2O_2 (Figure 4c) and O_3 (Figure 4d) can oxidize all halides at most environmental pH conditions, and these are the most favorable reactions for halide oxidation with ROS. Although H_2O_2 can oxidize the halides, these reactions become less favorable with increasing pH and are slow kinetically (*21*, *22*). This is a likely reason why plants have developed vanadium and iron dependent haloperoxidases (*21*, *22*), which are used to catalyze halide oxidation with subsequent formation of R-X compounds. Some of the haloperoxidases are not capable of oxidizing Cl^- and Br^-, which follows the thermodynamic pattern shown in Figure 4c.

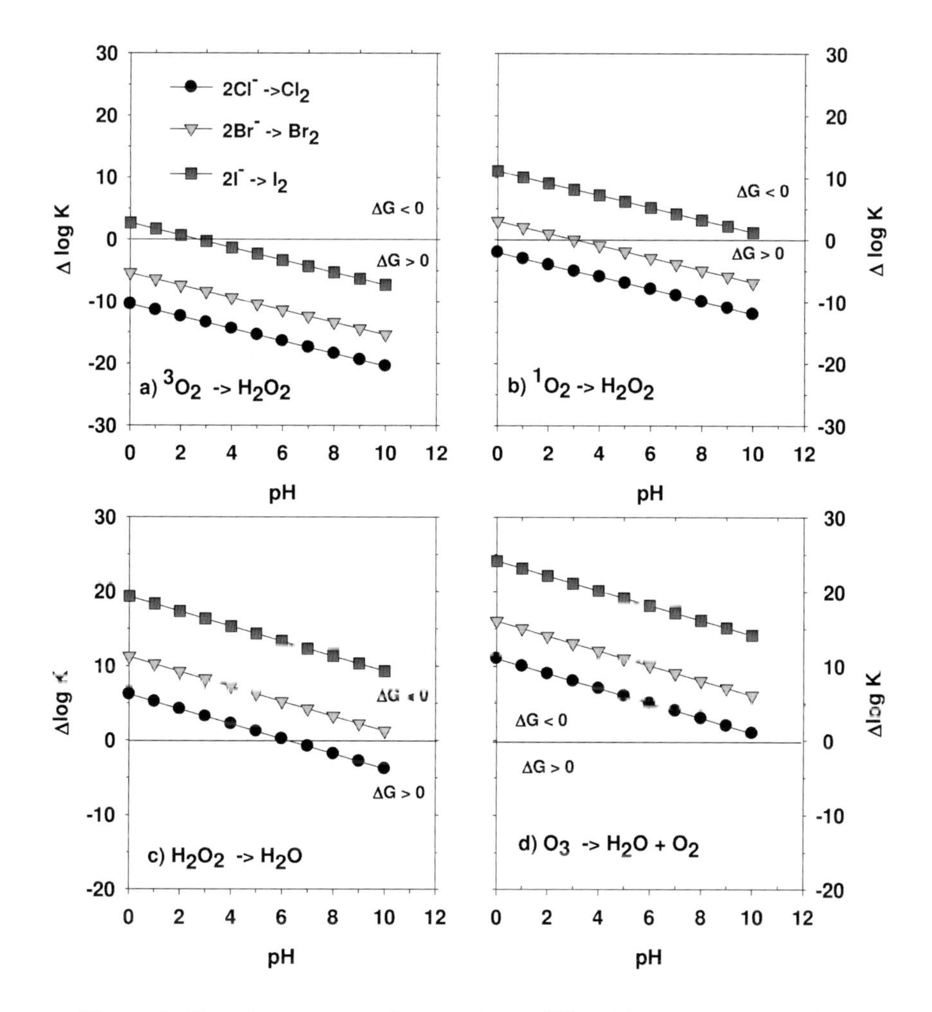

Figure 4. Two-electron transfer reactions of X^- with oxygen species from Appendix 1 to form X_2; (a) reaction O2 coupled with reactions CL1, Br1, Io1. (b) reaction O5 coupled with reactions CL1, Br1, Io1. (c) reaction O3 coupled with reactions CL1, Br1, Io1. (d) reaction O4 coupled with reactions CL1, Br1, Io1.

Figure 5 shows the thermodynamic calculations for the reaction of halides to form HOX species in two-electron transfer reactions. These reactions have similar thermodynamic trends but are not as favorable as those shown in Figure 4 to form X_2. $^3O_2$ (Figure 5a) again has a significant thermodynamic barrier to halide oxidation and H_2O_2 formation whereas $^1O_2$ (Figure 5b) can oxidize iodide. Formation of HOX appears more favorable than X_2 for H_2O_2 (Figure 5c) and O_3 (Figure 5d) oxidation of all halides and these reactions are likely O atom transfer processes. In water, direct formation of HOX may occur in peroxide reactions. However, I_2 has been shown to form from iodide oxidation by O_3 in brown algae, which were not submerged in water (*23*). Because O_3 does not penetrate the surface ocean microlayer from the atmosphere, its reactions are significant only in the surface microlayer (*24*).

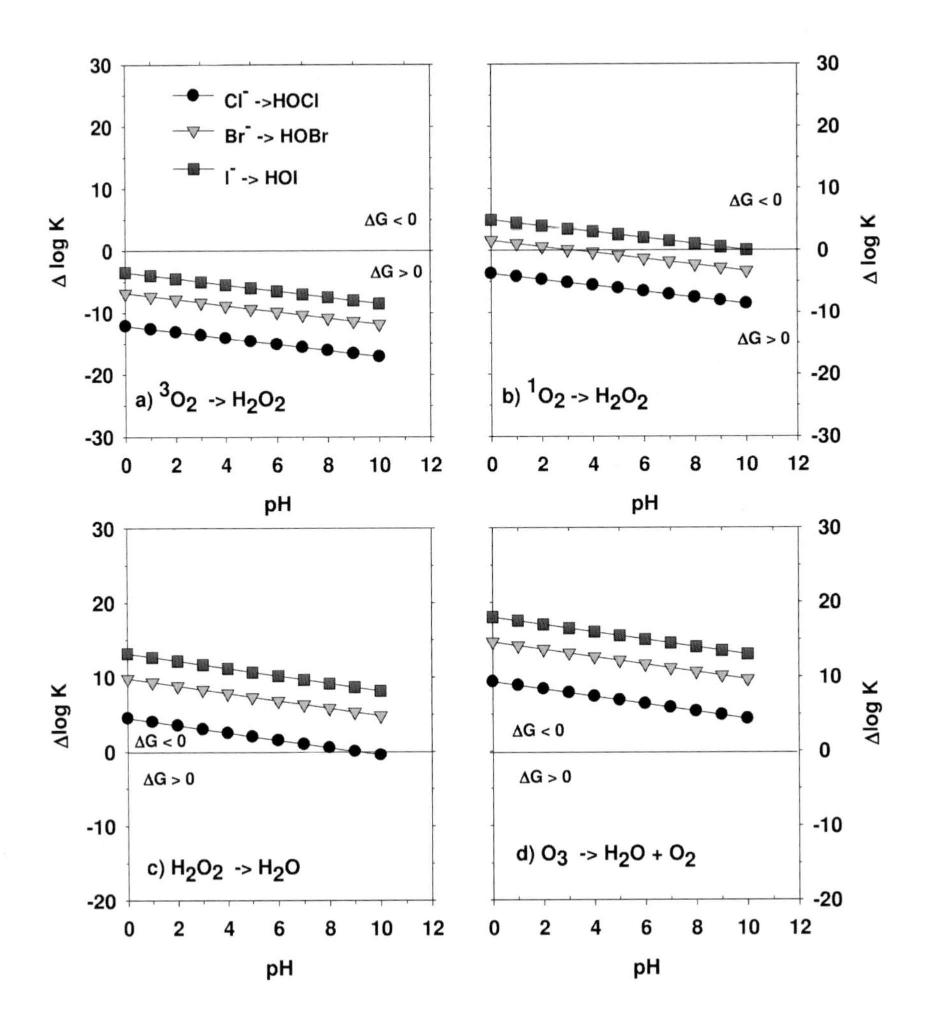

Figure 5. Two-electron transfer reactions of X^- with oxygen species from Appendix 1 to form HOX; (a) reaction O2 coupled with reactions CL2, Br2, Io2. (b) reaction O5 coupled with reactions CL2, Br2, Io2. (c) reaction O3 coupled with reactions CL2, Br2, Io2. (d) reaction O4 coupled with reactions CL2, Br2, Io2.

Oxidation of Halides to X· and X_2 by Fe(III) and Mn(III,IV) Solid Phases – Sedimentary Reactions

Other common environmental oxidants, especially in sediments, are the (oxy)hydroxides of Fe(III) and Mn(III,IV). Figure 6 shows that the one electron oxidation of halides to halide radicals (X·) for chloride (figure 6a) and bromide (figure 6b) are not favorable. Iodide oxidation (figure 6c) is only favorable for MnOOH and Mn_3O_4 at pH $\leq$ 2.

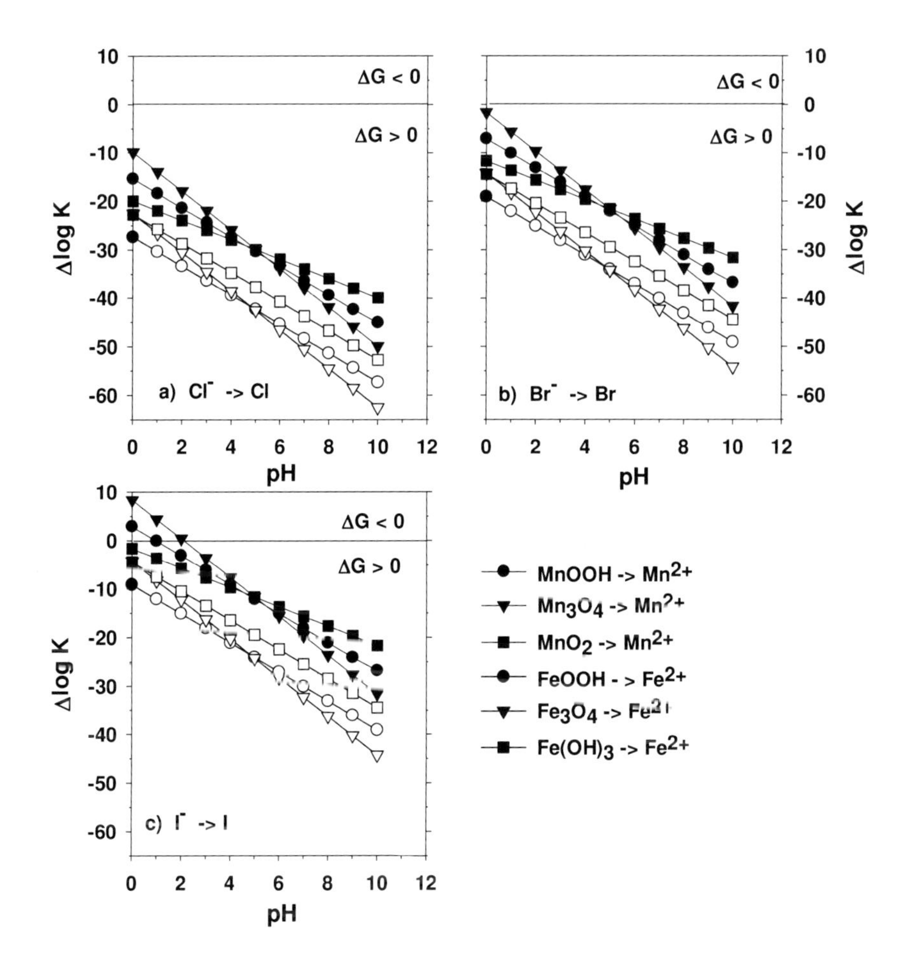

Figure 6. One-electron transfer reactions of X^- with oxidized metal species from Appendix 1 to form $X^{\cdot}$; (a) reaction CL3 coupled with reactions Mn1, Mn2, Mn3, Fe1, Fe2, Fe3. (b) reaction Br3 coupled with reactions Mn1, Mn2, Mn3, Fe1, Fe2, Fe3. (c) reaction Io3 coupled Mn1, Mn2, Mn3, Fe1, Fe2, Fe3.

However, Figure 7a-c show that the two electron halide oxidation reactions to form dihalogens, X_2, become favorable at low pH in the order chloride < bromide < iodide. These reactions are similar to the well-known reaction of MnOOH with iodide to form I_2 at low pH which is the basis for the Winkler oxygen titration (*25*). The kinetics of the reaction of Mn(IV) with iodide has recently been studied because MnO_2 can be a principal component of sediments (*26*) and a chemical agent to provide clean drinking water (*27*). In these two environmental settings, unwanted R-I transformations can occur, and in the case of sediments could promote the incorporation of radioactive iodine into a variety of compounds. At pH values above 5-7, the iodide oxidation reaction slows down or ceases (*26*, *27*) consistent with the thermodynamic data in Figure 7c.

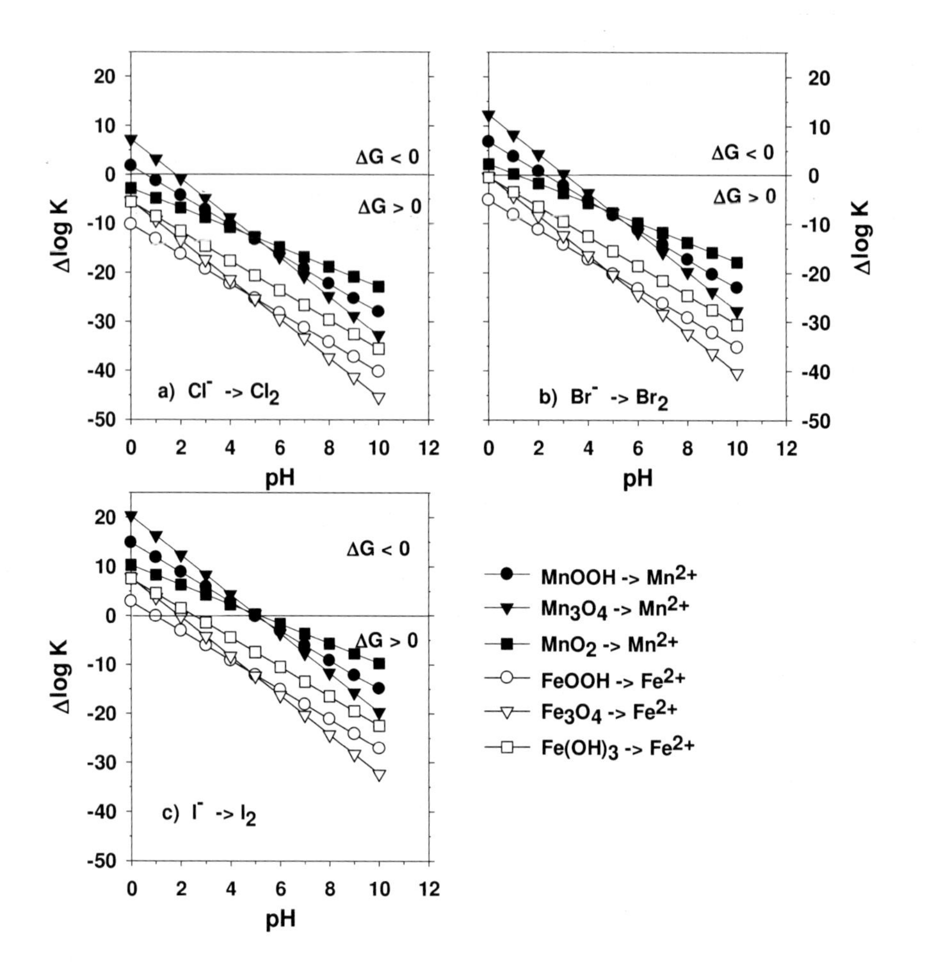

Figure 7. Two- electron transfer reactions of X^- with oxidized metal species from Appendix 1 to form X_2; (a) reaction CL1 coupled with reactions Mn1, Mn2, Mn3, Fe1, Fe2, Fe3. (b) reaction Br1 coupled with reactions Mn1, Mn2, Mn3, Fe1, Fe2, Fe3. (c) reaction Io1 coupled Mn1, Mn2, Mn3, Fe1, Fe2, Fe3.

Figure 7c shows that the two-electron reduction of Fe(III) and Mn(III,IV) phases by I^- to I_2 is favorable but decreases with increasing pH as observed by Fox et al (*26*) and von Gunten et al (*27*) for MnO_2. However, there is a thermodynamic barrier to Mn(III) reduction at a pH $\geq$ 6 and to Fe(III) reduction at pH $\geq$ 2. These Fe and Mn reactions are solid phase reactions and are surface controlled. The transition state also varies with increasing pH as the zero point of charge (*15*) of the metal oxide phases changes from positive to negative. Thus, two negative species that would repel each other are involved in the transition state at higher pH, and I^- does not adsorb to the MnO_2 surface at higher pH limiting its oxidation (*26*). For I^- to donate two electrons to an Fe(III) or Mn(III) phase would require that each I^- donate the electrons to a band containing several orbitals that are empty or partially occupied so that two Fe(III) or Mn(III) ions would accept the electrons. The resulting I^+ would react with another I^- to form I_2 or with water or OH^- to form HOI resulting in possible formation of IO_3^-. These calculations indicate that clusters or nanoparticles of oxidized metals would be important environmental oxidants.

Oxidation of Halides to HOX by Fe(III) and Mn(III,IV) Solid Phases

Figure 8 shows the reaction of Fe(III) and Mn(III,IV) phases with halides to form HOX species in two-electron transfer reactions. These reactions have similar thermodynamic trends as those shown in Figure 7. Figures 8a-c show that halide oxidation becomes favorable at low pH in the order chloride < bromide < iodide. Because the reaction of MnOOH with iodide to form I_2 at low pH is the basis for the Winkler oxygen tritration (*25*), HOI and other HOX species are not likely first products of halide oxidation with metal (oxy)hydroxide phases.

Iodate Formation

Recently, thermodynamic data for the iodous acid forms have become available (*19*) and the successive two electron or O atom transfer oxidation reactions of I^- to HOI to IO_2H to IO_3^- by H_2O_2 are favorable over all pH (data not shown). The corresponding reactions with triplet oxygen to form any of these species (e.g.; figures 2 and 4) are not favorable. Although iodide oxidation seems to be a likely way to form IO_3^-, disproportionation of HOI and IO_2H is possible and does not require O_2 or a reactive oxygen species (*7*, *8*).

Atmospheric Reactions Involving Halogen Atoms

Although X atoms do not readily form in solution, they are a common species in the atmosphere due to photochemical activation, and their formation and reactivity has been studied extensively (*28*, *29*). Atmospheric reactions are typically not complicated by water and H^+ transfers or formation. The formation of halogen atoms in the atmosphere appears related to three major natural processes involving plants and algae in marine and other natural waters, which facilitate volatile R-X formation and subsequent photochemical decomposition via homolytic bond cleavage of the C-X bond. In addition, photochemical decomposition of anthropogenic gaseous R-X compounds occurs.

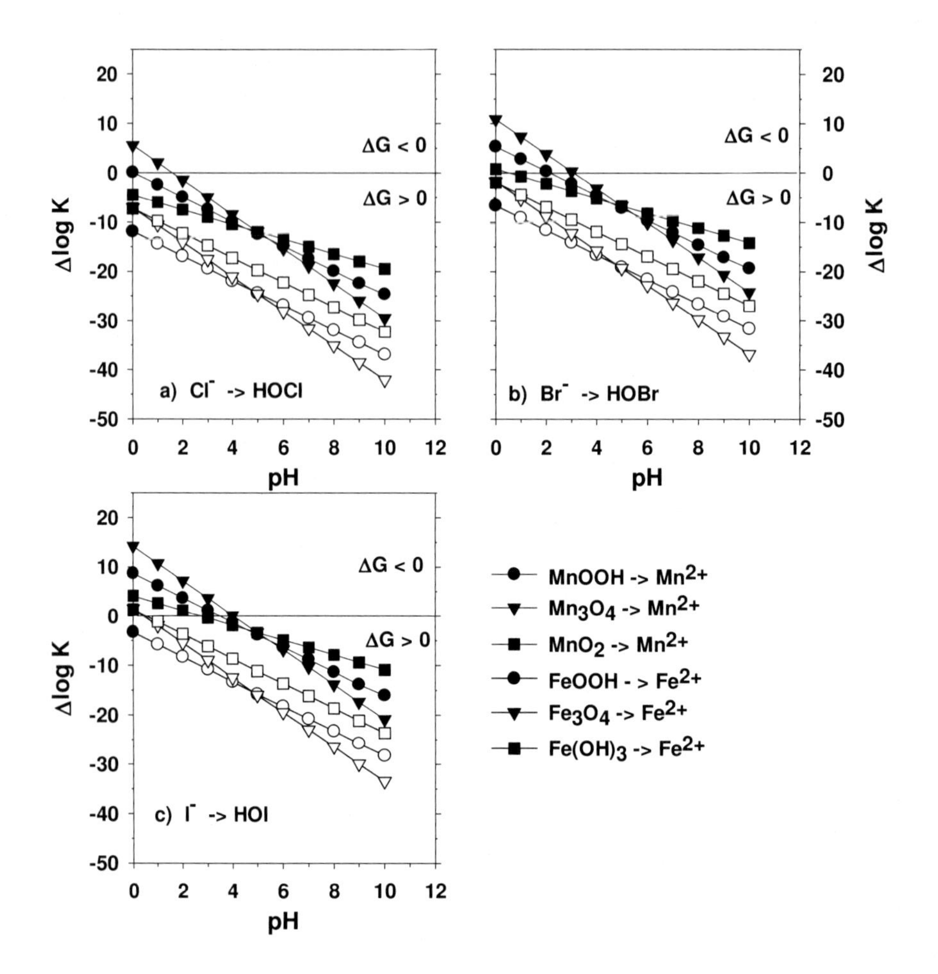

Figure 8. Two- electron transfer reactions of X⁻ with oxidized metal species from Appendix 1 to form HOX; (a) reaction CL2 coupled with reactions Mn1, Mn2, Mn3, Fe1, Fe2, Fe3. (b) reaction Br2 coupled with reactions Mn1, Mn2, Mn3, Fe1, Fe2, Fe3. (c) reaction Io2 coupled with reactions Mn1, Mn2, Mn3, Fe1, Fe2, Fe3.

First, O_3 can react with halides to form HOX and X_2, which can react with natural organic matter to form R-X compounds (*27*). Second macroalgae and microalgae produce R-X compounds presumably through haloperoxidases (*30–33*). Third, iodide in the interstitial fluids of the brown kelp, *Laminaria digitata*, and perhaps other plants can be released and used as an antioxidant to react with H_2O_2 and O_3 during oxidative stress (*23*). Oxidation of iodide above or out of the water leads to formation of I_2 rather than volatile organo halogen compounds, and I_2 formation can be three orders of magnitude larger than R-I and other R-X compound formation (*23*).

Relevant peroxide solution reactions to form HOX and X_2 from these processes are eqs. 2-3 whereas possible surface microlayer and lower atmospheric ozone reactions are eqs. 4-5, where X^- can come from sea salt spray. These latter reactions are most favorable for iodine (Figures 4 and 5), but are unwanted as they destroy O_3.

$$2X^- + H_2O_2 + 2H^+ \rightarrow X_2 + 2\,H_2O \qquad (2)$$

$$X^- + H_2O_2 + H^+ \rightarrow HOX + H_2O \qquad (3)$$

$$2X^- + O_3 + 2H^+ \rightarrow X_2 + O_2 + H_2O \qquad (4)$$

$$X^- + O_3 + H^+ \rightarrow HOX + O_2 \qquad (5)$$

Once X_2 and HOX are formed, they can react with organic matter to form R-X compounds (*4*), which are released to the atmosphere. X_2, HOX and R-X chemical species can be photochemically decomposed to X atoms, R radicals and OH radical as in eqs. 6-8. Because bonds of iodine with other atoms are weaker than corresponding Cl and Br analogs (*35*), iodine reactions are more facile.

$$I_2 + hv \rightarrow 2\,I \qquad (6)$$

$$RI + hv \rightarrow R + I \qquad (7)$$

$$HOI + hv \rightarrow OH + I \qquad (8)$$

Upon formation, X atoms can undergo a variety of thermodynamically favorable O atom transfer reactions with ROS and NO (eqs. 9-10) as discussed in Whalley et al (*34*). The destruction of O_3 is a main reaction that leads to XO radicals that can react with NO to form NO_2. Two IO molecules also react to form OIO (eq. 11), which further reacts with IO to form stable I_2O_3 (eq. 12) and I_2O_5 with eventual formation of nanoparticles or aerosols (*13*).

$$X + O_3 \rightarrow XO + O_2 \qquad (9)$$

$$XO + NO \rightarrow X + NO_2 \qquad (10)$$

$$IO + IO \rightarrow I + OIO \qquad (11)$$

$$IO + OIO \rightarrow I_2O_3 \qquad (12)$$

XO can also react with HO_2 to abstract a H· atom to form HOX (eq. 12), which is photochemically sensitive and regenerates OH (eq. 8).

$$XO + HO_2 \rightarrow HOX + O_2 \qquad (12)$$

Both OH and HO_2 also result in O_3 destruction (eqs. 13-14) in a similar manner to the halogen reactions above. Thus, decreasing halogen emissions to the atmosphere is of critical importance.

$$O_3 + OH \rightarrow HO_2 + O_2 \tag{13}$$

$$O_3 + HO_2 \rightarrow OH + 2O_2 \tag{14}$$

Conclusions

The thermodynamic data above show that halide oxidation in oxygenated waters is not a favorable process at any pH unless the reactive oxygen species, $O_2^{\cdot -}$, $\cdot OH$, $^1O_2$, H_2O_2 or O_3, can form via chemical, photochemical or enzymatic pathways. The best oxidants in aqueous solution appear to be the one electron oxidant, $\cdot OH$, and the two electron oxidants, $^1O_2$, H_2O_2 or O_3. However, thermodynamic favorability depends on the halide with iodide being the easiest to oxidize as well as on pH with thermodynamic favorability decreasing with increasing pH. Haloperoxidases are particularly efficient in oxidizing halides with H_2O_2. Because O_3 does not penetrate the ocean microlayer from the atmosphere, O_3 reactions appear significant only in the surface microlayer. Some plants, which have iodide in their pore fluids, can release iodide above the water to react with O_3 with I_2 formation (*23*). In sediments, Mn(III) (oxy)hydroxides are possible oxidants at near neutral pH for iodide; however, chloride and bromide reactions can only occur at pH values $\leq$ 2-3. In the atmosphere, halogen oxidation reactions are more likely due to photodecomposition of C-X bonds to X atoms, which react with O_3, NO, XO and other species by O atom transfer, a two electron transfer process.

Acknowledgments

This work was supported by grants from the Chemical Oceanography program of NSF (OCE-1031272) and the U.S. National Oceanic and Atmospheric Administration Sea Grant program (NA09OAR4170070). The author wishes to express his gratitude and appreciation to Don Macalady for his encouragement and friendship, and also wishes Don all the best as he embarks on the next phase of his career.

Appendix 1

Table 1. Reduction reactions for relevant oxygen, halide, Fe and Mn reactions normalized to one electron. All species are in aqueous form except for the oxidized Fe and Mn solids, and the reduced soluble metal species are the hexahydrate forms. Activities of all reactants other than H^+ are at unity

OXYGEN REACTIONS

four electron reaction normalized to one electron

$1/4\ O_2 + H^+ + e^- \rightarrow \frac{1}{2}\ H_2O$	$p\varepsilon = p\varepsilon^o - pH = 21.49\ -pH$	(O1)

two electron reactions normalized to one electron

$\frac{1}{2}\ O_2 + H^+ + e^- \rightarrow \frac{1}{2}\ H_2O_2$	$p\varepsilon = p\varepsilon^o - pH = 13.18\ -pH$	(O2)
$\frac{1}{2}\ H_2O_2 + H^+ + e^- \rightarrow 1\ H_2O$	$p\varepsilon = p\varepsilon^o - pH = 29.82\ -pH$	(O3)
$\frac{1}{2}\ O_3 + H^+ + e^- \rightarrow \frac{1}{2}\ O_{2(aq)} + \frac{1}{2}\ H_2O$	$p\varepsilon = p\varepsilon^o - pH = 34.64\ -pH$	(O4)
$\frac{1}{2}\ {}^1O_2 + H^+ + e^- \rightarrow \frac{1}{2}\ H_2O_2$	$p\varepsilon = p\varepsilon^o - pH = 21.57\ -pH$	(O5)

One electron transfer reactions only

$O_2 + e^- \rightarrow O_2^-$	$p\varepsilon = p\varepsilon^o = -2.72$	(O6)
$O_2^- + 2\ H^+ + e^- \rightarrow H_2O_2$	$p\varepsilon = p\varepsilon^o - 2pH = 29.08\ -2pH$	(O7)
$H_2O_2 + H^+ + e^- \rightarrow H_2O + OH\bullet$	$p\varepsilon = p\varepsilon^o - pH = 16.71\ -pH$	(O8)
$OH\bullet + e^- \rightarrow OH^-$	$p\varepsilon = p\varepsilon^o = 28.92 + pOH$	(O9a)
$OH\bullet + H^+ + e^- \rightarrow H_2O$	$p\varepsilon = p\varepsilon^o = 42.92 - pH$	(O9b)
${}^1O_2 + e^- \rightarrow O_2^-$	$p\varepsilon = p\varepsilon^o = 14.04$	(O10)
$O_3 + e^- \rightarrow O_3^-$	$p\varepsilon = p\varepsilon^o = 17.08$	(O11)

MANGANESE REACTIONS

two electron reactions normalized to one electron

$\frac{1}{2}\ MnO_{2(s)} + 2\ H^+ + e^- \rightarrow \frac{1}{2}\ Mn^{2+} + H_2O$	$p\varepsilon = p\varepsilon^o - 2pH = 20.80\ -2pH$	(Mn1)
$\frac{1}{2}\ Mn_3O_{4(s)} + 4\ H^+ + e^- \rightarrow 3/2\ Mn^{2+} + 2H_2O$	$p\varepsilon = p\varepsilon^o - 4pH = 30.82\ -4pH$	(Mn2)

One electron transfer reaction only

$MnOOH_{(s)} + 3\ H^+ + e^- \rightarrow Mn^{2+} + 2\ H_2O$	$p\varepsilon = p\varepsilon^o - 3pH = 25.35\ -3pH$	(Mn3)

Continued on next page.

Table 1. (Continued). Reduction reactions for relevant oxygen, halide, Fe and Mn reactions normalized to one electron. All species are in aqueous form except for the oxidized Fe and Mn solids, and the reduced soluble metal species are the hexahydrate forms. Activities of all reactants other than H^+ are at unity

IRON REACTIONS		
two electron reaction normalized to one electron		
$\frac{1}{2}\, Fe_3O_{4(s)} + 4\, H^+ + e^- \rightarrow 3/2\, Fe^{2+} + 2\, H_2O$	$p\varepsilon = p\varepsilon^o - 4pH = 18.20 - 4pH$	(Fe1)
One electron transfer reactions only		
$FeOOH_{(s)} + 3\, H^+ + e^- \rightarrow Fe^{2+} + 2\, H_2O$	$p\varepsilon = p\varepsilon^o - 3pH = 13.37 - 3pH$	(Fe2)
$Fe(OH)_{3(s)} + 3\, H^+ + e^- \rightarrow Fe^{2+} + 3\, H_2O$	$p\varepsilon = p\varepsilon^o - 3pH = 18.03 - 3pH$	(Fe3)
HALOGEN REACTIONS		
two electron reaction normalized to one electron		
$\frac{1}{2}\, Cl_2 + e^- \rightarrow Cl^-$	$p\varepsilon = p\varepsilon^o = 23.62$	(CL1)
$\frac{1}{2}\, Br_2 + e^- \rightarrow Br^-$	$p\varepsilon = p\varepsilon^o = 18.58$	(Br1)
$\frac{1}{2}\, I_2 + e^- \rightarrow I^-$	$p\varepsilon = p\varepsilon^o = 10.50$	(Io1)
$\frac{1}{2}\, HOCl + \frac{1}{2}\, H^+ + e^- \rightarrow \frac{1}{2}\, Cl^- + \frac{1}{2}\, H_2O$	$p\varepsilon = p\varepsilon^o - 0.5pH = 25.29 - 0.5pH$	(CL2)
$\frac{1}{2}\, HOBr + \frac{1}{2}\, H^+ + e^- \rightarrow \frac{1}{2}\, Br^- + \frac{1}{2}\, H_2O$	$p\varepsilon = p\varepsilon^o - 0.5pH = 20.00 - 0.5pH$	(Br2)
$\frac{1}{2}\, HOI + \frac{1}{2}\, H^+ + e^- \rightarrow \frac{1}{2}\, I^- + \frac{1}{2}\, H_2O$	$p\varepsilon = p\varepsilon^o - 0.5pH = 16.66 - 0.5pH$	(Io2)
One electron transfer reactions only		
$Cl\bullet + e^- \rightarrow Cl^-$	$p\varepsilon = p\varepsilon^o = 40.76$	(CL3)
$Br\bullet + e^- \rightarrow Br^-$	$p\varepsilon = p\varepsilon^o = 32.47$	(Br3)
$I\bullet + e^- \rightarrow I^-$	$p\varepsilon = p\varepsilon^o = 22.49$	(Io3)

References

1. Beal, E. J.; House, C. H.; Orphan, V. J. Manganese- and iron-dependent marine methane oxidation. *Science* **2009**, *325*, 184–187.
2. Luther, G. W., III. The role of one and two electron transfer reactions in forming thermodynamically unstable intermediates as barriers in multi-electron redox reactions. *Aquat. Geochem.* **2010**, *16*, 395–420.
3. Truesdale, V. W.; Luther, G. W., III; Canosa-mas, C. Molecular iodine reduction in seawater: an improved rate equation considering organic compounds. *Mar. Chem.* **1995**, *48*, 143–150.

4. Truesdale, V. W.; Luther, G. W., III. Molecular iodine reduction by natural and model organic substances in seawater. *Aquat. Geochem.* **1995**, *1*, 89–104.
5. Richardson, S. D.; Fasano, F.; Ellington, J. J.; Crumley, F. G.; Buettner, K. M.; Evans, J. J.; Blount, B. C.; Silva, L. K.; Waite, T. J.; Luther, G. W.; McKague, A. B.; Miltner, R. J.; Wagner, E. D.; Plewa, M. J. Occurrence and mammalian cell toxicity of iodinated disinfection byproducts in drinking water. *Environ. Sci. Technol.* **2008**, *42*, 8330–8338.
6. Gallard, H; Allard, S.; Nicolau, R.; von Gunten, U.; Croue, J. P. Formation of Iodinated Organic Compounds by Oxidation of Iodide-Containing Waters with Manganese Dioxide. *Environ. Sci. Technol.* **2009**, *43*, 7003–7009.
7. Truesdale, V. W.; Canosa-Mas, C. E.; Luther, G. W., III. On the disproportionation of molecular iodine added to seawater. *Mar. Chem.* **1995**, *51*, 55–60.
8. Truesdale, V. W.; Luther, G. W., III; Greenwood, J. E. The kinetics of iodine disproportionation: a system of parallel second-order reactions sustained by a multi-species pre-equilibrium. *Phys. Chem. Chem. Phys.* **2003**, *5*, 3428–3435.
9. Brookins, D. G. *Eh-pH diagrams for Geochemistry*. Springer-Verlag: Berlin, 1988; p176.
10. Luther, G. W., III; Cole, H. Iodine speciation in Chesapeake Bay waters. *Mar. Chem.* **1988**, *24*, 315–325.
11. Luther, G. W., III; Wu, J.; Cullen, J. Redox chemistry of iodine in seawater: frontier molecular orbital theory considerations. In *Aquatic Chemistry: Interfacial and interspecies processes*; Huang, C. P., O'Melia, C. R., Morgan, J. J., Eds.; Advances in Chemistry Series; American Chemical Society: Washington, DC, 1995; Vol. 244, 135−155.
12. Bloss, W. J.; Rowley, D. M.; Cox, R. A.; Jones, R. L. Kinetics and Products of the IO Self-Reaction. *J. Phys. Chem. A* **2001**, *105*, 7840–7854.
13. Kaltsoyanis, N.; Plane, J. M. C. Quantum chemical calculations on a selection of iodine-containing species (IO, OIO, INO_3, $(IO)_2$, I_2O_3, I_2O_4 and I_2O_5) of importance in the atmosphere. *Phys. Chem. Chem. Phys.* **2008**, *10*, 1723–1733.
14. Bylaska, E.; Salter-Blanc, A.; Tratnyek, P. One-electron reduction potentials from chemical structure theory calculations. In *Aquatic Redox Chemistry*, Tratnyek, P., Grundl, T. J., Haderlein, S. B., Eds.; American Chemical Society: Washington, DC, 2011; Vol. 1071, Chapter 3, pp 37−64.
15. Stumm W., Morgan J. J. *Aquatic Chemistry*, 3rd ed.; John Wiley: New York, 1996; p 1022.
16. Bard, A. J., Parsons, R., Jordan, J., Eds.; *Standard potentials in aqueous solution*, 1st ed.; M. Dekker: New York, 1985, p 834.
17. Stanbury, D. Reduction potentials involving inorganic free radicals in aqueous solution. In *Advances in Inorganic Chemistry*; Sykes, A. G., Ed.; Academic Press: New York, 1989; Vol. 33, pp 69−138.
18. Rickard, D.; Luther, G. W., III. Chemistry of Iron Sulfides. *Chem. Rev.* **2007**, *107*, 514–562.

19. Schmitz, G. Inorganic reactions of Iodine(III) in acidic solutions and free energy of iodous acid formation. *Int. J. Chem. Kinet.* **2008**, *40*, 647–652.
20. Anschutz, P.; Sundby, B.; Lefrançois, L.; Luther, G. W., III; Mucci, A. Interactions between metal oxides and species of nitrogen and iodine in bioturbated marine sediments. *Geochim. Cosmochim. Acta* **2000**, *64*, 2751–2763.
21. Butler, A.; Walker, J. V. Marine Haloperoxidases. *Chem. Rev.* **1993**, *93*, 1937–1944.
22. Blasiak, L. C.; Drennan, C. L. Structure perspective on enzymatic halogenations. *Acc. Chem. Res.* **2009**, *42*, 147–155.
23. Küpper, F. C.; Carpenter, L. J.; McFiggans, G. B.; Palmer, C. J.; Waite, T. J.; Boneberg, E.-M.; Woitsch, S.; Weiller, M.; Abela, R.; Grolimund, D.; Potin, P.; Butler, A.; Luther, G. W., III; Kroneck, P. M. H.; Meyer-Klaucke, W.; Feiters, M. C. Iodide accumulation provides kelp with an inorganic antioxidant impacting atmospheric chemistry. *Proc. Natl. Acad. Sci. U.S.A.* **2008**, *105*, 6954–6958.
24. Garland, J. A.; Elzerman, A. W.; Penkett, S. A. The mechanism for dry deposition of ozone to seawater surfaces. *J. Geophys. Res.* **1980**, *85*, 7488–7492.
25. Carpenter, J. H. The Chesapeake Bay Institute technique for the Winkler dissolved oxygen method. *Limnol. Oceanogr.* **1965**, *10*, 141–143.
26. Fox, P. M.; Davis, J. A.; Luther, G. W., III. The kinetics of iodide oxidation by the manganese oxide mineral birnessite. *Geochim. Cosmochim. Acta* **2009**, *73*, 2850–2861.
27. Allard, S.; von Gunten, U.; Sahli, E.; Nicolau, R.; Gallard, H. Oxidation of iodide and iodine on birnessite (delta-MnO2) in the pH range 4-8. *Water Res.* **2009**, *43*, 3417–3426.
28. Buxton, G. V.; Greenstock, C. L.; Helman, W. P.; Ross, A. B. Critical review of rate constants for reactions of hydrated electrons, hydrogen atoms and hydroxyl radicals in aqueous solution. *J. Phys. Chem. Ref. Data* **1988**, *17*, 513–886.
29. Yu, X.-Y. Critical Evaluation of Rate Constants and Equilibrium Constants of Hydrogen Peroxide Photolysis in Acidic Aqueous Solutions Containing Chloride Ions. *J. Phys. Chem. Ref. Data* **2004**, *33*, 747–763.
30. Martino, M.; Mills, G. P.; Woeltjen, J.; Liss, P. S. A new source of volatile organoiodine compounds in surface seawater. *Geophys. Res. Lett.* **2009**, *36*, L01609, doi:10.1029/2008GL036334.
31. Hughes, C.; Malin, G.; Nightingale, P. D.; Liss, P. S. The effect of light stress on the release of volatile iodocarbons by three species of marine microalgae. *Limnol. Oceanogr.* **2006**, *51*, 2849–2854.
32. Manley, S. L.; de la Cuesta, J. L. Methyl iodide production from marine phytoplankton cultures. *Limnol. Oceanogr.* **1997**, *42*, 142–147.
33. Nightingale, P. D.; Malin, G.; Liss, P. S. Production of chloroform and other low-molecular-weight halocarbons by some species of macroalgae. *Limnol. Oceanogr.* **1995**, *40*, 680–689.
34. Whaley, L. K.; Furneaux, K. L.; Goddard, A.; Lee, J. D.; Mahajan, A.; Oetjen, H.; Read, K. A.; Kaaden, N.; Carpenter, L. J.; Lewis, A. C.; Plane, J.

M. C.; Saltzman, E. S.; Wiedensohler, A.; Heard, D. E. The chemistry of OH and HO_2 radicals in the boundary layer over the tropical Atlantic Ocean. *Atmos. Chem. Phys.* **2010**, *10*, 1555–1576.

35. Huheey, J. E.; Keiter, E. A.; Keiter, R. L. *Inorganic Chemistry*, 4th ed.; Harper Collins; New York, NY, 1993; p A-30.

Chapter 3

One-Electron Reduction Potentials from Chemical Structure Theory Calculations

Eric J. Bylaska,[1] Alexandra J. Salter-Blanc,[2] and Paul G. Tratnyek[2,*]

[1]Environmental Molecular Sciences Laboratory, Pacific Northwest National Laboratory, P.O. Box 999, Richland, WA 99352
[2]Division of Environmental and Biomolecular Systems, Oregon Health & Science University, 20000 NW Walker Road, Beaverton, OR 97006
[*]tratnyek@ebs.ogi.edu

Many redox reactions of importance in aquatic chemistry involve elementary steps that occur by single-electron transfer (SET). This step is often the first and rate limiting step in redox reactions of environmental contaminants, so there has been a great deal of interest in the corresponding one-electron reduction potentials (E^1). Although E^1 can be obtained by experimental methods, calculation from first-principles chemical structure theory is becoming an increasingly attractive alternative. Sufficient data are now available to perform a critical assessment of these methods—and their results—for two types of contaminant degradation reactions: dehalogenation of chlorinated aliphatic compounds (CACs) and reduction of nitro aromatic compounds (NACs). Early datasets containing E^1's for dehalogenation of CACs by dissociative SET contained a variety of errors and inconsistencies, but the preferred datasets show good agreement between values calculated from thermodynamic data and quantum mechanical models. All of the datasets with E^1's for reduction of NACs by SET are relatively new, were calculated with similar methods, and yet yield a variety of systematic differences. Further analysis of these differences is likely to yield computational methods for E^1's of NAC nitro reduction that are similar in reliability to those for CAC dechlorination. However, comparison of the E^1 data compiled here with those calculated with a more universal predictive model (like SPARC) highlight a number of challenges with implementation of models for predicting properties over a wide range of chemical structures.

Introduction

The characteristic reactions of most major redox active species in the aquatic environment involve even-numbered electron transfers between closed-shell electron donors and acceptors. The standard potentials for these reactions are readily available, often in critically evaluated reviews (*1–4*), usually calculated from standard thermochemical data, but occasionally measured directly by electrochemical or other methods. Meta-analyses of these standard potentials have been performed to investigate various fundamental aspects of environmental redox processes, such as the role of thermodynamically unstable intermediates as barriers to multi-electron redox process in biogeochemistry (*5*, *6*); the viability of various pathways of microbial metabolism (*7–9*); feasibility of contaminant degradation pathways (*10*, *11*); and prospects for extraterrestrial life (*12*).

However, the mechanisms of most redox reactions are composed of elementary steps that involve single-electron transfers (SET), usually forming radical intermediates, which react further by various mechanisms to form stable products. The initial SET step is often thermodynamically unfavorable, providing the barrier that inhibits equilibration and controls the kinetics of multistep redox processes that are favorable overall (*5*, *6*). For overall redox reactions that are far from equilibrium under environmental conditions, such as those that result in degradation of organic contaminants, the initial SET step is particularly important in controlling the kinetics of contaminant transformation. This is well established for the two categories of organic contaminants whose reduction has been most extensively studied: chlorinated aliphatic compounds (CACs) (e.g., (*13–17*)) and nitro aromatics compounds (NACs) (e.g., (*18–21*)). A similar generalization can be made about the importance of SET in determining the kinetics of oxidation of several categories of contaminants, including substituted phenols and anilines (e.g., (*22–24*)).

The quantitative property that is most often used to describe the rate-limiting SET steps in these contaminant redox reactions is the one-electron reduction potential (E^1) for the corresponding half-reaction (*24*, *25*). Experimental values of E^1 for these half-reactions are relatively scarce, but there are a number of methods by which they can be obtained (*26*, *27*). E^1 data can be obtained from voltammetry, but aprotic solvents are usually necessary to stabilize the radical intermediates (*3*, *26*, *28*, *29*), or from pulse radiolysis, but this method usually requires the use of an intervening mediator compound (*19*, *26*, *30*). These complications make theoretical calculations of E^1 especially important for studies of redox reactions of organic contaminants. There is now a sufficient number of these datasets—with a sufficient variety of contrasting and complementary characteristics—to perform a meta-analysis of the results and critical assessment of the methods. The scope here will be restricted to dechlorination of CACs and nitro reduction of NACs, but the approach and some of the general conclusions about methods and results should be applicable to reduction of quinones and possibly azo compounds, oxidation of phenols and anilines, etc.

Background

The most basic strategy for determining E^1's is to calculate the free energy difference for the reaction, e.g.

$$\text{R-Cl}_{(aq)} + 1e^-_{(g)} \rightarrow \text{R}^\bullet_{(aq)} + \text{Cl}^-_{(aq)} \qquad [1]$$

$$\text{R-NO}_{2(aq)} + 1e^-_{(g)} \rightarrow \text{R-NO}^{\bullet-}_{2(aq)} \qquad [2]$$

and then convert ΔG to a potential using

$$E_{abs} = -\frac{\Delta G_{rxn}}{nF} \qquad [3]$$

where n is moles of electrons transferred and F is the Faraday constant. This type of potential is known in electrochemistry as an absolute potential (E_{abs}), which is difficult to measure directly from experiment. Instead, redox potentials are usually measured relative to an inert electrode, such as the standard hydrogen electrode (SHE), saturated calomel electrode (SCE), or silver/silver-chloride electrode. For comparisons among studies, it is conventional to report redox potentials relative to the SHE, which is sometimes also called the normal hydrogen electrode (NHE). This means that the free energy differences for the one-electron reduction half reactions given in eqs. 1-2 are usually reported in terms of the following overall reactions:

$$\text{R-Cl}_{(aq)} + \tfrac{1}{2}\text{H}_{2(g)} \rightarrow \text{R}^\bullet_{(aq)} + \text{Cl}^-_{(aq)} + \text{H}^+_{(aq)} \qquad [4]$$

$$\text{R - NO}_{2(aq)} + \tfrac{1}{2}\text{H}_{2(g)} \rightarrow \text{R - NO}^{\bullet-}_{2(aq)} + \text{H}^+_{(aq)} \qquad [5]$$

Converting between E_{abs} and SHE potentials (E_{SHE} or E_h) is given by

$$E_{SHE} = E_{abs} + E^0_h \qquad [6]$$

where E^0_h is potential associated with the absolute free energy of the hydrogen electrode reaction.

$$\tfrac{1}{2}\text{H}_{2(g)} \xrightarrow{\Delta G_{rxn}} \text{H}^+_{(aq)} + e^-_{(g)} \qquad E^0_h = -\frac{\Delta G_{rxn}}{F} \qquad [7]$$

Modern electronic structure methods and solvation models can be used to estimate SHE potentials. However, the development of a computational scheme that can accurately predict E^1 requires some care. A commonly used strategy that does not use any experimental data is to directly calculate the absolute potentials by using electronic structure methods and solvation models to calculate the free energy difference for the one-electron transfer reactions and then convert it to an SHE potential using eq. 7. For reactions in which the reaction energy difference is just an adiabatic electron affinity (such as eq. 5), this approach can provide reasonable estimates (errors less than 0.2 V) provided the solvation models are designed to correctly treat the radical anion product. Unfortunately, for reactions in which the one-electron transfer reaction involves the making or breaking of covalent bonds (such as eq. 4), this strategy is prone to large errors (i.e., greater than 0.5 V). This is because electronic structure methods need to include high-level treatments of the electronic correlation energy—which are very expensive to compute—in order to directly calculate bond energies.

Another strategy uses isodesmic reactions—where the types of chemical bonds broken in the reactants are the same as those formed in the products (see below)—to estimate the standard state aqueous free energies of formation of species in the reaction. An advantage of this approach is that lower level electronic structure methods can be used, but this requires that accurate thermodynamic data are available for similar species from experiment or high-level electronic structure calculations.

In addition to the issues associated with electronic structure methods, our objective (the prediction of redox potentials in solution) also requires the calculation of solvation energies. Several models exist for calculating solvation energies. Currently, the most computationally feasible models for estimating solvation energies are "continuum" reaction field solvation models (*31*). Despite the approximate treatment of solvation in these approaches, they have been shown to give hydration energies of many neutral molecules within 1 kcal/mol (0.05 V) when compared to experiment results (for charged species, errors are typically larger, on the order of 0.1 – 0.2 V).

Methods

In this study, different strategies were used to estimate the solution phase E^1's for CACs and NACs. For NACs, the absolute potentials were directly calculated from gas-phase reaction energy, entropy, and solvation energy differences using electronic structure calculations, gas-phase entropy estimates, and continuum solvation models. The absolute potentials were then converted to E^1's using eq. 6. A more involved approach was used to calculate reduction potentials of CACs—because eq. 2 involves breaking a covalent bond—and is summarized below.

First the gas phase enthalpies of formation, $\Delta_f H°$ (298.15 K, 1 bar), of all the unknown neutral and radical species in eq. 1 (i.e., RCl and $R^•$) were calculated using electronic structure calculations combined with isodesmic reaction strategies, followed by the calculation of their gas-phase entropies. These results were used to obtain the gas-phase free energy of formation, $\Delta_f G°$ (298.15 K, 1 bar). Then, the solvation energies of the R and $R^•$ species in the reaction were calculated. These solvation energies were used to obtain the solution-phase free energies of formation, $\Delta_f G°_{aq}$ (298.15 K, 1 M). The desired results—reaction energies for the one-electron transfer reactions in both the gas phase and solution phase—could then be estimated, because the necessary thermodynamic quantities are known either from experiment or obtained from the calculations described below. Moreover, since the standard states used for the gas-phase and solution-phase free energies of formation are defined in such a way as to make $\Delta_f G°(H_{2(g)}) = \Delta_f G°(H^+_{aq}) = 0$, the SHE potential can be calculated as

$$E^1 = \frac{\left(\Delta_f G^o\left(R^{\bullet}_{aq}\right) + \Delta_f G^o\left(Cl^-_{aq}\right) - \Delta_f G^o\left(RCl_{aq}\right)\right)}{23.06\frac{\text{kcal/mol}}{\text{V}}} \quad [8]$$

if the values for $\Delta_f G°(RCl_{aq})$, $\Delta_f G°(R^{\bullet}_{aq})$, and $\Delta_f G°(Cl^-_{aq})$ are given in kcal/mol.

The strategies outlined in this section have been used by the authors to estimate the thermodynamic parameters $\Delta_f H°$, $S°$, G_s, and $\Delta_f G°_{aq}$ for various CACs, including substituted chlorinated methanes, substituted chloromethyl radical and anions, 4,4′-DDT and it metabolites, and polychloroethylenyl radicals, anions, and radical anion complexes. Details of the methods and results from that work are given in manuscripts by Bylaska et al. (*13*, *32–37*). For the calculations on CACs reported in Table 1, an isodesmic strategy, based on DFT calculations at the B3LYP/6-311++G(2d,2p) level with solvation by the COSMO model, was used to determine the gas-phase free energies and solvation energies. The NWChem program suite (*38*) was used to perform these calculations.

Using Isodesmic Reactions to Estimate $\Delta_f H°$ (298 K, 1 bar)

It is difficult to obtain accurate estimates for $\Delta_f H°$ using electronic structure methods. This is because determining $\Delta_f H°$ by directly calculating the atomization energies and correcting for elemental standard states only works when very large basis sets (such as the correlation-consistent basis sets), high-level treatments of electronic correlation energy, and small correction factors (such as core-valence correlation energies and relativistic effects) are included in the electronic structure calculations. Unfortunately, these methods are extremely demanding, scaling at least as N^7 for N basis functions. Continuing improvements in the algorithms used for these calculations have made them faster (*39*), but currently only ~25 first row atoms can be calculated on today's petascale computers. In contrast, electronic structure methods with lower-levels of treatment for electronic correlation (i.e., Hartree-Fock, Density Functional Theory, MP2) are much more computationally efficient (scaling as N^3-N^5) and can be used to calculate a wide range of molecules containing hundreds of atoms on modest computational resources. Unfortunately, low-level methods often have large errors when they are used to calculate atomization energies, and thus they cannot be used to directly calculate $\Delta_f H°$.

An approach that can be used with low-level electronic structure methods for estimating $\Delta_f H°$ is an isodesmic reaction scheme (e.g., (*32*, *36*)). This strategy is computationally tractable for large molecules and is usually accurate to within a few kcal/mol. Isodesmic reactions are (hypothetical) chemical reactions in which there are an equal number of like bonds (of each formal type) on the left and right sides of the reaction (*31*). An example of isodesmic reactions that have been used to estimate thermochemical properties of several chlorinated hydrocarbons and degradation reaction intermediates are

$$CH_xCl_{3-x}L + CH_4 \rightarrow CH_{x+1}Cl_{3-x} + CH_3L \qquad [9]$$

where $L^- = F^-$, OH^-, SH^-, NO_3^-, or HCO_3^-, and $x = 0$, 1, or 2 (*36*).

Isodesmic reactions are designed to separate out the interactions between the additive functional groups and non-bonding electrons from the direct bonding interactions by having the direct bonding interactions largely cancel one another. Most first-principle methods give substantial errors when estimating direct bonding interactions due to the computational difficulties associated with electron pair correlation, whereas first-principle methods are expected to be more accurate for estimating neighboring interactions and long-range through-bond effects. In order to illustrate this strategy, we estimate $\Delta_f H°$ for the radical $CCl_3^•$ using an isodesmic reaction scheme. First, the reaction enthalpy for the following isodesmic reaction,

$$CHCl_3 + CH_3^• \xrightarrow{\Delta H_{rxn}^{isodesmic}} CCl_3^• + CH_4 \qquad [10]$$

is calculated from the electronic, thermal, and vibrational energy differences at 298.15 K and a consistent level theory. Computed values for $\Delta H^{isodesmic}{}_{rxn}$ at various levels of electronic structure theory are 17.40, 14.28, 12.30, and 12.47 kcal/mol respectively for the SVWN5/6-311++G(2d,2p), PBE96/6-311++G(2d,2p), B3LYP/6-311++G(2d,2p), and PBE0/6-311++G(2d,2p) levels. Given that $\Delta_f H°$, of the other three species are known from experiment ($\Delta_f H°(CHCl_3) = -24.65$ kcal/mol, $\Delta_f H°(CH_3^•) = 34.82$ kcal/mol, and $\Delta_f H°(CH_4)$ $= -17.88$ kcal/mol. $\Delta_f H°$ (298 K, 1 bar) of the unknown $CCl_3^•$ compound was then calculated with Hess's law.

$$\Delta_f H^o\left(CCl_3^•\right) = \Delta_f H^o\left(CHCl_3\right) + \Delta_f H^o\left(CH_3^•\right) - \Delta_f H^o\left(CH_4\right) + \Delta H_{rxn}^{isodesmic} \qquad [11]$$

This method is simple to apply as long as the selected enthalpies of formation of $CHCl_3$, $CH_3^•$, and CH_4 are known either from experiment or high quality first-principle estimates. The success of the isodesmic strategy is controlled by several factors including the accuracy of $\Delta_f H°$ for the reference species, level of the ab initio theory, size of the basis set used to calculate the electronic energy difference, and accuracy of the molecular vibration corrections. One should also bear in mind that it is often possible to use several different isodesmic reactions to estimate the enthalpy of formation of the same species. These different isodesmic reactions will give different results (hopefully small), and there is no a priori way to know which one gives more accurate results.

Estimating $S°$ and $\Delta_f G°$

Given an optimized structure and vibrational frequencies for a gas-phase polyatomic molecule, one can calculate thermodynamic properties using formulas derived from statistical mechanics (*40*). In many cases, results obtained with these formulas, and accurate structures and frequencies, can provide more accurate values than those determined by direct thermal measurements. Calculation of these formulas is straightforward and most electronic structure programs contain options for calculating them. However, for compounds that contain internal degrees of freedom (e.g., molecules that contain fragments such as $-NO_2$ rotors) that are not well described by normal vibrations, these formulas need to be corrected. Estimating accurate entropies in this situation is very demanding, and several strategies for this have been developed (*31*, *37*). The strategy that has been used by the authors is to explicitly solve for the energy levels of the rotational Schrödinger equation for each rotor and then use this as input in a canonical partition function to estimate its entropy (*35*, *37*).

The gas-phase free energy of formation $\Delta_f G°$ can be determined using the values of $\Delta_f H°$ and $S°$. This is done by calculating the entropy of formation, $\Delta_f S°$, found by subtracting the entropies of the atomic standard states from the virtual entropy of the compounds. For example, $\Delta_f S°(CCl_3^\bullet)$ and $\Delta_f G°(CCl_3^\bullet)$ are calculated from the following expressions:

$$\Delta_f S^o\left(CCl_3^\bullet\right) = S^o\left(CCl_3^\bullet\right) - \left[1S^o\left(\text{C-graphite}\right) + 3S^o\left(\tfrac{1}{2}Cl_2\right)\right] \quad [12]$$

$$\Delta_f G^o\left(CCl_3^\bullet\right) = \Delta_f H^o\left(CCl_3^\bullet\right) - T\Delta_f S^o\left(CCl_3^\bullet\right) \quad [13]$$

Estimating Solvation Energies

Solvent effects can be estimated using the self-consistent reaction field (SCRF) theories of Tomasi et al. (PCM) (*41–44*), Klampt and Schüürmann (COSMO and COSMO-RS) (*45*, *46*), Cramer SMx (*47*), or APBS (*48*). SCRF theory can be combined with a variety of ab initio electronic structure calculations, including DFT with the LDA, BP91, and B3LYP functionals, and MP2. Despite the approximate treatment of solvation in this approach, it (and others like it) has been shown to give hydration energies for many neutral molecules that are within a few kcal/mol of the experimental values.

In SCRF theory, the solvation energies for rigid solutes that do not react strongly with water are approximated as a sum of non-covalent electrostatic, cavitation, and dispersion energies. Several approaches have been proposed to calculate these contributions. For the electrostatic energy, the solvent is represented by an infinite homogeneous continuous medium having a dielectric constant (e.g., 78.3 for water), and the solute is represented by an empty cavity, inside which the solute's electrostatic charge distribution is placed. This approach self-consistently minimizes the electrostatic energy by optimizing the polarization of the continuous medium and charge distribution of the solute.

The cavitation and dispersion contributions to the solvation energy are less straightforward to handle because the interactions take place at short distances. There are several proposed ways to do this (*42*, *44*, *49–54*). One of the simplest approaches for estimating these terms is to use empirically derived expressions that depend only on the solvent accessible surface area. A widely used parameterization of this type has been given by Sitkoff et al. (*52*).

$$\Delta G_{\text{cav+disp}} = \gamma A + b \qquad [14]$$

where A is the solvent accessible surface area and γ and b are constants set to 5 cal/mol-Å^2 and 0.86 kcal/mol respectively. Sitkoff et al. fit the constants γ and b to the experimentally determined free energies of solvation of alkanes by using a least-squares fit (*55*). A shortcoming of this model is that it is not size extensive and cannot be used to study dissociative processes. Another accepted parameterization, which is size extensive, has been suggested by Honig et al. (*51*).

$$\Delta G_{\text{cav+disp}} = \gamma A \qquad [15]$$

where γ is a constant set to 25 cal/mol-Å^2. The solvent accessible surface area in eqs. 14 and 15 are defined by using a solvent probe with a radius of 1.4 Å rolled over the solute surface defined by van der Waal radii (i.e., H = 1.2 Å, C = 1.5 Å, O = 1.4 Å, Cl = 1.8 Å).

Adjustment to Environmentally Relevant Standard States

Calculated SCRF free energies of solvation cannot be compared directly to thermodynamic tables, because they use different standard states. The standard states for the SCRF model are 1 M at 298.15 K in the gas phase and 1 M at 298.15 K in the solution phase. Thermodynamics tables on the other hand define the standard state for the solute as 1 bar of pressure at 298.15 K in the gas phase and 1 M and 298.15 K in the solution phase. In order for the SCRF theory calculations to conform to the standard state of 1 bar of pressure at 298.15 K in the gas phase, a constant value of 1.90 kcal/mol must be added to the SCRF free energies of solvation.

For charged solutes, comparisons are less straightforward. Thermodynamic tables report free energies of formation for charged solutes or electrolytes in solution relative to H^+_{aq}, with the convention

$$\Delta_f H^o\left(\mathrm{H}^+_{(aq)}\right) = \Delta_f G^o\left(\mathrm{H}^+_{(aq)}\right) = S^o\left(\mathrm{H}^+_{(aq)}\right) = C^0_p\left(\mathrm{H}^+_{(aq)}\right) = 0 \qquad [16]$$

and the free energy of formation of ions in solutions are defined in terms of following redox reactions

$$\mathrm{M} + n\mathrm{H}^+_{(aq)} \rightarrow \mathrm{M}^{n+}_{(aq)} + \tfrac{n}{2}\mathrm{H}_{2(g)} \qquad [17]$$

$$\mathrm{X} + \tfrac{n}{2}\mathrm{H}_{2(g)} \rightarrow \mathrm{X}^{n-}_{(aq)} + n\mathrm{H}^+_{(aq)} \qquad [18]$$

This means that the absolute solvation free energy of a charged solute cannot be calculated using only thermodynamic tables. However, if the absolute free energy of the hydrogen electrode reaction

$$\tfrac{1}{2}\mathrm{H}_{2(g)} \rightarrow \mathrm{H}^{+}_{(aq)} + e^{-}_{(g)} \qquad \Delta G_{\mathrm{rxn}} = -E^{0}_{H} \qquad [19]$$

is known, then the solvation energy of a charged solute at 298.15 K can be found by subtracting the absolute free energy of the hydrogen electrode process, i.e.

$$\begin{aligned}\Delta G_{s}\left(\mathrm{X}^{-}\right) &= \Delta_{f}G^{o}\left(\mathrm{X}^{-}_{(aq)}\right) - \Delta_{f}G^{o}\left(\mathrm{X}^{-}_{(g)}\right) \\ &+\left\{-E^{0}_{H} - \left(\Delta_{f}G^{o}\left(\mathrm{H}^{+}_{(aq)}\right) + \Delta_{f}G^{o}\left(e^{-}_{(g)}\right) - \tfrac{1}{2}\Delta_{f}G^{o}\left(\mathrm{H}_{2(g)}\right)\right)\right\} \qquad [20] \\ &= \Delta_{f}G^{o}\left(\mathrm{X}^{-}_{(aq)}\right) - \Delta_{f}G^{o}\left(\mathrm{X}^{-}_{(g)}\right) - E^{0}_{H}\end{aligned}$$

Similarly, SCRF calculated solvation energies, $\Delta G_{\mathrm{SCRF}}(\mathrm{X}^{-})$, can be used to calculate the aqueous free energies of formation.

$$\begin{aligned}\Delta_{f}G^{o}\left(\mathrm{X}^{-}_{(aq)}\right) &= \Delta G_{SCRF}\left(\mathrm{X}^{-}\right) + \Delta_{f}G^{o}\left(\mathrm{X}^{-}_{(g)}\right) \\ &-\left\{-E^{0}_{H} - \left(\Delta_{f}G^{o}\left(\mathrm{H}^{+}_{(aq)}\right) + \Delta_{f}G^{o}\left(e^{-}_{(g)}\right) - \tfrac{1}{2}\Delta_{f}G^{o}\left(\mathrm{H}_{2(g)}\right)\right)\right\} \qquad [21] \\ &= \Delta G_{SCRF}\left(\mathrm{X}^{-}\right) + \Delta_{f}G^{o}\left(\mathrm{X}^{-}_{(g)}\right) + E^{0}_{H}\end{aligned}$$

The exact value for E°_{h} is still uncertain despite extensive experimental and computational efforts. However, Tissandier et al. (*56*) have recently reported a value of $\Delta G_{s}(\mathrm{H}^{+}) = -263.98$ kcal/mol, which can be used to approximate E°_{h}.

$$\begin{aligned}E^{0}_{h} &= \Delta G_{s}\left(\mathrm{H}^{+}\right) + \left(\Delta_{f}G^{o}\left(\mathrm{H}^{+}_{(g)}\right) - \Delta_{f}G^{o}\left(\mathrm{H}_{2(g)}\right)\right) \\ &= -263.98\ \text{kcal/mol} + 362.58\ \text{kcal/mol} \qquad [22] \\ &= 98.6\ \text{kcal/mol}\end{aligned}$$

Results

To demonstrate the application of the methods described above, we have calculated new values of E^{1} and have compared them to previously reported calculated and experimentally measured values below. For this purpose, we have limited our scope to the two most important, and extensively studied classes of organic contaminant degradation reactions: dechlorination and nitro reduction. Within these two groups, the results are arranged to reflect a combination of chronological and methodological considerations. Correlation analysis among the datasets for E^{1} is used to assess their consistency and accuracy. Other correlations—e.g., between E^{1} and vertical electron affinities (VEA) or energies of the highest occupied molecular orbitals (E_{LUMO})—would also be informative, but are not presented here.

Dechlorination

Dechlorination of CACs in the aquatic environment occurs by several pathways including hydrolysis, non-reductive elimination (dehydrohalogenation), reductive elimination (dihaloelimination), and hydrogenolysis (reductive dehalogenation). The latter two reductive pathways can occur by mechanisms that involve one-electron transfer, two-electron transfer, or atom transfer, but all these reduction mechanisms can—at least formally—be broken down into steps that involve SET, so E^1 for each SET can be useful for evaluating hypotheses about these mechanisms. Additionally, SET is known to be concerted with dissociation of the C-Cl bond in hydrogenolysis of many CACs (*13*, *57*, *58*), and these can be described by E^1 for the corresponding half-reaction (i.e., eqs. 1 and 4).

There are several relatively comprehensive sets of E^1 data for dissociative SET involving environmentally relevant CACs that ultimately are derived from experimental data. The first is by Curtis (*59*), who calculated $E^1{}_h$ (E^1's at standard conditions) for most of the chlorinated (and some brominated) methanes and ethanes using thermodynamic cycles and experimental gas-phase free-energy data from earlier literature. Next, the same approach was taken by Roberts et al. (*60*) to obtain a set of $E^{1\prime}{}_h$ (1-electron potentials at standard environmental conditions) for chlorinated ethylenes. These two datasets were combined in Scherer et al. (*61*) for use in correlation analysis with experimental rate constants for dechlorination of CACs. Later, Totten and Roberts (*62*) published a new set of $E^{1\prime}{}_h$ values for dechlorination of CACs (re)calculated using the same method as previous work (thermodynamic cycles using experimental gas-phase thermodynamic data from the literature), but with a much more thorough and critical analysis of the input data. Several types of errors in the E^1's calculated by Curtis (*59*) were noted and corrected. The combined E^1 dataset used by Scherer et al. (*61*) and the newer dataset from Totten and Roberts (*62*) are shown in Table 1 and are included in our correlation analysis of these results below.

Over the last decade, advances in computational chemistry have made it feasible to calculate accurate values of E^1 for CACs entirely from chemical structure theory (*37*, *63*). Cwiertny et al. (*64*) reported such a dataset for halogenated methanes and ethanes, obtained using a combination of G3MP2 methods for gas phase free energies and solvation with the SM5.42 model. We have previously reported thermodynamic data for one-electron reduction of chlorinated methanes (*36*) and ethenes (*13*, *34*) and one propane (*32*). We reported these only as standard state free energies ($\Delta G^{\circ}{}_{rxn}$), but here we have adjusted our previously reported results to $E^{1\prime}{}_h$ and added the results of calculations done using the same methods for other CACs. The result is the first dataset of E^1's for all the chlorinated methanes, ethanes, ethenes, and propanes, at standard conditions for aquatic chemistry, obtained by consistent methods that should be of high accuracy. Both theoretically-derived datasets—from Cwiertny et al. (*64*) and the new one reported here—are given in Table 1.

Table 1. One-Electron Reduction Potentials for Hydrogenolysis of Chlorinated Aliphatics

Name (Abbreviation)	Product	Scherer et al. [a] $E^1{}_h{}^b$ or $E^{1'}{}_h$ [c] (V)	Totten et al. [d] $E^{1'}{}_h$ (V)	Cwiertny et al. [e] $E^{1'}{}_h$ (V)	Bylaska et al. [f] $E^{1'}{}_h$ (V)
Tetrachloromethane (PCM)	$Cl_3C\bullet$	0.13 [b] (0.31[illegible]	0.085	0.182	-0.05
Trichloromethane (TCM)	$HCl_2C\bullet$	-0.23 [b] (-0.05[illegible])	-0.145	-0.089	-0.314
Dichloromethane (DCM)	$H_2ClC\bullet$	-0.50 [b] (-0.32)	-0.428	-0.323	-0.404
Chloromethane (CM)	$H_3C\bullet$	-0.88 [b] (-0.7[illegible])	-0.725		-0.605
Hexachloroethane (HCA)	$Cl_3CCl_2C\bullet$	0.33 [b] (0.51)	0.061	0.144	0.265
Pentachloroethane (PCA)	$Cl_3CClHC\bullet$	0.00 [b] (0.18)	-0.132	-0.132	-0.161
Pentachloroethane (PCA)	$Cl_2HCCl_2C\bullet$	0.31 [b] (0.49)	0.139	0.092	0.079
1,1,1,2-Tetrachloroethane (1112TeCA)	$Cl_3CH_2C\bullet$	-0.52 [b] (-0.34)	-0.545	-0.413	-0.415
1,1,1,2-Tetrachloroethane (1112TeCA)	$ClH_2CCl_2C\bullet$	-0.22 [b] (-0.043)	0.044	0.046 [g]	0.046
1,1,2,2-Tetrachloroethane (1122TeCA)	$Cl_2HCClHC\bullet$	-0.34 [b] (-0.16)	-0.257	-0.132 [g]	-0.187
1,1,1-Trichloroethane (111TCA)	$H_3CCl_2C\bullet$	-0.23 [b] (-0.053)	-0.020	-0.046	-0.178
1,1,2-Trichloroethane (112TCA)	$Cl_2HCH_2C\bullet$	-0.74 [b] (-0.56)	-0.589		-0.497
1,1,2-Trichloroethane (112TCA)	$ClH_2CClHC\bullet$	-0.42 [b] (-0.24)	-0.285	-0.213 [g]	-0.264
1,1-Dichloroethane (11DCA)	$H_3CClHC\bullet$	-0.40 [b] (-0.22)	-0.316	-0.320	-0.439
1,2-Dichloroethane (12DCA)	$ClH_2CH_2C\bullet$	-0.57 [b] (-0.39)	-0.577	-0.426 [g]	-0.513
Chloroethane (CA)	$H_3CH_2C\bullet$	-0.83 [b] (-0.65)	-0.690		-0.646

Continued on next page.

Table 1. (Continued). One-Electron Reduction Potentials for Hydrogenolysis of Chlorinated Aliphatics

Name (Abbreviation)	*Product*	*Scherer et al.* [a] $E^{1}{}_{h}{}^{b}$ *or* $E^{1\prime}{}_{h}$ [c] *(V)*	*Totten et al.* [d] $E^{1\prime}{}_{h}$ *(V)*	*Cwiertny et al.* [e] $E^{1\prime}{}_{h}$ *(V)*	*Bylaska et al.* [f] $E^{1\prime}{}_{h}$ *(V)*
Tetrachloroethene (PCE)	$Cl_2CClC\bullet$	-0.36 [c]	-0.598		-0.56
Trichloroethene (TCE)	$HClCClC\bullet$		-0.674		-0.838
Trichloroethene (TCE)	$Cl_2CHC\bullet$	-0.91 [c]	-0.998		-0.998
Trichloroethene (TCE)	$ClHCClC\bullet$	-0.62 [c]	-0.693		-0.803
1,1-Dichloroethene (11DCE)	$H_2C{=}ClC\bullet$	-0.72 [c]	-0.802		-0.816
c-1,2-Dichloroethene (c12DCE)	$HClC{=}HC\bullet$	-0.89 [c]	-1.012		-0.959
t-1,2-Dichloroethene (t12DCE)	$HClC{=}HC\bullet$	-0.85 [c]	-0.955		-0.946
Chloroethene (VC)	$H_2C{=}HC\bullet$	-0.95 [c]	-1.141		-1.085
1,1,1-Trichloropropane (111TCP)	$Cl_2C\bullet$-CH_2-CH_3				-0.192
1,1,2-Trichloropropane (112TCP)	Cl_2HC-$CH\bullet$-CH_3				-0.365
1,1,2-Trichloropropane (112TCP)	$ClHC\bullet$-$CHCl$-CH_3				-0.009
1,1,3-Trichloropropane (113TCP)	Cl_2HC-CH_2-$CH_2\bullet$				-0.608
1,1,3-Trichloropropane (113TCP)	$ClHC\bullet$-CH_2-CH_2Cl				-0.296
1,2,2-Trichloropropane (122TCP)	ClH_2C-$CCl\bullet$-CH_3				-0.157
1,2,2-Trichloropropane (122TCP)	$H_2C\bullet$-CCl_2-CH_3				-0.382
1,2,3-Trichloropropane (TCP)	ClH_2C-$CHCl$-$CH_2\bullet$				-0.491

Name (Abbreviation)	*Product*	*Scherer et al.* [c] $E^1{}_h$[b] *or* $E^{1\prime}{}_h$[c] *(V)*	*Totten et al.* [d] $E^{1\prime}{}_h$ *(V)*	*Cwiertny et al.* [e] $E^{1\prime}{}_h$ *(V)*	*Bylaska et al.* [f] $E^{1\prime}{}_h$ *(V)*
1,2,3-Trichloropropane (TCP)	ClH_2C-CH•-CH_2Cl				-0.443
1,1-Dichloropropane (11DCP)	ClHC•-CH_2-CH_3				-0.439
1,2-Dichloropropane (12DCP)	ClH_2C-CH•-CH_3				-0.595
1,2-Dichloropropane (12DCP)	H_2C•-CHCl-CH_3				-0.63
1,3-Dichloropropane (13DCP)	H_2C•-CH_2-CH_2Cl				-0.781
2,2-Dichloropropane (22DCP)	H_3C-CCl•-CH_3				-0.634
1-Chloropropane (1CP)	H_2C•-CH_2-CH_3				-0.66
2-Chloropropane (2CP)	H_3C-CH•-CH_3				-0.699

[a] Dataset compiled in (*61*) from [b] $E^1{}_h$ data in (*59*) and [c] $E^{1\prime}{}_h$ data in (*60*); calculated $E^{1\prime}{}_h$ values from the original data in (*59*) shown in parenthesis. [d] From (*62*) for $[Cl^-] = 10^{-3}$ M. [e] From (*64*) for $[Cl^-] = 10^{-3}$ M. [f] Calculated from free energies reported in previous work by Bylaska et al., for chlorinated methanes (*36*), ethenes (*13*, *34*), and propanes (*32*). [g] Boltzmann average energy for all possible conformers (i.e., *syn* and *anti*).

To evaluate the consistency and accuracy of the E^1 datasets in Table 1, we first consider the correlation between the two datasets that were derived from experimental data. Assuming the dataset from Totten and Roberts (*62*) is more reliable overall, we have assigned it to the abscissa in Fig. 1. The original dataset compiled by Scherer et al. (*61*) is shown as the ordinate in Fig. 1A. Overall, the correlation appears to be rather good, with uniform scatter about the (ordinary least squares) regression line, slope and intercept (0.94±0.09 and -0.03±0.05) that are not significantly different than for a perfect 1:1 correlation, no obvious outliers, and a root mean squared error (RMSE) of 0.15.

However, the apparent agreement between these two datasets is misleading because Scherer et al. (*61*) mixed E^1's from Curtis (*59*), which were calculated for all species at unit activity (denoted $E^1{}_h$ in Table 1), with E^1's from Roberts et al. (*60*), which were corrected for standard environmental conditions, in this case, $[Cl^-] = 10^{-3}$ M (denoted $E^{1\prime}{}_h$ in Table 1). Adjusting the former to $E^{1\prime}{}_h$ is straight-forward (*62*) and the resulting values are given in parenthesis in Table 1. Correlation of this adjusted version of the Scherer et al. dataset with the Totten and Roberts dataset is shown in Fig. 1B. The correlation still appears linear, and has slightly less overall scatter (RMSE = 0.14), but the slope is no longer 1:1 (the fitted slope and intercept are 1.10±0.08 and 0.16±0.05) because the adjusted data (open squares) now plot higher relative to the data from Roberts et al. (crossed squares).

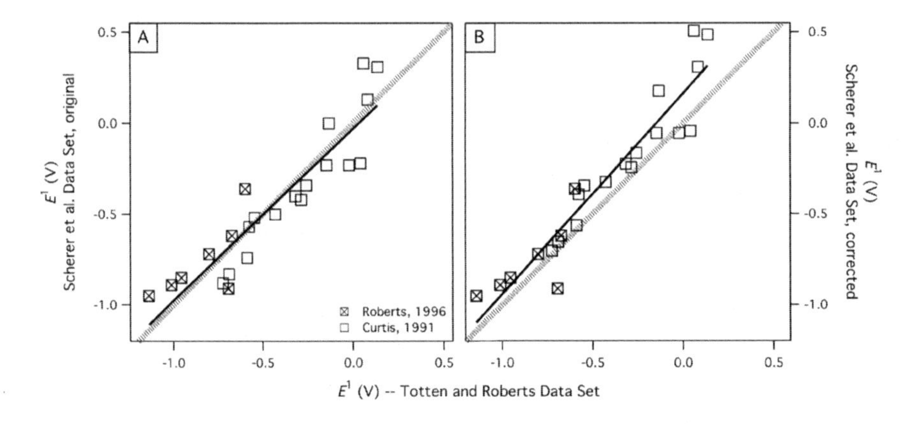

Figure 1. Comparison of E^1 datasets for dechlorination calculated from experimental thermodynamic data. Black lines are from ordinary least-squares regression; gray line is 1:1. The two types of symbols distinguish the source of data that were combined in Scherer et al. (61).

In addition to the adjustment described above, it should be noted that Totten and Roberts found errors in Curtis' calculations that resulted from using the gas phase standard entropy of the chlorine atom where it should have been chlorine gas (*62*). They determined that this error affected both the absolute magnitude of the E^1's as well as the relative magnitude, depending on the number of chlorines on the CAC (as seen in Fig. 1B). Coincidentally, these two sources of error (the incorrect standard entropy for chlorine and the lack of adjustment for chloride concentrations under aquatic conditions) largely obscure one another (c.f., Fig. 1B) so correcting for both does not suggest any substantive changes in correlation analysis of the kinetic data reported by Scherer et al. (*61*) (plots not shown).

The two new datasets of E^1's calculated from theory are compared to each other, and to the preferred experimentally-derived dataset, in Fig. 2. Data for structural classes common to each dataset (methanes and ethanes) have been fit using ordinary least-squares regression. Additional ethene data, when available, is shown in gray for comparison. Each of the three datasets considered was obtained by consistent methods and is reported as $E^{1'}_h$, so it is not surprising that they generally correlate well with each other. The correlation between the experimentally-derived data from Totten and Roberts (*62*) and the theory-derived dataset reported here is particularly good (Fig. 2B), with uniform scatter about the regression line, which has slope and intercept indistinguishable from 1 and 0, over the whole range of chlorinated methanes and ethanes (shown), as well as ethenes (not shown). (Note that two of the datasets discussed here —Totten and Roberts (*62*) and Cwiertny (*64*)—include brominated methanes and ethanes, but these were not considered in our correlation analysis. Also, the new dataset reported here is the only one to include chlorinated propanes, so the propanes do not appear in Fig. 2.)

The dataset from Cwiertny et al. (*64*) correlates very closely (RMSE = 0.05) with the experimentally-derived values from Totten and Roberts (*62*). The Bylaska dataset reported here correlates somewhat less well (RMSE = 0.1) mainly due to slightly high E^1's for tetra- and tri-chloromethane.

One peculiar characteristic of the data shown in Fig. 2 is evident only by comparison of the triangular-halves of the correlation matrix. Both correlations with Cwiertny's dataset on the abscissa (Fig. 2 A, F) have slopes that are not significantly different than one, but when Cwiertny's dataset is used as the ordinate (Fig. 2 C, D) the slopes are significantly different than one. Similarly, while a slope of one is observed when the Totten and Roberts dataset is plotted on the ordinate with the Bylaska dataset on the abscissa (Fig. 2 B), the slope deviates from one when the coordinates are reversed (Fig. 2 E). In each case, the orientation of the plot results in opposite conclusions due to the uneven distribution within each dataset. The use of orthogonal regression minimizes these effects. In this case, orthogonal regression resulted in only an ~0.02 difference in the deviation from a slope one between corresponding plots (results not shown). Without using orthogonal regression, considering only one half of the correlation matrix could have resulted in attributing physical significance to the differences in slope.

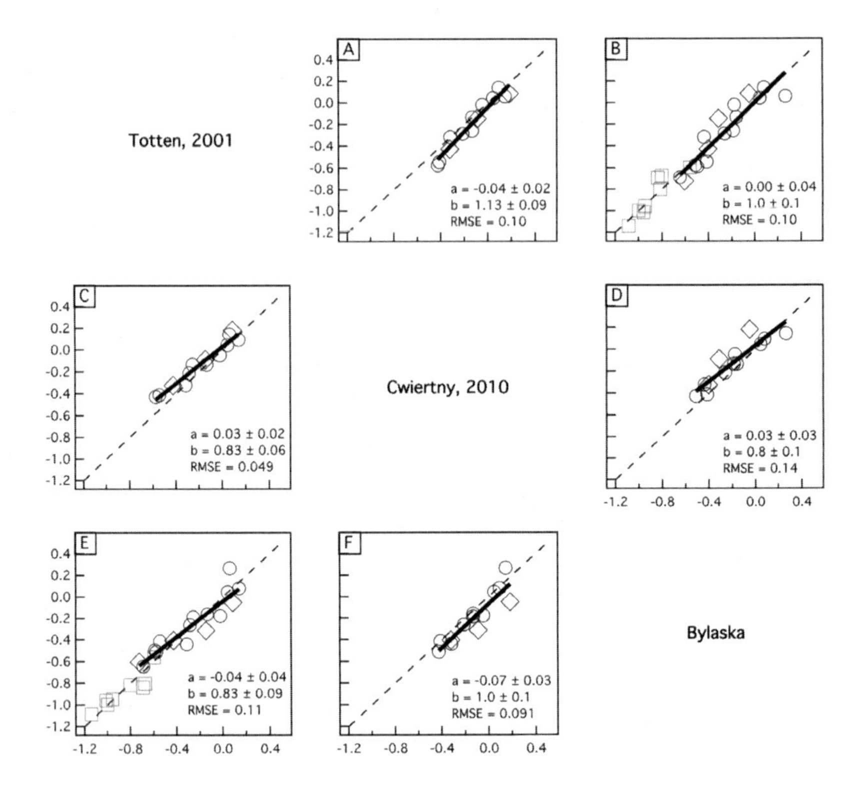

Figure 2. Scatter plot matrix of correlations between the three recommended E^1 datasets for dechlorination. Regression lines and fitting results are for ordinary least-squares regression ($y = a + bx$); dashed line is 1:1. Black diamond symbols = methanes; black circles = ethanes; and gray squares = ethenes (not included in regression).

It can also be noted that while the ethene data were not included in the regression analysis in order to maintain consistency throughout the correlation analysis, inclusion either has no effect or improves the affected comparisons. In both Totten and Roberts vs. Bylaska (Fig. 2 B) and Bylaska vs. Totten and Roberts (Fig. 2 E) the ethene data fall close to the one-to-one line and show little scatter. Inclusion of the ethene data in the regression analysis of Totten and Roberts vs. Bylaska (not shown) has no significant effect on either the slope, intercept, or RMSE. Inclusion of ethene data in the Bylaska vs. Totten and Roberts regression (not shown) improves the 1:1 correlation with a slope of 0.94 vs. 0.83 without ethenes; there is no significant change in the intercept or RMSE.

Nitro Reduction

Nitro reduction is usually interpreted as a series of SET and protonation steps, the exact order of which may vary with pH and the strength of the reductant (*65*, *66*). Under environmental conditions, the rate determining step is generally believed to be the first SET (*20*, *67*), which makes E^1's for the corresponding electron attachment (i.e., eqs. 2 and 5) a useful descriptor of the reaction. Alternatively, nitro reduction via $H^{\bullet}$ (as in catalytic hydrogenation) or H^{-} (as in some enzyme catalyzed reductions) is possible under environmental conditions and is implied by some recent data on nitro reduction by the cast iron used for groundwater remediation (*68*). However, as with dechlorination, these mechanisms involving concerted electron and atom transfers can be formulated as separate steps, so E^1 for the isolated SET still is useful in studies of these reactions.

The earliest dataset of E^1's for NACs of environmental interest was compiled by Schwarzenbach et al. (*18*). This dataset is comprised mostly of previously reported E^1's obtained from pulse radiolysis experiments (*69–72*). Also included were previously unreported values estimated using linear free energy relationships (LFERs) between E^1's and rate constants for nitro reduction measured in simple model systems. Hoffstetter et al. (*73*) later expanded this dataset, adding more values estimated from LFERs and additional values from the literature (*26*). We have used Hoffstetter's dataset to encompass all of this early work, and summarized it in Table 2.

Recently, Phillips et al. (*74*) compiled a set of purely measured values to validate their theoretical calculations (which are discussed below). Phillips et al. drew from the Schwarzenbach et al. dataset, and added additional data from Riefler and Smets (*75*) and others (*76–78*). For the analysis below, we have assumed that this is the most complete and critically evaluated, compiled set of E^1 data for reduction of NACs, and we have included it in Table 2.

As with dechlorination, it has recently become practical to calculate E^1's for nitro reduction from chemical structure theory. The first calculated E^1's for environmentally-significant NACs appear to have been reported by Zubatuk et al. (*79*), who calculated E^1's using density functional theory (DFT) and a variety of basis sets. Based on correlation with an experimental dataset, they concluded that the most accurate calculation of E^1 was obtained using the MPWB1K density functional and TZVP basis set with the PCM method for solvation. We have included these theoretically-derived E^1's in Table 2. However, Phillips et al. (*74*) noted two issues with this dataset. The first is the use of an out-of-date SHE potential, leading to an error of 0.08 V; the second is an apparent adjustment in the experimental dataset that was used for validation, the rationale for which was not reported.

Three approaches to calculating E^1's for explosives-related NACs were evaluated by Phillips et al. (*74*): (*i*) direct aqueous-phase DFT calculations, (*ii*) gas-phase DFT calculations with solvation corrections, and (*iii*) empirical correlation with electron affinity (EA) calculated using DFT methods. Neither purely theoretical approach was found to accurately predict E^1, and in the case of their second method, the source of the error was identified to be a lack of accuracy in the solvation calculation (errors using direct aqueous-phase calculations—the first method—were non-systematic). Their third approach—a hybrid of experimental and theoretical results, relying only upon gas-phase theoretical calculations—gave the best overall agreement to experimental data (dataset introduced above). This experimental dataset and the dataset obtained through correlation with EA are included in Table 2.

Table 2. One-Electron Reduction Potentials for Nitro Reduction

Name (Abbreviation)	*Hoffstetter et al.* [a] *$E^{1'}_h$ (V)*	*Phillips et al.* [b] *measured $E^{1'}_h$ (V)*	*Phillips et al.* [c] *corr. w/ EA $E^{1'}_h$ (V)*	*Zubatyuk et al.* [d] *E^{1}_h (V)*	*Bylaska et al. $E^{1'}_h$ (V)*
Nitrobenzene (NB)	-0.485	-0.486	-0.499	-0.88	-0.494
2-Methylnitrobenzene (2-CH_3-NB)	-0.590	-0.590	-0.519		-0.612
3- Methylnitrobenzene (3-CH_3-NB)	-0.475	-0.475	-0.507		-0.514
4- Methylnitrobenzene (4-CH_3-NB)	-0.500	-0.500	-0.516		-0.538
2-Chloronitrobenzene (2-Cl-NB)	-0.485		-0.470		-0.467
3-Chloronitrobenzene (3-Cl-NB)	-0.405	-0.405	-0.437		-0.391
4-Chloronitrobenzene (4-Cl-NB)	-0.450	-0.450	-0.447		-0.417
2-Acetylnitrobenzene (2-$COCH_3$-NB)	-0.470		-0.412		-0.445
3-Acetylnitrobenzene (3-$COCH_3$-NB)	-0.405	-0.437	-0.423		-0.406
4-Acetylnitrobenzene (4-$COCH_3$-NB)	-0.358	-0.356	-0.359		-0.255
2-4-6-Trinitrotoluene (TNT)	-0.300	-0.253	-0.245	-0.45	
2-Amino-4,6-dinitrotoluene (2-ADNT)	-0.390	-0.417	-0.386		
4-Amino-2,6-dinitrotoluene (4-ADNT)	-0.430	-0.449	-0.393	-0.85	
2,4-Diamino-6-nitrotoluene (2,4-DANT)	-0.515	-0.502	-0.557	-1.08	
2,6-Diamino-4-nitrotoluene (2,6-DANT)	-0.495				
2,4-Dinitrotoluene (2,4-DNT)		-0.397	-0.361		

Name (Abbreviation)	*Hoffstetter et al.* [a] $E^{1'}_h$ *(V)*	*Phillips et al.* [b] *measured* $E^{1'}_h$ *(V)*	*Phillips et al.* [c] *corr. w/ EA* $E^{1'}_h$ *(V)*	*Zubatyuk et al.* [d] E^{1}_h *(V)*	*Bylaska et al.* $E^{1'}_h$ *(V)*
2,6-Dinitrotoluene (2,6-DNT)		-0.402	-0.377		
2-Nitroanaline (2-NH_2-NB)	< -0.560		-0.533		
3-Nitroanaline (3-NH_2-NB)	-0.500		-0.520		
4-Nitroanaline (4-NH_2-NB)		-0.568	-0.569		
1,2-Dinitrobenzene (1,2-DNB)		-0.287	-0.335	-0.50	
1,3-Dinitrobenzene (1,3-DNB)		-0.345	-0.330	-0.68	
1,4-Dinitrobenzene (1,4-DNB)		-0.257	-0.248	-0.43	
2-Nitrobenzaldehyde (2-CHO)		-0.355	-0.365		
4-Nitrobenzaldehyde (4-CHO)		-0.322	-0.330	-0.60	
4-Nitrobenzyl alcohol (4-CH_2OH)		-0.478	-0.461	-0.90	

[a] As reported and compiled in (*73*). [b] Measured values compiled by and reported in the supporting information of (*74*). [c] As reported in (*74*)—calculated from correlation with EA. [d] As reported in (*79*).

We also include in Table 2 the results of preliminary calculations we performed for 10 NACs. For these calculations, the isodesmic strategy that was used with CACs was not employed. Instead, the E^1's were determined by directly taking free energy differences of eq. 2 at the B3LYP/6-311++G(2d,2p) level with solvation by the COSMO model and the effects of hindered rotors included in the calculation of gas-phase free energies and solvation energies. This strategy was similar to the one used by Zubatuk et al. (*79*)—and recently critiqued by Phillips et al. (*74*)—except that we included the effects of hindered rotors in our estimates. As pointed out in the background section, we expected that the main sources of error are in the calculations of the adiabatic electron affinity and the solvation energy of the radical anions. There errors are expected to be on the order of ~2 kcal/mol (0.1 V) and ~5 kcal/mol (0.2 V), respectively.

To evaluate the consistency and accuracy of the E^1 datasets in Table 2, we have correlated each combination of the five datasets in the matrix of scatter plots shown in Fig. 3. The two experimentally derived datasets—from Hofstetter et al. (*73*) and Phillips et al. (*74*)—include some identical data, so they correlate very closely (Fig. 3 A, E). Although the dataset represented by Hofstetter et al. (*73*) has been used to derive LFERs for nitro reduction rate constants in numerous studies (*25*, *80*, *81*), we will use the newer dataset compiled by Phillips et al. (*74*) for experimental data to compare with the theoretically-derived E^1's discussed below.

The correlation that lead Phillips et al. to select their third method of calculating E^1's as preferred is shown in Fig. 3 F and J. The slope and intercept of the corresponding (ordinary least-squares) regression lines for the calculated numbers vs. the measured numbers (Fig. 3 J) are not significantly different than 1 and 0, respectively, and the scatter about the line is modest (RMSE = 0.03). The other two purely theoretical methods of calculating E^1 that were used by Phillips et al. (*74*) gave correlations with considerably greater scatter, although the slope and intercept were still close to a 1:1 relationship (not shown here). In contrast, the theoretically calculated dataset of E^1 from Zubatyuk et al. (*79*), gives a correlation that is far from 1:1 (Fig. 3 N), but still linear with a small RMSE (0.04). This could be because the solvation energy of the radical anions were over stabilized, which would arise if the solvation cavity used in the continuum solvation energy calculations were chosen to be too small around the nitro groups.

The new dataset of E^1's for nitro reduction obtained in this work correlates to the experimental dataset with slope = 1.52±0.17, intercept = 0.25±0.08, and RMSE = 0.03 (Fig. 3 R). However, this fit is strongly influenced by the leverage of the point for 4-acetylnitrobenzene (which has the largest error relative to the experimental dataset: 0.10 V or ~2 kcal/mol) and absence of data for NACs that might balance this leverage (e.g., dinitrotoluenes). These results are reasonable given the expected errors in these types of calculations. Thus, it seems promising that further analysis using this approach will yield an improved method for calculating E^1's with purely-theoretical methods. Furthermore, the modest but systematic differences between our results and those obtained using similar methods by Zubatyuk et al. (*79*) and Phillips et al. (*74*) suggest that the estimation of E^1's is sensitive to the details of the computational methods used. Therefore, a more systematic evaluation of the source of errors in these calculations is needed before a fully satisfactory method for estimating E^1's for NACs can be selected.

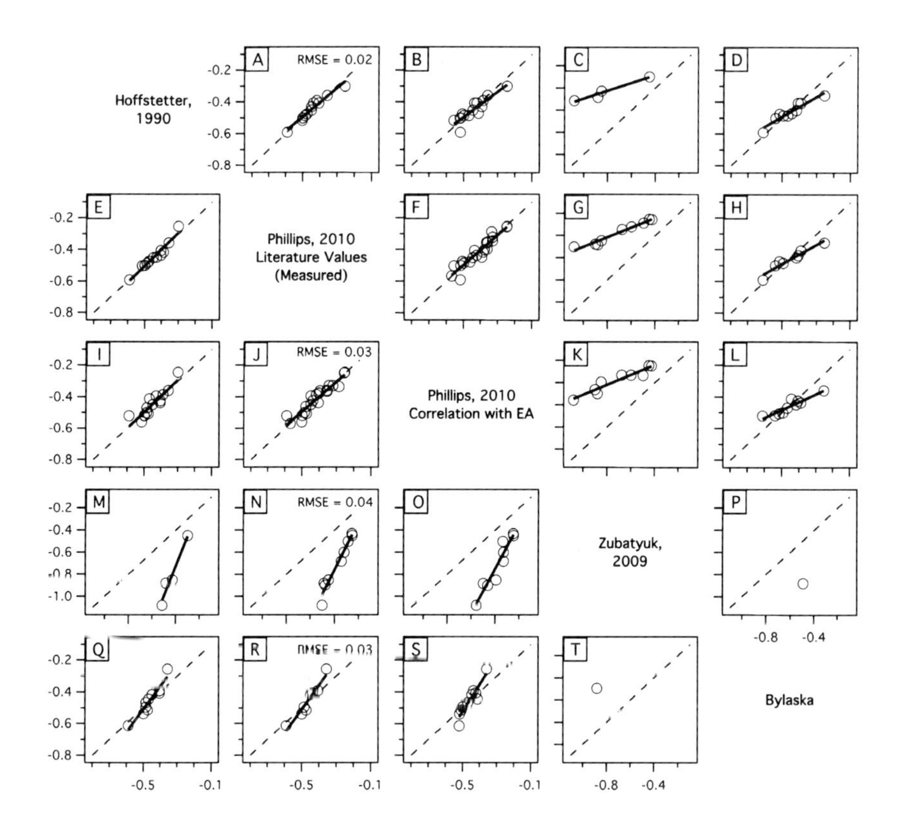

Figure 3. Scatter plot matrix of correlations between the five recommended E^1 datasets for nitro reduction. Regression lines and fitting results are for ordinary least-squares regression; dashed line is 1:1.

Other Organic Redox Reactions

Dechlorination of CACs and nitro reduction of NACs are the only types of reactions considered here, but there are other redox reactions of environmentally-relevant organics that should be amenable to similar treatment. For example, there are many data for E^1 of quinones (*26*), including some that are of interest as model compounds for the redox-active moieties associated with natural organic matter (*29*, *82*, *83*). Reduction of azo dyes is another redox reaction that plays a prominent role in determining contaminant fate, and E^1 data have been reported for this reaction (*84*).

As with reduction, oxidation of some contaminants can occur by SET, so the corresponding one-electron oxidation potentials are of interest in studies of the kinetics and mechanisms of these reactions. The methods used to calculate these E^1's from theory are similar to those described here for reduction. This has been reported in recent studies for polyphenols (*85*) and anilines (*63*, *86*).

Caveats and Future Prospects

The analysis provided here of E^1's for the two most thoroughly studied contaminant reduction reactions demonstrates that current theoretical calculations can produce datasets that agree well with experimental data. However, considerable effort is needed to determine what computational methods are appropriate for each type of SET, and to validate the accuracy of the results. Extrapolating these methods to additional compounds, and especially to other SET reactions, is not necessarily any more reliable by theoretical methods then it is for experimental methods. Therefore, a more general method of estimating E^1's would be of great value.

Currently, the most advanced effort to develop a method for predicting E^1's over a wide range of chemical types (and over a range of conditions such as solvent type) is SPARC (SPARC Performs Automated Reasoning in Chemistry (*87*)). The capability to calculate E^1's was added to SPARC relatively recently, and is based on previously developed and validated calculator for gas-phase electron affinities (*88*). SPARC's E^1 calculator has been validated against a set of experimental data, similar to those discussed here for dechlorination and nitro reduction, but for a much broader range of organic compounds and solvents (Hilal, personal communication). For nonaqueous solvents, SPARC's performance is good for the whole range of validation set compounds. However, for aqueous media, and some specific families of reactants—such as nitro reduction of NACs—agreement between SPARC and the other E^1 datasets presented here is weak. For reduction of CACs, SPARC calculates E^1 only for electron attachment not dissociative SET (e.g., eq. 1), and these two types of E^1 differ considerably. This illustrates the need for critical application of theoretical models to predicting chemical properties and one of the major obstacles to development of universal models for predicting contaminant fate.

Acknowledgments

We thank Drs. Said Halil and Eric Weber for providing us with unpublished details on the rationale and validation behind SPARC's algorithm for estimating E^1's. This work was supported by grants from the U.S. Department of Defense, Strategic Environmental Research and Development Program (SERDP Project No. ER-1735); and the U.S. Department of Energy (DOE), Division of Chemical Sciences, Geosciences, and BioSciences (DE-AC05-76RLO 1830, DE-FG07-02ER63485). Some of the calculations were performed on the Spokane and Chinook computing systems at the Environmental Molecular Sciences Laboratory (EMSL), a national scientific user facility sponsored by the DOE's Office of Biological and Environmental Research, and located at Pacific Northwest National Laboratory (PNNL). PNNL is operated by Battelle Memorial Institute. We also wish to thank the Scientific Computing Staff (Office of Energy Research) and DOE for a grant of computer time at the National Energy Research Scientific Computing Center (Berkeley, CA). This report has not been reviewed by any of these sponsors and therefore does not necessarily reflect their views and no official endorsement should be inferred.

References

1. Latimer, W. M. *Oxidation Potentials*; Prentice-Hall: New York, 1952.
2. Clark, W. M. *Oxidation-Reduction Potentials of Organic Systems*; Williams & Wilkins: Baltimore, 1960.
3. Meites, L.; Zuman, P. *CRC Handbook Series in Organic Electrochemistry*; CRC: Cleveland, OH, 1979.
4. Bard, A. J.; Parsons, R.; Jordan, J. *Standard Potentials in Aqueous Solutions*; Marcel Dekker: New York, 1985.
5. Luther, G. W., III. Thermodynamic redox calculations for one and two electron transfer steps: Implications for halide oxidation and halogen environmental cycling. In *Aquatic Redox Chemistry*; Tratnyek, P., Grundl, T. J., Haderlein, S. B., Eds.; American Chemical Society: Washington, DC, 2011; Vol. 1071, Chapter 2, pp 15−35.
6. Luther, G. W., III. The role of one- and two-electron transfer reactions in forming thermodynamically unstable intermediates as barriers in multi-electron redox reactions. *Aquat. Geochem.* **2010**, *16*, 395–420.
7. Dolfing, J. Thermodynamic considerations for dehalogenation. In *Dehalogenation: Microbial Processes and Environmental Applications*; Häggblom, M. M., Bossert, I. D., Bossert, I. D., Eds.; Kluwer: Boston, MA, 2003; pp 89−114.
8. Jin, Q.; Bethke, C. M. The thermodynamics and kinetics of microbial metabolism. *Am. J. Sci.* **2007**, *307*, 643–677.
9. Thauer, R. K.; Jungermann, K.; Decker, K. Energy conservation in chemotrophic anaerobic bacteria. *Bacteriol. Rev.* **1977**, *41*, 100–180.
10. Vogel, T. M.; Criddle, C. S.; McCarty, P. L. Transformations of halogenated aliphatic compounds. *Environ. Sci. Technol.* **1987**, *21*, 722–736.
11. Haas, J. R.; Shock, E. L. Halocarbons in the environment: Estimates of thermodynamic properties for aqueous chloroethylene species and their stabilities in natural settings. *Geochim. Cosmochim. Acta* **1999**, *63*, 3429–3441.
12. Majer, V.; Sedlbauer, J.; Wood, R. H. Calculation of standard thermodynamic properties of aqueous electrolytes and nonelectrolytes. In *Aqueous Systems at Elevated Temperatures and Pressures*; Palmer, D. A., Fernandez-Prini, R., Harvey, A. H., Eds.; Elsevier: 2004; pp 99−147.
13. Bylaska, E. J.; Dupuis, M.; Tratnyek, P. G. One-electron transfer reactions of polychlorinated ethylenes: Concerted versus stepwise cleavages. *J. Phys. Chem. A* **2008**, *112*, 3712–3721.
14. Wang, J.; Farrell, J. Investigating the role of atomic hydrogen on chloroethene reactions with iron using Tafel analysis and electrochemical impedance spectroscopy. *Environ. Sci. Technol.* **2003**, *37*, 3891–3896.
15. Arnold, W. A.; Winget, P.; Cramer, C. J. Reductive dechlorination of 1,1,2,2-tetrachloroethane. *Environ. Sci. Technol.* **2002**, *36*, 3536–3541.
16. Patterson, E. V.; Cramer, C. J.; Truhlar, D. G. Reductive dechlorination of hexachloroethane in the environment: Mechanistic studies via computational electrochemistry. *J. Am. Chem. Soc.* **2001**, *123*, 2025–2031.

17. Nonnenberg, C.; van der Donk, W. A.; Zipse, H. Reductive dechlorination of trichloroethylene: A computational study. *J. Phys. Chem. A* **2002**, *106*, 8708–8715.
18. Schwarzenbach, R. P.; Stierli, R.; Lanz, K.; Zeyer, J. Quinone and iron porphyrin mediated reduction of nitroaromatic compounds in homogeneous aqueous solution. *Environ. Sci. Technol.* **1990**, *24*, 1566–1574.
19. Riefler, R. G.; Smets, B. F. Enzymatic reduction of 2,4,6-trinitrotoluene and related nitroarenes: Kinetics linked to one-electron redox potentials. *Environ. Sci. Technol.* **2000**, *34*, 3900–3906.
20. Hartenbach, A. E.; Hofstetter, T. B.; Aeschbacher, M.; Sander, M.; Kim, D.; Strathmann, T. J.; Arnold, W. A.; Cramer, C. J.; Schwarzenbach, R. P. Variability of nitrogen isotope fractionation during the reduction of nitroaromatic compounds with dissolved reductants. *Environ. Sci. Technol.* **2008**, *42*, 8352–8359.
21. Barrows, S. E.; Cramer, C. J.; Truhlar, D. G.; Elovitz, M. S.; Weber, E. J. Factors contolling regioselectivity in the reduction of polynitroaromatics in aqueous solution. *Environ. Sci. Technol.* **1996**, *30*, 3028–3038.
22. Tratnyek, P. G.; Hoigné, J. Kinetics of reactions of chlorine dioxide (OClO) in water. II. Quantitative structure-activity relationships for phenolic compounds. *Water Res.* **1994**, *28*, 57–66.
23. Tratnyek, P. G. Correlation analysis of the environmental reactivity of organic substances. In *Perspectives in Environmental Chemistry*; Macalady, D. L., Ed.; Oxford: New York, 1998; pp 167−194.
24. Canonica, S.; Tratnyek, P. G. Quantitative structure-activity relationships for oxidation reactions of organic chemicals in water. *Environ. Toxicol. Chem.* **2003**, *22*, 1743–1754.
25. Tratnyek, P. G.; Weber, E. J.; Schwarzenbach, R. P. Quantitative structure-activity relationships for chemical reductions of organic contaminants. *Environ. Toxicol. Chem.* **2003**, *22*, 1733–1742.
26. Wardman, P. Reduction potentials of one-electron couples involving free radicals in aqueous solution. *J. Phys. Chem. Ref. Data* **1989**, *18*, 1637–1657.
27. Daasbjerg, K.; Pedersen, S. U.; Lund, H. Measurement and estimation of redox potentials of organic radicals. In *General Aspects of the Chemistry of Radicals*; Alfassi, Z. B., Ed.; Wiley: Chichester, 1999; pp 385−427.
28. Squella, J. A.; Bollo, S.; Nunez-Vergara, L. J. Recent developments in the electrochemistry of some nitro compounds of biological significance. *Curr. Org. Chem.* **2005**, *9*, 565–581.
29. Nurmi, J. T.; Tratnyek, P. G. Electrochemistry of natural organic matter. In *Aquatic Redox Chemistry*; Tratnyek, P., Grundl, T. J., Haderlein, S. B., Eds.; American Chemical Society: Washington, DC, 2011; Vol. 1071, Chapter 7, pp 129−151.
30. Meisel, D.; Czapski, G. One-electron transfer equilibriums and redox potentials of radicals studied by pulse radiolysis. *J. Phys. Chem.* **1975**, *79*, 1503–1509.
31. Cramer, C. J. *Essentials of Computational Chemistry: Theories and Models*; Wiley: 2005.

32. Bylaska, E. J.; Glaesemann, K. R.; Felmy, A. R.; Vasiliu, M.; Dixon, D. A.; Tratnyek, P. G. Free energies for degradation reactions of 1,2,3-trichloropropane from ab initio electronic structure theory. *J. Phys. Chem. A* **2010**, *114*, 12269–12282.
33. Valiev, M.; Bylaska, E.; Dupuis, M.; Tratnyek, P. G. Combined quantum mechanical and molecular mechanics studies of the electron transfer reactions involving carbon tetrachloride in solution. *J. Phys. Chem. A* **2008**, *112*, 2713–2720.
34. Bylaska, E. J.; Dupuis, M.; Tratnyek, P. G. Ab initio electronic structure study of one-electron reduction of polychlorinated ethylenes. *J. Phys. Chem. A* **2005**, *109*, 5905–5916.
35. Bylaska, E. J.; Dixon, D. A.; Felmy, A. R.; Apra, E.; Windus, T. L.; Zhan, C.-G.; Tratnyek, P. G. The energetics of the hydrogenolysis, dehydrohalogenation, and hydrolysis of 4,4′-dichloro-diphenyl-trichloroethane from ab initio electronic structure theory. *J. Phys. Chem. A* **2004**, *108*, 5883–5893.
36. Bylaska, E. J.; Dixon, D. A.; Felmy, A. R.; Tratnyek, P. G. One-electron reduction of substituted chlorinated methanes as determined from ab initio electronic structure theory. *J. Phys. Chem. A* **2002**, *106*, 11581–11593.
37. Bylaska, E. J. Estimating the thermodynamics and kinetics of chlorinated hydrocarbon degradation. *Theor. Chem. Acc.* **2006**, *116*, 281–296.
38. Valiev, M.; Bylaska, E. J.; Govind, N.; et al. NWChem: A comprehensive and scalable open-source solution for large scale molecular simulations. *Comput. Phys. Commun.* **2010**, *181*, 1477–1489.
39. de Jong, W. A.; Bylaska, E.; Govind, N.; et al. Utilizing high performance computing for chemistry: parallel computational chemistry. *Phys. Chem. Chem. Phys.* **2010**, *12*, 6896–6920.
40. McQuarrie, D. A. *Statistical Mechanics*; Harper & Row: New York, NY, 1973.
41. Cossi, M.; Barone, V.; Cammi, R.; Tomasi, J. Ab initio study of solvated molecules: a new implementation of the polarizable continuum model. *Chem. Phys. Lett.* **1996**, *255*, 327–335.
42. Floris, F. M.; Tomasi, J.; Pascual Ahuir, J. L. Dispersion and repulsion contributions to the solvation energy: Refinements to a simple computational model in the continuum approximation. *J. Comput. Chem.* **1991**, *12*, 784–791.
43. Miertus, S.; Scrocco, E.; Tomasi, J. Electrostatic interaction of a solute with a continuum. A direct utilization of ab initio molecular potentials for the prevision of solvent effects. *Chem. Phys.* **1981**, *55*, 117–129.
44. Tomasi, J.; Persico, M. Molecular interactions in solution: An overview of methods based on continuous distributions of the solvent. *Chem. Rev.* **1994**, *94*, 2027–2094.
45. Klamt, A.; Schüürmann, G. COSMO: A new approach to dielectric screening in solvents with explicit expressions for the screening energy and its gradient. *J. Chem. Soc., Perkin Trans. 2* **1993**, 799–803.

46. Klamt, A.; Eckert, F. COSMO-RS: A novel and efficient method for the a priori prediction of thermophysical data of liquids. *Fluid Phase Equilib.* **2000**, *172*, 43–72.
47. Cramer, C. J.; Truhlar, D. G. A universal approach to solvation modeling. *Acc. Chem. Res.* **2008**, *41*, 760–768.
48. Baker, N. A.; Sept, D.; Joseph, S.; Holst, M. J.; McCammon, J. A. Electrostatics of nanosystems: application to microtubules and the ribosome. *Proc. Natl. Acad. Sci.* **2001**, *98*, 10037–10041.
49. Pierotti, R. A. Aqueous solutions of nonpolar gases. *J. Phys. Chem.* **1965**, *69*, 281–288.
50. Huron, M. J.; Claverie, P. Calculation of the interaction energy of one molecule with its whole surrounding. II. Method of calculating electrostatic energy. *J. Phys. Chem.* **1974**, *78*, 1853–1861.
51. Honig, B.; Sharp, K. A.; Yang, A. Macroscopic models of aqueous solutions: Biological and chemical applications. *J. Phys. Chem.* **1993**, *97*, 1101–1109.
52. Sitkoff, D.; Sharp, K. A.; Honig, B. Accurate calculation of hydration free energies using macroscopic solvent models. *J. Phys. Chem.* **1994**, *98*, 1978–1988.
53. Cramer, C. J.; Truhlar, D. G. Implicit solvation models: Equilibrium, structure, spectra, and dynamics. *Chem. Rev.* **1999**, *99*, 2161–2200.
54. Eckert, F.; Klamt, A. Fast solvent screening via quantum chemistry: COSMO-RS approach. *AIChE J.* **2002**, *48*, 369–385.
55. Ben-Naim, A.; Marcus, Y. Solvation thermodynamics of nonionic solutes. *J. Chem. Phys.* **1984**, *81*, 2016–2027.
56. Tissandier, M. D.; Cowen, K. A.; Feng, W. Y.; Gundlach, E.; Cohen, M. H.; Earhart, A. D.; Coe, J. V.; Tuttle, T. R. The proton's absolute aqueous enthalpy and Gibbs free energy of solvation from cluster-ion solvation data. *J. Phys. Chem. A* **1998**, *102*, 7787–7794.
57. Eberson, L. Problems and prospects of the concerted dissociative electron transfer mechanism. *Acta Chem. Scand.* **1999**, *53*, 751–764.
58. Andrieux, C. P.; Savéant, J. M.; Su, K. B. Kinetics of dissociative electron transfer. Direct and mediated electrochemical reductive cleavage of the carbon-halogen bond. *J. Phys. Chem.* **1986**, *90*, 3815–3823.
59. Curtis, G. P. Reductive Dehalogenation of Hexachloroethane and Carbon Tetrachloride by Aquifer Sand and Humic Acid. Ph.D. Thesis, Stanford University, 1991.
60. Roberts, A. L.; Totten, L. A.; Arnold, W. A.; Burris, D. R.; Campbell, T. J. Reductive elimination of chlorinated ethylenes by zero-valent metals. *Environ. Sci. Technol.* **1996**, *30*, 2654–2659.
61. Scherer, M. M.; Balko, B. A.; Gallagher, D. A.; Tratnyek, P. G. Correlation analysis of rate constants for dechlorination by zero-valent iron. *Environ. Sci. Technol.* **1998**, *32*, 3026–3033.
62. Totten, L. A.; Roberts, A. L. Calculated one- and two-electron reduction potentials and related molecular descriptors for reduction of alkyl and vinyl halides in water. *Crit. Rev. Environ. Sci. Technol.* **2001**, *31*, 175–221.
63. Winget, P.; Cramer, C. J.; Truhlar, D. G. Computation of equilibrium oxidation and reduction potentials for reversible and dissociative

electron-transfer reactions in solution. *Theor. Chem. Acc.* **2004**, *112*, 217–227.

64. Cwiertny, D. M.; Arnold, W. A.; Kohn, T.; Rodenburg, L. A.; Roberts, A. L. Reactivity of alkyl polyhalides toward granular iron: Development of QSARs and reactivity cross correlations for reductive dehalogenation. *Environ. Sci. Technol.* **2010**, *44*, 7928–7936.
65. Laviron, E.; Vallat, A.; Meunier-Prest, R. The reduction mechanism of aromatic nitro compounds in aqueous medium. Part V. The reduction of nitrosobenzene between pH 0.4 and 13. *J. Electroanal. Chem.* **1994**, *379*, 427–435.
66. Lund, H. Cathodic reduction of nitro compounds. In *Organic Electrochemistry*; Marcel Dekker: New York, 1973; pp 315−345.
67. Hofstetter, T. B.; Neumann, A.; Arnold, W. A.; Hartenbach, A. E.; Bolotin, J.; Cramer, C. J.; Schwarzenbach, R. P. Substituent effects on nitrogen isotope fractionation during abiotic reduction of nitroaromatic compounds. *Environ. Sci. Technol.* **2008**, *42*, 1997–2003.
68. Oh, S. Y.; Cha, D. K.; Chiu, P. C. Graphite-mediated reduction of 2,4-dinitrotoluene with elemental iron. *Environ. Sci. Technol.* **2002**, *36*, 2178–2184.
69. Meisel, D.; Neta, P. One-electron redox potentials of nitro compounds and radiosensitizers. Correlation with spin densities of their radical anions. *J. Am. Chem. Soc.* **1975**, *97*, 5198–5203.
70. Neta, P.; Meisel, D. Substituent effects on nitroaromatic radical anions in aqueous solution. *J. Phys. Chem.* **1976**, *80*, 519–524.
71. Wardman, P. The use of nitroaromatic compounds as hypoxic cell radiosensitizers. *Curr. Top. Rad. Res. Q.* **1977**, *11*, 347–398.
72. Kemula, W.; Krygowski, T. M. Nitro compounds. In *Encyclopedia of Electrochemistry of the Elements*; Marcel Dekker: New York, 1979; Vol. 13; pp 77−130.
73. Hofstetter, T. B.; Heijman, C. G.; Haderlein, S. B.; Holliger, C.; Schwarzenbach, R. P. Complete reduction of TNT and other (poly)nitroaromatic compounds under iron-reducing subsurface conditions. *Environ. Sci. Technol.* **1999**, *33*, 1479–1487.
74. Phillips, K. L.; Sandler, S. I.; Chiu, P. C. A method to calculate the one-electron reduction potentials for nitroaromatic compounds based on gas-phase quantum mechanics. *J. Comput. Chem.* **2011**, *32*, 226–239.
75. Reifler, R. G.; Smets, B. F. Enzymatic reduction of 2,4,6-trinitrotoluene and related nitroarenes: Kinetics linked to one-electron redox potentials. *Environ. Sci. Technol.* **2000**, *34*, 3900–3906.
76. de Abreu, F. C.; de Paula, F. S.; dos Santos, A. F.; Sant'Ana, A. E. G.; de Almeida, M. V.; Cesar, E. T.; Trindade, M. N.; Goulart, M. O. F. Synthesis, electrochemistry, and molluscicidal activity of nitroaromatic compounds: Effects of substituents and the role of redox potentials. *Bioorg. Med. Chem.* **2001**, *9*, 659–664.
77. Sjoberg, L.; Eriksen, T. E. Nitrobenzenes: A comparison of pulse radiolytically determined one-electron reduction potentials and calculated electron affinities. *J. Chem. Soc., Faraday Trans. 1* **1980**, *76*, 1402–1408.

78. Adams, G. E.; Clarke, E. D.; Flockhart, I. R.; et al. Structure-activity relationships in the development of hypoxic cell radiosensitizers. I. Sensitization efficiency. *Int. J. Radiat. Biol. Relat. Stud. Phys., Chem. Med.* **1979**, *35*, 133–150.
79. Zubatyuk, R. I.; Gorb, L.; Shishkin, O. V.; Qasim, M.; Leszczynski, J. Exploration of density functional methods for one-electron reduction potential of nitrobenzenes. *J. Comput. Chem.* **2010**, *31*, 144–150.
80. Neumann, A.; Hofstetter, T. B.; Lussi, M.; Cirpka, O. A.; Petit, S.; Schwarzenbach, R. P. Assessing the redox reactivity of structural iron in smectites using nitroaromatic compounds as kinetic probes. *Environ. Sci. Technol.* **2008**, *42*, 8381–8387.
81. Gorski, C. A.; Nurmi, J. T.; Tratnyek, P. G.; Hofstetter, T. B.; Scherer, M. M. Redox behavior of magnetite: Implications for contaminant reduction. *Environ. Sci. Technol.* **2010**, *44*, 55–60.
82. Uchimiya, M.; Stone, A. T. Reversible redox chemistry of quinones: Impact on biogeochemical cycles. *Chemosphere* **2009**, *77*, 451–458.
83. Macalady Donald, L.; Walton-Day, K. Redox chemistry and natural organic matter (NOM). In *Aquatic Redox Chemistry*; Tratnyek, P., Grundl, T. J., Haderlein, S. B., Eds.; American Chemical Society: Washington, DC, 2011; Vol. 1071, Chapter 5, pp 85−111.
84. Sharma, K. K.; O'Neill, P.; Oakes, J.; Batchelor, S. N.; Rao, B. S. M. One-electron oxidation and reduction of different tautomeric forms of azo dyes: a pulse radiolysis study. *J. Phys. Chem. A* **2003**, *107*, 7619–7628.
85. Li, C.; Hoffman, M. Z. One-electron redox potentials of phenols in aqueous solution. *J. Phys. Chem. B* **1999**, *103*, 6653–6656.
86. Winget, P.; Weber, E. J.; Cramer, C. J.; Truhlar, D. G. Computational electrochemistry: aqueous one-electron oxidation potentials for substituted anilines. *Phys. Chem. Chem. Phys.* **2000**, *2*, 1231–1239.
87. Hilal, S. H.; Saravanaraj, A. N.; Whiteside, T.; Carreira, L. A. Calculating physical properties of organic compounds for environmental modeling from molecular structure. *J. Comput.-Aided Mol. Des.* **2007**, *21*, 693–708.
88. Hilal, S. H.; Carreira, L. A.; Karickhoff, S. W.; Melton, C. M. Estimation of electron affinity based on structure activity relationships. *Quant. Struct.-Act. Relat.* **1993**, *12*, 389–396.

Chapter 4

Thermodynamic Control on Terminal Electron Transfer and Methanogenesis

Christian Blodau*

School of Environmental Sciences, University of Guelph, N1G 2W1, Guelph, Canada
***christian.blodau@uni-muenster.de; present address: Hydrology Group, ILÖK, University of Münster, Germany**

Terminal electron accepting processes (TEAPs) control the fate of elements in anoxic environments. This study focuses on thermodynamic regulation of H_2-dependent TEAPs. H_2-dependent methanogenesis and sulfate reduction operate near free energy thresholds (ΔG_c) and can be inhibited by changes in thermodynamic conditions, whereas more 'potent' TEAPs occur far from their energy thresholds and lower H_2 concentrations to levels that exclude other TEAPs. Metabolic free energy thresholds depend on microbial physiology and occur when the energy conserved by ATP generation approaches the thermodynamic driving force. A model analysis for peat-sand mixtures suggests that acetoclastic methanogenesis can be inhibited by CH_4 and dissolved inorganic carbon (DIC) accumulation, lowering the free energy (ΔG_r) toward an energy threshold (ΔG_c), which was identified by inverse modeling near - 25 kJ mol^{-1}. Inhibition was sensitive to ΔG_c and acetate concentrations, so that $\Delta G_c \pm 5$ kJ mol^{-1} and a range of 1 to 100 µmol L^{-1} acetate lead to strongly differing steady state CH_4 concentrations in the model results.

Introduction and Review

Terminal Electron Transfer and the Role of Molecular Hydrogen

The distribution of chemical species in anoxic environments is governed by microbially mediated electron transfer, and in particular by terminal electron accepting processes and methanogenesis (*1*, *2*) (Table 1). Geochemists have recognized early that thermodynamic considerations may provide a framework for analyzing the occurrence of TEAPs in sediments (*3*). From such origins and the investigation of the bioenergetic regulation and limits of anerobic metabolism (*4–6*), the role of molecular hydrogen as regulator for TEAPs, methanogenesis, and syntrophic fermentation processes was elucidated. More recently, coherent and quantitative concepts addressing the regulation of TEAPs in anoxic environments have been developed (*7*). Efforts are also undertaken to integrate such concepts into reactive transport and diagenetic models (*8*). This study's objective is to summarize some of this progress and to analyze how constraints on energy and solute transport may control the kinetics of microbially mediated electron transfer processes in anoxic environments. The latter objective is also addressed based on a simple methanogenic, diffusion dominated system.

Anaerobic decomposition of both immobile and dissolved organic matter proceeds through the action of extracellular enzymes, fermentation, syntrophic processes, and TEAPs (*9*) (Fig. 1). Decomposition is initially controlled by the activity of extracellular enzymes, which are produced by fermenting bacteria, and hydrolyzed polymers that cannot pass the outer membrane of microorganisms (*10*). Among many decomposition products, acetate and molecular hydrogen (H_2) are primary substrates for methanogenic *Archaea* and a range of microorganisms mediating TEAPs (*11*). For simplicity, methanogenesis is included under the TEAP heading. Molecular H_2 plays a particularly important role in the anaerobic decomposition network. The microbial demand for H_2 is strong since the ability to oxidize H_2 through membrane-bound hydrogenases is phylogenetically widespread (*12*), linking H_2 to reduction of widespread electron acceptors such as NO_3^-, SO_4^{2-}, and CO_2, and to less common acceptors, such as As(V) and U(VI) (see also (*13*) and (*14*)). Thus, its concentration controls much of the energy available to TEA bacteria. H_2 typically has a half-life of less than a minute (*15*), which allows for rapid concentration adjustment with changing environmental conditions (*11*). The free energy of oxidation of H_2 is also highly dependent on H_2 activity because it involves only a transfer of two electrons. This network of processes involving H_2 transfer is subjected to constraints by transport of solutes and gases, microbial kinetics, stoichiometric limitations, and the availability of Gibbs free energy in form of electron donors and acceptors (Figure 1). The level of energy available results from fluxes of oxidants and reductants, and the rate at which mobile reactants are supplied, as analyzed by Kurtz and Peiffer (*16*) in this volume.

Lovley and Goodwin (*17*) recognized the role of H_2 in anaerobic decomposition networks and suggested its use as a redox indicator in studying competition between TEAPs. The authors originally argued that H_2 concentration under steady-state conditions should depend only on physiological characteristics of hydrogenotrophic microorganisms, i.e. the growth yield (Y) and the half-saturation constant (K) for H_2 uptake, which tend to increase and decrease, respectively, with increasing energy yield of a given TEAP. External factors, such as the H_2 supply rate, should not influence H_2 concentrations in steady-state systems (*18*). Levels of H_2 in anoxic environments, thus, decrease with increasing standard Gibbs free energy of the predominant TEAP. This way, energetically less potent TEAPs are excluded from substrate utilization as their energy thresholds are exceeded and their metabolism ceases. According to this concept, H_2 levels can be used as a redox indicator because levels adjust to ranges that are optimal and indicative for a particular predominating TEAP (*17*, *19*). For a compilation of these ranges the reader is referred to Heimann, et al. (*20*). The approach has been widely used to investigate the microbiology and redox chemistry of pristine and contaminated sediments (*21–24*). Similar observations were made for other low-molecular-weight intermediates (*25*) but concentration levels have been less clear than for H_2.

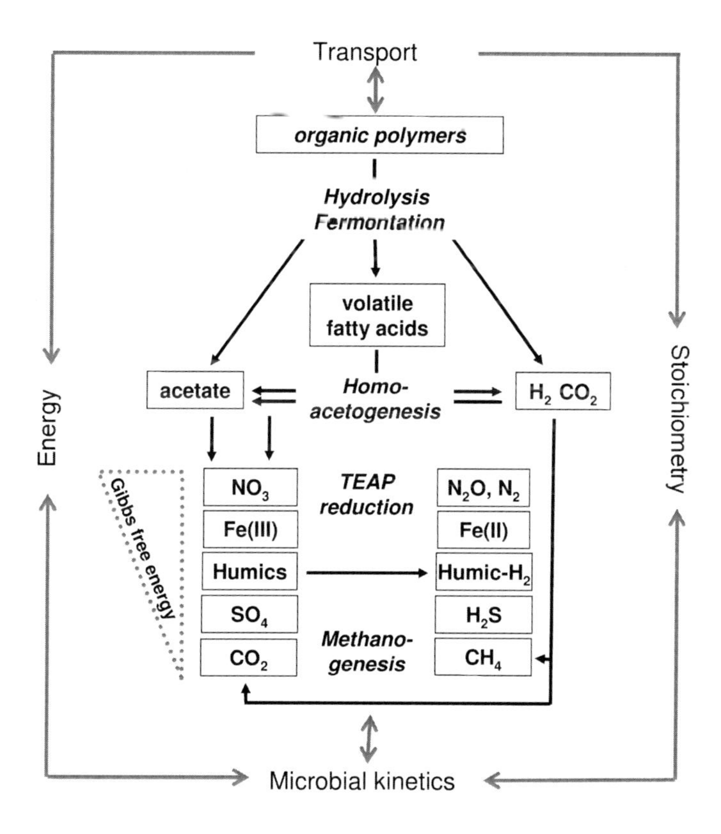

Figure 1. Simplified network of processes involved in anaerobic organic matter decomposition in anoxic aqueous systems and potential controls by transport, energy, microbial kinetics, and microorganism stoichiometry.

Table 1. Overview of important H_2-consuming TEAPs and their Gibbs free energies under standard (ΔG^0) and exemplary environmental (ΔG_r) conditions, respectively

Process	*Reaction stoichiometry*	*ΔG^0 (kJ mol^{-1})^a*	*ΔG_r (kJ mol^{-1})^b*
Oxic respiration	$1/2\ O_2 + H_2 \rightarrow H_2O$	-237.2	-206.7
Denitrification	$2/5\ NO_3^- + H_2 + 2/5\ H^+ \rightarrow 1/5\ N_2 + 6/5\ H_2O$	-240.1	-186.6
Iron reduction	$2\ FeOOH(a) + H_2 + 4\ H^+ \rightarrow 2\ Fe^{2+} + 4\ H_2O$	-182.5	-39.8
Arsenate reduction	$HAsO_4^{2-} + H_2 + 2\ H^+ \rightarrow H_3AsO_3 + H_2O$	-162.4	-53.9
Sulfate reduction	$1/4\ SO_4^{2-} + H_2 + 1/4\ H^+ \rightarrow 1/4\ HS^- + H_2O$	-48.0	-9.5
Hydrogenotrophic methanogenesis	$1/4\ HCO_3^- + H_2 + 1/4\ H^+ \rightarrow 1/4\ CH_4 + 3/4\ H_2O$	-43.9	-8.2
Homoacetogenesis	$1/2\ HCO_3^- + H_2 + 1/4\ H^+ \rightarrow 1/4\ Acetate^- + H_2O$	-36.1	+2.4

[a] Calculated from Gibbs free energies of formation (*26–28*); O_2, H_2, N_2, and CH_4 as gaseous species. [b] At the following conditions: T = 25 °C, $[O_2] = 0.21$, $[N_2] = 0.78$, $[CH_4] = [NO_3^-] = [Fe^{2+}] = [HAsO_4^{2-}] = [H_3AsO^3] = [SO_4^{2-}] = [HS^-] = [Acetate^-] = 10^{-4}$, $[HCO_3^-] = 10^{-2}$, $[H^+] = 10^{-7}$, $[H_2] = 10^{-5}$ (corresponding to an aqueous concentration of approximately 8 nmol L^{-1}). Square brackets indicate activities of aqueous species or fugacities of gaseous species. Data have previously been reported in (*29*) and (*20*).

The Energy Threshold – *in Situ* Energy Concept in Terminal Electron Transfer

Lovley and Goodwin's (*17*) H_2-based redox indicator concept did not always predict the occurrence of TEAPs well (*30–34*). Often TEAPs were also found to occur simultaneously (*35*, *36*). H_2 concentrations in real-world systems are not solely related to the physiology of the H_2-consuming microbes but are also influenced by the presence of other electron donors, the concentration levels of electron acceptors, transient environmental conditions, and micro-scale heterogeneity. Redox conditions and electron flow in anoxic systems have, thus, also been analyzed focussing on the *in situ* energy yields of individual processes (*30*, *37*, *38*). Energy yields can be calculated from the activities of substrates and products involved in a process at a given temperature and compared to a theoretical or empirical minimum energy requirement for the process, so called 'energy threshold'. This kind of analysis provides insight into the functioning of an individual electron transfer process. If *in situ* energies are smaller (more positive) than the energy threshold for a particular process, it cannot occur; if energies are much larger than usually observed while the process is ongoing, it is likely inhibited by some other constraint; and if the *in situ* energy of the process approaches its energy threshold it should become increasingly slow.

The expectation is that TEAPs operate at *in situ* energy levels that are close to but not at their specific energy thresholds. In agreement with this idea, ΔG_r for hydrogenotrophic methanogenesis was about -35 kJ mol^{-1} in several anaerobic freshwater systems during ongoing methanogenesis (*39*, *40*), whereas the theoretical energy threshold for the process is believed to be -20 to -25 kJ mol^{-1} under active growth (*5*).

Energy Thresholds

Energy thresholds have been derived from the ATP synthesis energy. The *in vivo* energy required for synthesis of one mol of ATP is around +50 kJ, depending on intracellular levels of ATP, ADP, phosphate, Mg^{2+}, and H^+ (*5*, *28*). Additional required thermodynamic driving force can be accounted for by simply adding an energy quantum for the process (*5*) and more generally by including a thermodynamic efficiency factor (*6*, *41*). Since the H^+/ATP ratio of the ATP synthase is typically in the range of 3-4 (*42*) the energy threshold equals the energy released by one proton returning into the cell through the ATP synthase. Based on this argument, the energy threshold should be around 1/3 or 1/4 of -70 kJ mol^{-1}, i.e. about -20 kJ mol^{-1} of substrate for methanogenesis (*5*). Under starvation conditions, smaller values of -10 to -15 kJ mol^{-1} of substrate may be found (*6*, *41*).

The H^+/ATP based energy threshold concept requires some flexibility with regard to methanogenic *Archaea*, which are particularly well-adapted to low energy levels (*43*). *Archaea* actively pump sodium ions across their membranes as a bioenergetic pathway, which may account for part of the ATP that is synthesized (*44*, *45*), and they are characterized by more variable and larger H^+/ATP ratios (*43*). Fermentation processes are not subject to the H^+/ATP energy threshold concept since ATP is synthesized via substrate level phosphorylation (*46*, *47*). Minimum energy requirements for these processes and acetogenesis were generally found to be smaller than in TEAPs (*46*, *47*). Compilations of minimum free energy yields for various TEAPs and fermentation processes were compiled by Hoehler (*6*) and Kleerebezem and Stams (*48*).

Differences in Energetic Control on TEAPs

The different levels of Gibbs free energies provided from TEAPs and methanogenesis (Table 1) have been found to correlate with the 'tightness' of thermodynamic control and the *in situ* energies that adjust during an ongoing process. This finding is reflected in the levels of energy thresholds that have been reported, and in the difference between these thresholds and *in situ* energies that were typically attained in anoxic environments. When Gibbs free energy is large, in particular with respect to denitrification, energy thresholds were found to be larger as well (*20*, *37*). Energy conservation per hydrogen consumed has been suggested to be twice or more that of sulfate reduction and methanogenesis, based on the energy conservation mechanisms; energy thresholds should accordingly be proportionately higher and attained when hydrogen levels are still farther from thermodynamic equilibrium (*7*). This being the case *in situ* energy yield did not reach energy thresholds in empirical studies because H_2 concentrations become minute when approaching the threshold (*37*). Enzyme kinetics and limits on diffusive flux probably control H_2 concentration and, thus, also *in situ* energy yields of such 'high energy' TEAPs (*37*).

For TEAPs falling into an intermediate range of energy yields, the relationship between energy thresholds and *in situ* energy yields is less uniform. Reduction of chlorinated organics, Cr(VI), As(V), and iron oxides represent examples of these TEAPs. During reductive dechlorination of ethenes, *in situ* energy yields encountered under field or laboratory conditions remained higher than is expected based on the energy thresholds concept (*49*). A similar conclusion has been reached with respect to As(V)-reduction with H_2 as electron donor (*29*). With respect to bacterial iron oxide reduction, no coherent picture has emerged so far, potentially caused by the phase transfer reaction and wide range of metastable phases and surface properties of the electron acceptor (*50*). In some cases, microbial reduction of goethite ceased at an energy threshold of around -23 kJ mol^{-1} (*51*), whereas in others experiments no such thresholds could be identified (*52*). An inherent difficulty in studies on iron oxide reduction is the variable nature of the mineral phase, which is altered by sorption, surface precipitation, and catalytic action of ferrous iron (*53*, *54*).

For processes at the low end of energy yields, empirical energy thresholds tended to be fairly consistent and small, and ranged from -20 to -28 kJ mol^{-1}, -9 to -50 kJ mol^{-1}, and -16 to -49 kJ mol^{-1} for hydrogenotrophic acetogenesis, methanogenesis, and sulfate reduction, respectively (when the reaction was written with the lowest possible integer coefficients) (*6*). The occurrence and kinetics of these processes is, thus, likely controlled by their *in situ* energy yields. The occurrence of these processes at very small ΔG_r has also been interpreted as an adaptation to survival under "substrate starvation" in depositional environments, where microbial communities are exposed to residual and increasingly recalcitrant organic matter (*41*).

Energy Thresholds and Transport – Methanogenesis as an Example

A consequence of energy thresholds is the potential for metabolic slowdown by accumulation of metabolic products. Such phenomena are well known from anaerobic bioreactors (*55*). Several closed incubation-type studies have also demonstrated that processes involved in anaerobic respiration slow down or cease when energy thresholds are approached (*41*, *46*, *56*). Extrapolating such experimental observations to peats, Beer et al. (*57*) proposed that small changes in ΔG_r of methanogenesis by accumulation of DIC and/or CH_4 may slow methanogenic decomposition in diffusion dominated methanogenic deposits. In this concept, the production of CH_4 reflects a system's limit to accommodate simultaneous constraints of transport, energy, and microbial kinetics.

To test this hypothesis, a simple column experiment was designed with water saturated and anaerobic peat-quartz sand mixtures containing 5%, 15%, and 50% of moderately decomposed peat. Differences in peat quality with depth were eliminated by homogenization and pore water profiles were allowed to approach steady-state in absence of vertical water flow. Under such conditions,

CH_4 production should diminish to similar and very small values deeper into the deposit regardless of C content, but increase proportionately with C content in the anaerobic layers near the column surface, where the energy constraints are alleviated by continuous removal of DIC and CH_4. Furthermore, the decrease in methanogenesis should be related to processes approaching their respective biological energy quantum.

Methods

Column Setup and Sampling

Commercially available ombrotrophic bog peat was homogeneously mixed with quartz sand and deionized water, and filled in PVC columns (140 cm x 20 cm) equipped with porewater peepers (*38*) for 120 cm of column length. These could be non-destructively retrieved from perforated and meshed frames embedded in the column filling. Columns were kept dark at 18°C with the water table slightly above the surface. Peepers were retrieved and replaced after 370 and 550 days. DIC, CH_4 and H_2 concentrations were quantified using headspead techniques, gas chromatography and a Trace Analytical TA 3000 hydrogen analyzer, and converted into dissolved concentration. Inorganic anions (ion chromatography), pH (potentiometry), and H_2S (amperometry) were analyzed as previously described (*38*), and acetate in selected samples was analyzed by HPLC with UV detection at 208 nm.

Modeling

To model production and transport of CH_4 a simple box model was implemented using STELLA© simulation software. The model encompassed 10 fully mixed vertical layers to simulate concentration profiles. Solute transport between layers was simulated using Fick's law with the diffusion coefficient of CH_4 taken from (*58*), adjusted to a temperature of 18°C and corrected for the effect of porosity (n) using a factor of n^2. The assumption that the system was diffusion dominated was checked by repeated static chamber measurements of CH_4 flux (*38*) at the end of the experiments. Ebullition (i.e. bubbling) of CH_4,detectable by rapid concentration increase in the chambers, was not observed. Concentrations of CH_4 at the upper and lower boundary of column were fixed at measured levels in the model. Porosity was determined from bulk density measurements in separate columns of analogous peat-sand mixtures, as well as estimates of specific density of peat (1.5 g cm^{-3}) and quartz components (2.65 g cm^{-3}). The production rate of methane in each depth layer was modelled using a modified Michalis-Menten kinetics (equation 1) with a maximum rate v_{max} (μmol L^{-1} d^{-1}) and half saturation constant k_s (μmol L^{-1}). The model further accounts for the free energy of the process by including the reaction quotient Q and an analogue for the equilibrium constant adjusted to a minimum energy requirement ΔG_c (kJ mol^{-1} (CH_4)). Originally this approach was presented by Hoh and Cord-Ruwisch (*55*) using the equilibrium constant instead. It represents a special case of the more general approach developed by Jin and Bethke (*7*). In the model, the production of methane equals zero when the $Q = K_{threshold}$, i.e. when the energy conserved in ATP generation equals the release of free energy from the overall chemical process. At this point, the thermodynamic driving force becomes zero.

$$-\frac{d[Ac]}{dt} = \frac{d[CH_4]}{dt} = \frac{v_{max}[Ac] \bullet (1 - \frac{Q}{K_{threshold}})}{k_s + [Ac] \bullet (1 + \frac{Q}{K_{threshold}})} \qquad \text{(equation 1)}$$

Gibbs free energy ΔG_r (kJ mol^{-1}) available for hydrogenotrophic ($4\ H_2(aq) + CO_2(aq) \rightarrow 2\ H_2O(l) + CH_4(aq)$) and acetoclastic ($CH_3COO^-(aq) + H^+(aq) \rightarrow CO_2(aq) + CH_4(aq)$) methanogenesis was calculated in the model for the smallest integer stoichiometry using the Nernst equation, as previously described (*38*):

$$\Delta G_r = \Delta G_r^0 + R \cdot T \cdot \ln Q \text{ with } Q = \frac{\prod_i (products)^{v_i}}{\prod_i (substrates)^{v_i}} \qquad \text{(equation 2)}$$

with ΔG_r^0 the standard Gibbs free energy of the reaction (kJ mol^{-1}) at *in situ* temperature of 18°C, R the gas constant (8.314×10^{-3} kJ mol^{-1} K^{-1}), T the absolute temperature (K), and ν the stoichiometric coefficients. The pressure dependency of ΔG_r and the effect of ionic strength on activities were neglected.

Hydrogenotrophic methanogenesis provided insufficient free energy for methanogenesis, at least at the spatial scale of solute sampling (data not shown). The model was thus restricted to the utilization of acetate and parameterized in agreement with literature ranges of v_{max}, k_s, and the critical energy converted to a $K_{threshold}$. Parameterization was further attempted with one consistent set of parameters for simulating CH_4 concentration profiles in 5 %, 15%, and 50% peat columns at 370 and 550 days, simultaneously. To account for an initial utilization of electron acceptors after flooding (*59*), a time lag of 100 days was assumed before methane production began and reached full production after 200 days. Acetate concentration was fixed at average levels of 10 µmol L^{-1}, or about half the detection limit of measurements. The DIC, here equivalent to dissolved CO_2, concentration profiles were simulated simultaneously using (1) first order kinetics of DIC production and decomposition constants of 0.0018 to 0.0028 d^{-1} to match the observed accumulation of DIC, and (2) an empirical linearly-increasing inhibition of DIC production at levels of 5000 to 10500 µmol L^{-1}, which was needed to match the evolution of DIC profiles in the experiments (data not shown).

The Michaelis-Menten model was also applied without the energy threshold to test if CH_4 profiles across the columns and over time could be created without a thermodynamic inhibition of acetoclastic methanogenesis. Finally, sensitivity analyses were carried out regarding v_{max}, k_s, ΔG_c, and concentrations of acetate.

Results and Discussion

Concentration Dynamics

Methane concentration in the peat columns increased with depth and time in all three peat sand mixtures (Figure 2). Concentrations reached 280 to 500 µmol L^{-1} (5% peat), 400 to 500 µmol L^{-1} (15 % peat), and 600 µmol L^{-1} (50% peat) in the columns. In the 50 % peat column concentrations hardly increased between 370 d and 550 d sampling; this column was near steady state by the end of the experiment, whereas concentration moderately increased in the 15%, and strongly increased in the 5% peat column.

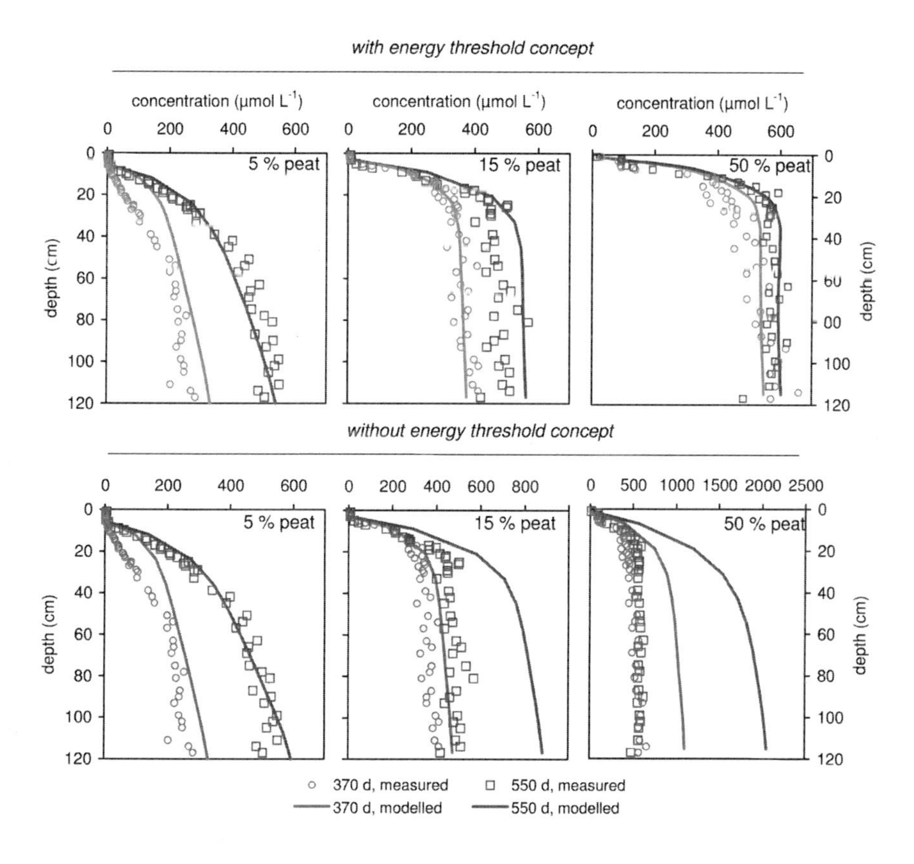

Figure 2. Measured and modelled CH_4 concentrations in experiments with 5% peat, 15% peat, and 50% peat (mass basis) in homogenized peat - quarz sand mixtures after 370 days and 550 days of incubation. Note different concentration scales. In the uppermost series of simulations, an energy threshold at -25 kJ mol^{-1} (CH_4) was implemented, resulting in slow down of concentration increase and the 'flattening' of the profiles observed with depth. In the lower panels, this constraint was removed.

Strikingly, the methane concentration increase with depth levelled off when the concentration profiles approached steady state, indicating an absence of methane production deeper into the peat sand mixtures. DIC concentrations increased in a similar but less steep manner to maximum levels of 5500 µmol L^{-1}, 8500 µmol L^{-1}, and 9500 µmol L^{-1}; H_2 concentrations varied from 0.1 to 2.7 nmol L^{-1}; and acetate concentration remained below the limit of detection. Values of pH were slightly to moderately acidic and ranged from 4.0 to 5.4, generally increasing with depth.

The methane dynamics in the columns could be adequately reproduced with the model that included an energy threshold of -25 kJ mol^{-1} (CH_4) and reasonably consistent V_{max} and k_s values that were in the broad range reported for peats and similar substrates (Figure 2). Values of k_s have been reported to be between 5 and 200 µmol L^{-1} (*60*); a value of 10 µmol L^{-1} was used in all simulations. A best fit could be reached when V_{max} increased from 6.5 µmol L^{-1} d^{-1} (5% peat), 11 µmol L^{-1} d^{-1} (15% peat) to 36.8 µmol L^{-1} d^{-1} (50% peat). Such values are well in line with measured *in situ* CH_4 production in similar bog peats during summer (*61*). Between the 15% and 50% treatments the increase in V_{max} was proportionate to the increase in organic matter content. Successful reproduction of the CH_4 profiles in the 5% treatment required a somewhat higher V_{max} than proportional to the differences in peat content. Acetate concentrations averaged 10 µmol L^{-1} in all model columns, but some adjustment of concentration levels with depth was required. In the 15 % and 50 % columns, a good fit was reached with a linear increase from 8.6 µmol L^{-1} to 11.4 µmol L^{-1}, whereas in the 5% column concentrations increased linearly from 5.5 to 14.5 µmol L^{-1}, resulting in faster rising CH_4 concentration deeper into the peat of this column (Figure 2). These assumptions are arbitrary but within the boundary stipulated by the acetate analyses.

Importantly, methane dynamics could not be reproduced across time and peat content without an energy threshold (ΔG_c) (Figure 2). In any particular column, a CH_4 profile could be adequately fitted by some parameter combination. In the 5% treatment, this exercise succeeded for both profiles at 370 d and 550 d (Figure 2). However, in the 15% and 50% treatments, a good fit of the 550 d profiles resulted in gross underestimates of concentrations after 370 d (data not shown). Using the general parameterization that produced a good fit in the 5% peat column, CH_4 concentrations in the other columns were strongly overestimated in absence of an energy threshold (Figure 2). While there is considerably degree of freedom in the model parameterization, these results suggest that a threshold to CH_4 production was present. This threshold was not yet reached in the 5%, but clearly reached in the 15% and particularly the 50% peat column. The nature of this threshold cannot be unequivocally ascertained, but the fact that it occurred when Gibbs free energies of acetoclastic methanogenesis reached -25 kJ mol^{-1} (which is close to theoretical energy thresholds (*5*)) suggest that methane production and increase in methane concentrations was inhibited thermodynamically and ultimately by the slowness of diffusive transport removing CH_4 and DIC.

Production Rates and Parameter Sensitivity

Some further insight into the methane production dynamics leading to this phenomenon can be gained by analyzing profiles of CH_4 production and Gibbs free energy in the model (Figure 3), as well as the sensitivity to processes to the parameters (Figure 4). In the 5% model, CH_4 production only slightly decreased over time and ΔG_r remained < -28 kJ mol^{-1} (Figure 3). In the 50% peat model, methane production remained strong in the uppermost layer under the same ΔG_r, whereas at larger depth production declined to zero. The 15% peat model was characterized by a more moderate inhibition of methanogenesis with time and depth. The decline of methanogenesis to zero in the 50% peat model was essentially caused by lower DIC concentration in the upper part of the column, which allowed for higher CH_4 concentrations at the energy threshold (see also equation (2)).

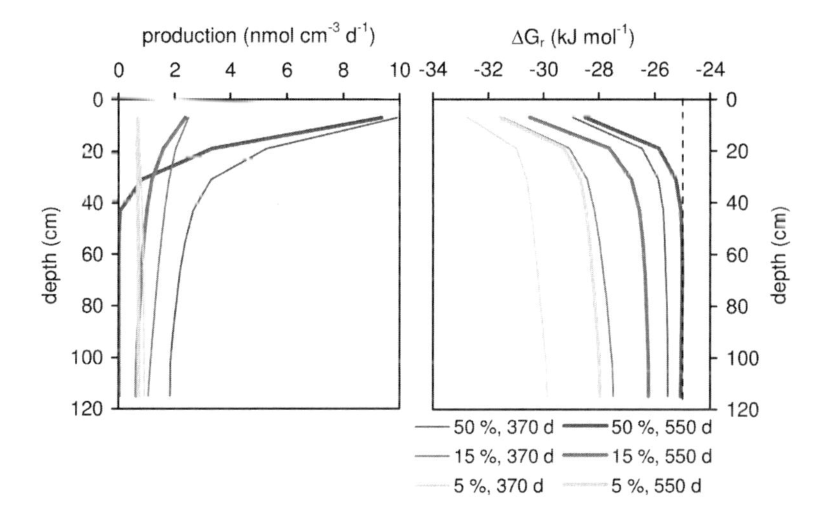

Figure 3. Modelled depth profiles of CH_4 production and respective ΔG_r of the processes at 370 and 550 days of incubation in experiments with 5% peat, 15% peat, and 50 % peat in homogenized peat - quarz sand mixtures.

As a result, CH_4 concentration slightly peaked at about 30 cm depth (Figure 2), where the gas diffused mostly upwards but also slightly downward, thus raising CH_4 concentration in deeper layers to levels that did not allow for any further production. This created the remarkably straight CH_4 concentration depth profiles in the model that were also observed in the empirical data (Figure 2).

The sensitivity analysis illustrates that concentration profiles were highly dependent on both the energy threshold and substrate concentration, and to a lesser degree on the kinetic V_{max} and k_s values. This assumes a reasonable range of these parameters in anoxic environments and confirms earlier studies (*55*). The finding is exemplified in the 50% peat column after 550 d (Figure 4). In each of the analyses, all parameters except one were kept constant.

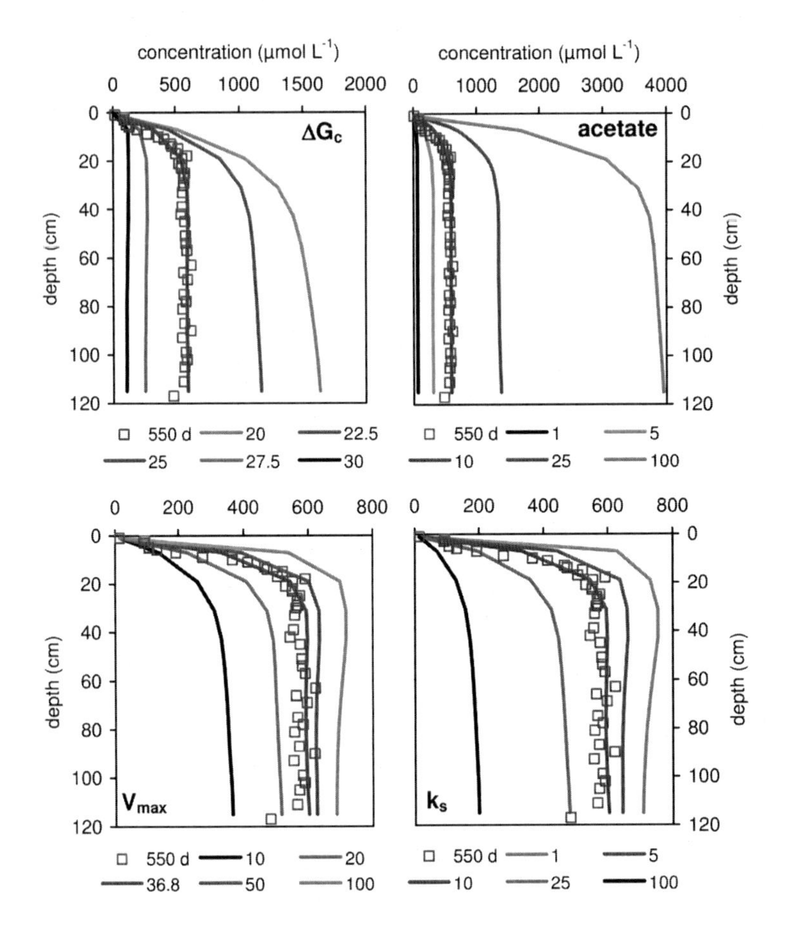

Figure 4. Sensitivity of modeled CH_4 concentrations in 50 % peat columns to changes in threshold energy ΔG_c, acetate concentration, maximum production rate V_{max}, and half saturation constant k_s, while keeping all other parameters constant in each simulation. When not varied in a simulation, ΔG_c= 25 kJ mol^{-1}, V_{max} =36.8 μmol L^{-1} d^{-1}, acetate = 10 μmol L^{-1}, and k_s = 0.0028 d^{-1}.

Moderate changes in the physiologically based energy threshold of ± 5 kJ mol^{-1}, which are smaller than widely reported and may reflect adaptations to the general availability of substrates (*6*), can thus lead to very different steady state CH_4 levels in methanogenic peat deposits, which may contribute to large variations that have reported from field sites (*57, 62*). A similarly strong influence of acetate concentration on the CH_4 concentrations was observed at an energy threshold of -25 kJ mol^{-1}. The parameters V_{max} and k_s primarily influenced the temporal dynamics in the model, i.e. the time required for CH_4 profiles to reach steady state near the energy threshold (Figure 4). Higher V_{max} and k_s also reinforce the already described phenomenon of increases of CH_4 concentrations beyond levels stipulated by ΔG_c in lower depths of this semi-closed soil system. This is caused by more vigorous CH_4 production in the upper layers, where DIC concentrations are lower than in larger depths, thus allowing for higher CH_4 concentrations at the energy threshold.

The effect of acetate concentration on CH_4 concentration levels stresses the importance of the dynamic equilibrium between acetate production and methanogenic consumption near energy thresholds. This point is further illustrated in Figure 5, which shows ΔG_r available for the fermenatation of propionate to acetate and its subsequent consumption by acetoclastic methanogenesis over a range of DIC, i.e. dissolved CO_2, and CH_4 concentrations. Whereas at low DIC and CH_4 concentrations both processes can occur over a wide range of acetate concentrations, in the example between 0.1 and 10 µmol L^{-1}, this range decreases with accumulation of DIC and CH_4 to a narrow band near 10 µmol L^{-1}. Deviation from this band would inhibit either process, and the effect of higher acetate concentration could only be compensated for by elevating concentrations of propionate. The 'energy signal' of DIC and CH_4 accumulation may be distributed through the preceding fermentation network, which could, for example, help to explain high levels of acetate and propionate in groundwater systems (*57*, *63*). Thus, temporary or local increases of acetate concentrations that have been widely reported for example in wetland (*64*, *65*) and aquifer systems (*63*, *66*) have the potential to ease thermodynamic constraints on acetoclastic methanogenesis, but, on the other hand, may inhibit preceding fermentation processes, as suggested in organic rich peat aquifers (*38*, *57*).

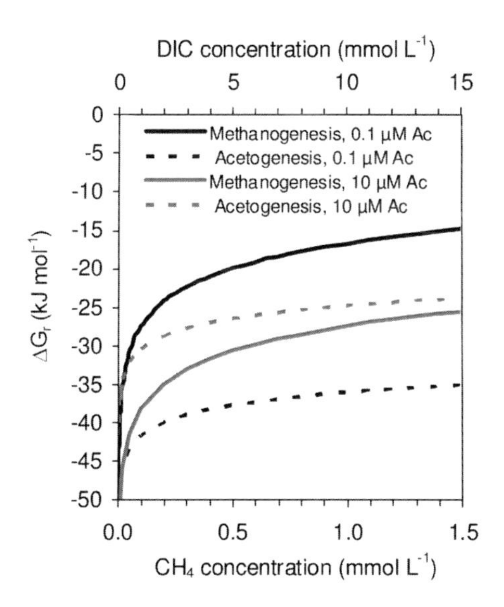

Figure 5. Gibbs free energy ΔG_r available from acetoclastic methanogenesis and syntrophic propionate fermentation to acetate (CH_3CH_3COOH (aq) + $2H_2O$ (l) $\Rightarrow$ CH_3COOH (aq) + CO_2 (aq) + 3 H_2 (aq)) with increasing DIC and CH_4 concentration. Assuming propionate concentrations of 1 µmol L^{-1} and either 0.1 or 10 µmol L^{-1} of acetate. DIC is assumed to be dissolved CO_2.

Conclusions

This review and case study demonstrate the usefulness of considering the energy threshold concept when interpreting the occurrence, rate and spatio-temporal distribution of H_2 and organic acid dependent TEAPs and methanogenesis in anoxic systems. H_2-dependent methanogenesis and sulfate reduction typically operate near free energy thresholds (ΔG_c) in soils and sediments, and can thus be inhibited by changes in geochemical conditions that are induced by constraints on transport of solutes and by provision of TEAPs that are energetically more potent. These TEAPS operate far from theoretical energy thresholds, despite higher physiological energy threshold levels, and can lower H_2 concentrations to levels excluding less potent TEAPs. Energy thresholds are dependent on microbial physiology and occur when the energy conserved by ATP generation comes close to the thermodynamic driving force. The case study illustrates that accumulation of methane in diffusion dominated and DIC rich systems may cause inhibition of acetoclastic methanogenesis near energy thresholds, and the high sensitivity of such a mechanism to small changes in acetate concentration and the energy threshold ΔG_c. The production and flux of CH_4, as the most reduced metabolic product of anaerobic organic matter decomposition, appeared to reflect the system's ability to simultaneously adjust to transport, energy, and microbial kinetic constraints on the redox process. Low energy, in particular methanogenic, systems may thus operate in a mode in which processes and constraints converge towards a characteristic steady state that reflects a specific capacity to process organic material and that is likely highly dependent on rates of solute transport.

Acknowledgments

I gratefully acknowledge the technical support by M. Heider, H. Zier, S. Hammer, L.Likke, and D. Burghardt, and contributions made by M. Siems and J. Beer to data analysis.

References

1. Lovley, D. R.; Coates, J. D. Novel forms of anaerobic respiration of environmental relevance. *Curr. Opin. Microbiol.* **2000**, *3* (3), 252–256.
2. Appelo, C. A. J.; Postma, D. *Geochemistry, groundwater and pollution*, 2nd ed.; A. A. Balkema Publishers: Leiden, The Netherlands, 2005; p 649.
3. Froehlich, P. N.; Klinkhammer, G. P.; Bender, M. L.; Luedke, N. A.; Heath, G. R.; Gullen, D.; Dauphin, P.; Hammond, D.; Hartman, B. Early oxidation of organic matter in pelagic sediments of the eastern equatorial Atlantic: suboxic diagenesis. *Geochim. Cosmochim. Acta* **1979**, *43*, 1075–1090.
4. Thauer, R. K.; Jungermann, K.; Decker, K. Energy-Conservation in Chemotropic Anaerobic Bacteria. *Bacteriol. Rev.* **1977**, *41* (1), 100–180.
5. Schink, B. Energetics of syntrophic cooperation in methanogenic degradation. *Microbiol. Mol. Biol. Rev.* **1997**, *61* (2), 262–280.

6. Hoehler, T. M. Biological energy requirements as quantitative boundary conditions for life in the subsurface. *Geobiology* **2004**, *2*, 205–215.
7. Jin, Q. S.; Bethke, C. M. Predicting the rate of microbial respiration in geochemical environments. *Geochim. Cosmochim. Acta* **2005**, *69* (5), 1133–1143.
8. Jakobsen, R. Redox microniches in groundwater: A model study on the geometric and kinetic conditions required for concomitant Fe oxide reduction, sulfate reduction, and methanogenesis. *Water Resour. Res.* **2007**, *43*, W12S12; doi:10.1029/2006WR005663.
9. Fenchel, T.; King, G. M.; Blackburn, T. H. *Bacterial Biogeochemistry*; Academic Press: London, 1998.
10. Decad, G. M.; Nikaido, H. Outer membrane of gram-negative bacteria. XII Molecular-sieving function of cell wall. *J. Bacteriol.* **1976**, *128*, 325–336.
11. Conrad, R. Contribution of hydrogen to methane production and control of hydrogen concentrations in methanogenic soils and sediments. *FEMS Microbiol. Ecol.* **1999**, *28* (3), 193–202.
12. Vignais, P. M.; Colbeau, A. Molecular biology of microbial hydrogenases. *Curr. Issues Mol. Biol.* **2004**, *6*, 159–188.
13. Hering, J.; Hug, S. J.; Farnsworth, C.; O'Day, P. A. Role of coupled redox transformations in the mobilization and sequestration of arsenic. In *Aquatic Redox Chemistry*, Tratnyek, P., Grundl, T. J., Haderlein, S. B., Eds.; American Chemical Society: Washington, DC, 2011; Vol. 1071, Chapter 21, pp 463−476.
14. O'Loughlin, E. J.; Boyanov, M. I.; Antonopoulos, D. A.; Kemner, K. M. Redox processes affecting the speciation of technetium, uranium, neptunium, and plutonium in aquatic and terrestrial environments. In *Aquatic Redox Chemistry*, Tratnyek, P., Grundl, T. J., Haderlein, S. B., Eds.; American Chemical Society: Washington, DC, 2011; Vol. 1071, Chapter 22, pp 477−517.
15. Conrad, R.; Goodwin, S.; Zeikus, J. G. Hydrogen metabolism in a mildly acidic lake sediment (Knaack Lake). *FEMS Microbiol. Ecol.* **1987**, *45* (4), 243–249.
16. Kurtz, W.; Peiffer, S., The role of transport in aquatic redox chemistry. In *Aquatic Redox Chemistry*, Tratnyek, P., Grundl, T. J., Haderlein, S. B., Eds.; American Chemical Society: Washington, DC, 2011; Vol. 1071, Chapter 25, pp 559−580.
17. Lovley, D. R.; Goodwin, S. Hydrogen concentrations as an indicator of the predominant terminal electron accepting reactions in aquatic sediments. *Geochim. Cosmochim. Acta* **1988**, *52*, 2993–3003.
18. Lovley, D. R.; Chapelle, F. H. Deep subsurface microbial processes. *Rev. Geophys.* **1995**, *33* (3), 365–381.
19. Chapelle, F. H.; Lovley, D. R. Competitive exclusion of sulfate reduction by Fe(III)-reducing bacteria: a mechanism for producing discrete zones of high-iron ground water. *Ground Water* **1992**, *30* (1), 29–36.
20. Heimann, A.; Jakobsen, R.; Blodau, C. Energetic constraints on H_2-dependent terminal electron accepting processes in anoxic environments

- a review of observations and model approaches. *Environ. Sci. Technol.* **2010**, *44* (1), 24–33.
21. Vroblesky, D. A.; Chapelle, F. H. Temporal and spatial changes of terminal electron-accepting processes in a petroleum hydrocarbon-contaminated aquifer and the significance for contaminant biodegradation. *Water Resour. Res.* **1994**, *30* (5), 1561–1570.
22. Chapelle, F. H.; McMahon, P. B.; Dubrovsky, N. M.; Fujii, R. F.; Oaksford, E. T.; Vroblesky, D. A. Deducing the distribution of terminal electron-accepting processes in hydrologically diverse groundwater systems. *Water Resour. Res.* **1995**, *31* (2), 359–371.
23. Christensen, T. H.; Bjerg, P. L.; Banwart, S. A.; Jakobsen, R.; Heron, G.; Albrechtsen, H. J. Characterization of redox conditions in groundwater contaminant plumes. *J. Contam. Hydrol.* **2000**, *45* (3−4), 165–241.
24. Gonsoulin, M. E.; Wilson, B. H.; Wilson, J. T. Biodegradation of PCE and TCE in landfill leachate predicted from concentrations of molecular hydrogen: a case study. *Biodegradation* **2004**, *15* (6), 475–485.
25. Lovley, D. R.; Phillips, E. J. P. Competitive mechanisms for inhibition of sulfate reduction and methane production in the zone of ferric iron reduction in sediments. *Appl. Environ. Microbiol.* **1987**, *53* (11), 2636–2641.
26. Stumm, W.; Morgan, J. J. *Aquatic chemistry - chemical equilibria and rates in natural waters*; Wiley-Interscience: New York, 1996.
27. Krauskopf, K. B.; Bird, D. K. *Introduction to Geochemistry*, 3rd ed.; McGraw-Hill: 1994; p 640.
28. Thauer, R. K.; Jungermann, K.; Decker, K. Energy conservation in chemotrophic anaerobic bacteria. *Bacteriol. Rev.* **1977**, *41* (1), 100–180.
29. Heimann, A.; Blodau, C.; Postma, D.; Larsen, F.; Viet, P. H.; Nhan, P. Q.; Jessen, S.; Duc, M. T.; Hue, N. T. M.; Jakobsen, R. Hydrogen thresholds and steady state concentrations associated with microbial arsenate respiration. *Environ. Sci. Technol.* **2007**, *41*, 2311–2317.
30. Jakobsen, R.; Albrechtsen, H. J.; Rasmussen, M.; Bay, H.; Bjerg, P. L.; Christensen, T. H. H_2 concentrations in a landfill leachate plume (Grindsted, Denmark): In situ energetics of terminal electron acceptor processes. *Environ. Sci. Technol.* **1998**, *32* (14), 2142–2148.
31. McGuire, J. T.; Smith, E. W.; Long, D. T.; Hyndman, D. W.; Haack, S. K.; Klug, M. J.; Velbel, M. A. Temporal variations in parameters reflecting terminal-electron-accepting processes in an aquifer contaminated with waste fuel and chlorinated solvents. *Chem. Geol.* **2000**, *169* (3−4), 471–485.
32. Cozzarelli, I. M.; Suflita, J. M.; Ulrich, G. A.; Harris, S. H.; Scholl, M. A.; Schlottmann, J. L.; Christenson, S. Geochemical and microbiological methods for evaluating anaerobic processes in an aquifer contaminated by landfill leachate. *Environ. Sci. Technol.* **2000**, *34* (18), 4025–4033.
33. Hansen, L. K.; Jakobsen, R.; Postma, D. Methanogenesis in a shallow sandy aquifer, Romo, Denmark. *Geochim. Cosmochim. Acta* **2001**, *65* (17), 2925–2935.
34. Watson, I. A.; Oswald, S. E.; Mayer, K. U.; Wu, Y. X.; Banwart, S. A. Modeling kinetic processes controlling hydrogen and acetate concentrations

in an aquifer-derived microcosm. *Environ. Sci. Technol.* **2003**, *37* (17), 3910–3919.

35. Broholm, M. M.; Crouzet, C.; Arvin, E.; Mouvet, C. Concurrent nitrate and Fe(III) reduction during anaerobic biodegradation of phenols in a sandstone aquifer. *J. Contam. Hydrol.* **2000**, *44* (3-4), 275–300.
36. Jakobsen, R.; Postma, D. Redox zoning, rates of sulfate reduction and interactions with Fe-reduction and methanogenesis in a shallow sandy aquifer, Romo, Denmark. *Geochim. Cosmochim. Acta* **1999**, *63* (1), 137–151.
37. Hoehler, T. M.; Alperin, M. J.; Albert, D. B.; Martens, C. S. Thermodynamic control on hydrogen concentrations in anoxic sediments. *Geochim. Cosmochim. Acta* **1998**, *62* (10), 1745–1756.
38. Beer, J.; Blodau, C. Transport and thermodynamics constrain belowground carbon turnover in a northern peatland. *Geochim. Cosmochim. Acta* **2007**, *71*, 2989–3002.
39. Rothfuss, F.; Conrad, R. Thermodynamics of methanogenic intermediary metabolism in littoral sediment of Lake Constance. *FEMS Microbiol. Ecol.* **1993**, *12* (4), 265–276.
40. Chin, K.-J.; Conrad, R. Intermediary metabolism in methanogenic paddy soil and the influence of temperature. *FEMS Microbiol. Ecol.* **1995**, *18* (2), 85–102.
41. Hoehler, T. M.; Alperin, M. J.; Albert, D. B.; Martens, C. S. Apparent minimum free energy requirements for methanogenic Archaea and sulfate-reducing bacteria in an anoxic marine sediment. *FEMS Microbiol. Ecol.* **2001**, *38* (1), 33–41.
42. Nicholls, D. G.; Ferguson, S. J. *Bioenergetics 3*, 3rd ed.; Academic Press, Elsevier Science: 2002.
43. Valentine, D. L. Adaptations to energy stress dictate the ecology and evolution of the Archaea. *Nat. Rev. Microbiol.* **2007**, *5* (4), 316–323.
44. Mulkidjanian, A. Y.; Dibrov, P.; Galperin, M. Y. The past and present of sodium energetics: May the sodium-motive force be with you. *Biochim. Biophys. Acta, Bioenerg.* **2008**, *1777*, 985–992.
45. Becher, B.; Muller, V. Delta-mu(Na+) drives the synthesis of ATP via an delta-mu(Na+)-translocating F1F0-ATP synthase in membrane-vesicles of the Archaeon *Methanosarcina mazei* Go1. *J. Bacteriol.* **1994**, *176* (9), 2543–2550.
46. Jackson, B. E.; McInerney, M. J. Anaerobic microbial metabolism can proceed close to thermodynamic limits. *Nature* **2002**, *415* (6870), 454–456.
47. Drake, H. L.; Küsel, K.; Mathiess, C. Acetogenic Prokaryotes. *Prokaryotes* **2006**, *2*, 355–420.
48. Kleerebezem, R.; Stams, A. J. M. Kinetics of syntrophic cultures: A theoretical treatise on butyrate fermentation. *Biotechnol. Bioeng.* **2000**, *67* (5), 529–543.
49. Heimann, A.; Jakobsen, R. Experimental evidence for a lack of thermodynamic control on hydrogen concentrations during anaerobic degradation of chlorinated ethenes. *Environ. Sci. Technol.* **2006**, *40* (11), 3501–3507.

50. Kocar, B.; Fendorf, S. Thermodynamic constraints on reductive reactions influencing the biogeochemistry of arsenic in soils and sediments. *Environ. Sci. Technol.* **2009**, *43*, 4871–4877.
51. Liu, C. X.; Kota, S.; Zachara, J. M.; Fredrickson, J. K.; Brinkman, C. K. Kinetic analysis of the bacterial reduction of goethite. *Environ. Sci. Technol.* **2001**, *35* (12), 2482–2490.
52. Roden, E. E. Diversion of electron flow from methanogenesis to crystalline Fe(III) oxide reduction in carbon-limited cultures of wetland sediment microorganisms. *Appl. Environ. Microbiol.* **2003**, *69* (9), 5702–5706.
53. Roden, E. E.; Urrutia, M. M. Influence of biogenic Fe(II) on bacterial crystalline Fe(III)oxide reduction. *Geomicrobiol. J.* **2002**, *19*, 209–251.
54. Pedersen, H. D.; Postma, D.; Jakobsen, R.; Larsen, O. Fast transformation of iron oxyhydroxides by the catalytic action of aqueous Fe(II). *Geochim. Cosmochim. Acta* **2005**, *69* (16), 3967–3977.
55. Hoh, C. Y.; Cord-Ruwisch, R. A practical kinetic model that considers endproduct inhibition in anaerobic digestion processes by including the equilibrium constant. *Biotechnol. Bioeng.* **1996**, *51* (5), 597–604.
56. Krylova, N. I.; Conrad, R. Thermodynamics of propionate degradation in methanogenic paddy soil. *FEMS Microbiol. Ecol.* **1998**, *26* (4), 281–288.
57. Beer, J.; Lee, K.; Whiticar, M.; Blodau, C. Geochemical controls on anaerobic organic matter decomposition in a northern peatland. *Limnol. Oceanogr.* **2008**, *53* (4), 1393–1407.
58. Lerman, A. *Geochemical Processes - Water and Sediment Environments*; Krieger: Malabar, FL, 1988.
59. Blodau, C.; Moore, T. R. Experimental response of peatland carbon dynamics to a water fluctuation. *Aquat. Sci.* **2003**, *65*, 47–62.
60. Segers, R. Methane production and methane consumption: a review of processes underlying wetland methane fluxes. *Biogeochemistry* **1998**, *41* (1), 23–51.
61. Blodau, C.; Roulet, N. T.; Heitmann, T.; Stewart, H.; Beer, J.; Lafleur, P.; Moore, T. R. Belowground carbon turnover in a temperate ombrotrophic bog. *Global Biogeochem. Cycles* **2007**, *21* (1), GB1021.
62. Chasar, L. S.; Chanton, J. P.; Glaser, P. H.; Siegel, D. I.; Rivers, J. S. Radiocarbon and stable carbon isotopic evidence for transport and transformation of dissolved organic carbon, dissolved inorganic carbon, and CH_4 in a northern Minnesota peatland. *Global Biogeochem. Cycles* **2000**, *14* (4), 1095–1108.
63. McMahon, P. B.; Chapelle, F. H. Microbial production of organic acids in aquitard sediments and its role in aquifer geochemistry. *Nature* **1991**, *349*, 233–235.
64. Duddleston, K. N.; Kinney, M. A.; Kiene, R. P.; Hines, M. E. Anaerobic microbial biogeochemistry in a northern bog: Acetate as a dominant metabolic end product. *Global Biogeochem. Cycles* **2002**, *16* (4), 1–9.
65. Shannon, R. D.; White, J. R. The effects of spatial and temporal variations in acetate and sulfate on methane cycling in two Michigan peatlands. *Limnol. Oceanogr.* **1996**, *41* (3), 435–443.

66. Chapelle, F. H.; Bradley, P. M. Microbial acetogenesis as a source of organic acids in ancient Atlantic Coastal Plain sediments. *Geology* **1996**, *24*, 925–928.

Chapter 5

Redox Chemistry and Natural Organic Matter (NOM): Geochemists' Dream, Analytical Chemists' Nightmare

Donald L. Macalady[1,*] and Katherine Walton-Day[2]

[1]Department of Chemistry and Geochemistry, Colorado School of Mines, Golden, CO 80401
[2]U.S. Geological Survey, Colorado Water Science Center, Denver Federal Center, Box 25046, MS 415, Denver, CO 80225
[*]dmacalad@mines.edu

Natural organic matter (NOM) is an inherently complex mixture of polyfunctional organic molecules. Because of their universality and chemical reversibility, oxidation/reductions (redox) reactions of NOM have an especially interesting and important role in geochemistry. Variabilities in NOM composition and chemistry make studies of its redox chemistry particularly challenging, and details of NOM-mediated redox reactions are only partially understood. This is in large part due to the analytical difficulties associated with NOM characterization and the wide range of reagents and experimental systems used to study NOM redox reactions. This chapter provides a summary of the ongoing efforts to provide a coherent comprehension of aqueous redox chemistry involving NOM and of techniques for chemical characterization of NOM. It also describes some attempts to confirm the roles of different structural moieties in redox reactions. In addition, we discuss some of the operational parameters used to describe NOM redox capacities and redox states, and describe nomenclature of NOM redox chemistry. Several relatively facile experimental methods applicable to predictions of the NOM redox activity and redox states of NOM samples are discussed, with special attention to the proposed use of fluorescence spectroscopy to predict relevant redox characteristics of NOM samples.

Introduction

Natural organic matter (NOM) is an inherently complex and inseparable group of molecules (*1*) primarily resulting from the partial decay of senescent plant materials and microorganisms. It is ubiquitous in all natural waters and soils. In addition to its redox chemistry, it plays significant roles in metal transport, microbial, photochemical and water treatment processes and in soil fertility. Because of its complexity and wide range of molecular sizes and physical properties, NOM has often been conceptually divided into separate classes of molecules. This classification has generally been based on solubility and pH properties of NOM samples, and a considerable collection of terms has been applied to operationally defined NOM fractions. Soil scientists define soil NOM as humic substances, which are composed of three fractions. One is water insoluble at all pH values (humin) and two are defined in terms of the base-solubilized portions of soil NOM. Humic acid is base soluble but re-precipitates at slightly acid pH. Fulvic acids are solubilized by base and remain in solution under acid conditions (*2*).

In aquatic science, solution-phase NOM is similarly divided into classes such as hydrophobic acids, hydrophilic acids, hydrophilic bases, and hydrophilic neutrals. The portions characterized as fulvic and humic acids (AHS, aquatic humic substances) reside primarily in the hydrophobic acid fraction. Frequently applied isolation techniques for aquatic NOM (after filtration to remove particulate fractions, generally using 0.45-μm membranes) include column chromatographic methods with non-ionic macroporous polymer sorbents composed of styrene divinylbenzene (e.g. XAD-2, XAD-4) or acrylic esters (e.g. XAD-7, XAD-8, DAX-8) (*3*). Weakly basic ion exchange resins (e.g. DEAE-cellulose) also have been successfully used to isolate AHS from water without the pH adjustment needed in the XAD-chromatography (*4*). Solid-phase extraction using modified styrene divinyl benzene polymer types of sorbents has also been applied to NOM isolation, especially from seawater (*5*). Tangential flow ultrafiltration methods have been successfully used to isolate dissolved NOM fractions based on nominal molecular mass fractions (*6, 7*). Finally, reverse osmosis combined with electrodialysis rounds out the rather long list of methods used to isolate NOM from natural waters (*8*).

Aquatic NOM in rivers and lakes is generally composed of 45-55% AHS (of which 80-90% is classified as fulvic acids), 20-30% hydrophilic acids, 15-20 % hydrophilic neutrals, and 1-5 % hydrophilic bases. For a detailed discussion of the fractional compositions of natural waters according to these classification schemes, refer to the book by Thurman (*9*). Standardized samples of AHS, including aquatic humic and fulvic acids have been made commercially available through the International Humic Substances Society. Less fully characterized commercial humic acids, such as Aldrich Humic Acid, are also commonly used in NOM research.

Standardized fractions of NOM are commonly used in research as a means of providing consistency to the many experimental procedures and approaches used to characterize NOM. One disadvantage of such practices is that they mask differences in the characteristics of NOM samples of different origins or from different water bodies. This can be especially important, as will be discussed in more detail below, for investigations of the redox reactions involving NOM. In addition, there is concern that NOM can be altered to varying extents during chemical or physical isolation, and the isolated fractions may not be representative of the material in its natural state. Nevertheless, most of the references contained in this chapter refer to NOM fractions rather than NOM. This is an inherent, but unavoidable, limitation to our review, and an inherent limitation to almost all of the research that has been reported for NOM. The disadvantages of using NOM fractions instead of NOM should be considered in evaluations of the validity of the research reported in this, or any other, discussion of NOM.

The objective of this chapter is to provide some context to and understanding of ongoing efforts to provide a coherent comprehension of aqueous oxidation/reduction (redox) chemistry involving natural organic matter (NOM). Initially, techniques for chemical characterization of NOM are presented to provide context for the subsequent discussions of redox-active constituents of NOM and redox chemistry of NOM. This is followed by discussions of those moieties in NOM that have been shown to be redox-active, and the methods and nomenclature used in investigations of the reversible redox behavior of NOM. Finally, methods applicable to predictions of NOM redox activity and redox state are outlined, with special consideration of the proposed use of statistical analyses of fluorescence spectral characteristics of NOM samples to predict the redox characteristics.

Chemical Characterization of NOM

Because this chapter focuses on redox reactions in aquatic environments, we are primarily interested in the nature of water soluble NOM, commonly called dissolved organic matter (DOM). Aquatic NOM commonly contains about 45-55% carbon, 35-45% oxygen, 3-5% hydrogen and 1-4% nitrogen by mass, with variable smaller amounts of sulfur (0.4-0.6 %) and phosphorus (0.1-0.45%). This elemental composition tells little, of course, about the chemical composition of aquatic NOM, but does provide some insight into the nature of the molecular components. For example, elemental ratios such as the C/H or C/N have been used to point to relative aromatic versus aliphatic content or the nature of biogenic precursors of NOM samples (*9–11*). Also, the analytical chemistry that defines the amount of NOM in a water sample invariably measures organic carbon, not organic matter. The elemental composition suggests that DOM, or aquatic NOM, content is close to twice the dissolved organic carbon (DOC) content.

Chemically, only a very small fraction of aquatic NOM contains identifiable organic compounds. These are mostly small organic acids such as acetic, citric, malic and other biogenic acids, including amino acids. Because these compounds are generally bioavailable, their presence is variable and transient. The more recalcitrant and enduring molecular components of NOM are primarily composed of complex organic acids with a variety of other functional group constituents. It is difficult to give precise descriptions of the functional group compositions of aquatic NOM, in part because almost all of the work to elucidate the nature and abundances of functional groups in NOM has been conducted on NOM fractions rather than on whole water NOM samples. However, a great deal of qualitative or semi-quantitative structural information has been learned through such investigations.

In the pursuit of the characterization of the structural features of NOM, a wide variety of spectroscopic and chromatographic techniques have been employed (e.g., (*12, 13*)), including those that derivatize, pyrolyze or otherwise alter the original moieties and molecules in NOM fractions. Prominent among the non-destructive methods are infrared (IR), ultraviolet-visible (UV-vis), fluorescence, and nuclear magnetic resonance (NMR) spectroscopies, mass spectrometry and all of its variations, X-ray techniques [e.g. X-ray fine-edge absorption spectroscopy (XFAS)], and chromatographic variants such as gel electrophoresis (e.g., (*14–18*)), plus many other references in this chapter). All of these techniques have their individual uses and limitations, advantages and disadvantages. The bottom line information about NOM chemical composition is that oxygen-containing functional groups are distributed among aliphatic and aromatic moieties, some of which may contain nitrogen or sulfur atoms in their structures. So, a preferred approach is to limit descriptions of the NOM in a given natural water sample in terms of dissolved organic carbon content (DOC) and the fraction of this carbon that exists as aromatic moieties or aliphatic structures. Further specification might involve the abundance of phenolic or ketone-like functionality in these fractions. The relative importance of sulfur and nitrogen functionalities can also be specified. In the context of this chapter, the important consideration is the relative abundance of functionalities that can be expected to be involved in reversible redox chemistry.

An additional aspect of NOM structure, ignored in most analytical investigations of NOM, is the state of complexation and/or chelation of NOM moieties with dissolved metal ions. NOM/metal ion complexes and/or chelates have been known and studied for decades (*19–21*), and represent an important component of many natural NOM samples. Complexation reactions have repeatedly been shown to primarily involve substituted aromatic constituents of NOM (*19*, *22–24*). As will be shown in the section on "Redox-Active NOM Constituents", these also are the NOM components thought to be most directly involved in redox transformations. Thus, the presence of metal complexes can be a critical structural feature in considerations of NOM redox chemistry.

Clearly, detailed information about the total functional group composition of a given NOM sample is beyond the capabilities of many analytical laboratories. One can know with reasonable assurance that certain functional groups are present, and that carbon-oxygen bonds dominate the functional group composition. In addition, several relatively facile laboratory procedures can provide useful information about important structural features of a given sample.

A procedure that can provide useful information about relative amounts of carboxylic and phenolic functional groups is a simple pH titration (*25*). Here, the sample is adjusted to pH ≈ 2 and subsequently titrated with a standard base solution. Titration to a pH of 7 deprotonates essentially all of the carboxylic acids in the sample. Further titration to a pH of 10 deprotonates the phenolic constituents in the sample. Normalization of these titrations to the organic carbon content gives a reliable indication of the relative amounts of carboxylate and phenolic moieties, for example, in moles of COOH per mole of carbon. Because the DOC concentrations of natural water samples are low, varying from <80 micromolar carbon for seawater to several millimolar carbon for certain surface waters or sediment pore waters, there is a serious limitation in the possible precision of such titrations (*26*). In addition, the accuracy of the procedure is limited somewhat by a slight overlap in pK_a values of phenols and carboxylic acids in NOM. Use of spectrophotometric titrations can alleviate some of this difficulty (*26*).

A second procedure is designed to give a relative indication of the aromatic contents of natural organic matter samples. The technique measures the specific ultraviolet absorption (SUVA) of a water sample, defined as the absorbance (1.0 cm cell) of the sample at 254 nm (or some other specified ultraviolet wavelength) divided by the organic carbon content of the sample in mg/L (*27*). The method is not reliable in the presence of substantial concentrations of transition metals, especially iron, or nitrogen compounds such as amino acids, nitrate or nitrite.

A third relatively simple procedure is the measurement of the intensities of infrared absorption bands by diffuse reflectance Fourier-transformed infrared spectroscopy (FTIR) (*28*). Though conceptually simple, caution must be used in the application of FTIR to NOM samples, as many of the observed peaks have pronounced pH dependence. However, at circumneutral pH values absorption at a characteristic wave number of 1725 cm^{-1} can be ascribed to C=O stretch of COOH and at 1620 cm^{-1} to aromatic C=C stretch/asymmetric $-COO^{-}$ stretch. An absorption peak/shoulder at 1270 cm^{-1} is due to C–OH stretch of phenolic –OH. Finally, absorption in the range of 1050 – 1200 cm^{-1} is due to mostly aliphatic –OH typical for carbohydrates. Infrared spectra of several NOM samples are shown in Fig. 1 (*29*) to illustrate that differences among NOM samples are clearly revealed by these IR spectra. The utility of such measurements in terms of redox chemistry will be discussed below.

Finally, fluorescence spectroscopic measurements have been proposed to be useful in assessing the redox properties of NOM. This also will be discussed in more detail in the section on "Predicting NOM Redox Activity".

Redox-Active NOM Constituents

It is important at this point to distinguish among several possible modes of participation of NOM in redox chemistry. It is of course true, that powerful oxidants can virtually destroy NOM and render it into simple inorganic oxides or small organic acids. We are not concerned here with these reactions, as the remarkable characteristic of the participation of NOM in aquatic redox reactions is their reversibility. So, we will limit our discussion to moieties in NOM that can reasonably be expected to participate in reversible redox reactions. Another limitation is that we are generally not interested in redox processes that occur outside of the limits of stability of water itself. In terms of oxidation/reduction potential, this means that we are interested in the electron potentials, expressed as E_H, in the range (at 298 K):

$$-0.0591\bullet(\mathrm{pH}) \leq E_H \leq +1.23 + 0.015\bullet\log[O_2] -0.0591\bullet(\mathrm{pH}),$$

where $[O_2]$ is the partial pressure of dissolved oxygen in atmospheres (*29*). Note that kinetic constraints sometimes allow redox couples to be significant even though they are outside of this range of thermodynamic stability for water.

These constraints limit the nature of the reversible redox reactions that can be expected with NOM as an oxidant or reductant. For example, a redox couple consisting of an aliphatic ketone and the corresponding alcohol cannot be expected to be part of the aquatic redox chemistry of NOM. So, what are the functional groups within NOM that can be involved in redox chemistry? As is discussed in detail below, the most widely accepted group of reversible redox-active moieties in NOM are quinones or quinone-like moieties. Fig. 2 illustrates the structures of some example quinones and an example of the reduced forms.

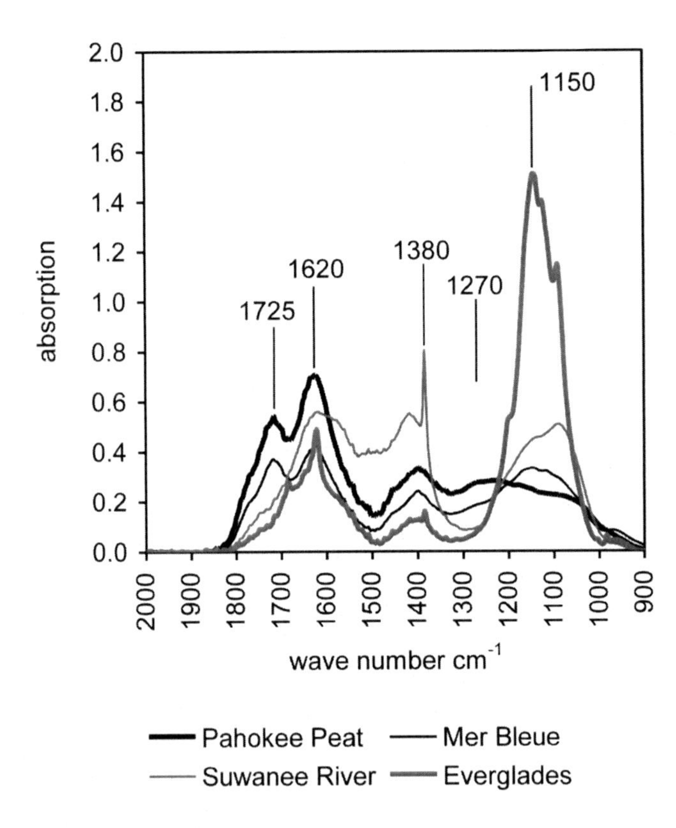

Figure 1. FTIR spectra of selected DOM samples. FTIR absorption intensity at 1725 cm^{-1} is ascribed to C=O stretch in COOH, 1620 cm^{-1} to aromatic C=C stretch, 1270 cm^{-1} to C–OH stretch of phenolic C, and 1150 cm^{-1} to C–OH stretch of aliphatic OH (29).

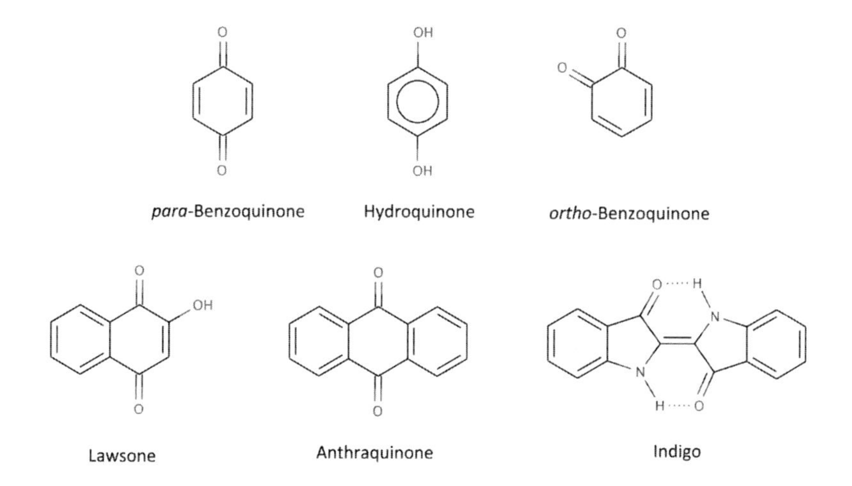

Figure 2. Structures of some quinones.

It is important to note that none of these molecules has been shown to be present in NOM. The structures are shown as examples only. There can be a considerable number of structural features imbedded within NOM molecules that exhibit quinone-like redox behavior, and redox transformations of these moieties can span a wide range of accessible electron potentials. The presence of a one-electron transfer reaction to semiquinone-type free radical intermediates further enhances the possible roles of such moieties in reversible redox transformations of NOM (Fig. 3).

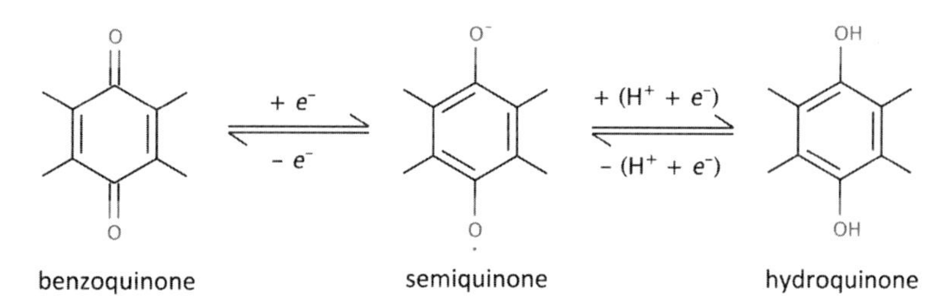

Figure 3. Structural diagram showing the redox reaction of a quinone, through the free-radical semiquinone intermediate to the fully reduced hydroquinone.

The presence of free radicals in NOM samples has been verified many times (recently, e.g. (*30*)). This reference reports the use of electron paramagnetic resonance (EPR) to show the presence of semiquinone-type radicals in aquatic NOM, along with smaller concentrations of carbon-centered "aromatic" radicals. At alkaline pH values, the semiquinone-type radicals dominate. Organic radical concentrations in NOM adjusted to pH 6.5 before freeze-drying are related to iron and aluminum contents.

Ultraviolet and visible irradiation of solid NOM can lead to more than a 10-fold increase of the concentration of organic radicals. The radicals are long-lived and have the same EPR properties as the original radicals. Similar effects were not observed with isolated humic- and fulvic acids, demonstrating the limited relationships of the environmental properties of these isolates to the original NOM in the samples (*30*). The roles of quinone-like free radicals in electron transfer reactions involving microorganisms also have been established (*31*). The importance of quinone-like molecules in the redox chemistry of natural organic matter is further discussed in elsewhere in this volume (*32*, *33*).

The discussion above illustrates that one of the most important "moieties" in the determination of NOM redox properties is the proton. The effects of pH on the properties of NOM, including metal chelation, molecular aggregation, optical properties and redox behavior, cannot be underestimated. Reference to almost any of the articles cited in the remainder of this chapter illustrates this importance. As one can infer from the data presented in reference (*30*), also cited above, redox activity of NOM generally increases with increasing pH (see also (*34*)).

Consideration of other possible moieties in NOM that are involved in reversible redox activity does not reveal many obvious choices. Among the most abundant structural features in NOM, only oxygen-containing structures are relevant. There are only a few literature references that report NOM redox activity that is not ascribed to aromatic, oxygenated moieties such as quinone-like or phenolic structures, and none that associate structures to such activity (*30*, *35*).

There are, however, some less common and less well investigated possibilities. In particular, sulfur species may be important in certain environments. Aqueous sulfide is known to form addition products with phenolic compounds (*36*) and fulvic acids (*34*), and these have been shown to be important in redox reactions of quinones (*37*) and fulvic acids (*38*). In addition, bisulfide linkages similar to those present in certain amino acids have been shown to reduce NOM (*39*) and also be minor constituents of native NOM samples. The possibility of redox activity by other sulfur-containing species in NOM cannot be discounted, but has not been investigated to any significant extent (*40*).

The possibility of redox activity by nitrogen-containing species, separate from aromatic nitrogen in quinone-like structures, cannot be eliminated, though there seem to be few likely candidates. Solid-state ^{13}C and ^{15}N NMR evidence shows that heterocyclic nitrogen moieties are present as very low fractions of the N content of NOM (*41*). Amines and amides are other possibilities, but their lability under environmental conditions has led some researchers to question their sustained presence in NOM. In any case, considerable disagreement seems to be present as to the dominant structural characteristics of nitrogen-containing NOM groups. A recent review (*42*) states that existing evidence supports amide N as the dominant chemical form of N in NOM, and that free amino acids are present as well. Heterocyclic N is a less significant contributor. Amide groups within NOM, which can have a net positive charge, whereas NOM has a net negative charge under typical environmental pH conditions (*43*), may have important impacts on NOM molecular conformation, interaction with mineral surfaces, and retention of nutrients or contaminants (*44*). However, little or no research has been reported as to the possible involvement of nitrogen-containing functional groups in the reversible redox chemistry of NOM.

Finally, the role of metal-organic species in NOM redox reactivity must be considered. As indicated above, metal complexation and chelation are important aspects of NOM chemistry. It is unreasonable to expect that the presence of redox-active metals such as iron within the structure of NOM molecules will not affect the redox properties of the system. Considerable research has confirmed that metal ions, particularly iron(II) and iron(III), have a strong influence on the redox chemistry of NOM. It is beyond the scope of this chapter to discuss the detailed nature of these effects, and the reader is referred to the considerable literature on this subject (see for example, (*45–48*)).

Redox Chemistry of NOM

The role of NOM in aquatic redox chemistry has most often been described in terms of its ability to act as an electron transfer mediator (*32*). Reduced NOM can be oxidized by the transfer of electrons to an electron acceptor and then be reduced again by a more abundant electron donor in the system. The details of this role are discussed in following chapters of this book (*32*, *33*), and need not be described here. One of the important variables to be quantified in assessing the ability of NOM to be involved in redox transformations is its capacity to accept (EAC) or donate (EDC) electrons in reversible reactions. Samples of NOM differ on their carbon normalized EACs and EDCs, depending upon the details of the chemical make-up of the NOM samples; that is, NOM has redox properties that depend on the source and history of the sample. Details as to the sources and causes of these variations in NOM properties are not well understood.

One problem with investigations of NOM redox chemistry is that the literature associated with determination of EACs and EDCs of NOM suffers from a lack of consistency in the terminology used to describe redox capacities and the methods used to measure them. In general, procedures to measure redox capacity involve the determination of the extent of reduction or oxidation of an external electron donating species and/or the reduction of an electron accepting species. That is:

Reduced NOM sample + Electron acceptor $\rightarrow$ Oxidized NOM; or
Oxidized NOM sample + Electron donor $\rightarrow$ Reduced NOM

Measurements of the consumption of electron donating or electron-accepting species can then be used to determine the capacity of the NOM sample to mediate redox reactions. There are several inherent conceptual difficulties with these important experiments. In principle, NOM can exist in a continuum of redox states within a range of redox potential (*33*), so discussion of oxidized or reduced NOM samples must be placed within a well-defined regime of redox potentials and system compositions. Within this range of conditions, NOM redox reactions are reversible and NOM degradation (mineralization) is minimal. Within this range of conditions, we can attempt to define a fully oxidized and a fully reduced state of NOM.

So, what is oxidized NOM? How can this state be described? The relevance of this question is perhaps most clearly illustrated by the fact that, for certain reactions, NOM samples that have been continuously exposed to light and atmospheric oxygen for long periods retain the ability to reduce certain species, for example As(V) (*49, 50*). Also, reaction of chemically reduced NOM with atmospheric oxygen and light for long periods does not remove the ability of the NOM to reduce iron(III) associated with the NOM molecules (*51*). Though redox stability in the presence of dioxygen is well known, this kinetic stability of reducing power under environmental conditions is rather remarkable. The causes and NOM structural features involved in this kinetic stability are unknown.

What is reduced NOM? Does the redox state of reduced NOM depend upon whether the sample was reduced abiotically or by microbial processes? What reactions and/or reagents should be used to determine redox capacities (the carbon-normalized number of moles of electrons to take an NOM sample from a fully oxidized to a fully reduced state)? The results clearly will depend on the redox potential of the chemical (or electrochemical) agent chosen to reduce the sample. These questions should be answered in a generally accepted manner in order to develop a comprehensive understanding of the redox capacities of NOM samples and the structural differences that cause observed differences.

To illustrate this confusion, consider the following. The amount (moles) of electrons transferred to an added oxidant by an unmodified NOM sample has been called "oxidation capacity" by Struyk and Sposito (*52*). Kappler et al. (*53*) used the name "reducing capacity" to describe the same process and "total reducing capacity" to describe electrons transferred to an oxidant from a chemically reduced NOM sample. Chen et al. (*54*) defined "oxidation capacity" to describe reaction of both microbially-reduced and non-reduced NOM samples.

Peretyazhko and Sposito (*55*) have recently proposed a unified nomenclature and a unified chemical approach to the determination of NOM redox capacities. They propose the generic term, "Reducing Capacity", to define the moles of electron charge per mole of carbon transferred by a NOM sample to an added oxidant, as observed over laboratory timescales. As special cases, they propose three new definitions: native reducing capacity (NRC) for samples with no reduction pretreatment in the laboratory; chemical reducing capacity (CRC) for samples that have been completely reduced chemically; and microbial reducing capacity (MRC) samples that have been completely reduced microbially.

They further proposed that chemical reduction be defined as reduction in the laboratory by gaseous hydrogen over a Pd/carbon catalyst. Microbial reduction capacity is assessed using an indigenous population of soil microorganisms, following the approach of Nevin and Lovley (*56*) and Kappler et al. (*53*), which is standard practice in studies of reductive transformations in natural soils (*57*). The reactions with oxidants were suggested to be standardized to the use of ferric citrate as the oxidant, by which the moles of reduction are measured as the moles of ferrous citrate produced. Using such standardized techniques, it is possible (at a given pH value) to determine the reduction capacities, and the native redox state of NOM samples in a well-defined manner, permitting comparison of NOM reducing capacities with structural parameters and with environmental conditions.

It is notable that recent experiments have been reported that provide an even more unambiguous method to measure electron donating and accepting capacities on NOM. It is somewhat obvious that the conceptually ideal method of measuring electron-accepting capacities is to directly use electrons for such measurements. Such a method has recently been reported by Aeschbacher et al. (*58*) and this method promises to provide new insights into the redox chemistry of NOM. This method, which uses straight-forward electrochemical techniques with added electron transfer mediators, can provide definitions of the electron accepting and electron donating capacities of NOM samples without the ambiguities associated with the choices of the reduction or oxidation agents employed in other techniques. It is, however, not a simple method, either conceptually or in terms of equipment and expertise required. It is our opinion that this technique, and developments that will almost certainly result from its application and improvement, represents a breakthrough in our ability to understand in detail the factors that control the redox chemistry of NOM. However, it will not replace the simple techniques described herein for routine, semi-quantitative techniques for estimation of NOM redox capacities that are usable in less well-equipped laboratories and in field analyses. It will also be limited to the types of information that can be provided by electrochemical titrations, as compared to spectrophotometric techniques.

Predicting NOM Redox Activity

We have now discussed in some detail the general nature of the chemistry of NOM, including the chemical structures that are important to redox chemistry. We also have detailed methods to define and determine the redox capacities (or reduction capacities, as suggested above) of different NOM samples. It would be helpful at this point if one could describe some reasonably accurate methods to predict the relative redox properties of NOM samples based on comparatively simple and straightforward observations and/or laboratory procedures.

The first helpful consideration is the origin of the NOM sample. As indicated near the beginning of this chapter, the origins of aquatic NOM include the decay products of senescent plant and microbial material. There is a commonly applied distinction among aquatic NOM samples relating to whether the NOM originates primarily from processes within the aqueous system under investigation (autochthonous NOM, (*59*)) or from sources within the catchment of the water body but not within the water body (allochthonous NOM, (*60*)). In fact, these distinctions are important to redox chemistry. To the extent that autochthonous NOM is derived primarily from microbial or non-woody aquatic plants (not containing lignin), and to the extent that it is not from an environment that has never been exposed to degradation processes that remove labile organic constituents such as amino acids, it contains much lower amounts of aromatic (vs. aliphatic) constituents and, therefore, has a much lower carbon-normalized reducing capacity (*61*). Allochthonous NOM, on the other hand, is dominated by the breakdown of woody plants and has a much higher aromatic content (*60*).

Allochthonous NOM is, therefore, much more redox active than autochthonous NOM (*51*).

The more aliphatic nature of autochthonous NOM means that it has a higher content of carboxylic protons relative to phenolic protons, suggesting that pH titrations, or spectrophotometric pH titrations (*26*) may provide indicators of redox activity. Samples with relatively low contents of phenolic vs. carboxylic protons can be expected to have relatively lower redox capacities. Thus one rather simple indicator of the relative redox capacities of NOM is the pH titration described earlier in this chapter.

Another possible indicator of redox activity is specific ultraviolet absorption (SUVA). This property is easily measured, as also described earlier in this chapter. In the absence of significant concentrations of iron, nitrate, nitrite or high levels of amino acids, NOM SUVA values reflect the relative content of aromatic moieties in the sample. This suggests a relationship between SUVA values and NOM redox activity, which is supported by some studies of NOM-mediated properties (*29*, *34*) but not others (*62*). Other spectral properties can be added to SUVA for a more comprehensive picture, but SUVA alone can often be helpful.

Infrared spectra also have been shown to have utility in the determination of NOM redox capacities. Blodau et al. (*29*) have recently shown that the ratios of the intensities of either the 1725, the 1620 or the 1270 cm^{-1} FTIR peaks to the intensity of the peak at 1150 cm^{-1} (see Fig. 1) correlate reasonably well with the variable EACs of NOM samples.

Finally, there has been a considerable amount of attention in the literature to the potential use of fluorescence spectroscopy to characterize NOM samples (*63*, *64*), especially to distinguish among allochthonous and autochthonous characteristics of NOM (*65*, *66*). This is made possible by the fluorescence of certain amino acids (e.g. tryptophan and tyrosine) that are residuals in autochthonous NOM sources. In addition, fluorescence spectroscopy can be useful to qualitatively differentiate not only NOM components of samples from varying origins, but also NOM subcomponents with varying compositions and functional properties. For example, polyphenolic-rich NOM fractions exhibit a much more intense fluorescence and a red shift of peak position in comparison to carbohydrate-rich NOM fractions (*67*).

More relevant to this chapter, several recent publications have reported an ability of statistical analyses of excitation emission matrixes (EEMs) produced from fluorescence spectroscopy to assess the redox state of NOM samples (*68–72*). These studies employed parallel factor analysis (PARAFAC) to resolve 3-dimensional fluorescence spectra (fluorescence intensity recorded over a range of emission and excitation wavelengths) into components that may represent fluorophores, or in complex mixtures such as NOM, may be "*approximations of the effects of other local processes (quenching or intra molecular charge transfer) occurring*" ((*73*), p. 574).

Miller et al. (*68*) presented a reducing index (RI), which is a metric calculated from the loadings of reduced and oxidized quinone-like components identified from the 13-component parallel factor analysis (PARAFAC) model of Cory and McKnight (*70*). Specifically, RI is a ratio of several PARAFAC components that were assigned to quinone-like NOM moieties: the sum of four reduced (one hydroquinone-like, HQ, and three semi-quinone-like, SQ1, 2, and 3) components to the sum of the loadings of the four reduced and three oxidized (quinone-like, Q1, 2, and 3) components identified using the PARAFAC model (*68*). Several recent publications have indicated variations in RI attributed to seasonal and source variations in DOM (*68*, *69*, *74*).

In contrast, Macalady and Walton-Day (*75*) showed that induced redox changes in NOM samples had no substantial effect on RI values, and concluded that the identification of reduced and oxidized quinone-like components in the PARAFAC model was questionable. These results are consistent with conclusions reported in another recent study that showed little change in fluorescence under electrochemical reduction and aerobic re-oxidation of a humic substance (*76*). They are also consistent with results from a third study that found optical properties of a suite of structurally diverse quinones/hydroquinones that are not compatible with the PARAFAC assignments of Cory and McKnight, and that large optical changes observed under borohydride reduction could not be assigned to quinones (*77*).

Miller et al. (*78*) suggested that inner filter effects caused by using samples having absorbance at 254 nm (A254) between 0.3 and 1.0 absorbance units might have obscured changes in the redox signature of the fluorescence spectra reported by Macalady and Walton-Day (*75*). To address this question of whether highly absorbing samples caused inner filter effects that obscured the redox signature of oxidized and reduced quinone-like components in the Macalady and Walton-Day study (*75*), we assessed the reducing index of samples having a range of A254 values between 0.2 and 0.7 absorbance units, and also conducted a series of redox experiments on samples having initial A254 values less than or equal to 0.3 absorbance units.

The experiments utilized four samples: two natural NOM samples collected as grab samples from the Black River in Michigan (location: 46.45121 N, 87.95093 W) on June 10 and October 27, 2010, and two extracts of black walnut husks created by soaking crushed walnut husks in deionized water for 10 weeks at 4 degrees C and then decanting the extract. Walnut husks are rich in juglone (5-hydroxynapthoquinone), a natural quinone (*79*). For the redox experiments, the NOM samples were first diluted to DOC concentrations such that A254 values were 0.3 absorbance units or less. These samples were then subjected to oxidation or reduction experiments designed to produce fully oxidized or reduced NOM. For oxidation, the samples were exposed in open containers to mid-day sunlight (in Golden, Colorado in early-March 2011) for periods of 4-9 hours (*80*, *81*). Two methods of reduction were used. In the first, de-oxygenated samples were exposed to metallic zinc for 24-48 hours (*29*, *75*), and were filtered and maintained under anaerobic conditions prior to UV-visible spectral and fluorescence analysis. In the second, 50-mL samples were spiked with approximately 1.0 g of either Pt/graphite or Pd/alumina catalyst and sparged with gaseous H_2 for 1-2 hours (*53*, *82*). These samples were maintained under the resulting H_2 atmosphere prior to filtration and UV-spectral and fluorescence analyses, which were performed on samples placed in sealed cuvettes for the appropriate spectral analysis in a glove box under an atmosphere of 5% H_2 in N_2.

Table 1. Sample type, dilution, redox treatment, reducing index (RI) and absorbance at 254 nm (A254) for Black River and Walnut Husk Extract samples

Dilution	Treatment	RI	RI New	A254	Proportion Quinone-like Components	Comments
Black River, 10 June 2010						
17.40%	None	0.70	0.70	0.28	0.45	
23.40%	None	0.70	0.70	0.37	0.45	
30.20%	None	0.69	0.69	0.48	0.46	
35.10%	None	0.68	0.68	0.54	0.45	
40.20%	None	0.70	0.70	0.62	0.46	
16.57%	5 hr sun	0.66	0.66	0.28	0.44	
12.35%	H_2	0.09	0.23	0.51	0.42	
12.35%	H_2	0.09	0.24	0.51	0.42	Lab Rerun
Black River, 27 October 2010						
16%	None	0.72	0.72	0.22	0.48	
24%	None	0.72	0.72	0.34	0.48	
16.44%	5 hr sun	0.64	0.65	0.29	0.44	
16.35%	Zn	0.73	0.73	0.21	0.56	
Walnut Husk Extract 1						
0.80%	None	0.78	0.78	0.22	0.50	
1.21%	None	0.78	0.78	0.32	0.50	
1.21%	None	0.79	0.79	0.32	0.53	Lab Rerun
2.00%	None	0.79	0.79	0.53	0.52	
2.51%	None	0.79	0.79	0.67	0.53	
1.03%	Zn	0.73	0.73	0.28	0.51	
1.03%	H_2	0.16	0.29	0.53	0.42	
Walnut Husk Extract 2						
1.00%	None	0.82	0.82	0.24	0.56	
1.00%	None	0.82	0.82	0.24	0.55	Lab Rerun
2.01%	None	0.81	0.81	0.47	0.55	
1.08%	4 hr sun	0.81	0.79	0.30	0.43	
0.40%	5 hr sun	0.80	0.79	0.30	0.41	
0.40%	5 hr sun	0.80	0.79	0.30	0.40	Lab Rerun
0.27%	9 hr sun	0.79	0.78	0.30	0.39	

Dilution is the proportion of the original sample. RI New is the reducing index recomputed using only non-negative component loadings. Proportion quinones is the sum of the proportions of all non-negative quinone components from the PARAFAC model (C2, C4, C5, C7, C9, C11, and C13); hr, hours; H_2, hydrogen treated. (Shaded rows indicate sample EEMs presented in Fig. 5).

Three-dimensional fluorescence spectra (emission wavelength, excitation wavelength, and fluorescence intensity in Raman Units) were obtained on a Fluoromax 4 (Horiba Jobin Yvon, Edison, NJ). Data were corrected for excitation and emission, were normalized to the area under the Raman curve, and a common inner filter correction was applied consistent with the methods of Cory and McKnight (*70*) and Macalady and Walton-Day (*75*). Corrected data were fitted to the Cory and McKnight (*70*) PARAFAC model by Kate Murphy, University of New South Wales. A set of samples (where data and correction factors were generated using a Fluoromax 3) that had been previously fitted to the Cory and McKnight model was included with these new samples and the model fits were identical, indicating consistency with previous data presented in Macalady and Walton-Day (*75*). Proportions of component loads were calculated by dividing the loading of each component by the sum of all component loadings. For a few samples, there were negative loadings for some of the components of the model. These negative loadings were excluded from the sums of loadings used to calculate the proportion of each component to the total load. As shown in Table 1, RI was calculated using the raw data and the new RI was calculated using the sum of component loadings that excluded negative component loadings.

Although a detailed discussion of the results of these experiments is beyond the scope of this chapter, the following limited discussion is presented. It proceeds under the assumption that the assignment of certain PARAFAC components to oxidized or reduced quinone-like moieties is correct. Recent research (*76*, *77*), in addition to the work of Macalady and Walton-Day (*75*), strongly suggests that these assignments of statistical components to quinone-like moieties is questionable. As shown below, our current results generally support this conclusion. We use the component assignments of Cory and McKnight (*70*) in this discussion: components C2 (Q2), C11 (Q1) and C12 (Q3) are assigned to fully oxidized quinone-like components; C5 (SQ1), C7 (SQ2) and C9 (SQ3) are assigned to semi-quinone-like components; and C4 (HG) to fully reduced hydroquinone-like components.

The results of these experiments are summarized in Table 1 and Figs. 4 and 5. First, the results shown in Table 1 establish that, for these samples, increases in UV absorbance at 254 nm (A254) have no substantial effect on the calculated RI values for these samples. In these trials, A254 values between 0.22 and 0.67 were observed in dilutions of untreated samples without changes in calculated RI values, so inner-filter effects did not have an appreciable effect on RI values and cannot be used as an explanation to negate the effects of sample treatments on RI values.

The Black River samples came from a small, shallow stream that is expected to be saturated or nearly saturated with O_2. The samples were not stored in an oxygen-free condition prior to analyses. It is, therefore, quite unexpected to find a model prediction that shows about 70% of the quinone-like moieties in semi-reduced or fully reduced forms. Exposure to very oxidizing conditions (open, shallow dishes exposed to bright sunlight) did indicate very slight decreases in the calculated RI (oxidation), but, not nearly so dramatic as those reported by Miller et al. (*68*) for winter to summer changes in RI values (changes from 0.5 to 0.3). The slight changes in RI are explained in the Black River June samples by a slight increase (up to 2 percent) in the C11Q1 and C12Q3 oxidized components and a slight decrease (up to one percent) in the C5SQ1 and C7SQ2 semi-quinone-like components, and in the Black River October samples by a slight increase (up to 2 percent) in the C11Q1 oxidized components and a slight decrease (up to 3 percent) in the C5SQ1 and C7SQ2 semi-quinone-like components, which is consistent with oxidation of a semi-quinone-like component to a quinone-like component.

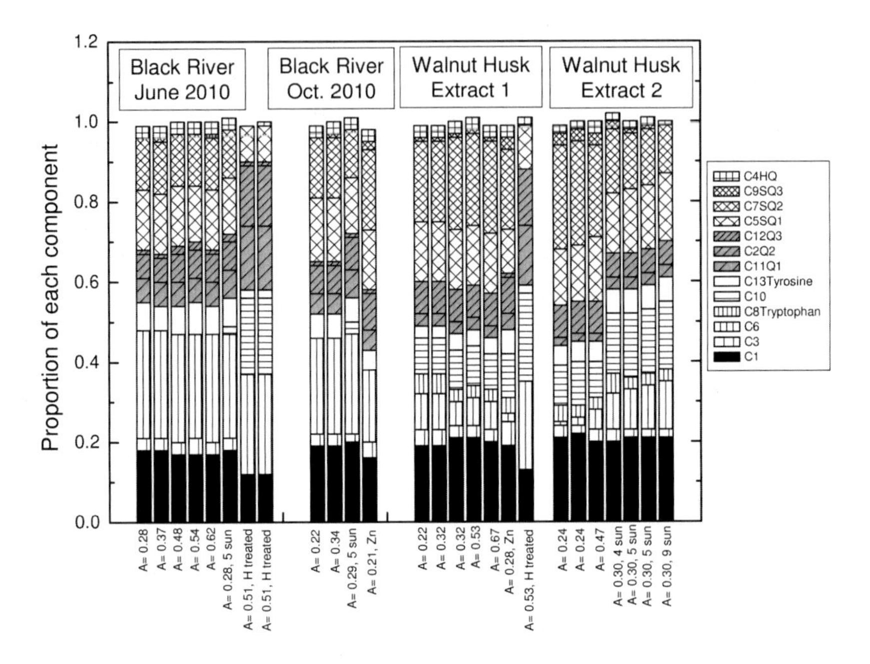

Figure 4. Stacked bar graph showing proportion of each component resulting from application of a 13-component PARAFAC model (70) to Black River and Walnut extract samples. Gray bars represent oxidized quinone-like components (Q1-Q3). Reduced quinone-like (semi (SQ1-SQ3)- and hydro-quinone(HQ)) components have hatched patterns and occur above the quinone-like components. Bar labels on x-axis indicate UV absorbance (A) at 254 nm and redox treatment applied to sample, where applicable.

However, as Fig. 4 illustrates, there were no concomitant decreases in the fully reduced quinone component. Furthermore, extreme reduction with H_2 and a noble metal catalyst showed a dramatic decrease in the calculated RI, indicating oxidation, not reduction. Reduction with zinc metal produced a calculated RI value that is not substantially different from the RI value before treatment. It must be concluded that the RI calculated from this PARAFAC model of fluorescence spectra and corresponding assignments of components to chemical moieties does not consistently indicate the redox status of the Black River samples.

The walnut husk extracts utilized in these experiments were prepared in brown-glass, sealed containers with little exposure to light or atmospheric oxygen. Although the samples were not protected from atmospheric exposure during filtration, dilution, and analyses, it might nevertheless be expected that these samples represent a more reduced state of NOM than the Black River samples. It is also expected that the walnut-husk extracts should be much richer in quinone and quinone-like components, because walnut husks are known to be rich in the water-soluble, substituted naphthoquinone, juglone (*79*).

a. Walnut Husk Extract 1, Untreated

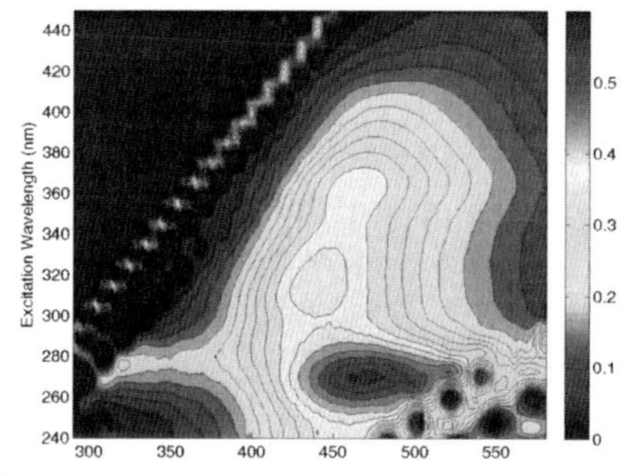

b. Walnut Husk Extract 1, Zn reduced

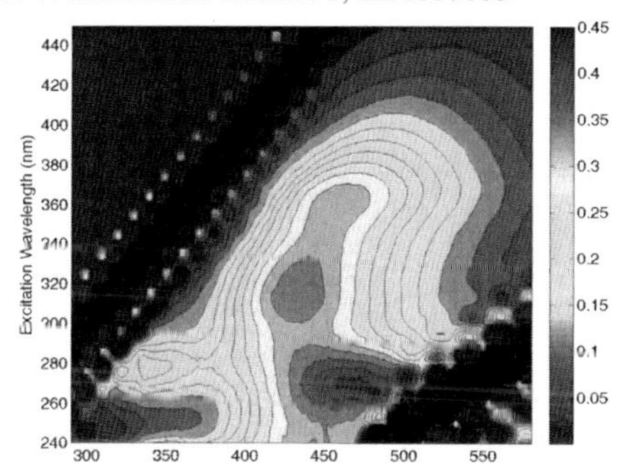

c. Walnut Husk Extract 1, H_2 reduced

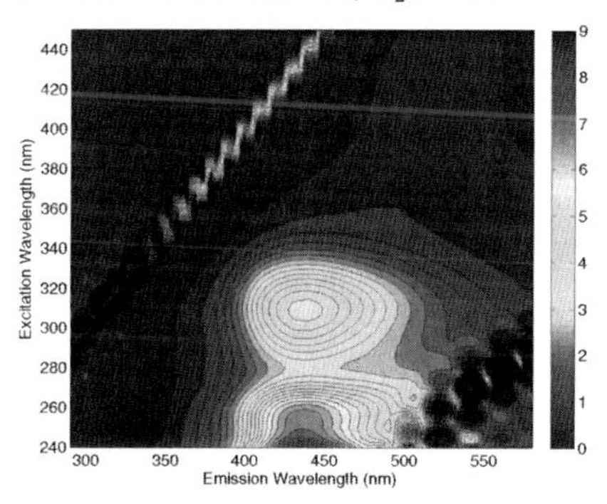

Figure 5. Corrected EEMs for (a) Walnut Husk Extract 1, untreated; (b) Walnut Husk Extract 1, Zn reduced, and (c) Walnut Husk Extract 1, H_2 reduced. Color scales indicate fluorescence intensity scale in Raman units.

In fact, one of the PARAFAC components for the untreated walnut-husk-extract samples, C7, which is identified as originating from semi-quinone-like moieties, is shown to be substantially enriched compared to the Black River samples (Fig. 4), in support of the conjecture that this component, C7, may be associated with quinone-like moieties. Also, the total proportion of quinones as measured by the sum of the 7 quinone-like components is slightly enriched in the walnut husk samples relative to the Black River samples (Table 1) supporting the idea that there are more quinones in the walnut husk samples. In addition, the RI values calculated for the untreated walnut-husk extracts are about 10% higher (Table 1) than those for the untreated Black River samples, supporting the expectation that these untreated samples were in a more reduced state than the untreated Black River samples.

However, these results from a cursory comparison of the EEMs for the Black River samples compared to the walnut samples, while first appearing to provide some support for the assignments of the components in the PARAFAC model (*70*), are not consistent with the results for oxygenated and reduced walnut-husk-extract samples. For the walnut samples, reduction by zinc or hydrogen reduced the calculated RI, indicating that oxidation, not reduction had occurred (Table 1).

EEMs (see online version for color figures) for untreated, zinc-reduced and hydrogen reduced walnut husk extract 1samples are shown in Fig. 5. The first impression is that these spectra are strikingly different from one another. So there is no question that these samples are chemically different. Yet the RI's calculated from the PARAFAC model of these EEMs show oxidation, not reduction (Table 1). It is notable that, for the zinc-treated sample, the EEMs show a slight change in shape and size of the peak in the protein (*83*) region (~Emission 340, Excitation 280) and the hydrogen-treated sample shows a blue shift of the two other major peaks. This blue shift with hydrogen reduction is consistent with fluorescence of samples occurring with borohydride reduction as reported by Ma and others (*77*).

Exposure to sunlight and O_2, which should have a dramatic oxidation effect if the samples start out in a reduced state, showed no substantial change in the calculated RI values. Thus, the walnut data also support the conclusions of Macalady and Walton-Day (*75*) and Ma et al. (*77*). These are that the PARAFAC model of Cory and McKnight (*70*) and their subsequent assignment of certain components to quinone-like moieties within NOM samples does not provide a reliable representation of NOM redox chemistry.

There are several possible explanations for these difficulties with the model assignments. Two categories of explanations are discussed briefly here. The first is the possibility that the model itself does not provide an accurate representation of the three-dimensional data in the EEMs spectra of these samples. Indeed, some studies have advised caution applying existing PARAFAC models to new datasets (*73*, *84*, *85*). First, the fact that the model was forced to provide negative values for the component loadings for some components and samples indicates problems with fitting the data to the model. Negative loadings, of course, have no chemical or physical significance. Also, although the residuals calculated for the goodness of fit of the data to the model were low (less than or equal to 10% of the fluorescence intensity, which is the same criterion used to judge good fit in other studies, e.g. (*71*, *74*)), non random structure observed within the model residuals indicated that the model did not provide an acceptable fit to the fluorescence data (*73*) (Kate Murphy, University of New South Wales, written communication, 2011).

Finally, the new data set was generated using a Fluoromax 4, and the Cory and McKnight model (*70*) was built using data generated on a Fluoromax 3. Although correction techniques applied to each dataset were consistent with manufacturer's recommendations and between data sets, there may be data biases related to the use of different instruments and laboratory procedures (*86*, *87*) that caused some of the observed lack of fit of our data with the Cory and McKnight PARAFAC model (*70*). These observations suggest that a different PARAFAC model for interpretation of EEMs spectral patterns built using these samples may provide the information we seek with respect to the chemical features of NOM samples.

The second category of explanation relates to ambiguities in the assignment of PARAFAC components to individual types of chemical species within the NOM sample. Here, we are suggesting that even if an appropriate PARAFAC model is available, assigning individual components to a specific chemical analyte is highly problematic. In this study, there are at least two alternative possibilities. The first is that the assignments of certain components to certain moieties within the system of quinone-like species is merely mixed up. The wrong components are assigned to one or more of the relevant species. If this is the case, then the model can be modified to provide the information we seek. For instance, inspection of Figure 4 shows that in the walnut husk extract 2 sample, component C6 increases with photochemical oxidation as component C7 decreases indicating that relative proportions of these two components are changing with oxidation and suggesting that these two components may represent redox-active constituents within the NOM. Is C6 the oxidized equivalent of C7? This might prove to be an avenue for additional investigation, but is not supported at this time. Indeed, Murphy et al. ((*84*), p. 2915) conclude that, PARAFAC spectra from OM (organic matter) data sets describe, "*the probabilistic distribution of an ensemble of individual spectra belonging to a range of spectrally similar chemical moieties from a range of sources, rather than exact chemical spectra*".

If, on the other hand, the fluorescence of one or more of the quinone-like moieties is very weak or otherwise swamped by the fluorescence of unrelated NOM chemical constituents and cannot be isolated from the overall EEMs pattern, then there is little or no hope of providing a means by which a meaningful index can be obtained from fluorescence data to indicate the redox state of quinone-like constituents in NOM samples. Unfortunately, this latter explanation is likely the correct one, as available data indicate that fluorescence of quinone moieties is far too weak to be isolated in fluorescence spectra of NOM samples (*77*, *88*).

Summary

Natural organic matter has been described in terms of its origins, operationally defined fractions and chemical composition. It is an inherently complex group of polyfunctional organic acids with composition that reflects its botanic and microbial origins and geochemical history. Aquatic NOM, the only form discussed in this chapter, refers to NOM that remains in solution in aqueous systems and has a ubiquitous presence in all natural waters. The oxidation/reduction, or redox, chemistry of aquatic NOM is especially interesting because of its reversibility and its potential participation in a wide range of environmental processes, from action as a mediator of redox reactions of pollutant organics to participation in electron transfer processes in microbial processes. Discussions of aquatic NOM chemistry are hampered somewhat by its chemical diversity, and also by its tendency to exhibit properties that depend significantly on sample origins and history.

Analytical determinations of chemical structures that represent the composition of aquatic NOM are generally unsatisfactory in terms of their completeness and ability to predict the relative behaviors of NOM samples in a given geochemical process.

The chemical structures within NOM that are active in environmental redox reactions are not completely quantified, though a major role of quinone-like structures is well established. The minor presence within NOM under predominant oxic conditions of incompletely oxidized moieties has been verified, but the structures responsible for this activity are unknown.

The redox activity of NOM is described in terms of its redox capacity, the carbon-normalized ability to accept or donate electrons, and its redox state, the relative proportions of oxidized and reduced redox-active components within the NOM system. The redox capacity of an NOM sample is related in a semi-quantitative way to its aromatic character (SUVA) and pH. Fourier-transform infrared spectroscopy has also been used to quantify the relative amounts of redox-active components in NOM samples. The redox capacity of NOM can be determined in a variety of ways, including redox titrations with a variety of oxidants and reductants, including direct electrochemical titration. Standardized methods for such procedures have not been developed, but progress has been made.

The redox state of an NOM sample refers to the relative extent that the redox-active functional groups are oxidized or reduced. Electrochemical titration is a recently developed method that shows promise as a standard method to determine the relative redox states of NOM samples (*58*, *76*). A recently proposed method for assessing the redox state of NOM samples which involves projecting 3-dimensional UV-visible fluorescence spectra of NOM samples upon a previously derived 13-component PARAFAC model has been evaluated experimentally. The observed failure of the method to consistently predict the redox state of our NOM samples may be due to a number of potential factors, including inadequacies in the original model and/or the fact the that model does not apply widely. Another possibility is that the underlying hypothesis that the redox state of quinones in NOM samples leads to predictable changes in fluorescence, is flawed. Thus, while there may be changes in EEMs of NOM samples that are associated with reversible changes in the redox status of the sample, current models are clearly incapable of exploiting these changes to enable understanding of the redox status. There is continuing discussion about the application of fluorescence to understanding the redox state of NOM samples, and new developments may bring new insights. The application of PARAFAC to this question would benefit from consistent and rigorous application of the technique as recommended in Stedmon and Bro (*73*) and Murphy et al. (*87*).

Acknowledgments

The authors gratefully acknowledge the assistance and manuscript review provided by Kate Murphy. Additional helpful reviews were provided by Neil Blough, Tony Ranalli, and four anonymous reviewers. Sara Gonzalez assisted compiling and formatting references. Any use of trade, firm, or product names is for descriptive purposes only and does not imply endorsement by the U.S. Government.

References

1. MacCarthy, P. The principles of humic substances. *Soil Sci.* **2001**, *166*, 738–751.
2. Eaton, A. D., Clesceri, L. S., Greenberg, A. E., Eds. *Humic Substances in Soil, Sediment, and Water: Geochemistry, Isolation and Characterization*; John Wiley and Sons: New York, 1985.
3. Leenheer, J. A. Comprehensive approach to preparative isolation and fractionation of dissolved organic carbon from natural waters and wastewaters. *Environ. Sci. Technol.* **1981**, *15*, 578–587.
4. Eaton, A. D., Clesceri, L. S., Greenberg, A. E., Eds. *Standard Methods for the Examination of Water and Wastewater*; American Public Health Association: Washington, DC, 1995.
5. Dittmar, T.; Koch, B.; Hertkorn, N.; Kattner, G. A simple and efficient method for the solid-phase extraction of dissolved organic matter (SPE-DOM) from seawater. *Limnol. Oceanogr.: Methods* **2008**, *6*, 230–235.
6. Everett, C. R.; Chin, Y. P.; Aiken, G. R. High-pressure size exclusion chromatography analysis of dissolved organic matter isolated by tangential-flow ultra filtration. *Limnol. Oceanogr.* **1999**, *44*, 1316–1322.
7. Benner, R.; Biddanda, B.; Black, B.; McCarthy, M. Abundance, size distribution, and stable carbon and nitrogen isotopic compositions of marine organic matter isolated by tangential-flow ultrafiltration. *Mar. Chem.* **1997**, *57*, 243–263.
8. Koprivnjak, J. F.; Perdue, E. M.; Pfromm, P. H. Coupling reverse osmosis with electrodialysis to isolate natural organic matter from fresh waters. *Water Res.* **2006**, *40*, 3385–3392.
9. Thurman, E. M. *Organic Geochemistry of Natural Waters*; Nijhoff, M.; Junk, W.: Dordrecht, Netherlands, 1985.
10. Rice, J. A.; MacCarthy, P. Statistical evaluation of the elemental composition of humic substances. *Org. Geochem.* **1991**, *17*, 635–648.
11. Reckhow, D. A.; Singer, P. C.; Malcolm, R. L. Chlorination of humic materials: Byproduct formation and chemical interpretation. *Environ. Sci. Technol.* **1990**, *24*, 1655–1664.
12. Fimmen, R. L.; Cory, R. M.; Chin, Y. P.; Trouts, T. D.; McKnight, D. M. Probing the oxidation-reduction properties of terrestrially and microbially derived dissolved organic matter. *Geochim. Cosmochim. Acta* **2007**, *71*, 3003–3015.
13. Abbt-Braun, G.; Lankes, U.; Frimmel, F. H. Structural characterization of aquatic humic substances: The need for a multiple method approach. *Aquat. Sci.* **2004**, *66*, 151–170.

14. Truong, H.; Lomnicki, S.; Dellinger, B. Potential for misidentification of environmentally persistent free radicals as molecular pollutants in particulate matter. *Environ. Sci. Technol.* **2010**, *44*, 1933–1939.
15. Fiorentino, G.; Spaccini, R.; Piccolo, A. Separation of molecular constituents from a humic acid by solid-phase extraction following a transesterification reaction. *Talanta* **2006**, *68*, 1135–1142.
16. Lehmann, J.; Solomon, D.; Brandes, J.; Fleckenstein, H.; Jacobsen, C.; Thieme, J. Synchrotron-based near-edge X-ray spectroscopy of natural organic matter in soils and sediments. In *Biophysico-Chemical Processes Involving Natural Nonliving Organic Matter in Environmental Systems*; Senesi, N., Xing, B., Huang, P. M., Eds.; John Wiley and Sons, Inc.: New York, 2009; pp 729−781.
17. Helal, A. A.; Murad, G. A. Characterization of different humic materials by various analytical techniques. *Arabian J. Chem.* **2011**, *4*, 51–54.
18. Redwood, P. S.; Lead, J. R.; Harrison, R. M.; Jones, I. P.; Stoll, S. Characterization of humic substances by environmental scanning electron microscopy. *Environ. Sci. Technol.* **2005**, *39*, 1962–1966.
19. Macalady, D.; Ritter, K.; Redman, A.; Skold, M. *Comparative characteristics of natural organic matter in the Fortymile River, Alaska*; Professional Paper 1685; U.S. Geological Survey: Reston, VA, 2004.
20. Pandey, A. K.; Pandey, S. D.; Misra, V. Stability constants of metal-humic acid complexes and its role in environmental detoxification. *Ecotoxicol. Environ. Saf.* **2000**, *47*, 195–200.
21. Musani, L. J.; Valenta, P.; Nurnberg, H. W.; Konrad, Z.; Branica, M. On the chelation of toxic trace metals by humic acid of marine origin. *Estuarine Coastal Mar. Sci.* **1980**, *11*, 639–649.
22. Leenheer, J. A.; Brown, G. K.; Maccarthy, P.; Cabaniss, S. E. Models of metal binding structures in fulvic acid from the Suwannee River, Georgia. *Environ. Sci. Technol.* **1998**, *32*, 2410–2416.
23. Manceau, A.; Matynia, A. The nature of Cu bonding to natural organic matter. *Geochim. Cosmochim. Acta* **2010**, *74*, 2556–2580.
24. Shin, H. S.; Rhee, S. W.; Lee, B. H.; Moon, C. H. Metal binding sites and partial structures of soil fulvic and humic acids compared: Aided by Eu(III) luminescence spectroscopy and DEPT/QUAT ^{13}C NMR pulse techniques. *Org. Geochem.* **1996**, *24*, 523–529.
25. Christl, I.; Kretzschmar, R. Relating ion binding by fulvic and humic acids to chemical composition and molecular size. 1. Proton binding. *Environ. Sci. Technol.* **2001**, *35*, 2505–2511.
26. Janot, N.; Reiller, P. E.; Korshin, G. V.; Benedetti, M. F. Using spectrophotometric titrations to characterize humic acid reactivity at environmental concentrations. *Environ. Sci. Technol.* **2010**, *44*, 6782–6788.
27. Weishaar, J. L.; Aiken, G. R.; Bergamaschi, B. A.; Fram, M. S.; Fujii, R.; Mopper, K. Evaluation of specific ultraviolet absorbance as an indicator of the chemical composition and reactivity of dissolved organic carbon. *Environ. Sci. Technol.* **2003**, *37*, 4702–4708.

28. Niemeyer, J.; Chen, Y.; Bollag, J. M. Characterization of humic acids, composts, and peat by diffuse reflectance Fourier-transform infrared spectroscopy. *Soil Sci. Soc. Am. J.* **1992**, *56*, 135–140.
29. Blodau, C.; Bauer, M.; Regenspurg, S.; Macalady, D. Electron accepting capacity of dissolved organic matter as determined by reaction with metallic zinc. *Chem. Geol.* **2009**, *260*, 186–195.
30. Paul, A.; Stösser, T. R.; Zehl, A.; Zwirnmann, E.; Vogt, R. D.; Steinberg, C. E. W. Nature and abundance of organic radicals in natural organic matter: Effect of pH and irradiation. *Environ. Sci. Technol.* **2006**, *40*, 5897–5903.
31. Scott, D. T.; McKnight, D. M.; Blunt-Harris, E. L.; Kolesar, S. E.; Lovley, D. R. Quinone moieties act as electron acceptors in the reduction of humic substances by humics-reducing microorganisms. *Environ. Sci. Technol.* **1998**, *32*, 2984–2989.
32. Sposito, G. Electron shuttling by natural organic matter: Twenty years after. In *Aquatic Redox Chemistry*; Tratnyek, P. G., Grundl, T. J., Haderlein, S. B., Eds.; ACS Symposium Series; American Chemical Society: Washington, DC, 2011; Vol. 1071, Chapter 6, pp 113−127.
33. Nurmi, J. T.; Tratnyek, P. G. Electrochemistry of natural organic matter. In *Aquatic Redox Chemistry*; Tratnyek, P. G., Grundl, T. J., Haderlein, S. B., Eds.; ACS Symposium Series; American Chemical Society: Washington, DC, 2011; Vol. 1071, Chapter 7, pp 129−151.
34. Dunnivant, F. M.; Schwarzenbach, R. P.; Macalady, D. L. Reduction of substituted nitrobenzenes in aqueous solutions containing natural organic matter. *Environ. Sci. Technol.* **1992**, *26*, 2133–2141.
35. Ratasuk, N.; Nanny, M. A. Characterization and quantification of reversible redox sites in humic substances. *Environ. Sci. Technol.* **2007**, *41*, 7844–7850.
36. Einsiedl, F.; Mayer, B.; Schäfer, T. Evidence for incorporation of H_2S in groundwater fulvic acids from stable isotope ratios and sulfur K-edge X-ray absorption near edge structure spectroscopy. *Environ. Sci. Technol.* **2008**, *42*, 2439–2444.
37. Perlinger, J. A.; Angst, W.; Schwarzenbach, R. P. Kinetics of the reduction of hexachloroethane by juglone in solutions containing hydrogen sulfide. *Environ. Sci. Technol.* **1996**, *30*, 3408–3417.
38. Guo, X.; Jans, U. Kinetics and mechanism of the degradation of methyl parathion in aqueous hydrogen sulfide solution: Investigation of natural organic matter effects. *Environ. Sci. Technol.* **2006**, *40*, 900–906.
39. Schwarzenbach, R. P.; Stierli, R.; Lanz, K.; Zeyer, J. Quinone and iron porphyrin mediated reduction of nitroaromatic compounds in homogeneous aqueous solution. *Environ. Sci. Technol.* **1990**, *24*, 1566–1574.
40. Buschmann, J.; Angst, W.; Schwarzenbach, R. P. Iron porphyrin and cysteine mediated reduction of ten polyhalogenated methanes in homogeneous aqueous solution: Product analyses and mechanistic considerations. *Environ. Sci. Technol.* **1999**, *33*, 1015–1020.
41. Knicker, H. Biogenic nitrogen in soils as revealed by solid-state carbon-13 and nitrogen-15 nuclear magnetic resonance spectroscopy. *J. Environ. Qual.* **2000**, *29*, 715–723.

42. Sutton, R.; Sposito, G. Molecular structure in soil humic substances: The new view. *Environ. Sci. Technol.* **2005**, *39*, 9009–9015.
43. Stevenson, F. J. *Humus Chemistry: Genesis, Composition, Reactions*; 2nd. ed.; John Wiley and Sons, Ltd.: New York, 1994.
44. Wershaw, R. L. *Evaluation of conceptual models of natural organic matter (humus) from a consideration of the chemical and biological processes of humification*; Scientific Investigations Report 2004-5121; U.S. Geological Survey: 2004.
45. Alberts, J. J.; Filip, Z. Metal binding in estuarine humic and fulvic acids: FTIR analysis of humic acid-metal complexes. *Environ. Sci. Technol.* **1998**, *19*, 923–931.
46. Hakala, J. A.; Chin, Y. P.; Weber, E. J. Influence of dissolved organic matter and Fe(II) on the abiotic reduction of pentachloronitrobenzene. *Environ. Sci. Technol.* **2007**, *41*, 7337–7342.
47. Perlinger, J. A.; Buschmann, J.; Angst, W.; Schwarzenbach, R. P. Iron porphyrin and mercaptojuglone mediated reduction of polyhalogenated methanes and ethanes in homogeneous aqueous solution. *Environ. Sci. Technol.* **1998**, *32*, 2431–2437.
48. Sachs, S.; Bernhard, G. Humic acid model substances with pronounced redox functionality for the study of environmentally relevant interaction processes of metal ions in the presence of humic acid. *Geoderma* **2011**, *162*, 132–140.
49. Redman, A. D.; Macalady, D. L.; Ahmann, D. Natural organic matter affects Arsenic speciation and sorption onto hematite. *Environ. Sci. Technol.* **2002**, *36*, 2889–2896.
50. Ritter, K.; Aiken, G. R.; Ranville, J. F.; Bauer, M.; Macalady, D. L. Evidence for the aquatic binding of arsenate by natural organic matter-suspended Fe(III). *Environ. Sci. Technol.* **2006**, *40*, 5380–5387.
51. Macalady, D. L.; Ranville, J. F. The chemistry and geochemistry of natural organic matter (NOM). In *Perspectives in Environmental Chemistry*; Macalady, D. L., Ed.; Oxford University Press: New York, 1999.
52. Struyk, Z.; Sposito, G. Redox properties of standard humic acids. *Geoderma* **2001**, *102*, 329–346.
53. Kappler, A.; Benz, M.; Schink, B.; Brune, A. Electron shuttling via humic acids in microbial iron(III) reduction in a freshwater sediment. *FEMS Microbiol. Biol.* **2004**, *47*, 85–92.
54. Chen, J.; Gu, B.; Royer, R. A.; Burgos, W. D. The roles of natural organic matter in chemical and microbial reduction of ferric iron. *Sci. Total Environ.* **2003**, *307*, 167–178.
55. Peretyazhko, T.; Sposito, G. Reducing capacity of terrestrial humic acids. *Geoderma* **2006**, *137*, 140–146.
56. Nevin, K. P.; Lovley, D. R. Potential for nonenzymatic reduction of Fe(III) via electron shuttling in subsurface sediments. *Environ. Sci. Technol.* **2000**, *34*, 2472–2478.
57. Yao, H.; Conrad, R.; Wassmann, R.; Neue, H. U. Effect of soil characteristics on sequential reduction and methane production in sixteen rice paddy soils from China, the Philippines, and Italy. *Biogeochemistry* **1999**, *47*, 269–295.

58. Aeschbacher, M.; Sander, M.; Schwarzenbach, R. P. Novel electrochemical approach to assess the redox properties of humic substances. *Environ. Sci. Technol.* **2010**, *44*, 87–93.
59. Christman, R. F.; Shi, J.; Wagoner, D.; Sharpless, C.; Fischer, E.; Schupbach, J. A new method for characterizing aquatic organic matter. In University of North Carolina: Chapel Hill, NC, 1998.
60. Tank, J. L.; Rosi-Marshall, E. J.; Griffiths, N. A.; Entrekin, S. A.; Stephen, M. L. A review of allochthonous organic matter dynamics and metabolism in streams. *J. N. Am. Benthol. Soc.* **2010**, *29*, 118–146.
61. Kordel, W.; Dassenakis, M.; Lintelmann, J.; Padberg, S. The importance of natural organic material for environmental processes in waters and soils. *Pure Appl. Chem.* **1997**, *69*, 1571–1600.
62. Nurmi, J. T.; Tratnyek, P. G. Electrochemical properties of natural organic matter (NOM), fractions of NOM, and model biogeochemical electron shuttles. *Environ. Sci. Technol.* **2002**, *36*, 617–624.
63. Chen, J.; Gu, B.; LeBoeuf, E. J.; Pan, H.; Dai, S. Spectroscopic characterization of the structural and functional properties of natural organic matter fractions. *Chemosphere* **2002**, *48*, 59–68.
64. Larsen, L. G.; Aiken, G. R.; Harvey, J. W.; Noe, G. B.; Crimaldi, J. P. Using fluorescence spectroscopy to trace seasonal DOM dynamics, disturbance effects, and hydrologic transport in the Florida Everglades. *J. Geophys. Res., [Biogeosci.]* **2010**, *115*.
65. Stedmon, C. A.; Markager, S. Tracing the production and degradation of autochthonous fractions of dissolved organic matter by fluorescence analysis. *Limnol. Oceanogr.* **2005**, *50*, 1415–1426.
66. Murphy, K. R.; Stedmon, C. A.; Waite, T. D.; Ruiz, G. M. Distinguishing between terrestrial and autochthonous organic matter sources in marine environments using fluorescence spectroscopy. *Mar. Chem.* **2008**, *108*, 40–58.
67. Chen, J.; LeBoeuf, E. J.; Dai, S.; Gu, B. Fluorescence spectroscopic studies of natural organic matter fractions. *Chemosphere* **2003**, *50*, 639–647.
68. Miller, M. P.; McKnight, D. M.; Cory, R. M.; Williams, M. W.; Runkel, R. L. Hyporheic exchange and fulvic acid redox reactions in an alpine stream/wetland ecosystem, Colorado front range. *Environ. Sci. Technol.* **2006**, *40*, 5943–5949.
69. Mladenov, N.; Zheng, Y.; Miller, M. P.; Nemergut, D. R.; Legg, T.; Simone, B.; Hageman, C.; Rahman, M. M.; Ahmed, K. M.; McKnight, D. M. Dissolved organic matter sources and consequences for iron and arsenic mobilization in Bangladesh aquifers. *Environ. Sci. Technol.* **2010**, *44*, 123–128.
70. Cory, R. M.; McKnight, D. M. Fluorescence spectroscopy reveals ubiquitous presence of oxidized and reduced quinones in dissolved organic matter. *Environ. Sci. Technol.* **2005**, *39*, 8142–8149.
71. Mladenov, N.; Huntsman-Mapila, P.; Wolski, P.; Masamba, W. R. L.; McKnight, D. M. Dissolved organic matter accumulation, reactivity, and redox state in ground water of a recharge wetland. *Wetlands* **2008**, *28*, 747–759.

72. Fulton, J. R.; McKnight, D. M.; Foreman, C. M.; Cory, R. M.; Stedmon, C.; Blunt, E. Changes in fulvic acid redox state through the oxycline of a permanently ice-covered Antarctic lake. *Aquat. Sci.* **2004**, *66*, 27–46.
73. Stedmon, C. A.; Bro, R. Characterizing dissolved organic matter fluorescence with parallel factor analysis: A tutorial. *Limnol. Oceanogr.: Methods* **2008**, *6*, 572–579.
74. Miller, M. P.; McKnight, D. M.; Chapra, S. C. Production of microbially-derived fulvic acid from photolysis of quinone-containing extracellular products of phytoplankton. *Aquat. Sci.* **2009**, *71*, 170–178.
75. Macalady, D. L.; Walton-Day, K. New light on a dark subject: On the use of fluorescence data to deduce redox states of natural organic matter (NOM). *Aquat. Sci.* **2009**, *71*, 135–143.
76. Maurer, F.; Christl, I.; Kretzschmar, R. Reduction and reoxidation of humic acid: Influence on spectroscopic properties and proton binding. *Environ. Sci. Technol.* **2010**, *44*, 5787–5792.
77. Ma, J.; Del Vecchio, R.; Golanoski, K. S.; Boyle, E. S.; Blough, N. V. Optical properties of humic substances and CDOM: Effects of borohydride reduction. *Environ. Sci. Technol.* **2010**, *44*, 5395–5402.
78. Miller, M. P.; Simone, B. E.; McKnight, D. M.; Cory, R. M.; Williams, M. W.; Boyer, E. W. New light on a dark subject: Comment. *Aquat. Sci.* **2010**, *72*, 269–275.
79. Dana, M. N.; Lerner, B. R. Black Walnut Toxicity, Purdue University, Department of Horticulture, Service, P. U. C. E.: West Lafayette, IN, HO-193, 1994.
80. Osburn, C. L.; Morris, D. P.; Thorn, K. A.; Moeller, R. E. Chemical and optical changes in freshwater dissolved organic matter exposed to solar radiation. *Biogeochemistry* **2001**, *54*, 251–278.
81. Sulzberger, B.; Durisch-Kaiser, E. Chemical characterization of dissolved organic matter (DOM): A prerequisite for understanding UV-induced changes of DOM absorption properties and bioavailability. *Aquat. Sci.* **2009**, *71*, 104–126.
82. Visser, S. A. Oxidation-reduction potentials and capillary activities of humic acids. *Nature* **1964**, *204*, 581.
83. Henderson, R. K.; Baker, A.; Murphy, K. R.; Hambly, A.; Stuetz, R. M.; Khan, S. J. Fluorescence as a potential monitoring tool for recycled water systems: A review. *Water Res.* **2009**, *43*, 863–881.
84. Murphy, K. R.; Hambly, A.; Singh, S.; Henderson, R. K.; Baker, A.; Stuetz, R.; Khan, S. J. Organic matter fluorescence in municipal water recycling schemes: Toward a unified PARAFAC model. *Environ. Sci. Technol.* **2011**, *45*, 2909–2916.
85. Fellman, J. B.; Miller, M. P.; Cory, R. M.; D'Amore, D. V.; White, D. Characterizing dissolved organic matter using PARAFAC modeling of fluorescence spectroscopy: A comparison of two models. *Environ. Sci. Technol.* **2009**, *43*, 6228–6234.
86. Cory, R. M.; Miller, M. P.; McKnight, D. M.; Guerard, J. J.; Miller, P. L. Effect of instrument-specific response on the analysis of fulvic acid fluorescence spectra. *Limnol. Oceanogr.: Methods* **2010**, *8*, 67–78.

87. Murphy, K. R.; Butler, K. D.; Spencer, R. G. M.; Stedmon, C. A.; Boehme, J. R.; Aiken, G. R. Measurement of dissolved organic matter fluorescence in aquatic environments: An interlaboratory comparison. *Environ. Sci. Technol.* **2010**, *44*, 9405–9412.
88. Blough, N. V.; Del Vecchio, R. Comment on "HILIC-NMR: Toward the identification of individual molecular components in dissolved organic matter". *Environ. Sci. Technol.* **2011**, *45*, 5908–5909.

Chapter 6

Electron Shuttling by Natural Organic Matter: Twenty Years After

Garrison Sposito*

Departments of Environmental Science, Policy & Management and Civil & Environmental Engineering, University of California, Berkeley, CA 94720
***gsposito@berkeley.edu**

The progress of science shows completeness and a logical development only in the settings provided by textbooks and review articles. Living science, like living human beings, invariably exhibits partial truths, tentativeness, trial-and-error, and subjectivity. The development of the concept of electron shuttling by natural organic matter is not an exception to this paradigm. In the following essay, the story of this development will be retold heuristically to endow it with a more logical structure, providing both a template for future research and a perspective on the contributions made by Donald Macalady.

The Beginnings

Our story begins, as is often true for innovative science, with a heuristic hypothesis, one concerning the mechanisms of reactions that degrade organic pollutants, such as pesticides, reductively—but abiotically—in soils, sediments, and natural waters. This hypothesis was put forth by Tratnyek and Macalady (*1*), who closed a decade-long debate about the dominant pathways of these reactions with a deductive insight that is also a model of scientific parsimony: "Speculation regarding the agents responsible for abiotic reduction of organic pollutants most commonly emphasizes ferrous iron or complexes of ferrous iron. However, it is a common generalization that natural organic matter is a strong reducing agent, and organic matter will reduce metals that can, in turn, reduce organic pollutants. Presumably, organic matter can also reduce organic pollutants directly. Studies to date have not identified the reducing site(s) on natural organic matter though they are likely to be associated with polyphenols, especially those with para orientation, i.e. hydroquinones". Thus, it was proposed that organic pollutants might be reductively degraded abiotically by electron transfer from reduced moieties in natural organic matter.

Proof-of-principle confirmation of the Tratnyek-Macalady hypothesis appeared soon afterward in a seminal paper by Dunnivant et al. (*2*). Using samples of aquatic organic matter extracted from 10 different natural waters, they quantitatively demonstrated its role as a mediator in the reductive degradation of substituted nitrobenzenes in aqueous solutions containing H_2S as an exogenous electron donor. They concluded that "under redox conditions typical for sulfate-reducing bacteria and, particularly, for methanogenic environments where reduction potentials well below −0.2 V have been measured, the abiotic reduction of nitro aromatic compounds by reduced [natural organic matter] constituents may be a significant transformation process. Various results of this study suggest that the reactive [natural organic matter] constituents may be quinone-type moieties which have recently been shown to be present in aqueous fulvic and humic acid samples."

Dunnivant et al. (*2*) summarized their results conceptually with a now-iconic diagram (see Scheme 1) depicting rapid electron transfer from a "Bulk Donor" [$H_2S(aq)$ in their experiments] to a "shuttle" (natural organic matter), which then transfers electrons at a moderate rate to an organic pollutant. With further prescient insight, they noted that "the reduction potential of natural systems is generally controlled by microorganisms, thus 'bulk' electron donors and reduced forms of electron-transfer mediators may also be replenished through microbial processes." In other words, natural organic matter, serving as a mediator of the reductive degradation of an organic pollutant, could be reduced, not by only by exogenous chemical electron donors, but also by microorganisms; and these same microorganisms could sustain the process by continually replenishing any electrons lost through subsequent reductive degradation of a pollutant. Accordingly, one could replace "Bulk Donor" in the two-step electron-transfer diagrammed by Dunnivant et al. (*2*) with "microbial consortium," then apply the diagram, thus broadened in scope, to the mediation of reductive degradation by any moiety in natural organic matter that is reducible by microbes.

A similar broadening of the electron-transfer scheme of Dunnivant et al. (*2*) is possible, if its second step, the reductive degradation of an organic pollutant, is replaced by *any* reduction reaction susceptible to mediation. This innovation was in fact already being explored independently by McBride and his coworkers (*3*, *4*) at the same time as Tratnyek and Macalady (*1*) were enlisting synthetic hydroquinones to mediate the reductive degradation of nitro aromatic pesticides. Kung and McBride (*3*), for example, demonstrated that synthetic hydroquinone can promote the reductive dissolution of the mixed-valence manganese oxide, hausmannite (Mn_3O_4), which is found commonly in soils. They noted that such quinone mediators can be expected from "the microbial breakdown of plant residues and humic substances". Their results were summarized in an electron-transfer diagram like that of Dunnivant et al. (*2*), but with a manganese oxide shown instead of an organic pollutant. Thus the terminal electron acceptor in the two-step scheme can be either organic or inorganic, contaminant or nutrient.

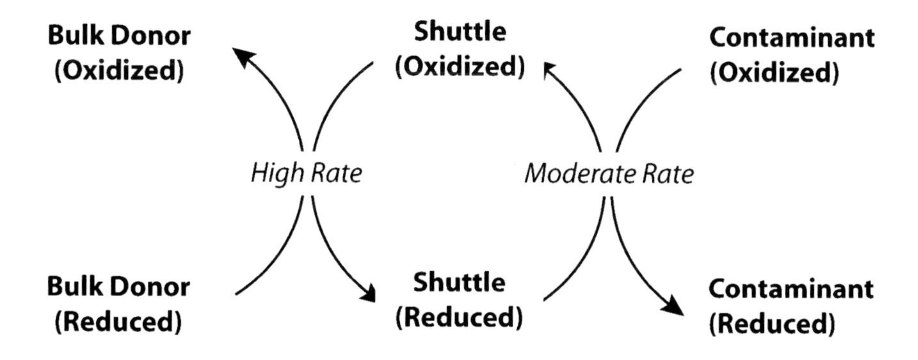

Figure 1. Conceptual scheme for shuttle- mediated electron transfer from a bulk donor to a contaminant in environmental systems. Adapted with permission from Dunnivant et al. (2). Copyright 1992 American Chemical Society.

Lovley et al. (*5*) utilized this broadened scope in a seminal investigation that fleshed out mechanistic details of the electron-transfer scheme suggested by Dunnivant et al. (*2*). Earlier work by Lovley's research group had discovered that the culturable bacterium, *Geobacter metallireducens*, can couple the complete oxidation of simple organic compounds (e.g., acetate) to the reduction of ferric iron. Lovley et al. (*5*) then "...speculated that if *G. metallireducens* could transfer electrons to humic substances, then humic substances might stimulate Fe(III) reduction in a two-stage process in which (1) *G. metallireducens* oxidizes acetate, with humic acids acting as the electron acceptor, and (2) reduced humic acids donate electrons to Fe(III)". Although "humic substances" is not a synonym for "natural organic matter," most of the latter is made up of the former, which refers to a dark-colored, biologically-refractory, heterogeneous family of compounds, partitioned by solubility criteria into humic acid and fulvic acid, that are the byproducts of microbial metabolism (*6*). They may account for up to 80 % of soil organic matter and up to half of that in natural waters, with the remainder being identifiable biomolecules.

Using a suspension comprising a poorly-crystalline Fe(III) oxide and commercially-available humic acid, Lovley et al. (*5*) demonstrated the greatly-enhanced production of Fe(II) products on timescales of hours when *G. metallireducens* and acetate were included; when they were not, or when they were but no humic acid was present, nothing happened. Additionally, in suspensions without Fe(III) oxide, acetate could not be oxidized by *G. metallireducens* when no humic acid was present to be the electron acceptor. And finally, humic acid reduced by *G. metallireducens* then placed in suspension with Fe(III) oxide promptly reduced it to Fe(II) products. Lovley et al. (*5*) concluded that "such electron shuttling may greatly facilitate the ability of Fe(III)-reducing bacteria to pass electrons to insoluble Fe(III) oxides ...," thereby putting an additional microbiological twist on the electron-transfer scheme of Dunnivant et al. (*2*) while coining a viral metaphor for the microbial mediation it invokes.

Is Electron Shuttling Ubiquitous?

The question posed directly above represents the first logical step to take following the hypothesis of electron shuttling by natural organic matter. It has several researchable facets. One of them is this: Are the microorganisms that can reduce natural organic matter ubiquitous? The answer to this essential question appears to be affirmative, according to a convincing number of laboratory and field studies that have been published over the past decade and a half (*7–15*). They indicate clearly that a broad variety of microorganisms thriving in anoxic environments (fermenters, iron-reducers, sulfate-reducers, methanogens, etc.) can do the job. This discovery of the widespread occurrence of microorganisms that can reduce natural organic matter has had ramifications in microbial ecology as well, since competitive advantages are conferred to microbes that use abundant natural organic matter as a terminal electron acceptor (*7*, *13–16*). Adding fascinating complexity to this issue, reduced organic matter can be "pirated" by microorganisms that are not themselves able to create it (*17*) and then used for electron shuttling. Of course, such piracy is not limited to microorganisms. Reduced organic matter prepared in the laboratory by incubation with aqueous extracts containing microbial consortia, or created abiotically by either electrochemical or wet-chemical methods, can then be used by humans to reduce both naturally-occurring or pollutant compounds (*18–21*).

It should be emphasized that the evidence for electron shuttling by natural organic matter presented thus far is circumstantial. This point can be illustrated by taking a closer look at the study of Kappler et al. (*12*) who measured profiles of Fe(II), pH, apparent electrode potential (E_H), and microbial population characteristics in a core of freshwater lake-bottom sediment exhibiting a strongly-delineated redox zonation. Predictably, they found the content of acid-extractable (1 *M* HCl) Fe(II) to increase with depth, from undetectable in the top centimeter of sediment ["oxygen (and probably nitrate) respiration zone"] to a convincing 80 % of the total Fe in a layer directly below and extending down to four centimeters ["iron-reduction zone"]. Below this layer, the content of extractable Fe(II) remained uniform.

The trend in pH observed by Kappler et al. (*12*) in their sediment core was a classic signature of reductive dissolution (*6*): a monotonic increase with depth throughout the upper 8 cm, beginning from about pH 6.7 to a plateau value of pH 7.3. The expected downward trend also was observed for E_H, which dropped monotonically from about +120 mV at the 1-cm depth to reach a plateau value near −200 mV at the 5-cm depth, the nominal boundary between the "iron-reduction zone" and the "sulfate-reduction zone," which was a 3-cm layer exhibiting black color and undetectable sulfate. On the microbial side, most-probable-number analyses of the indigenous bacterial populations in the sediment core revealed that humic-acid-reducing bacteria outnumbered iron-reducing bacteria by up to two orders of magnitude at any depth in the profile. The Fe(III)-oxide-reducing population was in fact non-negligible only in the "iron-reduction zone," and even there it was an order of magnitude smaller than the humic-acid-reducing population. The Fe(III)-oxide-reducers were identified by their ability to produce detectable Fe(II) in bicarbonate-buffered medium with both added acetate or lactate and poorly-crystalline Fe(III) oxide present. The humic-acid-reducers were identified by their ability to do the same when humic acid extracted from the sediment core also was added. These trends with depth provide strong circumstantial—but definitely not mechanistic—evidence that reducible natural organic matter in the lake sediment served as an electron shuttle for Fe(III) oxide bioreduction.

Now, if the answer to the question of ubiquity is to be strongly affirmative, then it should be possible to add *any* sort of natural organic matter to *any* sort of soil or sediment containing Fe(III) oxide minerals and find that under anoxic conditions, the reduction of these oxide minerals by native iron-reducers in the soil or sediment is significantly enhanced. Nevin and Lovley (*22*) apparently were the first to perform this crucial test, by placing off-the-shelf Aldrich humic acid into a suspension of contaminated aquifer sediments known to contain a high content of microbially-reducible Fe(III), incubating the sediments (with acetate added as an electron donor) under anoxic conditions for three months, then measuring acid-extractable (0.5 *M* HCl) Fe(II). Their results showed—unequivocally—that, although some measurable Fe(II) was generated in the absence of added organic matter, the amount produced in the presence of Aldrich humic acid was more than twice as large as in the control at the end of incubation. Similarly, Rakshit et al. (*23*) added commercial soil, peat, and aquatic humic substances to aqueous suspensions of an Ultisol soil that they incubated under anoxic conditions for one month without the benefit of an exogenous organic electron donor. Their results (Figure 1) also demonstrated a clear and substantial enhancement of soluble Fe(II) production by all of the added humic substances, regardless of their provenance.

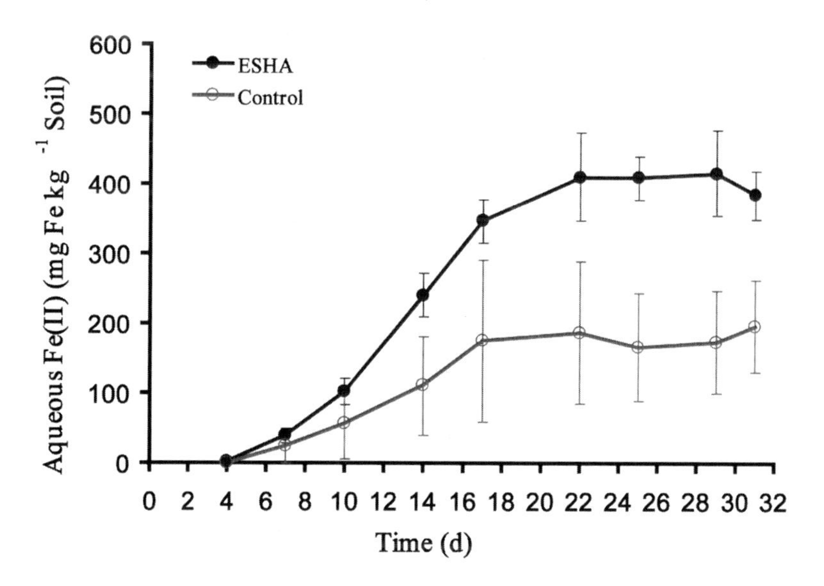

Figure 1. Soluble Fe(II) production during one month of anoxic incubation in the presence (blue filled symbols) or absence (red unfilled symbols) of Elliott soil humic acid ("ESHA") added to an aqueous soil suspension. [data from Rakshit et al. (23)]

Recently, Roden et al. (*24*) added significant breadth to these results by quantifying both the electrons transferred to solid-phase natural organic matter and the acid-extractable (0.5 *M* HCl) Fe(II) evolved during separate anoxic incubations of wetland sediment suspensions in the presence of dissimilatory iron-reducing bacteria. (Iron was removed from the sediments by extraction with citrate-dithionate-bicarbonate solution before the electron-transfer experiments were performed.) Electron transfer to solid-phase organic matter and Fe(II) evolution were found to be positively-correlated. Since only a relatively minor number of electrons was transferred in the absence of sediments (filtered suspensions), and no electrons were transferred in the absence of added bacteria, Roden et al. (*24*) concluded that direct electron transfer by the bacteria to solid-phase natural organic matter in the sediments was largely responsible for the Fe(III) bioreduction they had observed in unfiltered sediments. This important mechanistic insight cannot be obtained from the results of either Nevin and Lovley (*22*) or Rakshit et al. (*23*).

Quantitating the Shuttling Capacity

Given the ubiquity of electron shuttling by natural organic matter, the key question of how to quantify electron shuttling capacity in the laboratory can be addressed. Peretyazhko and Sposito (*25*) have reviewed much of the post-millennium literature seeking an operational definition of an electron-transfer capacity that can represent the second step in the electron-transfer scheme of Dunnivant et al. (*2*) while yielding an accessible quantitative measure of the shuttling capacity that has predictive value. Since the process at issue is the abiotic mediation of a *reductive* transformation, it makes sense, guided by a suggestion from A. Kappler (*A. Kappler, personal communication*), to identify the key shuttling property of the mediator as its *reducing capacity*. Quantitatively, reducing capacity is defined as the moles of electron charge, expressed per unit mass of organic matter (or per mole of carbon), that an electron shuttle makes available for transfer to an oxidant.

A measured value of reducing capacity will of course depend on the prior redox state of the electron shuttle. This state dependence has been illustrated in an especially clear way by Heitmann et al. (*13*), who differentiated "potential electron-donating capacity," the maximum value of a reducing capacity, from "*in situ* electron-donating capacity," the reducing capacity of an electron shuttle not maximally reduced. The positive difference between the two values of reducing capacity they termed "electron-accepting capacity" (*13*), which, in the present context, can be identified as an "oxidative capacity". Oxidative capacity refers to the first step in the electron-transfer scheme of Dunnivant et al. (*2*). It plays a key role in unraveling how special advantages accrue to microorganisms that are able to transfer electrons to natural organic matter (*13*).

Despite the quantitative dependence of reducing capacity on the prior redox state of an electron shuttle, its applicability should not depend on the details of how the prior redox state was attained. Consider, for example, the carefully-preserved sample of organic matter extracted from the "iron-reducing zone" in the lake-bottom sediments investigated by Kappler et al. (*12*). It is reasonable to refer to the reducing capacity of this sample as a *microbial reducing capacity* (*25*), with the additional adjective specifying the exogenous electron donor in the scheme of Dunnivant et al. (*2*). It is also reasonable to apply this term to the reducing capacity of the commercially-available humic acid sample utilized by Lovley et al. (*5*) for which the electron donor was the culturable bacterium, *G. metallireducens*. Microbial reducing capacity thus is applicable as a concept to natural organic matter that has been reduced either by a culturable microorganism (*5*, *7–11*) or by a microbial consortium in a sediment or soil (*12–14*, *22–25*).

Kappler et al. (*12*) extracted humic acid from lake-bottom sediments and then chemically reduced it by anoxic incubation for 24 hours under H_2(g) in the presence of a Pd catalyst, following a method devised many years ago by Visser (*26*). They did this in order to assess the prior redox state of the humic acid by comparison between its reducing capacities measured before and after exposure to H_2(g), i.e., by calculating its oxidative capacity in the lake-bottom sediments. The reducing capacity of their humic acid sample after reduction pretreatment can be termed its *chemical reducing capacity* (*25*), with the first adjective again referring to the exogenous electron donor in the scheme of Dunnivant et al. (*2*). Using the same line of reasoning as for microbial reducing capacity, one can conclude that the concept of chemical reducing capacity is applicable to natural organic matter that has been reduced in the laboratory by any one of several abiotic techniques, including H_2(g) with Pd catalyst (*7*, *12*, *25*, *26*), H_2S(aq) (*1*), or electrochemical cell (*18*). Adding an interesting microbial ecology dimension to this idea is the fact that chemical reductants such as H_2(g) and H_2S(aq) also can be produced by microorganisms in soils and sediments (*5–7*, *13*).

Laboratory measurement of either microbial or chemical reducing capacity necessarily involves use of an oxidant to harvest the electrons from a reduced organic matter shuttle. A variety of chemical oxidants has been used for this purpose, the most popular being ferric citrate [$FeC_6H_5O_7$ (*5*, *23*, *25*, *27*)] and potassium hexacyanoferrate [$K_3Fe(CN)_6$ (*7*, *12*, *18*, *28*, *29*)], both compounds exhibiting standard electrode potentials between 0.3 and 0.4 V for reduction to products containing Fe(II). An evident but often unacknowledged constraint on the choice of oxidant for measuring reducing capacity is that *it must react only with those moieties in natural organic matter that have been previously reduced microbially or chemically*, i.e., it should neither over-harvest nor under-harvest the electrons in a shuttle.

A clear example of over-harvesting was cited by Peretyazhko and Sposito (*25*), who found that using I_2(aq) as the oxidant (*30*) for reduced humic acids led to estimates of their chemical reducing capacity that were an order of magnitude larger than those obtained with either of the two Fe(III) oxidants mentioned above. The standard electrode potential for the reduction of I_2(aq) to I^-(aq) is about 1.0 V, which suggests that I_2(aq) was oxidizing moieties that are not involved in electron shuttling (*25*, *31*). An example of under-harvesting was provided by Bauer and Kappler (*31*), who examined Fe(III) oxyhydroxides as well as ferric citrate as oxidants, finding that well-crystallized mineral oxidants (hematite and goethite) produced chemical reducing capacities for Pahokee Peat humic acid that were less than one-third of the value found using ferric citrate.

The poorly-crystalline Fe(III) oxide, ferrihydrite, fared better, the overall trend being a monotonically-increasing chemical reducing capacity with increasing standard electrode potential for reduction of the oxidant to Fe(II) products (*31*, *32*). Bauer and Kappler (*31*) also noted that the reducing capacity of their humic acid prior to chemical reduction was not even detectable using well-crystallized Fe(III) minerals, whereas measurable values that compared well were found using either ferrihydrite or ferric citrate.

Peretyazhko and Sposito (*25*) and Jiang and Kappler (*33*) have compared the microbial reducing capacity with the chemical reducing capacity of humic substances. Peretyazhko and Sposito (*25*) used Pahokee Peat, Elliott Soil, and Leonardite humic acid in their study, inducing microbial reduction by anoxic incubation for two weeks in suspension with an indigenous population of soil microorganisms and chemical reduction by using H_2(g) over a Pd catalyst. Regardless of the mode of reduction, their samples were oxidized by ferric citrate. Measured values of the two reducing capacities were the same within experimental precision, leading them to conclude that "chemical reduction can be used as a convenient laboratory method to assess the [microbial reducing capacity] of [humic acid]". Jiang and Kappler (*33*) used Suwannee River fulvic and humic acids along with Aldrich humic acid in their experiments, inducing microbial reduction by anoxic incubation for 3 hours in suspension with *G.sulfurreducens* and chemical reduction by exposure to H_2(g) on a Pd catalyst. They oxidized their samples with potassium hexacyanoferrate. Again, within experimental precision, the two reducing capacities were the same, leading them to state (*33*) "...we can conclude that most redox-active functional groups in [humic substances] that are reduced chemically by H_2 are also bioreducible. We therefore conclude that chemically reduced [humic substances] can be used as representatives of microbially reduced natural humic substances".

We note that a reducing capacity determined with ferric citrate does not have the same value as that determined with potassium hexacyanoferrate, which is the more potent oxidant. For example, the chemical reducing capacities of Elliot soil and Pahokee Peat humic acid are equal to 0.561 ± 0.106 and 0.539 ± 0.047 mol_c/kg, respectively, if ferric citrate is used as the oxidant, whereas they are equal to 1.89 $\pm$ 0.04 and 2.37 ± 0.25 mol_c/kg, respectively, if hexacyanoferrate is the oxidant ((*25*) and *T. Peretazhko, unpublished data*). Similarly, the chemical reducing capacity of Suwannee River fulvic acid is 0.390 ± 0.008 mol_c/kg C if measured with ferric citrate (*23*), but equals 1.21 ± 0.13 mol_c/kg C when measured with hexacyanoferrate (*33*). These differences, a factor between 3 and 4, do not in principle affect direct comparison of microbial with chemical reducing capacity, but they certainly do affect the interpretation of reducing capacity as to which organic moieties are involved in electron shuttling.

Identifying the Shuttling Moieties

A third and final question to follow up the hypothesis of electron shuttling by natural organic matter is: which functional groups both contribute to the reducing capacity and are involved in electron shuttling? Royer et al. (*34*, *35*) took the first decisive steps toward answering this question through a series of clever experiments on a model system comprising the Fe(III) oxide, hematite, the iron-reducing bacterium, *Shewanella putrefaciens*, and a suite of natural organic matter samples of differing provenance and, therefore, with different chemical properties. The results of their experiments showed clearly that acid-extractable (0.5 *M* HCl) Fe(II) production after five days of anoxic incubation increased with both the concentration of natural organic matter added—thus confirming its putative role as an electron shuttle—and aromaticity, a trend that was noted as well by Chen et al. (*36*). No other chemical property of the organic matter samples showed a significant positive correlation with Fe(II) production.

They observed that adding anthraquinone-2,6-disulfonate, a synthetic quinone known to function as an electron shuttle for the bioreduction of Fe(III) in soils and sediments (*20*, *22*), enhanced Fe(II) production more after one day of incubation than it did after five days, whereas adding the strong Fe(II) complexing agent, ferrozine, did the exactly the reverse. Moreover, when the two compounds were present together, the resulting Fe(II) production could be predicted reliably by a linear combination of the effect of each compound alone. Anthraquinone-2,6-disulfonate engages in electron shuttling, but not Fe(II) complexation, and ferrozine engages in Fe(II) complexation but not electron shuttling (Figure 2), leading Royer et al. (*34*, *35*) to hypothesize from their results that electron shuttling mechanisms should be most effective during the initial stages of Fe(III) bioreduction, whereas Fe(II) complexation mechanisms (which lower the Fe^{2+} concentration) should be most effective later in Fe(III) bioreduction, after mass-transfer of Fe^{2+} away from the dissolving Fe(III) mineral had become rate-limiting.

Rakshit et al. (*23*) tested this hypothesis by monitoring the bioreduction of Fe(III) in soil using parallel incubation experiments designed to have the same initial chemical reducing capacity provided by two different samples of added natural organic matter, but with the initial total carboxyl content provided by the samples differing substantially (by a factor of 2) between the two experiments. The congruence of reducing capacity and the contrast in carboxyl content represented an effort to distinguish the role of Fe(II) complexation from that of pure electron shuttling. Their results showed that, in the presence of the organic matter sample with lower carboxyl content, soluble Fe(II) production rose to a plateau after about three weeks of incubation (see Figure 1). However, in the presence of the organic matter sample with larger carboxyl content, soluble Fe(II) production was essentially linear with incubation time, showing no tendency to plateau even after one month, except for an initial slow rise that was in fact congruent with that observed during incubation with the organic matter sample of lower carboxyl content, once adjustment of the results was made for the initial pH difference caused by the differing carboxyl content of the two samples. Rakshit et al. (*23*) concluded that, although electron shuttling dominated the early kinetics of Fe(III) bioreduction in their soil, Fe(II) complexation dominated the later kinetics.

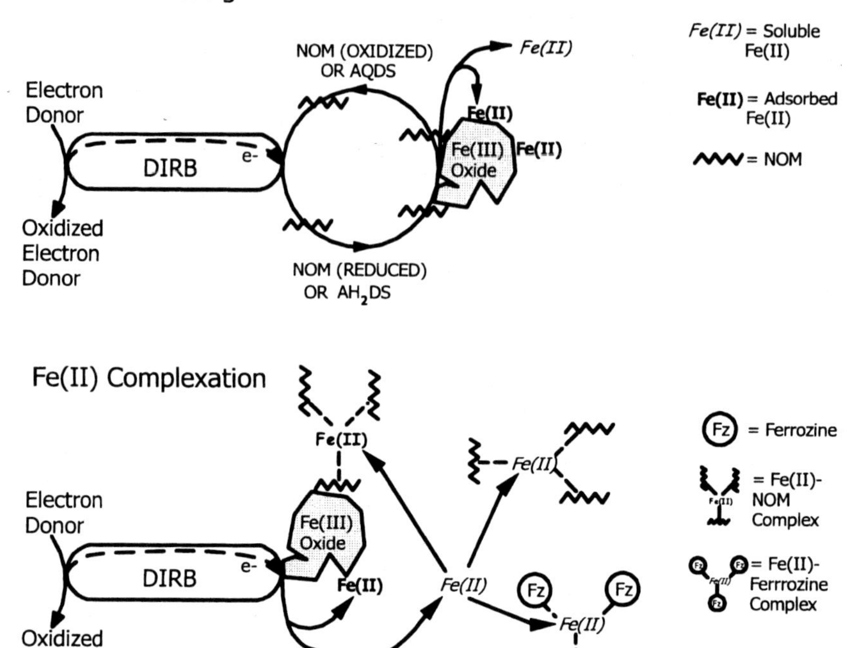

Figure 2. Mechanisms for the bioreduction of Fe(III) oxide in the presence of natural organic matter (NOM) based on experiments in which anthraquinone-2,6-disulfonate (AQDS) or ferrozine was added to suspensions containing dissimilatory iron-reducing bacteria (DIRB). Reprinted with permission from Royer et al. (34). Copyright 2002 American Chemical Society.

Having begun to clarify the ancillary role of carboxyl groups, one is still left to ask precisely which moieties in natural organic matter actually serve as electron shuttles. Tratnyek and Macalady (*1*) and Dunnivant et al. (*2*) proposed "quinone-type moieties" and indeed, adding anthraquinone-2,6-disulfonate to soils and sediments (*12*, *20*) enhances Fe(III) bioreduction similarly to adding natural organic matter (*22*, *23*), although not every synthetic quinone can do this (*34*). We also know that aromaticity is important to electron shuttling capability (*34–37*). Ratasuk and Nanny (*37*), however, raised questions concerning the quinone proposal by showing that humic substances reduced at pH 6.5 under H_2(g) on Pd catalyst supported by alumina apparently lose their quinone moieties (evidenced by loss of the aromatic ketone peak in their infrared spectra), but they do not lose their chemical reducing capacity (determined by reaction with ferric citrate). For example, Elliott soil humic acid and Pahokee Peat humic acid still retain about half of their chemical reducing capacities after reduction with H_2(g) over Pd/Al_2O_3 catalyst at pH 6.5 (*37*). Ratasuk and Nanny (*37*) concluded that their results "…suggest that electron-transfer processes involving [humic substances] could occur via several mechanisms besides that of the formation of semiquinone radicals".

Wolf et al. (*38*) added another piece to the puzzle in their systematic investigation of a model system realizing the schematic diagram in the upper part of Figure 2, with *G. Metallireducens* being the dissimilatory iron-reducing bacterium and ferrihydrite being the Fe(III) oxide. In respect to quinone electron shuttles, they examined not only anthraquinone-2,6-disulfonate, but seven other quinone compounds as well that were distinguished by varying molecular structure and, significantly, by their apparent electrode potential for reduction at pH 7. The results showed that the initial rate of acid-extractable (1 *M* HCl) Fe(II) production, normalized to that observed without quinone, exhibited a very strong maximum when plotted against the apparent electrode potential for reduction, this unimodal behavior reflecting the fact that some of the quinones simply could not enhance Fe(II) production. In particular, quinones having apparent electrode potentials for reduction at pH 7 in the range −225 to −137 mV were those that also greatly enhanced Fe(III) bioreduction (anthraquinone-2,6-disulfonate was best, with an apparent electrode potential equal to −184 mV at pH 7). Noting that the apparent electrode potential for the reduction of natural organic matter falls into the peak range they found for synthetic quinones, Wolf et al. (*38*) concluded that indeed quinone moieties are important to electron shuttling

The next piece of the puzzle was added by Aeschbacher et al. (*39*), who devised an ingenious electrochemical method for measuring oxidative capacity at pH 7 that also permitted inference as to which organic moieties contributed to it. Their flow-through electrochemical cell utilized a glassy carbon working electrode maintained at −0.49 V relative to the Standard Hydrogen Electrode, which is nearly at the bottom of the range of apparent electrode potentials for the reduction of synthetic quinones (*38*, *39*). An organic radical (the bipyridyl herbicide, diquat) was added to facilitate electron transfer between the natural organic matter sample being reduced and the working electrode. Quantitation of electron transfer in this cell was direct, by time-integration of the reductive current.

The resulting oxidative capacities, measured for a dozen commercial humic substances, were proportional to aromaticity [linear correlation (R^2 = 0.82), with a y-intercept not significantly different from 0]. This proportionality, as well as a strong linear correlation of oxidative capacity with the C/H molar ratio, and the range of apparent electrode potentials accessed by their electrochemical method led Aeschbacher et al. (*39*) to conclude "... that aromatic systems, likely quinone moieties, dominate the redox characteristics of [humic substances]". Most interesting for our quest, their values of oxidative capacity also were strongly and positively correlated with oxidative capacities reported by Ratasuk and Nanny (*37*) for the same humic substances chemically-reduced under H_2(g) on a Pd catalyst and oxidized with ferric citrate. Specifically, the electrochemical oxidative capacities measured by Aeschbacher et al. (*39*) were about three times larger than those measured by Ratasuk and Nanny (*37*). As noted above, this ratio is about the same as that which obtains for reducing capacities measured with potassium hexacyanoferrate vs. ferric citrate.

All of this is suggestive of quinone moieties being the seat of electron shuttling by natural organic matter. It would be very useful to develop a *direct* determination of the quinone content of humic substances to compare with the results of Aeschbacher et al. (*39*). Can we do this? Well, yes we can. Long ago Schnitzer and Riffaldi (*40*) published a wet-chemical method for quantitating quinones in humic substances. Basically, the method involves reaction of natural organic matter with $SnCl_2$ in a highly alkaline medium, with Sn^{2+} then reducing the quinone moieties, followed by potentiometric back titration with the strong oxidant, $K_2Cr_2O_7$. Schnitzer and Riffaldi (*40*) showed that their method was very accurate for synthetic quinones (and it gave the expected null result for aromatic compounds not possessing quinone moieties). Sudipta Rakshit, in the author's laboratory, has applied their method to anthraquinone-2,6-disulfonate, a compound which Schnitzer and Riffaldi (*40*) did not examine, and found a quinone content of 5.45 ± 0.21 mol/kg (*S. Rakshit, unpublished*), the same as the expected value, 5.46 mo/kg. When applied to Elliott soil humic acid and Pahokee Peat humic acid, the method yielded quinone contents of 1.38 and 1.25 mol/kg, respectively (*S. Rakshit, unpublished*). These results may be compared to the corresponding oxidative capacities of 1.96 and 1.62 mol_c/kg, reported by Aeschbacher et al. (*39*), and to the chemical reducing capacities of 1.89 and 2.37 mol_c/kg, cited above, which were obtained after reduction under H_2(g) on a Pd catalyst using potassium hexacyanoferrate as the oxidant (*T. Peretyazhko, unpublished*). It is difficult to believe that the close agreement among these independently-measured values is purely coincidental. It is more likely that the "Tratnyek-Macalady quinone hypothesis," is simply right.

Acknowledgments

Gratitude is expressed to the Andrew W. Mellon Foundation and to the National Science Foundation (Grant No. DEB-0543558) for supporting the author's research projects on electron shuttling by natural organic matter in soils. Thanks also to Whendee Silver for her leadership in securing this funding, and to Tanya Peretyazhko, Minori Uchimiya, and Sudipta Rakshit for their peerless laboratory experiments. And finally, thanks to Donald Macalady for changing forever the scientific study of natural organic matter with his research, which has always been both beautiful and true.

References

1. Tratnyek, P. G.; Macalady, D. L. Abiotic reduction of nitro aromatic pesticides in anaerobic laboratory systems. *J. Agric. Food Chem.* **1989**, *37*, 248–254.
2. Dunnivant, F. M.; Schwarzenbach, R. P.; Macalady, D. L. Reduction of substituted nitrobenzenes in aqueous solutions containing natural organic matter. *Environ. Sci. Technol.* **1992**, *26*, 2133–2141.
3. Kung, H. K.; McBride, M. B. Electron transfer processes between hydroquinone and hausmannite (Mn_3O_4). *Clays Clay Miner.* **1988**, *36*, 297–302.

4. Ukrainczyk, L.; McBride, M. B. Oxidation of phenol in acidic aqueous suspensions of manganese oxides. *Clays Clay Miner.* **1992**, *40*, 157–166.
5. Lovley, D. R.; Coates, J. D.; Blunt-Harris, E. L.; Phillips, E. J. P.; Woodward, J. C. Humic substances as electron acceptors for microbial respiration. *Nature* **1996**, *382*, 445–448.
6. Sposito, G. *The Chemistry of Soils,* 2nd ed.; Oxford: New York, 2008.
7. Benz, M.; Schink, B.; Brune, A. Humic acid reduction by *Propionibacterium freudenreichii* and other fermenting bacteria. *Appl. Environ. Microbiol.* **1998**, *64*, 4507–4512.
8. Coates, J. D.; Ellis, D. J.; Blunt-Harris, E. L.; Gaw, C. V.; Roden, E. E.; Lovley, D. R. Recovery of humic-reducing bacteria from a diversity of environments. *Appl. Environ. Microbiol.* **1998**, *64*, 1504–1509.
9. Lovley, D. R.; Fraga, J. L.; Blunt-Harris, E. L.; Hayes, L. A.; Phillips, E. J. P.; Coates, J. D. Humic substances as a mediator for microbially catalyzed metal reduction. *Acta Hydrochim. Hydrobiol.* **1998**, *26*, 152–157.
10. Lovley, D. R.; Kashefi, K.; Vargas, M.; Tor, J. M.; Blunt-Harris, E. L. Reduction of humic substances and Fe(III) by hyperthermophilic microorganisms. *Chem. Geol.* **2000**, *169*, 289–298.
11. Cervantes, F. J.; de Bok, F. A. M.; Duong-Dac, T.; Stams, A. J. M.; Lettinga, G.; Field, J. A. Reduction of humic substances by halorespiring, sulphate-reducing and methanogenic microorganisms. *Environ. Microbiol.* **2002**, *4*, 51–57.
12. Kappler, A.; Benz, M.; Schink, B.; Brune, A. Electron shuttling via humic acids in microbial iron(III) reduction in a freshwater sediment. *FEMS Microbiol. Ecol.* **2004**, *47*, 85–92.
13. Heitmann, T.; Goldhammer, T.; Beer, J.; Blodau, C. Electron transfer of dissolved organic matter and its potential significance for anaerobic respiration in a northern bog. *Global Change Biol.* **2007**, *13*, 1771–1785.
14. Keller, J. K.; Weisenhorn, P. B.; Megonigal, J. P. Humic acids as electron acceptors in wetland decomposition. *Soil Biol. Biochem.* **2009**, *41*, 1518–1522.
15. Lipson, D. A.; Jha, M.; Raab, T. K.; Oechel, W. C. Reduction of iron (III) and humic substances plays a major role in anaerobic respiration in an Arctic peat soil. *J. Geophys. Res., [Biogeosci.]* **2010**, *115*.
16. Cervantes, F. J.; van der Velde, S.; Lettinga, G.; Field, J. A. Competition between methanogenesis and quinone respiration for ecologically important substrates in anaerobic consortia. *FEMS Microbiol. Ecol.* **2000**, *34*, 161–171.
17. Lovley, D. R.; Fraga, J. L.; Coates, J. D.; Blunt-Harris, E. L. Humics as an electron donor for anaerobic respiration. *Environ. Microbiol.* **1999**, *1*, 89–98.
18. Kappler, A.; Haderlein, S. B. Natural organic matter as reductant for chlorinated aliphatic pollutants. *Environ. Sci. Technol.* **2003**, *37*, 2714–2719.
19. Petruzzelli, L.; Celi, L.; Ajmone-Marsan, F. Effect of soil organic fractions on iron oxide biodissolution under anaerobic conditions. *Soil Sci.* **2005**, *170*, 102–109.

20. Peretyazhko, T.; Sposito, G. Iron(III) reduction and phosphorous solubilization in humid tropical forest soils. *Geochim. Cosmochim. Acta* **2005**, *69*, 3643–3652.
21. Zhang, H.; Weber, E. J. Elucidating the role of electron shuttles in reductive transformations in anaerobic sediments. *Environ. Sci. Technol.* **2009**, *43*, 1042–1048.
22. Nevin, K. P.; Lovley, D. R. Potential for nonenzymatic reduction of Fe(III) via electron shuttling in subsurface sediments. *Environ. Sci. Technol.* **2000**, *34*, 2472–2478.
23. Rakshit, S.; Uchimiya, M.; Sposito, G. Iron(III) bioreduction in soil in the presence of added humic substances. *Soil Sci. Soc. Am. J.* **2009**, *73*, 65–71.
24. Roden, E. E.; Kappler, A.; Bauer, I.; Jiang, J.; Paul, A.; Stoesser, R.; Konishi, H.; Xu, H. Extracellular electron transfer through microbial reduction of solid-phase humic substances. *Nat. Geosci.* **2010**, *3*, 417–421.
25. Peretyazhko, T.; Sposito, G. Reducing capacity of terrestrial humic acids. *Geoderma* **2006**, *137*, 140–146.
26. Visser, S. A. Oxidation-reduction potentials and capillary activities of humic acids. *Nature* **1964**, *204*, 581.
27. Lovley, D. R.; Blunt-Harris, E. L. Role of humic-bound iron as an electron transfer agent in dissimilatory Fe(III) reduction. *Appl. Environ. Microbiol.* **1999**, *65*, 4252–4254.
28. Helburn, R. S.; MacCarthy, P. Determination of some redox properties of humic acid by alkaline ferricyanide titration. *Anal. Chim. Acta* **1994**, *295*, 263–272.
29. Matthiessen, A. Determining the redox capacity of humic substances. *Vom Vasser* **1995**, *84*, 229–235.
30. Matthiessen, A. Evaluating the redox capacity of humic substances by redox titrations. In *Humic Substances in the Global Environment*; Senesi, N., Miano, T. M., Eds.; Elsevier: New York, 1995; pp 187−192.
31. Bauer, I.; Kappler, A. Rates and extent of reduction of Fe(III) and O_2 by humic substances. *Environ. Sci. Technol.* **2009**, *43*, 4902–4908.
32. Bauer, M.; Heitmann, T.; Macalady, D. L.; Blodau, C. Electron transfer capacities and reaction kinetics of peat dissolved organic matter. *Environ. Sci. Technol.* **2007**, *41*, 139–145.
33. Jiang, J.; Kappler, A. Kinetics of microbial and chemical reduction of humic substances: Implications for electron shuttling. *Environ. Sci. Technol.* **2008**, *42*, 3563–3569.
34. Royer, R. A.; Burgos, W. D.; Fisher, A. S.; Unz, R. F.; Dempsey, B. A. Enhancement of biological reduction of hematite by electron shuttling and fe(ii) complexation. *Environ. Sci. Technol.* **2002**, *36*, 1939–1946.
35. Royer, R. A.; Burgos, W. D.; Fisher, A. S.; Jeon, B.-H.; Unz, R. F.; Dempsey, B. A. Enhancement of hematite bioreduction by natural organic matter. *Environ. Sci. Technol.* **2002**, *36*, 2897–2904.
36. Chen, J.; Gu, B.; Royer, R. A.; Burgos, W. D. The roles of natural organic matter in chemical and microbial reduction of ferric iron. *Sci. Total Environ.* **2003**, *307*, 167–178.

37. Ratasuk, N.; Nanny, M. A. Characterization and quantification of reversible redox sites in humic substances. *Environ. Sci. Technol.* **2007**, *41*, 7844–7850.
38. Wolf, M.; Kappler, A.; Jiang, J.; Meckenstock, R. U. Effects of humic substances and quinones at low concentrations on ferrihydrite reduction by *Geobacter metallireducens*. *Environ. Sci. Technol.* **2009**, *43*, 5679–5685.
39. Aeschbacher, M.; Sander, M.; Schwarzenbach, R. P. Novel electrochemical approach to assess the redox properties of humic substances. *Environ. Sci. Technol.* **2010**, *44*, 87–93.
40. Schnitzer, M.; Riffaldi, R. The determination of quinone groups in humic substances. *Soil Sci. Soc. Amer. Proc.* **1972**, *36*, 772–777.

Chapter 7

Electrochemistry of Natural Organic Matter

James T. Nurmi* and Paul G. Tratnyek

Division of Environmental and Biomolecular Systems, Oregon Health & Science University, 20000 NW Walker Road, Beaverton, OR 97006
***jnurmi@ebs.ogi.edu**

Natural organic matter (NOM) plays an important role in a variety of environmental redox processes, ranging from fueling the global carbon cycle to mediating microbial interactions with minerals. However, the complex and indeterminant composition of NOM makes characterization of its redox activity challenging. Approaches that have been taken to address these challenges include chemical probe reactions, potentiometric titrations, chronocoulometry, and voltammetry. Advantages of the latter include that it can be diagnostic and quantitative, but applying voltammetry to the characterization of NOM has been challenging. Improved results have been obtained recently by using aprotic solvents, microelectrodes, and various applied potential waveforms. Results obtained with several voltammetric methods and DMSO as the solvent strongly suggest that quinone-like moieties are the dominant redox active groups. Correcting the associated peak potentials for comparison with estimates of NOM redox potentials in aqueous solutions shows that the range of peak potentials resolved by voltammetry spans most of the redox potentials obtained by other means that have been reported for various types of NOM or NOM model compounds. The multiplicity of electron-transfer steps that are distinguishable by voltammetry, and the likelihood that there is a degree of redox-coupling among these moieties, suggests that the redox potential of NOM might best be modeled as a continuum of redox potentials. The kinetics of electron exchange along this continuum will vary with factors such as the complex tertiary structure of NOM. The kinetic limitations created by this tertiary structure may be overcome with organic solvents (which allow the structure to unravel) or electron shuttles (which can pass into the structure), which accounts for the improved resolution of methods that use these strategies in electrochemical characterization of NOM.

Introduction

Applications of electrochemistry to environmental science and engineering are of three general types: (*i*) engineering applications for treatment of contaminated water or sediment; (*ii*) analytical applications involving electrodes for analyte concentration and detection; and (*iii*) diagnostic applications to understanding fundamental aspects of redox processes in environmental systems. The engineering applications are well represented in reviews, some of which also include electrochemical aspects of green chemistry and energy capture/storage (*1–7*). Analytical applications of electrochemistry to environmental science are also sufficiently mature to have been reviewed many times (*8–13*). In contrast, application of electrochemical methods for the purpose of diagnosing fundamental aspects of environmental redox processes are dispersed throughout the literature on aquatic chemistry, biogeochemistry, etc.

Two areas where fundamental studies of environmental redox processes have relied heavily on electrochemical methods stand out because they illustrate some of the unique advantages of this approach. One particularly active area of research involves the use of microelectrodes to make spatially resolved, in situ measurements of redox-active species in structured systems like sediments and biofilms (*11, 14–17*). An area of environmental research where the electrochemical approach is comparatively novel involves direct characterization of reactions that occur on redox active solids—such as zerovalent metals, metal oxides and sulfides, and clays—by configuring these materials as electrodes (*18–22*). When these electrodes are made of fine-particulate materials, the structure and composition of the original particle surfaces can be preserved, thereby enabling investigation of the dynamics of surface property changes with other factors, such as applied overpotential or solution chemistry (*23*).

In addition to the approaches represented by the two examples given above, other applications of electrochemistry to mechanistic studies of aquatic or biogeochemical redox processes employ inert electrodes with a response that is mediated by a surface layer of substances whose properties are the main interest. Examples of this type of work involve microorganisms (*24*), cytochromes (*24–26*), other redox shuttles (*27*), clays (*26–30*), and humic substances.

This chapter concerns the redox properties of humic substances (HS)—or, more broadly, natural organic matter (NOM)—as determined by the electrochemical method commonly referred to as voltammetry. Other electrochemical methods, such as potentiometric titrations and coulometry have been used to study properties of NOM, such as the electron transfer capacity (e.g., (*31–33*)), but they are given only limited consideration in this review. We chose to focus on voltammetric techniques for three reasons: (*i*) the prevalence of data that currently exists is larger than for any other electrochemical technique, (*ii*) we have collected a lot of voltammetric data on various sources of NOM, and (*iii*) this is the only method that allows for the direct measurement of redox potentials at an electrode surface. The chapter starts with a review of the basic principles of the relevant voltammetric methods, followed by a summary of past work in this area, then presentation of some previously unpublished data, and finally a synthesis of the results and discussion of how electrochemical parameters such as redox potential and redox capacity relate to the biogeochemical redox state of environmental systems.

Background

Electrochemical Methods

In general, there are two categories of electrochemical methods, (*i*) static or equilibrium techniques, and (*ii*) dynamic or non-equilibrium methods. In the case of the static methods, no potential or current is applied to the working electrode so it equilibrates with the redox labile species in solution. The most precise term for this technique is zero-current chronopotentiometry, but it is more commonly referred to as open-circuit potential (OCP) or oxidation-reduction potential (ORP) measurement. In the second category of electrochemical measurements, a current or potential is applied to perturb conditions at the electrode-solution interface, and the system response is measured. The most basic dynamic electrochemical techniques are linear sweep voltammetry (LSV) and cyclic (linear sweep) voltammetry (CV), but there are many variations that have been developed for a wide range of purposes (*34–37*).

Dynamic voltammetric methods like LSV and CV involve applying a potential at an (usually) inert electrode surface and measuring the resulting current due to the transfer of charge to or from species in solution. In a typical voltammetry experiment, three electrodes are used. The potential is ramped between the working electrode and the reference electrode. When the potential is strong enough to either oxidize or reduce substances from the solution (e.g., NOM), current flows and is measured between the working electrode and an auxiliary (or counter) electrode.

Working electrodes are often made of noble metals, such as platinum or gold. Other materials have been used in environmental systems such as the hanging drop mercury electrode and glassy carbon electrodes. The various electrode materials offer different potential 'windows' in which experiments can be performed without reducing or oxidizing the solvent. Under aqueous conditions, this window also varies with the solvent and pH: for example, when using a Pt electrode in water at neutral pH, the window ranges from approximately -1.0 V to +1.0 V vs. Ag/AgCl. In DMSO, the window ranges from -1.5 V to +0.8 V vs. Ag/Ag^+. In this respect, water is more limiting (has a narrower potential window) than many alternative solvents.

Another important variable in voltammetry is the waveform of the applied potential. The simplest waveform used in voltammetric studies is called linear sweep voltammetry (LSV) and cyclic voltammetry (CV). In LSV, the applied potential is ramped linearly in one direction while sampling the current response.

CV works the same way, only that the potential is ramped back to the beginning value after reaching the end point (switching potential). Advantages of LSV and CV are that the data are comparatively easy to interpret and they can characterize redox activity over a wide potential region.

Another dynamic voltammetric method is differential pulse voltammetry (DPV), where the applied potential is ramped in pulses and the current is measured at the end of each pulse. In square wave voltammetry (SWV), the applied potential is ramped using square waves in which the potential pulse is applied in both the positive (anodic) and negative (cathodic) direction and the current is sampled at the end of each pulse. The net current is the total current response for both anodic and cathodic currents. DPV, and especially SWV, have higher sensitivity, reduce non-faradic currents (background currents), and scan times can be shorter than in LSV and CV.

In addition to solvent and waveform, many other variations on voltammetry are possible. One that is relevant to the results described below involves the use of micro-sized working electrodes. Microelectrodes offer three key advantages over macroelectrodes, (*i*) the thin diffusion layers on small electrodes favors diffusion limited currents, (*ii*) the small currents involved minimize perturbation of the solution chemistry, and (*iii*) the "ohmic drop" (voltage difference between the working electrode and the reference electrode) effect is minimized so less conductive solutions can be used.

A comparison of three representative types of voltammetric data is shown in Figure 1 for menadione (a model quinone, for comparison with data for NOM shown later). Conventional CV with a macro (mm-sized) working electrode gives two pairs of peaks (Figure 1A), labeled 1A, 1C, 2A, and 2C for anodic, cathodic, and first or second electron transfer (counting from the quinone form). The CV obtained with a micron-sized working electrode (Figure 1B) is simpler because the anodic and cathodic currents coincide and are diffusion limited, resulting in plateaus instead of peaks. SWV (with a microelectrode) gives sharp symmetrical faradaic peaks (Figure 1C) due to the way the current is sampled (removes all capacitive currents). The comparatively sharp definition of the peaks in Fig. 1C illustrates one of the advantages of SWV.

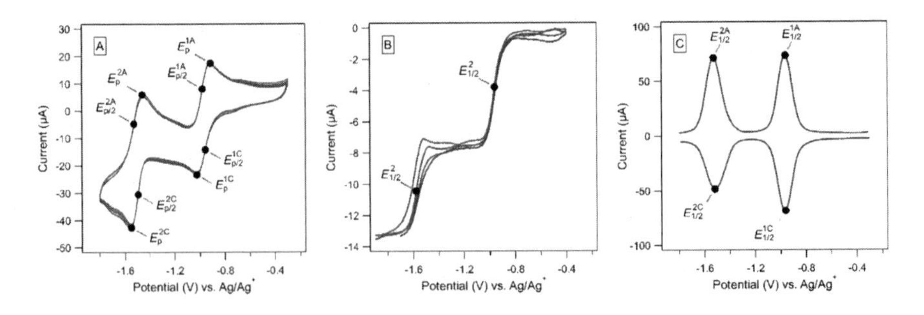

Figure 1. Typical electrochemical results for a quinone (1.0 mM menadione) in an aprotic solvent (DMSO). (A) CV with a 3-mm Pt working electrode. (B) CV with a 100 μm Pt working electrode. (C) SWV with a 100 μm Pt working electrode. E_p *and* $E_{p/2}$ *are peak and half-peak potentials, respectively.*

In all three cases shown in Fig. 1, the first and second electron transfer steps are well resolved in part because the data were obtained in an aprotic solvent. In water, the semiquinone intermediate would be less stable, resulting in one peak for both electron transfers. The effect of solvent on voltammetry of NOM is discussed further below.

Qualitative Voltammetry

To illustrate the qualitative interpretation of voltammetric data, consider a system that consists of a simple, reversible redox couple $A + e^- \rightleftharpoons D$. LSV on such a system will give a single peak in current (i) vs. potential (E). The peak is termed cathodic for reduction ($A \rightarrow D$) or anodic for oxidation ($D \rightarrow A$) depending on the scan direction. A CV will exhibit a pair of peaks, one anodic and the other cathodic, for this system. This behavior can be seen in Fig. 1A. Symmetrical pairs of anodic and cathodic peaks often are not observed because many redox couples are less than fully reversible and (or because) side reactions occur involving the species involved in the redox couple.

Voltammograms produce peaks even though the Butler-Volmer equation predicts that current should increase exponentially with increasing over-potential. This arises because the increase in net current measured by the working electrode is not only influenced by the potential dependence of interfacial charge transfer (the Butler-Volmer part), but also to the concentration of the redox labile species at the electrode surface. As the rate of interfacial charge transfer increases, electrolysis consumes A or D, which is only partially replenished by diffusion from bulk solution. Depletion of A or D in the interfacial region results in decreased current, producing the peak.

With more complex systems, voltammetry often produces multiple peaks, which indicates that multiple redox couples and/or multistep redox couples are contributing to the electrode response. To resolve these steps, experimental conditions such as solvent and sweep rate can be varied, as these factors influence the number, size, shape, and separation of peaks in voltammetry. These qualitative aspects of voltammogram interpretation are illustrated below, where we compare data obtained on various NOM samples to model compounds for likely NOM components.

Quantitative Voltammetry

There are several quantitative variables that can be determined from a voltammogram. The potential and current at which the peak occurs are described by the peak potential (E_p) and the peak current (i_p), respectively. In CV, the potential at half the peak height is called the half-peak potential ($E_{p/2}$), which is approximately equal to standard potential for reversible redox couples. The utility of this is limited, however, because any degree of irreversibility will cause $E_{p/2}$ to overestimate the standard potential. This issue is discussed further in Chapter 3 (*38*), as it relates to the determination of one-electron reduction potentials for contaminant degradation reactions. Quantitative analysis of data from DPV and SWV is similar, but with one major difference. Due to the way the current is sampled in SWV, E_p rather than $E_{p/2}$ is the estimator of the standard potential (for reversible couples).

Voltammetry can also be used to characterize the degree of reversibility of a redox couple. The peak potential (E_p) of a reversible redox couple is constant and independent of scan rate, but as a couple becomes less reversible, E_p shifts cathodically for reduction reactions and anodically for oxidation reactions.

Quantitatively, for the reversible process:

$$\left|E_p - E_{p/2}\right| = 2.20\left(\frac{\mathrm{R}T}{\mathrm{F}}\right) \qquad (1)$$

Whereas for irreversible electron transfer:

$$\left|E_p - E_{p/2}\right| = 1.86\left(\frac{\mathrm{R}T}{\alpha\mathrm{F}}\right) \qquad (2)$$

where α is the transfer coefficient or barrier symmetry, and R, T, and F have their usual meanings (gas constant, absolute temperature, and Faraday constant, respectively). Experimentally determined values of α are typically about 0.5, but they can vary between 0 and 1 and have been interpreted as a measure of progress on the reaction coordinate at charge transfer (*39*). The degree of reversibility also affects peak currents, with i_p for reversible couples tending to be larger than for irreversible couples (other factors, such as scan rate and concentration, being equal).

In CV, reversibility is also reflected in the difference between E_{ox} and E_{red}. Reversible couples give peak potentials that are separated by approximately 59 mV and this should be independent of scan rate. If n electrons are transferred in a reversible electrode process, the difference in peak potentials from CV is given by:

$$\left|E_p^{ox} - E_p^{red}\right| = 2.218\left(\frac{\mathrm{R}T}{n\mathrm{F}}\right) \qquad (3)$$

When a pair of peaks is observed by CV but the difference in peak potentials is < 59 mV, the couple is described as *quasi-reversible,* a condition that is fairly common and has been subjected to extensive interpretation (*34*).

Peak potentials and other quantitative properties obtained from voltammetry can be diagnostic of the species that dominate the electrode response, if other factors that influence the electrode response are constant or negligible. This strategy is used below in comparisons of electrochemical data obtained with NOM and NOM model compounds.

Voltammetry of Natural Organic Matter

It is well known that NOM can be redox active and this plays an important role in the biogeochemistry of environmental systems (*31*, *40–54*). There are three electrochemical aspects to this: (*i*) the *capacity* of NOM to accept or donate electrons, (*ii*) the *potential* at which charge transfers to or from NOM occurs, and (*iii*) the *kinetics* of these charge transfer reactions. The electron transfer capacity (ETC) of NOM to accept or donate electrons (EAC or EDC, respectively) makes it a significant redox buffer in environmental systems (*31*, *32*, *46*, *55–60*). However, EAC and EDC are properties of the aggregate material, and since NOM is composed of a complex mixture of redox labile compounds and moieties with different redox characteristics, this makes any measure of the ETC dependent on the potential and kinetics of the analysis. One response to this complication is to focus on isolating the effects of specific redox-active functional groups, which has been a prominent objective of many studies of NOM using electrochemical methods.

The redox activity of NOM is usually attributed mainly to the polyphenolic (especially hydroquinonoid) moieties (*50*, *54*, *61*); so much so that quinonoid-enriched NOM has been developed to provide enhanced redox properties (*60*). There are, however, other moieties that may contribute to the overall redox activity of NOM (e.g., complexed metals (*31*) and thiols (*62–64*)), and it is likely that the relative significance of these redox-active components varies with the source, type, and handling of the NOM. In principle, electrochemical methods can distinguish between the contributions of different redox active moieties—e.g., by peak potential, as described above—but this is only possible for a small number of relative distinct types of redox active moieties. While this ultimately limits the reductionist approach to interpreting electrochemical data for characterization of NOM, it has lead to several significant developments based on comparatively holistic interpretations of the data, which are discussed later.

Most early studies on the voltammetry of NOM (mainly NOM and fractions of NOM) involved simple voltammetric methods (LSV, CV) with a hanging drop mercury working electrode. It appears that there was consensus at the time that the humic fraction was not electrochemically reducible at the mercury drop (*65*), so the apparent electro-reduction of humic-like oxidation products from bituminous coal was attributed to nitro groups associated with the material (*66*). Later, it was shown that the effect of NOM and NOM fractions under these conditions was mainly due to their surfactancy, suppressing the electrode response by forming a redox inactive adsorbed layer (*67*). As a result of this, few of these early studies present sufficient electrochemical detail to allow further analysis of the results (e.g., (*67*, *68*)).

Subsequent studies of the voltammetry of NOM and NOM fractions have used solid-state working electrodes and explored the full range of potential waveforms summarized in the previous section (LSV, CV, DPV, and SWV). The evolution of this work is represented in Table 1, where the major works are summarized in chronological order (left to right), along with key experimental conditions that distinguish the studies (rows).

Table 1. Method details for key electrochemical studies of NOM redox properties

Properties	*Various Sources* [1]	*Yu, 1985 (73)*	*Ding, 1991 (69, 70)*	*Helburn, 1994 (75)*	*Motheo, 2000 (76)*	*Nurmi, 2002 (77, 78)*	*Aeschbacher, 2010 (33)*
Working electrode	Hg	GC	GC, CPE, CFbE	Au Wire	$Ti/Ir_nTi_nO_2$	Pt disk	GC
Counter electrode	Pt	Pt		Au Wire	Steel	Pt wire	Pt wire
Reference electrode	SCE	SCE	Ag/AgCl	SCE	SCE	Ag/Ag^+	Ag/AgCl
Potential Wave Form	LSV	LSV	DPS	LSV	LSV	LSV	Chrono-coulometry
Scan Rate			2 mV/s	500 mV/s	50 mV/s	10 mV/s	
Solvent	H_2O	H_2O	H_2O	H_2O	H_2O	DMSO	H_2O
Electrolyte	Unk [2]	Unk [2]	NH_4Ac	0.5 M KCl	0.1 M KCl	1.0 mM $NaClO_4$	0.1 M KCl
NOM conc.	Unk [2]	Unk [2]	100:5 [3]	3 g/L	30 mg/L	90 mg/L	2 g/L
NOM type	Various fractions of NOM	Unk [2]	Extracts of various tree Needles and leaves	Glenamoy-Ireland HA	Mogi Guacu River peat HA	Various NOM	Various NOM

[1] *(66, 79–82)*, [2] Unknown, [3] Ratio water to dry NOM.

The earliest of the studies shown in Table 1 focused on complex organic materials like the decomposition products of rice straw, manure, and rice paddy natural organic matter (*69*, *70*). In some cases these materials produced voltammograms with relatively good peak definition, but the data were not used to quantify redox potentials or identify the responsible compounds or moieties. When Yu et al. (*71–73*) studied the anaerobic decomposition products of rice straw and vetch by LSV with a (solid state) Pt electrode, only poorly defined peaks were observed. The lack of definite peaks prevented quantification of half-wave potentials, but comparison between these features showed that the quantity of redox active material was greatest at the stage of most vigorous decomposition.

In order to obtain better peak definition, DPS has been used to study various types of NOM ((*69*, *70*) and others cited therein). Using aqueous extracts of tree leaves, for example, a well-defined oxidation peak was observed at ~0.5 V vs. SCE, which was unaffected by passing the sample through a cation-exchange column, but completely removed after treatment by anion-exchange (*69*, *70*). Consistent with its apparent negative charge, this redox-active component of NOM was shown to adsorb strongly to various minerals and soil types (goethite, Rhodic Ferralsol, Ali-Haplic Acrisol, and Cambisol) (*74*). Subsequent work by the same group with solutions of decomposition products from pine needles characterized by DPS gave several moderately defined peaks between –0.05 and +0.69 mV vs. Ag/AgCl. After contact with various soils, the most cathodic of these peaks disappeared, but a large anodic peak appeared, which turned out to be Mn^{2+}, presumably from reduction of MnO_2 in the soil. Using DPS with a glassy carbon electrode, instead of Pt, gave somewhat improved peak definition in studies of solutions of pine needle and bamboo decomposition products (*69*), but the compounds or moieties responsible for these peaks were still not identified.

Rather than attempting to obtain interpretable voltammetry data from very complex, otherwise uncharacterized forms of NOM, most recent work has focused on relatively well characterized "standard" forms of NOM or fractions of NOM. This approach is exemplified by an electrochemical study of humic acid (HA) extracted from peat by Helburn and McCarthy (*75*). Using solutions of this HA prepared in water (and other conditions summarized in Table 1) they obtained CVs like the one shown in Figure 2A. At the switching potentials (where the sweep is reversed), the presence of HA resulted in more current, which Helburn and McCarthy ascribed to NOM-catalyzed oxidation and reduction of the solvent (in this case, water). Beyond this, their CVs are essentially featureless, so they provide no evidence for a direct redox reaction at the electrode involving the HA.

Another study on HA extract from peat reported CVs obtained in aqueous solution but performed with working electrodes made with mixtures of Ru, Ir, and Ti oxides deposited on a Ti substrate (*76*). Under the conditions of their study (Table 1), the CVs were also essentially featureless for a range of electrode compositions (Fig. 2B). The results do show, however, a significant effect of working electrode composition, with the Ir doped TiO_2 giving significantly greater total current. Still, there are no features in these CVs that can be ascribed to specific types of charge transfers between the electrode and the adsorbed organic matter.

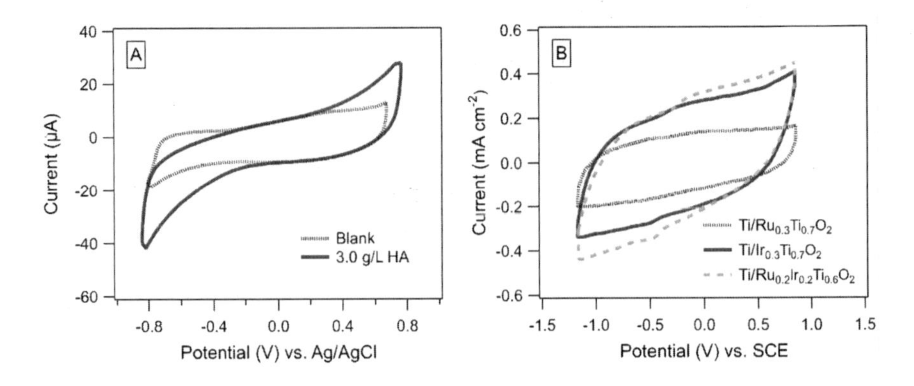

Figure 2. Examples of CV's for two different humic substances under aqueous conditions. (A) Replotted from data in (75); (B) replotted from data in (76).

Recently, the same basic electrochemical methods used in earlier work have been used with several critical refinements to achieve considerably greater success at characterizing the redox properties of NOM. Our approach has been to use an aprotic organic solvent, which greatly improves the resolution and sensitivity of voltammetry to direct electrode interactions involving NOM (*77, 78*). Below, we present some new data and further refinements for our approach and compare results with our method to those obtained by other electrochemical approaches such as chronocoulometry with direct electrochemical reduction or mediated electrochemical reduction with dissolved-phase electron-transfer mediators to measure the electron accepting capacities for various NOM fractions (*33*).

Methods

Table 1 provides a summary and comparison of materials and conditions used in this study vs. previously published attempts to use voltammetric methods to characterize the redox properties of NOM.

The main characteristics of our current method include Pt working and counter electrodes, dimethyl sulfoxide (DMSO) as the solvent, 1.0 mM $NaClO_4$ as the electrolyte, and a Ag/Ag^+ reference electrode. Further details and evidence for validation of this method are given in (*77, 78*). One new aspect of the setup used in this study is the comparison of two working electrode sizes: 1.6 mm dia. Pt disc vs. 125 µm dia. Pt wire.

The NOM, and fractions of NOM, used in this study were either provided by B. Gu (Oak Ridge National Laboratory) or were obtained from the International Humic Substance Society (IHSS). Samples obtained from B. Gu were isolated from raw concentrate obtained by reverse osmosis of brown water from a wetland pond in Georgetown, SC. This concentrate (NOM-GT) was purified with a column of cross-linked polyvinyl pyrrolidone polymer and a fraction that was enriched in polyphenol moieties (NOM-PP) or carbonhydrates (NOM-CH) was obtained. NOM-GT, NOM-PP, and other fractions of NOM-GT have been characterized by UV/Vis, IR, NMR, and EPR spectroscopy (*83–85*). For our work, stock solutions of NOM were prepared by dissolving the freeze-dried materials directly into DMSO.

Cyclic voltammetry (CV) was performed with the Pt microelectrode by starting at –1.8 V vs. Ag/Ag^+ for most of the NOM and model compounds. Square wave voltammetry (SWV) was also initiated at this potential and ramped toward more positive potentials. The frequency for the SWV was 100 Hz with a step potential of 2 mV with an amplitude of 10 mV.

Results

Electrochemistry in Nonaqueous Media

The method developed by Nurmi et al. (*77*)—CV with a Pt disk electrode in DMSO and minimal electrolyte—gave voltammograms with good definition for a variety of types of NOM and NOM model compounds. They were classified into six groups, and representative CVs for each group are shown in Fig. 3 (top row). A detailed interpretation of these groups is given in (*77*), but it is sufficient for the current purpose to recognize that they represent a progression from the electrochemical characteristics of pure, reversible quinone/hydroquinone type redox couples to complex materials containing a mixture of less than fully reversible or electrode active moieties. Further study with the method introduced in (*77*) was focused on NOM-PP (Category B); however, because it was the only type of NOM that gave CVs with all of the features of a model quinone (Category A).

In addition to the qualitative similarity between the CV of NOM-PP and model quinones, quantitative analysis of electrochemical data presented in (*77*) supports the conclusion that the redox properties of NOM-PP are dominated by quinone-like moieties. CVs for NOM-PP and model quinones both gave anodic/cathodic pairs of peaks with potentials (E_p) that were separated by slightly more than 0.059 V and peak currents (i_p) that were linearly related to the square root of the scan rate ($\nu^{0.5}$) and to analyte concentration (C). Equivalent results were obtained regardless of electrode rotation rate. These quantitative considerations indicate that the electrode reactions of NOM-PP and model quinones involve similar sequential pairs of one-electron, quasi-reversible, diffusion controlled, electron transfers (under the conditions of the method used in (*77*)).

A major limitation of the results obtained by the method used in (*77*) is that the less-fractionated samples gave relatively poorly defined CVs (Fig. 3, Groups D-F). To improve on this, we modified our method to take advantage of the properties of microelectrodes (still using the linear sweep waveform). The results obtained with NOM-PP were presented and discussed previously (*78*), and additional data for representatives of the other groups of materials are presented in the second row of Fig. 3. Group A and B materials show a clear shift from peaks to steps, as expected for mass transport control at microelectrodes. The Group C-F materials also gave CVs with steps, but the anodic and cathodic scans are still separate, suggesting less reversibility or slower electron transfer to the working electrode. Overall, the results obtained by CV with microelectrodes reinforced our interpretation of the data obtained by CV with conventional-size electrodes, but this change in method did not significantly improve the resolution of peaks for the unfractionated materials.

In a further effort to improve on our previous results, each representative of Groups A-F was characterized using a microelectrode and SWV (Fig. 3, 3rd Row). These data show the same number of peaks as the previous methods, but the resolution of the peaks is improved in such a way that E_p and i_p can be quantified with greater confidence. This effect is most notable for NOM-GT, where SWV revealed the largest number of peaks of any material tested. The apparent diversity of redox-active moieties associated with NOM-GT is unsurprising because this material comes from relatively young NOM with minimal purification. However, this result is notable in that it highlights a fundamental difficulty with methods of characterizing redox properties of NOM that give only a single, presumably aggregate, measure of potential.

The quantitative aspects of the data in Fig. 3 are summarized in Fig. 4. This representation shows peak and half-wave potentials obtained by CV with macroelectrodes (unfilled markers) and peak potentials obtained by SWV with microelectrodes (filled markers). For the data from CV, the line segments connect associated values of E_p and $E_{p/2}$, thereby highlighting the association between anodic and cathodic parts of each electron transfer peak. Note that the different fractions of NOM give a wide range of redox potentials and that the NOM sample with the least fractionation (NOM-GT) gives the largest range of potentials for all the samples.

Electrochemistry in Aqueous Media

One of the major impediments to using the results of electrochemical characterizations done with the methods described above is that they were done in nonaqueous solvent with a Ag/Ag^+ reference electrode (which is preferred in DMSO but not in water (*37*)). Adjusting these data for comparison with redox potentials under environmental conditions from first principles is not practical because it would require corrections for multiple factors including solvation of all the species involved in the electrode reaction, pKa's of protic species involved in the redox reaction, proton activity in the two solvent systems, etc. Instead, we can use the approach of calibrating the two systems using a redox couple that is well behaved in both systems. A redox couple that is often used for this purpose is ferricinium/ferricenium ion (Fc/Fc^+), which has a solvent independent redox potential of 0.692 vs. SHE (*86*).

The correction was performed by measuring the redox potential of the Fc/Fc^+ couple in DMSO vs. Ag/Ag^+ and in water vs. Ag/AgCl using CV in conditions similar to ones used to characterize the NOM samples. The results are summarized in Fig. 5 as two scale bars used to illustrate its application to our data for NOM and its fractions. The difference between the $E_{1/2}$ for ferrocene in DMSO and in water is 0.472 V. This difference is added to the potentials obtained in DMSO to reflect potentials in water vs. the SHE reference electrode.

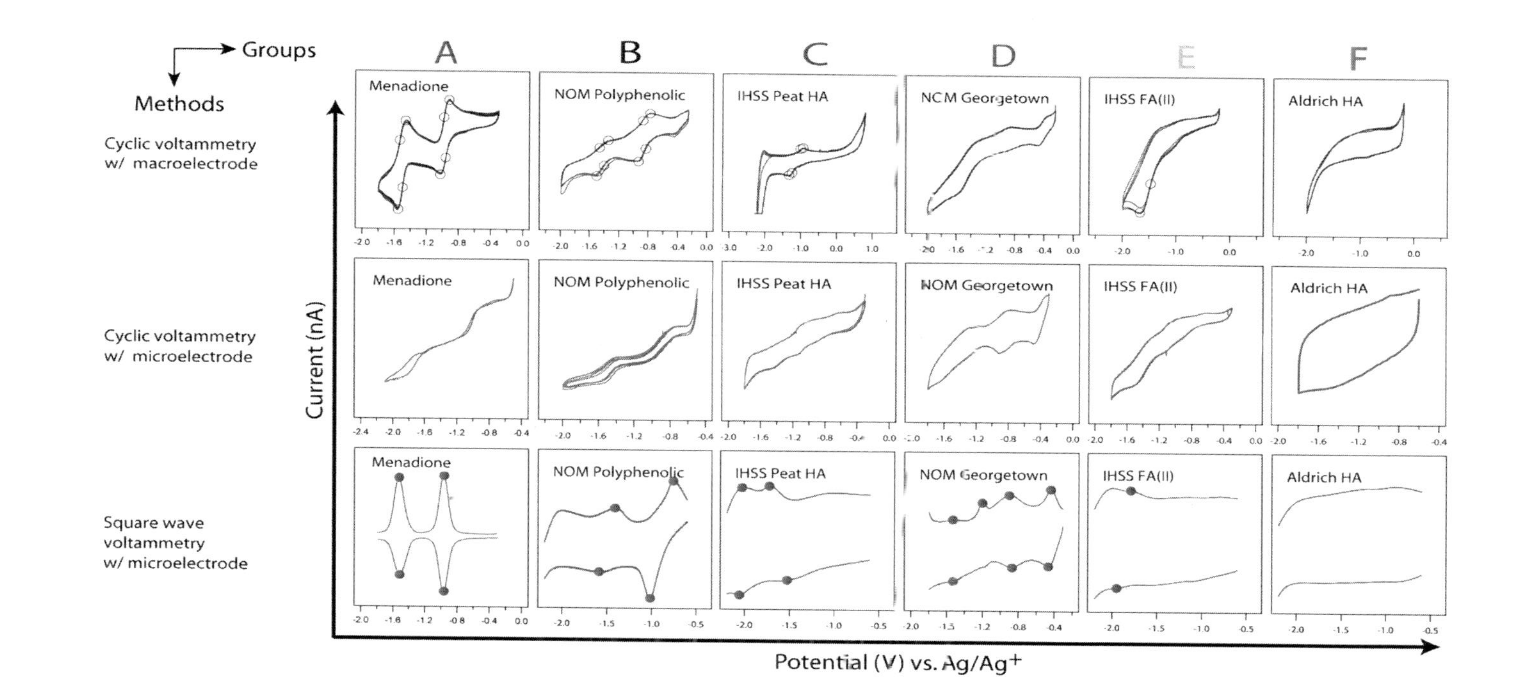

Figure 3. Comparison of voltammograms for representative types of NOM, fractions of NOM, and NOM model compounds (Columns A-F) obtained with three electrochemical methods (Rows). First row data from (77), second row data mostly from (78), third row data first reported here.

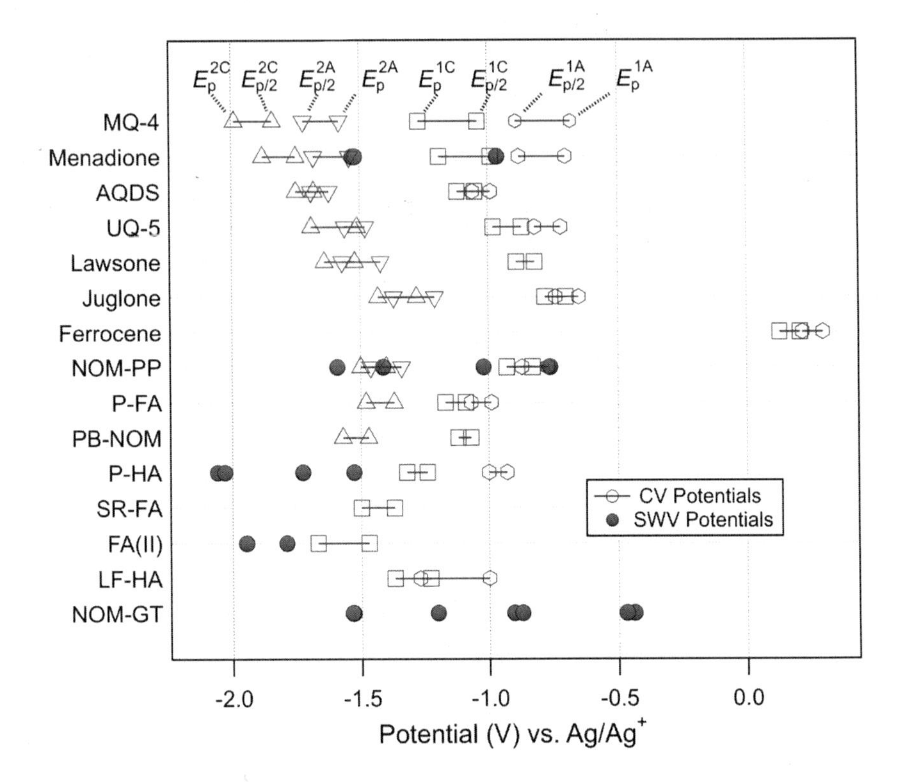

Figure 4. Comparison of potentials (E_p and $E_{p/2}$) obtained by CV and SWV for HS and HS model compounds in DMSO. Based on a figure in (77) with new data from (78) and this study. Definition of material abbreviations in (77).

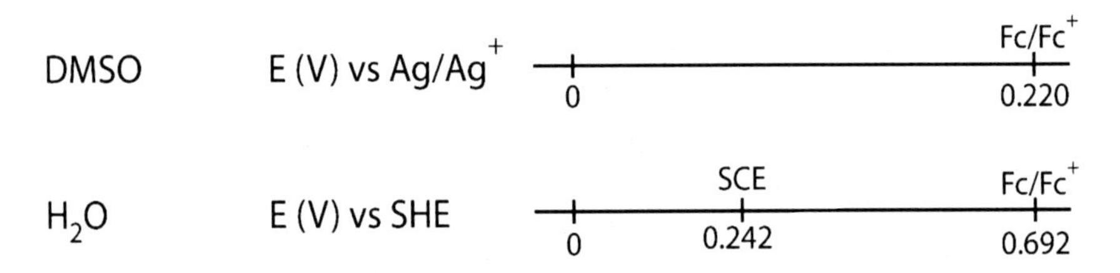

Figure 5. Converting redox potentials obtained in DMSO to environmentally relevant aqueous conditions.

Using this approach, we have adjusted the potentials shown Fig. 4 for approximate comparison to other data, including standard potentials for the major terminal electron accepting processes (TEAPs) and reduction reactions of representative environmental contaminants. The results are plotted as a redox ladder in Fig. 6, with TEAPs in the left column (to represent the process that determines the overall redox conditions in an environment), NOM and model quinones in the middle column (to represent its role as redox mediator), and contaminant reduction reactions in the right column (as ultimate electron acceptors of applied environmental interest). So, for example, Fig. 6 suggests sulfide formed under sulfate reducing conditions might reduce NOM-PP, and the reduced form of NOM-PP (labeled PPH_2 in Fig. 6) might reduce uranyl ion (U(VI)) to less soluble U(IV) species (*87*) or arsenate (As(V)) species to more soluble arsenite (As(III)) species (*88*).

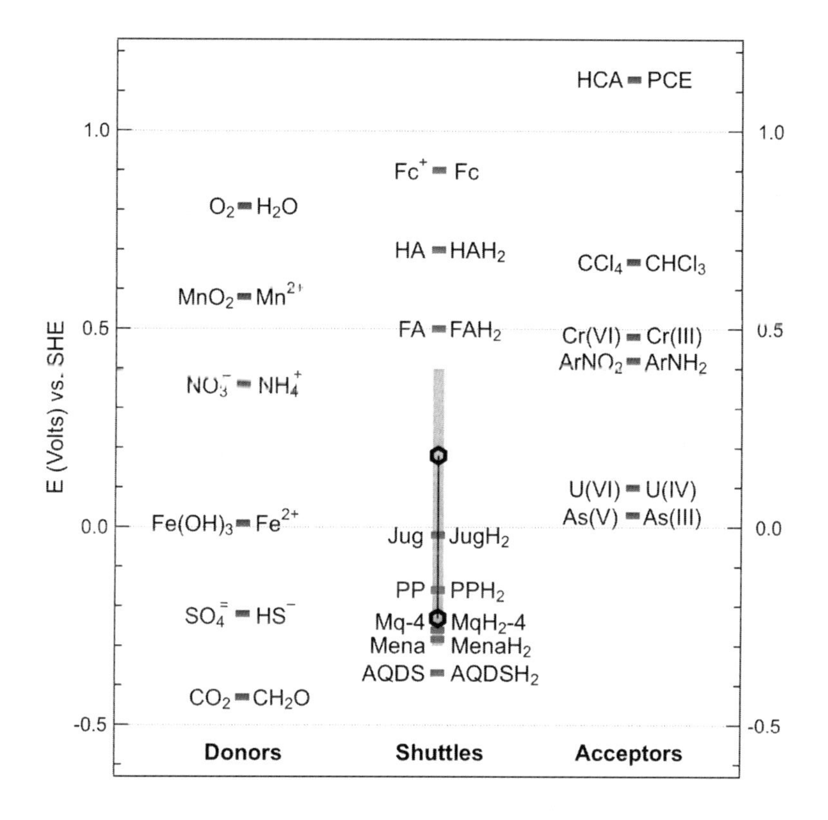

Figure 6. Redox ladder showing the major microbial terminal electron acceptor processes on the left, NOM and representative quinone electron-transfer mediators in the middle, and selected contaminant acceptors on the right. The grey bar represents the range of redox potentials obtained in this study for NOM-GT. The unfilled symbols represent the oxidation-reduction potential of the solution measured in Aeschbacher et al. (33).

The range of redox potentials for unfractionated NOM-GT, estimated with data from the voltammetric methods described here, is shown as the gray bar in the middle column of Fig. 6. It spans a broad range (>600 mV) centered a bit below 0 mV vs. SHE. The reducing end of the range (top) falls near the redox potentials for several model quinones (anthraquinone-2,6-disulfonate, mena-quinone, and menadione), while the middle of the range for NOM-GT aligns roughly with the potential for another model quinone, juglone. The redox potentials for the NOM fraction that is enriched in polyphenolic moieties (NOM-PP) plots between the groups of model quinone compounds.

Two additional sets of data included in Fig. 6 were obtained by methods that are based on potentiometric titration. This approach was used in a widely-cited early study that reported redox potentials for humic and fulvic acids (*89*). However, their results (labeled HA and FA in Fig. 6) fall significantly outside the range of more recent data, suggesting that the former are not representative. Recently, Aeschbacher et al. (*33*) have studied the electron accepting capacity of various fractions of NOM using chronocoulometry with both direct electrochemical reduction and with mediated electrochemical reduction by using electron-transfer mediators, but their results also provide estimates of NOM redox potential (for Leonardite humic acid after direct electrochemical reduction, shown as solid circles in the middle column in Fig. 6). The range of redox potentials represented by Aeschbacher's data fall within the range defined by our data for NOM-GT. Another recent study (*90*), reported redox potentials (not shown in Fig. 6) for similar types of NOM that fall inside the upper end of the range in Aeschbacher's data.

Conclusions

The application of voltammetric methods to study the redox properties of NOM has been variably successful, but most studies find evidence for multiple redox-active moieties (*69*, *70*, *73*, *77*). We have shown that voltammograms of NOM in DMSO have features consistent with quinone-like moieties being the dominant redox active groups, and this is consistent with structural, spectroscopic, and reactivity data reported by others (e.g., (*31*, *50*, *91–94*)). However, the exact structures of these quinone-like moieties could span a wide range of variations, and there are other possible redox active structures—such as organo-sulfur moieties and complexed metals—that could be important in some samples (*95*).

For such complex materials, it seems unlikely that their overall redox activity can be fully attributed to a single redox active compound or moiety and thus a single redox potential. Data that suggests such a singular response that apparently arises could simply be due to limited resolution under the method and conditions of characterization (e.g., as in potentiometric titrations and the use of single OCP measurements). Of greater interest is the possibility that distinct redox-active compounds or moieties are difficult to resolve in NOM because (redox) coupling among these species is characteristic of such polymeric materials. This coupling could result in a continuum of redox potentials, corresponding approximately to the gray band used in Fig. 5 to encompass the range of potentials measured for NOM-GT.

Precursory versions of this hypothesis have been articulated previously (*96*) but a thorough analysis of the direct evidence in support have not yet been reported. It is, however, consistent with the voltammetric data presented here.

Acknowledgments

This experimental work reported here was sponsored by the Natural and Accelerated Bioremediation Research (NABIR) Program, Office of Biological and Environmental Research, U.S. Department of Energy, under contract DE-AC05-00OR22725 with the Oak Ridge National Laboratory.

References

1. Kuhn, A. T. Role of electrochemistry in environmental control. *Mod. Aspects Electrochem.* **1972**, *8*, 273–340.
2. *Environmental Oriented Electrochemistry*; Sequeira, C. A. C., Ed. Elsevier: Amsterdam, The Netherlands; 1994.
3. Rajeshwar, K.; Ibanez, J. G. *Environmental Electrochemistry: Fundamentals and Applications in Pollution Abatement*; Academic Press: San Diego, 1997.
4. Simonsson, D. Electrochemistry for a cleaner environment. *Chem. Soc. Rev.* **1997**, *26*, 181–189.
5. MacDougall, B.; Bock, C.; Gattrell, M. Environmental electrochemistry. *Encycl. Electrochem.* **2007**, *5*, 855–883.
6. *Electrochemistry for the Environment*; Comninellis, C., Chen, G., Eds.; Springer: New York, 2009.
7. Avaca, L. A., Comminellis, C., Trasatti, S., Eds. Special issue: Electrochemistry for a healthy planet; Environmental analytical and engineering aspects. (Selection of Papers From the 6th ISE Spring Meeting 16-19 March 2008, Foz do Iguacu, Brazil). *Electrochim. Acta* **2009**, 54.
8. Vyskocil, V.; Barek, J. Mercury electrodes. Possibilities and limitations in environmental electroanalysis. *Crit. Rev. Anal. Chem.* **2009**, *39*, 173–188.
9. Wang, J. Stripping-based electrochemical metal sensors for environmental monitoring. *Comprehensive Analytical Chemistry*; Elsevier, 2007; Vol. 49, pp 131−141.
10. Hanrahan, G.; Patil, D. G.; Wang, J. Electrochemical sensors for environmental monitoring: design, development and applications. *J. Environ. Monit.* **2004**, *6*, 657–664.
11. *Environmental Electrochemistry: Analyses of Trace Element Bigeochemistry*; Taillefert, M., Rozan, T. F., Eds.; American Chemical Society: Washington, DC; 2002; Vol. 811.
12. Kalvoda, R. Environmental electroanalytical chemistry: Contemporary trends and prospects. *Crit. Rev. Anal. Chem.* **2000**, *30*, 31–35.
13. Fleet, B.; Gunasingham, H. Electrochemical sensors for monitoring environmental pollutants. *Talanta* **1992**, *39*, 1449–1457.
14. Luther, G. W., III; Glazer, B. T.; Ma, S.; et al. Use of voltammetric solid-state (micro)electrodes for studying biogeochemical processes: Laboratory measurements to real time measurements with an in situ electrochemical analyzer (ISEA). *Mar. Chem.* **2008**, *108*, 221–235.
15. Reimers, C. E. Applications of microelectrodes to problems in chemical oceanography. *Chem. Rev.* **2007**, *107*, 590–600.

16. Taillefert, M.; Luther, G. W., III; Nuzzio, D. B. The application of electrochemical tools for in situ measurements in aquatic systems. *Electroanalysis* **2000**, *12*, 401–412.
17. Revsbech, N. P.; Jørgensen, B. B. Microelectrodes: thier use in microbial ecology. *Adv. Microbial Ecol.* **1986**, *9*, 293–352.
18. Nurmi, J. T.; Bandstra, J. Z.; Tratnyek, P. G. Packed powder electrodes for characterizing the reactivity of granular iron in borate solutions. *J. Electrochem. Soc.* **2004**, *151*, B347–B353.
19. Grygar, T.; Marken, F.; Schroder, U.; Scholz, F. Electrochemical analysis of solids. A review. *Collect. Czech. Chem. Commun.* **2002**, *67*, 163–208.
20. Cachet-Vivier, C.; Vivier, V.; Cha, C. S.; Nedelec, J. Y.; Yu, L. T. Electrochemistry of powder material studied by means of the cavity microelectrode (CME). *Electrochim. Acta* **2001**, *47*, 181–189.
21. Lux, L.; Galova, M.; Hezelova, M.; Markusova, K. Investigation of the reactivity of powder surfaces by abrasive voltammetry. *J. Solid State Electrochem.* **1999**, *3*, 288–292.
22. Cha, C. S.; Li, C. M.; Yang, H. X.; Liu, P. F. Powder microelectrodes. *J. Electroanal. Chem.* **1994**, *368*, 47–54.
23. Tratnyek, P. G.; Salter-Blanc, A. J.; Nurmi, J. T.; Baer, D. R.; Amonette, J. E.; Liu, J.; Dohnalkova, A. Reactivity of zerovalent metals in aquatic media: Effects of organic surface coatings. In *Aquatic Redox Chemistry*; Tratnyek, P. G., Grundl, T. J., Haderlein, S. B., Eds.; ACS Symposium Series; American Chemical Society: Washington, DC, 2011; Vol. 1071, Chapter 18, pp 381−406.
24. Meitl, L. A.; Eggleston, C. M.; Colberg, P. J. S.; Khare, N.; Reardon, C. L.; Shi, L. Electrochemical interaction of *Shewanella oneidensis* MR-1 and its outer membrane cytochromes OmcA and MtrC with hematite electrodes. *Geochim. Cosmochim. Acta* **2009**, *73*, 5292–5307.
25. Khare, N.; Eggleston, C. M.; Lovelace, D. M. Sorption and direct electrochemistry of mitochondrial cytochrome C on hematite surfaces. *Clays Clay Miner.* **2005**, *53*, 564–571.
26. Sallez, Y.; Bianco, P.; Lojou, É. Electrochemical behavior of c-type cytochromes at clay-modified carbon electrodes: a model for the interaction between proteins and soils. *J. Electroanal. Chem.* **2000**, *493*, 37–49.
27. Borek, V.; Morra, M. J. Cyclic voltammetry of aquocobalamin on clay-modified electrodes. *Environ. Sci. Technol.* **1998**, *32*, 2149–2153.
28. Subramanian, P.; Fitch, A. Diffusional transport of solutes through clay: Use of clay-modified electrodes. *Environ. Sci. Technol.* **1992**, *26*, 1775–1779.
29. Villemure, G.; Bard, A. J. Clay modified electrodes: Part 9. Electrochemical studies of the electroactive fraction of adsorbed species in reduced-charge and preadsorbed clay films. *J. Electroanal. Chem. Interfacial Electrochem.* **1991**, *282*, 107–121.
30. Bard, A. J.; Mallouk, T. Electrodes modified with clays, zeolites, and related microporous solids. *Tech. Chem. (N.Y.)* **1992**, *22*, 271–312.
31. Struyk, Z.; Sposito, G. Redox properties of standard humic acids. *Geoderma* **2001**, *102*, 329–346.

32. Bauer, M.; Heitmann, T.; Macalady Donald, L.; Blodau, C. Electron transfer capacities and reaction kinetics of peat dissolved organic matter. *Environ. Sci. Technol.* **2007**, *41*, 139–145.
33. Aeschbacher, M.; Sander, M.; Schwarzenbach, R. P. Novel electrochemical approach to assess the redox properties of humic substances. *Environ. Sci. Technol.* **2010**, *44*, 87–93.
34. Bard, A. J.; Faulkner, L. R. *Electrochemical Methods. Fundamentals and Applications*; Wiley: New York, 2001.
35. Brett, C. M. A.; Brett, A. M. O. The solid metallic electrode: Some remarks. *Electrochemistry: Principles, Methods, and Applications*; Oxford University Press: Oxford, 1993; pp 56–65.
36. Bockris, J. O. M.; Reddy, A. K. N. *Modern Electrochemistry*; Plenum: New York, 1970.
37. Sawyer, D. T.; Sobkowiak, A.; Roberts, J. L. J. *Electrochemistry for Chemists*; Wiley: New York, 1995.
38. Bylaska Eric, J.; Salter-Blanc, A. J.; Tratnyek Paul, G. One-electron reduction potentials from chemical structure theory calculations. In *Aquatic Redox Chemistry*; Tratnyek, P. G., Grundl, T. J., Haderlein, S. B., Eds.; ACS Symposium Series; American Chemical Society: Washington, DC, 2011; Vol. 1071, Chapter 3, pp 37–64.
39. Bockris, J. O. M.; Khan, S. U. M. *Surface Electrochemistry. A Molecular Level Approach*; Plenum: New York, 1993.
40. Goodman, B. A. The characterization of iron complexes with soil organic matter. *Iron in Soils and Clay Minerals*; Reidel: Dordrecht, 1988; *NATO ASI Ser., Ser. C* **1988**, *217*, 677–687.
41. Macalady, D. L.; Tratnyek, P. G.; Wolfe, N. L. Influences of natural organic matter on the abiotic hydrolysis of organic contaminants in aqueous systems. In *Aquatic Humic Substances*; ACS Advances in Chemistry Series; Suffett, I. H., MacCarthy, P., Eds.; American Chemical Society: Washington, DC, 1989; Vol. 219, 323–332.
42. Schwarzenbach, R. P.; Stierli, R.; Lanz, K.; Zeyer, J. Quinone and iron porphyrin mediated reduction of nitroaromatic compounds in homogeneous aqueous solution. *Environ. Sci. Technol.* **1990**, *24*, 1566–1574.
43. Dunnivant, F. M.; Schwarzenbach, R. P.; Macalady, D. L. Reduction of substituted nitrobenzenes in aqueous solutions containing natural organic matter. *Environ. Sci. Technol.* **1992**, *26*, 2133–2141.
44. Glaus, M. A.; Heijman, C. G.; Schwarzenbach, R. P.; Zeyer, J. Reduction of nitroaromatic compounds mediated by *Streptomyces* sp. exudates. *Appl. Environ. Microbiol.* **1992**, *58*, 1945–1951.
45. Curtis, G. P.; Reinhard, M. Reductive dehalogenation of hexachloroethane, carbon tetrachloride, and bromoform by anthrahydroquinone disulfonate and humic acid. *Environ. Sci. Technol.* **1994**, *28*, 2393–2401.
46. Lovley, D. R.; Coates, J. D.; Blunt-Harris, E. L.; Phillips, E. J. P.; Woodward, J. C. Humic substances as electron acceptors for microbial respiration. *Nature* **1996**, *382*, 445–448.
47. Seeliger, S.; Cord-Ruwisch, R.; Schink, B. A periplasmic and extracellular c-type cytochrome of *Geobacter sulfurreducens* acts as a ferric iron

reductase and as an electron carrier to other acceptors or to partner bacteria. *J. Bacteriol.* **1998**, *180*, 3686–3691.

48. Collins, R.; Picardal, F. Enhanced anaerobic transformations of carbon tetrachloride by soil organic matter. *Environ. Toxicol. Chem.* **1999**, *18*, 2703–2710.
49. Macalady, D. L.; Ranville, J. F. The chemistry and geochemistry of natural organic matter (NOM). *Perspectives in Environmental Chemistry*; Oxford: New York, 1998; 94−137.
50. Scott, D. T.; McKnight, D. M.; Blunt-Harris, E. L.; Kolesar, S. E.; Lovley, D. R. Quinone moieties act as electron acceptors in the reduction of humic substances by humics-reducing microorganisms. *Environ. Sci. Technol.* **1998**, *32*, 2984–2989.
51. Lovley, D. R.; Fraga, J. L.; Blunt-Harris, E. L.; Hayes, L. A.; Phillips, E. J. P.; Coates, J. D. Humic substances as a mediator for microbially catalyzed metal reduction. *Acta Hydrochim. Hydrobiol.* **1998**, *26*, 152–157.
52. Lovley, D. R.; Fraga, J. L.; Coates, J. D.; Blunt-Harris, E. L. Humics as an electron donor for anaerobic respiration. *Environ. Microbiol.* **1999**, *1*, 89–98.
53. Buschmann, J.; Angst, W.; Schwarzenbach, R. P. Iron porphyrin and cysteine mediated reduction of ten polyhalogenated methanes in homogeneous aqueous solution: Product analyses and mechanistic considerations. *Environ. Sci. Technol.* **1999**, *33*, 1015–1020.
54. Macalady, D. L.; Walton-Day, K. Redox Chemistry and Natural Organic Matter (NOM): Geochemists' Dream, Analytical Chemists' Nightmare. In *Aquatic Redox Chemistry*; Tratnyek, P. G., Grundl, T. J., Haderlein, S. B., Eds.; ACS Symposium Series; American Chemical Society: Washington, DC, 2011; Vol. 1071, Chapter 5, pp 85−111.
55. Kappler, A.; Benz, M.; Schink, B.; Brune, A. Electron shuttling via humic acids in microbial iron(III) reduction in a freshwater sediment. *FEMS Microbiol. Ecol.* **2004**, *47*, 85–92.
56. Ratasuk, N.; Nanny, M. A. Characterization and quantification of reversible redox sites in humic substances. *Environ. Sci. Technol.* **2007**, *41*, 7844–7850.
57. Bauer, M.; Blodau, C. Arsenic distribution in the dissolved, colloidal and particulate size fraction of experimental solutions rich in dissolved organic matter and ferric iron. *Geochim. Cosmochim. Acta* **2009**, *73*, 529–542.
58. Peretyazhko, T.; Sposito, G. Reducing capacity of terrestrial humic acids. *Geoderma* **2006**, *137*, 140–146.
59. Maurer, F.; Christl, I.; Kretzschmar, R. Reduction and reoxidation of humic acid: Influence on spectroscopic properties and proton binding. *Environ. Sci. Technol.* **2010**, *44*, 5787–5792.
60. Perminova, I. V.; Kovalenko, A. N.; Schmitt-Kopplin, P.; Hatfield, K.; Hertkorn, N.; Belyaeva, E. Y.; Petrosyan, V. S. Design of quinonoid-enriched humic materials with enhanced redox properties. *Environ. Sci. Technol.* **2005**, *39*, 8518–8524.
61. Uchimiya, M.; Stone, A. T. Reversible redox chemistry of quinones: Impact on biogeochemical cycles. *Chemosphere* **2009**, *77*, 451–458.

62. Xia, K.; Weesner, F.; Bleam, W. F.; Bloom, P. R.; Skyllberg, U. L.; Helmke, P. A. XANES studies of oxidation states of sulfur in aquatic and soil humic substances. *Soil Sci. Soc. Am. J.* **1998**, *62*, 1240–1246.
63. Szulczewski, M. D.; Helmke, P. A.; Bleam, W. F. XANES spectroscopy studies of Cr(VI) reduction by thiols in organosulfur compounds and humic substances. *Environ. Sci. Technol.* **2001**, *35*, 1134–1141.
64. Gu, B.; Bian, Y.; Miller, C. L.; Dong, W.; Jiang, X.; Liang, L. Mercury reduction and complexation by natural organic matter in anoxic environments. *Proc. Natl. Acad. Sci. U.S.A.* **2011**, *108*, 1479–1483.
65. Kolthoff, I. M.; Lingane, J. J. *Polarography*; Wiley: New York, 1952.
66. Cody, A. F.; Milliken, S. R.; Kinney, C. R. Polarography of humic acid-like oxidation products of bituminous coal. *Anal. Chem.* **1955**, *27*.
67. Lucena-Conde, F.; Gonzalez-Crespo, A. The polarographic behavior of the three fractions of the organic matter of the soils. *7th International Congress of Soil Science*, Madison, WI, 1960; Vol. 2, pp 59−65.
68. Lindbeck, M. R.; Young, J. L. Polarography and coulometry, in dimethyl sulfoxide, on nitric acid oxidation products from soil humic acid. *Soil Sci.* **1966**, *101*, 366–372.
69. Ding, C.-P.; Wang, J. H. In *Electrochemical Methods in Soil and Water Research*; Yu, T.-R., Ji, G. L., Eds.; Pergammon Press: Oxford, 1993; pp 366−412.
70. Ding, C.-P.; Liu, Z.-G.; Yu, T.-R. Determination of reducing substances in soils by a voltammetric method. *Soil Sci.* **1982**, *134*, 252–257.
71. Yu, T.-R. Oxidation-reduction properties of paddy soils. In *Proceedings of Symposium on Paddy Soil*; Science Press: Beijing, 1981; pp 95−106.
72. Yu, T. *Physical Chemistry of Paddy Soils*; Science Press: Beijing, 1985.
73. Yu, T.-R.; Ji, G. L. *Electrochemical Methods in Soil and Water Research*; Pergamon: Tarrytown, NY, 1993.
74. Ding, C.-P.; Liu, Z.-G. Voltammetric determination of reducing substances in paddy soils. In *Proceedings of Symposium on Paddy Soil*; Science Press: Beijing, 1981; pp 251−254.
75. Helburn, R. S.; MacCarthy, P. Determination of some redox properties of humic acid by alkaline ferricyanide titration. *Anal. Chim. Acta* **1994**, *295*, 263–272.
76. Motheo, A. J.; Pinhedo, L. Electrochemical degradation of humic acid. *Sci. Total Environ.* **2000**, *256*, 67–76.
77. Nurmi, J. T.; Tratnyek, P. G. Electrochemical properties of natural organic matter (NOM), fractions of NOM, and model biogeochemical electron shuttles. *Environ. Sci. Technol.* **2002**, *36*, 617–624.
78. Nurmi, J. T.; Tratnyek, P. G. Voltammetric invesigations of natural organic matter. In *Proceedings of the 20th Anniversary Conference of the International Humic Substances Conference (IHSS), 21-26 July 2002, Northeastern University, Boston, MA*; 2002; pp 58−60.
79. Cominoli, A.; Buffle, J.; Haerdi, W. Voltammetric study of humic and fulvic substances. Part III. Comparison of the capabilities of the various polarographic techniques for the analysis of humic and fulvic substances. *J. Electroanal. Chem. Interfacial Electrochem.* **1980**, *110*, 259–275.

80. Buffle, J.; Cominoli, A. Voltammetric study of humic and fulvic substances. Part IV. Behavior of fulvic substances at the mercury-water interface. *J. Electroanal. Chem. Interfacial Electrochem.* **1981**, *121*, 273–299.
81. Mota, A. M.; Pinheiro, J. P.; Goncalves, M. L. Adsorption of humic acid on a mercury aqueous solution interface. *Water Res.* **1994**, *28*, 1285–1296.
82. Lindbeck, M. R.; Young, J. L. Polarography and coulometry, in dimethyl sulfoxide, on nitric acid oxidation products from soil humic acid. *Soil Sci.* **1966**, *101*, 360–372.
83. Chen, J.; Gu, B.; LeBoeuf, E. J.; Pan, H.; Dai, S. Spectroscopic characterization of structural and functional properties of natural organic matter fractions. *Chemosphere* **2002**, *48*, 59–68.
84. Chen, J.; Gu, B.; LeBoeuf, E. J. Fluorescence spectroscopic studies of natural organic matter fractions. *Chemosphere* **2003**, *50*, 639–647.
85. Chen, J.; Gu, B.; Royer, R. A.; Burgos, W. D. The roles of natural organic matter in chemical and microbial reduction of ferric iron. *Sci. Total Environ.* **2003**, *307*, 167–178.
86. Treimer, S. E.; Evans, D. H. Electrochemical reduction of acids in dimethyl sulfoxide. CE mechanisms and beyond. *J. Electroanal. Chem.* **1998**, *449*, 39.
87. O'Loughlin, E. J.; Boyanov, M. I.; Antonopoulos, D. A.; Kemner, K. M. Redox processes affecting the speciation of technetium, uranium, neptunium, and plutonium in aquatic and terrestrial environments.In *Aquatic Redox Chemistry*; Tratnyek, P. G., Grundl, T. J., Haderlein, S. B., Eds.; ACS Symposium Series; American Chemical Society: Washington, DC, 2011; Vol. 1071, Chapter 22, pp 477−517.
88. Palmer, N. E.; Freudenthal, J. H.; von Wandruszka, R. Reduction of arsenates by humic materials. *Environ. Chem.* **2006**, *3*, 131–136.
89. Szilágyi, M. The redox properties and the determination of the normal potential of the peat-water system. *Soil Sci.* **1973**, *115*, 434–437.
90. Palmer, N. E.; von Wandruszka, R. The influence of aggregation on the redox chemistry of humic substances. *Environ. Chem.* **2009**, *6*, 178–184.
91. Klapper, L.; McKnight, D. M.; Fulton, J. R.; Blunt-Harris, E. L.; Nevin, K. P.; Lovley, D. R.; Hatcher, P. G. Fulvic acid oxidation state detection using fluorescence spectroscopy. *Environ. Sci. Technol.* **2002**, *36*, 3170–3175.
92. Cory, R. M.; McKnight, D. M. Fluorescence spectroscopy reveals ubiquitous presence of oxidized and reduced quinones in dissolved organic matter. *Environ. Sci. Technol.* **2005**, *39*, 8142–8149.
93. Fimmen, R. L.; Cory, R. M.; Chin, Y.-P.; Trouts, T. D.; McKnight, D. M. Probing the oxidation-reduction properties of terrestrially and microbially derived dissolved organic matter. *Geochim. Cosmochim. Acta* **2007**, *71*, 3003–3015.
94. Macalady, D. L.; Walton-Day, K. New light on a dark subject: On the use of fluorescence data to deduce redox states of natural organic matter (NOM). *Aquat. Sci.* **2009**, *71*, 135–143.
95. Sutton, R.; Sposito, G. Molecular structure in soil humic substances: The new view. *Environ. Sci. Technol.* **2005**, *39*, 9009–9015.

96. Shea, D.; MacCrehan, W. A. Role of biogenic thiols in the solubility of sulfide minerals. *Sci. Total Environ.* **1988**, *73*, 135–141.

Chapter 8

Pathways Contributing to the Formation and Decay of Ferrous Iron in Sunlit Natural Waters

Shikha Garg,[1] Andrew L. Rose,[1,2] and T. David Waite[1,*]

[1]School of Civil and Environmental Engineering, The University of New South Wales, Sydney, NSW 2052, Australia
[2]Southern Cross GeoScience, Southern Cross University, Lismore NSW 2480, Australia
[*]d.waite@unsw.edu.au

Pathways contributing to the formation and decay of Fe(II) in sunlit natural waters are investigated in this chapter with insights drawn from both laboratory experiments and kinetic modelling. Our results support previous findings that superoxide-mediated iron reduction (SMIR) is the main pathway for photochemical reduction of Fe(III) at pH 8 while light-induced ligand to metal charge transfer (LMCT) is important in low pH environments. Furthermore, our work shows that triplet oxygen and photo-produced species are the main oxidants of Fe(II) while hydrogen peroxide, a relatively stable end product of SRFA photolysis, is not involved in Fe(II) oxidation. A kinetic model based on these observations is presented which provides an excellent description of the experimental results and is consistent with observations from a wide range of studies investigating the redox cycling of iron.

Introduction

The reduced (Fe(II)) form of iron is substantially more bioavailable than Fe(III) due to its much higher solubility and weaker affinity for organic ligands. As a result, researchers have given particular attention to reductive transformations of iron (*1–4*) and suggested that these processes can strongly influence iron availability in natural waters (*5–8*). While reductive dissolution of particulate iron(III) oxyhydroxides is likely to be important in some instances (*9–11*), it is from the dissolved pool that the bulk of the bioavailable iron is typically sourced.

Previous studies have shown that measurable concentrations of Fe(II) can exist in marine surface waters. While this in part may be attributable to biological activity (*5, 12*), the majority of Fe(II) is thought to arise from abiotic photochemical reactions (*13–15*). Three major pathways are considered to potentially account for photochemical reduction of dissolved Fe(III) in natural waters. These are: reduction of Fe(III) by photochemically-produced superoxide (O_2^-) (*1, 2*), i.e. superoxide-mediated iron reduction (SMIR); ligand-to-metal charge transfer (LMCT) in photoactive Fe(III) species (*16, 17*); and direct reduction of Fe(III) by organic radicals (e.g. semiquinone-type radicals) either initially present in natural organic matter (NOM) or formed during irradiation of NOM. O_2^- is known to readily reduce inorganic Fe(III) as well as a wide range of iron-organic complexes (*1, 18*). LMCT is known to occur in some iron-organic complexes (*16, 17*) and can also occur in dissolved inorganic Fe(III) at acidic pH (e.g. in $Fe(OH)_2^+$), but LMCT in inorganic Fe(III) is negligible around neutral pH (*19*). The role of organic radicals in reducing iron has not been well studied.

Although all three pathways are feasible, the relative importance of each under particular conditions is unclear. Based on measurements of O_2^- production kinetics during irradiation of the international standard NOM Suwannee River fulvic acid (SRFA) with simulated sunlight combined with kinetic calculations, Rose and Waite (*2*) showed that (SMIR) is probably more important than LMCT in marine waters and also concluded that reduction of iron by organic radicals is highly unlikely to play a significant role. It is likely however that the relative contribution of each of these three pathways to Fe(II) production depends on solution conditions such as the types of organic moieties present, the concentrations of these moieties, ionic strength and pH. For example, SMIR is likely to be more important at higher pH as the rapid rate of O_2^- disproportionation (self-reaction) at acidic pH (*20*) would be expected to render iron reduction insignificant by this pathway.

The persistence of photochemically produced Fe(II) in natural waters will also depend on its oxidation kinetics. Oxidation of Fe(II) in seawater is thought to occur primarily via its reactions with triplet dioxygen (O_2) and hydrogen peroxide (H_2O_2) if it is present in sufficient concentration (*2, 21, 22*). While other oxidants may include O_2^-, hydroxyl radical, and oxidizing organic radicals, these oxidants have not been shown to be important under conditions typically encountered in nature. Rose and Waite (*2*) showed that hydroxyl and oxidizing organic radicals play a minor role in Fe(II) oxidation, while O_2^- may play an important role in a LMCT dominated system due to its rapid reaction with the inorganic Fe(II) that would be produced (*23*).

Our recent study of reactive oxygen species (ROS) generation during photolysis of NOM showed that irradiation of SRFA at pH 8.1 with simulated sunlight resulted in production of nanomolar concentrations of O_2^- via reduction of O_2 (*24*). SRFA contains a redox-active chromophore which reduces O_2 to yield O_2^- upon photoexcitation. Based on our experimental data as well as past literature observations on ROS formation, we formulated a kinetic model capable of describing the results obtained. Key features of the model (which is shown in Table 1) are:

- near-instantaneous establishment of a steady-state concentration of singlet oxygen ($^1O_2$);
- a redox-active chromophore Q that, upon irradiation, facilitates donation of electrons from SRFA to O_2;
- a redox-active organic moiety A, which catalytically disproportionates O_2^- in the dark, but whose catalytic activity is inhibited during irradiation due to reaction of the reduced form of A with $^1O_2$; and
- a radical sink for O_2^-, R, that results in oxidation of O_2^- to O_2 during irradiation, along with its uncatalyzed disproportionation.

Table 1. Kinetic model for formation of ROS from irradiation of SRFA

No.	*Reaction*	*Model value*	*Published value*	*Reference*
1	$SRFA + h\nu \rightarrow SRFA^*$ $SRFA^* + {}^3O_2 \rightarrow SRFA + {}^1O_2$	Calculated $\Phi \sim 0.5\%$	- $\Phi \sim 0.5\%$	*(25)*
2	${}^1O_2 \xrightarrow{H_2O} {}^3O_2$	2.4×10^5 s^{-1}	2.4×10^5 s^{-1}	*(26)*
3	$Q + h\nu \rightarrow Q^-$	1.5×10^{-3} s^{-1} [a]	-	*(24)*
4	$Q^- + {}^3O_2 \rightarrow Q + O_2^-$	$k_5 \times 10^{-7}$ M^{-1} s^{-1}	-	*(24)*
5	$Q^- + {}^1O_2 \rightarrow Q + O_2^-$	$\sim 10^{10}$ M^{-1} s^{-1}	Diffusion limited	*(24)*
6	$A^- + O_2^- \xrightarrow{2H^+} A + H_2O_2$	$(6.6 \times 10^{-3})/(2[A]_T)$ M^{-1} s^{-1} [b]	-	*(24)*
7	$A + O_2^- \rightarrow A^- + O_2$	$\sim 10 \times k_6$		*(24)*
8	$A^- + {}^1O_2 \rightarrow A + O_2^-$	$>> 5.3 \times 10^3 \times k_6$		*(24)*
9	$O_2^- + O_2^- \xrightarrow{2H^+} O_2 + H_2O_2$	3.5×10^4 M^{-1} s^{-1} [c]	3.5×10^4 M^{-1} s^{-1}	*(20)*
10	$R + h\nu \rightarrow R^\bullet$	7.5×10^{-6} s^{-1} [d]	-	*(24)*
11	$R^\bullet + O_2^- \rightarrow R^- + O_2$	1×10^5 M^{-1} s^{-1}	10^4-10^9 M^{-1} s^{-1}	*(24)*
12	$R^\bullet + R^\bullet \rightarrow R_2$	1×10^3 M^{-1} s^{-1}	-	*(24)*

[a] Q represents the redox-active group responsible for O_2^- production [b] A represents the redox-active group responsible for catalyzed O_2^- disproportionation in dark but whose catalytic activity is inhibited during irradiation due to reaction of the reduced form of A with $^1O_2$ [c] a value of 1×10^7 M^{-1} s^{-1} was used at pH 4 as reported earlier *(20)* [d] R represents the redox-active group that results in oxidation of O_2^- during irradiation

Here we combine the proposed model for production of ROS with a model for the reactions of iron species with dioxygen and ROS and use the combined model to predict the relative importance of LMCT, SMIR and organic radical-mediated Fe(III) reduction to photochemical Fe(II) production under conditions typical of natural waters.

Experimental Methods

Reagents

Reagents were prepared using 18 MΩ cm resistivity Milli-Q water (MQ) unless stated otherwise. All experiments were performed in 2 mM $NaHCO_3$ and 10 mM NaCl solution (referred to as $NaHCO_3$/NaCl hereafter). All pH measurements were made on the NBS scale using a Hanna HI9025 pH meter calibrated daily using pH 7.01 and 10.01 buffers. Adjustment of pH was performed using 2% HCl and 2% NaOH prepared by dilution of high purity 30% w/v HCl (Sigma) and 30% w/v NaOH (Fluka puriss p.a plus) respectively. A 2 g L^{-1} stock solution of standard SRFA (International Humic Substances Society) was prepared in MQ. A 25 kU mL^{-1} superoxide dismutase (SOD) stock solution was prepared in MQ and stored at -85°C when not in use. A stock solution of 100 μM Rose Bengal (RB; Sigma) was prepared in MQ. A 0.1 M 3-(2-pyridyl)-5,6-diphenyl-1,2,4-triazine-4′,4″-disulfonic acid sodium salt (ferrozine; Fluka) stock solution was prepared in MQ and its pH adjusted to 8 with NaOH. Luminol reagent for total Fe(II) determination was prepared by dissolving 0.5 mM luminol (5-amino-2,3-dihydro-1,4-phthalazinedione; Fluka) in 1 M ammonia and adjusting the pH to 10.3 using 30% HCl. The solution was stored in a dark bottle for at least 24 hours prior to use to reach maximum efficiency by equilibration with atmospheric CO_2 (*27*). A stock solution of mixed reagent containing 200 μM Amplex Red (AR; Invitrogen) and 5000 kU L^{-1} horseradish peroxidase (HRP; Sigma) for H_2O_2 determination was prepared and stored as described previously (*28*).

A 0.5 mM Fe(III) working stock in 2 mM HCl was prepared three-monthly. The solution pH was sufficiently low to avoid hydrolysis of iron. A 4.0 mM Fe(II) stock solution in 0.2 M HCl was prepared for Fe(II) calibration. A working 4 μM Fe(II) stock in 0.2 mM HCl was prepared weekly by 1000-fold dilution of the 4 mM Fe(II) stock solution in MQ. The working stock pH was 3.5, which was sufficiently low to prevent significant Fe(II) oxidation during a week, but sufficiently high to prevent significant pH change when added to $NaHCO_3$/NaCl solution. Stock solutions of Fe(III)SRFA complex were prepared at pH 4 and 8 by mixing appropriate volumes of SRFA stock solution, $NaHCO_3$/NaCl and Fe(III) stock solution at various Fe:SRFA ratios, always ensuring that SRFA was sufficiently in excess of Fe(III) to avoid Fe(III) precipitation. The final pH was adjusted by addition of NaOH or HCl. SRFA concentrations investigated were 2.5 and 5.0 mg L^{-1} while total iron concentrations were 50, 100 and 200 nM.

All stock solutions were refrigerated in the dark at 4°C when not in use unless stated otherwise.

Experimental Setup

Photochemistry experiments in which concentrations of total Fe(II) (i.e. the sum of all inorganic and organically complexed Fe(II) species) and H_2O_2 were monitored in photolyzed Fe(III)SRFA solutions as a function of time were performed in a water-jacketed 1 L glass reactor equipped with a quartz side window. The reactor was covered with aluminum foil to exclude light and a gas-tight lid fitted to prevent gas exchange. The reaction solution was maintained at 25±0.5°C with a recirculating water bath. A ThermoOriel 150 W Xe lamp (equipped with AM1 filter to simulate the solar spectrum at the Earth's surface) was positioned horizontally adjacent to the quartz window to illuminate a 5 cm deep cross-section of sample.

For experiments in which the effect of SOD addition on H_2O_2 formation was examined, 3.5 mL solutions containing 5 mg L^{-1} SRFA and 100 nM total iron were irradiated in a quartz cuvette for 5 min. All other experiments were conducted in the 1 L glass reactor as described above.

Measurement of Total Fe(II) Production on Irradiation of Fe(III)SRFA

Concentrations of total Fe(II) were determined using the luminol chemiluminescence (CL) method. In this system, Fe(II) reacts rapidly with O_2 at the high pH of the CL reagent, facilitating oxidation of the reagent and resulting in emission of CL at a wavelength of 426 nm (*29*, *30*). Samples were withdrawn from the photochemistry reactor using a peristaltic pump, loaded into a 450 μL sample column, and injected into the flow cell every 2 min. The valve configuration and PMT settings were as described previously (*2*). Calibration was performed immediately after completion of experiments using the procedure described previously (*2*).

Ferrozine Trapping Experiments

To investigate the role of O_2^- in Fe(II) generation, we measured the rate of Fe(II) production from irradiation of 1 L of 100 nM total Fe(III) and 5 mg L^{-1} SRFA solution containing 1 mM ferrozine (FZ) in the absence and presence of 50 kU L^{-1} SOD. Ferrozine forms a stable complex ($Fe(FZ)_3$) with inorganic Fe(II) that absorbs strongly at 562 nm (*31*). The solution was continuously circulated through a 1 m pathlength type II liquid waveguide capillary cell (World Precision Instruments), and the absorbance of the solution measured at 562 nm, corrected for baseline drift by subtracting the absorbance at 690 nm (at which no components of the solution absorb significantly) using an Ocean Optics fiber optic spectrophotometry system. The system consisted of a broadband tungsten halogen lamp as the light source with a USB4000 spectrophotometer configured for use in the visible range. To account for $Fe(FZ)_3$ formation occurring due to processes other than light-mediated reduction of Fe(III)SRFA, the $Fe(FZ)_3$ formation rate was also measured in the absence of light and this value subtracted from the total measured rate.

Note that the concentration of Fe(II) measured using FZ would be expected to differ from that measured using luminol since FZ is present in the reactor and will prevent oxidation of inorganic Fe(II) or weakly complexed Fe(II) via the formation of the stable $Fe(FZ)_3$ complex, in contrast to the luminol CL method.

Measurement of Fe(II) Oxidation by Singlet Oxygen

Decay of 100 nM total Fe(II) in the presence of 1 μM Rose Bengal (RB), a photosensitizer that produces $^1O_2$ with a high yield, was measured under irradiation and in the dark at pH 4. The 150 W Xe lamp equipped with bandpass filters to eliminate light at wavelengths < 350 nm was used as the light source. Concentrations of total Fe(II) were determined using the luminol CL method as described above.

Measurement of H_2O_2 Production during Irradiation of Fe(III)SRFA

H_2O_2 concentrations during photolysis of Fe(III)SRFA were quantified fluorometrically using the Amplex Red method (*32*) in a Cary Eclipse spectrophotometer using settings and calibration procedures described previously (*28*). Samples for H_2O_2 determination at pH 8 were manually withdrawn from the reactor every 4 min then mixed in a quartz cuvette with mixed AR-HRP reagent solutionto yield final concentrations of 2 μM AR and 50 kU L^{-1} HRP. For H_2O_2 determination at pH 4, 1.5 mL of sample was manually withdrawn from the reactor every 4 min then mixed with 1.5 mL of 2 mM phosphate buffer at pH 7 in a quartz cuvette with AR-HRP reagent to yield final concentrations of 2 μM AR and 50 kU L^{-1} HRP.

Kinetic Modelling

Kinetic modeling was performed using ACUCHEM (*33*).

Results and Discussion

Formation of Total Fe(II) and H_2O_2 during Irradiation of Fe(III)SRFA at pH 8.1

Irradiation of the Fe(III)SRFA complex produced nanomolar concentrations of total Fe(II) and H_2O_2 (Figures 1 and 2). The Fe(II) concentration profile was similar to that observed for O_2^- in our prior work (in which iron was absent) with an initial peak in concentration followed by an approach to steady-state (*2*, *23*). Measured Fe(II) concentrations were greater with increasing total Fe(III) concentrations, however the relationship was non-linear. The peak Fe(II) concentration produced increased slightly with increasing SRFA concentration; however the steady-state Fe(II) concentration was similar at all SRFA concentrations, suggesting that SRFA affects both formation and oxidation rate of Fe(II). The H_2O_2 concentration increased with increasing SRFA concentration but did not vary with Fe(III) concentration and was similar in the presence and absence of iron.

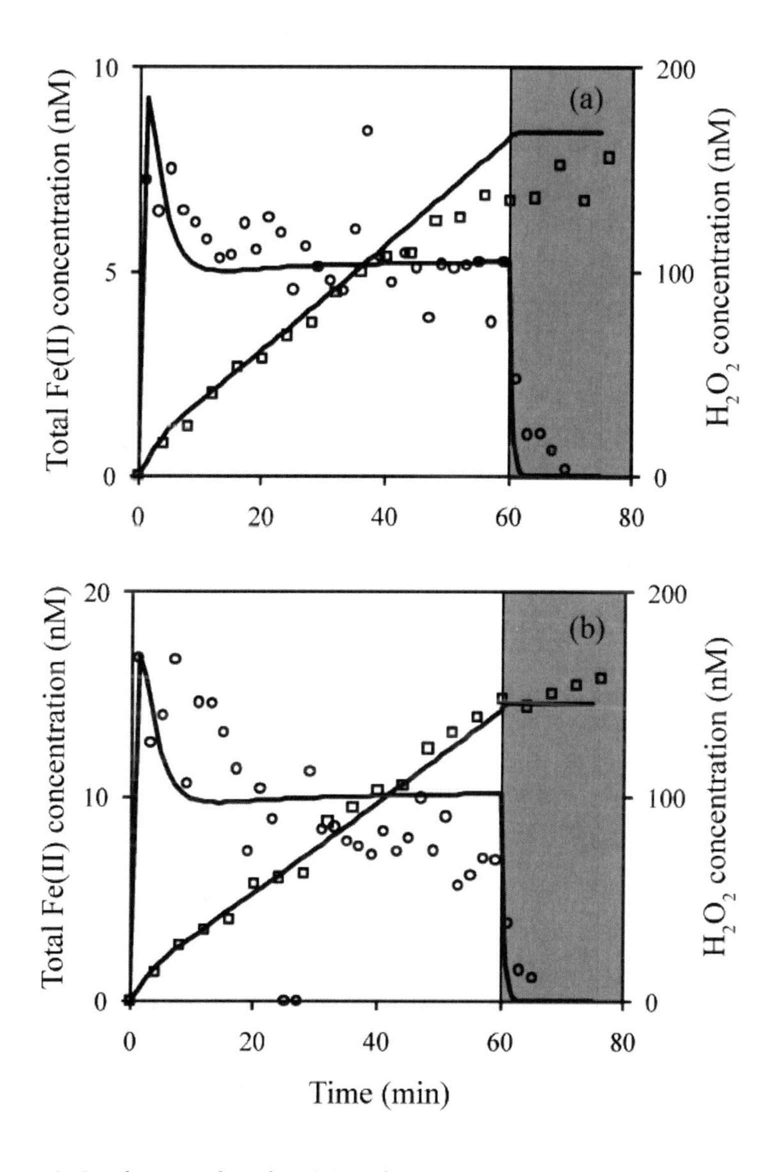

Figure 1. Production of total Fe(II) and H_2O_2 at pH 8.1 by irradiation of 5.0 mg L^{-1} SRFA containing Fe(III) at concentrations of (a) 100 nM and (b) 200 nM for 1 hr followed by 15 min in the dark after the lamp was extinguished (shaded region). Circles are experimentally measured total Fe(II) concentrations and squares are measured H_2O_2 concentrations. Lines represent model values. Data shown are the average from duplicate experiments.

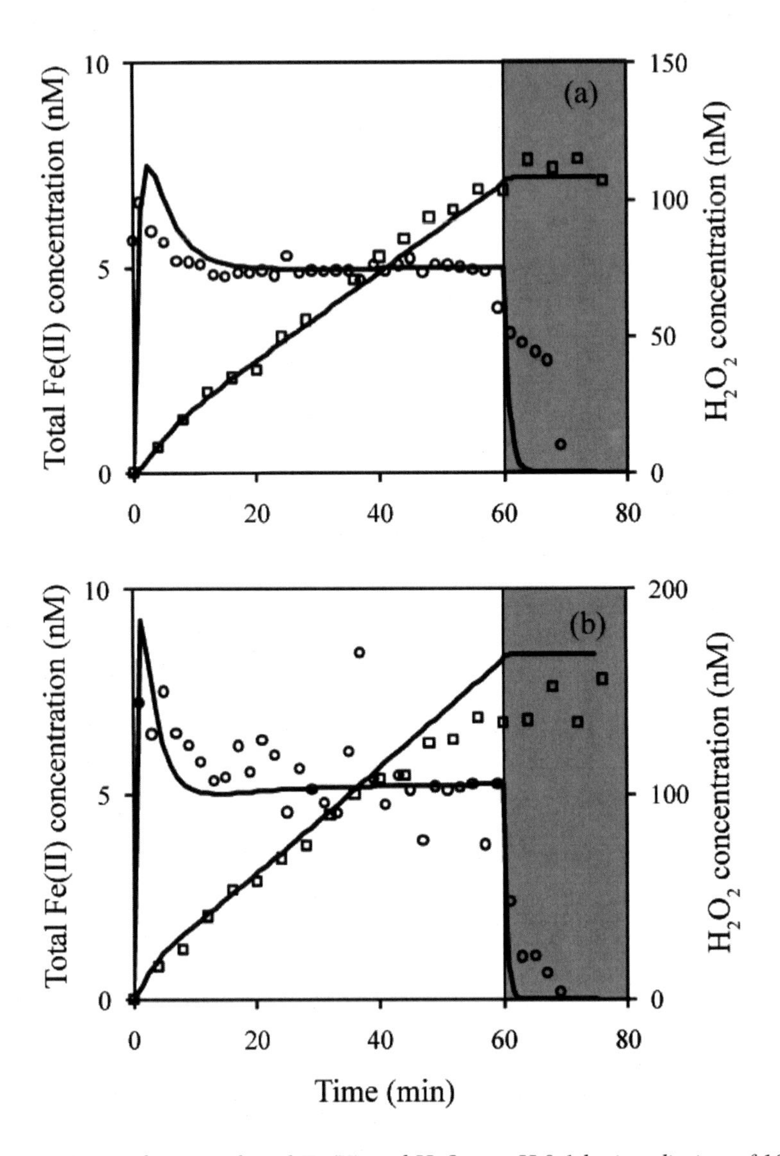

Figure 2. Production of total Fe(II) and H_2O_2 at pH 8.1 by irradiation of 100 nM total Fe(III) containing SRFA at concentrations of (a) 2.5 mg L^{-1} and (b) 5 mg L^{-1} for 1 hr followed by 15 min in the dark after the lamp was extinguished (shaded region). Circles are measured total Fe(II) concentrations and squares are measured H_2O_2 concentrations. Lines represent model values. Data shown are the average from duplicate experiments.

Mechanism of H_2O_2 Formation and Decay at pH 8.1

Addition of H_2O_2 had no affect on H_2O_2 production during irradiation, suggesting that H_2O_2 interacts negligibly with other entities in the system (data not shown). The presence of iron had no significant ($p < 0.05$, using an unpaired two tailed *t*-test) effect on the net rate of H_2O_2 production (Figure 1), confirming that iron plays a negligble role in production or consumption of H_2O_2 at pH 8.1.

We have previously shown that H_2O_2 production on irradiation of SRFA occurs almost entirely via uncatalyzed O_2^- disproportionation (*24*). This also appears to be the main pathway for H_2O_2 formation in the Fe-SRFA system when iron is present, given the negligible influence of iron on H_2O_2 concentrations. This was further confirmed by the observation that addition of SOD increased H_2O_2 production (Figure 3).

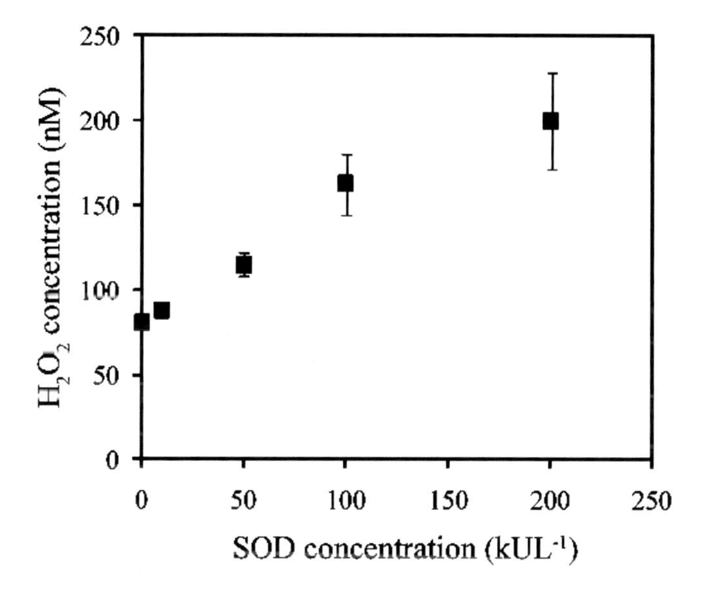

Figure 3. H_2O_2 concentration after irradiation of 3.5 mL of 5 mg L^{-1} SRFA and 100 nM Fe(III) solution in a quartz cuvette for 5 min in the presence of varying concentrations of SOD. Symbols are the mean and error bars are the standard deviation from duplicate experiments.

Mechanism of Fe(II) Formation at pH 8.1

Photochemical Fe(II) formation could potentially occur in our system via SMIR, LMCT or through photochemical production of Fe(III)-reducing organic radicals (*34*, *35*). Under oxygenated conditions, photogenerated organic radicals react mostly by addition of O_2 to form organic peroxyl radicals (*34*, *36*) at diffusion controlled rates. Recently, reduced organic radicals (represented hereafter as A^-) that are resistant to oxidation by O_2 (*37*) and which are capable of reducing Cu(II) to Cu(I) have also been shown to exist in SRFA in the dark (*24*). However, our recent work showed that these stable radicals (A^-) are oxidized by $^1O_2$ in irradiated solutions (*24*) and, as a result, would be unavailable to react with Fe(III). In view of the apparent insignificance of this pathway, we give particular attention below to assessing the importance of LMCT and SMIR pathways in the photochemical system.

To probe the importance of SMIR, we measured photochemical Fe(II) production in the presence of 50 kU L^{-1} SOD, an enzyme which catalyzes O_2^- decay to O_2 and H_2O_2. We used FZ trapping of Fe(II) as a probe for Fe(II) production in this case, because the luminol CL method involves O_2^- as an intermediate and is thus not suitable for determination of Fe(II) in the presence of SOD. Addition of SOD decreased $Fe(FZ)_3$ production (Figure 4), confirming that O_2^- is involved in Fe(II) production to some extent at least. FZ will not necessarily trap all Fe(II) produced since it binds only free Fe(II) that is able to dissociate from Fe(II)-organic complexes prior to oxidation of these complexes. Previous work indicates that FZ traps about 30% of Fe(II) produced from SMIR of Fe(III)SRFA (*1*), which suggests that the decrease in the initial total Fe(II) production in the presence of 50 kU L^{-1} SOD is about 55%. The decrease in total Fe(II) production in the presence of 50 kU L^{-1} SOD during the later stages of irradiation is about 48% only, possibly due to loss of SOD activity during irradiation, however this still suggests that SMIR is an important pathway for Fe(II) production during photolysis of Fe(III)SRFA.

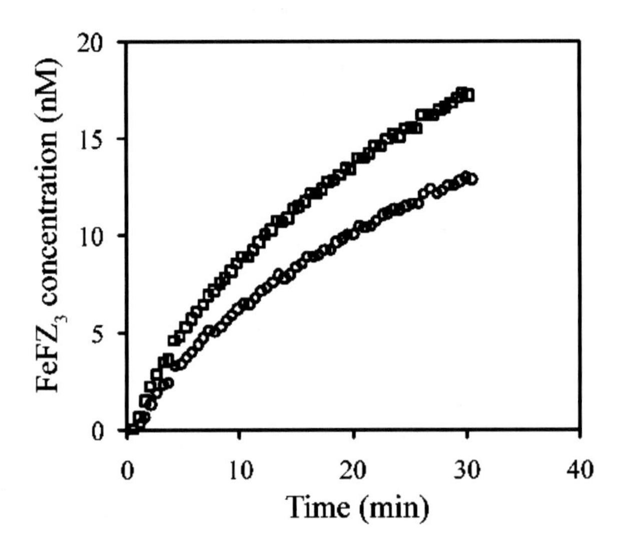

Figure 4. $Fe(FZ)_3$ concentration produced during irradiation of 100 nM total Fe(III) and 5 mg L^{-1} SRFA at pH 8.1 in the presence (circles) and absence (squares) of 50 kU L^{-1} SOD. Data shown are the average from duplicate experiments.

Furthermore, our earlier study (*24*) showed that > 200 kU L^{-1} SOD is required to outcompete all sinks of O_2^- produced during photolysis of SRFA, suggesting that 50 kU L^{-1} SOD may not be sufficient to outcompete the reaction of O_2^- with Fe(III). If this is the case then the contribution of SMIR to total Fe(II) production will be higher than the values calculated above. Thus, while we are unable to determine the exact contribution of SMIR based on these results, we still conclude that SMIR is the major pathway for Fe(II) production at pH 8.

Mechanism of Fe(II) Decay at pH 8.1

Fe(II) formed from photoreduction of Fe(III)SRFA can potentially be reoxidized in this system by dioxygen, hydroxyl radicals, O_2^-, H_2O_2 and oxidizing organic radicals. If we assume that triplet dioxygen (O_2) is the only important oxidant, then the rate law for total Fe(II) production by SMIR is:

$$\frac{d}{dt}[\mathrm{Fe(II)}] = k_r[\mathrm{Fe(III)}][O_2^-] - k_{ox}[\mathrm{Fe(II)}][O_2] \tag{1}$$

where k_r is the rate constant for reduction of total Fe(III) by O_2^- and k_{ox} is the rate constant for oxidation of Fe(II) by O_2. At steady-state, equating the derivative with respect to time to zero and applying the mass balance consideration [Fe(III)] + [Fe(II)] = $[\mathrm{Fe}]_T$:

$$[\mathrm{Fe(II)}]_{ss} = \frac{k_r[O_2^-]}{k_{ox}[O_2] + k_r[O_2^-]}[\mathrm{Fe}]_T \tag{2}$$

where the subscript ss denotes steady-state.

If we assume that total [Fe(III)] ≈ [Fe(III)SRFA], total [Fe(II)] ≈ [Fe(II)SRFA], $k_r = 1.5 \times 10^5$ M^{-1} s^{-1} (*1*) and $k_{ox} = 150$ M^{-1} s^{-1} (*38*), then substituting the [O_2] and [O_2^-], calculated using the measured H_2O_2 generation rate assuming that uncatalyzed [O_2^-] disproportionation is the main source of H_2O_2, in eq. (2) we obtain $[\mathrm{Fe(II)}]_{ss} \approx 14.2$ nM in the presence of 100 nM total Fe(III) and 5 mgL^{-1} SRFA. This is nearly 2-fold higher than the measured steady-state total Fe(II) concentration (Figure 1). Similar discrepancies between measured and calculated steady-state Fe(II) concentrations are obtained at other total Fe and SRFA concentrations. These calculations suggest that Fe(II) oxidants other than triplet dioxygen are present in the system. In calculating these values, we have assumed that all of the Fe(III) and Fe(II) is present in organically complexed form and we have ignored the contribution of LMCT to Fe(II) production. However, the total steady-state Fe(II) concentration would be even higher if these simplifications are invalid, given that k_r for inorganic Fe(III) (1.5×10^8 M^{-1} s^{-1} (*20*)) is higher than that for Fe(III)SRFA, k_{ox} for inorganic Fe(II) (13 M^{-1} s^{-1} (*39*)) is much lower than that for Fe(II)SRFA and increased production of Fe(II) due to LMCT will result in a higher steady-state Fe(II) concentration. Although we cannot calculate the exact contribution of these oxidants to the Fe(II) decay rate, these calculations show that these oxidants are important in the system under investigation. Below we assess the importance of other oxidants based on our experimental observations and reported rate constants for the reactions of Fe(II) with these oxidants.

Role of Hydroxyl Radical

Hydroxyl radicals can be immediately eliminated as a significant oxidant of Fe(II) in the system due to their rapid scavenging by organic molecules (*40*) and bicarbonate ions (*41*).

Role of H_2O_2

In order to test the role of H_2O_2 as an oxidant of Fe(II) in the system investigated, we measured the Fe(II) formation rate in the presence of added H_2O_2. This had no effect on the kinetics of Fe(II) formation (Figure 5), implying that H_2O_2 was a minor oxidant of both inorganic and organic Fe(II) at the concentrations present in our system. This is consistent with the previous finding that Fe(II)SRFA is oxidized slowly by H_2O_2 (*38*) and also in agreement with our observation that addition of iron did not alter the kinetics of H_2O_2 formation.

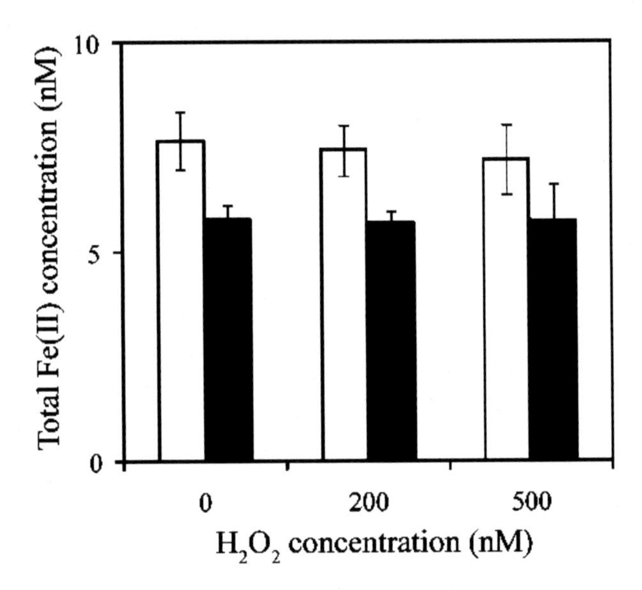

Figure 5. Peak concentration (open bars) and steady-state concentration (closed bars) of total Fe(II) at pH 8.1 produced during irradiation of 100 nM total Fe(III) and 5 mg L^{-1} SRFA in the presence of varying concentrations of H_2O_2. Peak concentrations correspond to maxima observed during the initial phase of irradiation, and steady-state concentrations correspond to the second phase of irradiation. Error bars represent the standard deviation of triplicate measurement.

Role of Superoxide

As shown earlier, O_2^- is an important reductant of Fe(III) and contributes at least 55% of the total Fe(II) produced during irradiation of 100 nM Fe(III) and 5 mg L^{-1} SRFA at pH 8.1. If O_2^- was also an important oxidant of Fe(II) in this system, we would expect faster O_2^- disproportionation kinetics in the presence of iron. However, this is inconsistent with the observation that H_2O_2 concentrations did not vary noticeably with iron concentrations (Figure 1), suggesting that O_2^- concentrations are similar in the presence and absence of iron. Thus, O_2^- mediated Fe(II) oxidation appears relatively small when SMIR is important, which is in agreement with the results of our previous study (*2*). Although O_2^- may oxidize Fe(II) formed via LMCT, the contribution of O_2^- to the overall oxidation would still be expected to be small given the small proportion of Fe(II) formed via LMCT at pH 8.1.

Role of Singlet Oxygen and/or Photo-Produced Organic Radicals

$^1O_2$ is produced during SRFA photolysis at substantial rates (*25, 42, 43*) through the quenching of photoexcited chromophoric organic moieties within NOM by O_2, and exists at elevated concentrations in the relatively hydrophobic microenvironment of NOM (*44*). Given that O_2 can oxidize Fe(II), $^1O_2$ should be able to oxidize Fe(II) even more readily since the reaction barrier imposed by the unusual electronic configuration of triplet oxygen will no longer be present. To probe the importance of $^1O_2$ in Fe(II) oxidation, we investigated the Fe(II) oxidation in the presence of $^1O_2$ at pH 4 where oxidation by O_2 is negligibly slow on the timescale of our experiments. $^1O_2$ was generated by photolysis of RB, a photosensitizer that produces $^1O_2$ with high yield. As shown in Figure 6, the Fe(II) concentration decreased in a solution containing RB during irradiation but not in the dark, suggesting that $^1O_2$ may oxidize Fe(II). Modifying eq. 2 to account for the additional oxidation of Fe(II) by $^1O_2$ gives:

$$[\mathrm{Fe(II)}]_{\mathrm{ss}} = \frac{k_{\mathrm{r}}[\mathrm{O_2^-}]}{k_{\mathrm{ox}}[\mathrm{O_2}] + k'_{\mathrm{ox}}[^1\mathrm{O_2}] + k_{\mathrm{r}}[\mathrm{O_2^-}]}[\mathrm{Fe}]_{\mathrm{T}} \tag{3}$$

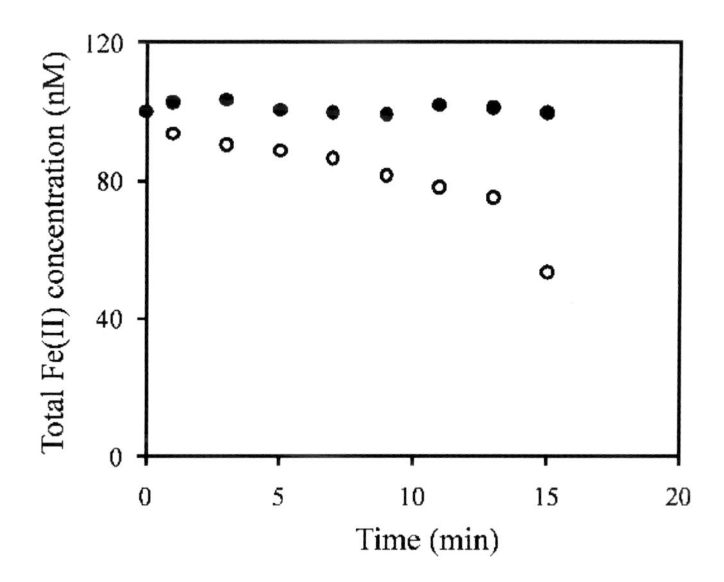

Figure 6. Decay of Fe(II) in presence of 1 μM RB at pH 4 in irradiated (open circles) and non-irradiated (closed circles) solution.

With an apparent $^1O_2$ concentration of $[^1O_2]_{app}$ = 3.5 pM, which is comparable to the value reported by Latch and McNeill (*44*) in photolysed NOM solution, and a second order rate constant for oxidation of Fe(II) close to the diffusion-controlled limit (~ 1 × 10^9 M^{-1} s^{-1}), we obtain $[Fe(II)]_{ss}$ comparable to the measured Fe(II) steady-state concentration at all concentrations of total Fe(III) and SRFA investigated here. Although, based on our experimental observations and kinetic calculation, Fe(II) oxidation by $^1O_2$ is a possibility, the assumed rate constant is very large with significantly more work required to justify such an assumption.

Another (more likely) possibility is that organic radicals that are produced on photolysis of SRFA may contribute to Fe(II) oxidation. Rose and Waite showed that Fe(II) oxidation by organic radicals is important at Fe(II) concentrations > 20 nM (*2*), which is close to the Fe(II) concentration observed here. The presence of semiquinone-quinone-hydroquinone type redox-active moieties have been demonstrated in SRFA and these radicals are known to oxidize Fe(II).

Thus, based on our experimental observations, we cannot identify precisely the nature of this additional Fe(II) oxidant but can conclude that this species is formed on photolysis of SRFA and is short-lived in the dark.

Formation of Total Fe(II) during Irradiation of Fe(III)SRFA at pH 4

In order to further probe the importance of SMIR under various pH conditions, we measured Fe(II) production at pH 4. As discussed previously, SMIR is likely to be less important at low pH given that O_2^- disproportionates rapidly under these conditions to yield H_2O_2 and O_2 (*20*). In contrast, we expect that production of Fe(II) via LMCT may be much less dependent on pH provided the concentration of photoactive FeSRFA complexes is independent of pH. Thus, measurement of Fe(II) production at low pH may provide some insight into the mechanism of Fe(II) production under different pH conditions.

As shown in Figure 7, irradiation of Fe(III)SRFA at pH 4 produced nanomolar concentrations of total Fe(II). The Fe(II) concentration profile at pH 4 is quite different from that observed at pH 8 using the same concentrations of Fe(III) and SRFA. This difference may be due to a change in Fe(II) production kinetics, a change in Fe(II) oxidation kinetics, or a combination of the two. Although Fe(II) oxidation by O_2 and H_2O_2 are negligible at pH 4 (data not shown), oxidation by $^1O_2$ and/or organic radicals may still be important (Figure 6). Assuming that $^1O_2$ is the main oxidant of Fe(II) at pH 4 and $[^1O_2]_{app}$ is the same as that measured at pH 8.1, we obtain a second-order rate constant of 5 × 10^8 M^{-1} s^{-1} for the reaction of $^1O_2$ with Fe(II), which is close to the diffusion-controlled limit. Further, we calculate Fe(II) production rates of 52 pM s^{-1} and 23 pM s^{-1} in the initial and steady-state phases of irradiation respectively in a solution containing 100 nM total Fe(III) and 5 mg L^{-1} SRFA. Based on the measured H_2O_2 production rates at pH 4 (Figure 5b), we calculate a steady-state concentrion of total O_2^- (i.e. including both O_2^- and $HO_2^•$) of 1.5 nM during irradiation of 100 nM Fe(III) and 5 mg L^{-1} SRFA assuming that uncatalyzed O_2^- disproportionation (second order disproportionation rate constant k_{disp} = 1 × 10^7 M^{-1} s^{-1} at pH 4 (*20*)) is the main source of H_2O_2. A steady-state concentration of 1.5 nM O_2^- yields a Fe(II) production rate of 23 pM s^{-1} (based on a rate constant of 1.5 × 10^5 M^{-1} s^{-1} for reduction of Fe(III)SRFA by O_2^- (*1*, *18*) measured at pH 8). This is equivalent to the measured Fe(II) production rate during the later phase of irradiation, but lower than the measured rate during the initial phase, which suggests that SMIR by itself is insufficient to explain the measured Fe(II) production rates at pH 4 and supports the hypothesis that LMCT is important at low pH.

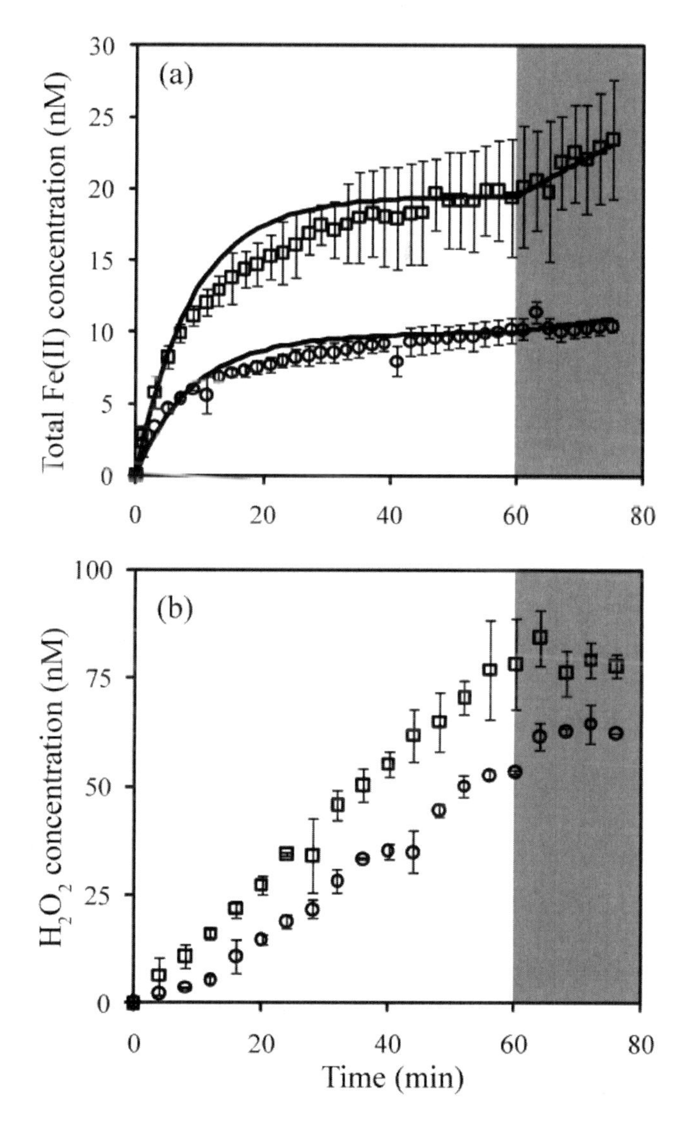

Figure 7. (a) Total Fe(II) concentration and (b) H_2O_2 concentration produced at pH 4 on irradiation of 100 nM total Fe(III) and 5 mg L^{-1} SRFA (open squares) and 50 nM total Fe(III) and 2.5 mg L^{-1} SRFA (open circles) for 1 hr followed by 15 min in the dark after the lamp was extinguished (shaded region). Symbols represent experimentally measured values while lines represent model values. Error bars represent the standard deviation of duplicate measurements.

Kinetic Model for Ferrous Iron Formation from Irradiation of Fe(III)SRFA

Here we couple our recently proposed model for ROS formation during irradiation of SRFA (*24*) with known reactions for redox cycling of iron in the presence of O_2^-. Based on the analysis presented in the previous sections, a kinetic model for Fe(II) formation during irradiation of Fe(III)SRFA must include the following features:

- SMIR of both inorganic and organically complexed Fe(III);
- a LMCT-mediated pathway for reduction of organically complexed Fe(III) whose relative contribution increases with decreasing pH;
- oxidation of inorganic and organically complexed Fe(II) by triplet oxygen and either photoproduced $^1O_2$ and/or organic radicals; and
- complexation and dissociation of both Fe(III) and Fe(II) in the presence of SRFA.

We now consider each of these processes in turn, with the overall kinetic model consisting of the reactions in Table 1 and the additional reactions presented in Tables 2 and 3.

Reduction of Fe(III) by O_2^-

Reactions 13 and 14 (Table 2) describe the reduction of organically-complexed and inorganic Fe(III) by O_2^-. The rate constants for reduction of inorganic Fe(III) and organic Fe(III) used are 1.5×10^8 M^{-1} s^{-1} and 1.0×10^5 M^{-1} s^{-1} respectively which are close to values reported earlier (*1*, *18*, *45*) at pH 8.1. The rate constant for reduction of organically-complexed by O_2^- at pH 4 (Reaction 25, Table 3) was assumed to be same as the value determined at pH 8.1. The rate constants for reduction of inorganic Fe(III) by O_2^- (Reaction 26, Table 3) at pH 4 used is same as that reported earlier (*45*).

Complexation-Dissociation of Fe(III)

Reaction 15 (Table 2) represents the formation of organically complexed Fe(III) and reaction 16 (Table 2) represents the dissociation of the complex to yield inorganic Fe(III). The rate constants used for these reactions are the same as reported earlier at pH 8.0 (*18*).

The rate constants for complexation (reaction 27, Table 3) and dissociation (reaction 28, Table 3) of organically complexed Fe(III) used at pH 4 are same as reported at pH 8. It is important to note here that the values assigned to these rate constants do not affect the model results much as long as most of the Fe(III) is present in organically-complexed form.

Table 2. Proposed kinetic model for Fe(III)SRFA reduction at pH 8.1

No.	*Reaction*	*Model value*	*Published value*	*Reference*
13	$Fe(III)L + O_2^- \rightarrow Fe(II)L + O_2$	$1.0 \times 10^5\ M^{-1}\ s^{-1}$	$(1.5\text{-}2.8) \times 10^5\ M^{-1}\ s^{-1}$	(*1*, *18*)
14	$Fe(III)' + O_2^- \rightarrow Fe(II)' + O_2$	$1.5 \times 10^8\ M^{-1}\ s^{-1}$	$1.5 \times 10^8\ M^{-1}\ s^{-1}$	(*45*)
15	$Fe(III)' + L \rightarrow Fe(III)L$	$6 \times 10^6\ M^{-1}\ s^{-1}$	$6 \times 10^6\ M^{-1}\ s^{-1}$	(*18*)
16	$Fe(III)L \rightarrow Fe(III)' + L$	$2 \times 10^{-6}\ s^{-1}$	$2 \times 10^{-6}\ s^{-1}$	(*18*)
17	$Fe(II)L + O_2 \rightarrow Fe(III)L + O_2^-$	$150\ M^{-1}\ s^{-1}$	$150\ M^{-1}\ s^{-1}$	(*38*)
18	$Fe(II)' + O_2 \rightarrow Fe(III)' + O_2^-$	$13\ M^{-1}\ s^{-1}$	$13\ M^{-1}\ s^{-1}$	(*39*)
19	$Fe(II)L + {}^1O_2 \rightarrow Fe(III)L + O_2^-$	$\sim 1 \times 10^9\ M^{-1}\ s^{-1}$	-	This work
20	$Fe(II)' + {}^1O_2 \rightarrow Fe(III)' + O_2^-$	$\sim 1 \times 10^9\ M^{-1}\ s^{-1}$ [a]		This work
21	$Fe(II)' + L \rightarrow Fe(II)L$	$1.4 \times 10^4\ M^{-1}\ s^{-1}$	$1.4 \times 10^4\ M^{-1}\ s^{-1}$	(*38*)
22	$Fe(II)L \rightarrow Fe(II)' + L$	$8 \times 10^{-4}\ s^{-1}$	$8 \times 10^{-4}\ s^{-1}$	(*18*)
23	$Fe(III)L \xrightarrow{h\nu} Fe(II)' + L_{ox}$	$3.0 \times 10^{-4}\ s^{-1}$ [b]	-	This work
24	$Fe(III)L + A^- \rightarrow Fe(II)L + A$	$7.0 \times 10^2\ M^{-1}\ s^{-1}$ [c]	-	This work

[a] Determined based on best-fit model results. [b] Determined based on best fit to Fe(II) concentration data observed at pH 4. [c] Determined based on Fe(II) production observed in the dark at pH 4 (data not shown).

Table 3. Proposed kinetic model for Fe(III)SRFA reduction at pH 4

No.	*Reaction*	*Model value*	*Published value*	*Reference*
25	$Fe(III)L + O_2^- \rightarrow Fe(II)L + O_2$	$1.0 \times 10^5\ M^{-1}\ s^{-1}$ [a]	-	
26	$Fe(III)' + O_2^- \rightarrow Fe(II)' + O_2$	$1.0 \times 10^7\ M^{-1}\ s^{-1}$	$1.0 \times 10^7\ M^{-1}\ s^{-1}$	(*45*)
27	$Fe(III)' + L \rightarrow Fe(III)L$	$6.0 \times 10^6\ M^{-1}\ s^{-1}$ [b]	-	(*18*)
28	$Fe(III)L \rightarrow Fe(III)' + L$	$2.0 \times 10^{-6}\ s^{-1}$ [b]	-	(*18*)

Continued on next page.

Table 3. (Continued). Proposed kinetic model for Fe(III)SRFA reduction at pH 4

No.	*Reaction*	*Model value*	*Published value*	*Reference*
29	$Fe(II)L + {}^1O_2 \rightarrow Fe(III)L + O_2^-$	~5.0 × 10^8 M^{-1} s^{-1}	-	This work
30	$Fe(II)' + {}^1O_2 \rightarrow Fe(III)' + O_2^-$	~5.0 × 10^8 M^{-1} s^{-1}	-	This work
31	$Fe(II)' + L \rightarrow Fe(II)L$	1.4 × 10^4 M^{-1} s^{-1} [b]	-	(*38*)
32	$Fe(II)L \rightarrow Fe(II)' + L$	8.0 × 10^{-4} s^{-1} [b]	-	(*18*)
33	$Fe(III)L \xrightarrow{h\nu} Fe(II)' + L_{ox}$	3.0 × 10^{-4} s^{-1} [c]	-	This work
34	$Fe(III)L + A^- \rightarrow Fe(II)L + A$	7.0×$10^2$$M^{-1}$ s^{-1} [d]	-	This work

[a] Determined based on best-fit to experimental data at pH 8.1. [b] Rate constant for formation and dissociation of Fe(III)L and Fe(II)L was assumed to be same as reported at pH 8.1. [c] Determined based on best fit to Fe(II) concentration observed at pH 4.
[d] Determined based on Fe(II) production observed in dark at pH 4 (data not shown).

Oxidation of Fe(II) by Triplet O_2

Reactions 17 and 18 (Table 2) represent oxidation of organically-complexed and inorganic Fe(II) by O_2. The rate constants for these reactions are based on those deduced previously at pH 8.0 (*38*). The rate constants for oxidation of organically-complexed and inorganic Fe(II) by O_2 at pH 4 were assumed to be negligibly small since no oxidation of Fe(II) was observed in the dark at pH 4 (data not shown).

Oxidation of Fe(II) by Photo-Produced Species

Due to uncertainity regarding the exact nature and production rates of these photo-produced species we have, for modeling purposes, assumed that ${}^1O_2$ (for which the production rates from SRFA on photolysis is known) contributes to Fe(II) oxidation. The possibility however that organic radicals rather than ${}^1O_2$ are key Fe(II) oxidants should be noted. Reactions 19 (Table 2) and 29 (Table 3) represent oxidation of organically-complexed Fe(II) by ${}^1O_2$, while reactions 20 (Table 2) and 30 (Table 3) represent oxidation of inorganic Fe(II) by ${}^1O_2$. The rate constant for oxidation of organic Fe(II) by ${}^1O_2$ at pH 8.1 was determined to be

~1 $\times 10^9$ M^{-1} s^{-1} based on best fit to our experimental data, which is close to the diffusion-controlled limit. The rate constant for oxidation of inorganic Fe(II) was assumed to be similar to the oxidation rate constant for the Fe(II)SRFA complex, however the model is highly insensitive to the value chosen, with similar fits obtained for almost any value of the rate constant. A value of 5×10^8 M^{-1} s^{-1} was used as the rate constant for oxidation of organically-complexed and inorganic Fe(II) by $^1O_2$ at pH 4 based on the best to fit to our experimental data. Although the rate constant assigned for oxidation of Fe(II) by $^1O_2$ at pH 4 is different to that assigned at pH 8.1, it is possible that the two rate constants are same if the $^1O_2$ concentrations are different at each pH. Both possibilities are equally consistent with our experimental data and, as such, the correct option cannot be determined from these data alone.

Complexation-Dissociation of Fe(II)

Reaction 21 (Table 2) and reaction 31 (Table 3) represent the formation of organically complexed Fe(II), while reaction 22 (Table 2) and reaction 32 (Table 3) represent the dissociation of the complex to form inorganic Fe(II). The rate constants used for these reactions at pH 8.1 are identical to those found to be appropriate previously (*18*, *38*). The same rate constants were used at pH 4 as the model was insensitive to these values at pH 4.

Fe(II) Formation via LMCT Pathway

Reaction 23 (Table 2) and Reaction 33 (Table 3) represent the formation of Fe(II) via LMCT at pH 8.1 and 4 respectively. The rate constant for the reduction of Fe(III)L by LMCT was determined from best fit of the kinetic model to experimental data at pH 4.

The model is able to successfully simulate the range of data collected in this study (Figures 1, 2 and 5). Based on our experimental data and the kinetic model, we also calculated the contribution of SMIR to Fe(II) production as function of pH (Figure 8). We did not attempt to model the measured H_2O_2 data at pH 4 since the rate constants for a number of the reactions involved in O_2^- production and decay are not known.

While the model accurately simulates our experimental data and is consistent with previous studies, uncertainty still remains with respect to some of the precise reaction details. Firstly, the pH dependence of Fe(II) production via LMCT, if any, is not determined here. Secondly, the involvement of organic radicals in Fe(II) oxidation and the nature of these radicals is unclear from our study.

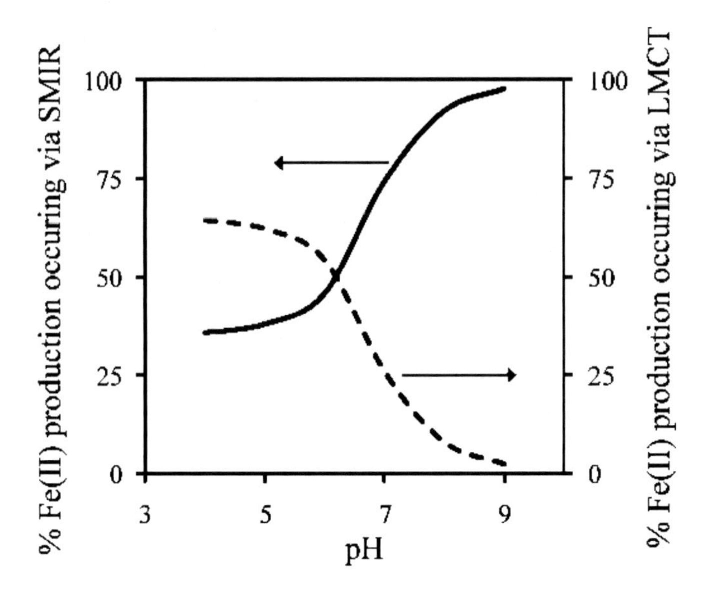

Figure 8. Effect of pH on the contribution of SMIR and LMCT to Fe(II) production in irradiated Fe(III)SRFA solution. It was assumed that the rate of Fe(II) production by LMCT is independent of pH and is equal to the rate determined at pH 4 from the experimental data. The Fe(II) production rate from SMIR is equal to $k_r[Fe(III)L[O_2^-]_{ss}$ where k_r is the reduction rate constant of Fe(III)SRFA by O_2^-. k_r was determined based on best-fit to the measured Fe(II) production rate at pH 4 and 8.1 and assumed to be constant with pH. $[O_2^-]_{ss}$ as function of pH was determined based on the measured H_2O_2 generation rate (calculated as $k_{disp}[O_2^-]_{ss}^2$; k_{disp} at each pH given by Bielski and co-workers (20); (data not shown) at various pH.

Conclusions

Photochemical reduction of Fe(III) in the presence of SRFA results in substantial production of Fe(II) and H_2O_2. The kinetics of Fe(II) production at pH 8.1 are characterized by a peak in Fe(II) concentration soon after commencement of irradiation, followed by a gradual decline over the remainder of the irradiation period. This is similar to the O_2^- concentration profile observed in the absence of Fe(III) and suggests that SMIR is more important than LMCT at this pH. This conclusion was supported by the observation that addition of SOD decreased total Fe(II) production from photolysis of Fe(III)SRFA. This does not mean that LMCT is never an important process; indeed, our data suggest that LMCT is an important pathway for Fe(II) formation at pH 4. However it does suggest that in many sunlit marine and freshwaters SMIR will be a major pathway for photochemical Fe(II) production. Reduction of iron by organic free radicals does not appear to be particularly important under the conditions investigated in this work, though this process may be important in the absence of dioxygen.

We have also shown that, over the range of iron and SRFA concentrations considered, O_2 and photo-produced species are the major oxidants of the photo-produced Fe(II).

We therefore suggest that at high pH where its disproportionation is slow, O_2^- is a critical intermediate controlling the speciation of iron in natural waters. However, the kinetics of photochemical O_2^- production is independent of iron concentration. Thus, sustained photochemical production of O_2^- will result in relatively constant and biologically significant Fe(II) production in marine waters. At low pH where the steady-state concentration of O_2^- is relatively low, LMCT processes most likely dominate the reduction of Fe(III).

References

1. Rose, A. L.; Waite, T. D. Reduction of organically complexed ferric iron by superoxide in a simulated natural water. *Environ. Sci. Technol.* **2005**, *39*, 2645–2650.
2. Rose, A. L.; Waite, T. D. Role of superoxide in photochemical reduction of iron in seawater. *Geochim. Cosmochim. Acta* **2006**, *70*, 3869–3882.
3. Öztürk, M.; Croot, P. L. C., Bertisson, S. B.; Abrahamssond, K.; Karlsone, B.; David, R. D.; Franssong, A. F.; Sakshauga, E. Iron enrichment and photoreduction of iron under UV and PAR in the presence of hydroxycarboxylic acid: implications for phytoplankton growth in the Southern Ocean. *Deep-Sea Res., Part II* **2004**, *51*, 2841–2856.
4. Voelker, B. M.; Sedlak, D. L. Iron reduction by photoproduced superoxide in seawater. *Mar. Chem.* **1995**, *50* (1-4), 93–102.
5. Rose, A. L.; Salmon, T. P.; Lukondeh, T.; Neilan, B. A.; Waite, T. D. Use of superoxide as an electron shuttle by the marine cyanobacterium *Lyngbya majuscula. Environ. Sci. Technol.* **2005**, *39*, 3708–3715.
6. Maldonado, M. T.; Price, N. M. Reduction and transport of organically bound iron by Thalassiosira oceanica (Bacillariophyceae). *J. Phycol.* **2001**, *37*, 298–309.
7. Shaked, Y.; Kustka, A. B.; Morel, F. M. M. A general kinetic model for iron acquisition by eukaryotic phytoplankton. *Limnol. Oceanogr.* **2005**, *50*, 872–882.
8. Salmon, T. P.; Rose, A. L.; Neilan, B. A.; Waite, T. D. The FeL model of iron acquisition: Non-dissociative reduction of ferric complexes in the marine environment. *Limnol. Oceanogr.* **2006**, *51*, 1744–1754.
9. Waite, T. D.; Morel, F. M. M. Photoreductive dissolution of colloidal iron oxides in natural waters. *Environ. Sci. Technol.* **1984**, *18*, 860–868.
10. Sulzberger, B.; Suter, D.; Siffert, C.; Banwart, S.; Stumm, W. Dissolution of Fe(III) (hydr)oxides in natural waters; laboratory assessment on the kinetics controlled by surface coordination. *Mar. Chem.* **1989**, *28*, 127–144.
11. Fujii, M.; Rose, A. L.; Waite, T. D.; Omura, T. Superoxide-Mediated Dissolution of Amorphous Ferric Oxyhydroxide in Seawater. *Environ. Sci. Technol.* **2006**, *40*, 880–887.

12. Cowart, R. E. Reduction of iron by extracellular iron reductases: implications for microbial iron acquisition. *Arch. Biochem. Biophys.* **2002**, *400*, 273–281.
13. Johnson, K. S.; Coale, K. H.; Elrod, V. A.; Tindale, N. W. Iron photochemistry in seawater from the equatorial Pacific. *Mar. Chem.* **1994**, *46* (4), 319–334.
14. Miller, W. L.; King, D. W.; Lin, J.; Kester, D. R. Photochemical redox cycling of iron in coastal seawater. *Mar. Chem.* **1995**, *50* (1−4), 63–77.
15. Waite, T. D.; Szymczak, R.; Espey, Q. I.; Furnas, M. J. Diel variations in iron speciation in northern Australian shelf waters. *Mar. Chem.* **1995**, *50* (1−4), 79–91.
16. Faust, B. C.; Zepp, R. G. Photochemistry of aqueous iron(III)-polycarboxylate complexes: roles in the chemistry of atmospheric and surface waters. *Environ. Sci. Technol.* **1993**, *27*, 2517–2522.
17. Sima, J.; Makánová, J. Photochemistry of iron(III) complexes. *Coord. Chem. Rev.* **1997**, *160*, 161–189.
18. Garg, S.; Rose, A. L.; Waite, T. D. Superoxide Mediated Reduction of Organically Complexed Iron(III): Comparison of Non-Dissociative and Dissociative Reduction Pathways. *Environ. Sci. Technol.* **2007**, *41* (9), 3205–3212.
19. King, D. W.; Aldrich, R. A.; Charnecki, S. E. Photochemical redox cycling of iron in NaCl solution. *Mar. Chem.* **1993**, *44*, 105–120.
20. Bielski, B. H. J.; Cabelli, D. E.; Arudi, R. L.; Ross, A. B. Reactivity of HO_2/O_2^- radicals in aqueous solution. *J. Phys. Chem. Ref. Data* **1985**, *14* (4), 1041–1100.
21. González-Dávila, M.; Santana-Casiano, J. M.; Millero, F. J. Oxidation of iron (II) nanomolar with H_2O_2 in seawater. *Geochim. Cosmochim. Acta* **2005**, *69* (1), 83–93.
22. King, D. W.; Farlow, R. Role of carbonate speciation on the oxidation of Fe(II) by H_2O_2. *Mar. Chem.* **2000**, *70*, 201–209.
23. Fujii, M.; Rose, A. L.; Waite, T. D.; Omura, T. Oxygen and superoxide-mediated redox kinetics of iron complexed by humic substances in coastal seawater. *Environ. Sci. Technol* **2010**, *44*, 9337–9342.
24. Garg, S.; Rose, A. L.; Waite, T. D. Photochemical Production of Superoxide and Hydrogen Peroxide from Natural Organic Matter. *Geochim. Cosmochim. Acta*, in press.
25. Paul, A.; Hackbarth, S.; Vogt, R. D.; Röder, B.; Burnison, B. K.; Steinberg, C. E. W. Photogeneration of singlet oxygen by humic substances: comparison of humic substances of aquatic and terrestrial origin. *Photochem. Photobiol. Sci.* **2004**, *3*, 273–280.
26. Dalrymple, R. M. Correlations between Dissolved Organic Matter Optical Properties and Quantum Yields of Singlet Oxygen and Hydrogen Peroxide. *Environ. Sci. Technol.* **2010**, *44*, 5824–5829.
27. Lan, Z.-H.; Mottola, H. A. Carbon dioxide-enhanced luminol chemiluminescence in the absence of added oxidant. *Analyst* **1996**, *121*, 211–218.

28. Garg, S.; Rose, A. L.; Waite, T. D. Production of Reactive Oxygen Species on Photolysis of Dilute Aqueous Solutions of Quinones. *Photochem. Photobiol.* **2007**, *83*, 904–913.
29. King, D. W.; Lounsbury, H. A.; Millero, F. J. Rates and mechanism of Fe(II) oxidation at nanomolar total iron concentrations. *Environ. Sci. Technol.* **1995**, *29*, 818–824.
30. Rose, A. L.; Waite, T. D. Chemiluminescence of luminol in the presence of iron(II) and oxygen: oxidation mechanism and implications for its analytical use. *Anal. Chem.* **2001**, *73*, 5909–5920.
31. Stookey, L. L. Ferrozine: a new spectrophotometric reagent for iron. *Anal. Chem.* **1970**, *42*, 779–781.
32. Zhou, M.; Diwu, Z.; Panchuk-Voloshina, N.; Haugland, R. P. A stable nonfluorescent derivative of resorufin for the fluorometric determination of trace hydrogen peroxide: Applications in detecting the activity of phagocyte NADPH oxidase and other oxidases. *Anal. Biochem.* **1997**, *253* (2), 162–168.
33. Braun, W.; Herron, J. T.; Kahaner, D. K. ACUCHEM: A computer program for modeling complex chemical reaction systems. *Int. J. Chem. Kinet.* **1988**, *20*, 51–60.
34. Blough, N. V. Electron paramagnetic resonance measurements of photochemical radical production in humic substances. 1. Effects of O_2 and charge on radical scavenging by nitroxide. *Environ. Sci. Technol.* **1988**, *22*, 77–82.
35. Zafiriou, O. C.; Blough, N. V.; Micinski, E.; Dister, B.; Kieber, D.; Moffett, J. W. Molecular probes for reactive transients in natural waters. *Mar. Chem.* **1990**, *30*, 45–70.
36. von Sonntag, C.; Dowideit, P.; Fang, X.; Mertens, R.; Pan, X.; Schuchmann, M. N.; Schuchmann, H.-P. The fate of peroxyl radicals in aqueous solution. *Water Sci. Technol.* **1997**, *35* (4), 9–15.
37. Aeschbacher, M.; Sander, M.; Schwarzenbach, R. P. Novel Electrochemical Approach to Assess the Redox Properties of Humic Substances. *Environ. Sci. Technol.* **2010**, *44*, 87–93.
38. Miller, C. J.; Rose, A. L.; Waite, T. D. Impact of natural organic matter on H2O2-mediated oxidation of Fe(II) in a simulated freshwater system. *Geochim. Cosmochim. Acta* **2009**, *73*, 2758–2768.
39. Rose, A. L.; Waite, T. D. Kinetic model for Fe(II) oxidation in seawater in the absence and presence of natural organic matter. *Environ. Sci. Technol.* **2002**, *36*, 433–444.
40. Goldstone, J. V.; Pullin, M. J.; Bertilsson, S.; Voelker, B. M. Reactions of hydroxyl radical with humic substances: bleaching, mineralization, and production of bioavailable carbon substrates. *Environ. Sci. Technol.* **2002**, *36*, 362–372.
41. Buxton, G. V.; Elliot, A. J. Rate constant for reaction of hydroxyl radicals with bicarbonate ions. *Int. J. Radiat. Appl. Instrum., Part C: Radiat. Phys. Chem.* **1986**, *27*, 241–243.

42. Haag, W. R.; Hoigné, J.; Gassman, E.; Braun, A. M. Singlet oxygen in surface waters – Part II: Quantum yields of its production by some natural humic materials as a function of wavelength. *Chemosphere* **1984**, *13*, 641–650.
43. Sandvik, S. L. H.; Bilski, P.; Pakulski, J. D.; Chignell, C. F.; Coffin, R. B. Photogeneration of singlet oxygen and free radicals in dissolved organic matter isolated from the Mississippi and Atchafalaya River plumes. *Mar. Chem.* **2000**, *69* (1−2), 139–152.
44. Latch, D. E.; McNeill, K. Microheterogeneity of Singlet Oxygen Distributions in Irradiated Humic Acid solutions. *Science* **2006**, *311*, 1743–1747.
45. Rush, J. D.; Bielski, B. H. J. Pulse radiolytic studies of the reactions of HO_2/O_2^- with Fe(II)/Fe(III) ions. The reactivity of HO_2/O_2^- with ferric ions and its implication on the occurrence of the Haber-Weiss reaction. *J. Phys. Chem.* **1985**, *89*, 5062–5066.

Chapter 9

The Role of Iron Coordination in the Production of Reactive Oxidants from Ferrous Iron Oxidation by Oxygen and Hydrogen Peroxide

Christina Keenan Remucal[1] and David L. Sedlak[2,*]

[1]Institute of Biogeochemistry and Pollutant Dynamics, ETH, Zürich, Switzerland
[2]Department of Civil and Environmental Engineering, University of California at Berkeley, Berkeley, California 94720
***sedlak@berkeley.edu**

A new picture of the Fenton reaction has emerged over the last two decades that extends our understanding beyond the acidic conditions studied previously. In the absence of ligands, the reaction produces hydroxyl radical under acidic conditions and a less reactive oxidant, presumed to be the ferryl ion (Fe[IV]), at circumneutral pH values. Formation of complexes between Fe(II) and organic ligands alters the reaction mechanism, resulting in production of hydroxyl radical over a wide pH range. As a result, iron coordination and pH determine the oxidants produced by the Fenton reaction. Consideration of the reactive oxidant produced by the Fenton reaction under environmentally- and biologically-relevant conditions is necessary to develop more effective treatment systems, to predict the fate of iron and carbon in natural waters, and to assess iron-mediated oxidative damage.

Introduction

Iron is the fourth most abundant element in the earth's crust by weight and is a physiological requirement for life. Redox cycling of iron, primarily between the ferrous (Fe[II]) and ferric (Fe[III]) oxidation states, allows iron to serve as an electron shuttle in numerous abiotic and biotic processes. The redox cycling of iron in aquatic, terrestrial, and atmospheric systems plays a critical role in a range of biogeochemical processes, including the oxidation of organic pollutants and natural organic matter (NOM) (*1*), mineral dissolution (*2*), and the regulation of iron bioavailability (*3*).

Under neutral and basic pH conditions, Fe(II) is thermodynamically unstable in the presence of oxygen (O_2) and is quickly oxidized to form sparingly soluble Fe(III) (hydr)oxides. Hydrogen peroxide (H_2O_2), which is often present in sunlit natural waters, also oxidizes Fe(II) quickly via the Fenton reaction. The reactions of Fe(II) with O_2 and H_2O_2 are important to iron redox cycling because they can lead to the production of reactive oxidants, such as hydroxyl radical ($OH^{\bullet}$). Reactive oxidants produced by Fe(II) oxidation are believed to be one of the main mechanisms through which organic compounds are oxidized in acidic waters such as cloudwater and acid-impacted streams (*1*, *2*, *4–6*).

Although $OH^{\bullet}$ is often considered to be the product of the reaction of Fe(II) and H_2O_2 under circumneutral and basic pH conditions, recent studies have provided additional evidence that Fe(II) oxidation may be more complex than previously believed and often does not result in $OH^{\bullet}$ production. To explain the oxidation reactions observed during Fe(II) oxidation, several researchers have proposed the formation of an alternate oxidant, such as the ferryl ion (Fe[IV]), which is more selective than $OH^{\bullet}$. Therefore, developing a better understanding of Fe(II) oxidation reactions under circumneutral and basic pH conditions is crucial for predicting how the iron redox cycle will affect other water constituents.

Although Fe(III) is usually the thermodynamically stable species in oxygen-containing waters, Fe(II) is frequently detected in sunlit waters (*7*, *8*) at concentrations up to approximately 10^{-11}, 10^{-8}, and 10^{-4} M in oceans (*9*), lakes (*10*, *11*), and atmospheric waters (*12*), respectively. Fe(III) can be reduced by direct photolysis of Fe(III)-complexes with electron-donating ligands, such as oxalate ($C_2O_4^{2-}$), as well as reactions with superoxide ($O_2^{-\bullet}$) and hydroperoxyl radical ($HO_2^{\bullet}$; Figure 1) (*4*, *6*, *12*, *13*). In the absence of sunlight, NOM and minerals, such as pyrite, are also able to reduce Fe(III) to Fe(II), resulting in oxidation of the reductants and low concentrations of Fe(II) (see Chapters 6-8) (*13–15*).

Iron redox cycling also plays a key role in certain contaminant treatment systems. For example, Fenton-based treatment systems rely on the production of reactive oxidants when Fe(II) is oxidized by H_2O_2 in water and soil (*16*). These systems typically employ high concentrations of H_2O_2 to facilitate reduction of Fe(III) (Figure 2) and are often conducted under acidic conditions or in the presence of iron-complexing ligands to limit Fe(III) precipitation (*17*, *18*). Variations on the traditional Fenton-based approach include photo-Fenton systems, where iron-complexes are reduced by light, and electro-Fenton systems, where Fe(II) is produced *in situ* with an electrode (*16*). The identification of the oxidant produced by the Fenton reaction under different solution conditions is necessary for selecting appropriate target contaminants and predicting their transformation products.

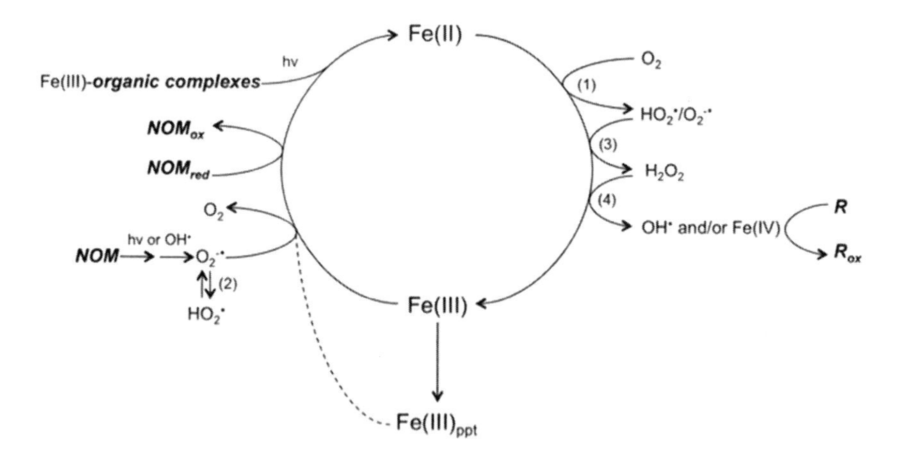

Figure 1. Schematic of iron redox cycling in natural sunlit waters. Bold text indicates processes where NOM may play a role in iron cycling. R indicates any species (organic or inorganic) present in water that is capable of reacting with OH• or Fe(IV). Numbers correspond to reactions in text.

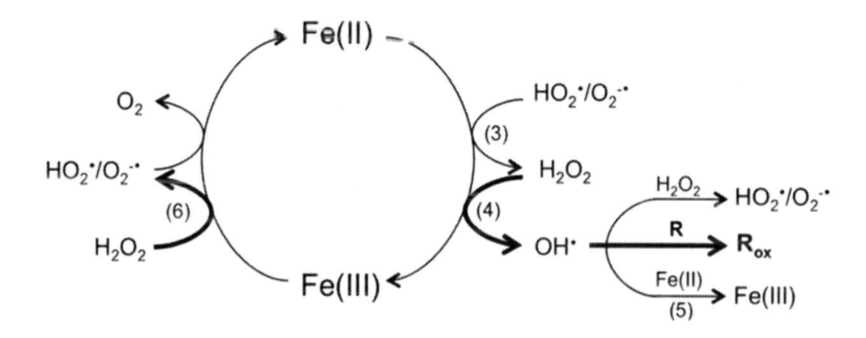

Figure 2. Schematic of iron redox cycling in a Fenton-based contaminant oxidation process, shown under acidic conditions in the dark. R represents the target contaminant(s) and darker arrows indicate the major reactions. Numbers correspond to reactions in the text.

In addition to its role in transforming NOM and contaminants, iron redox cycling contributes to oxidative damage in cells. Iron-containing aerosols or particulate matter can produce reactive oxidants *in vivo* when Fe(II) is oxidized by O_2 or H_2O_2 (*19*, *20*). Reductants released by cells also can facilitate Fe(III) reduction, followed by reactive oxidant production (*21*). A better understanding of the oxidant produced by the Fenton reaction *in vivo* and its reactivity with biomolecules is needed to predict oxidative stress.

In this chapter, we describe the production of reactive oxidants during the oxidation of Fe(II) by O_2 and H_2O_2 under conditions relevant to the situations described above. By considering the effect of complexation of Fe(II) with common ligands, we will gain an understanding of iron redox cycling, its role in the oxidation of NOM and contaminants, and its ability to induce oxidative damage.

Oxidation of Fe(II) by O₂

The oxidation of Fe(II) by oxygen occurs by a series of one-electron transfer reactions that were first described by Haber and Weiss (*22*, *23*). The initial reaction between Fe(II) and O_2 is the rate-limiting step (*24*, *25*):

$$\mathrm{Fe(II)} + \mathrm{O_2} \rightarrow \mathrm{Fe(III)} + \mathrm{O_2^{-\bullet}} \tag{1}$$

The superoxide radical anion rapidly equilibrates with hydroperoxyl radical:

$$\mathrm{HO_2^{\bullet}} \leftrightarrow \mathrm{H^+} + \mathrm{O_2^{-\bullet}} \qquad K_a = 10^{-4.8}\ (26) \tag{2}$$

The speciation of dissolved Fe(II) and Fe(III) depends upon pH and the presence of ligands, as discussed below. For simplicity, Fe(II) and Fe(III) will represent all dissolved ferrous and ferric iron species. Superoxide and $HO_2^{\bullet}$ react with Fe(II) via a second one-electron transfer to produce HO_2^-/O_2^{2-}, which is rapidly protonated to hydrogen peroxide (*27*):

$$\mathrm{Fe(II)} + \mathrm{HO_2^{\bullet}} \rightarrow \mathrm{Fe(III)} + \mathrm{HO_2^-} \xrightarrow{\mathrm{H^+}} \mathrm{Fe(III)} + \mathrm{H_2O_2} \tag{3a}$$

$$\mathrm{Fe(II)} + \mathrm{O_2^{-\bullet}} \rightarrow \mathrm{Fe(III)} + \mathrm{O_2^{2-}} \xrightarrow{\mathrm{2H^+}} \mathrm{Fe(III)} + \mathrm{H_2O_2} \tag{3b}$$

This reaction is followed by the oxidation of Fe(II) by H_2O_2, which is referred to as the Fenton reaction:

$$\mathrm{Fe(II)} + \mathrm{H_2O_2} \rightarrow \mathrm{Fe(III)} + \mathrm{OH^-} + \mathrm{OH^{\bullet}} \tag{4}$$

In the classic Haber-Weiss mechanism, which describes Fe(II) oxidation in the absence of $OH^{\bullet}$ scavengers (e.g., organic compounds, bicarbonate), $OH^{\bullet}$ produced by reaction 4 oxidizes Fe(II):

$$\mathrm{Fe(II)} + \mathrm{OH^{\bullet}} \rightarrow \mathrm{Fe(III)} + \mathrm{OH^-} \tag{5}$$

The overall stoichiometry of reactions 1-5 results in 4 moles of Fe(II) oxidized per mole of O_2 and has been confirmed experimentally in natural waters (*24*, *25*, *28*, *29*). Species present in natural waters (e.g., NOM) can compete with Fe(II) for $OH^{\bullet}$ (reaction 5), particularly in waters with lower iron concentrations, decreasing the number of moles of Fe(II) oxidized per O_2.

While $OH^{\bullet}$ is often assumed to be the dominant oxidant produced by the Fenton reaction (*30*), a number of researchers have invoked the production of a more selective oxidant to explain experimental observations under circumneutral pH conditions (*31–36*). Subsequent reactions of the oxidant typically result in the formation of Fe(III), H_2O, and OH^- without additional O_2 consumption. Thus, the overall stoichiometry of Fe(II) oxidation by O_2 is unaffected by changes in the mechanism of reaction 4. The identity of the oxidant produced by the Fenton reaction and the effect of solution conditions on the reaction mechanism are discussed below.

The rate of oxidation of Fe(II) by oxygen is strongly dependent on pH. Below pH 4, the reaction is very slow (e.g., $t_{1/2}$ ~ years in air-saturated water (*24*). Between pH 4.5 and 8, reaction 1 exhibits a second order dependence on OH^- (*24, 28, 29, 37, 38*). Assuming an air-saturated solution at 25° C (i.e., $[O_2]$ = 250 μM) and the absence of significant concentrations of Fe(II)-complexing ligands, the half-life of Fe(II) in reaction 1 is approximately 45 hours at pH 6 and 30 minutes at pH 7 in the absence of catalysts, such as surfaces or microbes (*25, 29*). The increase in reaction rate with pH is due to hydrolysis of Fe(II) (*28*) into $FeOH^+$ and $Fe(OH)_2^0$, which are extremely reactive and account for most of the loss of Fe(II) (Figure 4) (*29*) despite the fact that they account for a small fraction of the overall Fe(II) species in solution (Figure 3).

The high reactivity of hydrolyzed Fe(II) species is believed to be attributable to changes in reaction mechanisms with oxygen that occur upon hydrolysis. The reaction of the hexaquo species $Fe(H_2O)_6^{2+}$ with O_2 is described as an outer-sphere process based on molecular orbital theory arguments (*39*) and Marcus theory calculations (*40–42*). Although hydrolyzed Fe(II) species can also react via an outer-sphere process at neutral pH values (*39, 40*), recent studies suggest the oxidation of $FeOH^+$ and $Fe(OH)_2^0$ by O_2 occurs via an inner-sphere mechanism based on large differences between experimental rate constants and calculated outer-sphere rate constants (*41, 42*).

Oxidation of Fe(II) by H_2O_2

The oxidation of Fe(II) by H_2O_2 (the Fenton reaction) has been studied since the late 19th century (*45*). In addition to the oxidation of Fe(II) by H_2O_2 (reaction 4), H_2O_2 can also reduce Fe(III) (*46*):

$$\mathrm{Fe(III)} + \mathrm{H_2O_2} \rightarrow \mathrm{Fe(II)} + \mathrm{HO_2^{\bullet}} + \mathrm{H^+} \qquad (6)$$

The ability of H_2O_2 to both oxidize and reduce iron results in a catalytic cycle in which H_2O_2 is converted into H_2O through a series of reactions, which are referred to as Fenton-like reactions, involving $OH^{\bullet}$, H_2O_2 and $HO_2^{\bullet}$ (*16, 46*). Reaction 6 is several orders of magnitude slower than reaction 4 and thus serves as the rate-determining step in the loss of H_2O_2 (*16, 17*).

As with the oxidation of Fe(II) by O_2, the rate of the reaction of Fe(II) with H_2O_2 increases with pH. The reaction is independent of pH below pH 3 and increases with increasing pH above pH 3 (*43, 46, 47*). The linear increase in the rate of Fe(II) oxidation by H_2O_2 in the absence of Fe(II)-complexing ligands between pH 6 and 8 (*48*) has been attributed to the formation of the $FeOH^+$ and $Fe(OH)_2^0$ (Figures 4-5) (*43, 47*).

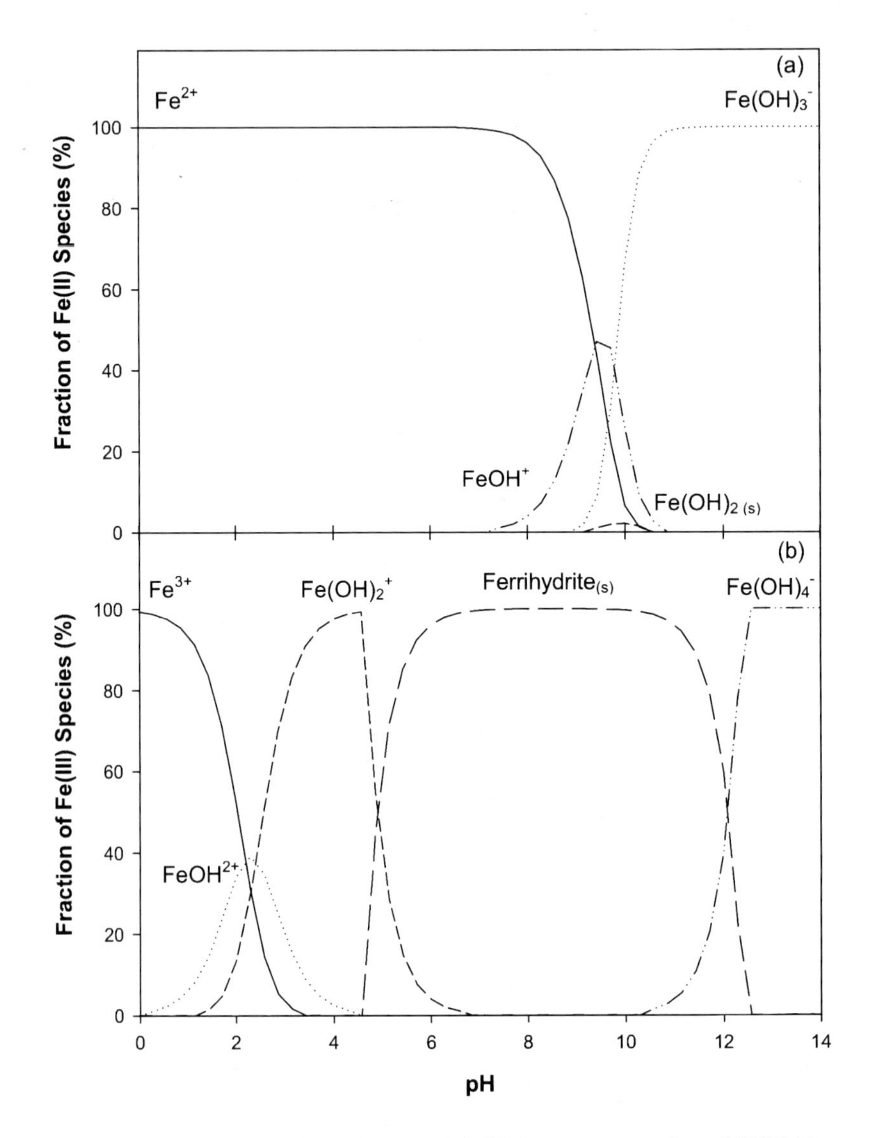

Figure 3. Speciation of (a) ferrous and (b) ferric species from MINEQL+ calculations based on 1 μM Fe in the absence of ligands with $Fe(OH)_{2(s)}$ and ferrihydrite considered.

Studies conducted under acidic conditions (e.g., pH <4) provide convincing experimental evidence that $OH^{\bullet}$ is produced by the Fenton reaction. Specifically, experiments under acidic conditions have yielded similar $OH^{\bullet}$ rate constants and target compound products as systems where $OH^{\bullet}$ was produced by pulse-radiolysis (*30*, *49*, *50*). The agreement of experimental data with kinetic models using reactions 1-6 provides additional evidence for $OH^{\bullet}$ production under acidic conditions (*18*, *51*, *52*).

Contaminant transformation rates and products in experiments conducted at higher pH values in the absence of Fe-complexing ligands or in the presence of

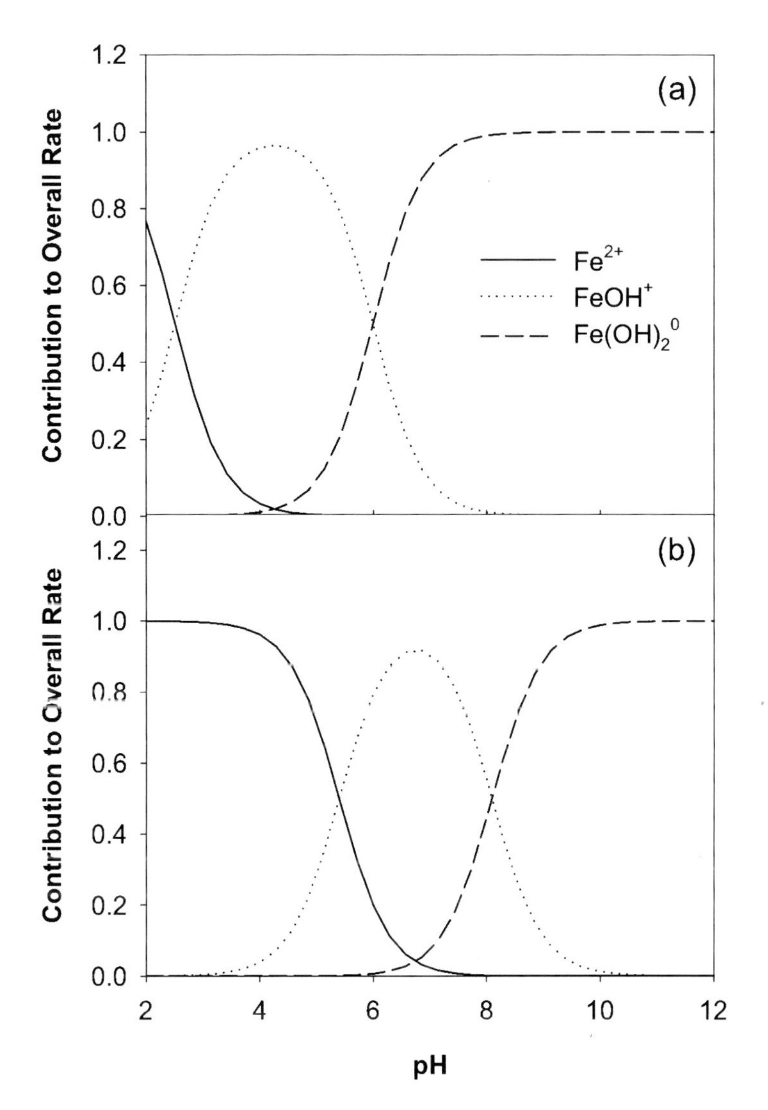

Figure 4. Fraction of contribution of Fe(II) species to total Fe(II) oxidation by (a) O_2 and (b) H_2O_2 using MINEQL+ speciation modeling and reaction rates from (37, 43, 44).

inorganic ligands, such as bicarbonate or phosphate, have yielded significantly different results from those expected for $OH^{\bullet}$. For example, the oxidation of arsenite (As[III]) in a thermal Fenton system was attributed to $OH^{\bullet}$ under acidic conditions, but high concentrations of $OH^{\bullet}$ scavengers (e.g., 2-propanol) were ineffective at preventing As(III) oxidation at circumnetural pH (*32*). Furthermore, kinetic models that invoked $OH^{\bullet}$ as the oxidant substantially over-predicted atrazine and 1,2,4-trichlorobenzene transformation at pH > 4 (*33*). Similar evidence against $OH^{\bullet}$ production during Fe(II) oxidation by O_2 (reactions 1-3) was obtained in experiments demonstrating that $OH^{\bullet}$ scavengers were able to prevent As(III) oxidation at pH 3-4, but not pH >5 (*32*). In a similar Fe(II)/O_2

system, a non-selective oxidant (e.g., $OH^{\bullet}$) was capable of oxidizing methanol, ethanol, benzoic acid, and 2-propanol at pH 3-5, whereas the oxidant produced at pH 6-9 was only able to oxidize methanol and ethanol (*34*, *35*). Additional evidence against production of $OH^{\bullet}$ under neutral conditions was obtained in a photo-Fenton system in which the yield of $OH^{\bullet}$ quantified using benzene as a probe compound at circumneutral pH. Under these conditions, the yield of phenol was much smaller than predicted based on the rate of the photo-Fenton reaction determined by monitoring H_2O_2 concentrations (*36*). Collectively, these studies indicate that the transformation of organic compounds and reduced metals cannot be predicted if $OH^{\bullet}$ is assumed to be the only product of the Fenton reaction at circumneutral pH values.

There are several possible explanations for the discrepancy between predicted and observed target compound transformation in Fenton systems under circumneutral pH conditions. One possibility is that the target compound transformation mechanism is pH-dependent. According to this explanation, the intermediates produced by the reaction of the compound with $OH^{\bullet}$ at circumneutral pH values re-form the parent compound in subsequent steps, rather than going on to form the oxidized product. Although the intermediates produced by the reaction of $OH^{\bullet}$ with aromatic compounds can be reduced (*53*), this pathway does not expain the observed pH-dependence of the oxidant produced in Fenton systems because experiments with compounds such as benzene and phenol in which $OH^{\bullet}$ is formed by H_2O_2 photolysis do not show decreased product yields at circumneutral pH values (*54–56*).

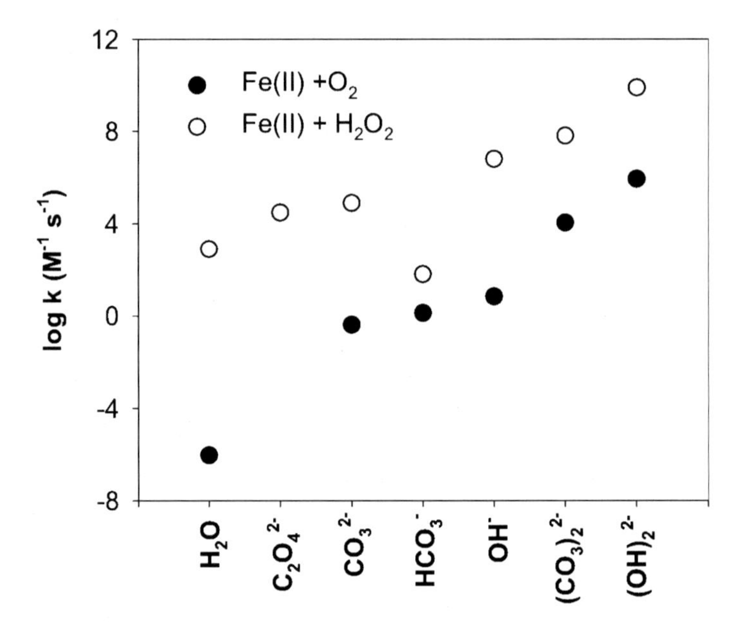

Figure 5. Fe(II) oxidation rate by O_2 and H_2O_2 when complexed with selected ligands (6, 43, 44).

A second explanation is that a carbon-centered radical ($C^{\bullet}$) produced when $OH^{\bullet}$ reacts with a target compound is subsequently reduced by a metal ion (e.g., Fe^{2+}) back to the parent compound (*30*, *49*). However, the relative rates of reactions of carbon-centered radicals with O_2 and Fe^{2+} (e.g., ~10^9 $M^{-1}s^{-1}$ and ~10^5 $M^{-1}s^{-1}$ with the radical produced by phenol oxidation, respectively; (*57*)) are too fast for the back-reaction with Fe^{2+} to be important in air-saturated waters. While it is likely that the Fe(II) hydrolysis species are more reactive with $C^{\bullet}$ than Fe^{2+}, the rate of reaction for $FeOH^+$ would need to be substantially greater than the diffusion-controlled maximum bimolecular rate constant (i.e., 10^{10} $M^{-1}s^{-1}$) for the hydrolyzed species to compete with O_2 for $C^{\bullet}$ (calculated at pH 7 with $[Fe]_{tot}$=100 μM). Similar rates of reaction have been reported between O_2 and carbon-centered radicals produced by the oxidation of other compounds, such as formate (*58*), suggesting that reduction of the radicals by Fe(II) is not responsible for the decreased oxidant yield at higher pH values.

A third possibility is that the reactive oxidant produced by the Fenton reaction depends on solution conditions. According to this explanation, a transient metal peroxide species forms as the first step in the reaction between Fe(II) and H_2O_2:

$$[Fe(OH)(H_2O)_5]^+ + H_2O_2 \rightarrow [Fe(OH)(H_2O_2)(H_2O)_4]^+ + H_2O \quad (7)$$

The existence of such a complex is supported by thermodynamic calculations indicating that most transition metal complexes (e.g., $FeOH^+$) react with H_2O_2 via an inner-sphere electron transfer mechanism (*31*, *59*, *60*), as well as spectroscopic data (*17*) and density functional theory calculations (*61*). In the next step, the peroxide dissociates to form $OH^{\bullet}$ (reaction 8) or an Fe(IV) species (reaction 9) (*59*):

$$[Fe(OH)(H_2O_2)(H_2O)_4]^+ \rightarrow [Fe(OH)(H_2O)_4]^{+2} + OH^- + OH^{\bullet} \quad (8)$$

$$[Fe(OH)(H_2O_2)(H_2O)_4]^+ \rightarrow [Fe(OH)(H_2O)_4]^{+3} + 2OH^- \quad (9)$$

According to this mechanism, the relative rates of reactions 8-9 determine which reactive oxidant is formed. Similar to the speciation-dependent reaction of Fe(II) with O_2 (*41*), it is possible that the inner-sphere reaction of $FeOH^+$ with H_2O_2 forms Fe(IV), whereas the outer-sphere reaction of uncomplexed Fe^{2+} with H_2O_2 forms $OH^{\bullet}$. The ferryl ion species formed in reaction 9 may also react with water to produce $OH^{\bullet}$ (Scheme 1) (*32*, *60*).

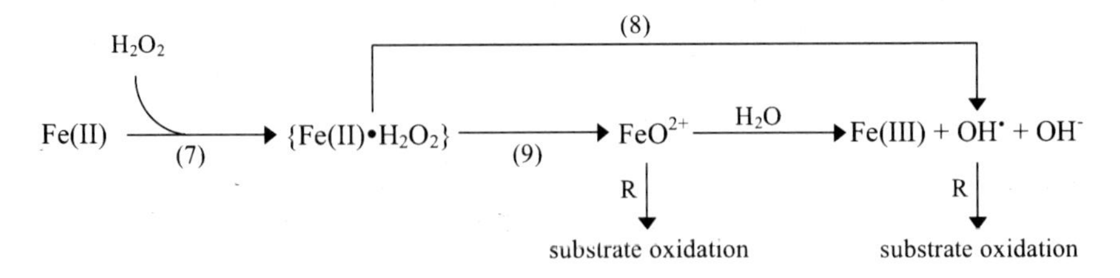

Scheme 1. Summary of possible reactions involved in the thermal Fenton reaction with simplified notations used for the iron complexes. Either the Fe(IV) species or the hydroxyl radical may oxidize a substrate, R. Adapted from (59, 60, 62).

The formation of Fe(IV) is noteworthy because it is a more selective oxidant than $OH^\bullet$. Hydroxyl radical reacts at near diffusion-controlled rates with many organic and inorganic compounds (*63*), giving it low selectivity and an extremely short lifetime in solution. The low selectivity of $OH^\bullet$ means that other compounds typically present in water, including NOM, bicarbonate, and Fe(II), can compete with contaminants for $OH^\bullet$. Fe(IV) is considered to be a weaker oxidant than $OH^\bullet$ based on standard reduction potentials (*64*). As a result, Fe(IV) has a much longer lifetime in solution (e.g., ~2 s in the absence of H_2O_2 compared to ~μs for $OH^\bullet$ under similar conditions; (*60*)). Unfortunately, few studies have been conducted on the reaction of Fe(IV) with organic compounds and the available measurements of rate constants for Fe(IV) reactions with organic compounds have been conducted under acidic conditions where the reaction of Fe(II) and ozone was used to produce Fe(IV) (*62*, *65*). While the lower reactivity of Fe(IV) limits its utility for the oxidation of many organic compounds, its selectivity could make Fe(IV) a more effective oxidant for As(III) and other contaminants with which it appears to react quickly (*32*, *66*).

Although it is difficult to distinguish between $OH^\bullet$ and Fe(IV) by direct observations (*16*, *31*, *49*), the inability to predict target compound oxidation based on reaction 4 and the related steps in the Haber-Weiss mechanism is noteworthy and implies that each molecule of H_2O_2 consumed does not always produce $OH^\bullet$ in Fenton systems. These observations have significant implications for interpreting the effect of Fe(II) oxidation on organic compound transformation under environmentally-relevant conditions and in remediation systems, as discussed below.

Role of Ligands and Surfaces

In Fenton-based oxidation systems used for contaminant oxidation, iron-complexing ligands are often added to increase Fe(III) solubility and enhance the rate of H_2O_2 activation via reactions 4 and 6 (*67*, *68*). Furthermore, some of the dissolved Fe(II) and Fe(III) in natural waters may be complexed by naturally-occurring ligands. In addition to increasing iron solubility, complexation changes the reduction potential of iron and can create a labile coordination position capable of forming an inner-sphere complex with O_2 or H_2O_2 (*69*, *70*). This shift in coordination can accelerate the rates of Fe(II) oxidation by

O_2 (reaction 1) (*70–72*) and H_2O_2 (reaction 4) (*6*, *68*, *69*). Chelates in which oxygen atoms serve as the ligand (e.g., $C_2O_4^{2-}$) stabilize Fe(III) and tend to accelerate Fe(II) oxidation (*72*), whereas chelates with nitrogen or sulfur atoms (e.g, porphyrins) often have the opposite effect.

Although the yield of transformation products produced by the reaction of H_2O_2 with iron in the absence of ligands (*33*, *34*, *60*) or in the presence of inorganic ligands, such as carbonate (*17*, *32*, *36*) and phosphate (*35*), is inconsistent with significant production of $OH^•$ at circumneutral pH, target compound oxidation in the presence of certain organic ligands is consistent with $OH^•$ production (*73*). For example, the yield of acetone from 2-propanol oxidation in the presence of Fe(II) and O_2 increased from $\ll$1% in the absence of ligands at pH 7 to 28.8% and 21.4% in the presence of oxalate and NTA, respectively (*34*, *73*, *74*), which is consistent with the expected stoichiometry of one $OH^•$ produced for every 3 Fe(II) oxidized (reactions 1-3). Furthermore, the relative oxidation rates of anisole and nitrobenzene in a photo-Fenton system containing oxalate or citrate were consistent with $OH^•$ production as predicted by rate constants measured by pulse radiolysis (*75*). Collectively, these studies provide evidence for the presence of $OH^•$ production via reactions 1-4 when iron is complexed by ligands such as oxalate, citrate, and NTA.

There are conflicting reports on the nature of the oxidant produced when Fe(II)-ethylenediaminetetraacetate (EDTA) complexes react with O_2 or H_2O_2. Some researchers have found evidence for the production of Fe(IV) at circumneutral pH values using probe compound transformation (*67*, *72*, *76*) and electron paramagnetic resonance (*77*). Others have observed reduced yields for target compound oxidation (*73*) and lower signals in spin-trapping studies (*78*), which they attributed to a mixture of $OH^•$ and Fe(IV). The oxidant produced by EDTA-chelated Fe(II) appears to be sensitive to numerous solution conditions, including pH, the ratio of EDTA:Fe, the presence of surfaces, and the concentrations of O_2 or H_2O_2.

NOM is heterogeneous and contains many different functional groups that could affect iron speciation and redox cycling. Its tendency to complex iron is most likely dominated by its numerous carboxylate groups (*13*, *79*). In general, terrestrially-derived NOM accelerates Fe(II) oxidation by O_2 in freshwater (*10*, *13*) and seawater (*80*), as expected for carboxylate ligands. The effect of NOM on reaction 4 is pH dependent; Suwannee River fulvic acid (SRFA) increases the rate of Fe(II) oxidation by H_2O_2 at pH 5 (*13*), has no effect at pH 7 (*36*), and decreases it at pH 8 (*81*). At these higher pH values, the rates of oxidation of the hydrolysis species (Figures 4-5) and other inorganic species (e.g., $FeCO_3$) are likely to dominate the observed rates, decreasing the importance of Fe-NOM complexes (*10*, *44*). Thus, NOM appears to accelerate the rates of Fe(II) oxidation primarily at pH values below 7.

Few studies have investigated the product of the Fenton reaction in the presence of NOM under conditions found in natural waters. Fe(II) complexed by carboxylate groups on NOM are expected to alter the product of the Fenton reaction at circumneutral pH in the same ways as oxalate and NTA (i.e., they would favor $OH^•$ production). However, in experiments with SRFA at pH 7, in which the oxidation of benzene to phenol was used to quantify the $OH^•$ production rate by reaction 4, the observed phenol production was only 26±13% of the value expected based on observed Fe(II) and H_2O_2 consumption rates (*36*), indicating the formation of an alternate oxidant. An EPR study with Fe-loaded humic acids at pH 7 found evidence for $OH^•$ production only at very high H_2O_2 concentrations (e.g., 1 mM) and evidence of an alternate oxidant, such as Fe(IV), when reactions 1-3 served as the source of H_2O_2 (*82*). While additional studies are needed to fully elucidate the effect of NOM on the Fenton reaction, the available data suggest that $OH^•$ is not formed stoichiometrically from the reaction of H_2O_2 with NOM-complexed Fe(II) under all conditions.

The presence of ferric iron has important implications for the oxidation of Fe(II) in oxygen-containing waters. Ferric iron undergoes hydrolysis and has very limited solubility at circumneutral pH values (Figure 3). Although Fe(II) is relatively soluble in homogenous solutions, ferrous iron can co-precipitate with Fe(III) oxy-hydroxides at pH values above 3 (*17*). The presence of surfaces, such as Fe(III) precipitates, also may accelerate the reaction of Fe(II) and O_2 (reaction 1) (*38*, *83*).

Most investigations into the reaction of H_2O_2 with surface-bound iron have been conducted in Fenton systems intended for contaminant remediation, where the concentrations of H_2O_2 are quite high (e.g., > mM). In such a ferrihydrite/H_2O_2 system at pH 4, the presence of $OH^•$ as the main oxidant was inferred by comparing relative decomposition rates of probe compounds (*18*, *84*). The efficiency of $OH^•$ production by heterogeneous Fenton-based processes at higher pH values is extremely low; generally the ratio of contaminant transformed to H_2O_2 consumed is $\ll$1% (*85–88*). One explanation for the observed inefficiency is that the majority of $OH^•$ reacts with the iron surface before diffusing into solution (*85*, *86*). An alternate explanation is that H_2O_2 is reduced directly to O_2 through a two electron transfer nonradical pathway on the iron surface (*18*, *87*), possibly accompanied by oxidation of Fe(II) to Fe(IV) (*88*, *89*). More research is needed to establish the H_2O_2 decomposition reaction mechanism in the presence of iron-containing surfaces, particularly under environmentally-relevant conditions where iron oxides could be important sinks for H_2O_2 (*85*).

Examples of Iron Redox Chemistry in Aerobic Systems

Sunlit Waters and Carbon Cycling

The photolysis of natural waters produces Fe(II) through ligand-to-metal charge transfer reactions of Fe(III)-complexed by polycarboxylate moieties on NOM and by reactions of Fe(III) with photoproduced reducants (e.g., $O_2^{-•}$; Figure 1) (*1*, *6*, *13*). In the absence of sunlight, NOM can slowly reduce Fe(III) through reactions with redox-active functional groups, such as quinone moieties (*13*, *14*, *50*), but the rates of these dark processes are substantially slower than those of the light-driven processes.

The reactions of photoproduced Fe(II) with O_2 and H_2O_2 can lead to oxidation of trace organic pollutants or NOM, provided that the oxidants produced during Fe(II) oxidation react with the organic compounds. As described above, the coordination of iron varies with pH (Figure 3) and therefore the production of oxidants is expected to shift as pH increases. As an example, consider sunlit waters with pH values of 5 and 7 and two different concentrations of NOM. Assuming that the rate of the photo-Fenton reaction (R_{Fenton}) is ~100 nM/hr (*36*) and that NOM is the major oxidant sink for $OH^•$ (*1*, *90*), the steady-state $OH^•$ concentration can be calculated according to (*36*):

$$[OH^•]_{ss} = (f_{OH^•})(R_{Fenton})/(k_{OH^•}^{NOM}[NOM]) \tag{10}$$

where $k_{OH^•}^{NOM}$ is 2.5 x 10^4 L $mg^{-1}s^{-1}$ (*91*) and $f_{OH^•}$ is the fraction of H_2O_2 converted to $OH^•$ in reaction 4. Assuming that the Fenton reaction produces $OH^•$ stoichiometrically at pH 5, $[OH^•]_{ss}$ is approximately 10^{-15} and 10^{-16} M in the surface of a sunlit water with 1 and 10 mg/L NOM, respectively. An organic contaminant (R) would have a half-life of 1 and 12 days due to Fenton-produced $OH^•$ under these conditions if $k_{OH^•}^{R}$ is 6 x 10^9 M^{-1} s^{-1} (*63*).

At pH 7, iron speciation has a significant effect on the formation of $OH^•$ by reaction 4. In NOM-rich water (10 mg/L), only about 25% of the photo-Fenton reaction appears to form $OH^•$ (*36*), giving $[OH^•]_{ss}$ of 10^{-17} M and a contaminant half-life of ~50 days if all of the Fe(II) is associated with NOM. Assuming that the yield of $OH^•$ from reaction 4 is <1% at lower NOM concentrations, the contaminant half-life due to Fenton-produced $OH^•$ is >120 days at circumnetural pH. The production of an alternate oxidant (e.g., Fe[IV]) by the photo-Fenton process could lead to significant contaminant transformation ($t_{1/2}$ <30 days), provided that the contaminant is capable of reaction with Fe(IV) at rates that are 104 times higher than NOM, where $k_{Fe(IV)}^{R}$ is in M^{-1} s^{-1}, $k_{Fe(IV)}^{NOM}$ is in L $mg^{-1}s^{-1}$, and NOM is 1 mg/L.

The effect of photo-Fenton processes on carbon cycling also will vary with pH and iron speciation. The photolysis of natural waters can lead to mineralization of NOM through nitrate photolysis, photo-Fenton reactions, and direct NOM photolysis (*90*, *92–94*). Under conditions where a significant fraction of the photo-Fenton reaction produces $OH^•$ (e.g., low pH or high NOM), the yield of $OH^•$ from this pathway is comparable to nitrate photolysis ($OH^•$ production by 0.1 mM nitrate = 70 nM/hr; (*1*)) and both mechanisms will be important for NOM oxidation. At circumneutral pH or in low-NOM waters, $OH^•$ production through Fe(II)-mediated processes will be negligible. The impact of other oxidants produced by Fe(II) oxidation, such as Fe(IV), on the mineralization of NOM is difficult to predict without a better understanding of the NOM transformation mechanisms.

H_2O_2-Based in Situ Chemical Oxidation

In situ chemical oxidation (ISCO) systems with H_2O_2 employ naturally present Fe-bearing minerals in soils to initiate Fenton-like production of reactive oxidants (reactions 4 and 6). However, only a small fraction of the H_2O_2 added is converted into oxidants that are capable of transforming recalcitrant contaminants at circumneutral pH (*86*, *88*, *95*). While oxidant scavenging by the iron oxide surfaces may partially account for the inefficiency of oxidation production (*85*, *86*), the production of a more selective oxidant by the Fenton reaction in ISCO systems is possible (*88*) and could limit the susceptibility of contaminants to remediation at neutral pH values. It is possible that altering the iron coordination environment can significantly increase $OH^•$ production. For example, $OH^•$ yields in silica-iron oxides and silica-alumina-iron oxides were ~20 and ~60 times higher than other Fe(III)-oxides at pH 7 (*88*). More research is needed to understand the role of iron complexation with silica and alumina in $OH^•$ production and to predict the efficacy of ISCO systems when different types of Fe-containing minerals are present. Such an approach could enable better selection of appropriate soils and contaminants for treatment by ISCO.

Biological Systems

The nature of the reactive oxidant produced by the Fenton reaction has important implications for oxidative tissue damage caused by iron-containing particles. For example, aerosol particles can contain considerable amounts of Fe(II), particularly under acidic conditions (*19*, *20*). Upon inhalation, the increase in pH that occurs when acidic aerosols encounter well-buffered lung fluid will result in the rapid oxidation of Fe(II) accompanied by a burst of oxidants (reactions 1-4). In addition, biomolecules, such as glutathione (*21*), contribute to the iron redox cycle by reducing Fe(III) to Fe(II). Although $OH^•$ is usually evoked as the oxidative stress-inducing oxidant produced by iron-containing particles (e.g., nanoparticles (*96*)), iron speciation is likely to be very different *in vivo* than in the environment and the effect of this speciation on the product of reaction 4 has not been fully considered. Furthermore, high-valent iron species, such as Fe(IV) are known to play a role in numerous enzymatic reactions (*97*). Damage to biomolecules and cells, including lung cells, has been attributed to Fe(IV) produced via reaction 4 (*35*, *98*, *99*), indicating that oxidants other than $OH^•$ can be responsible for oxidative damage.

Conclusions

The iron redox cycle is an important source of reactive oxidants in numerous systems, including sunlit natural waters, contaminant oxidation systems, and human lungs. Contrary to what is often assumed, the oxidation of Fe(II) by H_2O_2 does not always produce exclusively $OH^•$. The oxidant produced when H_2O_2 decomposes is sensitive to solution conditions, such as pH and iron coordination. As a result, the potential for oxidation of NOM, organic compounds, and biomolecules is frequently overestimated, especially under circumneutral conditions. One possible explanation for the lower-than-expected production of $OH^•$ is that outer-sphere reactions between Fe(II) and H_2O_2 yield $OH^•$, while inner-sphere reactions yield an alternate oxidant, such as the ferryl ion. While it is difficult to directly distinguish between $OH^•$ and Fe(IV) through direct observation, the discrepancies in H_2O_2 loss rates and target compound transformation yields implies that the oxidation of organic compounds cannot be predicted by assuming that the Fenton reaction produces stoichiometric yields of $OH^•$, especially under circumneutral pH conditions. Further research is needed to conclusively establish the identity of the oxidant produced by the Fenton reaction under circumneutral pH conditions, to identify factors influencing the yield, and to determine the reactivity of the oxidant with target contaminants and compounds present in natural waters, particularly NOM.

Acknowledgments

This work was supported by the U.S. National Institute of Environmental Health Services (Grant P42 ES004705).

References

1. Southworth, B. A.; Voelker, B. M. Hydroxyl radical production via the photo-Fenton reaction in the presence of fulvic acid. *Environ. Sci. Technol.* **2003**, *37*, 1130–1136.
2. Stumm, W.; Sulzberger, B. The cycling of iron in natural environments: Considerations based on laboratory studies of heterogeneous redox processes. *Geochim. Cosmochim. Acta* **1992**, *56*, 3233–3257.
3. Barbeau, K.; Rue, E.; Bruland, K.; Butler, A. Photochemical cycling of iron in the surface ocean mediated by microbial iron(III)-binding ligands. *Nature* **2001**, *413*, 409–413.
4. McKnight, D. M.; Kimball, B. A.; Bencala, K. E. Iron photoreduction and oxidation in an acidic moutain stream. *Science* **1988**, *240*, 637–640.
5. Zuo, Y.; Hoigné, J. Formation of hydrogen peroxide and depletion of oxalic acid in atmospheric water by photolysis of iron(III)-oxalato complexes. *Environ. Sci. Technol.* **1992**, *26*, 1014–1022.
6. Sedlak, D. L.; Hoigné, J. The role of copper and oxalate in the redox cycling of iron in atmospheric waters. *Atmos. Environ.* **1993**, *27A*, 2173–2185.
7. Sedlak, D. L.; Hoigné, J.; David, M. M.; Colvile, R. N.; Seyffer, E.; Acker, K.; Wiepercht, W.; Lind, J. A.; Fuzzi, S. The cloudwater chemistry of iron and copper at Great Dun Fell, UK. *Atmos. Environ.* **1997**, *31*, 2515–2526.
8. Deutsch, F.; Hoffmann, P.; Ortner, H. Field experimental investigations on the Fe(II)- and Fe(III)-content in cloudwater samples. *J. Atmos. Chem.* **2001**, *40*, 87–105.
9. Roy, E. G.; Wells, M. L.; King, D. W. Persistence of iron(II) in surface waters of the western subarctic Pacific. *Limnol. Oceanogr.* **2008**, *53*, 89–98.
10. Emmenegger, L.; King, D. W.; Sigg, L.; Sulzberger, B. Oxidation kinetics of Fe(II) in a eutrophic Swiss lake. *Environ. Sci. Technol.* **1998**, *32*, 2990–2996.
11. Aldrich, A. P.; van den Berg, C. M. G.; Thies, H.; Nickus, U. The redox speciation of iron in two lakes. *Mar. Freshwater Res.* **2001**, *52*, 885–890.
12. Behra, P.; Sigg, L. Evidence for redox cycling of iron in atmospheric water droplets. *Nature* **1990**, *344*, 419–421.
13. Voelker, B. M.; Sulzberger, B. Effects of fulvic acid on Fe(II) oxidation by hydrogen peroxide. *Environ. Sci. Technol.* **1996**, *30*, 1106–1114.

14. Pullin, M.; Cabaniss, S. The effects of pH, ionic strength, and iron-fulvic acid interactions on the kinetics of nonphotochemical iron transformations. II. The kinetics of thermal reduction. *Geochim. Cosmochim. Acta* **2003**, *67*, 4079–4089.
15. Cohn, C. A.; Mueller, S.; Wimmer, E.; Leifer, N.; Greenbaum, S.; Strongin, D. R.; Schoonen, M. A. A. Pyrite-induced hydroxyl radical formation and its effect on nucleic acids. *Geochem. T.* **2006**, *7*, 3.
16. Pignatello, J. J.; Oliveros, E.; MacKay, A. Advanced oxidation processes for organic contaminant destruction based on the Fenton reaction and related chemistry. *Crit. Rev. Environ. Sci. Technol.* **2006**, *36*, 1–84.
17. Pignatello, J. J.; Liu, D.; Huston, P. Evidence for an additional oxidant in the photoassisted Fenton reaction. *Environ. Sci. Technol.* **1999**, *33*, 1832–1839.
18. Kwan, W. P.; Voelker, B. M. Decomposition of hydrogen peroxide and organic compounds in the presence of dissolved iron and ferrihydrite. *Environ. Sci. Technol.* **2002**, *36*, 1467–1476.
19. Majestic, B. J.; Schauer, J. J.; Shafer, M. M. Application of synchrotron radiation for measurement of iron red-ox speciation in atmospherically processed aerosols. *Atmos. Chem. Phys.* **2007**, *7*, 2475–2487.
20. Cwiertny, D. M.; Young, M. A.; Grassian, V. H. Chemistry and photochemistry of mineral dust aerosol. *Annu. Rev. Phys. Chem.* **2008**, *59*, 27–51.
21. Valko, M.; Morris, H.; Cronin, M. T. D. Metals, toxicity and oxidative stress. *Curr. Med. Chem.* **2005**, *12*, 1161–1208.
22. Haber, F.; Weiss, J. Über die katalyse des hydroperoxydes. *Naturwissenschaften* **1932**, *20*, 948–950.
23. Weiss, J. Elektronenübergangsprozesse im Mechanismus von Oxydations- und Reduktionsreaktionen in Lösungen. *Naturwissenschaften* **1935**, *23*, 64–69.
24. Stumm, W.; Lee, G. F. Oxygenation of ferrous iron. *Ind. Eng. Chem.* **1961**, *53*, 143–146.
25. King, D. W.; Lounsbury, H. A.; Millero, F. J. Rates and mechanism of Fe(II) oxidation at nanomolar total iron concentrations. *Environ. Sci. Technol.* **1995**, *29*, 818–824.
26. Bielski, B. H. J.; Cabelli, D. E.; Arudi, R. L.; Ross, A. B. Reactivity of HO/O radicals in aqueous solution. *J. Phys. Chem. Ref. Data* **1985**, *14*, 1041–1100.
27. Rush, J. D.; Bielski, B. H. J. Pulse radiolytic studies of the reactions of HO_2/O_2^- with Fe(II)/Fe(III) ions. The reactivity of HO_2/O_2^- with ferric ions and its implication on the occurrence of the Haber-Weiss reaction. *J. Phys. Chem.* **1985**, *89*, 5062–5066.
28. Millero, F. J.; Sotolongo, S.; Izaguirre, M. The oxidation kinetics of Fe(II) in seawater. *Geochim. Cosmochim. Acta* **1987**, *51*, 793–801.
29. Millero, F. J.; Izaguirre, M. Effect of ionic strength and ionic interactions on the oxidation of Fe(II). *J. Solution Chem.* **1989**, *18*, 585–599.
30. Walling, C. Fenton's reagent revisited. *Acc. Chem. Res.* **1975**, *8*, 125–131.
31. Goldstein, S.; Meyerstein, D.; Czapski, G. The Fenton reagents. *Free Radical Biol. Med.* **1993**, *15*, 435–445.

32. Hug, S. J.; Leupin, O. Iron-catalyzed oxidation of arsenic(III) by oxygen and by hydrogen peroxide: pH-dependent formation of oxidants in the Fenton reaction. *Environ. Sci. Technol.* **2003**, *37*, 2734–2742.
33. Gallard, H.; de Laat, J.; Legube, B. Effect of pH on the oxidation rate of organic compounds by Fe-II/H_2O_2. Mechanisms and simulation. *New J. Chem.* **1998**, *22*, 263–268.
34. Keenan, C. R.; Sedlak, D. L. Factors affecting the yield of oxidants from the reaction of nanoparticulate zero-valent iron and oxygen. *Environ. Sci. Technol.* **2008**, *42*, 1262–1267.
35. Keenan, C. R.; Goth-Goldstein, R.; Lucas, D.; Sedlak, D. Oxidative stress induced by zero-valent iron nanoparticles and Fe(II) in human bronchial epithelial cells. *Environ. Sci. Technol.* **2009**, *43*, 4555–4560.
36. Vermilyea, A. W.; Voelker, B. M. Photo-Fenton reaction at near neutral pH. *Environ. Sci. Technol.* **2009**, *43*, 6927–6933.
37. Singer, P. C.; Stumm, W. Acidic mine drainage: The rate-determining step. *Science* **1970**, *167*, 1121–1123.
38. Sung, W.; Morgan, J. J. Kinetics and products of ferrous iron oxygenation in aqueous systems. *Environ. Sci. Technol.* **1980**, *14*, 561–568.
39. Luther, G. W. The Frontier-Molecular-Orbital Theory Approach in Geochemical Processes. In *Aquatic Chemical Kinetics*; Stumm, W., Ed.; Wiley-Interscience: New York, 1990; pp 173−198.
40. Wehrli, B., Redox Reactions of Metal Ions at Mineral Surfaces. In *Aquatic Chemical Kinetics*; Stumm, W., Ed.; Wiley-Interscience: New York, 1990; pp 311−336.
41. Rosso, K. M.; Morgan, J. J. Outer-sphere electron transfer kinetics of metal ion oxidation by molecular oxygen. *Geochim. Cosmochim. Acta* **2002**, *66*, 4223–4233.
42. Rosso, K. M.; Dupuis, M. Electron transfer in environmental systems: A frontier for theoretical chemistry. *Theor. Chem. Acc.* **2006**, *116*, 124–136.
43. González-Davila, M.; Santana-Casiano, J. M.; Millero, F. J. Oxidation of iron(II) nanomolar with H_2O_2 in seawater. *Geochim. Cosmochim. Acta* **2005**, *69*, 83–93.
44. King, D. W. Role of carbonate speciation on the oxidation rate of Fe(II) in aquatic systems. *Environ. Sci. Technol.* **1998**, *32*, 2997–3003.
45. Fenton, H. J. H. Oxidation of tartaric acid in the presence of iron. *J. Chem. Soc.* **1894**, *65*, 899–910.
46. Barb, W. G.; Baxendale, J. H.; George, P.; Hargrave, K. R. Reactions of ferrous and ferric ions with hydrogen peroxide. Part I.—The ferrous ion reaction. *Trans. Faraday Soc.* **1951**, *47*, 462–500.
47. Millero, F.; Sotolongo, S.; Stade, D.; Vega, C. Effect of ionic interactions on the oxidation of Fe(II) with H_2O_2 in aqueous solutions. *J. Solution Chem.* **1991**, *20*, 1079–1092.
48. Millero, F.; Sotolongo, S. The oxidation of Fe(II) with H_2O_2 in seawater. *Geochim. Cosmochim. Acta* **1989**, *53*, 1867–1873.
49. Walling, C. Intermediates in the reactions of Fenton type reagents. *Acc. Chem. Res.* **1998**, *31*, 155–157.

50. Duesterberg, C. K.; Waite, T. D. Kinetic modeling of the oxidation of p-hydroxybenzoic acid by Fenton's reagent: Implications of the role of quinones in the redox cycling of iron. *Environ. Sci. Technol.* **2007**, *41*, 4103–4110.
51. de Laat, J.; Gallard, H. Catalytic decomposition of hydrogen peroxide by Fe(III) in homogeneous aqueous solution: Mechanism and kinetic modeling. *Environ. Sci. Technol.* **1999**, *33*, 2726–2732.
52. Duesterberg, C. K.; Mylon, S. E.; Waite, T. D. pH effects on iron-catalyzed oxidation using Fenton's reagent. *Environ. Sci. Technol* **2008**, *42*, 8522–8527.
53. Kunai, A.; Hata, S.; Ito, S.; Sasaki, K. The role of oxygen in the hydroxylation reaction of benzene with Fenton's reagent. ^{18}O tracer study. *J. Am. Chem. Soc.* **1986**, *108*, 601–6016.
54. Castrantas, H. M.; Gibilisco, R. D. UV destruction of phenolic compounds under alkaline conditions. *ACS Symp. Ser.* **1990**, *422*, 77–99.
55. Kochany, J.; Bolton, J. R. Mechanism of photodegradation of aqueous organic pollutants. 2. Measurement of the primary rate constants for reaction of OH radicals with benzene and some halobenzenes using an EPR spin-trapping method following the photolysis of H_2O_2. *Environ. Sci. Technol.* **1992**, *26*, 262–265.
56. Shen, Y.; Lin, C. The effect of pH on the decomposition of hydrophenols in aqueous solutions by ultraviolet direct photolysis and the ultraviolet-hydrogen peroxide process. *Water Environ. Res.* **2003**, *75*, 54–60.
57. Chen, R.; Pignatello, J. Role of quinone intermediates as electron shuttles in Fenton and photoassisted Fenton oxidations of aromatic compounds. *Environ. Sci. Technol.* **1997**, *31*, 2399–2406.
58. Ilan, Y.; Rabani, J. On some fundamental reactions in radiation chemistry: Nanosecond pulse radiolysis. *Int. J. Radiat. Phys. Chem.* **1976**, *8*, 609–611.
59. Goldstein, S.; Meyerstein, D. Comments on the mechanism of the "Fenton like" reaction. *Accounts Chem. Res.* **1999**, *32*, 547–550.
60. Bossmann, S. H.; Oliveros, E.; Gob, S.; Siegwart, S.; Dahlen, E. P.; Payawan, L.; Straub, M.; Worner, M.; Braun, A. M. New evidence against hydroxyl radicals as reactive intermediates in the thermal and photochemically enhanced fenton reactions. *J. Phys. Chem. A* **1998**, *102*, 5542–5550.
61. Buda, F.; Ensing, B.; Gribnau, M. C. M.; Baerends, E. J. DFT study of the active intermediate in the Fenton reaction. *Chem.-Eur. J.* **2001**, *7*, 2775–2783.
62. Jacobsen, F.; Holcman, J.; Sehested, K. Reactions of the ferryl ion with some compounds found in cloud water. *Int. J. Chem. Kinet.* **1998**, *30*, 215–221.
63. Buxton, G. V.; Greenstock, C. L.; Helman, W. P.; Ross, A. B. Critical review of rate constants for reactions of hydrated electrons, hydrogen atoms and hydroxyl radicals in aqueous solution. *J. Phys. Chem. Ref. Data* **1988**, *17*, 513–886.
64. Koppenol, W. H.; Liebman, J. F. The oxidizing nature of the hydroxyl radical. A comparison with the ferryl ion (FeO^{2+}). *J. Phys. Chem.* **1984**, *88*, 99–101.

65. Mártire, D. O.; Caregnato, P.; Furlong, J.; Allegretti, P.; Gonzalez, M. C. Kinetic study of the reactions of oxoiron(IV) with aromatic substrates in aqueous solutions. *Int. J. Chem. Kinet.* **2002**, *34*, 488–493.
66. Katsoyiannis, I. A.; Ruettimann, T.; Hug, S. J. pH dependence of Fenton reagent generation and As(III) oxidation and removal by corrosion of zero valent iron in aerated water. *Environ. Sci. Technol.* **2008**, *42*, 7424–7430.
67. Sun, Y.; Pignatello, J. J. Chemical treatment of pesticide wastes. Evaluation of iron(III) chelates for catalytic hydrogen peroxide oxidation of 2,4-D at circumneutral pH. *J. Agric. Food Chem.* **1992**, *40*, 322–327.
68. Tachiev, G.; Roth, J. A.; Bowers, A. R. Kinetics of hydrogen peroxide decomposition with complexed and "free" iron catalysts. *Int. J. Chem. Kinet.* **2000**, *32*, 24–35.
69. Graf, E.; Mahoney, J. R.; Bryant, R. G.; Eaton, J. W. Iron-catalyzed hydroxyl radical formation. Stringent requirement for free iron coordination site. *J. Biol. Chem.* **1984**, *259*, 3620–3624.
70. Zang, V.; van Eldik, R. Kinetics and mechanisms of the autoxidation of iron(II) induced through chelation by ethylenediaminetetraacetate and related ligands. *Inorg. Chem.* **1990**, *29*, 1705–1711.
71. Kurimura, Y.; Ochiai, R.; Matsuura, N. Oxygen oxidation of ferrous ions induced by chelation. *Bull. Chem. Soc. Jpn.* **1968**, *41*, 2234–2239.
72. Welch, K. D.; Davis, T. Z.; Aust, S. D. Iron autoxidation and free radical generation: Effects of buffers, ligands, and chelators. *Arch. Biochem. Biophys.* **2002**, *397*, 360–369.
73. Keenan, C. R.; Sedlak, D. L. Ligand-enhanced reactive oxidant generation by nanoparticulate zero-valent iron and oxygen. *Environ. Sci. Technol.* **2008**, *42*, 6936–6941.
74. Lee, C.; Keenan, C. R.; Sedlak, D. L. Polyoxometalate-enhanced oxidation of organic compounds by nanoparticulate zero-valent iron and ferrous ion in the presence of oxygen. *Environ. Sci. Technol.* **2008**, *42*, 4921–4926.
75. Zepp, R. G.; Faust, B. C.; Hoigné, J. Hydroxyl radical formation in aqueoue reactions (pH 3-8) of iron(II) with hydrogen peroxide: The photo-Fenton reaction. *Environ. Sci. Technol.* **1992**, *26*, 313–319.
76. Rush, J. D.; Koppenol, W. H. Oxidizing intermediates in the reaction of ferrous EDTA with hydrogen peroxide. Reactions with oraganic molecules and ferrocytochrome-c. *J. Biol. Chem.* **1986**, *261*, 6730–6733.
77. Shiga, T.; Kishimoto, T.; Tomita, E. Aromatic hydroxylation catalyzed by Fenton's reagent. Electron paramagnetic resonance study. II. Benzoic acids. *J. Phys. Chem.* **1973**, *77*, 330–336.
78. Yamazaki, I.; Piette, L. H. EPR spin-trapping study on the oxidizing species formed in the reaction of the ferrous ion with hydrogen peroxide. *J. Am. Chem. Soc.* **1991**, *113*, 7588–7593.
79. Miles, C. J.; Brezonik, P. L. Oxygen consumption in humic-colored waters by a photochemical ferrous-ferric catalytic cycle. *Environ. Sci. Technol.* **1981**, *15*, 1089–1095.
80. Rose, A. L.; Waite, T. D. Effect of dissolved natural organic matter on the kinetics of ferrous iron oxygenation in seawater. *Environ. Sci. Technol.* **2003**, *37*, 4877–4886.

81. Miller, C. J.; Rose, A. L.; Waite, T. D. Impact of natural organic matter on H_2O_2-mediated oxidation of Fe(II) in a simulated freshwater system. *Geochim. Cosmochim. Acta* **2009**, *73*, 2758–2768.
82. Paciolla, M. D.; Davies, G.; Jansen, S. A. Generation of hydroxyl radicals from metal-loaded humic acids. *Environ. Sci. Technol.* **1999**, *33*, 1814–1818.
83. Tamura, H.; Goto, K.; Nagayama, M. Effect of ferric hydroxide on oxygenation of ferrous-ions in neutral solutions. *Corros. Sci.* **1976**, *16*, 197–207.
84. Kwan, W. P.; Voelker, B. M. Rates of hydoxyl radical generation and organic compound oxidation in mineral-catalyzed Fenton-like systems. *Environ. Sci. Technol.* **2003**, *37*, 1150–1158.
85. Lin, S.; Gurol, M. Catalytic decomposition of hydrogen peroxide on iron oxide: Kinetics, mechanism, and implications. *Environ. Sci. Technol.* **1998**, *32*, 1417–1423.
86. Valentine, R.; Wang, H. Iron oxide surface catalyzed oxidation of quinoline by hydrogen peroxide. *J. Environ. Eng.-ASCE* **1998**, *124*, 31–38.
87. Petigara, B.; Blough, N.; Mignerey, A. Mechanisms of hydrogen peroxide decomposition in soils. *Environ. Sci. Technol.* **2002**, *36*, 639–645.
88. Pham, A.; Lee, C.; Doyle, F.; Sedlak, D. A silica-supported iron oxide catalyst capable of activating hydrogen peroxide at neutral pH values. *Environ. Sci. Technol.* **2009**, *43*, 8930–8935.
89. Voegelin, A.; Hug, S. Catalyzed oxidation of arsenic(III) by hydrogen peroxide on the surface of ferrihydrite: An in situ ATR-FTIR study. *Environ. Sci. Technol.* **2003**, *37*, 972–978.
90. Brezonik, P.; Fulkerson-Brekken, J. Nitrate-induced photolysis in natural waters: Controls on concentrations of hydroxyl radical photo-intermediates by natural scavenging agents. *Environ. Sci. Technol.* **1998**, *32*, 3004–3010.
91. Haag, W.; Hoigne, J. Photosensitized oxidation in natural waters via OH radicals. *Chemosphere* **1985**, *14*, 1659–1671.
92. Miller, W. L.; Zepp, R. G. Photochemical production of dissolved inorganic carbon from terrestrial organic matter: Significance to the oceanic organic carbon cycle. *Geophys. Res. Lett.* **1995**, *22*, 417–420.
93. Vaughan, P.; Blough, N. Photochemical formation of hydroxyl radical by constituents of natural waters. *Environ. Sci. Technol.* **1998**, *32*, 2947–2953.
94. Pullin, M.; Bertilsson, S.; Goldstone, J.; Voelker, B. Effects of sunlight and hydroxyl radical on dissolved organic matter: Bacterial growth efficiency and production of carboxylic acids and other substrates. *Limnol. Oceanogr.* **2004**, *49*, 2011–2022.
95. Yeh, C. K.-J.; Chen, W.-S.; Chen, W.-Y. Production of hydroxyl radicals from the decomposition of hydrogen peroxide catalyzed by various iron oxides at pH 7. *Pract. Period. Hazard., Toxic, Radioact. Waste Manage.* **2004**, *8*, 161–165.
96. Nel, A.; Xia, T.; Madler, L.; Li, N. Toxic potential of materials at the nanolevel. *Science* **2006**, *311*, 622–627.
97. Groves, J. T. High-valent iron in chemical and biological oxidations. *J. Inorg. Biochem.* **2006**, *100*, 434–447.

98. Imlay, J. A.; Chin, S. M.; Linn, S. Toxic DNA damage by hydrogen peroxide through the Fenton reaction in vivo and in vitro. *Science* **1988**, *240*, 640–642.
99. Nieto-Juarez, J. I.; Pierzchla, K.; Sienkiewicz, A.; Kohn, T. Inactivation of MS2 coliphage in Fenton and Fenton-like systems: role of transition metals, hydrogen peroxide and sunlight. *Environ. Sci. Technol.* **2010**, *44*, 3351–3356.

Chapter 10

TiO_2 Photocatalysis for the Redox Conversion of Aquatic Pollutants

Jaesang Lee,[1] Jungwon Kim,[2] and Wonyong Choi*,[2]

[1]Water Environment Center, Environment Division, Korea Institute of Science and Technology (KIST), Hawolgok-dong, Seongbuk-gu, Seoul, Korea
[2]School of Environmental Science and Engineering, Pohang University of Science and Technology (POSTECH), Pohang 790-784, Korea
***wchoi@postech.edu**

Photo-induced redox chemical reactions occurring on irradiated semiconductor surfaces have been utilized for the purification of water contaminated with various inorganic and organic chemicals. Here, we focus on TiO_2 as the most popular photocatalyst and briefly describe its characteristics and applications mainly in relation with the photochemical redox conversion of aquatic pollutants. The photoexcitation of TiO_2 induces electron-hole pair formation and subsequent charge separation/migration/transfer leads to the production of highly reactive oxygen species (ROS) such as OH radical and superoxide on the surface of TiO_2. Aquatic organic pollutants subsequently react with ROS, holes, or electrons, and they undergo a series of redox chemical reactions, eventually leading to mineralization. The photo-induced ROS generation on TiO_2 is exploitable for bacterial/viral inactivation as well, while TiO_2 particles at the nano- and microscale possibly induce adverse biological effects in the absence of light. Photo-induced redox reactions on TiO_2 can also transform a variety of inorganic pollutants such as oxyanions (arsenite, chromate, bromate, etc.), ammonia, and metal ions. On the other hand, the photocatalytic degradation mechanism can be actively controlled by modifying the surface of TiO_2 to change the products. For example, the photocatalytic degradation of phenolic compounds can be accompanied by the simultaneous production of hydrogen when the surface of TiO_2 is modified with both platinum and fluoride. Finally, the photocatalytic activity of TiO_2 is highly dependent on the kind of substrates and the activity assessed with a specific test substrate is difficult to generalize. Therefore, the photocatalytic activities of TiO_2 should be assessed using multiple substrates to obtain balanced information.

Introduction

The photocatalytic activity of TiO_2 has been comprehensively explored and utilized in a variety of science and engineering sectors because it offers feasible ways to harness (solar) photon energy to drive various chemical redox reactions (*1–5*). TiO_2-mediated photocatalytic redox reactions have versatile applications, which include the degradation of pollutants in water and air (*6–8*), solar fuel production (*9–11*), lithography (*12*, *13*), metal ion recovery (*14*), corrosion protection of metals (*15*, *16*), organic photosynthesis (*17–19*), and inactivation of pathogenic microorganisms (*20–22*). In particular, much attention has been paid to the use of TiO_2 as an environmental photocatalyst that purifies contaminated water and air. The photo-induced production of reactive oxygen species (ROS) on TiO_2 enables the effective decontamination of hazardous substrates, which has shown great potential as an advanced oxidation process (AOP). The practical merits such as high oxidation power of holes, photochemical and chemical stability, abundance and easy availability, and low material cost have established TiO_2 as the most popular environmental photocatalyst among many semiconducting materials. While nanoparticles such as CdS and ZnO cause adverse biological effects, chemical inertness, photochemical stability, and non-toxicity of TiO_2 make it environmentally benign and practicable; otherwise unwanted release of TiO_2 would induce hazardous impact on the environment and human health. The use of TiO_2 is not limited to photocatalysis but is also widespread in many industrial sectors dealing with paints, paper, cosmetics, pharmaceuticals, optics, catalysis, and even foods (*23*, *24*).

The photo-induced excitation of electron-hole pairs and the subsequent hole migration to TiO_2 surface lead to the production of surface-trapped valence band (VB) holes or surface-bound OH radicals. Alternatively, conduction band (CB) electrons can migrate to the surface and react with dioxygen to produce superoxide (or hydroperoxide) radicals. Such a photochemical production of ROS with strong oxidizing power is the basis of remedial action of TiO_2 photocatalysis (*1*, *25*). An outstanding merit of TiO_2 photocatalysis is that the generation of ROS is enabled in ambient conditions and in any medium without the need of chemical oxidants except O_2 as long as photons are available. As a result, the photocatalytic remediation processes exhibit multi-phasic characteristics so that they can be applied to dry environments (air and solid) as well as aquatic environment (*26–30*). TiO_2 photocatalyst has been also successfully applied as self-cleaning and superhydrophilic material that continuously oxidizes and removes the organic contaminants deposited on its surface (*31*, *32*).

The superior properties of TiO_2 photocatalysis motivated the environmental research community to investigate it as a practical remediation technology. The number of research papers published on the subject of "TiO_2 photocatalysis" is continuously increasing every year although this field of study has been established with almost 30 years of research effort (*see* Figure 1). The photocatalytic remediation technology is continuously attracting researchers' interest from both academia and industry and still demands deeper understanding, improvements and breakthroughs in both science and engineering aspects. Research on TiO_2 photocatalysis is related to diverse subjects that include: kinetics and mechanisms for photocatalytic reactions; synthesis and characterization; modification and fabrication of photoactive composites; surface science and photoelectrochemistry; transient spectroscopic studies of charge recombination/transfer dynamics; photoreactor design and optimization of operation parameters. This chapter presents the use of TiO_2 photocatalysis in the redox conversion of aquatic pollutants.

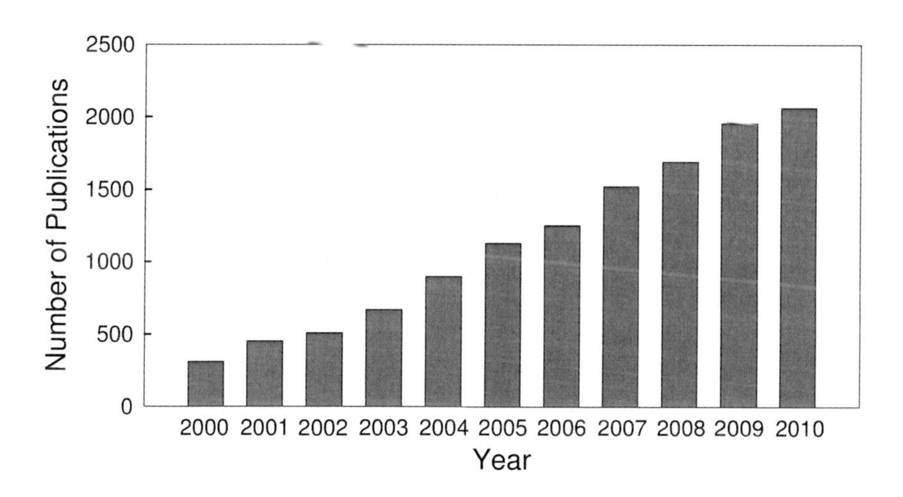

Figure 1. Annual number of papers published in the subject area of "TiO_2 photocatalysis". The literature search was carried out at the Scopus website (www.scopus.com) using the key word "TiO_2 photocatal".*

General Principle of TiO_2 Photocatalysis and the Redox Characteristics

TiO_2 can absorb a photon with energy that exceeds its bandgap energy (3.2 eV for anatase) to produce a charge pair consisting of a CB electron and a VB hole. The resulting charge separation recombines with a release of heat (or luminescence) or migrates to the surface to induce the interfacial charge transfer which triggers diverse redox reactions (*see* Figure 2). Such a photo-induced charge transfer can occur on any semiconductor surface as long as the charge transfer is energetically allowed. Figure 3 compares the bandgap and band edge position of various semiconductors along with the important reduction potentials in air-equilibrated water where the main electron acceptor and donor is O_2 and H_2O (or OH^-), respectively. It should be mentioned that the standard reduction potentials shown in Figure 3 are for the homogeneous solution. The exact reduction potentials in the presence of the semiconductor surface are unknown and can be different from the homogeneous counterpart. Therefore, the quantitative comparisons between the band edge potentials and the aquatic reduction potentials involving O_2 and H_2O should be made with caution. Nevertheless, the energy level diagram of Figure 3 should serve as a guideline for understanding the photo-induced charge transfer at the semiconductor interface. Since the spontaneous electron transfer takes place downward (from negative to positive direction) in the energy level diagram, the more negative CB position and the more positive VB position have the higher driving force for photo-induced (*following the bandgap excitation*) interfacial charge transfer.

In terms of the energetics, TiO_2 has CB/VB positions that enable both the electron transfer to O_2 and the hole transfer to H_2O, which makes TiO_2 suitable as an aquatic photocatalyst. Other metal-oxide semiconductors like ZnO, SnO_2, WO_3, and Fe_2O_3 can be compared with TiO_2 but they are less favorable as an aquatic photocatalyst. ZnO is very similar to TiO_2 in its bandgap and band position but is not stable enough in water and may undergo dissolution under acidic and irradiated conditions. The bandgap of SnO_2 is too wide to be activated by sunlight and its lower CB position does not allow the use of O_2 as an electron acceptor. WO_3 and Fe_2O_3 have smaller bandgaps which can absorb more solar light but their CB positions that are more positive than the reduction potential of O_2 make them unsuitable as an aquatic photocatalyst. In general, wide bandgap semiconductors like TiO_2 and ZnO have higher driving force for the photo-induced redox reactions but require the presence of UV instead of visible light. On the contrary, smaller bandgap semiconductors like CdS and Fe_2O_3 absorb more solar light but their photo-induced redox power is limited. The narrow-bandgap semiconductors are often quite unstable and suffer from photocorrosion in aquatic environments.

The successful performance of aquatic photocatalysts should depend on the ability to generate highly reactive radical species in water under light-irradiated conditions like other AOPs. The greatest merit of semiconductor photocatalysis as an AOP is that it produces the oxidizing radicals in the ambient condition without any extra chemicals. While other photochemical AOPs like H_2O/VUV (vacuum UV) (*33*), H_2O_2/UV (*34*, *35*), O_3/UV (*36–38*) require expensive photons (VUV or UVC) and/or the chemical oxidants (e.g., H_2O_2, O_3, persulfate) as precursors of ROS, TiO_2 photocatalysis can activate ambient H_2O and O_2 indirectly to generate ROS under sunlight or artificial UV light.

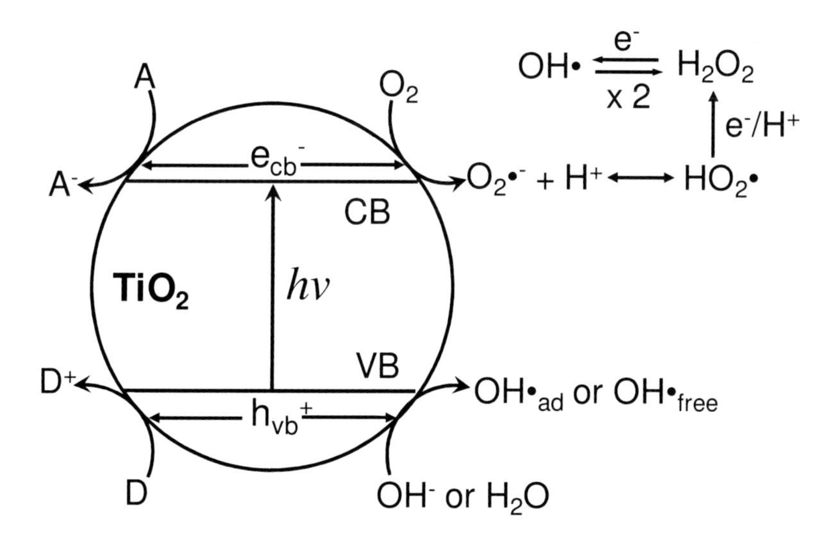

Figure 2. Photo-induced redox reactions and the generation of ROS occurring on the irradiated surface of TiO_2 photocatalyst.

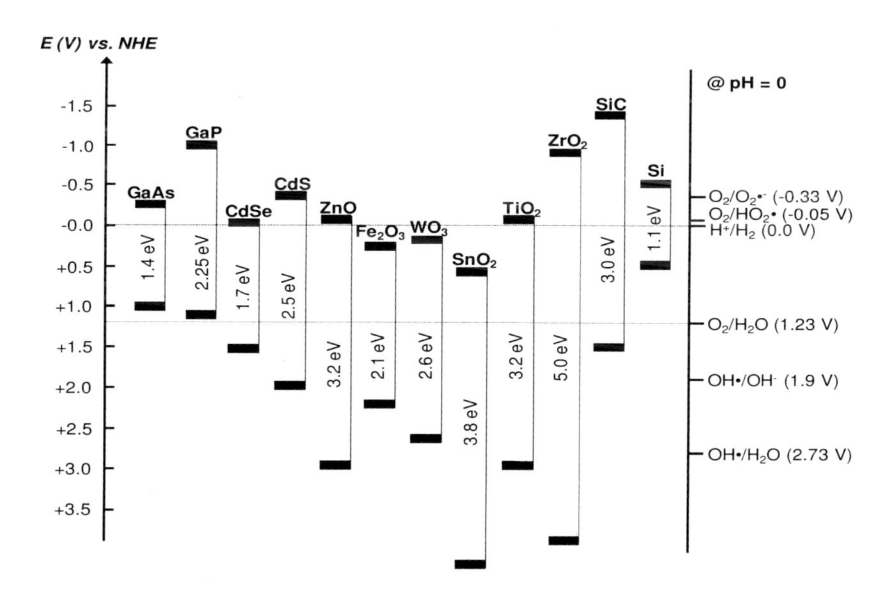

Figure 3. Energy-level diagram showing the bandgaps and CB/VB edge positions of various semiconductors and the selected aquatic redox potentials (at pH 0).

Figure 2 illustrates various photo-induced reaction pathways occurring on the TiO_2 surface. The VB holes can react with the surface-bound hydroxyl groups or adsorbed water molecules to generate OH radicals (reaction 1), which initiate the fast and non-selective oxidative degradation of organic pollutants. On the other hand, the CB edge positions are also critical in the generation of OH radicals because the CB electrons should be efficiently scavenged by O_2 (reaction 2) to retard the charge pair recombination. The further reduction of the superoxide by CB electron can provide an alternative pathway that leads to the generation of OH radical as Figure 2 shows.

$$h_{vb}^{+} + {>}Ti\text{-}OH\ (\textit{or}\ H_2O_{surf}) \rightarrow {>}Ti\text{-}(\bullet OH)_{surf} \quad (1)$$

$$e_{cb}^{-} + O_2 \rightarrow O_2^{-}\bullet\ (\text{or}\ HO_2\bullet) \quad (2)$$

The lower CB edge of TiO_2 at pH 7 is -0.5 V_{NHE} that does provide a sufficient potential to reduce O_2 [$E^0\ (O_2/O_2\bullet^-) = -0.33\ V_{NHE}$ and $E^0(O_2/HO_2\bullet) = -0.05\ V_{NHE}$]. The CB potential is pH-dependent and shifts to the negative direction by 59 mV with increasing one unit of pH (Nernstian behavior) (*39*). Therefore, the driving force of the CB electron transfer to the $O_2/O_2^{-}\bullet$ couple (with pH-independent potential) increases with pH while that to the $O_2/HO_2\bullet$ couple (with pH-dependent potential) is independent of pH. The resulting ROS (mainly hydroxyl and superoxide radicals) subsequently initiates the oxidation reactions of aquatic pollutants.

Figure 4 compares the UV/visible absorption spectra of various precursors of ROS (O_2, H_2O, H_2O_2, and O_3) with the absorption profile of TiO_2 and the terrestrial solar radiation spectrum. The direct photolysis of H_2O and O_2 that leads to the generation of ROS (e.g., OH• and O•) requires VUV photons ($\lambda < 200$ nm) that are expensive and not easily available (*33*). The addition of H_2O_2 or O_3 as an external precursor of ROS requires less energetic photons (UVC region: 200-280 nm) but even UVC light is completely absent in the solar spectrum. The UVC photolysis of H_2O_2 yields OH radicals with a quantum yield of 0.5 (*33–35*) and that of O_3 in aqueous media results in *in-situ* production of H_2O_2 which eventually converts to OH radicals (*33*, *37*, *38*, *40*). The homogeneous photochemical water treatment processes work in the short UV wavelength regions and are confronted with the high cost for maintenance and artificial light source development. On the other hand, TiO_2 photocatalysis is based on the indirect sensitization of H_2O and O_2 through bandgap excitation, which needs lower energy photons than the UVC-based processes. The photocatalytic activity of TiO_2 that generates OH radicals under sunlight (*see* the spectral overlap in Figure 4) makes it a cost-effective AOP. The overall efficiency of photocatalysis can be greatly enhanced if TiO_2 can absorb and utilize visible light which accounts for about a half of solar energy. The modification of TiO_2 by impurity dopants such as transition metal ions, nitrogen, and carbon has been established as a popular method for the development of visible light active photocatalysts (*41–43*). Other methods of visible light activation of TiO_2 include dye sensitization (*44–46*), surface complexation (*47*, *48*), and coupling with narrow bandgap semiconductors (*49–51*).

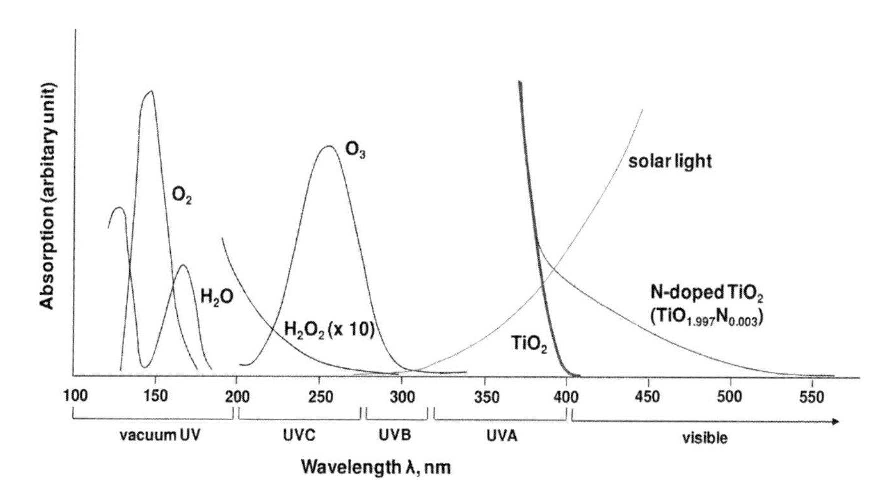

Figure 4. UV-visible absorption spectra of common precursors of ROS compared with the absorption of TiO_2 (and N-doped TiO_2 (e.g., $TiO_{1.997}N_{0.003}$)) photocatalyst and the terrestrial solar spectrum.

Photocatalytic Conversion of Organic Contaminants

The production of OH radical on the UV-illuminated TiO_2 surface initiates the oxidative degradation of organic compounds and the subsequent oxidation of intermediates leads to their mineralization to CO_2, H_2O, and inorganic ions (e.g., halide, sulfate, and nitrate). Being one of the most powerful oxidants, OH radicals react non-selectively with most organic substances. The reaction mode of OH radicals can be largely classified into three categories: (i) H-atom abstraction from a C-H bond, (ii) addition to a double bond, and (iii) addition to an aromatic ring, all of which lead to the generation of carbon-centered radicals that subsequently react with O_2 at a diffusion-controlled rate (reaction 3). The resulting peroxyl radicals further react and degrade into CO_2.

$$R_3C\bullet + O_2 \rightarrow R_3COO\bullet \tag{3}$$

Based on this photo-induced radical chemistry, TiO_2 photocatalytic oxidation can be applied to a variety of organic pollutants and a few selected examples are described below.

Tetramethylammonium hydroxide (TMA), used as a silicon etchant in semiconductor manufacturing process, is known to be very recalcitrant under the conventional water treatment processes (e.g., bioremediation and ozonation). TiO_2 photocatalysis successfully degraded TMA (*52*) along with the formation of $(CH_3)_3NH^+$, $(CH_3)_2NH_2^+$, $CH_3NH_3^+$, NH_4^+, NO_2^-, and NO_3^- as intermediates and products (Figure 5a). The total N-balance was satisfactorily met throughout the degradation process, which indicates that there were no major missing products. Figure 6 illustrates the proposed reaction pathways of TMA degradation. The degradation reaction is initiated via an H-atom abstraction from the methyl group by OH radical, and then undergoes stepwise demethylation as observed. The initial reaction of an OH radical with TMA molecule is rather slow but that of the demethylated products is faster (e.g., $k((CH_3)_4N^+ + \bullet OH) = 6.6 \times 10^6\ M^{-1}\ s^{-1}$ (*53*) *versus* $k((CH_3)_3NH^+ + \bullet OH) = 4 \times 10^8\ M^{-1}\ s^{-1}$ (*54*)). The photocatalytic degradation of aquatic pollutants proceeds to mineralization usually through generating several intermediates as this case. The identification and quantification of intermediates and products is the essential part in the study of aquatic pollutant degradation.

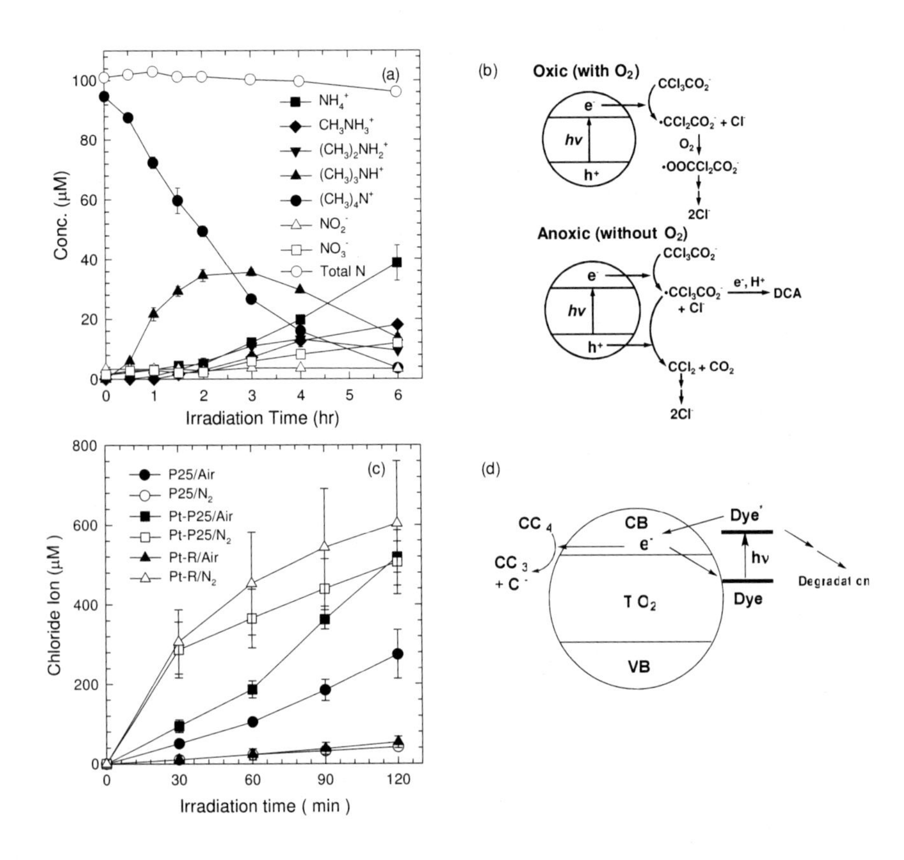

Figure 5. (a) Photocatalytic degradation of $(CH_3)_4N^+$ and the accompanying production of intermediates and products (52), (b) Two mechanistic paths of photocatalytic degradation of TCA on TiO_2 with or without dioxygen (58), (c) Evolution of chloride ions during the photocatalytic degradation of TCA on different photocatalysts including bare TiO_2 (P25), Pt/TiO_2 (P25), and Pt/TiO_2 (rutile) (58), (d) Schematic illustration of the dye-sensitized process occurring on TiO_2 under visible light (44).

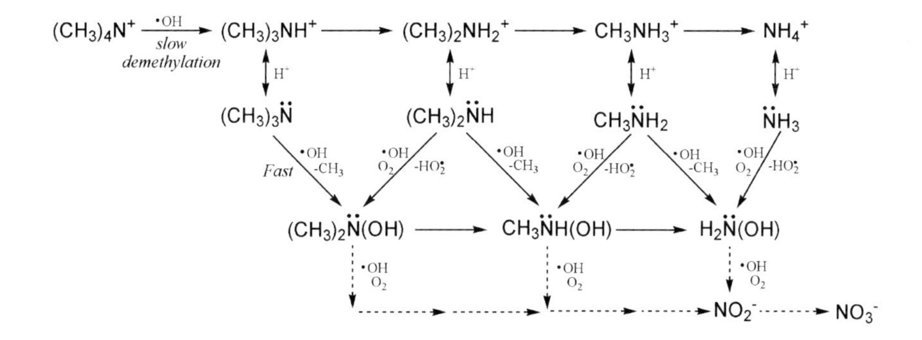

Figure 6. Schematic mechanism for the photocatalytic degradation of TMA through stepwise demethylation. (Adapted from ref. (52))

While OH radical generated on the photo-excited TiO_2 plays a dominant role in the degradation of organic pollutants (*25*, *52*, *55*, *56*), a series of CB electron and VB hole transfers are also essential in achieving the overall degradation. How the serial electron transfers occur determines the photocatalytic degradation mechanism. For example, the degradation of trichloroacetate (TCA), which has insignificant reactivity with OH radical (k(TCA + •OH) < 6.6×10^6 M^{-1} s^{-1}) (*55*)), is initiated by CB electron transfer and its subsequent reaction mechanism depends on reaction conditions (*57*). Figure 5b schematically compares the TCA degradation mechanism in oxic and anoxic conditions (*58*): Degradation pathways differ depending on the availability of oxygen. The anoxic mechanism proceeds through the formation of dichlorocarbene (CCl_2) intermediate, which requires the sequential transfer of a CB electron and a VB hole. Understanding the mechanistic pathways is critical in controlling the efficiency and selectivity of photocatalytic degradation. For instance, the stabilization of the dichlorocarbene intermediate by Pt nanoparticles deposited on TiO_2 kinetically enhances the anoxic degradation pathway of TCA, and causes oxygen to inhibit the degradation of TCA on Pt/TiO_2. On the contrary, the presence of oxygen accelerates TCA degradation on bare TiO_2 (Figure 5c). The photocatalytic degradation mechanism can sensitively depend on the surface properties of TiO_2 and can be actively controlled by modifying the surface properties.

Most TiO_2 photocatalytic reactions are carried out under UV irradiation because the bandgap is in the UV excitation region. Although TiO_2 is not activated by visible light, the degradation of dyes on TiO_2 under visible light is enabled through a dye-sensitization process in which dye is excited by absorbing visible light photons and immediately injects an electron into TiO_2 CB and initiates the degradation of dye (reactions 4-6).

$$Dye^*\text{-}TiO_2 \rightarrow Dye^{+\bullet}\text{-}TiO_2 + e_{cb}^- \quad (4)$$

$$e_{cb}^- + O_2 \rightarrow O_2^{-\bullet} \quad (5)$$

$$Dye^{+\bullet} + O_2/\,O_2^{-\bullet} \rightarrow \rightarrow \text{degradation} \quad (6)$$

This visible light-induced degradation of dyes on TiO_2 surface has been intensively investigated (*59–61*) and the treatment of dye wastewaters is one of the most frequently studied topics in AOPs (*62*). Such a dye-sensitized process can be also applied to redox conversion of aquatic pollutants (e.g., CCl_4) on TiO_2 under visible light if the reduction potentials are more positive than the TiO_2 CB edge (*44, 45, 63, 64*). Figure 5d illustrates the process of dye-sensitization occurring on dye/TiO_2 under visible light.

Photocatalytic Conversion of Inorganic Contaminants

TiO_2 photocatalysis can be also successfully applied to the conversion of inorganic pollutants such as nitrates (*65, 66*), ammonia (*67, 68*), cyanide (*69*), halides (*70–72*), chromate (*73–76*), arsenite (*77–82*) and heavy metal ions (*83, 84*). Although the photocatalytic degradation of organic pollutants largely depend on the oxidative power of VB holes and OH radicals, the conversion of the inorganic pollutants critically involves CB electrons as well. The oxidation states of inorganic elements such as nitrogen, halogen, and transition metals widely vary depending on their chemical forms. The reductive conversion of metal ions to lower oxidation state or zero-valent metallic state is one of the most common inorganic conversion reactions driven by photocatalysis. The followings are some examples.

$$Ag^+ + e_{cb}^- \rightarrow Ag^0 \quad (7)$$

$$3e_{cb}^- + 7H^+ + HCrO_4^- \rightarrow Cr^{3+} + 4H_2O \quad (8)$$

$$Fe^{3+} + e_{cb}^- \rightarrow Fe^{2+} \quad (9)$$

Such reductive conversion of metal ions has been commonly employed as a photocatalytic method of noble metal deposition on the surface of TiO_2 (*44, 85, 86*).

The dye-sensitization process (Figure 5d) can be also applied to the reductive conversion of metal ions. Figure 7a shows that the reductive deposition of Ag^+ to Ag^0 on TiO_2 could be achieved in the presence of dye (rhodamine B) under the visible light irradiation (i.e., reaction 4 followed by reaction 7). Such process can also achieve the simultaneous conversion of dyes and toxic heavy metal ions. Figure 7b shows that a ternary aquatic system that includes TiO_2, dye (acid orange 7), and chromate (Cr(VI)) synergistically enhanced the removal rate of the hexavalent chromium (reaction 8) (*87*). Although TiO_2 photocatalytic reduction of Cr(VI) was negligible under visible light illumination, the presence of dyes highly accelerated the reductive conversion.

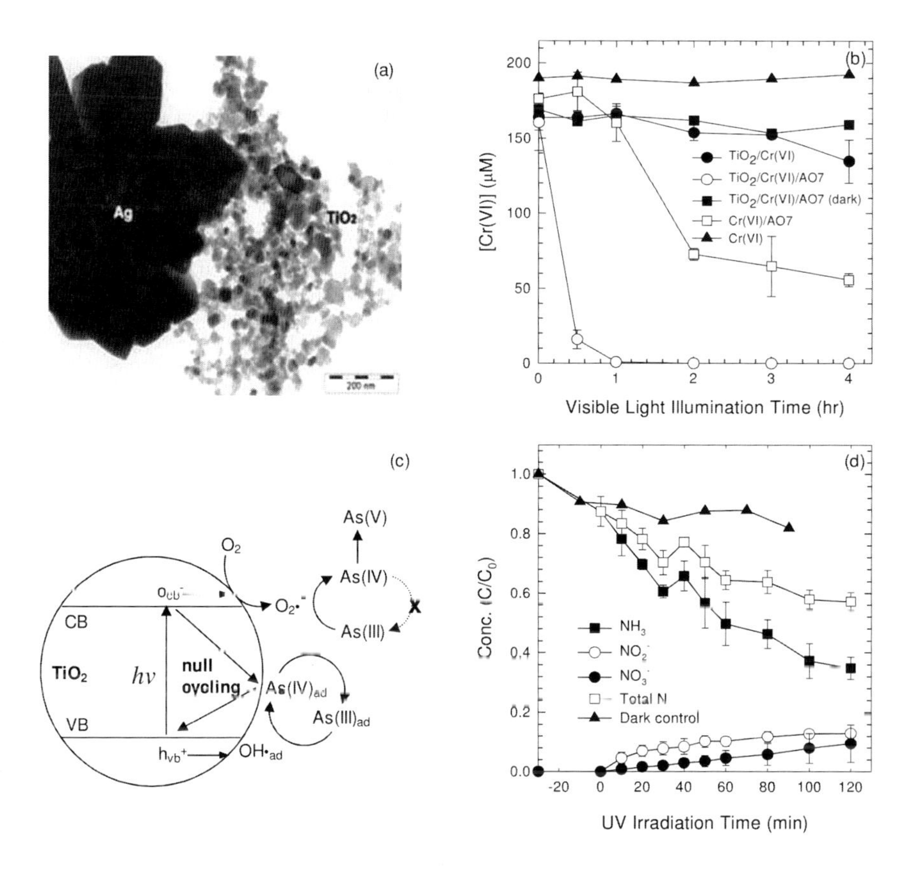

Figure 7. (a) Photo-reductive conversion of Ag^+ ions to silver particles in the dye/TiO_2 system (87), (b) Synergistic reductive conversion of Cr(VI) on TiO_2 in the presence of dye (AO7) under visible light irradiation (87), (c) Proposed photocatalytic mechanism of As(III) oxidation on TiO_2, which is induced mainly by superoxides (82), (d) Photocatalytic oxidation of NH_3 in the UV-illuminated aqueous suspension of Pt/TiO_2 (67).

The oxidative conversion of arsenite (As(III)) to arsenate (As(V)) as a pretreatment step promotes the efficacy in remediation of arsenic-contaminated water because As(V) is much less toxic and more easily adsorbed or coagulated. Being a highly oxidative photocatalyst, TiO_2 can rapidly oxidize As(III) to As(V) in the UV-irradiated aqueous suspension (*77, 78*). Interestingly, the photocatalytic oxidation of As(III) is not inhibited at all in the presence of excess amount of *tert*-butanol (scavenger of OH radicals). To explain why OH radicals do not seem to play the role of the primary oxidant in this specific photocatalytic oxidation, it has been hypothesized that the adsorbed As(III) on TiO_2 serves as an external charge-recombination center where the reaction of As(III) with an OH radical (or hole) is immediately followed by a CB electron transfer to make a null cycle (*see* Figure 7c). Transient spectroscopic and photoelectrochemical measurements in our recent study showed that the presence of As(III) accelerated the charge recombination in TiO_2 (*79*), supporting the proposed mechanism. A series of studies (*80–82*) have suggested that the photocatalytic oxidation of As(III) is largely mediated by superoxide/hydroperoxyl radicals ($O_2^{\bullet -}$/$HO_2^{\bullet}$) rather than OH radicals as illustrated in Figure 7c. The critical responsibility of superoxides for photocatalytic arsenite oxidation was also supported by the observation that a visible light-sensitized TiO_2 system that generates superoxides only, not OH radicals (because of the absence of bandgap excitation) was able to oxidize As(III) successfully (*79*, *81*). This case provides a unique example where the role of superoxide is emphasized whereas most TiO_2-mediated photocatalytic oxidation reactions are mainly driven by OH radicals or VB holes. The photocatalytic degradation mechanism often depends on the kind of substrate, is difficult to generalize, and therefore needs to be understood on a case-by-case basis. This aspect of photocatalytic activities of TiO_2 will be further discussed in the last section of this chapter.

The photocatalytic oxidation can be applied to the conversion of inorganic nitrogen compounds as well. For example, NH_3 can be oxidized to NO_2^-/NO_3^- in the irradiated suspension of TiO_2 under alkaline conditions where ammonia exists as a neutral (unprotonated) form which is highly susceptible to OH radical attack. However, NH_4^+ (protonated) has a very low reactivity with OH radical because of the absence of the lone electron pair. The photocatalytic oxidation of NH_3 to NO_2^-/NO_3^- proceeds stoichiometrically, which diminishes water quality by adding NO_2^- and NO_3^- which are more toxic than parent NH_3. The unwanted product formation could be controlled by modifying the surface of TiO_2 with Pt deposition. Pt/TiO_2 converts NH_3 to NO_2^-/NO_3^- with the imbalance of the total N (*see* Figure 7d), which implies the presence of missing products. The mass spectrometric analysis found that N_2 was evolved on Pt/TiO_2, but not on bare TiO_2. The presence of Pt catalyst on TiO_2 stabilized the transient intermediates (e.g., NH_x (x=0,1,2)), enabling the selective conversion of NH_3 [N(-III)] to N_2 [N(0)] while suppressing the complete oxidation to NO_2^-/NO_3^- ([N(+III)]/[N(+V)]) (*67*). The kinetics/mechanisms and intermediates/products distribution in TiO_2 photocatalytic reactions can be often changed by modifying the surface properties, as was the case for TCA degradation. The modification of surface properties of TiO_2 has been frequently investigated to control the photocatalytic reaction pathways (*88–92*).

Photocatalytic Degradation of Organic Contaminants with Simultaneous Production of Hydrogen

Photocatalytic reactions on TiO_2 can be applied not only to pollutant degradation but also hydrogen production (*93*, *94*). The two applications are very different and are usually carried out under different reaction conditions. However, a dual purpose photocatalysis that achieves the degradation of organic pollutants in water and the production of hydrogen simultaneously presents a methodology that recovers energy from wastewaters. For this purpose, the reaction of VB holes should favor the formation of OH radicals while that of CB electrons should lead to the production of hydrogen. This is possible by controlling the selectivity of TiO_2 photocatalysis through surface modification. A specific example is the simultaneous fluorination and platinization of TiO_2 (F-TiO_2/Pt), which enables the photocatalytic oxidation of phenolic pollutants (as sacrificial electron donors) in the absence of dioxygen (*91*). The F-TiO_2/Pt photocatalyst was successfully applied to the simultaneous degradation of phenolic compounds and the production of hydrogen (*92*). As Figure 8 shows, the dual purpose photocatalysis was only possible with F-TiO_2/Pt and not observed with any of bare TiO_2, F-TiO_2, and Pt/TiO_2. Such unique activity of F-TiO_2/Pt is ascribed to the combinative effects of the different surface modifications (fluorination and platinization). Platinum deposits on the TiO_2 surface hinder the charge recombination and accelerate the interfacial electron transfer to water and/or protons. On the other hand, surface fluorides inhibit the adsorption of phenolic substrates on TiO_2 but facilitate the generation of unbound OH radicals instead of surface bound OH radicals. This allows the photocatalytic degradation to proceed off the surface and retards the recombination of CB electrons with surface-bound OH radicals. Such selective photocatalysis of aquatic pollutants can be paired with solar conversion technology that achieves hydrogen production and water treatment simultaneously.

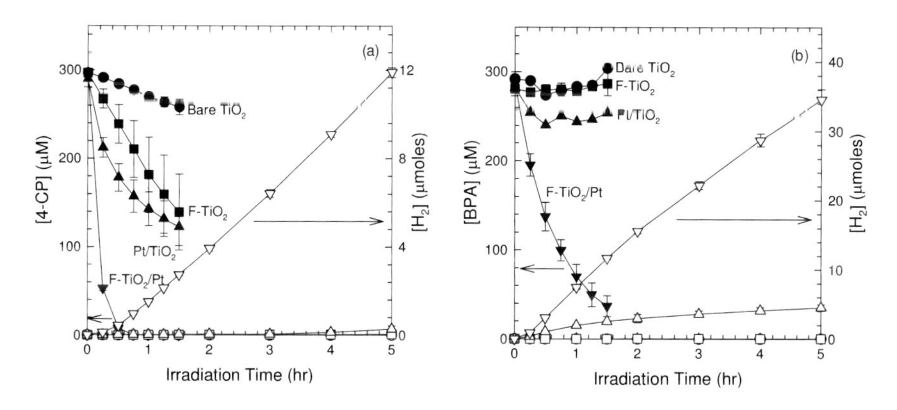

Figure 8. Simultaneous conversion of phenolic compounds and production of hydrogen in the presence of (a) 4-chlorophenol (4-CP) and (b) bisphenol A (BPA) in UV-irradiated suspensions of bare TiO_2, F-TiO_2, Pt/TiO_2, and F-TiO_2/Pt (filled symbols: the conversion of phenolic compound, open symbols: the accompanied production of hydrogen) (92). (Reproduced by permission of The Royal Society of Chemistry)

Photocatalytic Biocidal Activity and Inherent Toxicity

ROS produced during TiO_2 photoactivation can achieve effective inactivation of bacteria and virus, in that oxidants such as $\bullet OH$, $O_2\bullet^-$, and H_2O_2 induce oxidative damage on DNA and cell membrane (*95*). The comparison of CT values [CT refers to the residual concentration of a particular chemical disinfectant, C (mg/L), multiplied by the contact time between disinfectant and microorganism, T (min), and indicates how much disinfectant is required to achieve a desired level of disinfection] for a 2-log inactivation of *Escherichia coli* (*E.coli*) suggests that

OH radical has exceptionally powerful disinfection capacity compared to other chemical disinfectants such as chlorine, chlorine dioxide, and ozone (*20*, *96*). Matsunaga et al. (*97*) demonstrated that a VB hole produced by UV irradiation of TiO_2 triggered oxidative dimerization of coenzyme A inside *E.coli* cell, causing *E.coli* sterilization through enzyme deactivation and inhibition of cell respiration. The OH radical-induced oxidation is also critical in photocatalytic bactericidal property of TiO_2 since the addition of OH radical scavengers diminished the kinetics for *E.coli* inactivation (*98*, *99*). Figure 9a verifies that OH radical plays a predominant role in TiO_2-mediated photocatalytic bacterial disinfection by showing a linear correlation between the amount of photogenerated OH radical (quantified using *p*-chlorobenzoic acid (p-CBA) as an OH radical indicator) and the extent of *E.coli* inactivation (*20*). The photocatalytic inactivation mechanism seems to be different depending on the kind of microorganism. Addition of both *tert*-butanol and methanol at excess concentrations completely inhibited the photocatalytic inactivation of MS-2 bacteriophage (MS-2 phage) on the TiO_2 surface (Figure 9b), but not completely for *E.coli* (Figure 9c). *Tert*-butanol preferentially scavenges free OH radicals in the bulk phase whereas methanol consumes surface-bound as well as free OH radicals (Figure 9d). The different dependence of *E.coli* and MS-2 phage inactivation on two scavengers implies that MS-2 phage inactivation is mainly mediated by free OH radical while *E.coli* can be inactivated by both free and surface-bound OH radicals (*100*).

On the other hand, recent studies have continued to demonstrate the potential toxicological impacts of nanomaterials on environment and human health (*101*, *102*). In addition to the photo-induced cytotoxicity of TiO_2 nanoparticles, numerous research articles have reported the toxic potential of TiO_2 nanoparticles that do not involve photoactivation (*103–105*) although some still demonstrate that TiO_2 has negligible inherent toxicity through the comparison of toxicity between TiO_2 and other nanomaterials such as C_{60}, ZnO, and polystyrene nanopaticles (*106–108*). In the absence of light, ultrafine anatase TiO_2 particles (with diameters ranging from 10 to 20 nm) exhibit toxic activity toward human bronchial epithelial cells by inducing oxidative DNA damage and lipid peroxidation and by facilitating the formation of H_2O_2 and nitric oxide inside cells (*103*). On the other hand, such cytotoxicity vanishes as TiO_2 particle size increases up to 200 nm. P25 TiO_2 aggregates engulfed by brain microglia can stimulate the microglia to release ROS through *oxidative burst* (meaning the rapid ROS production from cells as a defense response when cells detect the presence of bacteria, fungi, or virus) (*104*). TiO_2 aggregates were also found potentially toxic to abalone (*Haliotis diversicolor supertexta*) embryos in the marine environment by hindering embryonic development, inhibiting hatching, and causing malfunctions (*105*). Based on a conservative assumption that TiO_2 particles cause hazardous effects on the aquatic environment, the application of TiO_2 photocatalyst for water treatment and disinfection processes should entail the reusability. For instance, TiO_2 immobilization (*32*) inhibits the unwanted discharge of TiO_2, mitigating the possible toxic effects on the aquatic ecosystems. Integration of TiO_2 photocatalysis with membrane filtration (*109*) enables the removal of catalyst from treated water. The fabrication of magnetic nanocomposites to achieve facile

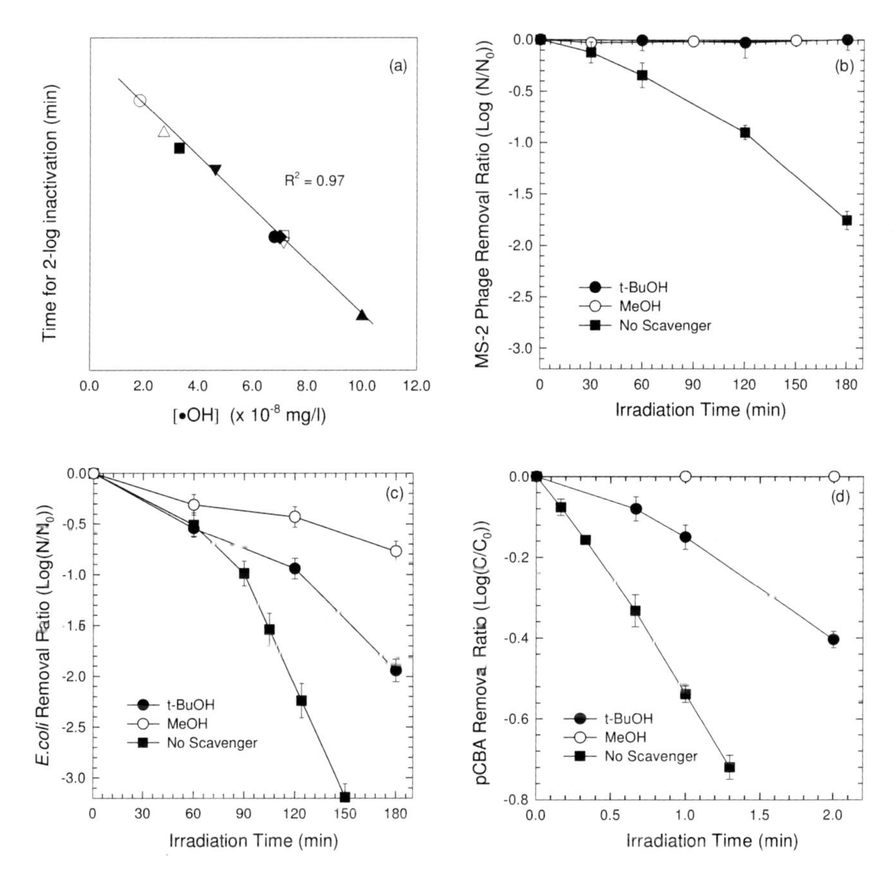

Figure 9. (a) Linear relationship between the amount of photogenerated OH radical and the degree of E.coli inactivation (20), TiO_2-mediated photodynamic inactivation of (b) MS-2 phage and (c) E.coli in the absence and presence of t-BuOH and MeOH, (d) TiO_2 photocatalytic degradation of p-CBA in the absence and presence of t-BuOH and MeOH (100). (Part (a) Reproduced by permission of Elsevier; parts (b−c) reproduced by permission of The American Society for Microbiology)

separation of TiO_2 (*110*) also can minimize the environmental release, eventually alleviating the possibility for TiO_2-induced secondary contamination.

Nature of Photocatalytic Activity

The activity assessment is an integral part in the studies of TiO_2 photocatalysis. However, how the activity of a given photocatalyst sample can be represented is not straightforward because the measured photocatalytic activity is usually substrate-specific. The photocatalytic activity is commonly quantified in terms of the degradation rate of a specific substrate but the measured activity cannot be generalized to other substrates. A recent study investigated the substrate-specific nature of TiO_2 photocatalysts in a systematic way (*111*).

Abridged results are givien in Table I where the photocatalytic activities of 6 commercial TiO_2 samples are compared for 10 test substrates. The measured activities exhibited a complex behavior that depended on the test substrate. Each TiO_2 sample showed the best activity for at least one test-substrate. For instance, Aldrich and Junsei TiO_2 samples exhibited the best activities for the degradation of 4-chlorophenol, while showing the least activity for the degradation of formic acid. Although many photocatalytic studies report the activity of a specific photocatalyst on the basis of a single substrate test and assume that the specific activity can be generalized to other substrates, such practice could be misleading. The single-substrate activity test shows only a part of the whole activity. To be more realistic, a multi-activity assessment is needed with employing multiple substrates instead of the single substrate.

Table I. Photocatalytic activities§ of six commercial TiO_2 samples measured with 10 test substrates. (Adapted from ref. (*111*))

	TiO_2 samples†					
Substrates	***D***	***H***	***J***	***A***	***M***	***I***
4-Chlorophenol	O	X	O	O		X
Formic acid			X	X		O
Methylamine	O		X		O	
Trichloroethylene	O		X			
$CHCl_3$	O				X	
Acid Orange 7	O	O	X			O
Methylene Blue		X		O		X
Methanol	O	X		X		
Cr(VI), Chromate	O		X	X		
Iodide		O	X	X		O

§ The most active TiO_2 samples for a given substrate (i.e., in the same row) are indicated by "O" and the least active ones by "X". † D: Degussa, H: Hombikat, J: Junsei, A: Aldrich, M: Millennium, I: Ishihara

Conclusions

A great number of research works have successfully demonstrated that TiO_2-mediated redox reactions are able to remediate diverse environmental media contaminated with organic and inorganic pollutants. The examples shown in this chapter demonstrate that the photocatalyzed redox reactions occurring on TiO_2 achieve effective destruction of a variety of pollutants, and offer a promising remediation strategy. The modification of TiO_2 in various ways boosts the efficacy or enables the visible light activation of TiO_2. The kinetics and mechanisms for TiO_2-photocatalyzed redox conversions should be comprehended on a case-by-case basis because the photocatalytic activity is specific to substrates and experimental conditions. Consequently, the photocatalytic activities of newly-synthesized or modified photocatalysts need to be assessed on the basis of multi-activity tests.

The focus of the published photocatalytic studies to date ranges from the fundamental science studies to the development of commercial products. Transient spectroscopic studies allow us to probe into the photophysical phenomenon relevant to charge generation, recombination and transfer dynamics in TiO_2. Investigation into the photocatalytic kinetics and mechanisms enabled the rational evaluation of TiO_2 activities and advanced understanding of the photocatalyzed chemical conversions. Research activities regarding TiO_2 coating and fabrication techniques have assisted in manufacture and commercialization of TiO_2-based photocatalytic systems such as water/air purifiers, deordorizers, and various self-cleaning products. In order to enable repeated use of TiO_2 and mitigate the unwanted environmental release, strategies to immobilize TiO_2 on various substrates or to readily recover TiO_2 particulates from treated water have attracted increasing attention. On the other hand, as nano-technologies can enhance morphological, physicochemical, and photochemical properties of TiO_2, many research activities are currently focused on the modification of TiO_2-based photocatalysts at the nanoscale, aiming to improve efficacy of pollutant removal and alleviate energy demand for photocatalyst activation. The development of cost-effective light sources such as light emitting diode and the optimization of photoreactors are also essential for the use of TiO_2 photocatalysis as a practical remediation technology.

Acknowledgments

This work was supported by KOSEF NRL program (No. R0A-2008-000-20068-0), KOSEF EPB center (Grant No. R11-2008-052-02002), and KCAP (Sogang Univ.) funded by MEST through NRF (NRF-2009-C1AAA001-2009-0093879).

References

1. Hoffmann, M. R.; Martin, S. T.; Choi, W.; Bahnemann, D. W. Environmental applications of semiconductor photocatalysis. *Chem. Rev.* **1995**, *95*, 69–96.
2. Thompson, T. L.; Yates, J. T., Jr. Surface science studies of the photoactivation of TiO_2-new photochemical processes. *Chem. Rev.* **2006**, *106*, 4428–4453.
3. Chen, X.; Mao, S. S. Titanium dioxide nanomaterials: synthesis, properties, modifications, and applications. *Chem. Rev.* **2007**, *107*, 2891–2959.
4. Grätzel, M. Photoelectrochemical cells. *Nature* **2001**, *414*, 338–344.
5. Tachikawa, T.; Majima, T. Single-molecule, single-particle fluorescence imaging of TiO_2-based photocatalytic reactions. *Chem. Soc. Rev.* **2010**, *39*, 4802–4819.

6. Mills, A.; Hunte, S. L. An overview of semiconductor photocatalysis. *J. Photochem. Photobiol., A* **1997**, *108*, 1–35.
7. Ollis, D. F.; Pelizzetti, E.; Serpone, N. Photocatalyzed destruction of water contaminants. *Environ. Sci. Technol.* **1991**, *25*, 1522–1529.
8. Choi, W. Pure and modified TiO_2 photocatalysts and their environmental applications. *Catal. Surv. Asia* **2006**, *10*, 16–28.
9. Fujishima, A.; Honda, K. Electrochemical photolysis of water at a semiconductor electrode. *Nature* **1972**, *238*, 37–38.
10. Chen, X.; Shen, S.; Guo, L.; Mao, S. S. Semiconductor-based photocatalytic hydrogen generation. *Chem. Rev.* **2010**, *110*, 6503–6570.
11. Kudo, A.; Miseki, Y. Heterogeneous photocatalyst materials for water splitting. *Chem. Soc. Rev.* **2009**, *38*, 253–278.
12. Kubo, W.; Tatsuma, T.; Fujishima, A.; Kobayashi, H. Mechanisms and resolution of photocatalytic lithography. *J. Phys. Chem. B* **2004**, *108*, 3005–3009.
13. Ishikawa, Y.; Matsumoto, Y.; Nishida, Y.; Taniguchi, S.; Watanabe, J. Surface treatment of silicon carbide using TiO_2(IV) photocatalyst. *J. Am. Chem. Soc.* **2003**, *125*, 6558–6562.
14. Herrmann, J.-M.; Disdier, J.; Pichat, P. Photocatalytic deposition of silver on powder titania: consequences for the recovery of silver. *J. Catal.* **1988**, *113*, 72–81.
15. Park, H.; Kim, K. Y.; Choi, W. A novel photoelectrochemical method of metal corrosion prevention using a TiO_2 solar panel. *Chem. Commun.* **2001**, 281–282.
16. Tatsuma, T.; Saitoh, S.; Ohko, Y.; Fujishima, A. TiO_2-WO_3 photoelectrochemical anticorrosion system with an energy storage ability. *Chem. Mater.* **2001**, *13*, 2838–2842.
17. Fujihira, M.; Satoh, Y.; Osa, T. Heterogeneous photocatalytic oxidation of aromatic-compounds on TiO_2. *Nature* **1981**, *293*, 206–208.
18. Yoshida, H.; Yuzawa, H.; Aoki, M.; Otake, K.; Itoh, H.; Hattori, T. Photocatalytic hydroxylation of aromatic ring by using water as an oxidant. *Chem. Commun.* **2008**, 4634–4636.
19. Palmisano, G.; García-López, E.; Marcí, G.; Loddo, V.; Yurdakal, S.; Augugliaro, V.; Palmisano, L. Advances in selective conversions by heterogeneous photocatalysis. *Chem. Commun.* **2010**, *46*, 7074–7089.
20. Cho, M.; Chung, H.; Choi, W.; Yoon, J. Linear correlation between inactivation of *E. coli* and OH radical concentration in TiO_2 photocatalytic disinfection. *Water Res.* **2004**, *38*, 1069–1077.
21. Kong, H.; Song, J.; Jang, J. Photocatalytic antibacterial capabilities of TiO_2-biocidal polymer nanocomposites synthesized by a surface-initiated photopolymerization. *Environ. Sci. Technol.* **2010**, *44*, 5672–5676.
22. Wu, P.; Xie, R.; Imlay, K.; Shang, J. K. Visible-light-induced bactericidal activity of titanium dioxide codoped with nitrogen and silver. *Environ. Sci. Technol.* **2010**, *44*, 6992–6997.
23. Fujishima, A.; Zhang, X.; Tryk, D. A. TiO_2 photocatalysis and related surface phenomena. *Surf. Sci. Rep.* **2008**, *63*, 515–582.

24. Fujishima, A.; Rao, T. N.; Tryk, D. A. Titanium dioxide photocatalysis. *J. Photochem. Photobiol., C* **2000**, *1*, 1–21.
25. Fox, M. A.; Dulay, M. T. Heterogeneous photocatalysis. *Chem. Rev.* **1993**, *93*, 341–357.
26. Cho, S.; Choi, W. Solid-phase photocatalytic degradation of PVC–TiO_2 polymer composites. *J. Photochem. Photobiol., A* **2001**, *143*, 221–228.
27. Lee, M. C.; Choi, W. Solid phase photocatalytic reaction on the soot/TiO_2 interface: the role of migrating OH radicals. *J. Phys. Chem. B* **2002**, *106*, 11818–11822.
28. Puddu, V.; Choi, H.; Dionysiou, D. D.; Puma, G. L. TiO_2 photocatalyst for indoor air remediation: influence of crystallinity, crystal phase, and UV radiation intensity on trichloroethylene degradation. *Appl. Catal., B* **2010**, *94*, 211–218.
29. Liu, Z.; Zhang, X.; Nishimoto, S.; Murakami, T.; Fujishima, A. Efficient photocatalytic degradation of gaseous acetaldehyde by highly ordered TiO_2 nanotube arrays. *Environ. Sci. Technol.* **2008**, *42*, 8547–8551.
30. Mills, A.; Crow, M.; Wang, J.; Parkin, I. P.; Boscher, N. Photocatalytic oxidation of deposited sulfur and gaseous sulfur dioxide by TiO_2 films. *J. Phys. Chem. C* **2007**, *111*, 5520–5525.
31. Wang, R.; Hashimoto, K.; Fujishima, A.; Chikuni, M.; Kojima, E.; Kitamura, A. Light-induced amphiphilic surfaces. *Nature* **1997**, *388*, 431–432.
32. Lee, C.-S.; Kim, J.; Son, J. Y.; Choi, W.; Kim, H. Photocatalytic functional coatings of TiO_2 thin films on polymer substrate by plasma enhanced atomic layer deposition. *Appl. Catal., B* **2009**, *91*, 628–633.
33. Legrini, O.; Oliveros, E.; Braun, A. M. Photochemical processes for water treatment. *Chem. Rev.* **1993**, *93*, 671–698.
34. Stefan, M. I.; Bolton, J. R. Mechanism of the degradation of 1,4-dioxane in dilute aqueous solution using the UV/hydrogen peroxide process. *Environ. Sci. Technol.* **1998**, *32*, 1588–1595.
35. Stefan, M. I.; Hoy, A. R.; Bolton, J. R. Kinetics and mechanism of the degradation and mineralization of acetone in dilute aqueous solution sensitized by the UV photolysis of hydrogen peroxide. *Environ. Sci. Technol.* **1996**, *30*, 2382–2390.
36. Glaze, W. H.; Peyton, G. R.; Lin, S.; Huang, R. Y.; Burleson, J. L. Destruction of pollutants in water with ozone in combination with ultraviolet radiation. 2. natural trihalomethane precursors. *Environ. Sci. Technol.* **1982**, *16*, 454–458.
37. Peyton, G. R.; Huang, F. Y.; Burleson, J. L.; Glaze, W. H. Destruction of pollutants in water with ozone in combination with ultraviolet radiation. 1. general principles and oxidation of tetrachloroethylene. *Environ. Sci. Technol.* **1982**, *16*, 448–453.
38. Peyton, G. R.; Glaze, W. H. Destruction of pollutants in water with ozone in combination with ultraviolet radiation. 3. photolysis of aqueous ozone. *Environ. Sci. Technol.* **1988**, *22*, 761–767.

39. Bolts, J. M.; Wrighton, M. S. Correlation of photocurrent-voltage curves with flat-band potential for stable photoelectrodes for the photoelectrolysis of water. *J. Phys. Chem.* **1976**, *80*, 2641–2645.
40. Staehelin, J.; Hoigné, J. Decomposition of ozone in water in the presence of organic solutes acting as promoters and inhibitors of radical chain reactions. *Environ. Sci. Technol.* **1985**, *19*, 1206–1213.
41. Kim, S.; Hwang, S.-J.; Choi, W. Visible light active platinum-ion-doped TiO_2 photocatalyst. *J. Phys. Chem. B* **2005**, *109*, 24260–24267.
42. Asahi, R.; Morikawa, T.; Ohwaki, T.; Aoki, K.; Taga, Y. Visible-light photocatalysis in nitrogen-doped titanium oxides. *Science* **2001**, *293*, 269–271.
43. Sakthivel, S.; Kisch, H. Daylight photocatalysis by carbon-modified titanium dioxide. *Angew. Chem., Int. Ed.* **2003**, *42*, 4908–4911.
44. Bae, E.; Choi, W. Highly enhanced photoreductive degradation of perchlorinated compounds on dye-sensitized metal/TiO_2 under visible light. *Environ. Sci. Technol.* **2003**, *37*, 147–152.
45. Park, Y.; Lee, S.-H.; Kang, S. O.; Choi, W. Organic dye-sensitized TiO_2 for the redox conversion of water pollutants under visible light. *Chem. Commun.* **2010**, *46*, 2477–2479.
46. Chen, C.; Ma, W.; Zhao, J. Semiconductor-mediated photodegradation of pollutants under visible-light irradiation. *Chem. Soc. Rev.* **2010**, *39*, 4206–4219.
47. Park, Y.; Singh, N. J.; Kim, K. S.; Tachikawa, T.; Majima, T.; Choi, W. Fullerol–titania charge-transfer-mediated photocatalysis working under visible light. *Chem.-Eur. J.* **2009**, *15*, 10843–10850.
48. Kim, G.; Choi, W. Charge-transfer surface complex of EDTA-TiO_2 and its effect on photocatalysis under visible light. *Appl. Catal., B* **2010**, *100*, 77–83.
49. Hou, Y.; Li, X.-Y.; Zhao, Q.-D.; Quan, X.; Chen, G.-H. Electrochemical method for synthesis of a $ZnFe_2O_4$/TiO_2 composite nanotube array modified electrode with enhanced photoelectrochemical activity. *Adv. Funct. Mater.* **2010**, *20*, 2165–2174.
50. Ratanatawanate, C.; Tao, Y.; Balkus, K. J., Jr. Photocatalytic activity of PbS quantum dot/TiO_2 nanotube composites. *J. Phys. Chem. C* **2009**, *113*, 10755–10760.
51. Bang, J. H.; Kamat, P. V. Quantum dot sensitized solar cells. a tale of two semiconductor nanocrystals: CdSe and CdTe. *ACS Nano* **2009**, *3*, 1467–1476.
52. Kim, S.; Choi, W. Kinetics and mechanisms of photocatalytic degradation of $(CH_3)_nNH_{4-n}^+$ ($0 \leq n \leq 4$) in TiO_2 suspension: the role of OH radicals. *Environ. Sci. Technol.* **2002**, *36*, 2019–2025.
53. Bobrowski, K. Pulse radiolysis studies concerning the reactions of hydrogen abstraction from tetraalkylammonium cations. *J. Phys. Chem.* **1980**, *84*, 3524–3529.
54. Simić, M.; Neta, P.; Hayon, E. Pulse radiolytic investigation of aliphatic amines in aqueous solution. *Int. J. Radiat. Phys. Chem.* **1971**, *3*, 309–320.
55. Mao, Y.; Schöneich, C.; Asmus, K.-D. Identification of organic acids and other intermediates in oxidative degradation of chlorinated ethanes on TiO_2

surfaces en route to mineralization. a combined photocatalytic and radiation chemical study. *J. Phys. Chem.* **1991**, *95*, 10080–10089.

56. Turchi, C. S.; Ollis, D. F. Photocatalytic degradation of organic water contaminants: mechanisms involving hydroxyl radical attack. *J. Catal.* **1990**, *122*, 178–192.
57. Choi, W.; Hoffmann, M. R. Novel photocatalytic mechanisms for $CHCl_3$, $CHBr_3$, and $CCl_3CO_2^-$ degradation and the fate of photogenerated trihalomethyl radicals on TiO_2. *Environ. Sci. Technol.* **1997**, *31*, 89–95.
58. Kim, S.; Choi, W. Dual photocatalytic pathways of trichloroacetate degradation on TiO_2: effects of nanosized platinum deposits on kinetics and mechanism. *J. Phys. Chem. B* **2002**, *106*, 13311–13317.
59. Liu, G.; Li, X.; Zhao, J.; Hidaka, H.; Serpone, N. Photooxidation pathway of sulforhodamine-B. dependence on the adsorption mode on TiO_2 exposed to visible light radiation. *Environ. Sci. Technol.* **2000**, *34*, 3982–3990.
60. Liu, G.; Wu, T.; Zhao, J.; Hidaka, H.; Serpone, N. Photoassisted degradation of dye pollutants. 8. irreversible degradation of alizarin red under visible light radiation in air-equilibrated aqueous TiO_2 dispersions. *Environ. Sci. Technol.* **1999**, *33*, 2081–2087.
61. Wu, T.; Liu, G.; Zhao, J.; Hidaka, H.; Serpone, N. Photoassisted degradation of dye pollutants. V. self-photosensitized oxidative transformation of *rhodamine B* under visible light irradiation in aqueous TiO_2 dispersions. *J. Phys. Chem. B* **1998**, *102*, 5845–5851.
62. Martínez-Huitle, C. A.; Brillas, E. Decontamination of wastewaters containing synthetic organic dyes by electrochemical methods: a general review. *Appl. Catal., B* **2009**, *87*, 105–145.
63. Kim, W.; Tachikawa, T.; Majima, T.; Choi, W. Photocatalysis of dye-sensitized TiO_2 nanoparticles with thin overcoat of Al_2O_3: enhanced activity for H_2 production and dechlorination of CCl_4. *J. Phys. Chem. C* **2009**, *113*, 10603–10609.
64. Füldner, S.; Mild, R.; Siegmund, H. I.; Schroeder, J. A.; Gruber, M.; König, B. Green-light photocatalytic reduction using dye-sensitized TiO_2 and transition metal nanoparticles. *Green Chem.* **2010**, *12*, 400–406.
65. Zhang, F.; Jin, R.; Chen, J.; Shao, C.; Gao, W.; Li, L.; Guan, N. High photocatalytic activity and selectivity for nitrogen in nitrate reduction on Ag/TiO_2 catalyst with fine silver clusters. *J. Catal.* **2005**, *232*, 424–431.
66. Kominami, H.; Nakaseko, T.; Shimada, Y.; Furusho, A.; Inoue, H.; Murakami, S.; Kera, Y.; Ohtani, B. Selective photocatalytic reduction of nitrate to nitrogen molecules in an aqueous suspension of metal-loaded titanium(IV) oxide particles. *Chem. Commun.* **2005**, 2933–2935.
67. Lee, J.; Park, H.; Choi, W. Selective photocatalytic oxidation of NH_3 to N_2 on platinized TiO_2 in water. *Environ. Sci. Technol.* **2002**, *36*, 5462–5468.
68. Zhu, X.; Castleberry, S. R.; Nanny, M. A.; Butler, E. C. Effects of pH and catalyst concentration on photocatalytic oxidation of aqueous ammonia and nitrite in titanium dioxide suspensions. *Environ. Sci. Technol.* **2005**, *39*, 3784–3791.

69. Frank, S. N.; Bard, A. J. Heterogeneous photocatalytic oxidation of cyanide ion in aqueous solutions at TiO_2 Powder. *J. Am. Chem. Soc.* **1977**, *99*, 303–304.
70. Draper, R. B.; Fox, M. A. Titanium dioxide photosensitized reactions studied by diffuse reflectance flash photolysis in aqueous suspensions of TiO_2 powder. *Langmuir* **1990**, *6*, 1396–1402.
71. Micic, O. I.; Zhang, Y.; Cromack, K. R.; Trifunac, A. D.; Thurnauer, M. C. Trapped holes on TiO_2 colloids studied by electron paramagnetic resonance. *J. Phys. Chem.* **1993**, *97*, 7277–7283.
72. Zhang, X.; Zhang, T.; Ng, J.; Pan, J. H.; Sun, D. D. Transformation of bromine species in TiO_2 photocatalytic system. *Environ. Sci. Technol.* **2010**, *44*, 439–444.
73. Yoneyama, H.; Yamashita, Y.; Tamura, H. Heterogeneous photocatalytic reduction of dichromate on n-type semiconductor catalysts. *Nature* **1979**, *282*, 817–818.
74. Chenthamarakshan, C. R.; Rajeshwar, K.; Wolfrum, E. J. Heterogeneous photocatalytic reduction of Cr(VI) in UV-irradiated titania suspensions: effect of protons, ammonium ions, and other interfacial aspects. *Langmuir* **2000**, *16*, 2715–2721.
75. Testa, J. J.; Grela, M. A.; Litter, M. I. Experimental evidence in favor of an initial one-electron-transfer process in the heterogeneous photocatalytic reduction of chromium(VI) over TiO_2. *Langmuir* **2001**, *17*, 3515–3517.
76. Testa, J. J.; Grela, M. A.; Litter, M. I. Heterogeneous photocatalytic reduction of chromium(VI) over TiO_2 particles in the presence of oxalate: involvement of Cr(V) species. *Environ. Sci. Technol.* **2004**, *38*, 1589–1594.
77. Yang, H.; Lin, W.-Y.; Rajeshwar, K. Homogeneous and heterogeneous photocatalytic reactions involving As(III) and As(V) species in aqueous media. *J. Photochem. Photobiol., A* **1999**, *123*, 137–143.
78. Bissen, M.; Vieillard-Baron, M.-M.; Schindelin, A. J.; Frimmel, F. H. TiO_2-catalyzed photooxidation of arsenite to arsenate in aqueous samples. *Chemosphere* **2001**, *44*, 751–757.
79. Choi, W.; Yeo, J.; Ryu, J.; Tachikawa, T.; Majima, T. Photocatalytic oxidation mechanism of As(III) on TiO_2: unique role of As(III) as a charge recombinant species. *Environ. Sci. Technol.* **2010**, *44*, 9099–9104.
80. Lee, H.; Choi, W. Photocatalytic oxidation of arsenite in TiO_2 suspension: kinetics and mechanisms. *Environ. Sci. Technol.* **2002**, *36*, 3872–3878.
81. Ryu, J.; Choi, W. Effects of TiO_2 surface modifications on photocatalytic oxidation of arsenite: the role of superoxides. *Environ. Sci. Technol.* **2004**, *38*, 2928–2933.
82. Ryu, J.; Choi, W. Photocatalytic oxidation of arsenite on TiO_2: understanding the controversial oxidation mechanism involving superoxides and the effect of alternative electron acceptors. *Environ. Sci. Technol.* **2006**, *40*, 7034–7039.
83. Wang, X.; Pehkonen, S. O.; Ray, A. K. Photocatalytic reduction of Hg(II) on two commercial TiO_2 catalysts. *Electrochim. Acta* **2004**, *49*, 1435–1444.
84. Litter, M. I. Heterogeneous photocatalysis transition metal ions in photocatalytic systems. *Appl. Catal., B* **1999**, *23*, 89–114.

85. Nguyen, V. N. H.; Amal, R.; Beydoun, D. Photodeposition of CdSe using Se-TiO_2 suspensions as photocatalysts. *J. Photochem. Photobiol., A* **2006**, *179*, 57–65.
86. Tada, H.; Ishida, T.; Takao, A.; Ito, S. Drastic enhancement of TiO_2-photocatalyzed reduction of nitrobenzene by loading Ag clusters. *Langmuir* **2004**, *20*, 7898–7900.
87. Kyung, H.; Lee, J.; Choi, W. Simultaneous and synergistic conversion of dyes and heavy metal ions in aqueous TiO_2 suspensions under visible-light illumination. *Environ. Sci. Technol.* **2005**, *39*, 2376–2382.
88. Park, H.; Choi, W. Photocatalytic reactivities of nafion-coated TiO_2 for the degradation of charged organic compounds under UV or visible light. *J. Phys. Chem. B* **2005**, *109*, 11667–11674.
89. Park, H.; Choi, W. Effects of TiO_2 surface fluorination on photocatalytic reactions and photoelectrochemical behaviors. *J. Phys. Chem. B* **2004**, *108*, 4086–4093.
90. Lee, J.; Choi, W. Effect of platinum deposits on TiO_2 on the anoxic photocatalytic degradation pathways of alkylamines in water: dealkylation and N-alkylation. *Environ. Sci. Technol.* **2004**, *38*, 4026–4033.
91. Kim, J.; Lee, J.; Choi, W. Synergic effect of simultaneous fluorination and platinization of TiO_2 surface on anoxic photocatalytic degradation of organic compounds. *Chem. Commun.* **2008**, 756–758.
92. Kim, J.; Choi, W. Hydrogen producing water treatment through solar photocatalysis. *Energy Environ. Sci.* **2010**, *3*, 1042–1045.
93. Ni, M.; Leung, M. K. H.; Leung, D. Y. C.; Sumathy, K. A review and recent developments in photocatalytic water-splitting using TiO_2 for hydrogen production. *Renewable Sustainable Energy Rev.* **2007**, *11*, 401–425.
94. Nada, A. A.; Barakat, M. H.; Hamed, H. A.; Mohamed, N. R.; Veziroglu, T. N. Studies on the photocatalytic hydrogen production using suspended modified TiO_2 photocatalysts. *Int. J. Hydrogen Energy* **2005**, *30*, 687–691.
95. Imlay, J. A.; Fridovich, I. Suppression of oxidative envelope damage by pseudoreversion of a superoxide dismutase-deficient mutant of *Escherichia coli*. *J. Bacteriol.* **1992**, *174*, 953–961.
96. Cho, M.; Kim, J.; Kim, J. Y.; Yoon, J.; Kim, J.-H. Mechanisms of *Escherichia coli* inactivation by several disinfectants. *Water Res.* **2010**, *44*, 3410–3418.
97. Matsunaga, T.; Tomoda, R.; Nakajima, T.; Nakamura, N.; Komine, T. Continuous-sterilization system that uses photosemiconductor powders. *Appl. Environ. Microbiol.* **1988**, *54*, 1330–1333.
98. Ireland, J. C.; Klostermann, P.; Rice, E. W.; Clark, R. M. Inactivation of *Escherichia coli* by titanium dioxide photocatalytic oxidation. *Appl. Environ. Microbiol.* **1993**, *59*, 1668–1670.
99. Wei, C.; Lin, W.-Y.; Zainal, Z.; Williams, N. E.; Zhu, K.; Kruzic, A. P.; Smith, R. L.; Rajeshwar, K. Bactericidal activity of TiO_2 photocatalyst in aqueous media: toward a solar-assisted water disinfection system. *Environ. Sci. Technol.* **1994**, *28*, 934–938.
100. Cho, M.; Chung, H.; Choi, W.; Yoon, J. Different inactivation behaviors of MS-2 phage and *Escherichia coli* in TiO_2 photocatalytic disinfection. *Appl. Environ. Microbiol.* **2005**, *71*, 270–275.

101. Colvin, V. L. The potential environmental impact of engineered nanomaterials. *Nature Biotechnol.* **2003**, *21*, 1166–1170.
102. Nel, A.; Xia, T.; Mädler, L.; Li, N. Toxic potential of materials at the nanolevel. *Science* **2006**, *311*, 622–627.
103. Gurr, J.-R.; Wang, A. S. S.; Chen, C.-H.; Jan, K.-Y. Ultrafine titanium dioxide particles in the absence of photoactivation can induce oxidative damage to human bronchial epithelial cells. *Toxicology* **2005**, *213*, 66–73.
104. Long, T. C.; Saleh, N.; Tilton, R. D.; Lowry, G. V.; Veronesi, B. Titanium dioxide (P25) produces reactive oxygen species in immortalized brain microglia (BV2): implications for nanoparticle neurotoxicity. *Environ. Sci. Technol.* **2006**, *40*, 4346–4352.
105. Zhu, X.; Zhou, J.; Cai, Z. TiO_2 nanoparticles in the marine environment: impact on the toxicity of tributyltin to abalone (*Haliotis diversicolor supertexta*) embryos. *Environ. Sci. Technol.* **2011**, *45*, 3753–3758.
106. Lovern, S. B.; Strickler, J. R.; Klaper, R. Behavioral and physiological changes in *daphnia magna* when exposed to nanoparticle suspensions (titanium dioxide, nano-C_{60}, and $C_{60}HxC_{70}Hx$). *Environ. Sci. Technol.* **2007**, *41*, 4465–4470.
107. Miller, R. J.; Lenihan, H. S.; Muller, E. B.; Tseng, N.; Hanna, S. K.; Keller, A. A. Impacts of metal oxide nanoparticles on marine phytoplankton. *Environ. Sci. Technol.* **2010**, *44*, 7329–7334.
108. Xia, T.; Kovochich, M.; Brant, J.; Hotze, M.; Sempf, J.; Oberley, T.; Sioutas, C.; Yeh, J. I.; Wiesner, M. R.; Nel, A. E. Comparison of the abilities of ambient and manufactured nanoparticles to induce cellular toxicity according to an oxidative stress paradigm. *Nano Lett.* **2006**, *6*, 1794–1807.
109. Ryu, J.; Choi, W.; Choo, K.-H. A pilot-scale photocatalyst-membrane hybrid reactor: performance and characterization. *Water Sci. Technol.* **2005**, *51*, 491–497.
110. Beydoun, D.; Amal, R.; Low, G. K.-C.; McEvoy, S. Novel photocatalyst: titania-coated magnetite. activity and photodissolution. *J. Phys. Chem. B* **2000**, *104*, 4387–4396.
111. Ryu, J.; Choi, W. Substrate-specific photocatalytic activities of TiO_2 and multiactivity test for water treatment application. *Environ. Sci. Technol.* **2008**, *42*, 294–300.

Chapter 11

Chlorine Based Oxidants for Water Purification and Disinfection

Gregory V. Korshin*

Department of Civil and Environmental Engineering, University of Washington, Seattle, WA 98195-2700
***korshin@u.washington.edu**

This chapter discusses the main aspects of chlorine and bromine speciation in systems with varying pHs, concentrations of bromide, chloride and total active chlorine. In the absence of ammonia, formal consideration of equilibria in solutions containing hypohalogenous acids, Cl_2, Br_2, BrCl and trihalogenide ions $Br_iCl_{3-i}^-$ can be carried out based on two reference species (OCl^-, Br^-). Formal constants necessary for implementing such an approach are presented in the paper. While haloamine formation constants can be determined based on the consideration of OCl^-, Br^- and NH_4^+ as reference species, this approach is deemed to be applicable only when monochloramine is prevalent. Properties of systems with halogens can be examined based on the electrochemical potential of the $HOCl/Cl^-$ couple.

Introduction

The use of chlorine-based species for oxidation and disinfection purposes has had a long and in some respects extraordinary history. For instance, the elimination of waterborne disease and ensuing notable increase of averaged life expectancy have been largely a result of use of chlorine to disinfect drinking water. While this fact is common knowledge, other aspects of water treatment processes that involve halogen-driven oxidations and related reactions remain a matter of continuing research and public debate. For instance, there are numerous issues related to further exploration of formation pathways, speciation and identification of disinfection by-products (DBPs) found in drinking water and their health effects (*1–3*). Formation and environmental effects of halogenated species generated in chlorinated wastewater or seawater used for desalination or cooling of nuclear reactors also need to be ascertained in more detail (*4*). Effects of chlorine and allied species on emerging trace-level organic species grouped into the operationally defined classes of endocrine disrupting chemicals, pharmaceuticals and personal care products (EDC/PPCP) have also attracted considerable attention (*5–10*). Roles of dissimilar halogen species in the deactivation of various microorganisms (e.g., (*11–13*)) and their influence on the kinetics of removal of trace-level organic contaminants are also subject of continuing research (*5–10*, *14–18*).

While effects of concentrations and speciation of aqueous halogen species on the viability of known and emerging pathogens, formation of DBPs caused by reactions of halogen species with natural and effluent organic matter found in surface waters and wastewater (NOM and EfOM, respectively), and the degradation of EDC/PPCP compounds are enormously important, the sheer complexity of these subjects preclude their review in this document. Likewise, this document will not cover issues related to the generation of iodine-containing oxidants, their reactions with NOM or EfOM and ensuing formation of iodinated DBPs. The intent of this chapter is to provide a reasonably detailed picture of the equilibrium chemistry of chlorine- and bromine-containing oxidants common in water treatment operation. This goal will be pursued based on a consistent description of reactions that involve diverse halogen compounds and define both the species that dominate in representative situations and also the nominal electrochemical potentials associated with the presence of these oxidants.

General Aspects of Halogen Concentration and Speciation

In most cases pertaining to water treatment operations, chlorine is added as chlorine gas Cl_2, or concentrated sodium hypochlorite NaOCl (bleach), or in some cases solid calcium hypochlorite $Ca(OCl)_2$. Monochloramine NH_2Cl is also frequently used for disinfection. The concentration of total active chlorine in water treatment operations does not normally exceed the threshold of 4 mg/L as Cl_2 (or $5.64\cdot10^{-5}$ M)

The introduction of chlorine can be accompanied by the formation of species of bromine (e.g., HOBr, OBr^- and others) generated via the relatively rapid oxidation of bromide ion (Br^-) by HOCl or OCl^- (*19*). The actual concentration and speciation of bromine species released as a result of the oxidation of Br^- depends on the bromide concentration in any particular water source and, as is discussed in more detail below, on the solution pH and chloride ion concentration. The concentration of bromide in surface waters is variable and can be from <10 μg/L to as high as 2 mg/L, for instance in Lake Kinneret (*20*, *21*). In contrast, the average concentration of bromide in seawater is $8.62\cdot10^{-4}$ M (68.9 mg/L) while the average concentration of Cl^- in this medium is 0.56 M (*22*). Surface waters or groundwater affected by seawater intrusion or exposed to geological strata formed in the presence of oceanic waters have notably higher bromide (and iodide) concentrations.

The speciation of chlorine and bromine species is well known to have a profound effect on the efficiency of disinfection and the kinetics of oxidative processes in drinking water and wastewater (*5–18*, *23*). While extremely important *per se*, the kinetics and mechanisms of interactions between halogen species and pollutants or microorganisms lie outside of the scope of this chapter, and only the relevant equilibria will be examined in the sections that follow.

Calculations of Electrochemical Potentials of Reactions Involving Halogen Species

The oxidative power of halogen species present in aquatic systems can be ascertained via calculations of electrochemical potentials corresponding to the reduction of HOCl to Cl^- ($HOCl + H^+ + 2e \rightarrow Cl^- + H_2O$) (*24*):

$$E_{HOCl/Cl^-}(V) = 1.428 + \frac{2.303RT}{2F}\log\left(\frac{\{HOCl\}\{H^+\}}{\{Cl^-\}}\right) \quad (1)$$

The potential calculated using expression (1) and discussed elsewhere are always quoted in this chapter vs. the normal hydrogen electrode (n.h.e.) Expression (1) can alternatively be written for the reduction of the deprotonated form OCl^- ($OCl^- + 2H^+ + 2e \rightarrow Cl^- + H_2O$)

$$E_{HOCl/Cl^-} = 1.428 + \frac{2.303RT}{2F}\log\left(\frac{\{OCl^-\}\{H^+\}^2}{K_{HOCl}\{Cl^-\}}\right) \quad (2)$$

where K_{HOCl} is the deprotonation constant of HOCl. Expressions (1) and (2) are identical and allow defining the electrochemical potential of the reduction of either HOCl or OCl^- to Cl^- at varying total concentrations of active chlorine, bromide, chloride, pH and ionic strengths.

The deprotonated form OCl^- will be used as the reference species in all calculations that follow. It is to be noted here that when the relevant equilibria that involve both chlorine and bromine species are taken into account, calculations of the electrochemical potential of the $HOBr/Br^-$ couple ($HOBr + H^+ + 2e \rightarrow Br^- + H_2O$) using equation (3) shown below yield the same values of the electrochemical potential of the system as those obtained with equation (1) or (2).

$$E_{HOBr/Br^-}(V) = 1.331 + \frac{2.303RT}{2F}\log\left(\frac{\{HOBr\}\{H^+\}}{\{Br^-\}}\right) \quad (3)$$

Calculations of electrochemical potentials of the $HOCl/Cl^-$ or OCl^-/Cl^- couples are not equivalent to the determination of actual redox potentials of aquatic systems that can be affected not only by halogen species but also by other oxidants (e.g., dissolved oxygen) and reducing species (e.g., NOM, EfOM, nitrite, etc.) present in the system. Still, given that the electrochemical potentials of the halogen species are considerably higher than that of the O_2/H_2O couple (*24*) and these species tend to be kinetically active, the electrochemical potential imposed by halogen species can be expected to be a good indicator of the redox status of aquatic systems containing chlorine and/or allied species.

Halogen Species Existing in Solutions Containing Species of Bromine, Chlorine and Chloride and Bromide Ions

This section addresses important features of systems containing chlorine and bromine species as well as Cl^- and Br^- ions in the absence of ammonia. The discussion below is largely based on chemical equilibria involving these compounds. The kinetics and other aspects of these reactions have been extensively explored in prior literature (*10*, *19*, *25–28*).

Three major groups of chlorine- and bromine-containing species are known to exist in aqueous solutions (Figure 1). The first and in many cases predominant of these groups comprises hypohalogenous acids HOX and their anionic forms OX^- (e.g., hypochlorous and hypobromous acids and their anions, $HOCl/OCl^-$ and $HOBr/OBr^-$). The second tier of species shown in Figure 1 comprises hydrated chlorine and bromine molecules Cl_2 and Br_2 as well as bromine chloride BrCl; all these species are volatile. The third tier of the halogen species is constituted by four trihalogenide ions with a general stoichiometry $(Br_{3-i}Cl_i)^-$. These anions include Cl_3^-, $BrCl_2^-$, Br_2Cl^- and Br_3^-. Other reactions involving chlorine and bromine species, for instance the formation of Cl_2O (*10*) and reactions hypothetically involving this compound (*18*) can be mentioned here. Calculations presented in the section that follows demonstrate that predicted concentration of Cl_2O are expected to be lower than those of hydrated chlorine Cl_2, but this does not exclude a potentially pronounced role of this compound in oxidations carried out in situations favoring the formation of Cl_2 and/or Cl_2O.

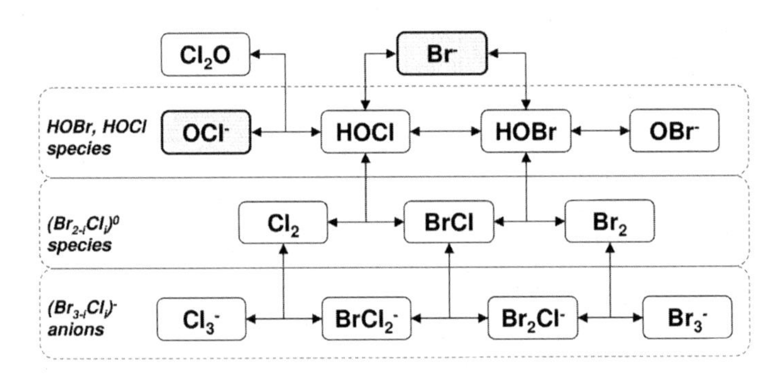

Figure 1. General scheme of formation of bromine and chlorine species in the absence of ammonia.

Examination of the published intrinsic rates of the formation of halogen species shown in Figure 1 indicates that, within time scales typical for water treatment, concentrations of these compounds reach a steady state nearly instantaneously (*19*). Accordingly, their speciation can be established using relevant equilibrium constants compiled by, for instance, Odeh et al. 2004 (*28*) (Table 1).

Table 1. Equilibrium constants of reactions governing the formation of chlorine- and bromine-containing species. Adopted from Odeh et al. 2004 (*28*)

Reaction	*Equilibrium constant*
(T1.1) $HOCl \leftrightarrow OCl^- + H^+$	pK = 7.47
(T1.2) $HOBr \leftrightarrow OBr^- + H^+$	pK = 8.59
(T1.3) $HOBr + Cl^- \leftrightarrow HOCl + Br^-$	$6.5 \cdot 10^{-6}$
(T1.4) $Cl_2 + H_2O \leftrightarrow HOCl + Cl^- + H^+$	$1.04 \cdot 10^{-3}$ M^2
(T1.5) $Br_2 + H_2O \leftrightarrow HOBr + Br^- + H^+$	$6.1 \cdot 10^{-9}$ M^2
(T1.6) $BrCl + H_2O \leftrightarrow HOBr + Cl^- + H^+$	$1.3 \cdot 10^{-4}$ M^2
(T1.7) $BrCl + H_2O \leftrightarrow HOCl + Br^- + H^+$	$8.7 \cdot 10^{-10}$ M^2
(T1.8) $2BrCl \leftrightarrow Cl_2 + Br_2$	$7.6 \cdot 10^{-3}$
(T1.9) $BrCl + Cl^- \leftrightarrow Cl_2 + Br^-$	$9.1 \cdot 10^{-7}$
(T1.10) $Cl_2 + Cl^- \leftrightarrow Cl_3^-$	0.18 M^{-1}
(T1.11) $Cl_2 + Br^- \leftrightarrow BrCl_2^-$	$4.2 \cdot 10^{6}$ M^{-1}
(T1.12) $BrCl + Cl^- \leftrightarrow BrCl_2^-$	3.8 M^{-1}
(T1.13) $Br_2 + Cl^- \leftrightarrow Br_2Cl^-$	1.3 M^{-1}
(T1.14) $BrCl + Br^- \leftrightarrow Br_2Cl^-$	$1.8 \cdot 10^{4}$ M^{-1}
(T1.15) $Br_2 + Br^- \leftrightarrow Br_3^-$	16.1 M^{-1}

Further analysis of these reactions involves a short discussion concerning the origin of bromine-containing species in typical water treatment situations. In the context of this paper, bromine species listed in Figure 1 and Table 1 are deemed to be formed solely as a result of the oxidation of the bromide ion (assumed to be present in the water due to local geochemical conditions rather than introduced as a treatment reagent) by HOCl or related species (e.g., reaction T1.3 in Table 1).

Correspondingly, chemical equilibria in these systems can be described based on two reference components (e.g., OCl^- and Br^-) while all the other species are assumed to be generated via interactions between these components and other solution components, notably H^+, Cl^- and Br^-. The introduction of these *a priori*-defined reference species allows for a more consistent and unambiguous interpretation of the equilibria of aquatic halogen compounds, with each reaction written to define explicitly the stoichiometries and formation constants of reactions that always start from the reference species, as opposed to the equations compiled in Table 1 that do not necessarily involve the reference species but rather compounds formed as a result on their interactions.

That is, this approach allows redefining reaction stoichiometries pertaining to the reactions shown in Table 1 and, using the data summarized in (*10*, *28*), calculating their modified equilibria constants. These constants that apply to the equilibria between OCl^-, Br^-, H^+, Cl^- and all the other halogen species formed as a result of their interactions are compiled in Table 2. Explicit equations relating activities of the species listed in Tables 1 and 2 are presented below.

Protonated form of hypochlorous acid

$$K_{HOCl} = \frac{\{HOCl\}}{\{OCl^-\}\{H^+\}} \tag{4}$$

Molecular chlorine

$$K_{Cl_2} = \frac{\{Cl_2\}}{\{OCl^-\}\{Cl^-\}\{H^+\}^2} \tag{5}$$

Trichloride ion

$$K_{Cl_3^-} = \frac{\{Cl_2\}}{\{OCl^-\}\{Cl^-\}^2\{H^+\}^2} \tag{6}$$

Dichlorine oxide

$$K_{Cl_2O} = \frac{\{Cl_2O\}}{\{OCl^-\}^2\{H^+\}^2} \tag{7}$$

Deprotonated form of hypobromous acid

$$K_{OBr^-} = \frac{\{OBr^-\}\{Cl^-\}}{\{OCl^-\}\{Br^-\}} \tag{8}$$

Protonated form of hypobromous acid

$$K_{HOBr} = \frac{\{HOBr\}\{Cl^-\}}{\{OCl^-\}\{Br^-\}\{H^+\}} \quad (9)$$

Molecular bromine

$$K_{Br_2} = \frac{\{Br_2\}\{Cl^-\}}{\{OCl^-\}\{Br^-\}^2\{H^+\}^2} \quad (10)$$

Bromine chloride

$$K_{BrCl} = \frac{\{BrCl\}}{\{OCl^-\}\{Br^-\}\{H^+\}^2} \quad (11)$$

Bromodichloride ion

$$K_{BrCl_2^-} = \frac{\{BrCl_2^-\}}{\{OCl^-\}\{Cl^-\}\{Br^-\}\{H^+\}^2} \quad (12)$$

Dibromochloride ion

$$K_{Br_2Cl^-} = \frac{\{Br_2Cl^-\}}{\{OCl^-\}\{Br^-\}^2\{H^+\}^2} \quad (13)$$

Tribromide ion

$$K_{Br_3^-} = \frac{\{Br_3^-\}\{Cl^-\}}{\{OCl^-\}\{Br^-\}^3\{H^+\}^2} \quad (14)$$

Equilibrium constants compiled in Table 2 can be employed in calculations made with chemical equilibria programs, for instance MINEQL+ or Visual MINTEQ (*29*, *30*), to ascertain effects of solution composition on the speciation of halogen species. Because not all equilibria considered in this chapter are included in the default databases of these programs, this document provides sufficient information for the user to modify these or similar databases. Results of such calculations for selected representative aquatic systems are described in the sections that follow.

Table 2. Equilibrium constants for halogen species formation reactions based on OCl^- and Br^- reference species. All constants are derived from on the data presented in refs. (*10*) and (*28*) (Table 1)

Species	*Formal reaction stoichiometry*	*LogK$_i$*
HOCl	(T2.1) $OCl^- + H^+ \leftrightarrow HOCl$	7.47
Cl_2	(T2.2) $OCl^- + Cl^- + 2H^+ \leftrightarrow Cl_2 + H_2O$	10.5
Cl_3^-	(T2.3) $OCl^- + 2Cl^- + 2H^+ \leftrightarrow Cl_3^- + H_2O$	9.7
Cl_2O	(T2.4) $2OCl^- + 2H^+ \leftrightarrow Cl_2O + H_2O$	12.9
OBr^-	(T2.5) $OCl^- + Br^- \leftrightarrow OBr^- + Cl^-$	4.1
HOBr	(T2.6) $OCl^- + Br^- + H^+ \leftrightarrow HOBr + Cl^-$	12.7
Br_2	(T2.7) $OCl^- + 2Br^- + 2H^+ \leftrightarrow Br_2 + H_2O + Cl^-$	20.9
BrCl	(T2.8) $Br^- + OCl^- + 2H^+ \leftrightarrow BrCl + H_2O$	16.5
$BrCl_2^-$	(T2.9) $OCl^- + Cl^- + Br^- + 2H^+ \leftrightarrow BrCl_2^- + H_2O$	16.3
Br_2Cl^-	(T2.10) $OCl^- + 2Br^- + 2H^+ \leftrightarrow Br_2Cl^- + 2H_2O$	21.0
Br_3^-	(T2.11) $OCl^- + 3Br^- + 2H^+ \leftrightarrow Br_3^- + Cl^- + H_2O$	22.1

General Features of Chlorine and Bromine Speciation in Typical Surface Waters

Surface waters are normally characterized by low ionic strengths (e.g., 10^{-3} to 10^{-2} M), low to moderate chloride concentration (e.g., from <10 to 250 mg/L, or $2.8 \cdot 10^{-4}$ to $7.0 \cdot 10^{-3}$ M) and Br^- levels < 1 mg/L (<$1.25 \cdot 10^{-5}$ M). Calculations of the speciation of active chlorine and bromine species in these conditions yield well-known and expected results. That is, the speciation of active chlorine is dominated at HOCl and OCl^- at pH < 7.4 and >7.6, respectively (Figure 2A). For a 150 mg/L Cl^- concentration, the formation of Cl_2 becomes noticeable at pH < 3 while the presence of the Cl_3^- anion remains insignificant at all pHs. The speciation of active bromine species has features similar to those shown in Figure 2A, with HOBr and OBr^- predominating at pH < 8.4 and 8.7, respectively and the presence of bromine chloride BrCl becoming important at pH < 3.5 (Figure 2B).

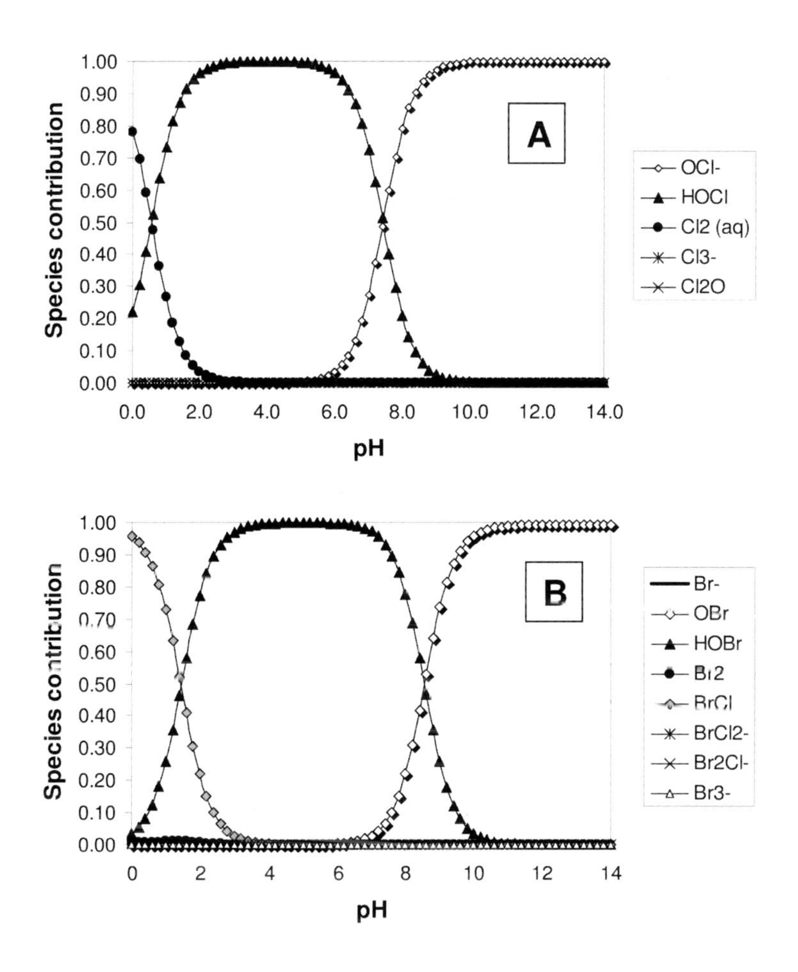

Figure 2. Speciation of (A) chlorine in the absence of bromide and (B) bromine species in a systems containing 4 mg/L total active chlorine, 1 mg/L bromide, 150 mg/L chloride, ionic strength 0.01 M.

Other details of the speciation of chlorine compounds represented in Figure 2 can be examined based on a logC representation of the concentrations of relevant compounds (Figure 3). These data show that at low to medium chloride concentrations and ionic strengths typical for a majority of surface waters, concentrations of Cl_2, Cl_3^- and Cl_2O are always low. Despite their low concentrations, these species and especially Cl_2 and possibly Cl_2O can play an important role in the oxidations of many PPCPs, notably those with phenolic functional groups in their structures since Cl_2 appears to have higher intrinsic rates of the oxidative attack on the phenolic structures (*5*, *8*, *10*, *18*) compared with HOCl and even more so with OCl^-.

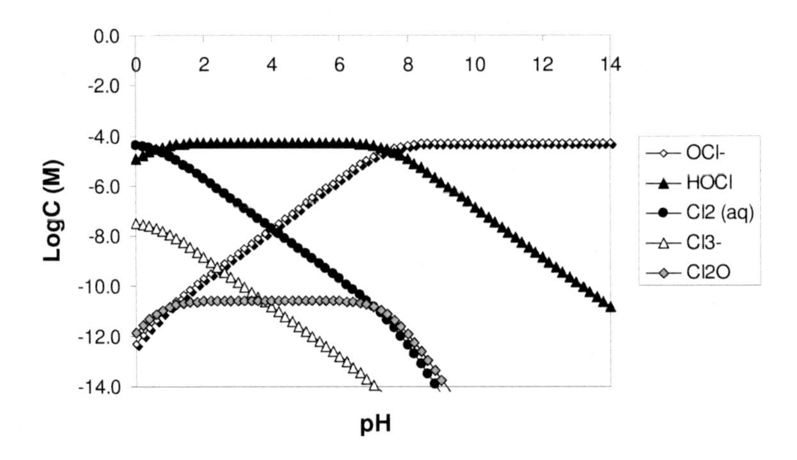

Figure 3. LogC plot for chlorine species in a solution containing 4 mg/L total chlorine, 150 mg/ Cl^-, no bromide, ionic strength 0.01 M.

In systems having low- to medium chloride concentrations and ionic strengths, the speciation of bromine species differs prominently from that shown in Figure 2. While effects of variations of background chloride concentrations will not be discussed in detail here, data generated for a representative situation (ionic strength 0.01 M, chloride concentration 150 mg/L) demonstrate that a small fraction of Br^- ion (that is subject to oxidation by chlorine) remains in the system due to the reversibility of reaction T1.3 in Table 1 (Figure 4). The concentration of bromine chloride BrCl prominent at pH< 3 is predicted to exceed considerably that of Br_2 and other species that can affect rates of the oxidation of trace-level organic species at low pHs (*5*, *8*, *10*). On the other hand, concentrations of $BrCl_2^-$, Br_2Cl^- and Br3- are predicted to remain low in all conditions.

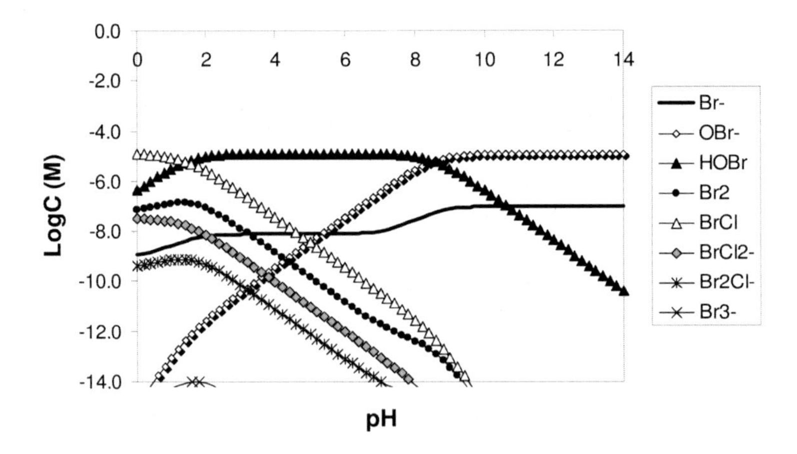

Figure 4. LogC plot for halogen species in a system containing 4 mg/L total chlorine, 1 mg/L bromide, 150 mg/L Cl^-, ionic strength 0.01 M.

Speciation of Halogen Compounds in Seawater

Seawater has average chloride and bromide concentrations of 0.56 M and $8.62 \cdot 10^{-4}$ M, respectively, while its ionic strength is close to 0.7 M. Results of numeric modeling on the speciation of halogen species generated as a result of addition of $5.6 \cdot 10^{-5}$ M (4 mg/L) active chlorine to seawater is shown in Figure 5. It demonstrates that at pH <5 and in the absence of ammonia, halogen species are dominated by molecular bromine Br_2, dibromochloride ion Br_2Cl^- and to a lesser extent by BrCl (Figure 5A).

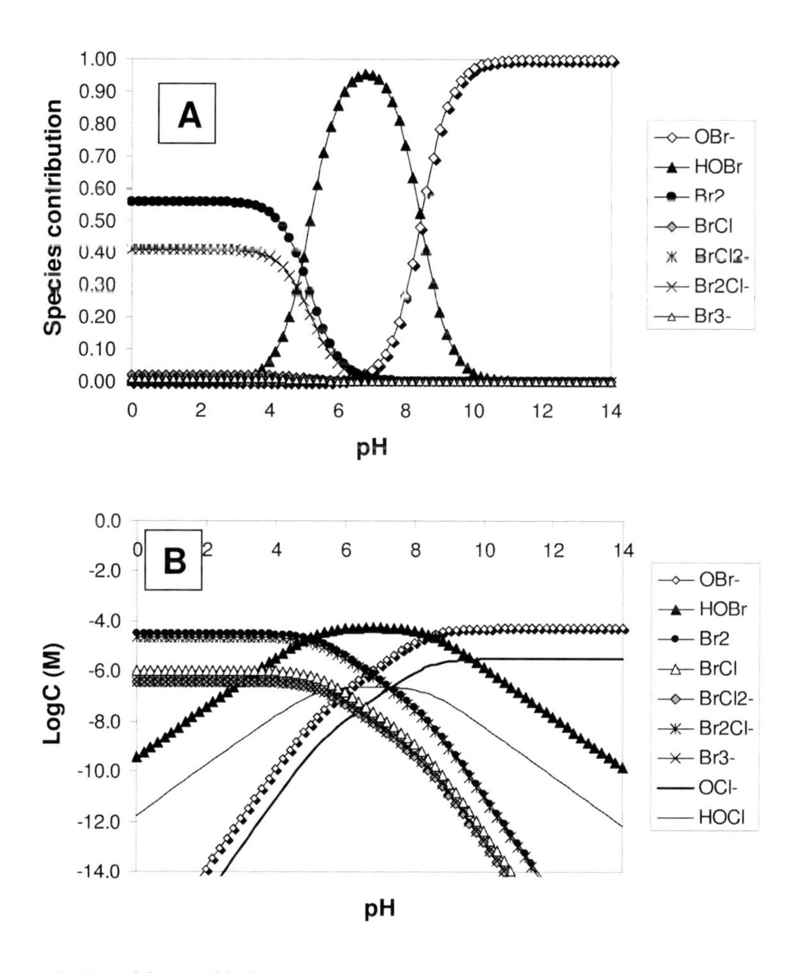

Figure 5. Equilibria of halogen species in seawater (ionic strength 0.7 M, Cl^- and Br^- concentrations 0.56 and $8.62 \cdot 10^{-4}$ M). (A) chlorine and bromine speciation; (B) logC plot for HOCl, OCl^- and bromine species.

Hypobromous acid HOBr tends to dominate in the range of pH from 5 to 8 while at pH > 8.6 the anionic form OBr^- becomes dominant, with a notable presence of the OCl^- anion. LogC plots shown in Figure 5B complement this observation indicating that at low pHs the system is dominated by bromine species while in the circumneutral pHs and especially at pH > 8.6 contributions of HOCl and OCl^- becoming considerably more important.

Examination of electrochemical potentials of the $HOCl/Cl^-$ couple ($E_0 = 1.482$ V (*24*)) shows that the oxidative power of halogen species in seawater is expected to be lower compared with that in surface waters (Figure 6).

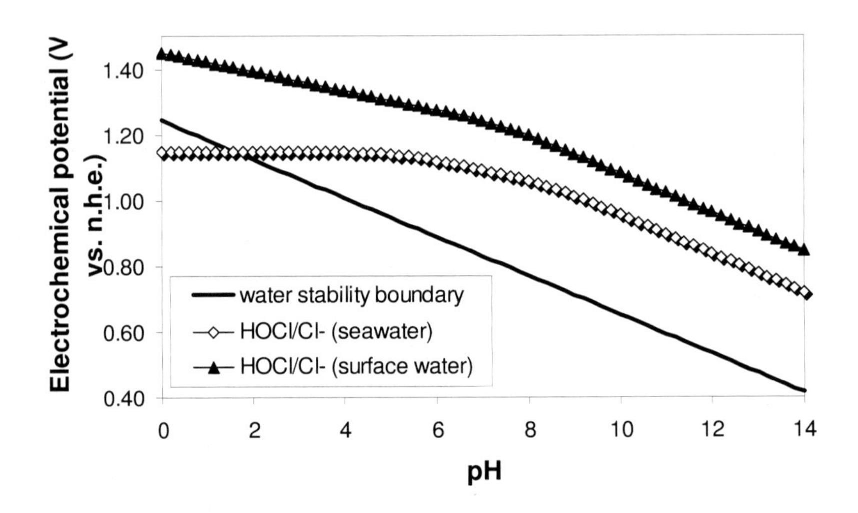

Figure 6. Electrochemical potentials of the $HOCl/Cl^-$ couple calculated for a typical surface water (Cl^- concentration $4.2 \cdot 10^{-4}$ M, traces of bromide, ionic strength 0.001 M) and seawater (Cl^- and Br^- concentrations 0.56 and $8.62 \cdot 10^{-4}$ M, respectively, ionic strength 0.7 M). Total active chlorine 4 mg/L as Cl_2.

In the latter case, electrochemical potentials of the $HOCl/Cl^-$ couple are always above the electrochemical potential of the O_2/H_2O couple (this corresponds to the nominal water stability boundary). In seawater, the electrochemical potential of the $HOCl/Cl^-$ couple is lowered due to the formation of trihalogenide ions and less oxidation-active species of bromine combined with a much higher concentration of Cl^- ion.

In fact, at pH<2, $E_{HOCl/Cl-}$ values in seawater are predicted to be lower than those of the O_2/H_2O couple while in the circumneutral range of pHs, for instance for pHs between 6 and 8, the $HOCl/Cl^-$ electrochemical potentials change from 1.12 to 1.06 V in seawater, while this potential changes from 1.27 to 1.20 V vs. in a typical surface water.

Formation of Haloamines

Elucidation of the kinetics, equilibria and speciation of haloamines has been a challenging issue over last several decades. While tremendous progress has been made in determining the intrinsic rates of chloramines and bromamines generation and their reactivity (*31–40*), numerous questions concerning the speciation and persistence of haloamines, especially bromamines in environmental systems remain. While this chapter can not address these complex issues in sufficient detail, it will examine general features pertinent to haloamine speciation.

A general scheme of the incorporation of chlorine and bromine into haloamines is presented in Figure 7. That figures includes mono-, di- and tri-chloro- and bromamines as well as mixed haloamines such NHBrCl, $HBrCl_2$ and NBr_2Cl. The existence of the mixed haloamine species has been documented in literature (*41–43*), but the kinetics and equilibria of their formation have not been explored as extensively as those of chloro- and bromamines.

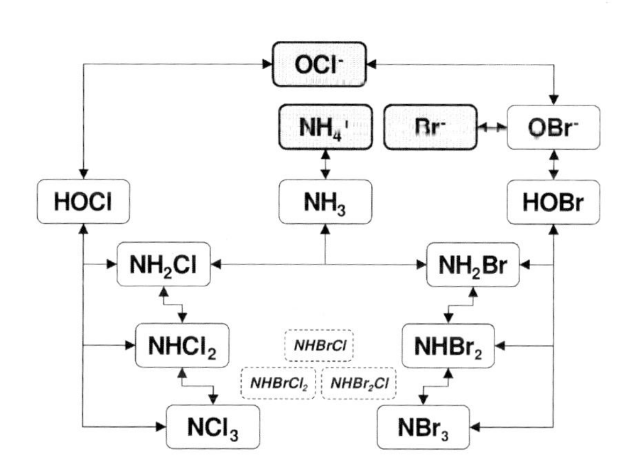

Figure 7. General scheme of haloamines formation in a system containing ammonia, chlorine and bromine formed via the oxidation of bromide.

Prior literature provides extensive evidence that reactions controlling the concentrations and speciation of haloamines are kinetically controlled. This appears to be especially true for reactions involving bromamines that tend to undergo both rapid formation and breakdown (*36*, *37*, *44*, *45*).

As a result, examination of properties of aquatic systems containing haloamines should be primarily based on the kinetics of relevant reactions. Nonetheless, some of the major aspects of the chemistry of chloramines can be traced via examination of the apparent equilibria related to these reactions.

A detailed sequence of NH_2Cl, $NHCl_2$ and NCl_3 formation and breakdown reactions ultimately accompanied by the reduction of OCl^- to Cl^- and oxidation of NH_3 to N_2 and NO_3^- was presented by Valentine et al. in refs. (*33–35*) and ensuing publications that also demonstrate the formation of yet unidentified products of haloamine breakdown (*46, 47*). Rates of the most important forward and reverse reactions involving chloramines are complied in Table 3. While representative, this table is incomplete because it does not reflect effects of NOM on the degradation of haloamines (*35*).

Table 3. Rates of forward and reverse reactions associated with the formation and breakdown of chloramines (based on the compilation presented in refs. (*33–35*)

Formal reaction stoichiometry	*Rate constant at 25°C*
(T3.1f) $HOCl + NH_3 \xrightarrow{k_{1f}} NH_2Cl + H_2O$	$k_{1f} = 1.5 \cdot 10^{10}$ $M^{-1}h^{-1}$
(T3.1r) $NH_2Cl \xrightarrow{k_{1r}} HOCl + NH_3$	$k_{1r} = 7.6 \cdot 10^{-2}$ h^{-1}
(T2.2f) $HOCl + NH_2Cl \xrightarrow{k_{1f}} NHCl_2 + H_2O$	$k_{1f} = 1.0 \cdot 10^{6}$ $M^{-1}h^{-1}$
(T3.2r) $NHCl_2 \xrightarrow{k_{1r}} HOCl + NH_2Cl$	$k_{2r} = 2.3 \cdot 10^{-3}$ h^{-1}
(T3.3f) $NH_2Cl + NH_2Cl \xrightarrow{k_{3f}} NHCl_2 + NH_3$	See footnote a
(T3.3r) $NHCl_2 + NH_3 \xrightarrow{k_{3r}} NH_2Cl + NH_2Cl$	$k_{3r} = 2.16 \cdot 10^{8}$ $M^{-2}h^{-1}$
(T3.4) $NHCl_2 + H_2O \xrightarrow{k_4} Intermediate$	$k_4 = 4.0 \cdot 10^{5}$ $M^{-1}h^{-1}$
(T3.5) $Intermediate + NHCl_2 \xrightarrow{k_5} HOCl + products$	$k_5 = 1.0 \cdot 10^{8}$ $M^{-1}h^{-1}$
(T3.6) $Intermediate + NH_2Cl \xrightarrow{k_6} products$	$k_6 = 3.0 \cdot 10^{7}$ $M^{-1}h^{-1}$
(T3.7) $NH_2Cl + NHCl_2 \xrightarrow{k_7} products$	$k_7 = 55$ $M^{-2}h^{-1}$
(T3.8) $HOCl + NHCl_2 \xrightarrow{k_8} NCl_3 + H_2O$	See footnote b
(T3.9) $NHCl_2 + NCl_3 + 2H_2O \xrightarrow{k_9} 2HOCl + products$	$k_9 = 2.0 \cdot 10^{14}$ $M^{-2}h^{-1}$

Continued on next page.

Table 3. (Continued). Rates of forward and reverse reactions associated with the formation and breakdown of chloramines (based on the compilation presented in refs. (*33–35*)

Formal reaction stoichiometry	*Rate constant at 25°C*
(T3.10) $NHCl_2 + NCl_3 + H_2O \xrightarrow{k_{10}} 2HOCl + products$	$k_{10} = 5.0 \cdot 10^{12}$ $M^{-2}h^{-1}$
(T3.11) $NHCl_2 + 2HOCl + H_2O \xrightarrow{k_{10}} NO_3^- + 5H^+ + 4Cl^-$	$k_{11} = 8.3 \cdot 10^5$ $M^{-1}h^{-1}$

[a] The rate of reaction (T3.3f) depends on the pH as this reaction is deemed to be catalyzed by the proton, carbonate and phosphate species:

$$k_{3f} = k_{3f\_H^+}[H^+] + k_{3f\_H_2CO_3}[H_2CO_3] + k_{3f\_HCO_3^-}[HCO_3^-] + k_{3f\_H_3PO_4}[H_3PO_4] + k_{3f\_H_3PO_4^-}[H_2PO_4^-]$$

The values of k_{3f_H+}, k_{3f_H2CO3}, k_{3f_HCO3-}, k_{3f_H3PO4} and k_{3f_H2PO4-} are $2.5 \cdot 10^7$, $2.7 \cdot 10^3$, 7.2, $3.2 \cdot 10^6$ and $1.3 \cdot 10^3$ $M^{-2}h^{-1}$, respectively. [b] The rate of reaction (T3.8) is a function of pH as a result of catalytic action of hydroxyl-ion, chloride and carbonate:

$$k_8 = k_{OCl^-}[OCl^-] + k_{OH^-}[OH^-] + k_{CO_3^{2-}}[CO_3^{2-}]$$

The values of k_{OCl-}, k_{OH-} and k_{CO32-} in the above expression are $3.24 \cdot 10^8$ $M^{-2}h^{-1}$, $1.18 \cdot 10^{13}$ $M^{-2}h^{-1}$ and $2.16 \cdot 10^{10}$ $M^{-2}h^{-1}$, respectively.

General Aspects of Haloamine Formation Equilibria

In principle, the speciation of haloamines can be examined based on the kinetic rates of reactions of step-wise incorporation of halogen atoms into NH_3 and ensuing formation of mono-, di- and tri- haloamines denoted as NH_2X, NHX_2 and NHX_3, respectively, where X is Cl or Br. Excluding for the purposes of this discussion the formation of mixed haloamines (e.g., NHBrCl), equilibria of each step of these reactions can be defined as shown in equations (15), (16) and (17) below:

$$K_{NH_2X} = \frac{\{NH_2X\}}{\{HOX\}\{NH_3\}} \qquad (15)$$

$$K_{NHX_2} = \frac{\{NHX_2\}}{\{HOX\}\{NH_2X\}} \qquad (16)$$

$$K_{NX_3} = \frac{\{NX_3\}}{\{HOX\}\{NHX_2\}} \qquad (17)$$

Because the system was assumed to be governed by reactions originating from three major reference species (e.g., OX^-, Br^- and NH_4^+) that, in accord with the approach employed here are considered to be the progenitors of all the other compounds in the system, equilibria written in equations (15), (16) and (17) need to be redefined to account for the references status of OX^-, Br^- and NH_4^+. As a result, the following formal expressions can be generated and applied to the formation of chloramines (equations 18, 19 and 20):

Monochloramine

$$\{NH_2Cl\} = K_{NH_2Cl}\{HOCl\}\{NH_3\} = \frac{K_{NH_2cl}K_{NH_4^+}}{K_{HOCl}}\{OCl^-\}\{NH_4^+\} = \tag{18}$$
$$= K^*_{NH2Cl}\{OCl^-\}\{NH_4^+\}$$

Dichloramine

$$\{NHCl_2\} = K_{NHCl2}\{HOCl\}\{NH_2Cl\} = \frac{K_{NHCl2}K^*_{NH2Cl}}{K_{HOCl}}\{H^+\}\{OCl^-\}^2\{NH_4^+\} = \tag{19}$$
$$= K^*_{NHCl2}\{H^+\}\{OCl^-\}^2\{NH_4^+\}$$

Trichloramine

$$\{NCl_3\} = K_{NCl3}\{HOCl\}\{NHCl_2\} = \frac{K_{NCl3}K^*_{NHCl2}}{K_{HOCl}}\{H^+\}^2\{OCl^-\}^3\{NH_4^+\} = \tag{20}$$
$$= K^*_{NCl3}\{H^+\}^2\{OCl^-\}^3\{NH_4^+\}$$

Similar expressions can be applied to the formation of bromamines. However, because the concentration of OBr^- in these reactions is assumed to be controlled by the fast oxidation of Br^- by HOCl and/or OCl^- (reaction T13 in Table 1), the concentration of OBr^- in the expressions accounting for the step-wise incorporation of bromine into NH_3, NH_2Br and $NHBr_2$ molecules is defined as a function of the activities of OCl^-, Cl^- and Br^- in the system:

$$\{OBr^-\} = \frac{K_{OBr^-}\{OCl^-\}\{Br^-\}}{\{Cl^-\}} \tag{21}$$

where the value of K_{OBr-} is given in Table 2.

Application of the above expression results in the following nominal formation expressions for bromamines NH_2Br, $NHBr_2$ and NBr_3 (equations 22, 23 and 24):

Monobromamine

$$\{NH_2Br\} = K^{*}_{NH2Br}\{OBr^-\}\{NH_4^+\} = K^{*}_{NH2Br}K_{OBr^-}\frac{\{NH_4^+\}\{OCl^-\}\{Br^-\}}{\{Cl^-\}} =$$
$$= K^{**}_{NH2Br}\frac{\{NH_4^+\}\{OCl^-\}\{Br^-\}}{\{Cl^-\}} \tag{22}$$

Dibromamine

$$\{NHBr_2\} = K^{*}_{NHBr2}\{H^+\}\{OBr^-\}^2\{NH_4^+\} =$$
$$= K^{*}_{NHBr2}\left(K_{OBr^-}\right)^2\frac{\{H^+\}\{NH_4^+\}\{OCl^-\}^2\{Br^-\}^2}{\{Cl^-\}^2} = K^{**}_{NHBr2}\frac{\{H^+\}\{NH_4^+\}\{OCl^-\}^2\{Br^-\}^2}{\{Cl^-\}^2} \tag{23}$$

Tribromamine

$$\{NBr_3\} = K^{*}_{NBr3}\{H^+\}^2\{OBr^-\}^3\{NH_4^+\} = K^{*}_{NBr3}\left(K_{OBr^-}\right)^3\frac{\{H^+\}^2\{OCl^-\}^3\{Br^-\}^3\{NH_4^+\}}{\{Cl^-\}^3} =$$
$$= K^{**}_{NBr3}\frac{\{H^+\}^2\{OCl^-\}^3\{Br^-\}^3\{NH_4^+\}}{\{Cl^-\}^3} \tag{24}$$

Limited Applicability of the Conversion of Kinetic Rates of Haloamine Formation to Apparent Equilibrium Constants

Apparent equilibrium constants of monochloramine and dichloramine formation can be determined as the ratio of the rates of forward and reverse NH_2Cl and $NHCl_2$ generation reactions, e.g., (T3.1f)/(T3.1r) and (T3.2f)/(T3.2r) shown in Table 3.

For monochloramine, that ratio yields a $\log K_{NH2Cl}$ value of 11.3 (Table 4), which is close to that utilized in recent publications concerned with effects of chloramines on metal release (*38*, *46–48*). Similarly, determination of the ratios of the forward and reverse kinetic rates of dichloramine formation (reactions T3.2f and T3.3r in Table 3) yield a $\log K_{NHCl2}$ estimate of 8.6. These and other apparent formation constants obtained for chloramines and bromamines (based on the data presented in ref. (*36*)) are compiled in Table 4.

Table 4. Formal haloamine formation constants estimated based on the rates of reactions complied in Table 3 and in ref. (*36*)

Species	$\log K_{NH_{3-i}Cl_i}$	$\log K^{*}_{NH_{3-i}Cl_i}$	$\log K_{NH_{3-i}Br_i}$	$\log K^{*}_{NH_{3-i}Br_i}$	$\log K^{**}_{NH_{3-i}Br_i}$
	Chloramines		*Bromamines*		
NH_2X	11.3	9.5	10.5	8.9	13.0
NHX_2	8.6	25.6	8.7	26.2	34.4
NX_3	a	a	6.7	44.5	56.8

[a] Apparent equilibrium constant corresponding to NCl_3 formation from $NHCl_2$ remains to be determined. Estimates show that the nominal value of $logK^{*}_{NH3\text{-}iCli}$ is likely to be close to 6.0. This corresponds to a 39.1 value of $logK^{*}_{NH3\text{-}iCli}$.

While the latter table presents estimates of nominal haloamine formation constants based on the available kinetic data and application of equations (15) to (24), it should be stated that the applicability of these estimates to modeling actual aquatic systems containing haloamines is likely to be quite limited.

Detailed discussion of these limitations goes beyond the scope of this document. In a general case, a rigorous determination of concentrations and speciation of haloamines must be done via kinetic modeling that includes forward and reverse reactions listed in Table 3 and elsewhere (*32–36*). However, MINEQL+ calculations indicate that there is a range of conditions where monochloramine is prevalent. For instance, at pH >7, NH_2Cl dominates in most conditions relevant to water treatment, while at lower pHs the formation of dichloramine, and in some conditions of trichloramine, is expected to become important.

The effects of monochloramine formation on oxidation processes in treated surface waters can be evaluated based on the changes of the electrochemical potential of the $HOCl/Cl^-$ couple in the presence of ammonia. Result of these calculations (that assumed the formation of monochloramine only at pH>7) are shown in Figure 8.

The data of numerical modeling indicate that NH_2Cl formation will cause a significant decrease in the redox conditions at practically all pHs relevant to water treatment. Calculations for varying ammonia levels show that the electrochemical potential of the $HOCl/Cl^-$ couple at a fixed pH typical for water treatment processes (e.g., pH 8) is expected to decrease when the molar concentration of ammonia approaches that of the total chlorine (Figure 9).

At equimolar concentrations of ammonia and chlorine, the decrease of the electrochemical potential at pH 8 is expected to be close to 0.075 V. This decrease in the electrochemical potential of the system has been shown to be significant enough to affect NOM oxidation and disinfection processes and destabilize some of the solids, for instance PbO_2, formed in drinking water distribution systems in the presence of chlorine (*38*, *46–48*).

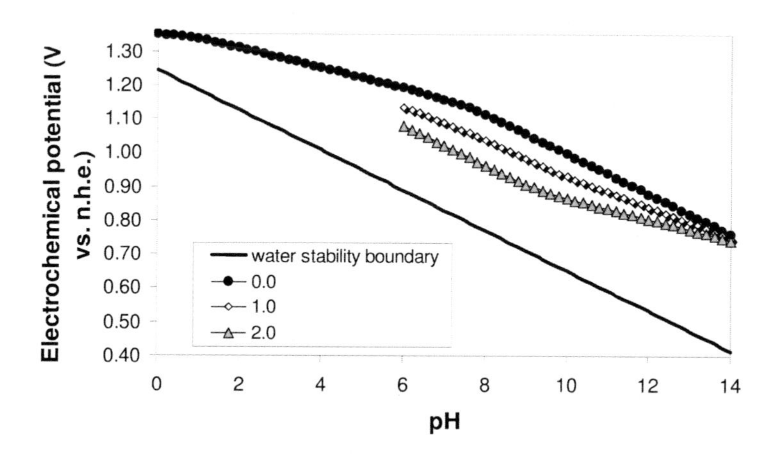

Figure 8. Effects of varying ammonia/total chlorine molar ratios on the electrochemical potential of the HOCl/Cl- couple. 150 mg/L chloride, no bromide, ionic strength 0.01 M. Total active chlorine concentration 4 mg/L as Cl_2.

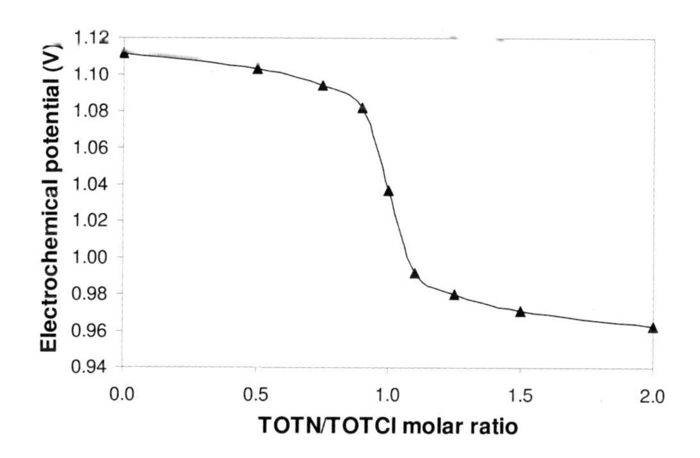

Figure 9. Change of the electrochemical potentials of the HOCl/Cl- couple at varying ammonia/total chlorine molar ratios at pH 8. 150 mg/L chloride, no bromide, ionic strength 0.01 M. Total active chlorine concentration 4 mg/L as Cl_2. NOM is assumed to be absent.

Results shown in Figure 8 and Figure 9 are likely to underestimate effects of haloamine formation on the redox potential of water since these calculations do not account neither for the formation of $NHCl_2$ and NCl_3 nor the effects of NOM on the equilibrium concentration of HOCl that ultimately defines the electrochemical potential of systems with halogens.

The formation of di- and trichloramine as well as the engagement of HOCl and OCl^- in reactions with amine groups (and other reactive functionalities) present in NOM will undoubtedly decrease the concentration of these oxidants. These effects can be quantified via detailed examination and subsequent interpretation of the redox potential of waters containing varying levels of chlorine, ammonia and NOM. These measurements have not been done with adequate consistency and need to be carried out in the future.

Conclusions

Major aspects of the speciation of chlorine and bromine in aquatic systems with widely varying pHs, concentrations of bromide, chloride and total active chlorine are discussed in this chapter. In the absence of ammonia, formal consideration of equilibria involving three main types of halogen species (hypohalogenous acids HOX and their anionic forms, molecular forms of halogens Cl_2, Br_2 and BrCl and trihalogenide ions $Br_iCl_{3-i}^-$) can be carried out based on two major reference species (OCl^-, Br^-) and accounting for effects of free Br^-, Cl^-, pH and ionic strength of the relevant reactions. Formal equilibria constants that are necessary for implementing such an approach were calculated in the paper and compiled in Table 2. Formal haloamine formation constants can also be defined based on the consideration of OCl^-, Br^- and NH_4^+ as reference species (Table 4) but this approach is deemed to have a limited applicability, primarily to conditions when monochloramine is expected to be prevalent. Detailed kinetic modeling that takes into account forward and reverse reactions associated with haloamine formation as well as redox transformation yielding Cl^-, N_2, NO_3^- and other species needs to be employed to model actual aquatic systems containing both ammonia and halogens. Redox conditions in such systems can be examined based on the electrochemical potential of the $HOCl/Cl^-$ couple that be modeled numerically and quantified experimentally.

Acknowledgments

This work was partially supported by the Chemical, Bioengineering, Environmental and Transport Systems Program (CBET) of the National Science Foundation (project # 0931676) and WateReuse Foundation (project # WRF09-10).

References

1. Hua, G.; Reckhow, D. A.; Kim, J. Effect of bromide and iodide ions on the formation and speciation of disinfection by-products during chlorination. *Environ. Sci. Technol.* **2006**, *42* (9), 3050–3056.
2. Smith, E. M.; Plewa, M. J.; Lindell, C. L.; Richardson, S. D.; Mitch, W. A. Comparison of byproduct formation in waters treated with chlorine and iodine: relevance to point-of-use treatment. *Environ. Sci. Technol.* **2010**, *44* (22), 8446–8452.

3. Bichsel, Y; von Gunten, U. Formation of iodo-trihalomethanes during disinfection and oxidation of iodide containing waters. *Environ. Sci. Technol.* **2000**, *34* (13), 2784–2791.
4. Latteman, S.; Hoepner, T. *Seawater Desalination. Impacts of Brine and Chemical Discharge on the Marine Environment*; Balaban Desalination Publications: 2003.
5. Deborde, M.; Rabouan, S.; Gallard, H.; Legube, B. Aqueous chlorination kinetics of some endocrine disruptors. *Environ. Sci. Technol.* **2004**, *38* (21), 5577–5583.
6. Gallard, H.; Leclercq, A.; Croué, J. P. Chlorination of bisphenol A: kinetics and by-products formation. *Chemosphere* **2004**, *56* (6), 465–473.
7. Westerhoff, P.; Yoon, Y.; Snyder, S.; Wert, E. Fate of endocrine-disruptor, pharmaceutical, and personal care product chemicals during simulated drinking water treatment processes. *Environ. Sci. Technol.* **2005**, *39* (17), 6649–6663.
8. Lu, J.; Korshin, G. V. A spectroscopic study of the bromination of the endocrine disruptor ethynyl estradiol. *Chemosphere* **2008**, *72* (3), 504–508.
9. Sharma, V. K. Oxidative transformations of environmental pharmaceuticals by Cl_2, ClO_2, O_3 and Fe(VI): kinetics assessment. *Chemosphere* **2008**, *73* (9), 1379–1386.
10. Deborde, M.; von Gunten, U. Reactions of chlorine with inorganic and organic compounds during water treatment - Kinetic and mechanisms: a critical review. *Water Res.* **2008**, *42* (1), 13–51.
11. Sirikanchana, K.; Shisler, J. L.; Marinas, B. J. Inactivation kinetics of adenovirus serotype 2 with monochloramine. *Water Res.* **2008**, *42* (6-7), 1467–1474.
12. Page, M. A.; Shisler, J. L.; Marinas, B. J. Kinetics of adenovirus type 2 inactivation with free chlorine. *Water Res.* **2009**, *43* (11), 2916–2926.
13. Amiri, F.; Mesquita, M. M. F.; Andrews, S. A. Disinfection effectiveness of organic chloramines, investigating the effect of pH. *Water Res.* **2010**, *44* (3), 845–853.
14. Dodd, M. C.; Huang, C. H. Transformation of the antibacterial agent sulfamethoxazole in reactions with chlorine: Kinetics mechanisms, and pathways. *Environ. Sci. Technol.* **2004**, *38* (21), 5607–5615.
15. Dodd, M. C.; Shah, A. D.; Von Gunten, U.; Huang, C. H. Interactions of fluoroquinolone antibacterial agents with aqueous chlorine: Reaction kinetics, mechanisms, and transformation pathways. *Environ. Sci. Technol.* **2005**, *39* (18), 7065–7076.
16. Dodd, M. C.; Huang, C. H. Aqueous chlorination of the antibacterial agent trimethoprim: Reaction kinetics and pathways. *Water Res.* **2007**, *41* (3), 647–655.
17. Acero, J. L.; Benitez, F. J.; Real, F. J.; Roldan, G. Kinetics of aqueous chlorination of some pharmaceuticals and their elimination from water matrices. *Water Res.* **2010**, *44* (14), 4158–4170.
18. Sivey, J. D.; McCullough, C. E.; Roberts, A. L. Chlorine monoxide (Cl_2O) and molecular chlorine (Cl_2) as active chlorinating agents in reaction of

dimethenamid with aqueous free chlorine. *Environ. Sci. Technol.* **2010**, *44* (9), 3357–3362.

19. Kumar, K.; Margerum, D. W. Kinetics and mechanisms of general acid-assisted oxidation of bromide by hypochlorite and hypochlorous acid. *Inorg. Chem.* **1987**, *26* (16), 2706–2711.
20. Heller-Grossman, L.; Idin, A.; Limoni-Relis, B.; Rebhun, M. Formation of cyanogen bromide and other volatile DBPs in the disinfection of bromide-rich lake water. *Environ. Sci. Technol.* **1999**, *33* (6), 932–937.
21. Richardson, S. D.; Thruston, A. D.; Rav-Acha, C.; Groisman, L.; Popilevsky, I.; Juraev, O.; Glezer, V.; McKague, A. B.; Plewa, M. J.; Wagner, E. D. Tribromopyrrole, brominated acids, and other disinfection byproducts produced by disinfection of drinking water rich in bromide. *Environ. Sci. Technol.* **2003**, *37* (17), 3782–3793.
22. Haag, W. R.; Hoigne, J. Ozonation of bromide-containing waters: kinetics of formation of hypobromous acid and bromate. *Environ. Sci. Technol.* **1983**, *17* (5), 261–267.
23. Faust, S. D.; Ali, O. M. *Chemistry of Water Treatment*; Ann Arbor Press: 1998.
24. *CRC Handbook on Chemistry and Physics*, 71st ed.; Lide, D. R., Ed.; CRC Press: Boca Raton, 1991.
25. Beckwith, R. C; Margerum, D. W. Kinetics of hypobromous acid disproportionation. *Inorg. Chem.* **1997**, *36* (17), 3754–3760.
26. Beckwith, R. C.; Wang, T. X.; Margerum, D. W. Equilibrium and kinetics of bromine hydrolysis. *Inorg. Chem.* **1996**, *35* (4), 995–1000.
27. Liu, Q.; Margerum, D. W. Equilibrium and kinetics of bromine chloride hydrolysis. *Environ. Sci. Technol.* **2001**, *35* (6), 1127–1133.
28. Odeh, I. N.; Nicolson, J. S.; Hartz, K. E. H.; Margerum, D. W. Kinetics and mechanisms of bromine chloride reactions with bromite and chlorite. *Inorg. Chem.* **2004**, *43* (23), 7412–7420.
29. Schecher, W. D.; McAvoy, D. C. MINEQL+. *A Chemical Equilibrium Modeling System.* Environmental Research Software, 1998 (updated versions at http://www.mineql.com/).
30. Visual MINTEQ Version 3.0 (http://www2.lwr.kth.se/English/OurSoftware/vminteq).
31. Valentine, R. L.; Jafvert, C. T.; Leung, S. W. Evaluation of a chloramine decomposition model incorporating general acid catalysis. *Water Res.* **1988**, *22* (9), 1147–1153.
32. Yiin, B. S.; Margerum, D. W. Non-metal redox kinetics: reactions of trichloramine with ammonia and with dichloramine. *Inorg. Chem.* **1990**, *29* (11), 2135–2141.
33. Jafvert, C. T.; Valentine, R. L. Reaction scheme for the chlorination of ammoniacal water. *Environ. Sci. Technol.* **1992**, *26* (3), 577–586.
34. Vikesland, P. J.; Ozekin, K.; Valentine, R. L. Effect of natural organic matter on monochloramine decomposition: Pathway elucidation through the use of mass and redox balances. *Environ. Sci. Technol.* **1998**, *32* (10), 1409–1416.
35. Vikesland, P. J.; Ozekin, K.; Valentine, R. L. Monochloramine decay in model and distribution system waters. *Water Res.* **2001**, *35* (7), 1766–1776.

36. Hofmann, R.; Andrews, R. C. Ammoniacal bromamines: a review of their influence on bromate formation during ozonation. *Water Res.* **2001**, *35* (3), 599–604.
37. Von Gunten, U.; Hoigné, J. Bromate formation during ozonation of bromide-contaning waters: interactions of ozone and hydroxyl radical reactions. *Environ. Sci. Technol.* **1994**, *28* (7), 1234–1242.
38. Rajasekharan, V. V.; Clark, B. N.; Boonsalee, S.; Switzer, J. A. Electrochemistry of free chlorine and monochloramine and its relevance to the presence of Pb in drinking water. *Environ. Sci. Technol.* **2007**, *41* (12), 4252–4257.
39. Lin, Y. P.; Valentine, R. L. Release of Pb(II) from monochloramine-mediated reduction of lead oxide (PbO_2). *Environ. Sci. Technol.* **2008**, *42* (24), 9137–9143.
40. Lin, Y. P.; Valentine, R. L. Reductive dissolution of lead dioxide (PbO_2) in acidic bromide solution. *Environ. Sci. Technol.* **2010**, *44* (10), 3895–3900.
41. Valentine, R. L. Bromodichloramine oxidation of N,N′-diethyl-p-phenylenediamine in the presence of monochloramine. *Environ. Sci. Technol.* **1986**, *20* (2), 166–170.
42. Yamamoto, K.; Fukushima, M. Disappearance rates of chlorine-induced combined oxidant in estuarine water. *Water Res.* **1992**, *26* (8), 1105–1109.
43. Gazda, M.; Dejarme, L. E.; Choudhury, T. K.; Cooks, R. G.; Margerum, D. W. Mass spectrometric evidence for the formation of bromochloramine and N-bromo-N-chloromethylamine in aqueous solution. *Environ. Sci. Technol.* **1993**, *27* (3), 557–561.
44. Lei, H. X.; Marinas, B. J.; Minear, R. A. Bromamine decomposition kinetics in aqueous solutions. *Environ. Sci. Technol.* **2004**, *38* (7), 2111–2119.
45. Lei, H. X.; Minear, R. A.; Marinas, B. J. Cyanogen bromide formation from the reactions monobromamine and dibromamine with cyanide ion. *Environ. Sci. Technol.* **2006**, *40* (8), 2559–2564.
46. Leung, S. W.; Valentine, R. L. An unidentified chloramine decomposition product. 1. Chemistry and characteristics. *Water Res.* **1994**, *28* (6), 1475–1483.
47. Leung, S. W.; Valentine, R. L. An unidentified chloramine decomposition product. 1. A proposed formation mechanism. *Water Res.* **1994**, *28* (6), 1485–1495.
48. Boyd, G. R.; Dewis, K. M.; Korshin, G. V.; Reiber, S. H.; Sandvig, A. M.; Giani, R. Effects of changing disinfectants on lead and copper release in distribution systems – a review. *J. - Am. Water Works Assoc.* **2008**, *100* (11), 75–84.

Chapter 12

Remediation of Chemically-Contaminated Waters Using Sulfate Radical Reactions: Kinetic Studies

Stephen P. Mezyk,[1,*] Kimberly A. Rickman,[1] Garrett McKay,[1] Charlotte M. Hirsch,[1] Xuexiang He,[2] and Dionysios D. Dionysiou[2]

[1]Department of Chemistry and Biochemistry, California State University Long Beach, 1250 Bellflower Blvd., Long Beach, CA 90840
[2]School of Energy, Environment, Biological and Medical Engineering, University of Cincinnati, Cincinnati, OH 45221-0012
*Phone: 562-985-4649, Fax: 562-985-8557, Email: smezyk@csulb.edu

The quantitative removal of chemical contaminants in water is one of the most pressing problems facing water utilities today. To augment traditional water treatments that are usually based on adsorptive and chemical-physical processes, radical-based advanced oxidation and reduction processes (AO/RPs) are now being considered. While most AO/RPs utilize the hydroxyl radical in treatment the use of oxidizing sulfate radicals is also gaining interest. To help assess the applicability of sulfate radical based AO/RPs in remediating contaminated waters, here we have determined absolute rate constants and reaction mechanisms for $SO_4^{-\bullet}$ reaction with four β-lactam antibiotics (amoxicillin, penicillin-G, piperacillin, tircarcillin), three estrogenic steroids (ethynylestradiol, estradiol, and progesterone) and one personal care product (isoborneol). For the four antibiotics of this study the relatively fast rate constant values suggests that the majority of the $SO_4^{-\bullet}$ oxidation occurs at the sulfur atom in the ring adjacent to the β-lactam moiety, as opposed to the hydroxyl radical reaction which occurs at peripheral aromatic rings. The measured sulfate radical rate constants for estradiol and progesterone are identical, with the slightly faster value for ethynylestradiol suggesting significant oxidation occurring at its ethynyl moiety. For isoborneol, the sulfate radical reactivity was slightly lower, but still fast enough that AO/RP treatment utilizing this radical might be feasible at large-scale. Piperacillin was also chosen for a detailed investigation of its degradation by both $SO_4^{-\bullet}$ and $^{\bullet}OH$ in a laboratory scale homogeneous UV photochemical system. It was found that although the absolute reaction rate constant for piperacillin reaction with $SO_4^{-\bullet}$ was lower than for $^{\bullet}OH$, the overall removal of this antibiotic was more effective when using $UV/S_2O_8^{2-}$ than UV/H_2O_2. For an initial oxidant dose of 1 mM and an antibiotic concentration of 50 μM, percentage removals of 65.2% and 33.0%, respectively, at a UV fluence of 320 mJ/cm^2 were obtained. This difference was attributed to the higher quantum yield of sulfate radical production from persulfate under UV 254 nm irradiation.

Introduction

The adverse ecological impacts of endocrine-disrupting compounds, personal care products, antibiotics, and pesticides or herbicides in water supplies (*1–6*) and wastewater effluents are causing concern amongst regulatory groups and the public. Traditional water treatment relies primarily upon adsorptive and chemical-physical processes to remove or transform these unwanted organic contaminants. However, these treatment processes may sometimes not be sufficient (*6*), as quantitative removal of low (ng L^{-1}) concentrations of dissolved chemicals may be complicated by the presence of much higher levels of other water constituents such as dissolved organic matter (DOM) and carbonate.

In order to augment traditional water treatment processes, the use of *in-situ* generated radical species, such as the oxidizing hydroxyl radical ($^{\bullet}OH$) and/or reducing electron (e_{aq}^-) and hydrogen atoms ($H^{\bullet}$), to react with and destroy trace contaminants following standard water treatment processes could be a viable approach. These additional treatments are generally referred to as advanced oxidation/reduction processes (AO/RPs) (*7–11*). These radicals can be created using a variety of techniques (*12*); for example, the hydroxyl radical ($^{\bullet}OH$) can be generated through using a combination of O_3/H_2O_2, O_3/UV-C, or H_2O_2/UV-C, and mixtures of $^{\bullet}OH$, e_{aq}^-, and $H^{\bullet}$ are produced from the UV irradiation of titanium dioxide, sonolysis, or the irradiation of water via electron beams or γ rays.

The utilization of reducing radicals to destroy chemical contaminants in real-world waters is problematic due to the presence of dissolved oxygen, which preferentially scavenges these radicals to create the much less reactive superoxide radical, $O_2^{-\bullet}$ ($[O_2] \sim 2.5 \times 10^{-4}$ M, $k = 1.9 \times 10^{10}$ M^{-1} s^{-1} (*13*)). Therefore, the most widespread AO/RPs are based on only the $^{\bullet}OH$ radical production. However, another AO/RP that is gaining interest utilizes sulfate radical ($SO_4^{-\bullet}$) reactions (*14*). The sulfate radical is also strongly oxidizing ($E^o = 2.3$ V, (*15*)) which means that it can react with most organic chemical contaminants. It typically reacts by electron abstraction from electron-rich centers in molecules, in contrast to hydroxyl radical based oxidations that mainly occur by hydrogen atom abstraction and/or addition to aromatic moieties.

Sulfate radicals can be readily formed through persulfate ($S_2O_8^{2-}$) decomposition induced by UV light (UV/ $S_2O_8^{2-}$), the presence of a catalyst, or higher temperatures. In addition, they can be formed in water by the addition of $S_2O_8^{2-}$ to selected AO/RPs, where the reducing radicals will react with this species according to (*13*):

$$S_2O_8^{2-} + e_{aq}^- \rightarrow SO_4^{2-} + SO_4^{-\bullet} \qquad k_1 = 1.2 \times 10^{10}\ M^{-1}\ s^{-1} \qquad (1)$$

$$S_2O_8^{2-} + H^{\bullet} \rightarrow H^+ + SO_4^{2-} + SO_4^{-\bullet} \qquad k_2 = 1.4 \times 10^{7}\ M^{-1}\ s^{-1} \qquad (2)$$

Persulfate addition itself has been shown to be effective in treating subsurface soils contaminated with chemicals such as diesel (*16*), PCBs (*17*), PAHs (*18*, *19*), chlorinated hydrocarbons (*20*, *21*) and VOCs (*22*). These remediation processes are considerably enhanced by persulfate activation, which increases its rate of decomposition to form sulfate radicals. However, much less investigation of sulfate radical use in chemically-contaminated water treatment has been reported. The sulfate radical reaction has previously been shown to have high efficiency in degrading model organic chemicals in water (*14*), but little is known about its chemistry with contaminants of higher molecular weight in real-world waters.

The optimal, quantitative, removal of water contaminants through the use of AO/RPs requires a thorough understanding of the redox chemistry occurring between free radicals and the contaminant chemicals of concern under the conditions of use. This can be accomplished through kinetic computer models, that give the most information and provide the best test of the proposed treatment (*23*) as all the chemistry in the system is considered. A critical component for kinetic modeling of any free radical based process is the full understanding of the kinetics and mechanisms of the radical reactions occurring. These fundamental data allow for quantitative computer modeling of AO/RP systems to establish the feasibility and large-scale efficiency of using radicals for specific contaminant removal under real-world conditions.

Therefore, in this work we describe absolute rate constant measurements for the sulfate radical reaction with four typical antibiotics, three representative estrogenic steroids, and one personal care compound isoborneol (see Figure 1) measured using an electron pulse radiolysis system. In addition, the radical-induced degradation of one specific antibiotic, pipericillin, was investigated in a laboratory scale homogeneous photochemical system using UV 254 nm /$S_2O_8^{2-}$ and UV 254 nm/H_2O_2 to provide further insight into the radical chemistry occurring.

Experimental

Kinetic Studies

Chemicals used in this study were purchased from Sigma-Aldrich Chemical Company at the highest purity available (steroids, >98%, KSCN, 99%, $K_2S_2O_8$, 99%, $Na_2S_2O_8$, 98%, β-lactam antibiotics >98%, isoborneol >98%). m-Toluic acid (3-methylbenzoic acid) was purchased from Fisher Scientific (99%). All were used as received.

There are many possible methods of producing AO/RP radicals (*12*) but the use of an electron beam for kinetic studies is optimal, as its energy deposition quantitatively generates a mixture of $^{\bullet}OH$, e_{aq}^- and $H^{\bullet}$ directlyfrom breaking bonds and ionizing water molecules (*13*) according to the stoichiometry:

$$H_2O \text{ -/\!\!\!\!\/\!\!\!\/\!\!\!\/\!\!\rightarrow } [0.28]\,^{\bullet}OH + [0.27]e_{aq}^- + [0.06]H^{\bullet} + [0.07]H_2O_2 + [0.05]H_2 + [0.27]H_3O^+ \quad (3)$$

The numbers preceding each species in Equation (3) are their absolute yields (G-value) in units of μmol J^{-1}. These yields are constant for solutions containing relatively low concentrations (< 0.10 M) of solutes irradiated in the pH range 3-10. Electron pulse radiolysis allows generation of all these species in nanoseconds, while the secondary reaction between the produced radicals and any added solute molecule typically occurs on a microsecond timescale. Hydrogen peroxide reactions occur at much longer times (milliseconds or greater), and so do not interfere with radical kinetic measurements.

The study of sulfate radical kinetic measurements by this technique requires the prior removal of hydroxyl radicals in order to isolate the reducing species. Therefore, these kinetic experiments were conducted using a constant high concentration of *tert*-butanol, $(CH_3)_3COH$, as a co-solvent (0.5-2.0 M), which immediately scavenges the radiolytically produced hydroxyl radicals and most hydrogen atoms (Equations 4 & 5) to produce the relatively inert $^{\bullet}CH_2(CH_3)_2COH$ alcohol radical (*13*):

$$(CH_3)_3COH + {}^{\bullet}OH \rightarrow {}^{\bullet}CH_2(CH_3)_2COH + H_2O \quad k_4 = 6.6 \times 10^8\ M^{-1}\ s^{-1} \quad (4)$$

$$(CH_3)_3COH + H^{\bullet} \rightarrow {}^{\bullet}CH_2(CH_3)_2COH + H_2 \quad k_5 = 1.7 \times 10^5\ M^{-1}\ s^{-1} \quad (5)$$

The isolated hydrated electron quantitatively reacts with added persulfate (in our experiments 5.0 mM) to give the oxidizing sulfate radical. The sulfate radical will also slowly react with the added *tert*-butanol (*13*),

$$(CH_3)_3COH + SO_4^{-\bullet} \rightarrow SO_4^{2-} + {}^{\bullet}CH_2(CH_3)_2COH + H^+ \quad k_6 = 8.4 \times 10^5\ M^{-1}\ s^{-1} \quad (6)$$

but by careful selection of added concentrations a significant fraction of sulfate radicals will react with the added chemical solute.

All rate constant data were collected using the linear accelerator facilities at the Radiation Laboratory, University of Notre Dame. This irradiation and transient absorption detection system has been described in full detail previously (*24*). Absolute radical concentrations (dosimetry) were determined using the hydroxyl radical oxidation of N_2O-saturated 1.0 x 10^{-2} M thiocyanate (KSCN) solutions at natural pH, whose efficiency has been previously established (*25*). These measurements were performed daily.

The presence of the high alcohol concentration means that significant intra-spur radical scavenging will occur, which will increase the initial hydrated electron

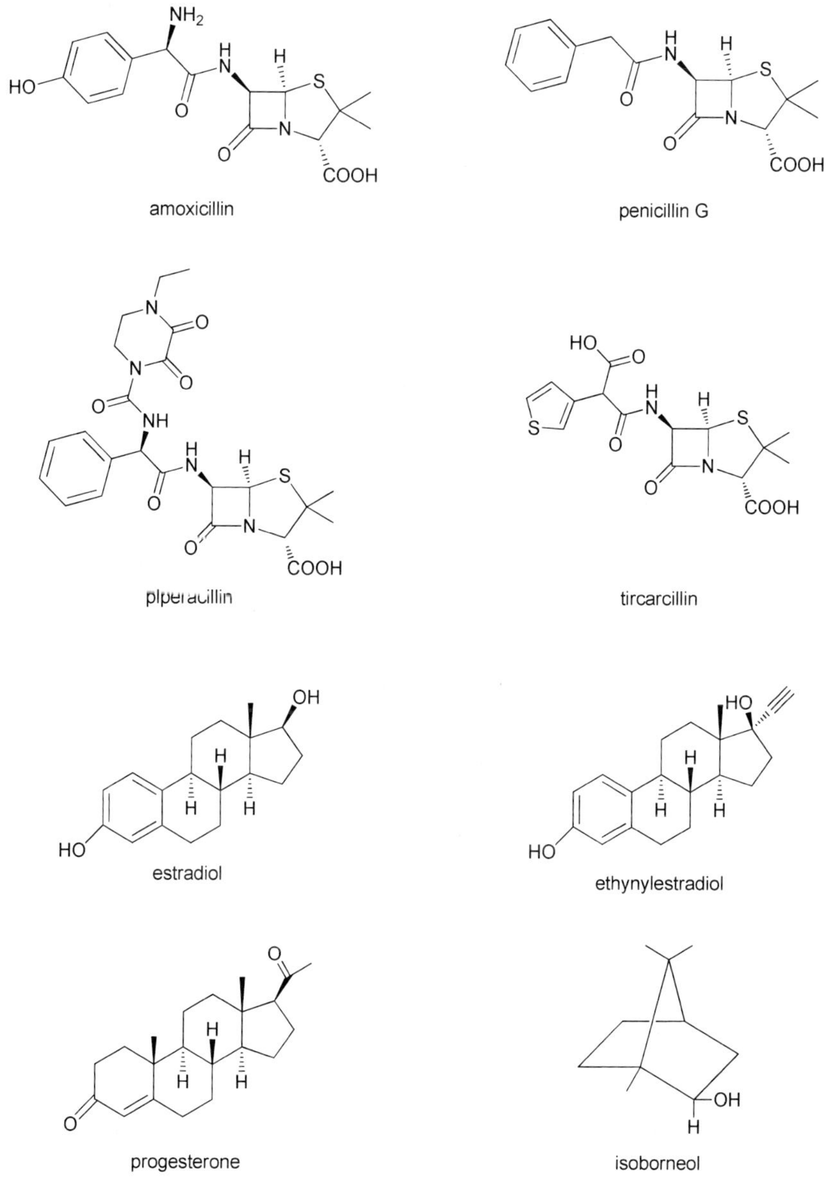

amoxicillin

penicillin G

piperacillin

tircarcillin

estradiol

ethynylestradiol

progesterone

isoborneol

Figure 1. Structures of four antibiotics (amoxicillin, penicillin-G, piperacillin, tircarcillin), three estrogenic steroids (estradiol, ethynylestradiol, progesterone) and one personal care compound (isoborneol) of interest in this study.

and sulfate radical yields (*26*). However, this mixed solvent solution also allowed higher concentrations of the steroids to be dissolved (isoborneol and the β-lactam antibiotics were sufficiently soluble in water) which meant that good pseudo-first-order conditions ([Solute]:[$SO_4^{-\bullet}$] > 20:1) were maintained. This considerably simplified the data analysis.

Solution flow rates were adjusted so that each pulse irradiation was performed on a fresh sample, and multiple traces (5-15) were averaged to produce a single kinetic trace. Typically, 3-6 ns pulses of 8 MeV electrons generating radical concentrations of 2-10 μM $SO_4^{-\bullet}$ per pulse were used in these experiments. All of these experiments were conducted at ambient room temperature (20 ± 2°C) with the temperature variation in any given experiment being less than ± 0.3°C. Rate constant error limits reported here are the combination of experimental precision and compound purities.

Homogeneous Photochemical Degradation of Piperacillin

In addition to the directly measured kinetic parameters of this study, one antibiotic, pipericillin, was also chosen for further investigation into its radical-induced degradation using a homogeneous UV 254 nm photochemical system as an example of the application of AO/RPs to remediate chemically contaminated waters. Two different radicals, $SO_4^{-\bullet}$ and $^{\bullet}OH$, were generated by UV irradiation (15 W low-pressure UV lamps by Cole-Parmer, λ_{max} = 254 nm) of added $S_2O_8^{2-}$ and H_2O_2, respectively. The experiment was conducted in a collimated system made according to Bolton and Linden (*27*). The irradiance was determined by three different methods, iodide/iodate actinometry (*28*), ferrioxalate actinometry (*29*), and a radiometer. Before each experiment, the lamps were allowed to warm up for at least 30 minutes. A Pyrex® glass petri dish (60 mm × 15 mm) with a quartz cover was used as the reactor, to which a solution of 10 mL was added and mildly mixed with a magnetic stirrer bar. During the experiment, 0.1 mL samples were taken at specific fluence levels, and then mixed with 0.1 mL methanol to quench all of the radical reactions occurring.

The solution concentration of piperacillin was determined by HPLC. An Agilent 1100 Series quaternary LC and a Nova-Pak C18 Waters 5-μm (3.9 mm × 150 mm) column was used with the photodiode array detector set at 238 nm. The mobile phase was the combination of 0.1% acetic acid in Milli-Q water (A) and acetonitrile (B) with a gradient mode of 95% A and 5% B as the initial, gradually changing to 65% A and 35% B in 12 minutes, 85% and 15% in the following 6 minutes, and return to the original combination at 20 minutes. The flow rate was set at 0.5 mL/min, the injection volume was 20 μL, and the temperature of the column was 25 °C.

The rate constant of piperacillin with $SO_4^{-\bullet}$ was determined in this homogeneous UV 254 photochemical system using a competition approach with m-Toluic acid as the standard. The initial concentrations of pipericillin, Toluic acid, sodium persulfate, *tert*-butanol, and phosphate buffer (pH=7.4) were 50 μM, 50 μM, 10 mM, 500 mM, and 5 mM, respectively. The high concentration of persulfate was to ensure the sufficient production of $SO_4^{-\bullet}$ while the high concentration of *tert*-butanol was used again to quench any hydroxyl radical produced under UV irradiation of the reaction solution.

Results and Discussion

Kinetic Studies

The sulfate radical has a broad absorption spectrum, with a maximum near 450 nm (*30*). The reactions of sulfate radicals with four β-lactam antibiotics, amoxicillin, penicillin-G, piperacillin and ticarcillin, were investigated in this work. These experiments were conducted in solutions containing 0.5 M *tert*-butanol which were adjusted to a pH of 7.4 using 5.00 mM phosphate buffer. Maintaining a constant near-neutral pH was necessary for reactivity comparison as a pH dependence has been previously reported for sulfate radical reaction with carboxylic acids and TCE in aqueous solution (*31*, *32*). Typical antibiotic concentrations were 100-500 μM. From the rate of change of the first-order decay kinetics observed with varying antibiotic concentration (see Figure 2a) second order reaction rate constants could be readily determined (Figure 2b). Our measured rate constants are summarized in Table 1.

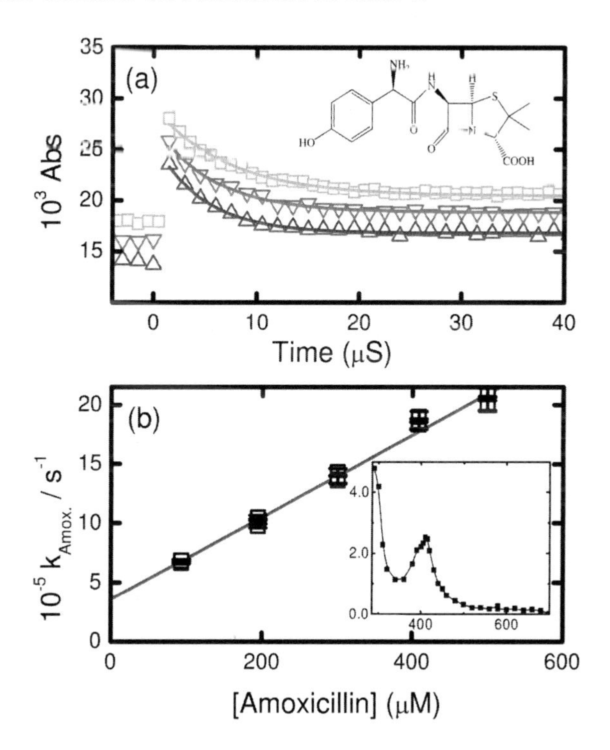

Figure 2. a) Decay of $SO_4^{-\bullet}$ radical at 450 nm for 92.7 (□), 300.0 (▿) and 500.0 μM (▵) added amoxicillin. Kinetic curves are offset in order to aid visibility. Solid lines through data points are fitted first-order kinetics, with pseudo-first-order rate constants of (7.06 ± 0.12) × 10^5, (1.30 ± 0.03) × 10^6 and (2.09 ± 0.05) × 10^6 s^{-1}, respectively. b) Second order transformation of first-order fitted values plotted against amoxicillin concentration. Solid line corresponds to second order rate constant, k = (3.48 ± 0.05) × 10^9 M^{-1} s^{-1}. Inset: Oxidized amoxicillin product species absorbance in arbitrary units taken at end of sulfate radical reaction.

Table 1. Summary of sulfate radical reaction rate constants determined in this study

Compound	*Measured* $k_{SO_4^{\cdot-}}$ [a] $M^{-1}\ s^{-1}$	$k_{\cdot OH}$ $M^{-1}\ s^{-1}$
amoxicillin	2.9×10^9	$(6.94 \pm 0.44) \times 10^9$ (*39*)
penicillin G	1.4×10^9 (*30*)	$(8.70 \pm 0.32) \times 10^9$ (*40*)
piperacillin	$(1.74 \pm 0.11) \times 10^9$ (*30*) [b] 1.2×10^9 (*30*) 1.85×10^9 [b] 1.16×10^9	$(7.84 \pm 0.49) \times 10^9$ (*40*)
ticarcillin	8.0×10^8 (*30*)	$(8.18 \pm 0.99) \times 10^9$ (*40*)
EE2	$(3.01 \pm 0.28) \times 10^9$ (*38*)	$(1.52 \pm 0.23) \times 10^{10}$ (*38*)
estradiol	$(1.21 \pm 0.16) \times 10^9$ (*38*)	$(1.15 \pm 0.28) \times 10^{10}$ (*38*)
progesterone	$(1.19 \pm 0.16) \times 10^9$ (*38*)	$(8.5 \pm 0.9) \times 10^8$ (*38*)
isoborneol	$(5.28 \pm 0.13) \times 10^8$	———

[a] Rate constants for antibiotics are zero ionic strength corrected values. [b] Rate constants for antibiotics with $SO_4^{\cdot-}$ are not corrected for zero ionic strength.

It is important to note that at this pH these antibiotics are negatively charged, like the sulfate radical, and so the measured reaction rate constants will be dependent upon the solution total ionic strength. As our stock solution had a relatively high ionic strength of 0.025 M, we corrected our measured values to zero ionic strength using the standard equation (*33*):

$$\log k = \log k^o + z_1 z_2 I^{1/2} \tag{7}$$

where k is the measured second-order rate constant, k^o is the corresponding zero ionic strength value, z_1 and z_2 are the charges of the sulfate radical (-1) and antibiotic, respectively, and I is the solution ionic strength. The antibiotic charges at pH 7.4 were calculated based on literature pK_a values (*34–36*). For amoxicillin, with its reported pK_a values of 2.4 and 7.49 (*37*), this reduces the $SO_4^{\cdot-}$ reaction rate constant from its measured value of 3.48×10^9 to 2.96×10^9 $M^{-1}\ s^{-1}$. Previously (*38*) we had used an incorrect high pK_{a2} value of 9.6 for this correction which resulted in an erroneously high $SO_4^{\cdot-}$ rate constant for this antibiotic. The zero ionic strength rate constants for penicillin G, piperacillin, and tircarcillin were found to be 1.44×10^9, 1.17×10^9, and 0.80×10^9 $M^{-1}\ s^{-1}$, respectively (see Table 1).

While amoxicillin has the fastest sulfate radical reaction rate constant, the other three antibiotics show effectively the same value within experimental error. Moreover, the transient absorption spectra (see for example Figure 2b, Inset) obtained for the initial species produced in sulfate radical oxidation of all these antibiotics as well as their parent (+)-6-aminopenicillanic acid compound (*38*) are similar, which implies a consistent reaction mechanism for these antibiotics. Based on these data we infer that the predominant oxidizing $SO_4^{-\bullet}$ reaction occurs at the S atom in the five-member ring immediately adjacent to the β-lactam ring in these antibiotics. In contrast, it has previously been reported (*39*, *40*) that the corresponding hydroxyl radical reactions for these antibiotic species occur predominately at peripheral aromatic rings, producing hydroxylated species with an intact β-lactam core. While the hydroxyl radical oxidations are considerably faster than those of the sulfate radical (see Table 1) the closer site of reactivity of the latter (Figure 3) implies that it could be more efficient in destroying antibiotic activity than the corresponding hydroxyl radical reaction.

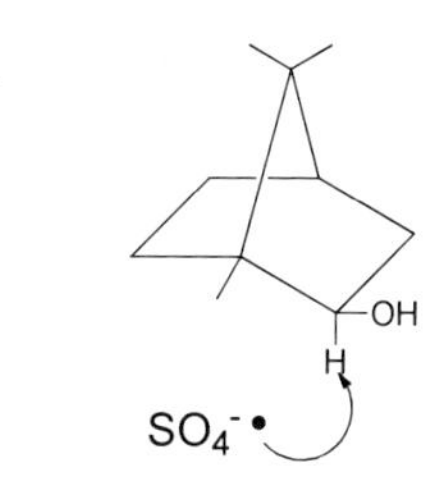

Figure 3. Suggested initial reaction mechanisms of sulfate radical oxidation of amoxicillin, EE2, and isoborneol.

$SO_4^{-\bullet}$ Reaction with Estrogenic Steroids

The sulfate radical reactivity with three typical contaminant estrogenic steroids (Figure 1) was also determined in this work. The considerably decreased aqueous solubility of these chemicals meant that much lower steroid concentrations had to be used. However, following the same methodology as for the antibiotics, sulfate radical reaction rate constants were obtained (*38*) for ethynylestradiol (EE2), estradiol and progesterone as 3.01×10^9, 1.21×10^9 and 1.19×10^9 M^{-1} s^{-1}, respectively (see Table 1). These second-order kinetic data are shown in Figures 4(a-c). For both progesterone and estradiol the $SO_4^{-\bullet}$ rate constant was slower than for EE2, implying that different reaction mechanisms were occurring. While these three rate constants do not allow the specific mechanism of oxidation for progesterone and estradiol to be quantitatively determined, the significantly faster rate constant for EE2 suggests that significant $SO_4^{-\bullet}$ oxidation occurs at the ethinyl bond in this molecule (Figure 3).

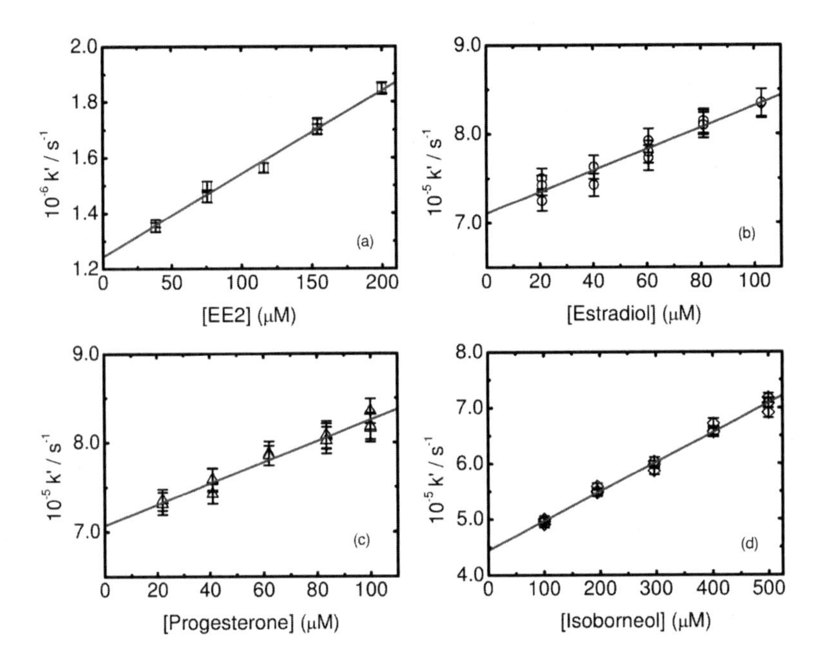

Figure 4. Second order kinetic plots for $SO_4^{-\bullet}$ radical reaction with a) ethynylestradiol (EE2), b) estradiol, c) progesterone, and d) isoborneol. Solid lines are weighted linear fits, corresponding to rate constants of $(3.01 \pm 0.28) \times 10^9$ M^{-1} s^{-1} and $(1.21 \pm 0.16) \times 10^9$ M^{-1} s^{-1}, $(1.19 \pm 0.16) \times 10^9$ M^{-1} s^{-1}, and $(5.28 \pm 0.13) \times 10^8$ M^{-1} s^{-1}, respectively.

All these sulfate radical oxidations are slower than measured for the hydroxyl radical, (see Table 1) indicating that the initial reaction mechanisms for these two oxidizing radicals differ. As observed for the antibiotic oxidations (*38*) the $^{\bullet}OH$ reaction is anticipated to add to the constituent phenyl ring in these steroids. This would give dihydroxy stable product species, and for estradiol these dihydroxy species have been shown to have comparable steroidal activity and higher aqueous solubility (*41*). Therefore, the use of sulfate radicals which preferentially reacts at other electron-rich centers in these molecules may prove advantageous for the total removal of estrogenic activity in the treatment of large-scale real-world waters.

$SO_4^{-\bullet}$ Reaction with Isoborneol

We have also investigated the sulfate radical reaction of a common personal care chemical, isoborneol (Figure 1). This aliphatic ring based chemical is a perfume agent, food additive, and moth repellant, and can be a skin, eye, and respiratory irritant (*42*). While isoborneol itself is not a major contaminant of concern, this compound was chosen for study because it is a saturated monoterpene that is more soluble in water than 2-methyl isoborneol, an analogous contaminant that has a major water quality impact due to its low odor threshold in the ppb range.

The second-order reaction rate constant for the sulfate radical with isoborneol (Figure 3) was found to be $(5.28 \pm 0.13) \times 10^8$ M^{-1} s^{-1}. While its reactivity is slower than for the estrogenic steroids and antibiotics of this study (Table 1), it was still surprisingly fast when compared to simple aliphatic alcohols such as *tert*-butanol (Equation 6, $k_6 = 8.4 \times 10^5$ M^{-1} s^{-1}). A previous, systematic, investigation of other sulfate radical reaction rate constants with other (non-ring) aliphatic alcohols in water (*43*) has established a relationship with number and type of C-H bonds in these molecules. Based on the principle of additive reactivity, individual rate constants for primary, secondary, and tertiary C-H bond were calculated as 7.4×10^5, 1.8×10^7 and 9.8×10^7 M^{-1} s^{-1}, respectively. Using these values for isoborneol, an overall sulfate radical rate constant of 3.1×10^8 M^{-1} s^{-1} is predicted, only slightly lower than its measured value. In addition, this analysis suggests that the predominant oxidation occurs by hydrogen atom abstraction from a tertiary carbon atom, as shown in Figure 3. Unfortunately, no hydroxyl radical rate constant for isoborneol could be found in the literature, precluding any further quantitative comparison between these two oxidizing radicals. However, it would be expected that $^\bullet OH$ reaction would also occur by hydrogen atom abstraction, with a rate constant in the 10^8 - 10^9 M^{-1} s^{-1} range (*13*). Similar values would be expected for the 2-methyl isoborneol compound as based on their close structures. Overall, these sulfate radical data suggest that treatment of aqueous waste streams containing isoborneols by AO/RPs could be a viable treatment method, with both the oxidizing hydroxyl and sulfate radicals assisting in the total removal of these chemicals.

Homogeneous Photochemical Degradation of Piperacillin

Persulfate is a strong inorganic oxidant; however, it can be activated to form sulfate radicals (*44–47*) by use of a transition metal, UV irradiation or by increasing the temperature. In this study the application of sulfate radicals was shown by the degradation of a β-lactam antibiotic, piperacillin, in a laboratory scale homogeneous UV photochemical system (Figure 5). Initial concentrations of pipercillin and persulfate were 50 μM and 1 mM, respectively. At a UV fluence of 320 mJ/cm^2 the degradation of piperacillin was 4.7%, 33.0% and 65.2% for UV, UV/$S_2O_8^{2-}$ and UV/H_2O_2, respectively. Under dark conditions, there was no significant removal of pipericillin either by $S_2O_8^{2-}$ or by H_2O_2 during the same time interval. Previously, the second order rate constants for $SO_4^{-\bullet}$ and $^\bullet OH$ radicals with pipericillin were determined to be 1.2×10^9 M^{-1} s^{-1} (*30*) and $(7.84 \pm 0.49) \times 10^9$ M^{-1} s^{-1} (*40*), respectively, (the sulfate radical value was corrected to

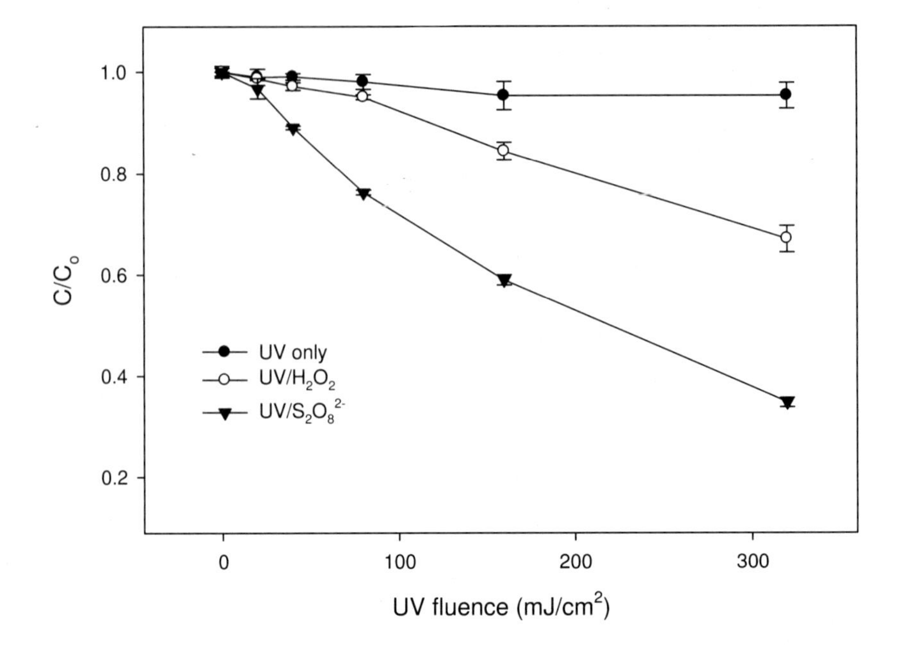

Figure 5. Measured degradation of piperacillin in the homogeneous photochemical system, with the initial concentration of the antibiotic and the oxidant to be 50 μM and 1 mM, respectively, in Milli-Q water.

zero ionic strength using Equation (7)). The slightly faster degradation of PIP by the UV/$S_2O_8^{2-}$ AO/RP in comparison to the UV/H_2O_2 was therefore attributed to the rate of formation of the radicals, which was shown in the difference in the quantum yield of the oxidants under UV-254 nm irradiation (Equations 8 & 9, (*48–50*))

$$H_2O_2 \rightarrow 2{\cdot}OH \qquad \Phi=1.0 \tag{8}$$

$$S_2O_8^{2-} \rightarrow 2SO_4^{-}{\cdot} \qquad \Phi=2 \tag{9}$$

In this study, a chemical competition kinetics approach was also evaluated to determine the second order rate constant for sulfate radical reaction with pipericillin. The competitor, m-Toluic acid, was chosen because of its similar rate constant to pipericillin with $SO_4^{-\bullet}$, 2×10^9 M^{-1} s^{-1} (*51*). As shown in Figure 6, neither pipericillin nor m-Toluic acid underwent directly photolysis; however, both of them followed UV fluence-based pseudo-first-order reaction kinetics, with $k_{obs(PIP)} = 1.74 \times 10^{-3}$ cm^2/mJ and $k_{obs(TA)} = 1.88 \times 10^{-3}$ cm^2/mJ (m-Toluic acid) when sulfate radicals were generated in the AO/RP. The absolute second order rate constant could thus be determined by equation (10)

$$k_{SO_4^{-\bullet}(PIP)} = \frac{k_{obs(PIP)}}{k_{obs(TA)}} \times k_{SO_4^{-\bullet}(TA)} \tag{10}$$

to be $k_{SO4^{-\bullet}(PIP)} = 1.85 \times 10^9$ M^{-1} s^{-1}. Correcting this value for ionic strength effects (the pK_a of pipericillin was 4.14 and the total ionic strength at pH 7.4 was 0.041 M) according to equation (7) gave a limiting value of $k_{SO4^{-\bullet}(PIP)} = 1.16 \times 10^9$ M^{-1} s^{-1}, in excellent agreement with our previous result (*30*).

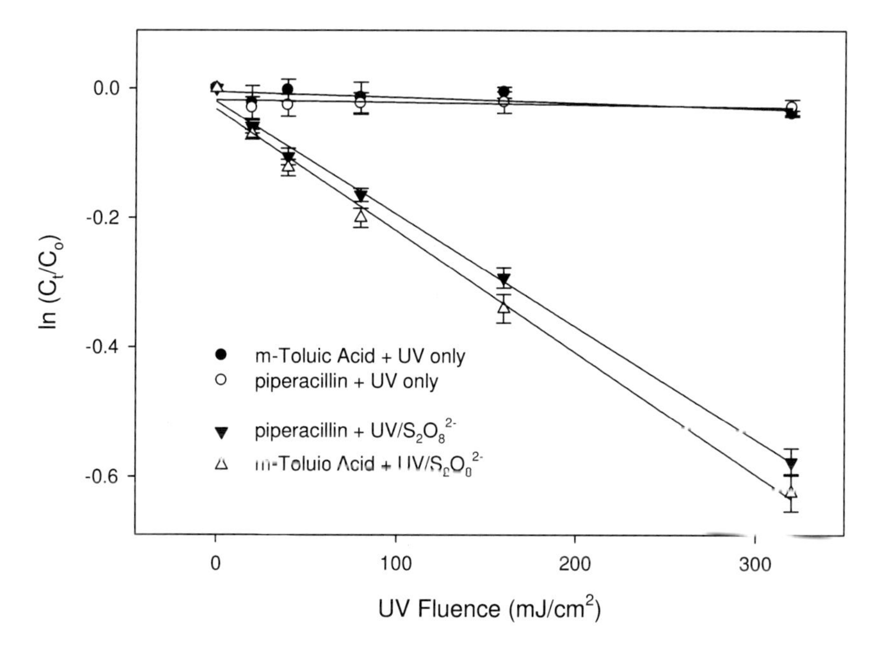

Figure 6. Determination of second order rate constant of piperacillin with $SO_4^{-\bullet}$ through m-Toluic acid competition kinetics. For the $UV/S_2O_8^{2-}$ experiment the initial concentrations of piperacillin, m-Toluic acid, sodium persulfate, tert-butanol, and phosphate buffer (pH=7.4) were 50 μM, 50 μM, 10 mM, 500 mM, and 5 mM, respectively. The chemical concentrations were the same for the UV-only experiment in the absence of persulfate.

Conclusions

Rate constants for the reactions of oxidizing sulfate radicals have been determined for four common antibiotics, three estrogenic steroids, and one personal care product in water. The relatively fast rate constants extrapolated to zero ionic strength for $SO_4^{-\bullet}$ reaction with amoxicillin (2.9×10^9 M^{-1} s^{-1}), penicillin-G (1.4×10^9 M^{-1} s^{-1}), piperacillin (1.2×10^9 M^{-1} s^{-1}), and tircarcillin (8.7×10^8 M^{-1} s^{-1}), suggests a consistent mechanism, believed to be electron abstraction from the sulfur atom in the five-member ring adjacent to the β-lactam ring in these compounds. For the three estrogenic steroids, ethynylestradiol (($3.01 \pm 0.28) \times 10^{10}$ M^{-1} s^{-1}), estradiol (($1.21 \pm 0.16) \times 10^{10}$ M^{-1} s^{-1}, and progesterone (($1.19 \pm 0.16) \times 10^9$ M^{-1} s^{-1}) the lower values for the latter two suggests that significant sulfate radical reaction at the triple bond occurs for EE2. For the saturated aliphatic isoborneol (($5.28 \pm 0.13) \times 10^8$ M^{-1} s^{-1}), the sulfate radical reaction is still relatively fast suggesting that AO/RP treatment utilizing this radical may be feasible at large-scale. Although the absolute rate constant of piperacillin with $SO_4^{-\bullet}$ was lower than that for $^{\bullet}OH$, the removal of the compound was faster by $UV/S_2O_8^{2-}$ than by UV/H_2O_2, with a percentage loss of 65.2% and 33.0%, respectively, at UV fluence of 320 mJ/cm^2. This finding is attributed to the higher quantum yield of persulfate under UV 254 nm irradiation.

Acknowledgments

Rate constant measurements were performed at the Radiation Laboratory, University of Notre Dame, which is supported by the Office of Basic Energy Sciences, U.S. Department of Energy.

References

1. Ternes, T. A. Occurrence of drugs in German sewage treatment plans and rivers. *Water Res.* **1998**, *32*, 3245–3260.
2. Synder, S. A.; Westerhoff, P.; Yoon, Y.; Sedlak, D. L. Pharmaceuticals, personal care products, and endocrine disruptors in water: implications for the water industry. *Environ. Eng. Sci.* **2003**, *20*, 449–469.
3. Westerhoff, P.; Yoon, Y.; Snyder, S. A.; Wert, E. Fate of endocrine-disruptor, pharmaceutical, and personal care product chemicals during simulated drinking water treatment processes. *Environ. Sci. Technol.* **2005**, *39*, 6649–6663.
4. Kummerer, K. Antibiotics in the aquatic environment: A review Part I. *Chemosphere* **2009**, *75*, 417–434.
5. Kummerer, K. Antibiotics in the aquatic environment: A review Part II. *Chemosphere* **2009**, *75*, 435–441.
6. Kolpin, D. W.; Furlong, E. T.; Meyer, M. T.; Thurman, E. M.; Zaugg, S. D.; Barber, L. B.; Buxton, H. T. Pharmaceuticals, Hormones, and Other Organic Wastewater Contaminants in U.S. Streams, 1999−2000: A National Reconnaissance. *Environ. Sci. Technol.* **2002**, *36*, 1202–1211.
7. Ikehata, K.; Nagashkar, N. J.; El-Din, M. C. Degradation of aqueous pharmaceuticals by ozonation and advanced oxidation processes: A review. *Ozone: Sci. Eng.* **2006**, *28*, 353–414.
8. Burbano, A. A.; Dionysiou, D. D.; Suidan, M. T. Effect of oxidant-to-substrate ratios on the degradation of MTBE with Fenton reagent. *Water Res.* **2008**, *42*, 3225–3239.
9. Ning, B.; Graham, N.; Zhang, Y. P.; Nakonechny, M.; El-Din, M. G. Degradation of endocrine disrupting chemicals by ozone/AOPs. *Ozone: Sci. Eng.* **2007**, *29*, 153–176.
10. Lee, Y.; Escher, B. I.; von Gunten, U. Efficient removal of estrogenic activity during oxidative treatment of waters containing steroid estrogens. *Environ. Sci. Technol.* **2008**, *42*, 6333–6339.
11. Huber, M. M.; Ternes, T. A.; von Gunten, U. Removal of estrogenic activity and formation of oxidation products during ozonation of 17 alpha-ethinylestradiol. *Environ. Sci. Technol.* **2004**, *38*, 5177–5186.

12. Cooper, W. J.; Cramer, C. J.; Martin, N. H.; Mezyk, S. P.; O'Shea, K. E.; von Sonntag, C. Free radical mechanisms for the treatment of methyl tert-butyl ether (MTBE) via advanced oxidation/reductive processes in aqueous solution. *Chem. Rev.* **2009**, *109*, 1302–1345.
13. Buxton, G. V.; Greenstock, C. L.; Helman, W. P.; Ross, A. B. Critical review of rate constants for reactions of hydrated electrons, hydrogen atom and hydroxyl radicals ($^{\bullet}OH/^{\bullet}O^{-}$) in aqueous solutions. *J. Phys. Chem. Ref. Data* **1988**, *17*, 513–886.
14. Anipsitakis, G. P.; Dionysiou, D. D. Transition metal/UV-base advanced oxidation technologies for water decontamination. *Appl. Catal. B* **2004**, *54*, 155–163.
15. Huie, R. E.; Clifton, C. L.; Neta, P. Electron transfer rates and equilibria of the carbonate and sulfate radical anions. *Int. J. Radiat. Phys. Chem.* **1991**, *5*, 477–481.
16. Do, S.-H.; Kwon, Y.-J.; Kong, S.-H. Effect of metal oxides on the reactivity of persulfate/Fe(II) in the remediation of diesel-contaminated soil and sand. *J. Hazard. Mater.* **2010**, *182*, 933–936.
17. Yukselen-Aksoy, Y.; Khodadoust, A. P.; Reddy, K. R. Destruction of PCB 44 in spiked subsurface soils using activated persulfate oxidation. *Water, Air, Soil Pollut.* **2010**, *209*, 419–427.
18. Andreottola, G.; Bonomo, L.; De Gioannis, G.; Ferrarese, E.; Muntoni, A.; Polettini, A.; Pomi, R.; Saponaro, S. Lab-scale feasibility tests for sediment treatment using different physico-chemical techniques. *J. Soils Sediments* **2010**, *10*, 142–150.
19. Tsai, T. T.; Kao, C. M.; Hong, A. Treatment of tetrachloroethylene-contaminated groundwater by surfactant-enhanced persulfate/BOF slag oxidation – a laboratory feasibility study. *J. Hazard. Mater.* **2009**, *171*, 571–576.
20. Teel, A. L.; Cutler, LM.; Watts, R. J. Effect of sorption on contaminant oxidation in activated persulfate systems. *J. Environ. Sci. Health, Part A* **2009**, *44*, 1098–1103.
21. Liang, C.; Lin, Y.-T.; Shih, W.-H. Treatment of trichloroethylene by adsorption and persulfate oxidation in batch studies. *Ind. Eng. Chem. Res.* **2009**, *48*, 8373–8380.
22. Huang, K.-C.; Zhao, Z.; Hoag, G. E.; Dahmani, A.; Block, P. A. Degradation of volatile organic compounds with thermally activated persulfate oxidation. *Chemosphere* **2005**, *61*, 551–560.
23. Crittenden, J. C.; Hu, S.; Hand, D. W.; Green, S. A. A kinetic model for H_2O_2/UV process in a completely mixed batch reactor. *Water Res.* **1999**, *33*, 2315–2328.
24. Whitman, K.; Lyons, S.; Miller, R.; Nett, D.; Treas, P.; Zante, A.; Fessenden, R. W.; Thomas, M. D.; Wang, Y. Linear accelerator for radiation chemistry research at Notre Dame 1995. *Proceedings of the '95 Particle Accelerator Conference & International Conference of High Energy Accelerators*, Dallas, TX, 1996.
25. Buxton, G. V.; Stuart, C. R. Re-evaluation of the thiocyanate dosimeter for pulse radiolysis. *J. Chem. Soc. Faraday Trans.* **1995**, *91*, 279–282.

26. LaVerne, J. A.; Pimblott, S. M. Yields of hydroxyl radical and hydrated electron scavenging reactions in aqueous solutions of biological interest. *Rad. Res.* **1993**, *135*, 16–23.
27. Bolton, J. R.; Linden, K. G. Standardization of methods for fluence (UV dose) determination in bench-scale UV experiments. *J. Environ. Eng.* **2003**, *129* (3), 209–215.
28. Rahn, R. O. Potassium iodide as a chemical actinometer of 254 nm radiation: use of iodate as an electron scavenger. *Photochem. Photobiol.* **1997**, *66*, 450–455.
29. Murov, S. L.; Carmichael, I.; Hug, G. L. *Handbook of photochemistry*, 2nd ed.; Marcel Dekker: New York, 1993; pp 330−336.
30. Rickman, K. A.; Mezyk, S. P. Kinetics and mechanisms of sulfate radical oxidation of β-lactam antibiotics in water. *Chemosphere* **2010**, *81*, 359–365.
31. Criquet, J.; Nebout, P.; Karpel Vel Leitner, N. Enhancement of carboxylic acid degradation with sulfate radical generated by persulfate activation. *Water Sci. Technol.* **2010**, *61*, 1221–1226.
32. Liang, C.; Zih-Sin, W.; Bruell, C. J. Influence of pH on persulfate oxidation of TCE at ambient temperatures. *Chemosphere* **2007**, *66*, 106–113.
33. Connors, K. A. *Chemical kinetics: The study of reaction rates in solution*; VCH Publishers Inc.: New York, NY, 1990.
34. Florey, E., Ed. *Analytical profiles of drug substances*; Academic Press: New York, 1978.
35. Hanch, C., Sammas, P. G., Taylor, J. B., Eds. *Comprehensive medicinal chemistry*; Pergamon Press: Oxford, 1990; Vol. 6.
36. Alkseev, V. G.; Volkova, I. A. Acid-base properties of some penicillins. *Russ. J. Gen. Chem.* **1991**, *73*, 1616–1618.
37. Andreozzi, R.; Canterino, M.; Marotta, R.; Paxeus, N. Antibiotic removal from wastewaters: The ozonation of amoxicillin. *J. Hazard. Mater.* **2005**, *122*, 243–250.
38. Mezyk, S. P.; Abud, E. M.; Swancutt, K. L.; McKay, G.; Dionysiou, D. D. Removing steroids from contaminated waters using radical reactions; *Contaminants of Emerging Concern in the Environment: Ecological and Human Health Considerations*; ACS Symposium Series; American Chemical Society: Washington, DC, 2010; Vol. 1048, Chapter 9, pp 213−225.
39. Song, W.; Weisang, C.; Cooper, W. J.; Greaves, J.; Miller, G. E. Free radical destruction of β-lactam antibiotics in aqueous solution. *J. Phys. Chem. A.* **2008**, *112*, 7411–7417.
40. Dail, M. K.; Mezyk, S. P. Hydroxyl-radical-induced degradative oxidation of β-lactam antibiotics in water: Absolute rate constant measurements. *J. Phys. Chem. A* **2010**, *114*, 8391–8395.
41. Zhao, Z.; Kosinska, W.; Khmelnitsky, M.; Cavalieri, E. L.; Rogan, E. G.; Chakravarti, D.; Sacks, P. G.; Guttenplan, J. B. Mutagenic activity of 4-hydroxyestradiol, but not 2-hydroxyestradiol, in BB Rat2 embryonic cells, and the mutational spectrum of 4-hydroxyestradiol. *Chem. Res. Toxicol.* **2006**, *19*, 475–479.
42. MSDS sheet. http://msds.chem.ox.ac.uk/IS/isoborneol.html.

43. Huie, R. E.; Clifton, C. L. Rate constants for hydrogen atom abstraction reactions of the sulfate radical, SO_4^{-}. Alkanes and ethers. *Int. J. Chem. Kinet.* **1989**, *21*, 611–619.
44. Anipsitakis, G. P.; Dionysiou, D. D. Degradation of organic contaminants in water with sulfate radicals generated by the conjunction of peroxymonosulfate with cobalt. *Environ. Sci. Technol.* **2003**, *37* (20), 4790–4797.
45. Bandala, E. R.; Peláez, M. A.; Dionysiou, D. D.; Gelover, S.; Garcia, J.; Macías, D. Degradation of 2,4-dichlorophenoxyacetic acid (2,4-D) using cobalt-peroxymonosulfate in Fenton-like process. *J. Photochem. Photobiol., A* **2007**, *186* (2–3), 357–363.
46. Waldemer, R. H.; Tratnyek, P. G.; Johnson, R. L.; Nurmi, J. T. Oxidation of chlorinated ethenes by heat-activated persulfate: kinetics and products. *Environ. Sci. Technol.* **2007**, *41*, 1010–1015.
47. Kronholm, J.; Metsälä, H.; Hartonen, K.; Riekkola, M.-L. Oxidation of 4-chloro-3-methylphenol in pressurized hot water/supercritical water with potassium persulfate as oxidant. *Environ. Sci. Technol.* **2001**, *35*, 3247–3251.
48. Baxendale, J. H.; Wilson, J. A. The photolysis of hydrogen peroxide at high light intensities. *Trans. Faraday Soc.* **1957**, *53*, 344–356.
49. Yu, X.-Y.; Bao, Z.-C.; Barker, J. R. Free radical reactions involving $Cl^{\bullet}$, $Cl_2^{-\bullet}$, and $SO_4^{-\bullet}$ in the 248 nm photolysis of aqueous solutions containing $S_2O_8^{2-}$ and Cl^{-}. *J. Phys. Chem. A* **2004**, *108*, 295–308.
50. Hori, H.; Yamamoto, A.; Koike, K.; Kutsuna, S.; Osaka, I.; Arakawa, R. Perfulfate-induced photochemical decomposition of a fluorotelomer unsaturated carboxylic acid in water. *Water Res.* **2007**, *41*, 2962–2968.
51. Neta, P; Madhavan, V.; Zemel, H.; Fessenden, R. W. Rate constants and mechanism of reaction of $SO_4^{-\bullet}$ with aromatic compounds. *J. Am. Chem. Soc.* **1977**, *99* (1), 163–164.

Chapter 13

Voltammetry of Sulfide Nanoparticles and the FeS(aq) Problem

G. R. Helz,[*,1] I. Ciglenečki,[2] D. Krznarić,[2] and E. Bura-Nakić[2]

[1]Department of Chemistry and Biochemistry, University of Maryland, College Park, MD 20742, USA
[2]Center for Marine and Environmental Research, Ruđer Bošković Institute, Bijenička 54, 1000 Zagreb, Croatia
[*]helz@umd.edu

Voltammetry at Hg drop electrodes is a promising method for detecting sulfide nanoparticles in natural waters. Recent research suggests that such nanoparticles might affect organisms in unforeseen ways. Sulfide nanoparticles diffusing to Hg surfaces are captured selectively even from unfiltered waters that contain larger amounts of other nanoscale materials, such as organic macromolecules or clay minerals. Optimum size sensitivity for capture is roughly 5-100 nm at Hg drop electrodes. Sulfide nanoparticles are stabilized at Hg surfaces by transformation to adlayers, whose accumulation can be quantified electrochemically. Study of FeS adlayers has led to new insights regarding the puzzling -1.1 V vs. Ag/AgCl signal observed in sulfidic natural waters. This signal has been attributed previously to Fe sulfide clusters or complexes. New evidence shows that it arises from reduction of Fe^{2+} at FeS adlayers formed by sorption of FeS nanoparticles on Hg electrodes. Partial coverage of Hg with FeS creates in essence two electrodes. These reduce Fe^{2+} at separate potentials.

Introduction

Aquatic and marine chemists by convention have relied on filtration to divide samples of natural waters into dissolved and particulate fractions prior to analysis. For many decades, 0.45 μm pore size filters were preferred, although

0.2 μm or smaller pore sizes have been also in common use. It has always been recognized that this approach provides a less than ideal way of classifying analytes. The filtration process itself is procedure-dependent and problematic (*1*). More importantly, solutes in the so-called dissolved (filterable) fraction consist not solely of hydrated ions and molecules, but include nanoparticles. Recently, nanoparticles have become the focus of increasing interest.

Exactly what constitutes a nanoparticle has received much discussion. Interested readers are referred to a review by Lead and Wilkinson (*2*). Some physicists and chemists define nanoparticles as objects so small (usually < 5 nm) that they possess optical properties influenced by quantum confinement. In natural waters, this definition is operationally impractical because of the presence of optically interfering substances, especially macromolecular organic materials. Additionally, the relevant optical properties are well-defined only for nanoparticles having narrow ranges of composition and size, an unlikely situation in nature. Many environmental scientists now prefer a size-based definition that nanoparticles simply have at least one dimension in the 1 to 100 nm range (*2*). Although neither filtration nor ultrafiltration achieve precise size-based separations, sizes of 1 to 100 nm are roughly consistent with an operation-based definition that nanoparticles pass filters with nominal pore sizes in the tenths of a micron range but are retained by ultrafiltration membranes with cutoffs of a few kDa (*3*).

Some nanoparticles may possess unique toxicological properties that differ from those of the same components as bulk solids or as ions or molecules in solution (*4–6*). A subclass of nanoparticles, the chalcogenide quantum dots, has attracted particular attention. These are sulfide-, selenide- or telluride-containing semiconductors in the quantum confinement size range. Some quantum dots appear to be able to enter cells and cause damage that differs from that caused by the same components as ions in solution (*7–10*). Whether this behavior is a general property of chalcogenide quantum dots is not yet clear (*11*). Filterable (0.2 μm) HgS nanoparticles are sufficiently hydrophobic to be extracted into octanol and may be responsible for the biological activity previously attributed to HgS^0 complexes or clusters (*12*).

As discussed in the next section, good evidence indicates that sulfide nanoparticles of natural origin exist in the environment. The possibility that such nanoparticles might also exhibit unusual biological behavior creates a need for analytical methods to characterize and quantify them in natural waters. The needed methods should be able to detect sulfide nanoparticles in the presence of more abundant and common nanoscale materials, including organic macro-molecules, Fe and Mn oxyhydroxides and clay minerals.

In this chapter, we first review evidence that metal sulfide nanoparticles probably exist in natural waters, possibly even in the presence of O_2. Then we describe recent research at the Ruđer Bošković Institute aimed at characterizing their voltammetric behavior. The strong affinity of Hg for sulfide makes the Hg drop electrode an attractive device for selectively capturing sulfide particles from samples containing an abundance of other nanoparticulate materials. Finally we present some preliminary results to show that this approach might prove useful for direct determination of sulfide nanoparticles in natural waters.

Sulfide Nanoparticles in Nature

Expectations from Thermodynamics

Figure 1 shows thermodynamic speciation calculations for two sulfide-reactive metals, Cu and Fe. Saturation with respect to elemental S is assumed, resulting in redox control by the S(-II)/S(0) couple. Notice that even quite low sulfide concentrations cause the principal dissolved complexes in both cases to be in lower oxidation states at equilibrium.

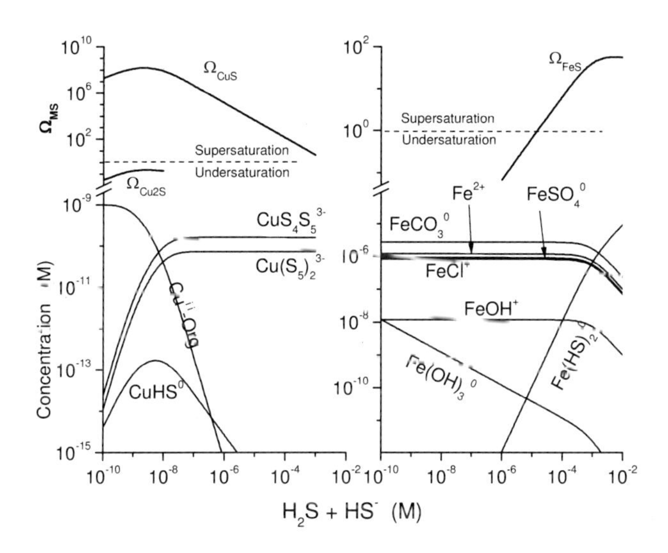

Figure 1. Equilibrium speciation and degree of saturation (Ω) in waters containing 10^{-9}M total Cu (left panel) and 10^{-5}M total Fe (right panel) as a function of free sulfide (H_2S+HS^-)concentration. Metal concentrations are representative for suboxic or anoxic natural waters. $\Omega = [M^{2+}][HS^-]/10^{-pH}/K_{sp}$. Assumed conditions: pH 7.5; major inorganic anions at seawater concentrations; saturation with $S_{rhombic}$. (Data: (13–17)).

The species distributions assume that no sulfide phases precipitate, but the figure indicates that very high degrees of supersaturation would exist if this were actually possible. To the contrary, huge supersaturations near the maxima in the Ω curves imply negligible barriers to sulfide phase nucleation; rates of nucleation should occur nearly at diffusion limits. Rapid nucleation produces numerous small primary particles, usually having nanoscale dimensions (*18*). For example, FeS precipitation in the laboratory produces primary particles < 10 nm (*19–22*). Copper sulfide precipitates are similar (*23*, *24*).

In nature, huge degrees of supersaturation that would drive nanoparticle precipitation can be created wherever sharp redox gradients exist. Examples of such places include pycnoclines above euxinic water columns and sediment-water interfaces. At the microscale, sharp gradients occur around reducing microniches in what might appear to be homogeneously oxidized media (*25*).

Figure 1 shows that huge supersaturations with respect to CuS (covellite) exist already at the lowest sulfide concentrations shown. In this range, dissolved Cu(I) complexes involving sulfide and polysulfide are not yet competitive with Cu(II)-organic matter complexes, but Cu sulfide particles are nonetheless stable. Thus if Cu is involved in preserving traces of filterable sulfide in oxic natural waters, as has been proposed, then Cu sulfide nanoparticles rather than dissolved Cu complexes are most likely the agent (*26*).

As sulfide increases in Figure 1, the degree of supersaturation passes through a maximum and then decreases as Cu(I) sulfide and polysulfide complexes start to dominate in the aqueous phase. Under S^0-saturation conditions, Cu polysulfide complexes are most abundant, but they would be replaced by sulfide complexes or bidentate thioanion complexes (*27*, *28*) at lower activity of zero-valent sulfur. For Cu, the greatest driving force for precipitation occurs near 0.01 μM sulfide.

In the case of Fe, sulfide complexes do not become significant until sulfide concentrations exceed those normally found in euxinic waters; this prediction is consistent with field evidence (*29*, *30*). The supersaturation curve for Fe is qualitatively similar to that of Cu, but the greatest driving force for FeS (mackinawite) precipitation is centered at much higher sulfide.

A very important implication is that Cu and Fe sulfide nanoparticles are most likely to be found in different sulfide concentration regimes in nature. Remarkably, thermodynamics suggest that Cu sulfide nanoparticles are most likely to be found in waters that would be judged non-sulfidic when assessed by conventional sulfide analytical methods, which have detection limits near 10^{-6} M.

Field Evidence

To date, evidence that filterable (submicron) FeS particles indeed exist in sulfidic natural waters is indirect. Based on sampling with size exclusion columns, as much as a third of the filterable Fe below the chemocline in the Black Sea consists of particles that are less than 50 nm in size (*31*). This material is soluble at pH 5.5 and is inferred to be FeS. By comparing colorimetrically and electrochemically determined concentrations of filterable Fe(II) and S(-II) in Lake Bret, Buffle et al. (*32*) established that both components occur in submicron colloids. Curiously, colloidal Fe exceeded colloidal S by about 2-fold on a molar basis. If this is not an analytical artifact related to the FeSaq problem (see below), then mackinawite cannot be the only Fe-bearing nanoscale precipitate. Bura-Nakić et al. (*33*) concluded that roughly half the methylene blue determined S(-II) in Lake Pavin resides in a filterable form that is not free sulfide. This form is most likely FeS based on thermodynamic arguments as well as analytical constraints.

Similar evidence from natural waters supports existence of sulfide nanoparticles containing other metals. Skei et al. (*34*) captured aggregates of ZnS and CuS at the top of the sulfidic water column of Framvaren Fjord by 0.4 μm filtration. The aggregates themselves exceeded 1 μm in size, but consisted of smaller particles. Spherical aggregates consisting of 1-5 nm primary ZnS particles have been observed in biofilms in a flooded Pb-Zn mine (*35*, *36*). Similar aggregates have been found in H_2S-rich pore waters in wetlands (*37*). Copper-rich sulfide particles < 50 nm in diameter, as well as hollow spheres 50 – 150 nm, were produced in experiments with flooded soils (*38*). On the other hand, the highly sulfidic Black Sea water column contains copper mainly in an anionic form, not as nanoparticles (*31*). This supports the inference from thermodynamics that Cu sulfide and polysulfide anionic complexes are more likely than nanoparticles at higher sulfide concentrations (Figure 1).

Even in oxic environments, tentative evidence exists for sulfide nanoparticles. Wastewaters contain filterable forms of Ag having properties consistent with Ag_2S nanoparticles, including retention by ultrafiltration and resistance to oxidation (*39*). Nanoscale Ag_2S (5-20 nm) now actually has been imaged by transmission electron microscopy in wastewater sludge (*40*). Oxic surface waters and sewage treatment plant effluents contain filterable forms of bound S(-II) that are partly retained by ultrafilters; these behave analytically like Cu and Zn sulfide particles (*41*). Counter-intuitively, Sukola et al. (*26*) demonstrate that Cu, Zn and Cd sulfide species that are probably nanoparticles persist in oxic water for weeks.

A shortcoming of this evidence is that filtration is necessary in most cases to qualify analytes as nanoparticulate. Experiments with recovery of synthesized sulfide nanoparticles suggest that they readily adhere to surfaces and therefore are likely to be underdetermined in filtered samples (*26*, *42*, *43*). Another shortcoming is that the evidence is largely indirect. Nanoparticles are considered to be what is left after other analytes have been accounted for. In many cases, nanoparticles themselves are not being observed nor are their properties being measured.

Behavior of Sulfide Nanoparticles at Hg Electrode Surfaces

Sulfide macro- and nanoparticles readily sorb to Hg surfaces owing to the great affinity of Hg^0 for reduced S in almost any form (*44*, *45*). As a consequence, metallic Hg is an effective and selective sample collection device. For example, we have shown that an Hg drop exposed to stirred, unfiltered Adriatic Sea water can be used as a sulfide nanoparticle collector (*44*).

Sulfide Adlayers at Hg^0 Surfaces

To understand the voltammetry of nanoparticles, it is necessary to understand the adlayers that form on Hg electrode surfaces. *Adlayer* is a portmanteau word that simply means adsorbed layer; the word is used throughout surface science, not simply with regard to electrode surfaces. In metallurgy, oxide adlayers are commonly used to impart corrosion resistance to metal surfaces; bluing of gunmetal and anodizing of aluminum are familiar examples.

It has been known for more than a century that HgS adlayers accumulate spontaneously under certain conditions on Hg electrodes that are exposed to sulfide solutions (*45*). For this to happen the electrode potential must be held within a window of approximately +0.15 V to -0.65 V. (All potentials quoted in this paper are vs. a Ag/AgCl, 3 M KCl reference electrode; for electrochemical reactions that are H^+- or HS^--dependent, potentials are approximate and assume near-neutral pH, $\sim 10^5$ M HS^-, which are common conditions in sulfidic natural waters.)

Outside this window, HgS adlayers are unstable. At more positive potentials, HgS in the adlayer is transformed to HgO (or calomel in chloride solutions); at more negative potentials, Hg^{II} is reduced to Hg^0 with release of dissolved sulfide (*45*). The reduction reaction ($HgS + H^+ + 2e^- \rightarrow Hg^0 + HS^-$) is often exploited in the voltammetric determination of dissolved sulfide. To reduce completely an HgS monolayer, a charge of about 180 microcoulombs per square centimeter of electrode surface must be provided (*45*). When HgS adlayers exceed monolayer thicknesses, the necessary reduction charge per layer is greater and the reduction potential shifts to more negative values. This implies that multilayers are more dense and thermodynamically stable than monolayers.

When metal sulfide (MS) nanoparticles are present in sulfidic solution, MS adlayers as well as HgS adlayers are a possibility on Hg electrodes. We have explored CuS and FeS adlayers in some detail (*24*, *46*, *47*), and explored adlayers of a few other metals to a limited extent (*48*).

The potential window for CuS adlayer stability reaches from an anodic limit of about -0.3 V to a cathodic limit of about -0.95 V. The anodic reaction involves oxidation of Hg^0 to HgS with release of Cu^{2+} or a Cu^{II} complex, depending on solution composition (*46*). Notice that this limit does not involve oxidation of CuS, itself, which occurs at much more positive potentials (*49*). The cathodic limit, around -0.95 V, is nominally due to $CuS + H^+ + 2e^- \rightarrow Cu^0 + HS^-$, but Cu^0 in this case represents an amalgam or intermetallic Cu-Hg compound on the electrode surface. As in the case of HgS, the reduction potential of CuS shifts in the negative direction as the adlayer becomes thicker (*24*).

The potential window for stable FeS adlayers is broader, reaching from about -0.45 V to beyond -1.50 V. The anodic limit is established by $FeS + Hg^0 \rightarrow HgS + Fe^{2+} + 2e^-$ (*47*). The cathodic side of this window, which has not been investigated carefully, presumably is established by reduction of FeS to Fe^0 with release of sulfide to solution. The cathodic limit is hundreds of millivolts more negative than the thermodynamic reduction potential of FeS, which is estimated to be -1.2 V (*50*). This overpotential is very similar to the known overpotential for $Fe^{2+} + 2e^- \rightarrow Fe^0$ at Hg electrodes (*51*). Both overpotentials are related to the difficulty of nucleating Fe^0, a non-amalgamating metal, on Hg electrodes. It is important to note that the stability windows for FeS and CuS adlayers overlap partly, but not entirely, with the stability window of HgS adlayers.

The key to voltammetric determination of MS nanoparticles is to accumulate them in the form of MS adlayers on an electrode. This must be done at potentials where no interfering HgS adlayer can form. After accumulation, the amount of MS adlayer can be quantified by scanning past either the anodic or cathodic limit of stability of the MS adlayer and measuring the current as the adlayer decomposes electrochemically. The strategy is analogous to the time-honored one of determining dissolved sulfide by accumulating an HgS adlayer and then scanning past the potential where that adlayer is reduced.

Sulfide Nanoparticle Interactions with Hg^0 Electrodes

Figure 2 presents a conceptualization of what happens if an Hg surface is exposed to FeS nanoparticles in suspension. It is assumed that the electrode potential is being held at a value that lies beyond the negative limit of HgS adlayer stability.

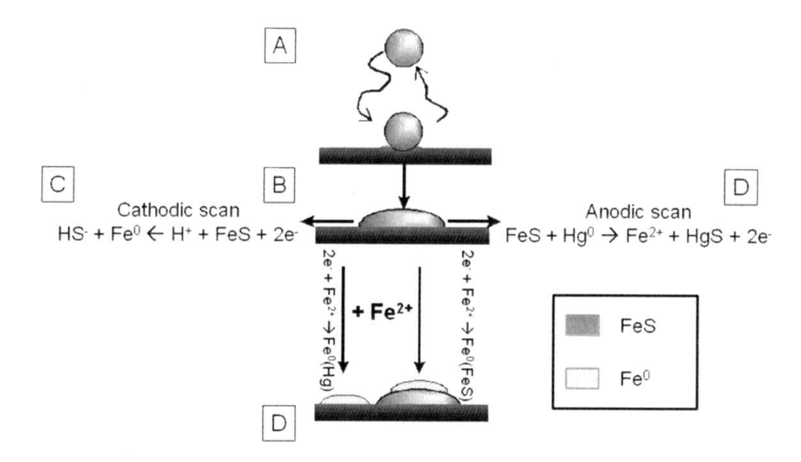

Figure 2. Schematic of the interaction of FeS nanoparticles with a mercury electrode surface.

The FeS nanoparticles diffuse by Brownian motion and bind to Hg on the surface (A in Figure 2). The electrode has a negative surface charge at this potential. Sulfide particles usually have negative zeta potentials at near-neutral pH if kept out of contact with O_2 (*52*, *53*). In this situation, the incoming particles must overcome an electrostatic potential energy barrier in order to reach the electrode. Particle collection at the Hg drop electrode then is analogous to coagulation; collection efficiency is sensitive to cation charge and concentration in the electrolyte as predicted by the Schulze-Hardy Rule (*24*). If the electrode and particles have opposite surface charges, then particle collection is facilitated by electrostatics, rather than impeded. Collection efficiency becomes comparatively insensitive to electrolyte composition (*24*).

This colloid-like behavior during accumulation of an adlayer demonstrates that the nanoparticles, themselves, are transporting the analyte to the electrode surface. Transport is not due to a minor dissolved intermediate.

Coagulation-like behavior discriminates against very small nanoparticles (less than a few nm). They are less able to penetrate electrostatic energy barriers and more likely to escape back to solution (*54*). In some cases small particles also may be lost by reduction during the accumulation period and thus not assayed during the subsequent scan (*24*).

The mass accumulation rate of spherical particles by Brownian diffusion is proportional to radius^{-1}, resulting in discrimination against the largest particles (i.e. particle radii on the order of 100 nm or larger). Although we have shown that particles in the micron size range, such as found in powder slurries, will accumulate readily on Hg electrodes (*44*, *48*) the accumulation process is most sensitive to particles in the 5-100 nm diameter range.

This size selection effect is easily observed by repeatedly analyzing an equimolar metal-sulfide mixture (*46*, *48*). Typically, no metal sulfide deposit is formed on an Hg electrode immediately after preparing the mixture. At this stage, particle sizes usually are in the 1-10 nm range (*24*), and most particles are too small to be captured efficiently. As assays continue, however, signals due to metal sulfide adlayers gradually grow because the primary particles coagulate and recrystallize, becoming more efficiently captured by Hg^0. Eventually though, the signals decay as the particles pass out of the optimum size range for capture.

Nanoparticles that adhere to Hg surfaces apparently transform spontaneously to adlayers (A→B in Figure 2). The thermodynamic drive for this involves maximizing the exergonic sulfide-Hg^0 interfacial interaction. In the case of CuS, a large shift in reduction potential occurs during this transformation, implying that the adlayer is much more stable than the adsorbed particle (*24*).

Once formed, the adlayer can be destabilized and its mass quantified by either anodic (B→D) or cathodic (B→C) scans. In the case of FeS, the cathodic reaction lies at extremely negative potentials where many interfering reactions are possible; therefore the anodic reaction is best for quantification. On the other hand, CuS is more conveniently quantified by its cathodic reaction.

Within the stability window of FeS adlayers, cathodic reactions with Fe^{2+} and its labile complexes ($FeCl^+$, $FeSO_4^0$ etc.) occur (B→D in Figure 2). As discussed in the next section, if the electrode surface consists partly of an FeS adlayer and partly of bare Hg^0, then Fe^{2+} can be reduced at both surfaces, but at different potentials.

An alternate mechanism of producing an FeS adlayer on Hg involves first accumulating an HgS adlayer at a relatively positive potential, such as -0.2 V (i.e. within the HgS stability window but outside the FeS stability window). FeS nanoparticles diffusing to the electrode will oxidize Hg^0 to HgS with release of Fe^{2+} rather than accumulating as an FeS adlayer. Then, if a cathodic scan is made in Fe^{2+}-containing solution, the HgS adlayer can be transformed to an FeS adlayer (D→B) (*47*). By the same kind of replacement reaction, HgS can produce CuS adlayers (*24*, *44*, *46*). Thus an analyst who wishes to detect nanoparticles by the adlayer they deposit on Hg^0 must avoid accumulation potentials where HgS adlayers form.

The FeSaq Problem

Our study of FeS adlayers has led to an interesting new explanation for the mysterious analyte that is commonly designated FeSaq (*50*, *55*). This putative analyte yields a reduction peak, which is sometimes a doublet, near -1.1 V in sulfidic natural waters and in synthetic solutions containing Fe and sulfide. Davison et al. (*55*) presented a detailed summary of the properties of FeSaq but were unable to deduce its actual composition. They did show that it has a diffusion coefficient similar to Fe^{2+} and therefore must have a low molecular mass. However it also seemed to display coagulation behavior, disappearing at ionic strengths above 0.1 M. Other authors have proposed that FeSaq is a charge-neutral iron sulfide cluster of uncertain composition (*50*). The FeSaq signal appears only in solutions that are near saturation with respect to FeS precipitates (*33*, *47*, *50*, *55*). Such solutions can be suspected of containing FeS nanoparticles.

The electrochemical landscape associated with the putative FeSaq signal is illustrated in Figure 3. This figure shows both anodic and cathodic scans starting from an accumulation potential of -0.75 V in solutions containing mixtures of Fe(II) and sulfide. The mixtures were initially supersaturated with mackinawite and contain suspended nanoparticulate FeS (diameter <50 nm) as determined by dynamic light scattering measurements. At this accumulation potential, FeS nanoparticles can deposit at the Hg surface and then transform to FeS adlayers. In the cathodic direction, an FeSaq peak (i.e. -1.1 V peak labeled C1) is encountered first, followed by a C4 peak. The C4 peak is agreed to be due to $Fe^{2+} + 2e^- \rightarrow Fe^0$. Note that both C1 and C4 increase proportionately with increasing Fe(II)/sulfide ratio. In the anodic direction, an A3 peak occurs where FeS adlayers decompose by oxidizing Hg^0 (B→D in Figure 2).

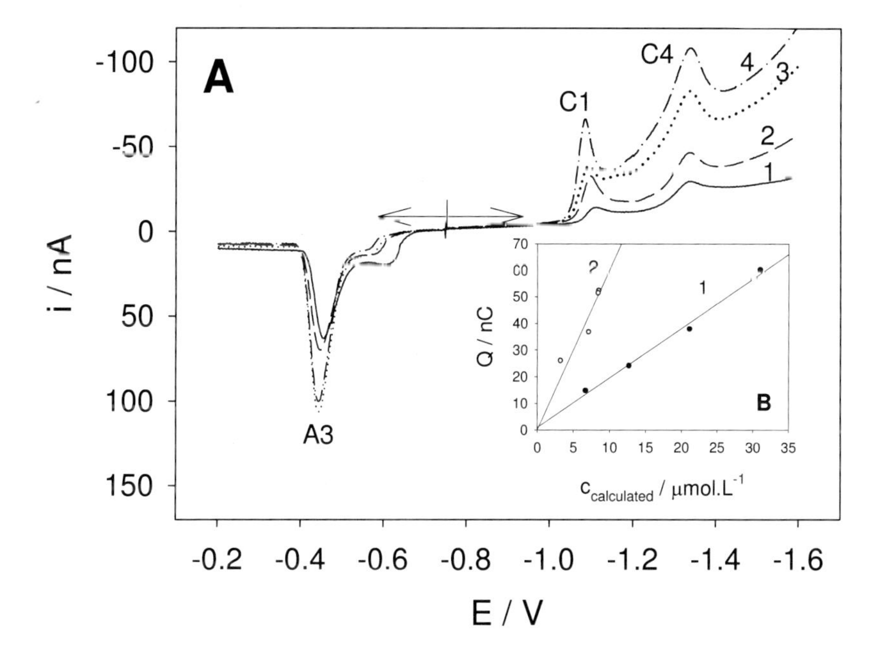

Figure 3. Voltammetric curves for 2.2 x 10^{-5} M Na_2S, 0.01 M NaCl and (1) 1.1 x 10^{-5} M, (2) 2.2 x 10^{-5} M, (3) 3.3 x 10^{-5} M, (4) 4.4 x 10^{-5} M Fe(II). Deposition with stirring was at -0.75 V (Ag/AgCl reference; deposition time 10 s; Hg electrode area 0.0054 cm^2). First, the potential was scanned towards more negative values and then the measurement was repeated by scanning towards more positive values. (Inset, B) Dependence of (1) A3 peak charge vs. theoretically calculated concentration of precipitated FeS nanoparticles and (2) sum of C1 and C4 peak charges vs. theoretically calculated dissolved Fe(II) in the solutions given for part A of the figure.

A thermodynamic model was used to calculate the equilibrium concentrations of Fe^{2+} and its labile complexes, as well as the amount nanoparticulate FeS that would be precipitated if the solution equilibrated with mackinawite (*47*). In the inset to Figure 3, we show that the sum of charges for C1+C4 is proportional to the sum of dissolved Fe(II) species, whereas the peak charge for A3 is proportional to the calculated amount of FeS precipitate.

Our interpretation of these observations is illustrated in part D of Figure 2. During the accumulation period (10 s in this case), nanoparticulate FeS has sorbed to the electrode and partially covered its surface with FeS adlayers. During the subsequent cathodic scan, Fe^{2+} is reduced to Fe^0 at -1.1 V on the FeS-covered portion of the electrode and at -1.4 V on the bare Hg portion. In other words, the overpotential known to be associated with this reduction reaction is greater on Hg^0 than on FeS. As a consequence, the one analyte gives two signals; a second analyte, FeSaq, is not required.

Corroboration of this interpretation is supplied by two observations. First, if a scan is reversed after the C1 peak has been recorded, an anodic peak due to oxidation of Fe^0 is observed (*47*). In the scientific literature, this peak has been reported after Fe^{2+} was reduced on solid state electrodes but not after it was reduced on Hg^0 electrodes. Fe^0 is wetted and absorbed into Hg^0 (*56*). Thus it is removed from the electrode surface and cannot be oxidized during an anodic scan. Appearance of an Fe^0 oxidation peak after an FeS adlayer has been deposited on an Hg^0 electrode shows that some Fe^0 must reside on the adlayer, as depicted in Figure 2.

Second, and more compellingly, our interpretation is corroborated by a demonstration that the relative charges of the C1 and C4 peaks can be manipulated simply by changing the amount of FeS adlayer deposited prior to a cathodic scan (see Figure 7 in (*47*)). Increasing an electrode's FeS adlayer coverage by increasing the accumulation time increases the C1 peak at the expense of the C4 peak. When adlayer coverage becomes complete, the C4 peak disappears even though the concentration of dissolved Fe^{2+} has not changed.

This new interpretation of the C1 signal explains some previously puzzling findings in natural waters. For example, several workers have observed what they interpret as an FeSaq complex or cluster in waters that appear to contain no Fe^{2+}, as measured by C4 at -1.4 V. If the signal arose from an FeSaq cluster or complex, why would this species not disassociate, releasing Fe^{2+}, until an equilibrium was reached? Alternatively, if the FeSaq cluster or complex were so stable that the equilibrium Fe^{2+} were immeasurably small in the presence of dissolved sulfide, then why in other samples are both Fe^{2+} and FeSaq (C4 and C1), as well as free sulfide, seemingly found together (*51*, *57*)? If the C1 peak were an analytical signal from a dissolved FeSaq complex or cluster, it should be possible to find a mass action law relationship between the sizes of the C1 and C4 signals in sulfidic waters; no such relationship has been reported. In the laboratory, iron sulfide solutions are observed to equilibrate on time scales from a few seconds (*50*) to an hour or so (*20*, *55*), implying that slow kinetics probably cannot be invoked. On the other hand, according to the interpretation presented here, a C1 peak with no C4 peak simply means that the electrode has become completely covered with an FeS adlayer and all dissolved Fe^{2+} is being reduced at C1.

Our interpretation of the C1 peak suggests that voltammetry will underdetermine Fe^{2+} when the -1.4 V (C4) peak alone is measured. Thus the $Fe_{colloid}/S_{colloid}$ ratio of ~2, rather than 1, in the colloids of Lake Bret (*32*) might have arisen because Fe^{2+} was underdetermined, resulting in overestimation of $Fe_{colloid}$, which was obtained by difference: $Fe_{colloid} = Fe(II)_{total} - Fe^{2+}$.

Identifying Metal Sulfide Nanoparticles

Several metal-sulfide adlayers can be reduced to metal amalgam plus dissolved sulfide on Hg electrodes. Potentials of their reduction peaks are given in Table 1. For the cases where reduction peaks have been measured, two peaks are observed (*48*). The ratios of these peaks change with experimental conditions, including accumulation time and nanoparticle ageing time. Reasons for two peaks remain to be fully explored, but a contributing factor may be formation of metal sulfide adlayers having different thicknesses. Thicker adlayers are produced by longer accumulation times, but also might result when larger, aged particles sorb to an electrode; larger particles create localized high concentrations of metal sulfide on the electrode surface.

Table 1. Sulfide nanoparticle particle reduction potentials (*24, 44, 46, 48*)

Investigated sulfide species	*Peak potentials / V*	
	C1(1)	*C1(2)*
FeS	< -1.5	
CuS	–0.95	-1.15
HgS	–0.89	–1.01
PbS	–1.18	–1.33

Notice that reduction potentials of Cu, Hg and Pb sulfides are rather similar. Additionally they fall in the same range as the putative FeSaq peak. Thus reduction potential appears not promising for identifying compositions of sulfide nanoparticles.

To overcome this problem, additional tests can be conducted. Table 2 shows how nanoparticles of the four metal sulfides in Table 1 respond to treatment with EDTA or with dilute HCl. For example after addition of excess EDTA to HgS or CuS nanoparticle solutions, reduction peaks remain. Both of these sulfides are too insoluble to be appreciably dissolved by $\sim10^{-4}$ M EDTA. On the other hand, in the case of PbS or FeS, reduction peaks completely disappear when EDTA is added. In addition to such chemical tests, geochemical abundance and thermodynamic criteria (e.g. Figure 1) can be helpful in assessing what metal sulfide might be responsible for an observed reduction peak.

Table 2. Responses of C1 reduction peaks to EDTA addition or acidification (*33*, *48*)

Nanoparticulate phase	*C1 peaks observed after 10^{-4} M EDTA addition*	*C1 peaks observed after acidification with HCl (pH=2) and purging with N_2*
PbS	No	No
HgS	Yes	Yes
CuS	Yes	Yes
FeS	No	No

Preliminary Application to Natural Waters

Lake Pavin, in the Massif Central of France is permanently sulfidic below 60 m (*33*, *58*). Sulfide, measured by the methylene blue colorimetric method (designated $S(-II)_{MB}$), reaches ~12 μM in the deeper waters. These waters are dominated by dissolved Fe(II), which exceeds S(-II) by ~100-fold. A maximum in the $S(-II)_{MB}$ concentration at about 65 m attests to saturation of the lake waters with FeS at this depth and below (*33*). This is affirmed by thermodynamic calculations. As shown in Figure 4, a -1.1 V peak is observed in samples from the sulfidic water column of this lake.

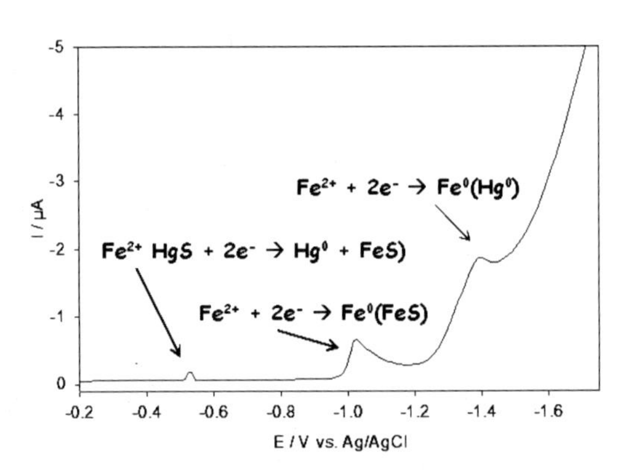

Figure 4. Cathodic scan in a sample collected from the iron-rich but sulfidic water column of Lake Pavin at a depth of 71 m. Measurements were performed ex situ under diffusion control (no stirring) at a starting potential of -0.2 V, scan rate (v) of 0.1 V s^{-1}. The middle peak has previously been attributed to reduction of an iron sulfide complex or cluster of unknown composition (FeSaq) but in this work is interpreted as due to reduction of Fe^{2+} and its labile complexes on FeS adlayers.

Previously, we found that sulfide measured by voltammetry in unfiltered samples (designated $S(-II)_V$) was much lower than $S(-II)_{MB}$, which had been measured in filtered samples (*33*). The difference, typically ~9 μM, was attributed to FeS, which is known to be included in $S(-II)_{MB}$ but not $S(-II)_V$. Because the samples analyzed by the methylene blue method had been filtered (0.2 μm), the extra sulfide found by methylene blue was inferred to be nanoscale FeS. Based on extensive prior work on trace elements in this lake (*58*), all likely metal sulfides other than FeS would be too scarce to contribute significantly to the 9 μM difference.

The -1.1 V peak in Figure 4 is evidence that the electrode has acquired an FeS adlayer. However, this experiment alone does not demonstrate that FeS nanoparticles exist in Lake Pavin. An HgS adlayer would have formed during the early part of the scan in this experiment. This HgS would be transformed to an FeS adlayer by the cathodic reaction near -0.5 V. Thus conditions for reduction of Fe^{2+} to Fe^0 on an FeS adlayer (i.e. the -1.1 V peak) might have arisen without involvement of FeS nanoparticles.

On the other hand, direct evidence for FeS nanoparticles in Lake Pavin deep waters is shown in Figure 5. This voltammetric scan was obtained after poising a fresh Hg drop in unfiltered Lake Pavin water for 30 s at -0.75 V. At this potential, no HgS adlayer can form (*45, 47*). After the accumulation period, a scan in the positive direction produced a robust peak, which is due to oxidation of Hg^0 by FeS adlayers (B→D in Figure 2). To confirm the nature of the observed oxidation peak, we treated the sample with EDTA as described above. The oxidation peak disappeared completely after this treatment. At the same time, the HgS (C3) peak increased due to the increment of sulfide released when FeS dissolved (*33*).

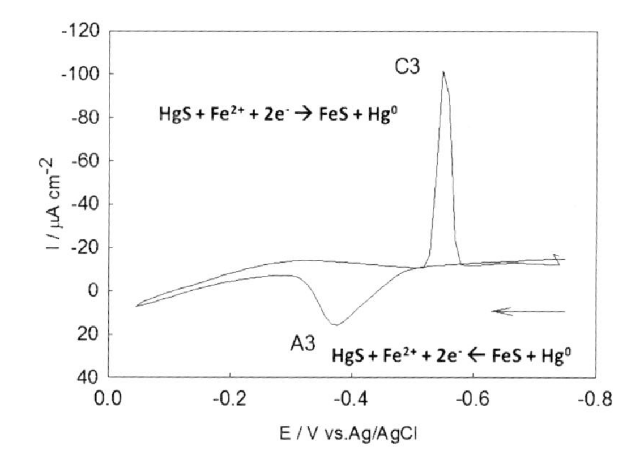

Figure 5. Cyclic voltammetric (CV) curve for a sulfidic Pavin Lake sample taken at 71 m depth. The A3 peak is a measure of FeS nanoparticles and is recorded after accumulation with stirring (30s) at an accumulation potential of -0.75 V; scan rate (v) of 0.1 V s^{-1}. In the return scan, the reduction peak is due to reduction of HgS produced by the oxidation reaction as well as accumulation of HS^- during the scan. In these Fe-rich waters, HgS is reduced at about -0.55 V by reaction with Fe^{2+} rather than at -0.65 V, where HgS is reduced in the absence of dissolved Fe^{2+}.

As Figure 5 shows, at the high concentrations of Fe attained in anoxic waters, FeS nanoparticles can produce robust signals. On the other hand, we have not observed a similar FeS signal in saline (38‰) Rogoznica Lake, even though its deep waters are saturated with respect to mackinawite (*59*). Probably the residence time of FeS nanoparticles is much shorter in saltwater (Rogoznica) relative to freshwater (Pavin), owing to coagulation and settling. Shorter residence times would contribute to lower steady state concentrations of nanoparticles.

Whether voltammetry will be useful for metals other than Fe remains to be determined. Preliminary efforts to identify CuS nanoparticles in Rogoznica Lake have been unproductive. Possibly this simply confirms that the high sulfide concentrations in Rogoznica (*60*) lie outside the optimum range for Cu nanoparticles, as predicted by Figure 1. Nonetheless, when 10^{-7} M Cu^{2+} was added to Rogoznica waters, a CuS nanoparticle signal appeared immediately. The same has been seen in North Adriatic Sea water (*44*). In this case, ambient sulfide concentrations were submicromolar.

How to calibrate voltammetric peaks so as to obtain quantitative information is a critical remaining problem. Integration of an A3 peak, such as the one in Figure 5, gives a measure of the amount of FeS adlayer accumulated on the electrode, but the relationship of that to the amount of nanoparticulate FeS in the water is not clear. The particle size dependence of the electrode's collection efficiency is a key problem. An accurate determination of the mass concentration of a nanoparticulate analyte might require standards containing particles of the same composition, degree of crystallinity, reactive surface area and size. It may be necessary to pair voltammetric measurements with other size-dependent techniques, such as dynamic light scattering or field flow fractionation, in order to overcome this obstacle. Until progress is made on this problem, the voltammetric determination of nanoparticles will be useful mainly for qualitative analysis.

Even without quantification, the evidence in Figure 5 provides important direct support for the previous conclusion from indirect evidence that Lake Pavin contains FeS nanoparticles. A tool to study how varying chemical and physical properties in natural waters affect rates of formation and loss of these particles now exists. A key transient in the biogeochemical cycles of iron and sulfur is now directly accessible to experiment.

References

1. Horowitz, A. J.; Lum, K. R.; Garbarino, J. R.; Hall, G. E. M.; LeMieux, C; Demas, C. R. Problems associated with using filtration to define dissolved trace element concentrations in natural water samples. *Environ. Sci Technol.* **1996**, *30*, 954–963.
2. Lead, J. R.; Wilkinson, K. J. Environmental colloids and particles: Current knowledge and future developments. In *Environmental Colloids and Particles*; Wilkinson, K. J., Lead, J. R., Eds.; IUPAC Series, Analytical and Physical Chemistry; John Wiley & Sons: Chichester, 2007; Vol 10, pp 1−15.
3. Guo, L.; Santschi, P. H. Ultrafiltration and its applications to sampling and characterization of aquatic colloids. In *Environmental Colloids and Particles*; Wilkinson, K. J., Lead, J. R., Eds.; IUPAC Series, Analytical and Physical Chemistry; John Wiley & Sons: Chichester, 2007; Vol 10, pp 159−221.

4. Nel, A.; Xia, T.; Mädler, L.; Li, N. Toxic potential of materials at the nanolevel. *Science* **2006**, *311*, 622–627.
5. Oberdörster, G; Oberdörster, E.; Oberdörster, J. Nanotoxicology: An emerging discipline evolving from studies of ultrafine particles. *Environ. Health Perspect.* **2005**, *113*, 823–839.
6. Nowack, B.; Bucheli, T. D. Occurrence, behavior and effects of nanoparticles in the environment. *Environ. Pollut.* **2007**, *150*, 5–22.
7. Lovrić, J.; Cho, S. J.; Winnik, F. M.; Maysinger, D. Unmodified cadmium telluride quantum dots induce reactive oxygen species formation leading to multiple organelle damage and cell death. *Chem. Biol.* **2005**, *12*, 1227–1234.
8. Lin, P.; Chen, J.-W.; Chang, L. W.; Wu, J.-P.; Redding, L.; Chang, H.; Yeh, T.-K.; Yang, C. S.; Tsai, M.-H.; Wang, H.-J.; Kuo, Y.-C.; Yang, R. S. H. Computational and ultrastructural toxicology of a nanoparticle, Quantum Dot 705, in mice. *Environ. Sci. Technol.* **2008**, *42*, 6264–6270.
9. Priester, J. H.; Stoimenov, P. K.; Mielke, R. E.; Webb, S. M.; Ehrhardt, C.; Zhang, J. P.; Stucky, G. D.; Holdern, P. A. Effects of Soluble Cadmium Salts Versus CdSe Quantum Dots on the Growth of Planktonic Pseudomonas aeruginosa. *Environ. Sci. Technol.* **2009**, *43*, 2589–2594.
10. Su, Y; Hu, M.; Fan, C.; He, Y.; Li, Q.; Li, W.; Wang, L.-H.; Shen, P.; Huang, Q. The cytotoxicity of CdTe quantum dots and the relative contributions from released cadmium ions and nanoparticle properties. *Biomaterials* **2010**, *31*, 4829–4834.
11. Aruguete, D. M.; Guest, J. S.; Yu, W. W.; Love, N. G.; Hochella, M. F. Interaction of CdSe/CdS core-shell quantum dots and *Pseudomonas aeruginosa*. *Environ. Chem.* **2009**, *7*, 28–35.
12. Deonarine, A.; Hsu-Kim, H. Precipitation of mercuric sulfide nanoparticles in NOM-containing water: implications for the natural environment. *Environ. Sci. Technol.* **2009**, *43*, 2368–2373.
13. Shea, D.; Helz, G. R. The solubility of copper in sulfidic waters: Sulfide and polysulfide complexes in equilibrium with covellite. *Geochim. Cosmochim. Acta* **1988**, *52*, 1815–1825.
14. Thompson, R. A.; Helz, G. R. Copper speciation in sulfidic solutions at low sulfur activity. Further evidence of cluster complexes? *Geochim. Cosmochim. Acta* **1994**, *58*, 2971–2983.
15. Davison, W.; Phillips, N.; Tabner, B. J. Soluble iron sulfide species in natural waters: reappraisal of their stoichiometry and stability constants. *Aquat. Sci.* **1999**, *61*, 23–43.
16. Mountain, B. W.; Seward, T. W. The hydrosulphide sulphide complexes of copper(I): Experimental determination of stoichiometry and stability at 22 degrees C and reassessment of high temperature data. *Geochim. Cosmochim. Acta* **1999**, *63*, 11–29.
17. King, D. W.; Farlow, R. Role of carbonate speciation on the oxidation of Fe(II) by H_2O_2. *Mar. Chem.* **2000**, *70*, 201–209.
18. Matijević, E. Nanosize precursors as building blocks for monodispersed colloids. *Colloid J.* **2007**, *69*, 29–38.
19. Michel, F. M.; Antao, S. M.; Chupas, P. J.; Lee, P. L.; Parise, J. B.; Schoonen, M. A. A. Short- to medium-range atomic order and crystallite

size of the initial FeS precipitate from pair distribution function analysis. *Chem. Mater.* **2005**, *17*, 6246–6255.
20. Wothers, M.; Charlet, L.; van der Linde, P. R.; Rickard, D.; van der Weijden, C. H. Surface chemistry of disordered mackinawite (FeS). *Geochim. Cosmochim. Acta* **2005**, *69*, 3469–3481.
21. Ohfuji, H.; Rickard, D. High resolution transmission electron microscopic study of synthetic nanocrystalline mackinawite. *Earth Planet. Sci. Lett.* **2006**, *241*, 277–233.
22. Jeong, H. Y; Lee, J. H.; Hayes, K. F. Characterization of synthetic nanocrystalline mackinawite: crystal structure, particle size, and specific surface area. *Geochim. Cosmochim. Acta* **2008**, *72*, 493–505.
23. Shea, D.; Helz, G. R. Kinetics of inhibited crystal growth: Precipitation of CuS from solutions containing chelated copper (II). *J. Colloid Interface Sci.* **1987**, *116*, 373–383.
24. Krnarić, D.; Helz, G. R.; Bura-Nakić, E.; Jurašin, D. Accumulation mechanism for metal chalcogenide nanoparticles at Hg^0 electrodes: Copper sulfide example. *Anal. Chem.* **2008**, *80*, 742–749.
25. Widerlund, A.; Davison, W. Size and density distribution of sulfide-producing microniches in lake sediments. *Environ. Sci. Technol.* **2007**, *41*, 8044–8049.
26. Sukola, K.; Wang, F.; Tessier, A. Metal sulfide species in oxic waters. *Anal. Chim. Acta* **2005**, *528*, 183–195.
27. Clarke, M. B.; Helz, G. R. Metal-thiometalate transport of biologically active trace elements in sulfidic environments. 1. Experimental evidence for copper thioarsenite. *Environ. Sci. Technol.* **2000**, *34*, 1477–1482.
28. Helz, G. R.; Erickson, B. E. Extraordinary stability of copper(I)-tetrathiomolybdate complexes: Possible implications for aquatic ecosystems. *Environ. Toxicol. Chem.* **2011**, *30*, 97–102.
29. Landing, W. M.; Westerlund, S. The solution chemistry of iron(II) in framvaren fjord. *Mar. Chem.* **1988**, *23*, 329–343.
30. Dierssen, H.; Balzer, W.; Landing, W. M. Simplified synthesis of an 8-hydroxyquinoline chelating resin and a study of trace metal profiles from Jellyfish Lake, Palau. *Mar. Chem.* **2001**, *73*, 173–192.
31. Lewis, B. L.; Landing, W. M. The biogeochemistry of iron and manganese in the Black Sea. *Deep-Sea Res.* **1991**, *38* (Suppl. 2A), S773–S803.
32. Buffle, J.; Zali, O.; Zumstein, J.; DeVitre, R. Analytical methods for the direct determination of inorganic and organic species: Seasonal changes of iron, sulfur, and pedogenic and aquogenic organic constituents in the eutrophic Lake Bret, Switzerland. *Sci. Total Environ.* **1987**, *64*, 41–59.
33. Bura-Nakić, E.; Viollier, E.; Jézéquel, D.; Thiam, A.; Ciglenečki, I. Reduced sulfur and iron species in anoxic water column of meromictic Lake Pavin (Massif Central, France). *Chem. Geol.* **2009**, *266*, 311–317.
34. Skei, J. M.; Loring, D. H.; Rantala, R. T. T. Trace metals in suspended particulate matter and in sediment trap material from a permanently anoxic fjord—Framvaren, South Norway. *Aquat. Geochem.* **1996**, *2*, 131–147.
35. Labrenz, M.; et al. Formation of sphalerite (ZnS) deposits in natural biofilms of sulfate-reducing bacteria. *Science* **2000**, *290*, 1744–1747.

36. Moreau, J. W.; Webb, R. I.; Banfield, J. F. Ultrastructure, aggregation-state, and crystal growth of biogenic nanocrystalline sphalerite and wurtzite. *Am. Mineral.* **2004**, *89*, 950–960.
37. Gammons, C. H.; Frandsen, A. K. Fate amd transport of metals in H_2S-rich waters at a treatment wetland. *Geochem. Trans.* 2001, DOI: 10.1039/b008234.
38. Weber, F.-A.; Voegelin, A.; Kaegi, R.; Kretzschmar, R. Contaminant mobilization by metallic copper and metal sulphide colloids in flodded soil. *Nat. Geosci.* **2009**, *2*, 267–271.
39. Adams, N. W. H.; Kramer, J. R. Silver speciation in wastewater effluent, surface waters and pore waters. *Environ. Toxicol. Chem.* **1999**, *18*, 2667–2573.
40. Kim, B.; Park, C.-S.; Murayama, M.; Hochella, R. F., Jr. Discovery and characterization of silver sulfide nanoparticles in final sewage sludge products. *Environ. Sci. Technol.* **2010**, *44*, 7509–7514.
41. Rozan, T. F.; Benoit, G.; Luther, G. W., III. Measuring metal sulfide complexes in oxic river waters with square wave voltammetry. *Environ. Sci. Technol.* **1999**, *33*, 3021–3026.
42. Simpson, S. L.; Apte, S. C.; Batley, G. E. Sample storage artifacts affecting the measurement of dissolved copper in sulfidic waters. *Anal. Chem.* **1998**, *70*, 4202–4205.
43. Bowles, K. C.; Bell, R. A.; Ernste, M. J.; Kramer, J. R.; Manolopoulos, H.; Ogden, N. Synthesis and characterization of metal sulfide clusters for toxicological studies. *Environ. Toxicol. Chem.* **2002**, *21*, 693–699.
44. Ciglenečki, I.; Krznarić, D.; Helz, G. R. Voltammetry of copper sulfide particles and nanoparticles: Investigation of the cluster hypothesis. *Environ. Sci. Technol.* **2005**, *39*, 7492–7498.
45. Krznarić, D.; Ciglenečki-Jušič, I. Electrochemcial processes of sulfide in NaCl electrolyte solutions on mercury electrode. *Electroanalysis* **2005**, *17*, 1317–1324.
46. Krznarić, D.; Helz, G. R.; Ciglenečki, I. Prospect for determining copper sulfide nanoparticles by voltammetry: A potential artifact in supersaturated solutions. *J. Electroanal. Chem.* **2006**, *590*, 201–214.
47. Bura-Nakić, E.; Krznarić, D.; Helz, G. R.; Ciglenečki, I. Characterization of iron sulfide species in model solutions by cyclic voltammetry. Revisiting an old problem. *Electroanalysis* **2011**, *23*, 1376–1382.
48. Bura-Nakić, E.; Krznarić, D.; Jurašin, D.; Helz, G. R.; Ciglenečki, I. Voltammetric characterization of metal sulfide particles and nanoparticles in model solutions and natural waters. *Anal. Chim. Acta* **2007**, *594*, 44–51.
49. Yin, Q.; Vaughan, D. J.; England, K. E. R.; Kelsall, G. H. Electrochemical oxidation of covellite (CuS) in alkaline-solution. *J. Colloid Interface Sci.* **1994**, *166*, 133–142.
50. Theberge, S. M.; Luther, G. W., III. Determination of the electrochemical properties of a soluble aqueous FeS species present in sulfidic solutions. *Aquat. Geochem.* **1997**, *3*, 191–211.
51. Buffle, J.; DeVitre, R. R.; Perret, D.; Leppard, G. G. Combining field measurements for speciation in non perturbable water samples. In *Metal*

Speciation. Theory, Analysis and Applications; Kramer, J. R., Allen, H. E., Eds.; Lewis Publishers: Chelsea, MI, 1988; pp 99–124.
52. Bebie, J.; Schoonen, M. A. A.; Fuhrmann, M.; Strongin, D. R. Surface charge development on transition metal sulfides: An electrokinetic study. *Geochim. Cosmochim. Acta* **1998**, *62*, 633–642.
53. Fullston, D.; Fornasiero, D.; Ralston, J. Zeta potential study of the oxidation of copper sulfide minerals. *Colloids Surf., A* **1999**, *146*, 113–121.
54. Higashitani, K.; Kondo, M.; Hatade, S. Effect of particle size on coagulation rate of ultrafine colloidal particles. *J. Colloid Interface Sci.* **1991**, *142*, 204–213.
55. Davison, W.; Buffle, J.; DeVitre, R. Voltammetric characterization of a dissolved iron sulphide species by laboratory and field studies. *Anal. Chim. Acta* **1998**, *377*, 193–203.
56. Winkler, K.; Krogulec, T.; Galus, Z. Formation of FeS and its effect on the electrode-reactions of the Fe(II)/Fe system in thiocyanate solutions at mercury-electrodes. *Electrochim. Acta* **1985**, *30*, 1055–1062.
57. Hebert, A. B.; Morse, J. W. Microscale effects of light on H_2S and Fe^{2+} in vegetated (*Zostera marina*) sediments. *Mar. Chem.* **2003**, *81*, 1–9.
58. Viollier, E.; Jézéquel, D.; Michard, G.; Pèpe, M.; Sarazin, G.; Albéric, P. Geochemical study of a crater lake (Pavin Lake, France) - trace-element behavior in the monimolimnion. *Chem. Geol.* **1995**, *125*, 61–72.
59. Helz, G. R.; Bura-Nakić, E.; Mikac, N.; Ciglenečki, I. New model for molybdenum behavior in euxinic waters. *Chem. Geol.* **2011**, *284*, 323–332.
60. Bura-Nakić, E.; Helz, G. R.; Ciglenečki, I.; Ćosović, B. Reduced sulfur species in a stratified seawater lake (Rogoznica Lake, Croatia); seasonal variations and argument for organic carriers of reactive sulfur. *Geochim. Cosmochim. Acta* **2009**, *73*, 3738–3751.

Chapter 14

Redox Reactivity of Organically Complexed Iron(II) Species with Aquatic Contaminants

Timothy J. Strathmann*

Department of Civil and Environmental Engineering, University of Illinois at Urbana-Champaign, Urbana, IL 61801 USA
***strthmnn@illinois.edu**

Extracellular organic ligands and ligand functional groups within macromolecular natural organic matter can significantly influence the speciation and kinetic redox reactivity of Fe(II) with aquatic contaminants. Fe(II) complexation by Fe(III)-stabilizing ligands (e.g., carboxylate, catecholate, thiol) leads to formation of Fe(II) species with low standard reduction potentials (E_H^0) and enhanced reactivity with reducible contaminants (e.g., nitroaromatics and halogenated alkanes). Rates of contaminant reduction by Fe(II) are highly variable and dependent upon the identity and concentration of specific organic ligands as well as environmental conditions that affect the extent of complex formation. Linear free energy relationships have been developed to predict the aqueous reactivity of individual Fe(II) species with contaminants. Studies on the reactivity of Fe(II) complexes with model ligands also provide mechanistic insights into the potential mechanisms responsible for contaminant transformations observed in more complex aquatic systems where Fe(II) co-accumulates with more poorly defined natural organic matter.

Introduction

Abiotic reductive transformation processes represent important environmental sinks for many contaminants (*1*). The dominant microbial respiration processes in anoxic environments generate reduced extracellular products that can act as chemical reductants for contaminants possessing electron-poor functional groups in their structures. Thus, knowledge of important chemical reductants and the factors controlling their reactivity with aquatic contaminants is needed to predict the natural attenuation of contaminants in anoxic and suboxic environments and to develop effective remedial actions.

Evidence from field and laboratory investigations indicates that ferrous iron is abundant and mediates the reductive transformation of many classes of contaminants in soils, sediments, and aquifers. In the absence of oxygen, Fe(III) is often the most abundant terminal electron acceptor available to subsurface microorganisms, and the ubiquity of dissimilatory Fe(III)-reducing bacteria (e.g., *Geobacter metallireducens and Shewanella oneidensis*) is well documented (*2*). Their activity leads to elevated concentrations of Fe(II) in many soil, sediment, and groundwater environments (*3–5*). Contaminants that can be reduced by Fe(II) include halogenated solvents (*6–11*), nitroaromatic compounds (*8*, *12–18*), *N*-heterocyclic nitramine explosives (*19–22*), oxime carbamate pesticides (*23–26*), nitrite (*27*), nitrate (*28*), selenate (*29*, *30*), and several metal contaminants (*31–36*).

Fe(II) is a versatile reductant that can exhibit a wide range of standard one-electron reduction potentials, E_H^0, in aquatic systems depending on the metal's speciation (Figure 1) (*31*, *37*, *38*). Most laboratory studies have focused on the characterizing contaminant reactions with Fe(II) species that are associated with mineral surfaces, either adsorbed or incorporated into the mineral surface structure (*8*, *9*, *16*, *20*, *25*, *26*, *32*, *39–43*). These species are abundant and exhibit low E_H^0 values in comparison to dissolved Fe^{2+}(aq), making them quantitatively important reductants for many contaminants in heterogeneous aquatic systems. It has been commonly assumed that dissolved Fe(II) species are unreactive with most contaminants, but recent work demonstrates that dissolved Fe(II) complexes with some classes of organic ligands are also potent chemical reductants of aquatic contaminants (*10*, *13–15*, *21*, *24*, *31*, *44*). In addition to being of potential environmental relevance, dissolved Fe(II) complexes with low molecular weight organic ligands are very useful as probes for studying molecular-scale factors that control homogeneous and heterogeneous Fe(II) redox reactivity since they have more well-defined structures and properties than the mixture of Fe(II) species present at aqueous/mineral interfaces. Finally, similar Fe(II)-organic complexes may also form on the surfaces of soil minerals that are coated with organic matter, and such complexes may be important contributors to contaminant attenuation observed in soil systems (*45*).

This contribution summarizes efforts to characterize the redox reactivity of organically complexed Fe(II) species, with emphasis on kinetic aspects of redox reactions with aquatic contaminants. Important factors and properties governing the formation and redox reactions of Fe(II)-organic complexes are described. Models quantitatively linking reaction rates with equilibrium Fe(II) speciation and reactant molecular properties are also described. Finally, recent work examining the influence of humic substances on Fe(II) reactions with aquatic contaminants is discussed and results are interpreted in light of mechanistic information gained from reports on Fe(II) complexes with model ligands.

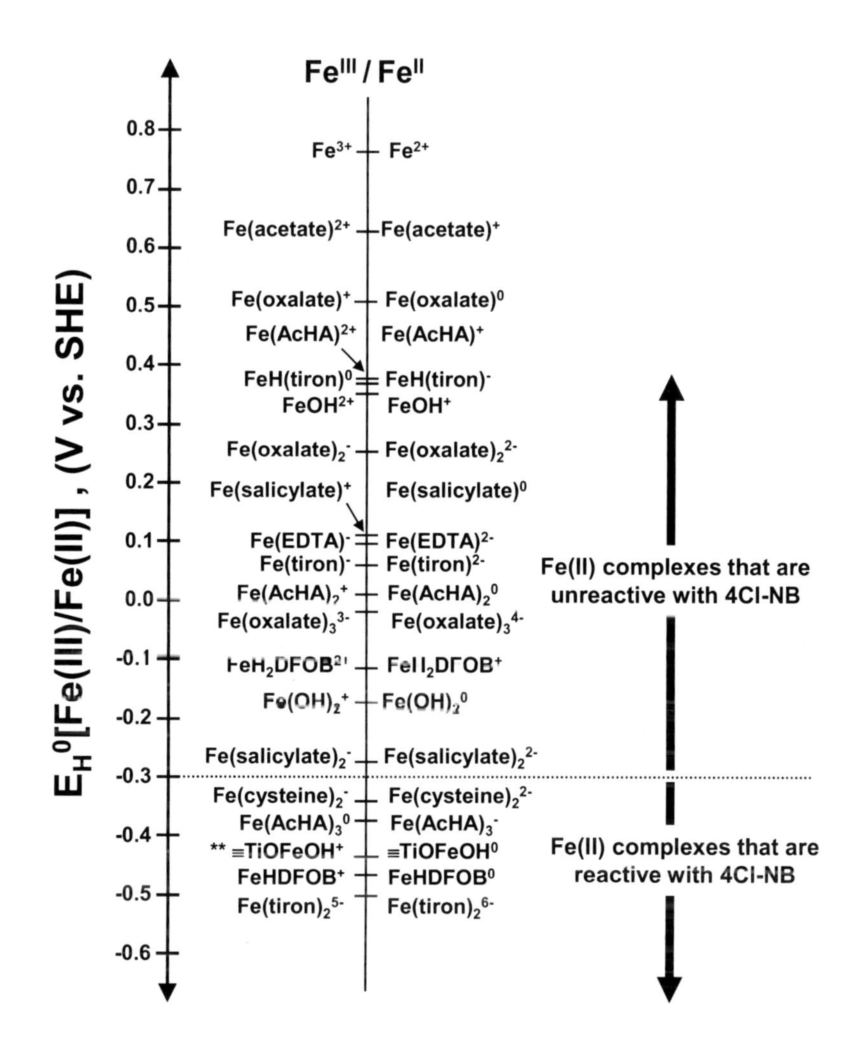

*Figure 1. E_H^0[Fe(III)/Fe(II)] values for selected dissolved and mineral-surface-complexed Fe(II) species. **E_H^0 value for hydroxylated Fe(II) surface complex on $TiO_{2(s)}$ estimated using a linear free energy relationship reported in (15) and kinetic data from (26). Abbreviations: AcHA = acetohydroxamate; DFOB = desferrioxamine B; EDTA = ethylenediaminetetraacetate; tiron = 4,5-dihydroxy-1,3-disulfonate; 4Cl-NB = 4-chloronitrobenzene. (Adapted with permission from (15). Copyright 2009 Elsevier).*

Organic Ligands in Soil and Aquatic Environments

Co-accumulation of Fe(II) and dissolved organic carbon (DOC) has been reported in reducing environments such as wetland soils and sedimentary pore water (*5*, *46*), and molecular interactions between the two constituents have been documented (*47*). Natural organic matter (NOM) is the broad term used to encompass both specific low molecular weight biochemicals and more poorly defined macromolecules like humic substances (*48*). The wide diversity of organic molecules that make up NOM originates from both the non-specific decay of plant and animal materials and specific release from microorganisms and plants into extracellular environments (*48*). Increasingly, inputs of synthetic organic chemicals also need to be considered as sources of dissolved organic carbon (DOC) in aquatic systems, especially in contaminated environments.

Humic substances and low molecular weight natural organic compounds contain a range of ligand functional groups (also called Lewis bases) that possess free electron pairs that can be donated to iron and other metal ions (Lewis acids) to form coordination complexes. Important Fe-complexing ligands present in biological molecules include several oxygen-, nitrogen-, and sulfur-containing functional groups (Figure 2A). Oxygen-containing ligands, including carboxylate and phenolate groups, are most abundant. The widely used Suwannee River NOM reference material available from the International Humic Substances Society (IHSS) is reported to contain 9.9 mmole carboxylate and 3.9 mmole phenolate groups per gram of carbon (*49*). Estimated carboxylate and phenolate contents for other IHSS humic materials range from 7.1-15.2 mmole/gc and 1.8-4.2 mmole/gc, respectively (*49*). Individual low molecular weight carboxylate compounds (e.g., lactate, citrate, oxalate, and formate) are also reported in soils at appreciable concentrations, sometimes as high as 1 mM (*50*). These compounds are byproducts from the breakdown of more complex organic materials and are excreted by plants and microorganisms. The high phenolic and aromatic acid content in the humic fraction of NOM results from high inputs of lignin and other aromatic precursors to terrestrial environments (*48*, *51*).

Although less abundant, nitrogen- and sulfur-based ligands are also important for Fe(II) and Fe(III) complexation in aquatic environments. Modern synchrotron-based spectroscopic tools show that amide and heterocyclic nitrogen groups (e.g., pyridine, imidazole, pyrazole) are important *N*-containing functional groups in humics and other natural organic materials, likely derived from biological precursors (e.g., amino acids, nucleic acids, porphyrins) (*52*, *53*). Sulfur-based ligands are particularly important in reducing environments. X-ray absorption spectroscopy measurements indicate that a significant fraction of sulfur atoms within humic substances are present in reduced oxidation states (e.g., thio, thiol, sulfide) (*54*), and several low molecular weight organosulfur compounds have been detected in sediments (*55*–*57*). Organosulfur compounds can form by specific biochemical pathways as well as extracellular abiotic reactions between inorganic sulfide species and electrophilic organic molecules (*58*).

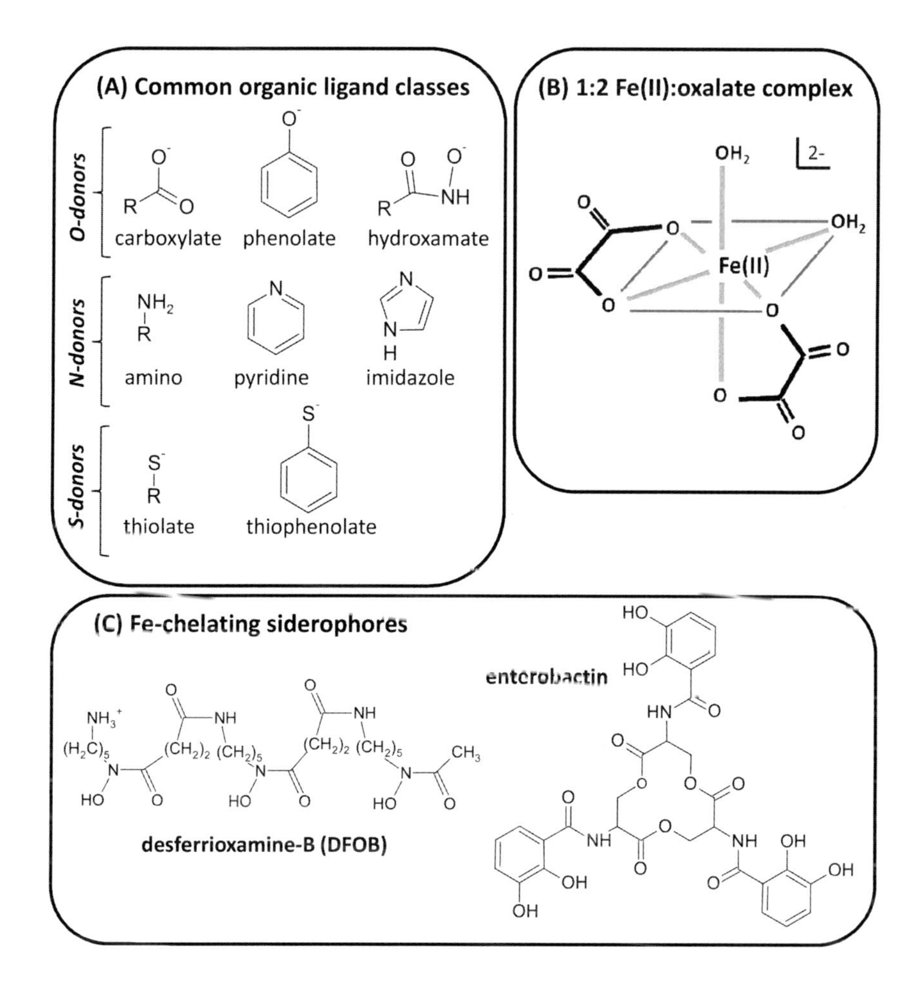

Figure 2. (A) Common naturally occurring organic ligand donor functional groups. (B) Structure of the Fe(II) chelation complex between Fe(II) and two oxalate ligands. (C) Siderophore ligands that form highly stable complexes with iron.

Organic molecules containing multiple Lewis base donor atoms in their structure can form multidentate complexes with Fe ions, which are typically more stable than monodentate complexes (*59*, *60*). Ligands capable of forming highly stable multidentate complexes are sometimes referred to as chelating agents. Five-member chelation complexes like those shown in Figure 2B are exceptionally stable (*60*, *61*). Low molecular weight polycarboxylate chelating agents (e.g., oxalate, citrate, malonate) are commonly released by organisms to help solubilize Fe and other sparingly soluble metal ions (*62*). Bacteria, fungi, and graminaceous plants (grasses) have also been found to excrete a number of Fe-specific chelating agents, siderophores, into their extracellular environments to solubilize iron for subsequent uptake (*62–64*). Structures of two siderophores are provided in Figure 2C. The exceptional ability of siderophores to chelate Fe (e.g., log β > 30) results from the presence of multiple Lewis base structures (typically hydroxamate, catecholate, and/or carboxylate groups with less common occurrence of amine, hydroxyphenyloxazolone, and hydroxypyridinone moieties (*65*)) spatially arranged in a manner that leads to formation of highly stable hexadentate complexes that saturate the metal's inner coordination shell (*64*). To date, over 500 siderophores have been identified (*66*). Reports of siderophore concentrations in the environment are sparse. Hydroxamate siderophore concentrations in bulk soil solutions have been estimated between 10^{-7} and 10^{-8} M, but concentrations in microenvironments like biofilms and the rhizosphere have been estimated to be many orders-of-magnitude greater (*63*, *67*, *68*). Synthetic chelating agents, most notably aminocarboxylates (e.g., EDTA, NTA) and phosphonates, are also increasingly ubiquitous in aquatic environments due to widespread use in consumer and industrial products and recalcitrance to conventional wastewater treatment technologies (*69*, *70*).

Influence of Ligands on Fe Speciation and Redox Properties

Fe(II) and Fe(III) Speciation

Fe(II) and Fe(III) are transition metal ions that typically adopt octahedral coordination geometries in aqueous environments, forming coordinate covalent bonds with six ligand donor atoms in the inner coordination sphere (see Figure 2B); seven-coordinate geometries are also reported in some rare cases (e.g., Fe complexes with the synthetic chelating agent EDTA) (*71*). The identity of the Fe-coordinated donor groups depends on solution conditions and the identity (Figure 2A) and concentration of ligands present, with water molecules serving as the default ligands (~55.6 M concentration in freshwater systems) in hexaquo "free ion" complexes ($Fe(H_2O)_6^{2+}$ and $Fe(H_2O)_6^{3+}$). In the absence of other Fe-complexing ligands, Fe(II)- and Fe(III)-hydroxo complexes of varying stoichiometry form when increasing pH due to metal ion-induced hydrolysis of coordinated water molecules. Equilibrium expressions and thermodynamic stability constants are available to predict formation of several Fe(II)-hydroxo complexes in aqueous solution (*72*):

$$Fe(H_2O)_6^{2+} \rightleftharpoons FeOH(H_2O)_5^{+} + H^{+} \qquad \log K_{FeOH} = -9.40 \qquad (1)$$

$$Fe(H_2O)_6^{2+} \rightleftharpoons Fe(OH)_2(H_2O)_4^{0} + 2\,H^{+} \qquad \log K_{Fe(OH)2} = -20.49 \qquad (2)$$

$$Fe(H_2O)_6^{2+} \rightleftharpoons Fe(OH)_3(H_2O)_3^{-} + 3\,H^{+} \qquad \log K_{Fe(OH)3} = -28.99 \qquad (3)$$

Organic ligands, especially those capable of chelating metals, significantly alter both Fe(II) and Fe(III) speciation by exchanging with coordinated water molecules at one or more positions in the inner coordination shell. The nature and extent of Fe ion complexation is highly dependent upon solution conditions, most notably pH and the ligand-to-metal concentration ratio. As an example, consider equilibrium Fe(II) speciation in solution containing the catecholate ligand tiron (L^{4-}). In addition to the hexaquo-coordinated Fe^{2+} species and hydrolyzed complexes thereof, three Fe(II)-tiron complexes of varying stoichiometry and protonation state ($Fe(II)H_xL_y^{2+x-4y}$) can form (Figure 3A; coordinated H_2O molecules omitted for clarity) (*14*). Formation of these complexes can be described by the following equilibrium expressions and stability constants (*14*):

$$Fe^{2+} + L^{4-} \rightleftharpoons FeL^{2-} \quad \log K_{FeL} = 10.63 \tag{4}$$

$$Fe^{2+} + H^{+} + L^{4-} \rightleftharpoons FeHL^{-} \quad \log K_{FeHL} = 17.85 \tag{5}$$

$$Fe^{2+} + 2\,L^{4-} \rightleftharpoons FeL_2^{6-} \quad \log K_{FeL2} = 15.33 \tag{6}$$

As shown in Figure 3B-C, the extent of Fe(II) complexation by tiron and the distribution of different Fe(II) species are heavily influenced by pH and ligand concentration. Tiron also forms complexes with any Fe(III) present in solution (*72*), and similar equilibrium species distribution diagrams can be calculated using available equilibrium constants (*72*). Organic complexation significantly alters the molecular properties of Fe(II) and Fe(III) ions, including, as will be shown, their aqueous redox reactivity with many contaminants.

Distinction between E_H^0 and E_H

Complexation of Fe(II) and Fe(III) by organic ligands can significantly alter the thermodynamic favorability of the Fe(III)/Fe(II) redox couple (*31*):

$$Fe(III) + e^{-} \rightleftharpoons Fe(II) \tag{7}$$

When discussing the effects of organic ligands on the energetics of the Fe redox couple, it is very important to make a distinction between the effects on the standard-state one-electron reduction potentials of individual Fe species (E_{H0}) and the overall "condition-dependent" reduction potential of the Fe(III)/Fe(II) redox couple (E_H) (*73*). E_H^0 values represent a molecular property for the energy associated with transferring an electron from or to individual Fe(II) or Fe(III) species, respectively. We will show below that E_H^0 values are useful molecular property descriptors for predicting the kinetic reactivity of individual Fe species. In contrast, the overall E_H value is a measure of the favorability or unfavorability of an iron redox reaction at specific environmental conditions, and is useful for predicting the extent of Fe(II) oxidation or Fe(III) reduction that will occur before equilibrium is achieved.

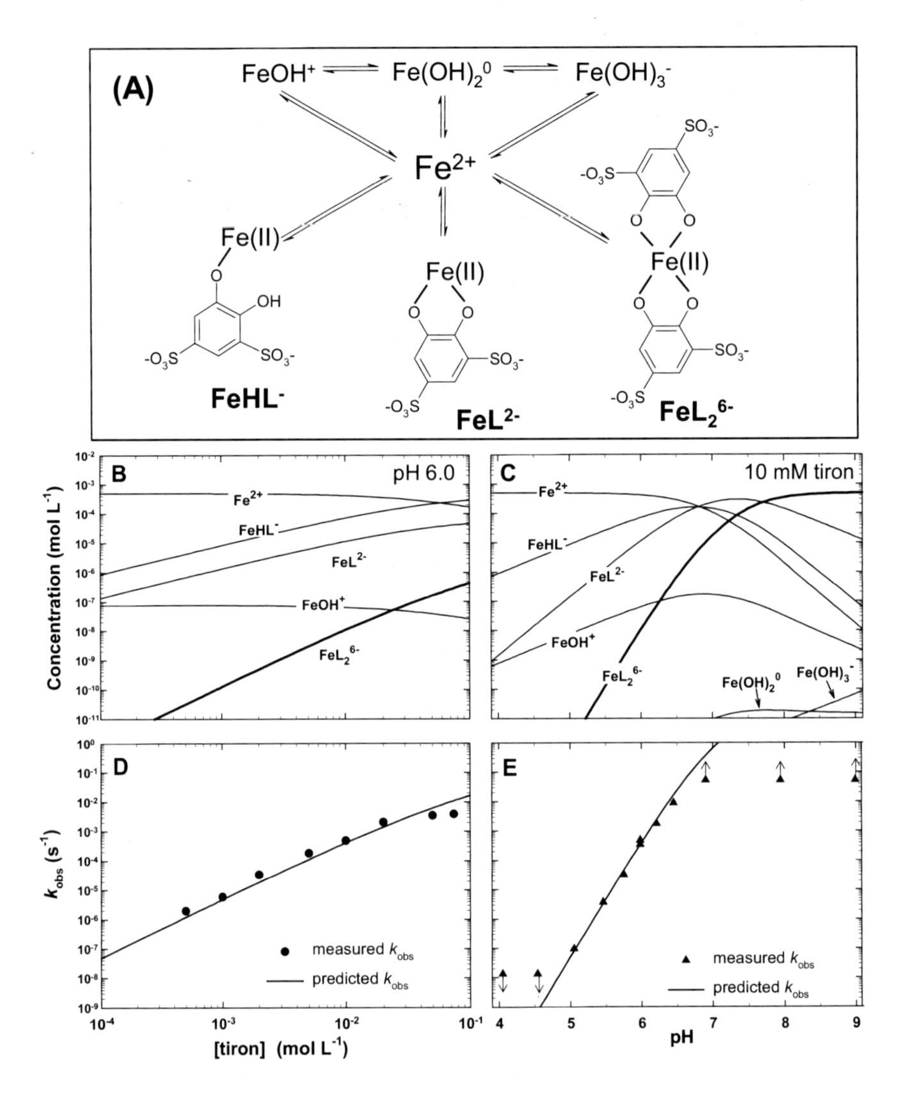

Figure 3. (A) Simplified depiction of Fe(II) speciation in the presence of the catecholate ligand tiron. (B-C) Effect of tiron concentration and pH on the speciation of 0.5 mM Fe(II). (D-E) Effect of tiron concentration and pH on k_{obs} for reduction of 25 μM 4-chloronitrobenzene in solutions described by panels B-C. (Figure Panels B-E adapted from reference (14). Copyright 2006 American Chemical Society.)

Values of E_H^0 for individual Fe species can be calculated whenever equilibrium constants are available for formation of equivalent Fe(II)- and Fe(III)-ligand complexes (i.e., complexes that differ only in oxidation state). This is accomplished as follows. Equation 8 describes the half reaction and E_H^0 value of the redox couple involving the uncomplexed (i.e., hexaquo-coordinated) Fe(III) and Fe(II) species (*74*):

$$Fe^{3+} + e^- \rightleftharpoons Fe^{2+} \quad E_H^0[Fe^{3+}/Fe^{2+}] = -\frac{\Delta G^0_{Fe(3+)/Fe(2+)}}{F} = +0.77\,V \quad (8)$$

where ΔG^0 is the standard state Gibbs free energy change of the half reaction and F is Faraday's constant. The half reaction and associated standard potential for a redox couple involving a Fe-ligand complex is then obtained by combining eq 8 with equilibrium expressions for the formation of equivalent Fe(II)- and Fe(III)-ligand complexes (*23*):

$$Fe^{2+} + xH^+ + yL \rightleftharpoons Fe(II)H_xL_y \quad \Delta G^0_{Fe(II)L} = -RT\ln K_{Fe(II)L} \quad (9)$$

$$Fe^{3+} + xH^+ + yL \rightleftharpoons Fe(III)H_xL_y \quad \Delta G^0_{Fe(III)L} = -RT\ln K_{Fe(III)L} \quad (10)$$

where x and y are the stoichiometric coefficients for protons and ligand molecules (L), respectively, R is the universal gas constant, T is the absolute temperature, and $K_{Fe(II)L}$ and $K_{Fe(III)L}$ are equilibrium constants for Fe(II) and Fe(III) complexation, respectively. The three reactions depicted in eqs 8-10 can then be combined to arrive at a half reaction for the Fe(III)-ligand/Fe(II)-ligand redox couple:

$$Fe(III)H_xL_y + e^- \rightleftharpoons Fe(II)H_xL_y$$

$$\Delta G^0_{Fe(III)L/Fe(II)L} = \Delta G^0_{Fe(3+)/Fe(2+)} + \Delta G^0_{Fe(II)L} - \Delta G^0_{Fe(III)L} \quad (11)$$

Substituting into eq 11 expressions for each ΔG^0 term from eqs 8-10, E_H^0 of the Fe(III)-ligand/Fe(II)-ligand redox couple can then be expressed in terms of equilibrium constants for formation of the Fe(II) and Fe(III) complexes in question:

$$E_H^0[Fe(III)L/Fe(II)L] = +0.77\,V - \frac{RT}{F}\ln\left(\frac{K_{Fe(III)L}}{K_{Fe(II)L}}\right) \quad (12)$$

Figure 1 shows E_H^0 values calculated for several Fe-organic complexes, illustrating the wide range in E_H^0 values spanning >1 V. Most naturally occurring organic ligands complex more strongly with Fe(III) than Fe(II) (i.e., $K_{Fe(III)L} >> K_{Fe(II)L}$) (*60*), and it follows from eq 12 that these complexes exhibit E_H^0 values lower than +0.77 V. In terms of thermodynamics, oxidation of such Fe(II)-ligand complexes will be more favorable than Fe^{2+}, whereas reduction of the corresponding Fe(III)-ligand complexes will be less favorable than Fe^{3+}.

Although related in nature, the effects of ligand complexation on the overall E_H value of the Fe(III)/Fe(II) redox couple is different from the effects on E_H^0 values of individual Fe-ligand complexes and is dependent on the solution conditions (*62*, *73*, *75*). E_H values for the reversible Fe(III)/Fe(II) couple can be calculated at different solution conditions and reactant concentrations using the Nernst equation:

$$E_H = +0.77 - \frac{RT}{F}\ln\left(\frac{\{Fe^{2+}\}}{\{Fe^{3+}\}}\right) \quad (13)$$

where +0.77 V is the standard state potential for the redox couple involving the uncomplexed Fe species (eq 8), and {i} represent thermodynamic activities, which are equal to the molar concentration [i] multiplied by an activity coefficient (γ_i) that can be calculated for ionic solutes using the Davies equation or other expressions (*37*). Note that the Nernst equation can alternatively be written in terms of E_H^0[Fe(III)L/Fe(II)L] determined in eq 12 and the ratio for activities of the Fe(III)L and Fe(II)L complexes at solution conditions of interest. However, using the formulation presented in eq 13 helps to illustrate the point that the presence of organic ligands affect E_H by suppressing the activities of uncomplexed Fe^{2+} and Fe^{3+} to differing degrees. As mentioned already, most naturally occurring organic ligands complex more strongly with Fe(III) than Fe(II). Thus, they will tend to suppress Fe^{3+} to a greater degree than Fe^{2+}, thereby increasing the magnitude of the $\{Fe^{2+}\}/\{Fe^{3+}\}$ term in eq 13 and lowering E_H. Figure 4 shows the effects of pH and concentration of the hydroxamate siderophore desferrioxamine-B (DFOB) on E_H of the Fe(III)/Fe(II) redox couple. For the conditions shown, E_H decreases dramatically with increasing pH and DFOB concentration since both factors promote greater suppression of $\{Fe^{3+}\}$ than $\{Fe^{2+}\}$. For the plots shown in Figure 4, calculations assumed fixed and equal total concentrations of Fe(II) and Fe(III). Changes in relative amounts of total Fe(II) and Fe(III) also affect E_H values (*73*), as does presence of different Fe(III)- and Fe(II)-solubility limiting mineral phases (*62*). Thus, there are a number of assumptions that go into any calculation of E_H. Most often, it is assumed that the total Fe(III) and total Fe(II) concentrations are equal (as in Figure 4), but this condition is not reflective of many environments where one oxidation state predominates (e.g., Fe(III) predominates in aerobic soils).

Contaminant Reactions with Fe(II)-Organic Complexes

As already mentioned, it has commonly been assumed that dissolved Fe(II) species are unreactive with many classes of persistent contaminants, and that contaminant reduction only occurs via Fe(II) species associated with mineral surfaces. This stems from the lack of contaminant reactivity often observed in solutions prepared from simple inorganic salts of Fe(II) (e.g., $FeCl_2$, $FeSO_4$). These salts dissociate and the resulting aquo-coordinated Fe^{2+} species is unreactive or exhibits only limited reactivity with many important classes of contaminants, including nitro- and halogenated compounds (*9*, *20*, *39*) (some classes of contaminants such as Cr(VI) and oxime carbamate pesticides are reduced by dissolved Fe^{2+} at appreciable rates (*23*, *76*)). The limited ability of Fe^{2+} to act as a reductant for aquatic contaminants is attributed to the high E_H^0 value of the associated Fe^{3+}/Fe^{2+} redox couple (+0.77 volts) (Figure 1) (*38*).

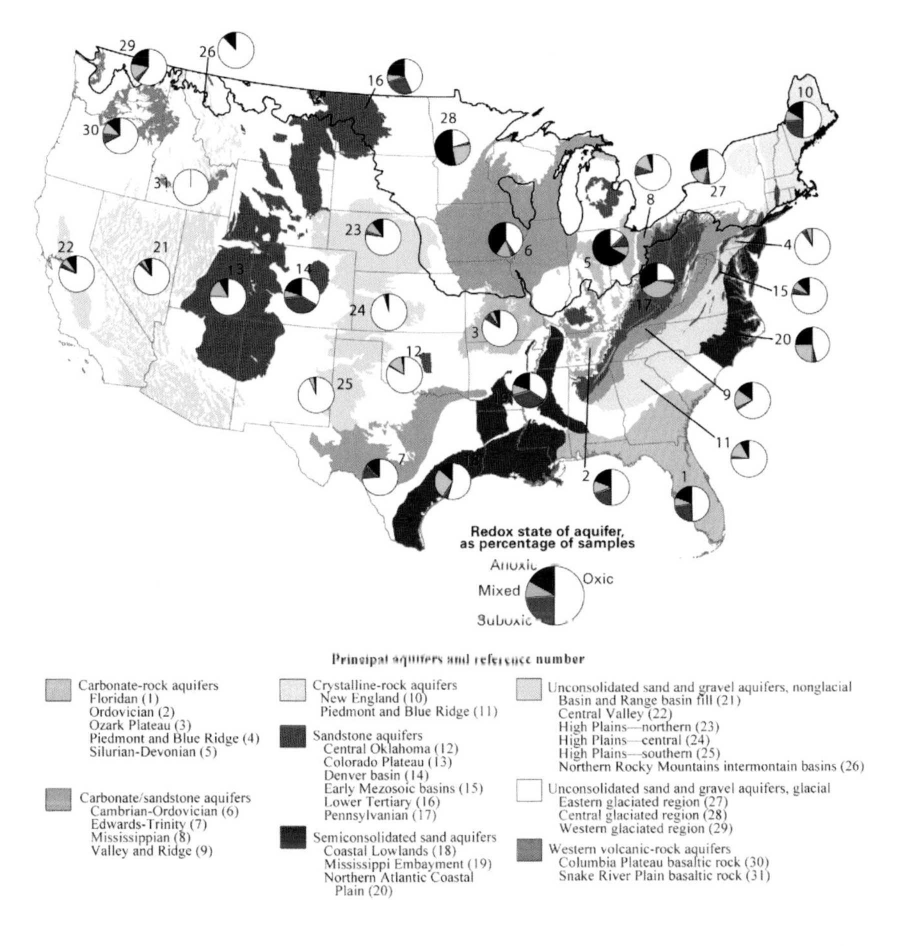

Figure 8. Pie diagrams indicating the percentages of domestic well samples that were oxic, suboxic, anoxic, or diagnostic of mixed redox processes in selected regional aquifers of the United States. (modified from reference (58))

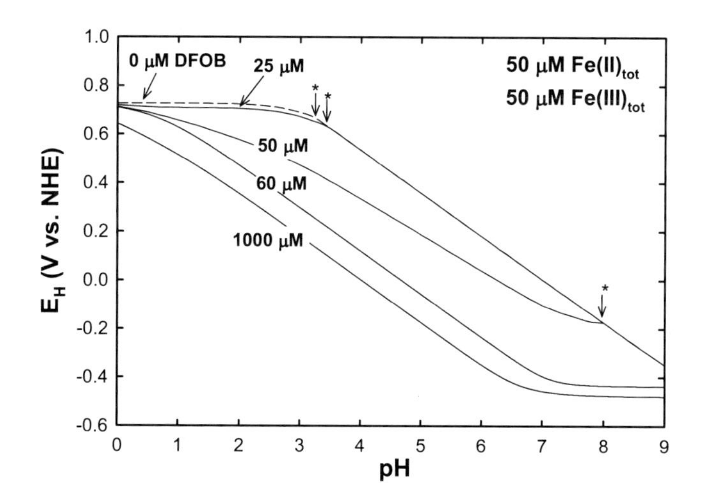

Figure 4. Influence of pH and DFOB ligand concentration on E_H for the Fe(III)/Fe(II) redox couple calculated by eq 13. Asterisks indicate pH conditions where onset of $Fe(OH)_{3(s)}$ precipitation is predicted to occur for 0, 25, and 50 μM DFOB concentrations, respectively

It has been known for many years that specialized Fe(II)-porphyrin complexes, representative of electron transfer centers in hemoproteins, are capable of rapidly reducing a number of organic functional groups. More than 30 years ago, Castro and co-workers demonstrated that reduction of alkyl halides, quinones, and nitro and nitroso compounds can be coupled to the oxidation of Fe(II)-porphyrin complexes in homogeneous non-aqueous solutions (*77–79*). The rapid oxidation of Fe(II)-porphyrin centers by alkylhalides and the consequential disruption of the respiratory chain was even suggested as a possible mode of alkylhalide toxicity (*80*). These complexes have also been proposed as electron-transfer mediators, rapidly cycling between Fe(III) and Fe(II) redox states and providing a lower activation energy pathway for transfer of electrons between bulk reductants (e.g., cysteine) and target contaminants (*6*, *12*).

More recently, model organic ligands which are representative of ligands that are abundant in extracellular aquatic and soil environments have been found to significantly enhance Fe(II) reaction kinetics with aquatic contaminants (*13–15*, *24*, *31*). For example, Figure 5 shows the effects of Fe(II) complexation by tiron (a model catecholate ligand) and desferrioxamine B (a hydroxamate siderophore) on rates of abiotic reduction for 4-chloronitrobenzene (4Cl-NB), a nitroaromatic compound. A six-electron reduction of 4Cl-NB to the corresponding aniline (4Cl-An) is coupled to the one-electron oxidation of six Fe(II) ions (*17*):

$$\underset{\text{4Cl-NB}}{NO_2\text{-}C_6H_4\text{-}Cl} \xrightarrow[H_2O]{2e^- + 2H^+} \underset{\text{4Cl-NsB}}{NO\text{-}C_6H_4\text{-}Cl} \xrightarrow{2e^- + 2H^+} \underset{\text{4Cl-HA}}{NHOH\text{-}C_6H_4\text{-}Cl} \xrightarrow[H_2O]{2e^- + 2H^+} \underset{\text{4Cl-An}}{NH_2\text{-}C_6H_4\text{-}Cl} \qquad (14)$$

Organic ligand classes found to promote Fe(II) reactions with nitroaromatic contaminants include catecholate, hydroxamate, and organothiol ligands capable of forming Fe(II) complexes with very low $E_H{}^0$ values (Figure 1) (*13–15*). The first two classes are common ligand donor groups in siderophores excreted by microorganisms to increase iron bioavailability (*63*). Organothiols are produced by biochemical routes (e.g., cysteine, glutathione) as well as extracellular abiotic routes (e.g., Michael addition reactions) and are abundant in many sulfate-reducing aquatic environments (*55–58*). The same classes of ligands have also been shown to promote Fe(II) reactions with *N*-heterocyclic nitramine explosives (e.g., RDX, HMX) and polyhalogenated alkane contaminants (*10*, *21*). Although other common classes of natural organic ligands, including carboxylates and aminocarboxylates, do not promote significant Fe(II) reactions with these contaminants, they have been shown to significantly accelerate Fe(II) reactions with Cr(VI) (*31*), oxime carbamate pesticides (*24*), and naturally occurring oxidant species (O_2, H_2O_2, Fe(III) oxyhydroxide minerals) (*81–87*). Fe(II) complexation by these ligands also lowers $E_H{}^0$ of the Fe(III)/Fe(II) redox couple, but generally to a lesser degree than the catecholate, hydroxamate, and organothiol ligands that induce Fe(II) reactions with nitroaromatic compounds (see Figure 1).

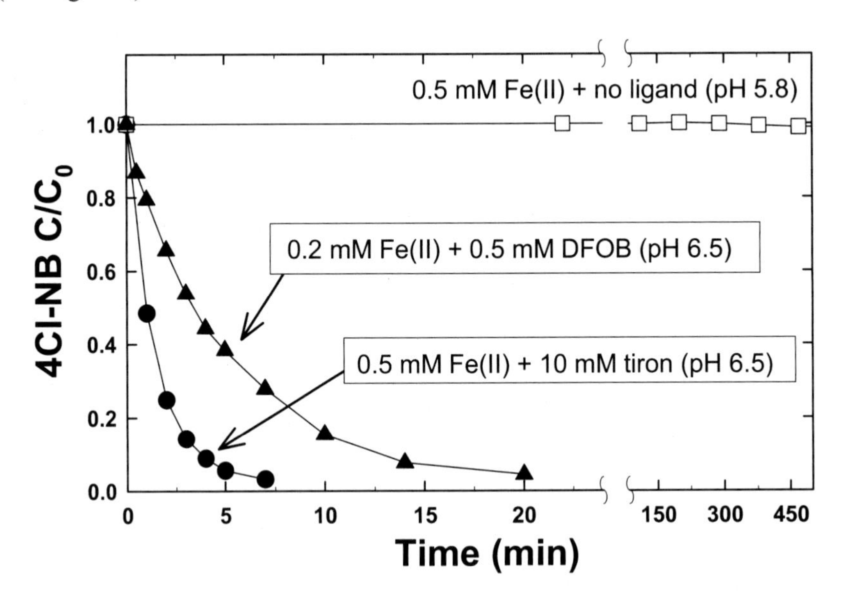

Figure 5. Timecourses for the reduction of 4Cl-NB by Fe(II) in the absence and presence of organic ligands. Data from references (14, 15).

Effects of Changing Fe(II) Speciation on Reaction Rates

The influence that organic ligands have on the kinetics of Fe(II) reactions with contaminants can be explained by comparing the effects that tiron has on Fe(II) speciation (Figure 3B-C) with the effects that the ligand has on pseudo-first-order rate constants for 4Cl-NB reduction (k_{obs}; s^{-1}) by Fe(II) (Figure 3D-E). Fe(II) speciation trends were calculated using eqs 1-6 together with equilibrium expressions for tiron acid-base chemistry. From a redox standpoint, individual Fe(II) species can be considered as different reagents with distinct redox properties and kinetic reactivity. Thus, contaminant reduction can be viewed as occurring along several parallel pathways involving different Fe(II) species. For tiron containing solutions, the following possible parallel reactions and associated second-order rate constants (k_i; $M^{-1}s^{-1}$) can be written for Fe(II) reactions with 4Cl-NB,

$$Fe^{2+} + 4Cl\text{-}NB \rightarrow \text{products} \qquad k_{Fe(2+)} \tag{15a}$$

$$FeOH^{+} + 4Cl\text{-}NB \rightarrow \text{products} \qquad k_{FeOH(+)} \tag{15b}$$

$$Fe(OH)_2^{0} + 4Cl\text{-}NB \rightarrow \text{products} \qquad k_{Fe(OH)2(0)} \tag{15c}$$

$$Fe(OH)_3^{-} + 4Cl\text{-}NB \rightarrow \text{products} \qquad k_{Fe(OH)3(-)} \tag{15d}$$

$$FeHL^{-} + 4Cl\text{-}NB \rightarrow \text{products} \qquad k_{FeHL(-)} \tag{15e}$$

$$FeL^{2-} + 4Cl\text{-}NB \rightarrow \text{products} \qquad k_{FeL(2-)} \tag{15f}$$

$$FeL_2^{6-} + 4Cl\text{-}NB \rightarrow \text{products} \qquad k_{FeL2(6-)} \tag{15g}$$

If equilibrium between Fe(II) species is assumed to be maintained, the values of k_{obs} measured in different batch reactions represent the weighted sum contribution of individual reactions listed in eqs 15a-g,

$$\begin{aligned} k_{obs} = {} & k_{Fe(2+)}[Fe^{2+}] + k_{FeOH(+)}[FeOH^{+}] + k_{Fe(OH)2(0)}[Fe(OH)_2^0] + \\ & + k_{Fe(OH)3(-)}[Fe(OH)_3^-] + k_{FeHL(-)}[FeHL^-] + k_{FeL(2-)}[FeL^{2-}] + k_{FeL2(6-)}[FeL_2^{6-}] \end{aligned} \tag{16}$$

or in an more generalized form,

$$k_{obs} = \sum_i k_i [Fe^{II}{}_i] \tag{17}$$

Although eq 16 contains several terms, often many be ignored if the contributing Fe(II) species are known to be unreactive or if evidence is presented that other terms dominate. For example, the first 4 terms in eq 16 (corresponding to eqs 15a-d) can be neglected because no significant 4Cl-NB reduction occurs in tiron-free solutions (Figure 5) (*13*, *14*, *26*, *39*) (note that reports of 4Cl-NB reduction by dissolved Fe(II)-hydroxo complexes (*18*) have been attributed to artifacts likely resulting from the presence of fine colloidal Fe hydroxide precipitates that form reactive sorbed Fe(II) surface complexes (*40*)). Furthermore, comparison of kinetic trends (Figure 3D-E) with the corresponding trends in Fe(II) speciation (Figure 3B-C) indicate that eq 16 can be reduced to a single term involving the 1:2 Fe(II)-tiron complex:

$$k_{obs} = k_{FeL2(6-)}[FeL_2^{6-}] \quad (18)$$

In the tiron system, good linear correlations between $[FeL_2^{6-}]$ and k_{obs} measurements that span several orders-of-magnitude were found for 4Cl-NB as well as model *N*-heterocyclic nitramine (RDX) and polyhalogenated alkane (1,1,1-trichloroethane) contaminants (Figure 6). Fit-derived values of $k_{Fel2(6-)}$ determined for the three model contaminants were 3.8×10^4 M^{-1} s^{-1}, 7.3×10^2 M^{-1} s^{-1}, and 9.5×10^{-2} M^{-1} s^{-1}, respectively (*10, 14, 21*). The controlling influence of FeL_2^{6-} on contaminant reduction kinetics is consistent with its very low E_H^0 value in comparison to the other Fe(II)-tiron complexes (-0.509 V for FeL_2^{6-} vs. +0.045 V for FeL^{2-} and +0.352 V for $FeHL^-$) (*14*).

Whereas a strong correlation between measured k_{obs} values and the concentration of a single Fe(II) species was noted for reactions with these contaminants, this is not always the case. For example, describing the influence of organic ligands on Fe(II) reactions with Cr(VI) or oxime carbamate pesticides requires additive contributions from multiple reactive Fe(II) species (*24, 31*).

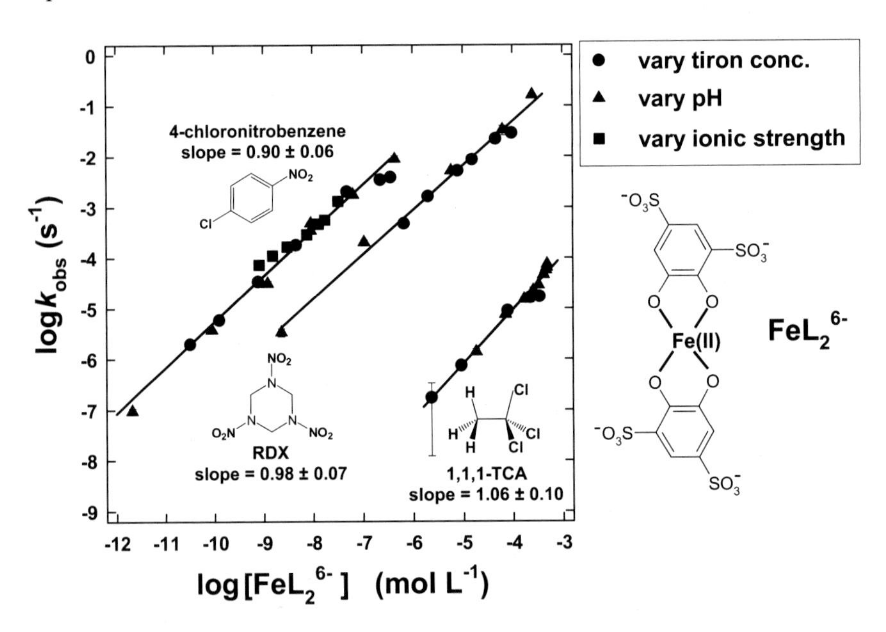

Figure 6. Apparent reaction order plots showing linear correlations between the pseudo-first-order rate constants for reduction of three different organic contaminants and predicted concentrations of the 1:2 Fe(II)-tiron complex (FeL_2^{6-}) at varying solution conditions. (Reproduced from reference (10). Copyright 2007 American Chemical Society.)

Linear Free Energy Relationships

The reactivity of contaminants with only selected Fe(II)-organic complexes is attributed to their low E_H^0 values. For example, 4Cl-NB is reduced only by Fe(II) complexes that possess E_H^0 values lower than -0.3 V (Figure 1) (*15*).

The relative reactivity of different Fe(II) complexes can sometimes be described by linear free energy relationships (LFERs) that correlate kinetic reactivity with overall reaction energetics (*38*). LFERs can be derived whenever there is a correlation between the effects that structural changes have on the overall free energy of the reaction (or other molecular property descriptor) and the effects that the changes have on the activation energy of the rate-determining step in the reaction mechanism (*88*). Figure 7A shows a plot of the logk_i values for 4Cl-NB reacting with different Fe(II)-organic complexes versus the E_H^0 values of the complexes calculated using eq 12 (*15*). Although the data set is small, it can be fit with a LFER of the following form:

$$\log k_i = a\frac{E_H^0}{0.059\ \mathrm{V}} + b \tag{19}$$

As expected, reactivity is greatest for complexes exhibiting the lowest E_H^0 values. Regression of the data in Figure 7A yields values of 1.6(±1.1) and -9.6(±8.3) for the slope and intercept (uncertainties provided at 95% CI), respectively. Similar LFERs can be derived for aqueous Fe(II) complexes reacting with Cr(VI) (a = -0.53±0.13), oxamyl (a = -0.37±0.09), O_2 (a = -0.72±0.19), and H_2O_2 (a = -0.47±0.26) (*24*, *31*, *81–85*, *89*, *90*) (Figures 7B-D). The value of the LFER slopes differ considerably, indicating large differences in apparent sensitivity of individual reaction mechanisms to the free energy change associated with the Fe(III)/Fe(II) redox couple. The elevated sensitivity of 4Cl-NB is consistent with the fact this compound only reacts with a small number of complexes exhibiting very low E_H^0 values, whereas the other four contaminants/oxidants react with a much wider range of Fe(II) species. The larger slope value observed for 4Cl-NB is also consistent with expectations for reactions in which rates are controlled predominantly by electron transfer (*92*), but a large nitrogen isotope fractionation occurs in 4Cl-NB when reacting with Fe(II)-tiron complexes (*93*), consistent with a more complex mechanism wherein the rate of 4Cl-NB reduction depends on a rate-limiting irreversible N−O bond cleavage step as well as several reversible pre-equilibrium electron and proton transfer steps (*94*). For oxamyl and H_2O_2, some negative outliers to the LFERs are also noted. These points correspond to Fe(II) complexes that are coordinately saturated by the complexing organic ligands (i.e., six inner-sphere coordination positions are occupied), and the suppressed reactivity of these species is attributed to the ligands blocking contaminant/oxidant access to the Fe(II) center (*24*, *81*). Thus, for these target compounds, inner-sphere electron transfer mechanisms are believed to be more rapid than outer-sphere pathways that do not involve direct bonding between the electron donor and acceptor prior to electron transfer (*92*).

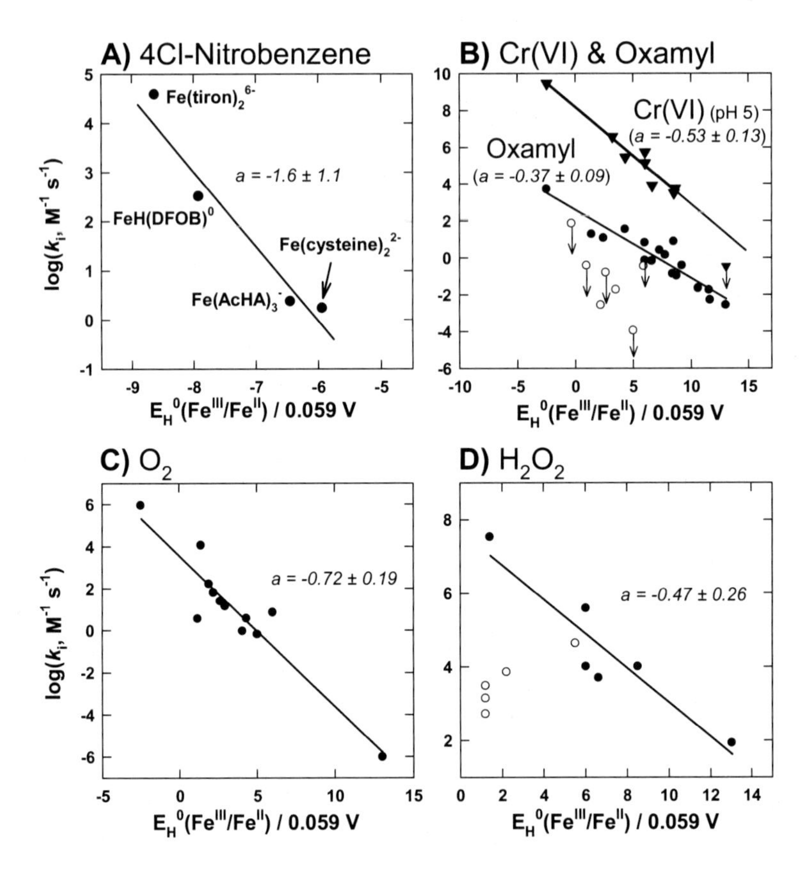

Figure 7. Correlation between the second-order rate constants measured for reactions of Fe(II) complexes with different oxidizing species and E_H^0[Fe(III)L/Fe(II)L] values. Open symbols for reactions of oxamyl and H_2O_2 indicate coordinatively saturated Fe(II) species considered to be outliers to the LFERs. Symbols with downward arrows indicate upper-limit estimates of the rate constants. (Panel A adapted with permission from reference (15). Copyright 2009 Elsevier.) (Data for panels B-D from references (24, 31, 81–85, 89–91)).

Within individual classes of contaminants, the relative reactivities of different analogues with an individual Fe(II)-organic complex can also be described by LFERs. For example, LFERs of the type shown in eq 19 have been demonstrated for substituted nitrobenzenes reacting with various Fe(II)-organic complexes (Figure 8A) (*12–15*). These LFERs indicate increased reactivity for nitrobenzenes exhibiting larger apparent one-electron reduction potentials, $E_H^{1\prime}(ArNO_2)$, which occurs when electron- withdrawing ring substituents (e.g., -Cl, -acetyl) are introduced in place of electron-donating substituents (e.g., $-CH_3$) (*38*). LFERs of this type have also been reported for reduction of nitrobenzenes in a variety of homogeneous and heterogeneous aqueous systems (*17*). Originally, the slopes obtained from such LFERs were thought to be diagnostic of whether or not transfer of the first electron to the $R\text{-}NO_2$ group is rate determining

(*38*). However, as already mentioned, results from nitrogen stable isotope fractionation studies indicate that the rate of nitrobenzene reduction depends on a rate-limiting irreversible N–O bond cleavage step as well as several reversible pre-equilibrium electron and proton transfer steps (*94*). Thus, the slopes obtained for the LFERs represent the collective effects of aromatic ring substituents on all the pre-equilibrium steps relative the substituent effects on $E_H^{1'}(ArNO_2)$ values (*93–95*).

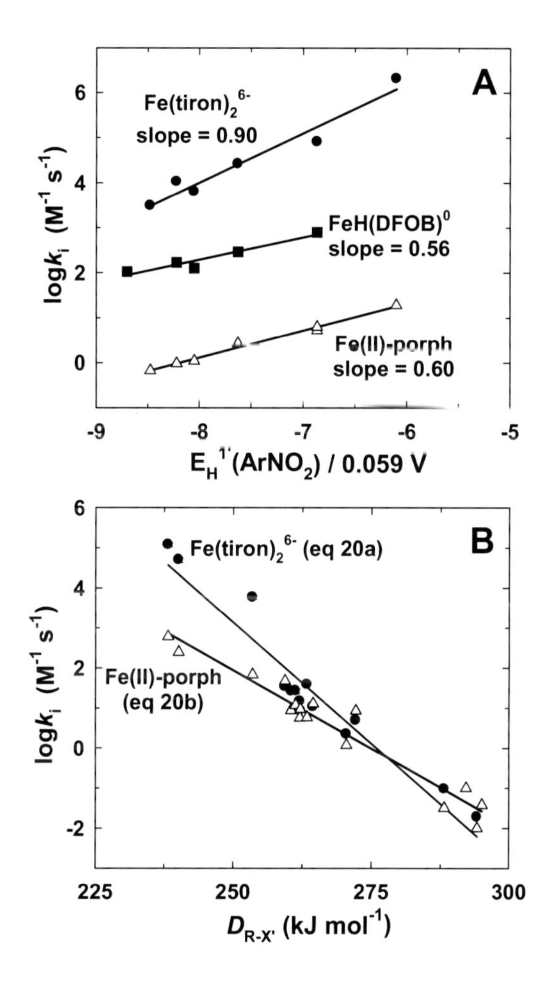

Figure 8. Linear free energy relationships for predicting second-order rate constants for reaction of aqueous Fe(II)-organic complexes with (A) nitroaromatic compounds and (B) polyhalogenated alkanes. Data from references (10, 12, 14, 15, 96).

The relative reactivity of different polyhalogenated alkane analogues with both Fe(II)-tiron complexes and Fe(II)-porphyrin complexes can best be described by LFERs that use the theoretical bond dissociation energies of the weakest carbon-halogen bond ($D_{R\text{-}X'}$) as the molecular descriptor (Figure 8B):

$$\text{Fe(II)-tiron } (10)\text{: } \log k_{FeL_2^{6-}} = -0.121(\pm 0.022)D_{\text{R-X'}} + 33.4(\pm 5.7) \quad (20a)$$

$$\text{Fe(II)-porph } (96)\text{: } \log k_{FeP} = -0.0781(\pm 0.0112)D_{\text{R-X'}} + 21.5(\pm 3.0) \quad (20b)$$

These correlations have been interpreted as evidence that a dissociative one-electron transfer process occurs during the rate-limiting step in the reaction mechanisms.

Using E_H To Predict Equilibrium Conditions

In an earlier section, a distinction was made between the system-independent E_H^0 values and the system- and condition-dependent E_H values. Whereas the former have been used as molecular descriptors in LFERs for predicting the relative kinetic reactivity of different Fe(II) complexes, a direct correlation between observed rate constants for aquatic redox reactions and E_H values is difficult to make (*75*). However, for reversible redox reactions, E_H can be useful for predicting the equilibrium ratio of [Fe(III)]:[Fe(II)] and the oxidized-to-reduced forms of the its reacting partner (i.e., the equilibrium distribution of products and reactants). A generalized reversible reaction between iron and a reagent capable of reversible redox conversion (involving "n" electrons) between oxidized (C_{OX}) and reduced (C_{RED}) forms can be depicted as follows:

$$n\,\text{Fe(II)} + C_{\text{OX}} \rightleftharpoons n\,\text{Fe(III)} + C_{\text{RED}} \quad (21)$$

E_H values for each of the participating redox half reactions can be calculated by appropriate Nernst equations:

$$\text{Fe(III)} + e^- \leftrightarrow \text{Fe(II)}: \quad \text{E}_\text{H} = +0.77 - \frac{RT}{F}\ln\left(\frac{\{\text{Fe}^{2+}\}}{\{\text{Fe}^{3+}\}}\right) \quad (13)$$

$$C_{\text{OX}} + ne^- \leftrightarrow C_{\text{RED}}: \quad \text{E}_\text{H} = E_H^0[C_{\text{OX}}/C_{\text{RED}}] - \frac{RT}{nF}\ln\left(\frac{\{C_{\text{RED}}\}}{\{C_{\text{OX}}\}}\right) \quad (22)$$

The E_H values for each half reaction will shift in opposite directions as the reaction proceeds and the ratios of [Fe(II)]/[Fe(III)] and [C_{RED}]/[C_{OX}] vary (e.g., [Fe(II)]/[Fe(III)] decreases and [C_{RED}]/[C_{OX}] increases if the reaction is proceeding in the forward direction for eq 21). An equilibrium will then occur when the ratios of [Fe(II)]/[Fe(III)] and [C_{RED}]/[C_{OX}] shift to the point where E_H values calculated by eqs 13 and 22 reach the same value (where the electrode potential for the overall redox reaction is then zero). If the E_H value for one of the two redox couples is "buffered", e.g., because [C_{OX}] and [C_{RED}] are much larger than the changes in concentrations (Δ[C_{OX}] and Δ[C_{RED}]) that occur when reacting with Fe(II) or Fe(III), then the equilibrium ratio of the other redox couple (i.e., [Fe(II)]/[Fe(III)]) can be calculated by assuming that E_H for its half reaction at equilibrium will be equal to the E_H value of the buffered redox couple.

In a recent contribution (*73*), it was shown that E_H values were accurate predictors of the equilibrium solution composition for the reversible redox reaction between flavin mononucleotide (FMN) and Fe-DFOB complexes,

$$2\,Fe(II)\text{-}DFOB + FMN_{OX} + 2H^+ \rightleftharpoons 2\,Fe(III)\text{-}DFOB + FMN_{RED} \quad (23)$$

where FMN_{OX} and FMN_{RED} represent the oxidized and reduced forms of FMN, respectively. Figure 9 illustrates the known reversible redox chemistry and acid/base chemistry of FMN (*97*). Using this information, a Nernst equation can be written for the FMN_{OX}/FMN_{RED} redox couple:

$$E_H = +0.194 - \frac{RT}{2F}\ln\left(\frac{\{FMN_{HQ(1)}\}}{\{FMN_{OX(1)}\}\{H^+\}^2}\right) \quad (24)$$

where +0.194 V (*97*) is the $E_H{}^0$ value determined at standard state conditions (pH 0, 1 M concentration of all species in the reaction quotient term, 25° C, 1 atm). The subscripts RED(1) and OX(1) refer to the fully protonated species for each oxidation state, which correspond to the reacting species at standard state pH conditions where $E_H{}^0$ is valid. The reaction quotient term also includes H^+ since the FMN_{OX}/FMN_{RED} redox half reaction involves 2 protons (Figure 9).

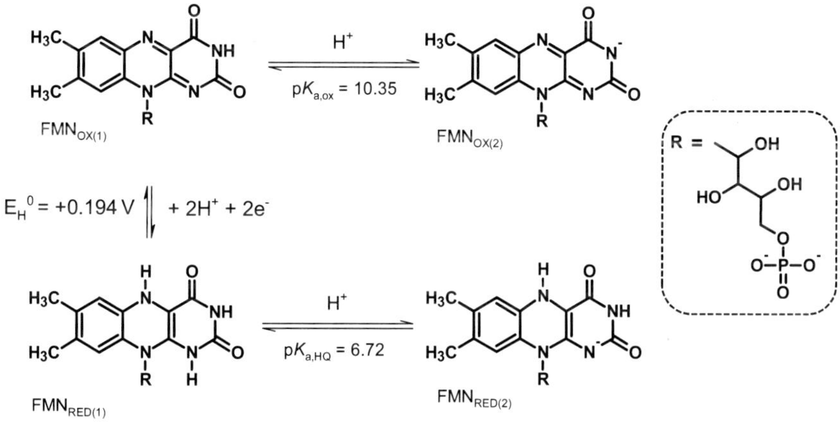

Figure 9. Flavin mononucleotide (FMN) structure in the reduced (FMN_{RED}) and oxidized (FMN_{OX}) states, along with the reported pK_a values and the midpoint reduction potential derived for standard state conditions. Numbers in parentheses within subscripts refer to different protonation levels.

Figure 10A compares the effect of pH on the calculated E_H value of the FMN redox couple (eq 24) where $[FMN_{OX}]_{tot}$ = $[FMN_{RED}]_{tot}$ (i.e., so-called midpoint potential) with E_H values for the iron redox couple calculated by eq 13 for different ratios of $[Fe(III)]_{tot}$:$[Fe(II)]_{tot}$. Where the FMN midpoint potential line intersects each of the lines calculated for the iron redox couple represents the pH condition where the $[Fe(III)]_{tot}$:$[Fe(II)]_{tot}$ ratio in question is predicted to be in equilibrium with a 50:50 ratio of FMN_{OX}/FMN_{RED}. For example, a 25:75 [Fe(III)]:[Fe(II)] ratio is predicted to be in equilibrium with the 50:50 $[FMN_{OX}]:[FMN_{RED}]$ mixture at pH ~4.7. According to this analysis, if the FMN_{OX}/FMN_{RED} ratio is buffered, the equilibrium distribution of iron species will shift from predominantly Fe(II) at pH < 4 to predominantly Fe(III) at pH > 5.75, with intermediate ratios of the two Fe oxidation states being predicted in between. These predictions agree very closely with experimental measurements shown in Figure 10B. In the experiment shown, either 0.1 mM Fe(II)-DFOB or 0.1 mM Fe(III)-DFOB was initially spiked into a solution containing FMN_{OX} and FMN_{RED} in significant excess and equal concentrations (i.e., $[FMN_{OX}]/[FMN_{RED}]$ buffered at ~1.0). After reacting, the measured [Fe(III)]:[Fe(II)] ratios were found to be independent of the initial oxidation state of Fe added to solution and agree closely with Nernst equation predictions.

For many contaminants, reduction by Fe(II) or other reductants leads to irreversible transformation (e.g., dehalogenation), so equilibrium ratios of the products and reactants needed for the Nernst equation formulation cannot be readily calculated. Instead, if a reaction is favorable, it will generally lead to complete depletion of the limiting reagent. However, in these situations, E_H calculations may still be useful for illustrating the limits on the favorability of reactions between Fe(II)-organic complexes and contaminants. As an interesting example, Uchimya (*44*) compared E_H-pH curves for the Fe(III)/Fe(II) redox couple calculated with eq 13 for solutions containing Fe(II) and Fe(III) plus different organic ligands with the E_H value for the initial reversible one-electron reduction of 2,4-dinitrotoluene ($Ar\text{-}NO_2$):

$$ArNO_2 + e^- \rightleftharpoons ArNO_2^{\bullet-} \tag{25}$$

estimated using a Nernst equation for the half reaction (*44*),

$$E_H = -0.38 - \frac{RT}{F}\ln\left(\frac{\{ArNO_2^{\bullet-}\}}{\{ArNO_2\}}\right) \tag{26}$$

and assuming that $ArNO_2^{\bullet-}$ is an unstable transient species that reaches a very low steady-state concentration (1 nM compared to 25 μM $ArNO_2$) during the overall six-electron reduction of each nitro group (see eq 14) in the structure of 2,4-dinitrotoluene. The calculations show that the catecholate ligand tiron lowers E_H much more dramatically than carboxylate-based ligands, suggesting that 2,4-dinitrotoluene reduction by Fe(II) is thermodynamically favorable over a much wider pH range in solutions containing tiron than in solutions containing carboxylate ligands.

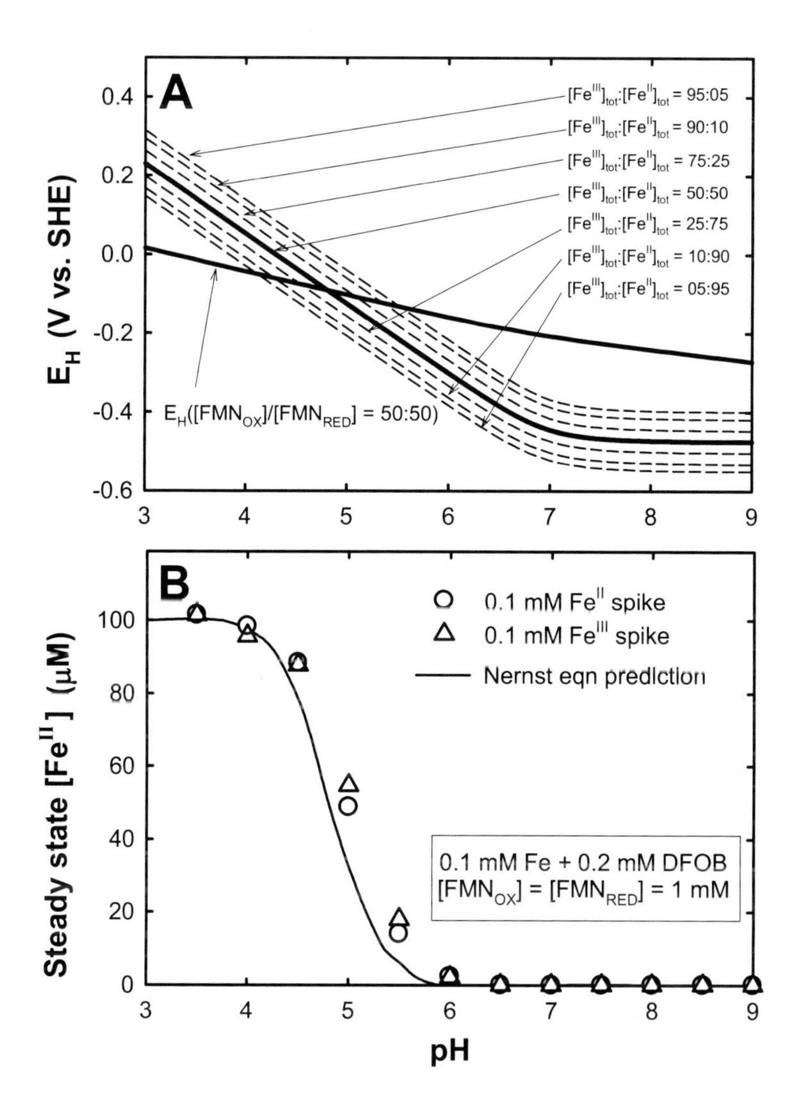

Figure 10. (A) Comparison of the effects of pH on the midpoint potential of the FMX_{OX}/FMN_{RED} redox couple with the E_H of the Fe(III)/Fe(II) redox couple calculated for different $[Fe(III)]_{tot}:[Fe(II)]_{tot}$ ratios (0.1 mM Fe_{tot}, 2 mM DFOB). (B) Steady-state Fe(II) concentrations measured and predicted at different pH values in solutions containing DFOB and excess concentrations of both FMN_{OX} and FMN_{RED}. (Adapted with permission from reference (73). Copyright 2010 Elsevier.)

Contaminant Reduction in the Presence of Fe(II) and NOM

A few studies have also examined the reduction of aquatic contaminants in solutions containing Fe(II) and more poorly defined NOM sources (*24*, *31*, *44*, *45*, *98–102*). Results from these studies indicate that, depending on geochemical conditions and the target contaminant, interactions with NOM can enhance, inhibit, or have little effect on Fe(II) redox reactivity. These contradictory reports can be attributed to a complex set of environmental factors as well as the experimental procedures employed by investigators (*99–101*). First, as mentioned above, NOM derived from different sources contain some of the same ligand functional groups shown to complex and enhance Fe(II) redox reactivity, including carboxylate, catecholate, and thiol groups (*49*, *52–57*). Second, the structure of NOM includes redox-active functional groups (e.g., quinones) that can react with Fe(II) and act as electron transfer intermediates in reactions with contaminants (*12*, *103*); the net effect on reactions with contaminants then depends on the relative rates of contaminant reduction by Fe(II) complexes versus the organic electron transfer intermediates. Finally, the presence of high levels of NOM can act to slow contaminant reduction by inhibiting formation of colloidal Fe(III) oxyhydroxide solids (*100*) that act autocatalytically on Fe(II) reactions with contaminants (by promoting formation of highly reactive sorbed Fe(II) species (*8*, *9*, *16*, *20*, *25*, *26*, *32*, *39–43*)). If these factors are controlling, then the net effect of NOM on contaminant reactions by NOM is dependent on the relative importance of rate-enhancing versus rate-inhibiting factors for different target contaminants and different experimental systems that have been reported.

Some studies have reported that elevated concentrations of different NOM sources enhance Fe(II) reaction rates with polyhalogenated alkanes (*45*), Cr(VI) (*31*, *99*), and oxime carbamate pesticides (*24*). Little or no contaminant reduction is observed in NOM-only controls, so reactivity is attributed to either formation of reactive Fe(II)-organic complexes or to quinones and other NOM functional groups acting as electron transfer mediators in reactions (*12*, *75*, *104*)). Buerge and Hug (*31*) found that NOM extracted from an organic soil horizon collectively had a similar effect on Fe(II) reactivity with Cr(VI) as the carboxylate ligand tartrate (on a per mass basis). Curtis and Reinhard (*45*) reported that addition of 25 mg/L DOC from different NOM sources increased the rate of hexachloroethane reduction by Fe(II) by an order of magnitude. Similarly, Strathmann and Stone (*24*) reported a 63-fold increase in the rate of oxamyl reduction when DOC from reconstituted Great Dismal Swamp Water was increased from 0 to 75 mg/L while keeping Fe(II) concentration relatively unchanged.

In contrast to the above studies, some reports indicate minimal effects, or even inhibitory effects of NOM on Fe(II) reactions with contaminants (*44*, *98*). For example, Colón and coworkers (*98*) reported observing no reduction of *p*-cyanonitrobenzene (CNNB) in aqueous solutions containing Fe(II) and Suwannee River humic acid (SRHA), and found that SRHA actually lowered CNNB reduction rates in heterogeneous systems containing Fe(II) and the mineral goethite (α-$FeOOH_{(s)}$). Inhibition likely results from either SRHA blocking contaminant access to reactive Fe(II) species sorbed to the goethite surface or to suppressed formation of goethite-sorbed Fe(II) species that are more reactive than the NOM-complexed Fe(II) species formed in their place.

Hakala, Chin and coworkers (*99–101*) have reported both enhanced and inhibitory effects of NOM on Fe(II) reactions with contaminants. Whereas enhanced rates of pentachloronitrobenzene (PCNB) reduction are observed when NOM is added to filtered homogeneous Fe(II) solutions, NOM addition slows PCNB reduction in comparable solutions that are not filtered prior to initiating reactions (*100*), a finding the authors attribute to NOM suppressing the formation of colloidal autocatalytic Fe(III) oxyhydroxides that form in the absence of NOM. Interestingly, these investigators also showed that the apparent reactivity of contaminants in a filtered sediment pore water rich in both Fe(II) and NOM was also significantly affected by methods used to preserve the pore water samples prior to experimentation. PCNB reduction was found to be much slower in pore water stored at its native pH prior to reaction than in pore water that was acidified for preservation and then re-adjusted to the native pH just before initiating the reaction (*101*). The investigators hypothesized that this results from release of Fe(III) from highly Fe(III)-stabilizing ligands (e.g., catecholates) upon acidification, making the ligands available for formation of reactive Fe(II) complexes upon pH re-adjustment.

Ultimately, it is difficult to characterize or predict the exact structures of NOM-complexed Fe(II) species, and so it is also challenging to obtain direct evidence of the mechanisms responsible for changes in apparent Fe(II) redox reactivity with contaminants when NOM is present. However, combining results from NOM experiments with studies conducted using model ligands and redox-active model compounds (e.g., model quinones) provides insights into the potential mechanisms that are operative in NOM-rich aquatic matrices.

Conclusions and Implications

A growing body of research indicates that organically complexed iron species play critical, but previously unrecognized, roles in iron redox cycles and reactions that transform and attenuate many aquatic contaminants, particularly in environments where dissolved iron and natural organic matter co-accumulate (e.g., sediment pore water). Rates of Fe(II) reactions with contaminants are highly variable and depend on a variety of factors, including solution conditions and the concentration and identity of Fe(II)-complexing ligands present. Work with model organic ligands representative of Lewis base donor groups present in natural organic matter and microbial/plant exudates demonstrate that enhanced Fe(II) reactivity can result from formation of complexes with Fe(III)-stabilizing ligands that lower the standard one-electron reduction potential of the Fe(III)/Fe(II) redox couple. It follows that reports of both enhanced and inhibited rates of contaminant reduction in solutions containing elevated Fe(II) and natural organic matter may be attributed, in part, to Fe(II) complexation by Fe(III)-stabilizing versus Fe(II)-stabilizing ligand groups present in different natural organic matter sources. Current tools for analyzing Fe speciation in natural waters containing dissolved organic matter is limited to bulk characteristics (e.g., uncomplexed versus complexed (*47*)) which are not sufficient for predicting Fe(II) redox reactivity with contaminants. However, research has identified specific Lewis base functional groups most likely to promote Fe(II) redox reactivity (i.e., phenolic, thiol, hydroxamic), which, if quantified in dissolved organic matter, may serve as useful indicators of Fe(II) redox reactivity in such systems. It is also worth noting that mechanistic insights obtained from experiments conducted in model systems with well-defined Fe(II)-organic complex structures can provide general insights into molecular properties that control Fe(II) redox reactivity in more complex and ill-defined heterogeneous systems. For example, studies with model organic ligands indicate that the speciation-dependent standard reduction potential of the Fe(II)/Fe(III) redox couple and oxidant access to iron's inner coordination shell are important factors contributing to the metal's apparent redox reactivity with different chemicals in aquatic environments. Such factors can help to explain differences in reactivity of Fe(II) sorbed to different mineral surfaces (e.g., Al_2O_3 versus TiO_2) (*25*, *32*).

Findings from the work summarized in this report may also support the development of novel subsurface remediation strategies. Soluble Fe(II) complexes with select ligands, either pre-formed or formed in situ in Fe(II)-rich groundwater, can be introduced as reactive amendments to promote reductive transformation of target contaminants. A number of research teams are investigating the injection of colloidal and nanoparticulate zerovalent iron materials into subsurface zones of contamination (*105–107*), but injection and transport of particles through porous media is challenging and may be limited to shallower zones of contamination. Aqueous reductants, including Fe(II)-organic complexes, may be more suitable injection amendments for sites where contamination is deeper or more widespread. In addition to direct reactions with amenable contaminants, Fe(II)-organic complexes may indirectly promote microbial reductive transformation processes by scavenging dissolved oxygen (*85*) and promoting formation of more reducing conditions that favor growth of contaminant-degrading anaerobic microbial populations (e.g., *Dehalococcoides* (*108*)).

Although research summarized in this chapter has advanced understanding of the mechanistic factors contributing to the redox reactivity of Fe(II)-organic complexes with aquatic contaminants, many issues continue to remain unresolved and a number of promising applications require further development. Thus, further research is suggested in a number of directions, including developing a better understanding of the link between the molecular structure of Fe(II) species and redox reactivity with important classes of contaminants and environmental oxidants, establishing general rules for predicting Fe(II)-organic complex reactivity, bridging the gap between well-defined model Fe(II)-organic complex structures and Fe(II) complexes with more ill-defined natural organic matter, assessing the potential role of Fe(II)-organic complex centers in natural organic matter as electron transfer mediators, and development and testing of remedial strategies involving reactive Fe(II)-organic complexes.

Acknowledgments

Funding support was provided by the National Science Foundation (CBET CAREER 07-46453). Several students, collaborators, and mentors who contributed significantly to my thinking on this subject are acknowledged, including Dongwook Kim, Daisuke Naka, Genevieve Nano, Adam Bussan, Owen Duckworth, Thomas Hofstetter, Richard Carbonaro, and Alan Stone.

References

1. Macalady, D. L.; Tratnyek, P. G.; Grundl, T. J. Abiotic Reduction Reactions of Anthropogenic Organic Chemicals in Anaerobic Systems: A Critical Review. *J. Contam. Hydrol.* **1986**, *1*, 1–28.
2. Lovley, D. R.; Holmes, D. E.; Nevin, K. P. Dissimilatory Fe(III) and Mn(IV) Reduction. *Adv. Microb. Physiol.* **2004**, *49*, 219–286.
3. Schlesinger, W. H. *Biogeochemistry - An Analysis of Global Change*, 2nd ed.; Academic Press: San Diego, CA, 1997; p 588.
4. Hering, J. G.; Stumm, W. Oxidative and Reductive Dissolution of Minerals. In *Mineral-Water Interface Geochemistry*; Hochella, M. F. J., White, A. F., Eds.; American Mineralogical Society: Washington, DC, USA, 1990; Vol. 23, pp 427–465.
5. O'Loughlin, E. J.; Chin, Y.-P. Quantification and Characterization of Dissolved Organic Carbon and Iron in Sedimentary Pore Water from Green Bay, WI, USA. *Biogeochemistry* **2004**, *71*, 371–386.
6. Perlinger, J. A.; Buschmann, J.; Angst, W.; Schwarzenbach, R. P. Iron Porphyrin and Mercaptojuglone Mediated Reduction of Polyhalogenated Methanes and Ethanes in Homogeneous Aqueous Solution. *Environ. Sci. Technol.* **1998**, *32*, 2431–2437.
7. McCormick, M. L.; Adriaens, P. Carbon Tetrachloride Transformation on the Surface of Nanoscale Biogenic Magnetite Particles. *Environ. Sci. Technol.* **2004**, *38*, 1045–1053.
8. Elsner, M.; Schwarzenbach, R. P.; Haderlein, S. B. Reactivity of Fe(II)-Bearing Minerals towards Reductive Transformation of Organic Contaminants. *Environ. Sci. Technol.* **2004**, *38*, 799–807.
9. Pecher, K.; Haderlein, S. B.; Schwarzenbach, R. P. Reduction of Polyhalogenated Methanes by Surface-Bound Fe(II) in Aqueous Suspensions of Iron Oxides. *Environ. Sci. Technol.* **2002**, *36*, 1734–1741.
10. Bussan, A. L.; Strathmann, T. J. Influence of Organic Ligands on the Reduction of Polyhalogenated Alkanes by Iron(II). *Environ. Sci. Technol.* **2007**, *41*, 6740–6747.
11. Amonette, J. E.; Workman, D. J.; Kennedy, D. W.; Fruchter, J. S.; Gorby, Y. A. Dechlorination of Carbon Tetrachloride by Fe(II) Associated with Goethite. *Environ. Sci. Technol.* **2000**, *34*, 4606–4613.
12. Schwarzenbach, R. P.; Stierli, R.; Lanz, K.; Zeyer, J. Quinone and Iron Porphyrin Mediated Reduction of Nitroaromatic Compounds in Homogeneous Aqueous Solution. *Environ. Sci. Technol.* **1990**, *24*, 1566–1574.

13. Naka, D.; Kim, D.; Strathmann, T. J. Abiotic Reduction of Nitroaromatic Contaminants by Iron(II) Complexes with Organothiol Ligands. *Environ. Toxicol. Chem.* **2008**, *27*, 1257–1266.
14. Naka, D.; Kim, D.; Strathmann, T. J. Reduction of Nitroaromatic Contaminants by Aqueous Iron(II)-Catechol Complexes. *Environ. Sci. Technol.* **2006**, *40*, 3006–3012.
15. Kim, D.; Duckworth, O. W.; Strathmann, T. J. Hydroxamate Siderophore-Promoted Reactions between Iron(II) and Nitroaromatic Groundwater Contaminants. *Geochim. Cosmochim. Acta* **2009**, *73*, 1297–1311.
16. Hofstetter, T. B.; Schwarzenbach, R. P.; Haderlein, S. B. Reactivity of Fe(II) Species Associated with Clay Minerals. *Environ. Sci. Technol.* **2003**, *37*, 519–528.
17. Haderlein, S. B.; Schwarzenbach, R. P. Environmental Processes Influencing the Rate of Abiotic Reduction of Nitroaromatic Compounds in the Subsurface. In *Biodegradation of Nitroaromatic Compounds*; Spain, J. C., Ed.; Plenum Press: New York, 1995; pp 199−225.
18. Schultz, C. A.; Grundl, T. J. pH Dependence on Reduction Rate of 4-Cl-Nitrobenzene by Fe(II)/Montmorillonite Systems. *Environ. Sci. Technol.* **2000**, *34*, 3641–3648.
19. Williams, A. G. B.; Gregory, K. B.; Parkin, G. F.; Scherer, M. M. Hexahydro-1,3,5-Trinitro-1,3,5-Triazine Transformation by Biologically Reduced Ferrihydrite: Evolution of Fe Mineralogy, Surface Area, and Reaction Rates. *Environ. Sci. Technol.* **2005**, *39*, 5183–5189.
20. Gregory, K. B.; Larese-Casanova, P.; Parkin, G. F.; Scherer, M. M. Abiotic Transformation of Hexahydro-1,3,5-trinitro-1,3,5-triazine by Fe^{II} Bound to Magnetite. *Environ. Sci. Technol.* **2004**, *38*, 1408–1414.
21. Kim, D.; Strathmann, T. J. Role of Organically Complexed Iron(II) Species in the Reductive Transformation of RDX in Anoxic Environments. *Environ. Sci. Technol.* **2007**, *41*, 1257–1264.
22. Alessi, D. S.; Grundl, T. J. Reduction of 2,4,6-Trinitrotoluene and Hexahydro-1,3,5-Trinitro-1,3,5-triazine by Hydroxyl-Complexed Fe(II). *J. Environ. Eng.* **2008**, *134*, 937–943.
23. Strathmann, T. J.; Stone, A. T. Reduction of the Pesticides Oxamyl and Methomyl by Fe^{II}: Effect of pH and Inorganic Ligands. *Environ. Sci. Technol.* **2002**, *36*, 653–661.
24. Strathmann, T. J.; Stone, A. T. Reduction of Oxamyl and Related Pesticides by Fe^{II}: Influence of Organic Ligands and Natural Organic Matter. *Environ. Sci. Technol.* **2002**, *36*, 5172–5183.
25. Strathmann, T. J.; Stone, A. T. Mineral Surface Catalysis of Reactions between Fe(II) and Oxime Carbamate Pesticides. *Geochim. Cosmochim. Acta* **2003**, *67*, 2775–2791.
26. Nano, G. V.; Strathmann, T. J. Application of Surface Complexation Modeling to the Reactivity of Iron(II) with Nitroaromatic and Oxime Carbamate Contaminants. *J. Colloid Interface Sci.* **2008**, *321*, 350–359.
27. Sørensen, J.; Thorling, L. Stimulation by Lepidocrocite (γ-FeOOH) of Fe(II)-Dependent Nitrite Reduction. *Geochim. Cosmochim. Acta* **1991**, *55*, 1289–1294.

28. Hansen, H. C. B.; Koch, C. B.; Nancke-Krogh, H.; Borggaard, O. K.; Sorensen, J. Abiotic Nitrate Reduction to Ammonium: Key Role of Green Rust. *Environ. Sci. Technol.* **1996**, *30*, 2053–2056.
29. Myneni, S. C. B.; Tokunaga, T. K.; Brown, G. E. Abiotic Selenium Redox Transformations in the Presence of Fe(II,III) Oxides. *Science* **1997**, *278*, 1106–1109.
30. Refait, P.; Simon, L.; Génin, J.-M. R. Reduction of SeO_4^{2-} Anions and Anoxic Formation of Iron(II)-Iron(III) Hydroxy-Selenate Green Rust. *Environ. Sci. Technol.* **2000**, *34* (5), 819–825.
31. Buerge, I. J.; Hug, S. J. Influence of Organic Ligands on Chromium(VI) Reduction by Iron(II). *Environ. Sci. Technol.* **1998**, *32*, 2092–2099.
32. Buerge, I. J.; Hug, S. J. Influence of Mineral Surfaces on Chromium(VI) Reduction by Iron(II). *Environ. Sci. Technol.* **1999**, *33*, 4285–4291.
33. Liger, E.; Charlet, L.; VanCappellen, P. Surface Catalysis of Uranium(VI) Reduction by Iron(II). *Geochim. Cosmochim. Acta* **1999**, *63*, 2939–2956.
34. Sedlak, D. L.; Chan, P. G. Reduction of Hexavalent Chromium by Ferrous Iron. *Geochim. Cosmochim. Acta* **1997**, *61*, 2185–2192.
35. Cui, D.; Eriksen, T. E. Reduction of Pertechnetate by Ferrous Iron in Solution: Influence of Sorbed and Precipitated Fe(II). *Environ. Sci. Technol.* **1996**, *30*, 2259–2262.
36. O'Loughlin, E. J.; Kelly, S. D.; Kemner, K. M.; Csencsits, R.; Cook, R. E. Reduction of Ag^{I}, Au^{III}, Cu^{II}, and Hg^{II} by Fe^{II}/Fe^{III} Hydroxysulfate Green Rust. *Chemosphere* **2003**, *53*, 437–446.
37. Stumm, W.; Morgan, J. J. *Aquatic Chemistry: Chemical Equilibria and Rates in Natural Waters*, 3rd ed.; Wiley and Sons: New York, 1996; p 1022.
38. Schwarzenbach, R. P.; Gschwend, P. M.; Imboden, D. M. *Environmental Organic Chemistry*, 2nd ed.; Wiley Interscience: Hoboken, NJ, 2003.
39. Klausen, J.; Tröber, S. P.; Haderlein, S. B.; Schwarzenbach, R. P. Reduction of Substituted Nitrobenzenes by Fe(II) in Aqueous Mineral Suspensions. *Environ. Sci. Technol.* **1995**, *29*, 2396–2404.
40. Klupinski, T. P.; Chin, Y.-P.; Traina, S. J. Abiotic Degradation of Pentachloronitrobenzene by Fe(II): Reactions on Goethite and Iron Oxide Nanoparticles. *Environ. Sci. Technol.* **2004**, *38*, 4353–4360.
41. Williams, A. G. B.; Scherer, M. M. Kinetics of Cr(VI) Reduction by Carbonate Green Rust. *Environ. Sci. Technol.* **2001**, *35*, 3488–3494.
42. Butler, E. C.; Hayes, K. F. Effects of Solution Composition on the Reductive Dechlorination of Hexachloroethane by Iron Sulfide. *Environ. Sci. Technol.* **1998**, *32*, 1276–1284.
43. Danielsen, K. M.; Gland, J. L.; Hayes, K. pH Dependence of Carbon Tetrachloride Reductive Dechlorination by Magnetite. *Environ. Sci. Technol.* **2004**, *38*, 4745–4752.
44. Uchimiya, M. Reductive Transformation of 2,4-Dinitrotoluene: Roles of Iron and Natural Organic Matter. *Aquat. Geochem.* **2010**, *16*, 547–562.
45. Curtis, G. P.; Reinhard, M. Reductive Dehalogenation of Hexachloroethane, Carbon Tetrachloride, and Bromoform by Anthraquinone Disulfonate and Humic Acid. *Environ. Sci. Technol.* **1994**, *28*, 2393–2401.

46. Chin, Y.-P.; Traina, S. J.; Swank, C. R.; Backhus, D. Abundance and Properties of Dissolved Organic Matter in Pore Waters of a Freshwater Wetland. *Limnol. Oceanogr.* **1998**, *43*, 1287–1296.
47. Luther, G. W.; Shellenbarger, P. A.; Brendel, P. J. Dissolved Organic Fe(III) and Fe(II) Complexes in Salt Marsh Porewaters. *Geochim. Cosmochim. Acta* **1996**, *60*, 951–960.
48. Stevenson, F. J. *Humus Chemistry: Genesis, Composition, and Reactions*, 2nd ed.; Wiley Interscience: New York, NY, 1994.
49. Ritchie, J. D.; Perdue, E. M. Proton-Binding Study of Standard and Reference Fulvic Acids, Humic Acids, and Natural Organic Matter. *Geochim. Cosmochim. Acta* **2003**, *67*, 85–96.
50. Fox, T. R.; Comerford, N. B. Low Molecular Weight Organic Acids in Selected Forest Soils of Southeastern USA. *Soil Sci. Soc. Am. J.* **1990**, *54*, 1139–1144.
51. Stone, A. T.; Godtfredsen, K. L.; Deng, B., Sources and Reactivity of Reductants Encountered in Aquatic Environments. In *Chemistry of Aquatic Systems: Local and Global Perspectives*; Bidoglio, G., Stumm, W., Eds.; Kluwer: Dordrecht, The Netherlands, 1994; pp 337−374.
52. Vairavamurthy, A.; Wang, A. S. Organic Nitrogen in Geomacromolecules: Insights on Speciation and Transformation with K-edge XANES Spectroscopy. *Environ. Sci. Technol.* **2002**, *36*, 3050.
53. Leinweber, P.; Walley, F.; Kruse, J.; Jandl, G.; Eckhardt, K.-W.; Blyth, R. I. R.; Regier, T. Cultivation Affects Soil Organic Nitrogen: Pyrolysis-Mass Spectrometry and Nitrogen K-edge XANES Spectroscopy Evidence. *Soil Sci. Soc. Am. J.* **2009**, *73*, 82–92.
54. Xia, K.; Weesner, F.; Bleam, W. F.; Bloom, P. R.; Skyllberg, U. L.; Helmke, P. A. XANES Studies of Oxidation States of Sulfur in Aquatic and Soil Humic Substances. *Soil Sci. Soc. Am. J.* **1998**, *62*, 1240–1246.
55. Vairavamurthy, A.; Mopper, K. Geochemical formation of organosulphur compounds (thiols) by addition of H_2S to sedimentary organic matter. *Nature* **1987**, *329*, 623–625.
56. Ferdelman, T. G.; Church, T. M.; Luther, G. W. Sulfur Enrichment of Humic Substances in a Delaware Salt Marsh Sediment Core. *Geochim. Cosmochim. Acta* **1991**, *55*, 979–988.
57. Urban, N. R.; Ernst, K.; Bernasconi, S. Addition of Sulfur to Organic Matter during Early Diagenesis of Lake Sediments. *Geochim. Cosmochim. Acta* **1999**, *63*, 837–853.
58. Perlinger, J. A.; Kalluri, V. M.; Venkatapathy, R.; Angst, W. Addition of Hydrogen Sulfide to Juglone. *Environ. Sci. Technol.* **2002**, *36*, 2663–2669.
59. Martell, A. E.; Hancock, R. D. *Metal Complexes in Aqueous Solutions*; Plenum: New York, NY, 1996; p 253.
60. Martell, A. E.; Motekaitis, R. J.; Chen, D.; Hancock, R. D.; McManus, D. Selection of New Fe(III)/Fe(II) Chelating Agents as Catalysts for the Oxidation of Hydrogen Sulfide to Sulfur by Air. *Can. J. Chem.* **1996**, *74*, 1872–1879.
61. Bell, C. F. *Principles and Applications of Metal Chelation*; Clarendon Press: Oxford, U.K., 1977; p 147.

62. Stone, A. T. Reactions of Extracellular Organic Ligands with Dissolved Metal Ions and Mineral Surfaces. In *Geomicrobiology: Interactions between Microbes and Minerals*; Banfield, J. F., Nealson, K. H., Eds.; Mineralogical Society of America: Washington, DC, 1997; Vol. 35, pp 309−344.
63. Kraemer, S. M. Iron Oxide Dissolution and Solubility in the Presence of Siderophores. *Aquat. Sci.* **2004**, *66*, 3–18.
64. Dhungana, S.; Crumbliss, A. L. Coordination Chemistry and Redox Processes in Siderophore-Mediate Iron Transport. *Geomicrobiol. J.* **2005**, *22*, 87–98.
65. Hider, R. C.; Kong, X. L. Chemistry and Biology of Siderophores. *Nat. Prod. Rep.* **2010**, *27*, 637–657.
66. Ratledge, C. Iron Metabolism and Infection. *Food Nutr. Bull.* **2007**, *28*, S515–S523.
67. Römheld, V. The Role of Phytosiderophores in Acquisition of Iron and Other Micronutrients in Graminaceous Species: An Ecological Approach. *Plant Soil* **1991**, *130*, 127–134.
68. Powell, P. E.; Cline, G. R.; Reid, C. P. P.; Szaniszlo, P. J. Occurrence of Hydroxamate Siderophore Iron Chelators in Soils. *Nature* **1980**, *287*, 833–834.
69. Nowack, B. Environmental Chemistry of Aminopolycarboxylate Chelating Agents. *Environ. Sci. Technol.* **2002**, *36*, 4009–4016.
70. Nowack, B. Environmental Chemistry of Phosphonates. *Water Res.* **2003**, *37*, 2533–2546.
71. Mizuta, T.; Wang, J.; Miyoshi, K. A 7-Coordinate Structure of Iron(II)-Ethylenediamine-N,N,N′,N′-tetraaceto complex as Determined by X-Ray Crystal Analysis. *Bull. Chem. Soc. Jpn.* **1993**, *66*, 2547–2551.
72. Martell, A. E.; Smith, R. M.; Motekaitis, R. J. *Critically Selected Stability Constants of Metal Complexes Database*, Version 4.0; U.S. Department of Commerce, National Institute of Standards and Technology: Gaithersbug, MD, 1997.
73. Kim, D.; Duckworth, O. W.; Strathmann, T. J. Reactions of Aqueous Iron-DFOB (Desferrioxamine B) Complexes with Flavin Mononucleotide in the Absence of Strong Iron(II) Chelators. *Geochim. Cosmochim. Acta* **2010**, *74*, 1513–1529.
74. Bard, A. J.; Faulkner, L. R. *Electrochemical Methods*; Wiley: New York, 1980; p 718.
75. Uchimiya, M.; Stone, A. T. Redox Reactions between Iron and Quinones: Thermodynamic Constraints. *Geochim. Cosmochim. Acta* **2006**, *70*, 1388–1401.
76. Buerge, I. J.; Hug, S. J. Kinetics and pH Dependence of Chromium(VI) Reduction by Iron(II). *Environ. Sci. Technol.* **1997**, *31*, 1426–1432.
77. Ong, J. H.; Castro, C. E. Oxidation of Iron(II) Porphrins and Hemoproteins by Nitro Aromatics. *J. Am. Chem. Soc.* **1977**, *99*, 6740–6745.
78. Wade, R. S.; Castro, C. E. Oxidation of Iron(II) Porphyrins by Alkyl Halides. *J. Am. Chem. Soc.* **1973**, *95*, 226–230.
79. Wade, R. S.; Havlin, R.; Castro, C. E. The Oxidation of Iron(II) Porphyrins by Organic Molecules. *J. Am. Chem. Soc.* **1969**, *91*, 7530–7530.

80. Castro, C. E. Rapid Oxidation of Iron(II) Porphyrins by Alkyl Halides Possible Mode of Intoxication of Organisms by Alkyl Halides. *J. Am. Chem. Soc.* **1964**, *86*, 2310.
81. Rush, J. D.; Koppenol, W. H. The Reaction between Ferrous Polyaminocarboxylate Complexes and Hydrogen Peroxide: An Investigation of the Reaction Intermediates by Stopped Flow Spectrophotometry. *J. Inorg. Biochem.* **1987**, *29*, 199–215.
82. Rush, J. D.; Maskos, Z.; Koppenol, W. H. Reactions of Iron(II) Nucleotide Complexes with Hydrogen Peroxide. *FEBS Lett.* **1990**, *261*, 121–123.
83. Croft, S.; Gilbert, B. C.; Smith, J. R. L.; Stell, J. K.; Sanderson, W. R. Mechanisms of Peroxide Stabilization. An Investigation of Some Reactions of Hydrogen Peroxide in the Presence of Aminophosphonic Acids. *J. Chem. Soc., Perkin Trans. 2* **1992**, 153–160.
84. Park, J. S. B.; Wood, P. M.; Davies, M. J.; Gilbert, B. C.; Whitwood, A. C. A Kinetic and ESR Investigation of Iron(II) Oxalate Oxidation by Hydrogen Peroxide and Dioxygen as a Source of Hydroxyl Radicals. *Free Radical Res.* **1997**, *27*, 447–458.
85. Brown, E. R.; Mazzarella, J. D. Mechanism of Oxidation of Ferrous Polydentate Complexes by Dioxygen. *J. Electroanal. Chem.* **1987**, *222*, 173–192.
86. Borghi, E. B.; Regazzoni, A. E.; Maroto, A. J. G.; Blesa, M. A. Reductive Dissolution of Magnetite by Solutions Containing EDTA and Fe^{II}. *J. Colloid Interface Sci.* **1989**, *130*, 299–310.
87. Blesa, M. A.; Marinovich, H. A.; Baumgartner, E. C.; Maroto, A. J. G. Mechanism of Dissolution of Magnetite by Oxalic Acid-Ferrous Ion Solutions. *Inorg. Chem.* **1987**, *26*, 3713–3717.
88. Tratnyek, P. G.; Weber, E. J.; Schwarzenbach, R. P. Quantitative Structure-Activity Relationships for Chemical Reductions of Organic Contaminants. *Environ. Toxicol. Chem.* **2003**, *22*, 1733–1742.
89. King, D. W. Role of Carbonate Speciation on the Oxidation Rate of Fe(II) in Aquatic Systems. *Environ. Sci. Technol.* **1998**, *32*, 2997–3003.
90. King, D. W.; Farlow, R. Role of Carbonate Speciation on the Oxidation of Fe(II) by H_2O_2. *Mar. Chem.* **2000**, *70*, 201–209.
91. Rahhal, S.; Richter, H. W. Reduction of Hydrogen Peroxide by the Ferrous Iron Chelate of Diethylenetriamine-*N,N,N',N'',N''*-Pentaacetate. *J. Am. Chem. Soc.* **1988**, *110*, 3126–3133.
92. Eberson, L. E. *Electron Transfer Reactions in Organic Chemistry*; Springer-Verlag: Berlin, Germany, 1987; p 234.
93. Hartenbach, A. E.; Hofstetter, T. B.; Aeschbacher, M.; Sander, M.; Kim, D.; Strathmann, T. J.; Arnold, W. A.; Cramer, C. J.; Schwarzenbach, R. P. Variability of Nitrogen Isotope Fractionation during the Reduction of Nitroaromatic Compounds with Dissolved Reductants. *Environ. Sci. Technol.* **2008**, *42*, 8352–8359.
94. Hartenbach, A.; Hofstetter, T. B.; Berg, M.; Bolotin, J.; Schwarzenbach, R. P. Using Nitrogen Isotope Fractionation To Assess Abiotic Reduction of Nitroaromatic Compounds. *Environ. Sci. Technol.* **2006**, *40*, 7710–7716.

95. Hofstetter, T. B.; Neumann, A.; Arnold, W. A.; Hartenbach, A. E.; Bolotin, J.; Cramer, C. J.; Schwarzenbach, R. P. Substituent Effects on Nitrogen Isotope Fractionation During Abiotic Reduction of Nitroaromatic Compounds. *Environ. Sci. Technol.* **2008**, *42*, 1997–2003.
96. Perlinger, J. A.; Venkatapathy, R. Linear Free Energy Relationships for Polyhalogenated Alkane Transformation by Electron-Transfer Mediators in Model Aqueous Systems. *J. Phys. Chem. A* **2000**, *104*, 2752–2763.
97. Mayhew, S. G. The Effects of pH and Semiquinone Formation on the Oxidation-Reduction Potentials of Flavin Mononucleotide. *Eur. J. Biochem.* **1999**, *265*, 698–702.
98. Colón, D.; Weber, E. J.; Anderson, J. L. Effect of Natural Organic Matter on the Reduction of Nitroaromatics by Fe(II) Species. *Environ. Sci. Technol.* **2008**, *42*, 6538–6543.
99. Agrawal, S. G.; Fimmen, R. L.; Chin, Y.-P. Reduction of Cr(VI) to Cr(III) by Fe(II) in the Presence of Fulvic Acids and in Lacustrine Pore Water. *Chem. Geol.* **2009**, *262*, 328–335.
100. Hakala, J. A.; Chin, Y.-P.; Weber, E. J. Influence of Dissolved Organic Matter and Fe(II) on the Abiotic Reduction of Pentachloronitrobenzene. *Environ. Sci. Technol.* **2007**, *41*, 7337–7342.
101. Hakala, J. A.; Fimmen, R. L.; Chin, Y.-P.; Agrawal, S. G.; Ward, C. P. Assessment of the Geochemical Reactivity of Fe-DOM Complexes in Wetland Sediment Pore Waters using a Nitroaromatic Probe Compound. *Geochim. Cosmochim. Acta* **2009**, *73*, 1382–1393.
102. Koons, B. W.; Baeseman, J. L.; Novak, P. J. Investigation of Cell Exudates Active in Carbon Tetrachloride and Chloroform Degradation. *Biotechnol. Bioeng.* **2001**, *74*, 12–17.
103. Dunnivant, F. M.; Schwarzenbach, R. P.; Macalady, D. L. Reduction of Substituted Nitrobenzenes in Aqueous Solutions Containing Natural Organic Matter. *Environ. Sci. Technol.* **1992**, *26*, 2133–2141.
104. Nevin, K. P.; Lovley, D. R. Potential for Nonenzymatic Reduction of Fe(III) via Electron Shuttling in Subsurface Sediments. *Environ. Sci. Technol.* **2000**, *34*, 2472–2478.
105. Tratnyek, P. G.; Johnson, R. L. Nanotechnologies for Environmental Cleanup. *Nano Today* **2006**, *1*, 44–48.
106. Saleh, N.; Kim, H. J.; Phenrat, T.; Matyjaszewski, K.; Tilton, R. D.; Lowry, G. V. Ionic Strength and Composition Affect the Mobility of Surface-Modified Fe(0) Nanoparticles in Water-Saturated Sand Columns. *Environ. Sci. Technol.* **2008**, *42*, 3349–3355.
107. He, F.; Zhao, D. Y. Preparation and Characterization of a New Class of Starch-Stabilized Bimetallic Nanoparticles for Degradation of Chlorinated Hydrocarbons in Water. *Environ. Sci. Technol.* **2005**, *39*, 3314–3320.
108. Cupples, A. M.; Spormann, A. M.; McCarty, P. L. Growth of a Dehalococcoides-Like Microorganism on Vinyl Chloride and cis-Dichloroethene as Electron Acceptors as Determined by Competitive PCR. *Appl. Environ. Microbiol.* **2003**, *69*, 953–959.

Chapter 15

Fe^{2+} Sorption at the Fe Oxide-Water Interface: A Revised Conceptual Framework

Christopher A. Gorski[1] and Michelle M. Scherer[2,*]

[1]Environmental Chemistry, Eawag, Swiss Federal Institute of Aquatic Science and Technology, 8600 Duebendorf, Switzerland
[2]Civil and Environmental Engineering, University of Iowa, Iowa City, IA, 52242, USA
***michelle-scherer@uiowa.edu**

Sorption of aqueous Fe^{2+} at the Fe oxide-water interface has traditionally been viewed in the classic framework of sorption at static oxide surface sites as formulated in surface complexation models (SCMs). Significant experimental and theoretical evidence has accumulated, however, to indicate that the reaction of aqueous Fe^{2+} with Fe^{3+} oxides is much more complex and is comprised of sorption, electron transfer, conduction, dissolution, and, in some cases, atom exchange and/or transformation to secondary minerals. Here, we provide a brief historical review of Fe^{2+} sorption on Fe oxides and present a revised conceptual model based on the semiconducting properties of Fe oxides that incorporates recent experimental evidence for Fe^{2+} - Fe^{3+}_{oxide} electron transfer, bulk electron conduction, and Fe atom exchange. We also discuss the implications of this revised conceptual model for important environmental processes, such as trace metal cycling and contaminant fate.

Introduction

Iron (Fe) (oxyhydr)oxides are ubiquitous. They are present in the soil beneath your feet, the rust on your car, the hard drive in your computer, and the rocks on Earth and Mars. Essentially anywhere that Fe is exposed to oxygen, Fe oxides form. These tiny, often nanoscale, particles are responsible for most of the red, yellow, green, and black color you observe around you, and they profoundly influence the quality of our water, air, and soil through the biologically-driven redox cycling between oxidized ferric (Fe^{3+}) and reduced ferrous (Fe^{2+}) iron. The mechanisms of how these Fe oxides react in the environment are a fascinating subject, and research in the field of Fe biogeochemistry has exploded in the last few decades (e.g., (*1–8*)).

The critical role of the Fe^{2+} - Fe^{3+} redox couple in understanding soil and groundwater chemistry has been recognized since the 1960s (*9–11*). Much of this earlier work was driven from a practical agricultural perspective of optimizing rice yields. Today, redox cycling of Fe is linked to several diverse environmental processes, including cycling of both macro (e.g., (*12*, *13*)) and trace (e.g., (*11*, *14*)) biological nutrients; mediating biological and abiotic reactions under Fe-reducing conditions (e.g., (*6*)); and mobility and fate of groundwater pollutants (some early key studies include: (*15–18*)). The Fe^{2+} - Fe^{3+} redox couple has also been shown to significantly impact the dissolution, secondary mineral precipitation, and bioavailability of Fe^{3+} oxides (e.g., (*19–24*)).

Despite the importance of Fe^{2+} - Fe^{3+} cycling in these processes, it has been difficult to study the heterogeneous reaction of aqueous Fe^{2+} with Fe^{3+} oxides due to the very small quantity of Fe^{2+} at the interface. Because of this challenge, much of the earlier work on Fe oxide reactivity was based on macroscopic observations, such as aqueous Fe^{2+} measurements, with changes in Fe oxide behavior interpreted within the framework of changes in surface energy and surface site availability due to complexation reactions at the oxide surface (e.g., (*25–28*)). Within the framework of these models, reaction mechanisms of anions and cations, such as Fe^{2+}, with Fe oxide particles were constrained primarily to forming static complexes with surface Fe^{3+} atoms that could potentially catalyze a variety of environmentally important redox reactions.

Advances in spectroscopic and microscopic techniques (e.g., (*29–34*)), stable Fe isotope measurements (e..g, (*35–40*)), and theoretical calculations of mineral electronic structures (e.g., (*41–47*)), however, have led to a clear paradigm shift in our understanding of the heterogeneous reaction of aqueous Fe^{2+} with Fe^{3+} oxides. From these studies, a new conceptual framework for the heterogeneous reaction between aqueous Fe^{2+} and Fe^{3+} oxides is beginning to emerge that considers the well-known semiconducting properties of Fe oxides. Fundamental processes contributing to this paradigm shift include: (*i*) electron transfer between aqueous Fe^{2+} and structural Fe^{3+}, (*ii*) secondary mineral phase transformations, (*iii*) bulk electron conduction, and (*iv*) Fe^{2+} - Fe^{3+}_{oxide} atom exchange. Although early research studies examining Fe^{2+} on Fe oxides proposed these reactions (*14*, *28*, *48–53*), and the reactions have generally been supported by theoretical calculations (*54*, *55*), only recently has direct experimental confirmation for each process been obtained.

In this chapter, we review the literature and current understanding of the reaction of aqueous Fe^{2+} with Fe oxides. Our goals are to provide some historical perspective on the reaction of aqueous Fe^{2+} with Fe oxides and to summarize the key experiments and concepts that have required us to frame a revised conceptual model for understanding the reaction of Fe^{2+} with Fe oxides based on the well-known semiconducting properties of these particles.

A Historical Perspective on Fe^{2+} Sorption on Fe Oxides

Research as early as the 1960s suggested that Fe oxides were a dominant sorbent for cations (typically Mn^{2+}, Co^{2+}, Ni^{2+}, Cu^{2+}, Cr^{3+}, and Pb^{2+}) in natural environments due to their ubiquity and high sorptive capacities (*11*, *56*, *57*). Cation sorption on Fe oxides has been traditionally thought to occur at surface hydroxyl sites via both chemisorption and physisorption mechanisms (e.g., (*25*, *58–64*) and refs. therein). Numerous studies of cation sorption on Fe oxides were conducted in the latter half of the 20th century and resulted in the general conclusion that the extent of cation sorption on Fe oxides is significant at environmentally relevant pH values and that it is an important reaction to consider when examining the fate and transport of metals (e.g., (*14*, *25*, *52*, *58*, *65–67*)). More recent studies have also examined the sorption behavior for several radionucleotides including U^{6+} (*68–71*), Th^{4+} (*69*), and Tc^{7+} (*72*) due to their relevance in radioactive water contamination and long-term engineered radioactive waste storage.

When these studies were conducted, techniques were not yet available that could discern the surface speciation of sorbed cations (*61*, *73–75*). As a result, initial characterization of sorbed species was limited to describing cation loss from solution (i.e., uptake) using sorption isotherms, such as Langmuir, Freundlich, and linear isotherms (e.g., (*56*, *57*)). In the 1970s, surface complexation models (SCMs) were developed to provide a more quantitative and mechanistic approach to describing sorption (e.g., (*63*, *76–78*)). SCMs typically contained four components: (*i*) acid-base protonation/hydroxylation reactions that occur at the particle surface, (*ii*) an equilibrium expression for sorption, (*iii*) mass balance equations for surface sites and the sorbent, and (*iv*) capacitance terms to account for the electrostatic charge between the particle and aqueous phase. Note that the most commonly varied parameter between different models is the capacitance term (ψ), which is modeled with respect to distance from the oxide surface (*75*, *79*, *80*). If charge accumulation was not considered (i.e., capacitance was ignored), then a Langmuir isotherm could be derived from these equations (*73*, *81*, *82*).

While SCMs, such as the constant capacitance model (CCM), double-layer model (DLM), triple-layer model (TLM), and, more recently, the CD-MUSIC model, have significantly improved our ability to describe and predict the macroscopic uptake of ions from solution (e.g., (*83–86*)), more direct evidence from spectroscopy is often needed to discern surface speciation (e.g., (*87*)). The complex behavior observed for cations uptake has, in particular, been difficult to model with SCMs, and often additional reactions, such as surface precipitation, have had to be invoked to adequately capture trends in cation uptake behavior (*73*, *88*, *89*). Throughout the 1980s, the need for spectroscopic evidence to validate surface species proposed in SCMs was recognized and discussed at length in several excellent reviews (e.g., (*61*, *75*, *79*, *80*)).

Environmental studies specifically measuring Fe^{2+} sorption on Fe oxides via Fe^{2+} uptake from solution began to appear in the literature in the early 1990s (e.g., (*14*, *16*, *17*, *90*, *91*)). Several earlier works had reacted Fe^{2+} with Fe oxides, but they were largely concerned with secondary mineral transformation products and Fe^0 corrosion reactions (e.g., (*52*, *92*, *93*)). The increased interest in Fe^{2+} sorption experiments over the past two decades is most likely due to the implication of sorbed Fe^{2+} as a potentially important reductant for environmental contaminants (e.g., (*16*, *17*)) and the discovery that microbes could respire directly on Fe^{3+} oxides (i.e., dissimilatory iron reducing bacteria (DIRB)) (e.g., (*94–97*)).

The primary focus of these early Fe^{2+} sorption studies was to understand contaminant reduction by sorbed Fe^{2+} based on Fe^{2+} sorption isotherms. Several studies found that contaminant reduction rates (k_{obs}) for radionuclides, nitroaromatics, and carbon tetrachloride (CCl_4) could be expressed as a linear function of the amount of Fe^{2+} sorbed on the mineral surface (e.g., (*17*, *98–100*)). The increased reactivity of sorbed Fe^{2+} relative to aqueous Fe^{2+} was attributed to hydroxyl ligands at the oxide surface acting as sigma donor ligands, which increased the electron density of the Fe^{2+} atoms (*64*, *101*). Others used SCMs to describe Fe^{2+} uptake and proposed specific surface species, such as $\equiv Fe^{3+}OFe^{2+}OH^0$, to be responsible for contaminant reduction (*83*, *102*). While the trends in Fe^{2+} sorption and contaminant reduction rates were compelling, most studies lacked the spectroscopic measurements necessary to confirm the presence of the proposed surface species.

At the same time, another avenue of research was being explored that investigated Fe^{2+} sorption on Fe oxides in relation to the sorption of other cations, such as Co^{2+} and Ni^{2+}. One of the earliest, and perhaps most notable, studies was done by Tronc et al., who compared the uptake of Co^{2+}, Ni^{2+}, and Fe^{2+} on magnetite (Fe_3O_4) (*52*). In this work, they found that Fe^{2+} uptake was three times greater than Co^{2+} or Ni^{2+} despite the similar atomic sizes of the cations, and that the uptake kinetics for Co^{2+} and Ni^{2+} was a rapid, one-step process, whereas for Fe^{2+}, the uptake behavior contained several kinetic domains stretching over several days. A similar observation was later made by Jeon and co-workers with hematite (α-Fe_2O_3), Fe^{2+}, and other metals (*103*, *104*). In experiments that focused on the competition of metals for sorption sites on goethite (α-FeOOH), it was noted that Fe^{2+} did not appear to compete with the other metal cations, but surprisingly, the presence of aqueous Fe^{2+} *increased* Cu^{2+}, Co^{2+}, and Ni^{2+} uptake (*14*). A lack of competition for oxide surface sites from aqueous Fe^{2+} was also observed for hematite in the presence of a wide variety of metal cations (*103–105*).

From these studies and others, the idea that an electrochemical reaction was occurring between sorbed Fe^{2+} and the underlying Fe^{3+} oxides emerged (*14*, *52*, *101*, *105*). Within this context, Coughlin and Stone proposed the following equations to illustrate the redox reaction between sorbed Fe^{2+} and the underlying Fe oxide (eqs. 1, 2) (*14*):

$$\equiv Fe^{3+}\text{-}OH + Fe^{2+} + H_2O \leftrightarrow \equiv Fe^{3+}OFe^{2+}OH + 2H^+ \qquad (1)$$

$$\equiv Fe^{3+}OFe^{2+}OH \rightarrow \equiv Fe^{2+}OFe^{3+}OH \qquad (2)$$

In these studies, the semiconducting properties of Fe oxides were often invoked, and the need to consider the bulk electronic properties of the oxide was also proposed (in contrast to SCMs, which considered only surface reactions) (*41*, *52*, *101*, *105*). The semiconducting properties of Fe oxides had already been invoked to describe other redox reactions involving Fe oxides and Fe sulfides, such as the photoelectrolysis of water by illuminated hematite (e.g., (*106–108*)); the photochemical reductive dissolution of Fe oxides by oxalate (*101*, *109*, *110*) and bisulfide (*111*); the photooxidation of pyrite (FeS_2) in the absence of oxygen (*112*); and several other reactions (e.g., (*113*) and refs. therein).

As our knowledge of the sorbed Fe^{2+} - Fe^{3+}_{oxide} reaction has progressed over the past half-century, studies have evolved from largely empirical observations interpreted through sorption isotherms and SCMs to where we are now, which is several compelling, but mostly unlinked, experimental observations, and an ever-increasing awareness of the electrochemical aspect of the reaction. In the next two sections, we discuss two of the compelling observations that our group and others have been focused on for the last several years: (*i*) electron transfer between aqueous Fe^{2+} and Fe oxides and (*ii*) Fe atom exchange between aqueous Fe^{2+} and Fe oxides. In the last section, we build on our experimental findings, as well as the experimental and theoretical work of many others, to propose a revised conceptual framework for describing the reaction of aqueous Fe^{2+} with Fe oxides.

Fe^{2+} - Fe^{3+}_{oxide} Electron Transfer

Oxidation of Sorbed Fe^{2+}

Although there was ample indication in the earlier literature that electron transfer was likely occurring between aqueous Fe^{2+} and Fe oxides (*14*, *49*, *50*, *52*, *101*, *105*), direct observations of Fe^{2+} - Fe^{3+}_{oxide} electron transfer remained elusive. Spectroscopic measurements of Fe^{2+} - Fe^{3+}_{oxide} electron transfer were difficult to make because of the very small quantity of Fe^{2+} at the interface, interference from the bulk structural Fe^{3+}, and the sensitivity of Fe^{2+} species to oxidation by oxygen (*103*, *114*). Other spectroscopic techniques that have been used to characterize sorbed surface species, such as electron paramagnetic resonance (EPR) and X-ray absorption spectroscopy (e.g., extended X-ray absorption fine structure (EXAFS) and X-ray absorption near edge structure (XANES)), but these bulk techniques that could not distinguish sorbed Fe from underlying structural Fe. Surface-sensitive techniques, such as X-ray photoelectron spectroscopy (XPS), detected only the surface Fe atoms, but they operated at high vacuum and required dry samples that are not representative of hydrated mineral surfaces found in natural environments.

Over the last several years, our group has taken advantage of the isotope specificity of ^{57}Fe Mössbauer spectroscopy to overcome these challenges ((*30*, *115–118*); for an introduction to Mössbauer spectroscopy, see: (*119–122*)). ^{57}Fe Mössbauer spectroscopy detects *only* the ^{57}Fe isotope. We capitalized on this specificity to isolate the sorbed Fe signal from the structural Fe^{3+} in the bulk oxide by reacting aqueous $^{57}Fe^{2+}$ with oxides synthesized from ^{56}Fe. Since ^{56}Fe is transparent to ^{57}Fe Mössbauer spectroscopy, oxides made from ^{56}Fe, including ^{56}goethite, ^{56}hematite, and ^{56}magnetite, are transparent and result in a negligible Mössbauer signal, as shown in Figure 1. Mössbauer spectra of the ^{56}oxides after reaction with aqueous $^{57}Fe^{2+}$ revealed that the $^{57}Fe^{2+}$ became oxidized and formed phases similar to the underlying oxide. Fe^{3+} and Fe^{2+} phases are easily distinguishable with Mössbauer spectroscopy based on the center shift (CS) derived from spectral fitting. The CS is measured relative to ^{57}Fe in α-Fe metal (which is operatively assigned a CS value of $v = 0$ mm/s). The shift in velocity, based on the middle of the Mössbauer spectrum, from $v = 0$ mm/s provides information on the electron density at the Fe nucleus; with one less electron, Fe^{3+} shifts about 0.3 to 0.4 mm/s which is much lower than the 1.0 to 1.3 mm/s shift typically observed for Fe^{2+} (*123*). CS values for the spectra of $^{57}Fe^{2+}$ after reaction with ^{56}hematite, ^{56}goethite, and ^{56}magnetite range from 0.35 to 0.44 mm/s, providing direct spectroscopic evidence for oxidation of the $^{57}Fe^{2+}$ (*30, 115, 116*).

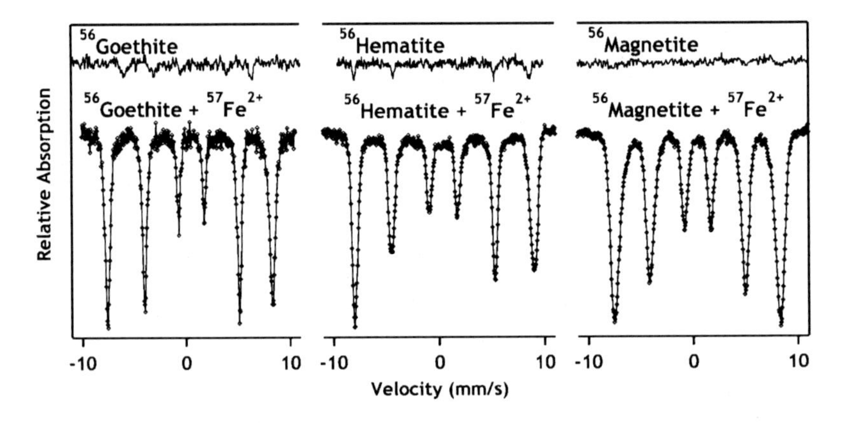

Figure 1. Mössbauer spectra of ^{56}goethite, ^{56}hematite, and ^{56}magnetite before and after reaction with $^{57}Fe^{2+}$. Fe oxides were synthesized from ^{56}Fe and are transparent to ^{57}Fe Mössbauer spectroscopy (top row). After reaction with $^{57}Fe^{2+}$, sextets with parameters consistent with Fe^{3+} are observed (bottom row) indicating that the $^{57}Fe^{2+}$ was oxidized and that electron transfer had occurred between adsorbed Fe^{2+} and structural Fe^{2+} in goethite (30), hematite (116), and magnetite (115). Other oxidants, such as oxygen, were ruled out by complete recovery of the Fe^{2+} upon complete dissolution of the Fe oxides.

Additionally, the presence of six peaks, instead of two, indicates that ^{57}Fe is magnetically ordered (i.e., the electron spins are aligned), which is characteristic of a mineral phase with significant Fe-Fe interactions (*119–122*). The hyperfine parameters of the spectra indicate that the ^{57}Fe spectra are all characteristic of the underlying oxide phases, demonstrating mineral growth. In essence, a type of homoepitaxy has occurred, where magnetite has formed on magnetite, hematite on hematite, and goethite on goethite. Note that magnetite is more complex than the other mineral phases in that it contains structural Fe^{2+} and the Mössbauer spectra are more challenging to interpret (*115*, *124*). The spectrum shown does indicate that most of the $^{57}Fe^{2+}$ has become oxidized (≈90%), although a portion has become Fe^{2+} structurally incorporated into magnetite (≈10%) (*115*). At the present time, we have used this approach to examine several Fe oxide phases, including goethite (*30*), hematite (*30, 116*), ferrihydrite (*30*), and magnetite (*115*), as well as an Fe-bearing clay mineral (*118*), all of which provide compelling evidence that sorbed Fe^{2+} is oxidized in the presence of these Fe^{3+}-bearing mineral phases.

In addition to the Mössbauer work of our group and others (*125*), additional techniques have emerged that have been able to measure oxidation of Fe^{2+} by Fe oxides. A recent study conducted by Yanina and Rosso exposed a large, single crystal of hematite to aqueous Fe^{2+} and oxalate under mildly acidic conditions (*32*). Under these conditions, Fe^{2+} preferentially sorbed at the (001) crystal face forming pyramids that could be seen using microscopy, while pitting occurred at the (hk0) face. This data was interpreted to mean that oxidative sorption occurred at one crystalline face (001), while reductive dissolution occurred at another (hk0). A recent study using X-ray reflectivity measurements has confirmed that face-dependent sorption and dissolution occurs with Fe^{2+} and a single crystal of hematite under more environmentally relevant conditions (i.e., pH 7, no oxalate) (*34*). Similarly, Tanwar and co-workers have used crystal truncation rod (CTR) diffraction to show that Fe^{2+} reacted with hematite is oxidized and results in hematite growth (*31*).

Fate of Injected Electron ("Reduction of Fe Oxide")

With strong evidence for the oxidation of Fe^{2+} by Fe oxides, the compelling question that arose was: What is the fate of the injected electron? Several photoelectrolysis studies, including the work of Eggleston and co-workers, have conclusively demonstrated that electrons are mobile within Fe oxides using scanning tunneling microscopy (STM) (e.g., (*126–129*)) and electrochemical techniques (e.g., (*107*, *108*, *130*), and refs. therein). In these experiments, however, the Fe oxides were often saturated with light or current, whereas Fe^{2+} - Fe^{3+}_{oxide} electron transfer usually involves a relatively small number of transferred electrons. In the 1980s, Mulvaney and co-workers performed experiments using pulse radiolysis where, similar to the Fe^{2+} - Fe^{3+}_{oxide} electron transfer situation, a relatively small number of electrons were added to aqueous hematite suspensions (*49*, *50*, *131*, *132*). By reducing hematite and measuring the fraction of recovered electrons by acidic dissolution, they proposed three potential sinks for the injected electron: (*i*) as a sorbed species at the hematite surface, (*ii*) as a localized, trapped electron in a structural defect of the hematite, and (*iii*) as a dissolved Fe^{2+} species resulting from reductive dissolution (*50*). The criterion they used for distinguishing between sorbed and trapped electrons was whether or not the electrons were recoverable in the acidic extraction (*50*). Results from their work suggest that a combination of localized and delocalized electrons can co-exist, with the relative abundances being dependent on the structural properties of the oxide (e.g., purity, particle size, etc.), the extent of reduction, and the solution conditions. Subsequent theoretical work has slightly revised the three possibilities for the fate of the transferred electron to include: (*i*) a delocalized, conducting electron; (*ii*) a localized, trapped electron; or (*iii*) a dissolved Fe^{2+} atom formed from reductive dissolution (*41*, *131*).

Our group has attempted to probe the fate of the injected electron by continuing to capitalize on the isotopic specificity of Mössbauer spectroscopy. To address the fate of the injected electron, we switched the isotopes and reacted aqueous $^{56}Fe^{2+}$, which is transparent in Mössbauer spectroscopy, with either ^{57}Fe oxides or isotopically normal oxides (which contain approximately 2.1% ^{57}Fe). We could then observe any changes to the oxide after the $^{56}Fe^{2+}$ was oxidized and the electron was injected into the oxide. The changes have, not surprisingly, varied dramatically among ferrihydrite (*30*), hematite (*116*), goethite, and magnetite (*115*). For ferrihydrite, low temperature spectra indicated localized paramagnetic Fe^{2+}, similar to spectra previously collected from co-precipitated Fe^{2+} and Fe^{3+} at high pH values (*133*). Hematite reacted with Fe^{2+} showed quite different behavior, with a subtle change in the magnetic behavior of the particles as evidenced by a partial repression of the Morin transition (i.e., a temperature-dependent magnetic transition) which could be caused by either localized or delocalized electrons (*134*). For goethite, no significant change could be observed in the spectrum after exposure to $^{56}Fe^{2+}$ (our group, unpublished data).

Of the Fe oxides we have reacted with $^{56}Fe^{2+}$, we observed the most distinct change in magnetite. Magnetite can exist with different amounts of structural Fe^{2+}, with the Fe^{2+}/Fe^{3+} ratio ranging from 0.5 (stoichiometric magnetite) to 0 (completely oxidized magnetite or maghemite, γ-$Fe_{2.67}O_4$). Magnetite with intermediate Fe^{2+}/Fe^{3+} ratios is known as non-stoichiometric or partially-oxidized magnetite. When we added $^{56}Fe^{2+}$ to ^{57}Fe-containing non-stoichiometric magnetite (Fe^{2+}/Fe^{3+} = 0.31), we saw the underlying magnetite convert to stoichiometric magnetite based on the Mössbauer spectrum (Fe^{2+}/Fe^{3+} = 0.48; Figure 2) (*115*). After reaction with $^{56}Fe^{2+}$, the magnetite spectrum shows a decrease in the octahedral Fe^{3+} spectral area coupled to an increase in the octahedral $Fe^{2.5+}$ phase area (with the mixed-valent phase observed because the electron hopping rate between octahedral sites in magnetite is faster than the characteristic time of Mössbauer spectroscopy) (*124*). Unlike the other Fe oxides, with magnetite we could for the first time distinguish the fate of the electrons within the oxide structure (that is, the electron reduces an octahedral Fe^{3+} atom to produce an octahedral Fe^{2+} atom).

Fe Atom Exchange

Above we discussed how the electron injected into the Fe oxide (from oxidative sorption of Fe^{2+}) might become trapped in the oxide structure as either a localized or delocalized electron. Alternatively, the electron might not remain within the oxide, but instead may be released into solution as an Fe^{2+} atom via reductive dissolution. A vast literature exists on the reductive dissolution of Fe oxides by other reductants, such as ascorbic acid, oxalate, and sulfide (e.g., (*101*, *111*, *135*, *136*)), but there is little discussion on whether sorbed Fe^{2+} was capable of promoting reductive dissolution of an Fe oxide.

To study the movement of Fe atoms between the solid and aqueous phases, we, and others, have used Fe isotopes as tracers. Fe has four stable isotopes (natural abundances: ^{54}Fe (5.8%), ^{56}Fe (91.8%), ^{57}Fe (2.1%), and ^{58}Fe (0.3%) (*137*)) and several radioactive isotopes (most notably ^{55}Fe; half-life = 1004 days (*138*)). Measuring the relative abundance of isotopes with the precision necessary for tracking exchange reactions is, however, difficult and requires high

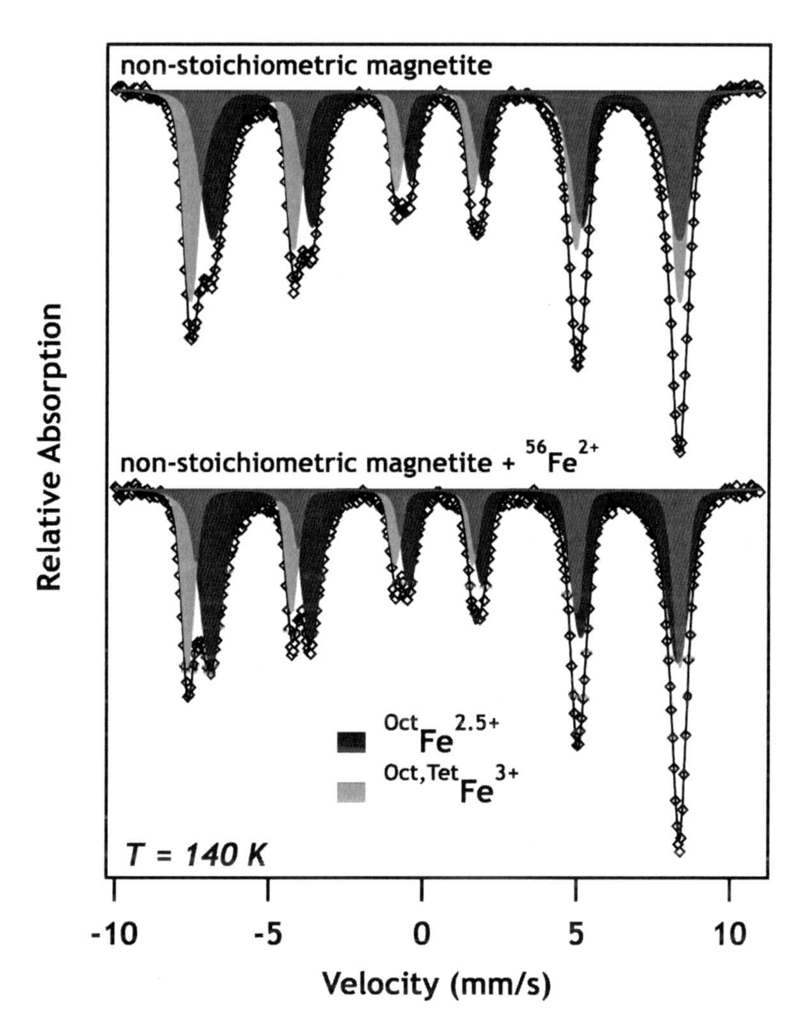

Figure 2. Mössbauer spectrum of non-stoichiometric magnetite (Fe^{2+}/Fe^{3+} = 0.31) before and after (Fe^{2+}/Fe^{3+} = 0.48) reaction with $^{56}Fe^{2+}$. Initial concentration of aqueous $^{56}Fe^{2+}$ was 3 mM. After 24 hours, 1.5 mM was taken up. Open markers represent the observed spectrum, with the total fit shown as a solid line. (Reproduced with permission from reference (115). Copyright 2009 ACS)

precision mass spectrometers for stable isotope measurements (multi-collector inductively-coupled plasma mass spectrometry or MC-ICP-MS) (*139–141*), or scintillation counters for radioactivity measurements (*22*, *142*). Many of the earliest works using stable Fe isotopes to examine the interaction of aqueous Fe^{2+} with Fe oxides were focused on determining sorption-induced changes in Fe isotope fractionations in order to gain insight into natural fractionation processes (e.g., (*36*, *37*, *143–148*)).

One of the first studies to use Fe isotopes to track the exchange of Fe between aqueous Fe^{2+} and Fe oxides used ^{55}Fe as a tracer (*22*). Pedersen et al. synthesized four Fe oxides enriched with ^{55}Fe and observed the release of ^{55}Fe

into solution upon exposure to aqueous Fe^{2+} over the period of a month. Release of ^{55}Fe into solution was observed for ferrihydrite, lepidocrocite (γ-FeOOH), and goethite, but not for hematite (Figure 3, circles). The amount of ^{55}Fe released into solution approached isotopic equilibrium for ferrihydrite over a matter of days, demonstrating that nearly all the structural Fe in ferrihydrite exchanged with the aqueous Fe^{2+} phase. Partial exchange was observed for two different-sized lepidocrocites and goethite, indicating that only a fraction of the structural Fe participated in exchange or reductive dissolution reactions. These findings showed, at least for ferrihydrite, lepidocrocite, and goethite, that aqueous Fe^{2+} catalyzed extensive interfacial Fe atom exchange. Jones et al. more recently used ^{55}Fe-enriched Fe oxides to measure the amount of isotope exchange between aqueous Fe^{2+} and ferrihydrite, jarosite, lepidocrocite, and schwertmannite with and without the presence of silica and natural organic matter (NOM) (*40*). Significant exchange was observed for all the oxides, but only ferrihydrite showed complete exchange (Figure 3, triangles). When silica and NOM were present, the extent of exchange decreased, which was attributed to their blocking reductive dissolution sites (*40*, *149*, *150*).

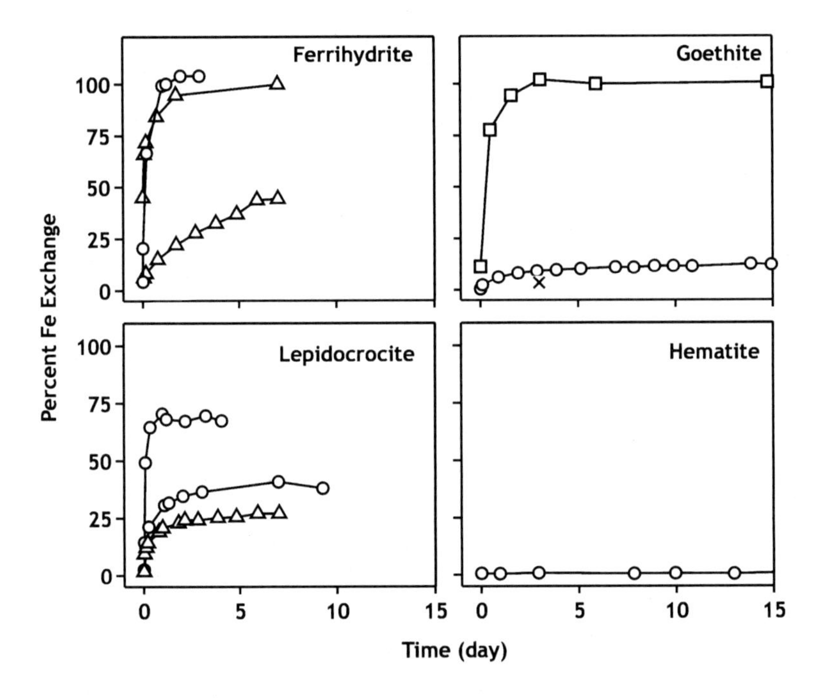

Figure 3. Kinetics of Fe atom exchange between aqueous Fe^{2+} and Fe oxides (circles (22), triangles (40), squares (39), and X (38)). Complete exchange (100%) indicates complete equilibration of isotopes in each phase, while negligible exchange (≈0%) means no atomic exchange has occurred. Partial exchange is observed when only a fraction of the structural Fe exchanges with the aqueous Fe^{2+}.

Fe atom exchange between aqueous Fe^{2+} and Fe oxides has also been quantified using natural Fe isotope abundances (e.g., (*36–38*, *147*, *151*, *152*)). For example, Mikutta et al. measured $\delta^{56/54}Fe$ values for aqueous Fe^{2+} reacted with goethite and goethite-coated sand (*38*); based on mass-balance estimates, they determined that approximately a single-surface atom layer of Fe was exchanged ($\approx$ 4%), which was significantly less than that observed for the ^{55}Fe goethite studied by Pedersen et al. (Figure 3). Both Jang et al. and Crosby et al. observed minor exchange between aqueous Fe^{2+} and goethite (*36*, *37*) or hematite (*36*), but quantifying the extent of exchange was difficult due to only minor initial isotopic differences between the aqueous and solid phase Fe. Additionally, many of these studies tracked the isotopic composition of only the aqueous phase, making mass and isotope balance calculations more difficult to perform.

Compelled by the findings from Pedersen et al. and others, we used a ^{57}Fe-enriched tracer approach to simultaneously measure changes in the isotope composition of both the aqueous Fe and the goethite Fe (Figure 4) (*39*, *153*). We enriched the aqueous Fe^{2+} with ^{57}Fe atoms ($\delta^{57/56}$ Fe = +840.43‰), while the goethite remained naturally abundant in Fe ($\delta^{57/56}Fe$ = -0.12‰). By tracking the $\delta^{57/56}Fe$ value for the aqueous and solid phases over time, we were able to determine the extent of mixing by measuring changes in both Fe reservoirs (see (*39*) for calculations). Over the course of 30 days, the $\delta^{57/56}Fe$ value of aqueous Fe^{2+} decreased from +840.43 to +39.94‰ and the solid phase $\delta^{57/56}Fe$ value increased from -0.12 to +38.75‰ (Figure 4). The decrease in the aqueous Fe^{2+} $\delta^{57/56}Fe$ value and the increase in the solid phase $\delta^{57/56}Fe$ value both indicate that Fe atom exchange occurred between the two phases. The convergence of the two values to the mass balance average indicates that nearly 100% of the structural Fe in the goethite exchanged with aqueous Fe atoms (Figure 3, squares).

As shown in Figure 3, estimates of the extent of exchangeable Fe^{3+} in Fe oxides after reaction with Fe^{2+} vary significantly. As one might expect, the rate and extent of mixing vary significantly for each Fe oxide, with the least stable Fe oxide (ferrihydrite) exhibiting more exchange than the more stable Fe oxides. There is also significant variation between studies for the same oxide, particularly for goethite, where nearly complete exchange is observed after a week in our work (*39*), and much less exchange is observed ($\leq$ 10%) in two other works over longer time periods (*22*, *38*). The reason for these variations is, at this time, unclear. Relative to the experimental conditions of Pedersen et al., our goethite particles had a greater surface area (110 vs. 37 m^2g^{-1}), and we worked at a higher solution pH (7.5 vs. 6.5) and higher goethite concentrations (2 g/L vs. ~50 mg/L). Mikutta et al. used goethite-coated quartz (as opposed to goethite particles) and a flow-through column (instead of batch reactors) to estimate the extent of exchange (*38*). These differences, as well as other experimental variables such as pH buffer and background electrolyte, may have contributed to the discrepancies in amount of Fe exchanged among the goethite experiments. What controls the extent of exchangeable Fe^{3+} in Fe oxides after reaction with Fe^{2+} is a critical question that remains to be addressed.

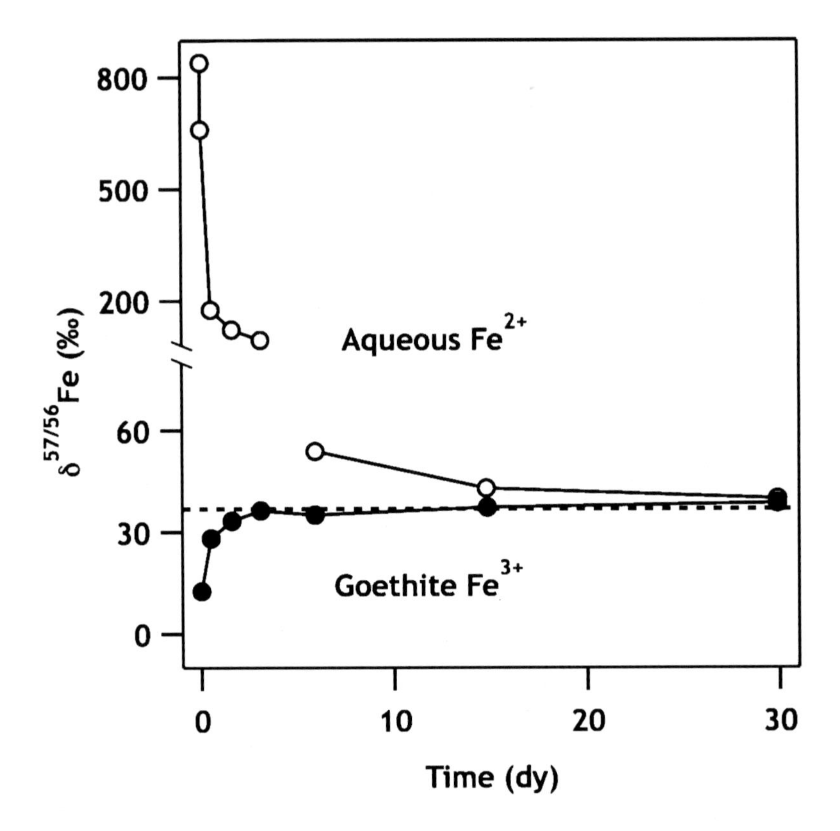

Figure 4. Measured $\delta^{57/56}Fe$ values of aqueous (○) and goethite-Fe (●) over time. The dashed line is the calculated completely mixed $\delta^{57/56}Fe$ value of 37.6. Because the initial molar amount of aqueous Fe^{2+} was much smaller than the amount of Fe initially present within the goethite particles (15.85 mmole vs. 337.66 mmole, respectively), the completely mixed $\delta^{57/56}Fe$ value is much closer to the initial isotopic composition of the goethite solids than the aqueous Fe^{2+}. Each data point represents the average of triplicate reactors. Standard deviations of replicates are contained within markers. (Reproduced with permission from reference (39). Copyright 2009 ACS)

Despite the differences in the extent of atom exchange observed in these studies, the works collectively suggest that atom exchange is preceded by oxidative sorption of Fe^{2+} by structural Fe^{3+}. Based on the observations that (*i*) sorbed Fe^{2+} oxidizes at Fe oxide surface and (*ii*) significant reductive dissolution of structural Fe atoms occurs, we proposed a conceptual model to describe the dynamics of aqueous Fe^{2+} - Fe^{3+}_{oxide} electron transfer and atom exchange (Figure 5). We hypothesized that five sequential steps (i.e., sorption – electron transfer – crystal growth – conduction – dissolution) are occurring in what can be envisioned as a redox-driven conveyor belt (Figure 5) (*39*). This cyclic reaction offers a potential explanation for how significant isotope exchange can occur between phases without significant change observed in the particle dimensions or aqueous Fe^{2+} concentration over the course of the experiment. Although the energetic driving force of this reaction mechanism remains unclear, recent measurements with hematite have revealed that a potential gradient exists between crystalline faces when the oxide is exposed to aqueous Fe^{2+} (*32*). In this work, the authors invoked the semiconducting nature of hematite (i.e., the ability of the oxide to conduct electrons through the crystal lattice) to explain how spatially separated sites on hematite could be coupled in oxidative sorption and reductive dissolution reactions (i.e., the proximity effect: (*41*)).

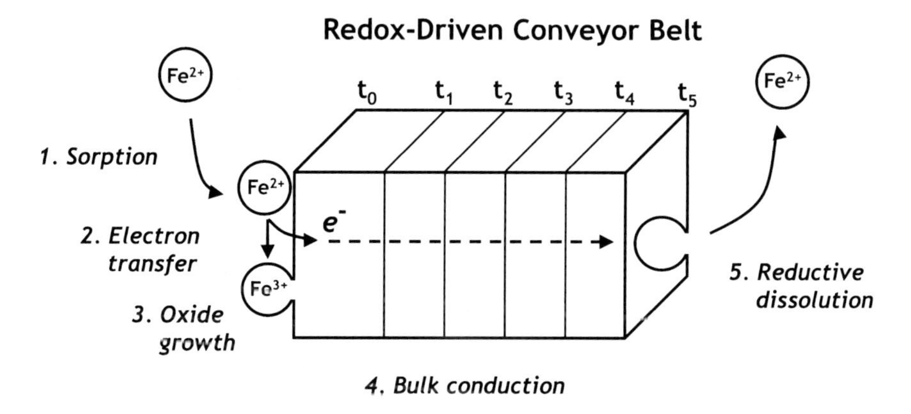

Figure 5. Conceptual model for the five steps associated with the redox-driven conveyor belt mechanism to explain how bulk goethite Fe^{3+} atoms and aqueous Fe^{2+} can become completely mixed via growth and dissolution at separate goethite surface sites. The left surface may be considered a reference plane in the original goethite crystal at the start of the process (t_0), and through growth on the left and dissolution on the right, this reference plane will migrate over time ($t_0 \rightarrow t_5$) until time t_5, at which point 100% atom exchange has occurred. (Reproduced with permission from reference (39). Copyright 2009 ACS)

Moving Toward a Revised Conceptual Model for Fe^{2+} Uptake

From the discussed studies, compelling evidence has emerged showing that aqueous Fe^{2+} - Fe^{3+}_{oxide} electron transfer occurs and, in some cases, atom exchange and/or bulk conduction through the oxide may occur after the injection of the electron. Many of these observations, however, are with different oxides under different experimental conditions, and it is difficult to interpret these observations without a clear conceptual framework (e.g., (*154*)). Here, we attempt to combine the experimental evidence for Fe^{2+} - Fe^{3+}_{oxide} electron transfer, bulk electron conduction, and Fe atom exchange into a revised conceptual framework that builds on the well-known semiconducting properties of Fe oxides.

The transfer of electrons into semiconducting oxide particles has long been studied due to its use in photocatalytic reactions (i.e., the conversion of light into chemical energy) (some classic studies and reviews include: (*49*, *107*, *108*, *130*, *131*, *155–160*)). In these works, several approaches have utilized semiconducting nanoparticles to convert visible light into chemical energy (for a historical perspective, see (*158*)). In studies analogous to Fe^{2+} sorption on Fe oxides, this field of work has examined how electrons are transferred to and from semiconducting nanoparticles (e.g., (*49, 50, 131, 156, 157, 159–165*)). Reviewing this extensive literature provides an excellent starting point for considering how Fe oxides can be studied and thought of as electrochemically-active semiconductors.

In a solid, the molecular orbital energy levels become smeared into energy bands due to their similar energy levels and the Pauli exclusion principle (for a more thorough review, see (*131, 155, 158, 166, 167*)). The gap between the highest energy states occupied by electrons (i.e., the valence band, VB) and the lowest unoccupied energy states (i.e., the conduction band, CB) is known as the band gap (E_g), and is the amount of energy required to excite an electron from the VB to CB (0.5-3.0 eV for semiconductors) (Figure 6). In an ideal, flawless oxide particle, no energy levels exist between the VB and CB, thus an electron added to the oxide goes into the CB. In real oxides, however, atomic impurities, crystalline defects, and surface effects alter the ideal structure and create sites where electrons can stabilize at energy levels between the VB and CB (Figure 6). The nature of the energy levels of these impurities and defects can dictate whether a semiconductor is negative- or *n*-type, where excess electrons transfer charge in the CB, or *p*-type, where positively charged holes in the VB act as charge carriers (e.g., (*166–172*), and refs. therein). The common Fe oxides have E_g values ranging from approximately 1.9 to 2.2 eV, with the exception of magnetite, which has an E_g of 0.1 eV, making it closer to a conductor (*150*). Note that a 2 eV band gap corresponds to the absorption of light with a wavelength shorter than ~620 nm, which results in the transmission of only red and orange visible light and provides the Fe oxides with their characteristic colors.

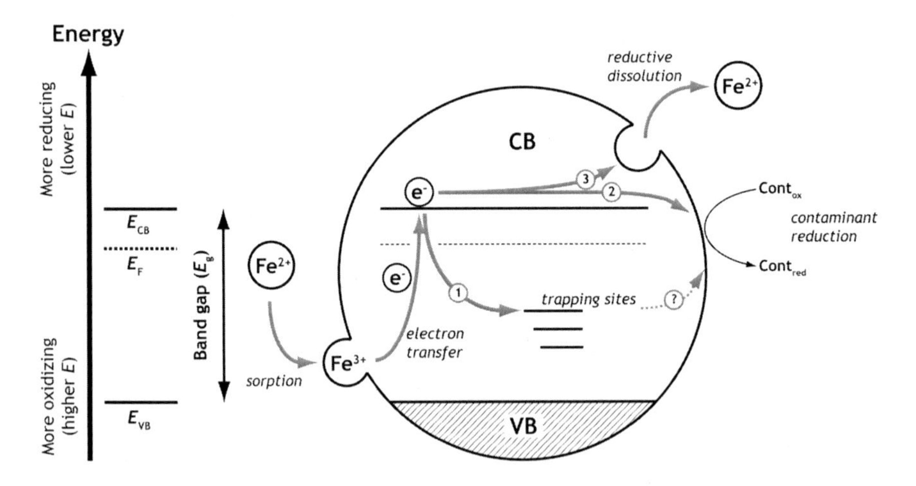

Figure 6. Conceptual schematic representation of how aqueous Fe^{2+} interacts with Fe oxides with environmental reactions in mind. Upon oxidative sorption of Fe^{2+}, the electron is transferred to the CB. From the conduction band, the electron has three possible fates: (1) it can be localized in a trapping site; (2) it can reduce an oxidant in solution, such as an environmental contaminant or a biological electron mediator; or (3) it can partake in reductive dissolution.

By coupling models previously devised in the literature (*39*, *41*, *54*, *131*, *156–159*) with experiments investigating Fe^{2+} uptake by Fe oxides (*22*, *30*, *32*, *39*, *49*, *50*, *52*, *115*), we propose a revised conceptual framework for interpreting the reaction of aqueous Fe^{2+} with Fe oxides (Figure 6). In this model, an Fe^{2+} atom sorbs to an Fe oxide particle and becomes oxidized to Fe^{3+} with the electron transferred into the CB. As described by Mulvaney and co-workers and others (*41*, *50*, *131*, *157*, *159*, *160*), this electron has three potential fates which are illustrated in Figure 6: (*i*) the electron may become immobilized in a trapping site; (*ii*) the electron may stay in the conduction band and react with a redox-active aqueous substituent, such as an environmental contaminant; or (*iii*) the electron may migrate near the surface and cause another Fe^{3+} atom to participate in reductive dissolution. For Fe oxides, reductive dissolution likely occurs at a reduction potential more positive (lower on the y-axis in Figure 6) than that for the reduction of H_2O to H_2 in the absence of light (*131*).

Interfacial redox reactions occurring between semiconducting oxide particles and redox-active aqueous species have been well established in the literature; in fact, it has become common in the physical chemistry literature to use redox-active probe compounds to determine the reduction potential (E) of an oxide by assuming equilibrium between the phases (i.e., $E_{oxide} = E_{redox\ probe}$) (e.g., (*49*, *50*, *131*, *156*, *157*, *159*, *160*)). The potential of the oxide (E_{oxide}) is defined as the Fermi level (E_F), which is the energy level in the oxide that has a ½ probability of being occupied by an electron (E_F, arbitrarily placed between the CB and VB in Figure 6). As an oxide becomes more reduced or electron-doped, the E_F will become more negative as higher energy orbitals are occupied (higher on the y-axis in Figure 6).

As the E_F becomes more negative, the oxide becomes a stronger reductant, and vice versa. Fe^{2+} removal from solution by the oxide would lead to a lowering of the oxide E_F and an increase in the potential of the Fe^{2+} in solution (E_{soln}). In light of the semiconducting properties of Fe oxides, several researchers have speculated that the extent of Fe^{2+} uptake by Fe oxides may be controlled by a redox-driven process, such as E_F-E_{soln} equilibration between the Fe oxide and the $Fe^{2+}/Fe^{3+}{}_{aq}$ redox couple(s) (*24*, *32*, *49*, *125*, *131*, *160*). Calculating the extent of expected Fe^{2+} uptake using this framework is still not a trivial calculation, however, as several co-existing variables can influence E_F and band localities and capacities, such as the abudance of trapping sites and structural defects (e.g., (*49*, *50*, *131*)), inadvertant or intentional structural doping (e.g., (*130*, *172*)), the relative abundance of crystalline faces (e.g., (*32*, *34*)), secondary-mineralization pathways and irreversible reactions (e.g., (*19–24*)), and VB and CB band-bending effects and E_F-pinning (e.g., (*130*, *173*, *174*)).

Reconsidering Fe^{2+} sorption on Fe^{3+} oxides in light of the semiconducting framework illustrated in Figure 6 provides some thought-provoking insights to consider. For example, the extent of Fe^{2+} sorption should be controlled by E_F equilibration, not surface site availability. This is consistent with our group's recent finding that Fe^{2+} uptake by magnetite is controlled by a bulk property, rather than particle surface area (*115*). More specifically, Fe^{2+} sorption isotherms were measured for magnetite batches with varying stoichiometries ($x = Fe^{2+}/Fe^{3+}$). The specific surface areas (SSA) of the oxides were carefully controlled in synthesis to minimize any differences (BET SSA = 62 ± 8 m^2g^{-1} ($\pm\sigma$)). Fe^{2+}

uptake by the magnetite was strongly dependent on the initial stoichiometry (x), with more oxidized magnetite batches (lower x) exhibiting significantly more Fe^{2+} uptake from solution (Figure 7, left panel). When the particles were dissolved to measure the total Fe^{2+}/Fe^{3+} ratio after Fe^{2+} sorption, the stoichiometries of all the magnetite batches plateaued at approximately $x = 0.5$, the value corresponding to stoichiometric magnetite ($x = 0.5$: $^{Tet}Fe^{3+}[^{Oct}Fe^{2+}Fe^{3+}]O_4$) (Figure 7, right panel). Similar observations have also been made based on Fe^{2+} titrations with magnetite (*52*).

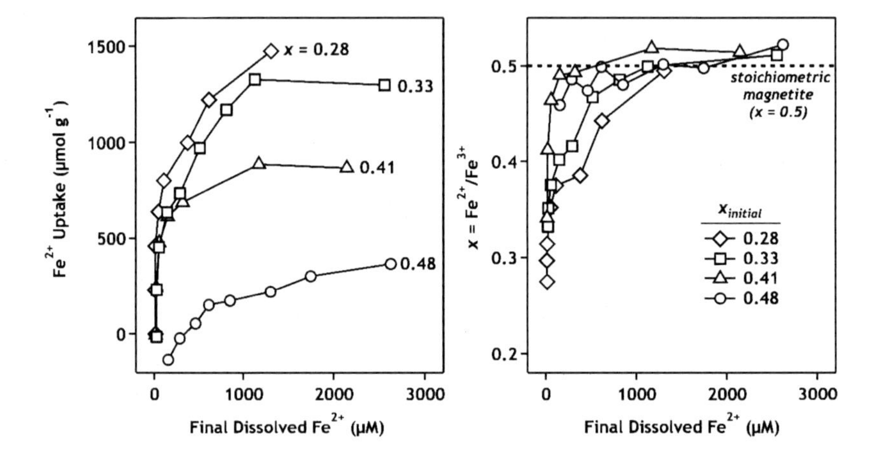

Figure 7. (Left) Effect of initial magnetite stoichiometry ($x = Fe^{2+}/Fe^{3+}$) on Fe^{2+} uptake from solution (presented as sorption isotherms). The average specific surface areas for the magnetite batches were: 62 ± 8 m^2g^{-1}. (Right) Magnetite stoichiometry (x) measured by acidic dissolution after reaction with aqueous Fe^{2+}. The data in each panel is taken from the same experiment. (Reproduced with permission from reference (115). Copyright 2009 ACS)

Observations regarding contaminant reduction by sorbed Fe^{2+} are also interesting to revisit. For example, several studies report negligible reduction of a model contaminant (i.e., NO_2^-, O_2, and nitrobenzene) when an Fe oxide is exposed to aqueous Fe^{2+} and either the residual aqueous Fe^{2+} is removed or all Fe^{2+} sorbs. In the presence of aqueous Fe^{2+}, however, rapid reduction occurs (*30, 175, 176*). Using the above model, removal of Fe^{2+}_{aq} would raise E_{soln} to a more positive potential, which consequently raises the E_F of the solid. A significant portion of the Fe^{2+} associated with the particle may also be present in localized trapping sites, which may or may not be capable of reducing an aqueous oxidized species. Note that in these works, the reaction kinetics were only measured for relatively short time periods making it difficult to assess if the reaction was simply very slow (indicative of a more positive E_F) or completely stopped (suggesting significant or complete trapping). Future contaminant fate studies should consider the possibility of trapped electrons, especially when evaluating the extent of a reaction.

One final consideration is how Fe^{2+} catalyzes Fe oxide transformation reactions (e.g., ferrihydrite + Fe^{2+} → goethite or magnetite; lepidocrocite + Fe^{2+} → magnetite) (e.g., (*19–24*)). A topic of discussion has been whether Fe^{2+} promotes mineral transformations by topotactic reformation (i.e., an internal reorganization) or by nucleation and recrystallization (i.e., a breakdown of the mineral and reconstruction of a new phase) (e.g., (*20–22*, *24*)). Recently, Yang et al. argued that both topotactic reformation and nucleation can occur, with the rate of electron-doping (i.e., Fe^{2+} uptake) dictating which mechanism occurred. They argue that at low Fe^{2+} concentrations, reductive dissolution and oxidative sorption would be in equilibrium, with nucleation and recrystallization being the dominant mineral transformation mechanism. At high Fe^{2+} concentrations, however, the rate of Fe^{2+} uptake could be greater than the rate of reductive dissolution (e.g., (*177*)). In this case, the significant charge accumulation within the Fe oxide could lead to topotactic reformation. In the Yang et al. model, the extent of electron doping into the Fe oxide dictates which mechanism, and likely which products, are predominant. Further developing an understanding of these reaction mechanisms may also offer insights into how foreign cations, such as trace nutrients (e.g., Mg^{2+}, Ca^{2+}, Cu, Mn, Zn) and heavy metals (e.g., U, As, Cr), can become and/or remain structurally incorporated in Fe oxides (e.g., (*14*, *104*, *178–181*)).

We end this chapter by presenting a few examples of how the semiconducting nature of Fe oxides may also be used advantageously in engineered remediation efforts. Several studies have demonstrated that the particle size of Fe oxides can be finely tuned down to the nanometer scale, which can dramatically influence reactivity beyond what would be predicted by only considering the surface area (e.g., reductive dissolution (*182*, *183*), CCl_4 reduction (*184*), As sorption (*185*), and Mn^{2+} oxidation (*186*)). Doping of Fe oxides with foreign cations may also be utilized to promote otherwise unfavorable reactions: reduced hematite (i.e., electron-doped) and Mg^{2+}-doped hematite (i.e., *p*-doped) can reduce H_2O to H_2 and N_2 to NH_3 (*106*, *187*, *188*), while Sn^{4+}- and Ti^{4+}-doped hematite (i.e., *n*-doped) can reduce dissolved O_2 (*172*). While Fe oxides have long been utilized as semiconductors in other fields, such as photocatalytic energy production (e.g., (*106–108*, *168–170*, *189*)), their electronic potential in engineered remediation remains to be fully explored (e.g., (*113*, *190*)).

Acknowledgments

We thank Phil Larese-Casanova, Michael Sander, and two anonymous reviewers for helpful and insightful comments in reviewing the manuscript.

References

1. Lovely, D. R. Microbial Fe(III) reduction in subsurface environments. *FEMS Microbiol. Rev.* **1997**, *20*, 305–313.
2. Brown, G. E.; Henrich, V. E.; Casey, W. H.; Clark, D. L.; Eggleston, C.; Felmy, A.; Goodman, D. W.; Gratzel, M.; Maciel, G.; McCarthy, M. I.;

Nealson, K. H.; Sverjensky, D. A.; Toney, M. F.; Zachara, J. M. Metal oxide surfaces and their interactions with aqueous solutions and microbial organisms. *Chem. Rev.* **1999**, *99*, 77–174.

3. Hochella, M. F. There's plenty of room at the bottom: Nanoscience in geochemistry. *Geochim. Cosmochim. Acta* **2002**, *66*, 735–743.
4. Sarthou, G.; Timmermans, K. R.; Blain, S.; Treguer, P. Growth physiology and fate of diatoms in the ocean: A review. *J. Sea Res.* **2005**, *53*, 25–42.
5. Borch, T.; Campbell, K.; Kretzschmar, R. How electron flow controls contaminant dynamics. *Environ. Sci. Technol.* **2009**, *44*, 3–6.
6. Borch, T.; Kretzschmar, R.; Kappler, A.; Van Cappellen, P.; Ginder-Vogel, M.; Voegelin, A.; Campbell, K. Biogeochemical redox processes and their impact on contaminant dynamics. *Environ. Sci. Technol.* **2010**, *44*, 15–23.
7. Schmidt, C.; Behrens, S.; Kappler, A. Ecosystem functioning from a geomicrobiological perspective - A conceptual framework for biogeochemical iron cycling. *Environ. Chem.* **2010**, *7*, 399–405.
8. Konhauser, K. O.; Kappler, A.; Roden, E. E. Iron in microbial metabolisms. *Elements* **2011**, *7*, 89–93.
9. Takai, Y.; Kamura, T. The mechanism of reduction in waterlogged paddy soil. *Folia Microbiol.* **1966**, *11*, 304–313.
10. Ponnamperuma, F. N.; Tianco, E. M.; Loy, T. Redox equilibria in flooded soils: I. The iron hydroxide systems. *Soil Sci.* **1967**, *103*, 374–382.
11. Jenne, E. A. Controls on Mn, Fe, Co, Ni, Cu, and Zn concentration in soils and water - Significant role of hydrous Mn and Fe oxides. *Adv. Chem. Ser.* **1968**, *73*, 337–387.
12. Lack, J. G.; Chaudhuri, S. K.; Chakraborty, R.; Achenbach, L. A.; Coates, J. D. Anaerobic biooxidation of Fe(II) by Dechlorosoma suillum. *Microb. Ecol.* **2002**, *43*, 424–431.
13. Murray, G. C.; Hesterberg, D. Iron and phosphate dissolution during abiotic reduction of ferrihydrite-boehmite mixtures. *Soil Sci. Soc. Am. J.* **2006**, *70*, 1318–1327.
14. Coughlin, B. R.; Stone, A. T. Nonreversible adsorption of divalent metal-ions (Mn^{II}, Co^{II} Ni^{II} Cu^{II} and Pb^{II}) onto goethite - Effects of acidification, Fe^{II} addition, and picolinic-acid addition. *Environ. Sci. Technol.* **1995**, *29*, 2445–2455.
15. Macalady, D. L.; Tratnyek, P. G.; Grundl, T. J. Abiotic reduction reactions of anthropogenic organic chemicals in anaerobic systems. *J. Contam. Hydrol.* **1986**, *1*, 1–28.
16. Sorensen, J.; Thorling, L. Stimulation by lepidocrocite (γ-FeOOH) of Fe(II)-dependent nitrite reduction. *Geochim. Cosmochim. Acta* **1991**, *55*, 1289–1294.
17. Klausen, J.; Trober, S. P.; Haderlein, S. B.; Schwarzenbach, R. P. Reduction of substituted nitrobenzenes by Fe(II) in aqueous mineral suspensions. *Environ. Sci. Technol.* **1995**, *29*, 2396–2404.
18. Hofstetter, T. B.; Heijman, C. G.; Haderlein, S. B.; Holliger, C.; Schwarzenbach, R. P. Complete reduction of TNT and other

(poly)nitroaromatic compounds under iron-reducting subsurface conditions. *Environ. Sci. Technol.* **1999**, *33*, 1479–1487.

19. Roden, E. E.; Urrutia, M. M. Influence of biogenic Fe(II) on bacterial crystalline Fe(III) oxide reduction. *Geomicrobiol. J.* **2002**, *19*, 209–251.
20. Hansel, C. M.; Benner, S. G.; Neiss, J.; Dohnalkova, A.; Kukkadapu, R.; Fendorf, S. Secondary mineralization pathways induced by dissmilatory iron reduction of ferrihydrate under advective flow. *Geochim. Cosmochim. Acta* **2003**, *67*, 2977–2992.
21. Hansel, C. M.; Benner, S. G.; Fendorf, S. Competing Fe(II)-induced mineralization pathways of ferrihydrite. *Environ. Sci. Technol.* **2005**, *39*, 7147–7153.
22. Pedersen, H. D.; Postma, D.; Jakobsen, R.; Larsen, O. Fast transformation of iron oxyhydroxides by the catalytic action of aqueous Fe(II). *Geochim. Cosmochim. Acta* **2005**, *69*, 3967–3977.
23. Cutting, R. S.; Coker, V. S.; Fellowes, J. W.; Lloyd, J. R.; Vaughan, D. J. Mineralogical and morphological constraints on the reduction of Fe(III) minerals by Geobacter sulfurreducens. *Geochim. Cosmochim. Acta* **2009**, *73*, 4004–4022.
24. Yang, L.; Steefel, C. I.; Marcus, M. A.; Bargar, J. R. Kinetics of Fe(II)-catalyzed transformation of 6-line ferrihydrite under anaerobic flow conditions. *Environ. Sci. Technol.* **2010**, *44*, 5469–5475.
25. Stumm, W.; Kummert, R.; Sigg, L. A ligand exchange model for the adsorption of inorganic and organic ligands at hydrous oxide interfaces. *Croat. Chem. Acta* **1980**, *53*, 291–312.
26. Zinder, B.; Furrer, G.; Stumm, W. The coordination chemistry of weathering: II. Dissolution of Fe(III) oxides. *Geochim. Cosmochim. Acta* **1986**, *50*, 1861–1869.
27. Dzombak, D. A.; Morel, F. M. M. *Surface Complexation Modeling: Hydrous Ferric Oxide*; John Wiley & Sons: New York, 1990; p 393.
28. Wehrli, B. Redox reactions of metal ions at mineral surfaces. In *Aquatic Chemical Kinetics: Reaction Rates of Processes in Natural Waters*; Stumm, W., Ed.; Wiley-Interscience: New York, 1990; pp 311−336.
29. Fenter, P. A. X-ray reflectivity as a probe of mineral-fluid interfaces: A user guide. In *Applications of Synchrotron Radiation in Low-Temperature Geochemistry and Environmental Sciences*; Fenter, P. A., Rivers, M. L., Sturchio, N. C., Sutton, S. R., Eds.; Mineralogical Society of America: Washington, DC, 2002; Vol. 49, pp 149−220.
30. Williams, A. G. B.; Scherer, M. M. Spectroscopic evidence for Fe(II)-Fe(III) electron transfer at the Fe oxide-water interface. *Environ. Sci. Technol.* **2004**, *38*, 4782–4790.
31. Tanwar, K. S.; Petitto, S. C.; Ghose, S. K.; Eng, P. J.; Trainor, T. P. Strucutral study of Fe(II) adsorption on hematite (11(bar)02). *Geochim. Cosmochim. Acta* **2008**, *72*, 3311–3325.
32. Yanina, S. V.; Rosso, K. M. Linked reactivity at mineral-water interfaces through bulk crystal conduction. *Science* **2008**, *320*, 218–222.

33. Catalano, J. G.; Fenter, P.; Park, C. Water ordering and surface relaxations at the hematite (110)-water interface. *Geochim. Cosmochim. Acta* **2009**, *73*, 2242–2251.
34. Catalano, J. G.; Fenter, P.; Park, C.; Zhang, Z.; Rosso, K. M. Structure and oxidation state of hematite surfaces reacted with aqueous Fe(II) at acidic and neutral pH. *Geochim. Cosmochim. Acta* **2010**, *74*, 1498–1512.
35. Beard, B. L.; Johnson, C. M.; Cox, L.; Sun, H.; Nealson, K. H.; Aguilar, C. Iron isotope biosignatures. *Science* **1999**, *285*, 1889–1892.
36. Crosby, H. A.; Roden, E. E.; Johnson, C. M.; Beard, B. L. The mechanisms of iron isotope fractionation produced during dissimilatory Fe(III) reduction by Shewanella putrefaciens and Geobacter sulfurreducens. *Geobiology* **2007**, *5*, 169–189.
37. Jang, J. H.; Mathur, R.; Liermann, L. J.; Ruebush, S.; Brantley, S. L. An iron isotope signature related to electron transfer between aqueous ferrous iron and goethite. *Chem. Geol.* **2008**, *250*, 40–48.
38. Mikutta, C.; Wiederhold, J. G.; Cirpka, O. A.; Hofstetter, T. B.; Bourdon, B.; Von Gunten, U. Iron isotope fractionation and atom exchange during sorption of ferrous iron to mineral surfaces. *Geochim. Cosmochim. Acta* **2009**, *73*, 1795–1812.
39. Handler, R. M.; Beard, B. L.; Johnson, C. M.; Scherer, M. M. Atom exchange between aqueous Fe(II) and goethite: An Fe isotope tracer study. *Environ. Sci. Technol.* **2009**, *43*, 1102–1107.
40. Jones, A. M.; Collins, R. N.; Rose, J.; Waite, T. D. The effect of silica and natural organic matter on the Fe(II)-catalysed transformation and reactivity of Fe(III) minerals. *Geochim. Cosmochim. Acta* **2009**, *73*, 4409–4422.
41. Becker, U.; Rosso, K. M.; Hochella, M. F. The proximity effect on semiconducting mineral surfaces: A new aspect of mineral surface reactivity and surface complexation theory? *Geochim. Cosmochim. Acta* **2001**, *65*, 2641–2649.
42. Kerisit, S.; Rosso, K. M. Kinetic Monte Carlo model of charge transport in hematite (α-Fe_2O_3). *J. Chem. Phys.* **2007**, *127*, 124706.
43. Russell, B.; Payne, M.; Ciacchi, L. C. Density functional theory study of Fe(II) adsorption and oxidation on goethite surfaces. *Phys. Rev. B* **2009**, *79*, 165101.
44. Wigginton, N. S.; Rosso, K. M.; Stack, A. G.; Hochella, M. F. Long-range electron transfer across cytochrome-hematite (α-Fe_2O_3) interfaces. *J. Phys. Chem. C* **2009**, *113*, 2096–2103.
45. Rosso, K.; Zachara, J. M.; Fredrickson, J. K.; Gorby, Y. A.; Smith, S. C. Nonlocal bacterial electron transfer to hematite surfaces. *Geochim. Cosmochim. Acta* **2003**, *67*, 1081–1087.
46. Kerisit, S.; Rosso, K. M. Charge transfer in FeO: A combined molecular-dynamics and ab initio study. *J. Chem. Phys.* **2005**, *123*, 224712.
47. Kerisit, S.; Rosso, K. M. Computer simulation of electron transfer at hematite surfaces. *Geochim. Cosmochim. Acta* **2006**, *70*, 1888–1903.
48. Stimming, U.; Schultze, J. W. A semiconductor model of the passive layer on iron electrodes and its application to electrochemical reactions. *Electrochim. Acta* **1979**, *24*, 858–869.

49. Mulvaney, P.; Cooper, R.; Grieser, F.; Meisel, D. Charge trapping in the reductive dissolution of colloidal suspensions of iron(III) oxides. *Langmuir* **1988**, *4*, 1206–1211.
50. Mulvaney, P.; Swayambunathan, V.; Grieser, F.; Meisel, D. Dynamics of interfacial charge-transfer in iron(III) oxide colloids. *J. Phys. Chem.* **1988**, *92*, 6732–6740.
51. Tronc, E.; Belleville, P.; Jolivet, J. P.; Livage, J. Transformation of ferric hydroxide into spinel by Fe(II) adsorption. *Langmuir* **1992**, *8*, 313–319.
52. Tronc, E.; Jolivet, J.-P.; Lefebvre, J.; Massart, R. Ion adsorption and electron transfer in spinel-like iron oxide colloids. *J. Chem. Soc., Faraday Trans.* **1984**, *80*, 2619–2629.
53. Tronc, E.; Jolivet, J. P.; Belleville, P.; Livage, J. Redox phenomena in spinel iron-oxide colloids induced by adsorption. *Hyperfine Interact.* **1989**, *46*, 637–643.
54. Rosso, K. M.; Smith, D. M. A.; Dupuis, M. An *ab initio* model of electron transport in hematite (α-Fe_2O_3) basal planes. *J. Chem. Phys.* **2003**, *118*, 6455–6466.
55. Wang, J. W.; Rustad, J. R. A simple model for the effect of hydration on the distribution of ferrous iron at reduced hematite (012) surfaces. *Geochim. Cosmochim. Acta* **2006**, *70*, 5285–5292.
56. Ahrland, S.; Grenthe, I.; Noren, B. The ion exchange properties of silica gel .1. The sorption of Na^+, Ca^{2+}, Ba^{2+}, UO_2^{2+}, Gd^{3+}, Zr(IV) + NB, U(IV), and Pu(IV). *Acta Chem. Scand.* **1960**, *14*, 1059–1076.
57. Stanton, J.; Maatman, R. W. Reaction between aqueous uranyl ion and surface of silica gel. *J. Colloid Interface Sci.* **1963**, *18*, 132–146.
58. Ambe, S.; Iwamoto, M.; Maeda, H.; Ambe, F. Multitracer study on adsorption of metal ions on α-Fe_2O_3. *J. Radioanal. Nucl. Chem.* **1996**, *205*, 269–275.
59. Ambe, S.; Ambe, F. Mossbauer study of ferric ions adsorbed at alpha-ferric oxide/aqueous solution interface. *Langmuir* **1990**, *6*, 644–649.
60. Davis, J. A.; Hayes, K. F. Geochemical processes at mineral surfaces: An overview. In *Geochemical Processes at Mineral Surfaces*; Davis, J. A.; Hayes, K. F., Eds.; American Chemical Society: Washington, DC, 1986; Vol. 323, pp 2−18.
61. Westall, J. C.; Hohl, H. A comparison of electrostatic models for the oxide/solution interface. *Adv. Colloid Interface Sci.* **1980**, *12*, 265–294.
62. Schindler, P. W.; Stumm, W. The surface chemistry of oxides, hydroxides, and oxide minerals. In *Aquatic Surface Chemistry*; Stumm, W., Ed.; John Wiley & Sons: New York, 1987; pp 83−110.
63. Stumm, W.; Hohl, H.; Dalang, F. Interaction of metal-ions with hydrous oxide surfaces. *Croat. Chem. Acta* **1976**, *48*, 491–504.
64. Stumm, W. *Aquatic Surface Chemistry: Chemical Processes at the Particle-Water Interface*; Wiley: New York, 1987.
65. Schultz, T. W. Relative toxicity of para-substituted phenols: log k_{OW} and pK_a-depedent structure-activity relationships. *Bull. Environ. Contam. Toxicol.* **1987**, *38*, 994–999.

66. Bruemmer, G. W.; Gerth, J.; Tiller, K. G. Reaction kinetics of the adsorption and desorption of nickel, zinc and cadminum by goethite. I. Adsorption and diffusion of metals. *J. Soil Sci.* **1988**, *39*, 37–52.
67. Ainsworth, C. C.; Pilon, J. L.; Gassman, P. L.; Vandersluys, W. G. Cobalt, cadmium, and lead sorption to hydrous iron-oxide - Residence time effect. *Soil Sci. Soc. Am. J.* **1994**, *58*, 1615–1623.
68. Hsi, C. D.; Langmuir, D. Adsorption of uranyl onto ferric oxyhydroxides: Application of the surface complexation site-binding model. *Geochim. Cosmochim. Acta* **1985**, *49*, 1931–1941.
69. Murphy, R. J.; Lenhart, J. J.; Honeyman, B. D. The sorption of thorium (IV) and uranium (VI) to hematite in the presence of natural organic matter. *Colloids Surf., A* **1999**, *157*, 47–62.
70. Moyes, L. N.; Parkman, R. H.; Charnock, J. M.; Vaughan, D. J.; Livens, F. R.; Hughes, C. R.; Braithwaite, A. Uranium uptake from aqueous solution by interaction with goethite, lepidocrocite, muscovite, and mackinawite: An X-ray absorption spectroscopy study. *Environ. Sci. Technol.* **2000**, *34*, 1062–1068.
71. Duff, M. C.; Coughlin, J. U.; Hunter, D. B. Uranium co-precipitation with iron oxide minerals. *Geochim. Cosmochim. Acta* **2002**, *66*, 3533–3547.
72. Wakoff, B.; Nagy, K. L. Perrhenate uptake by iron and aluminum oxyhydroxides: An analogue for pertechnetate incorporation in Hanford waste tank sludges. *Environ. Sci. Technol.* **2004**, *38*, 1765–1771.
73. Farley, K. J.; Dzombak, D. A.; Morel, F. M. M. A surface precipitation model for the sorption of cations on metal oxides. *J. Colloid Interface Sci.* **1985**, *106*, 226–242.
74. Davis, J. A.; Kent, D. B. Surface complexation modeling in aqueous geochemistry. In *Mineral-Water Interface Geochemistry*; Hochella, J. M. F.; White, A. F., Eds.; Mineralogical Society of America: 1990; Vol. 23, pp 177−260.
75. Hayes, K. F.; Redden, G.; Ela, W.; Leckie, J. O. Surface complexation models: An evaluation of model parameter estimations using FITEQL and oxide mineral titration data. *J. Colloid Interface Sci.* **1991**, *142*, 448–469.
76. Huang, C. P.; Stumm, W. Specific adsorption of cations on hydrous γ-Al_2O_3. *J. Colloid Interface Sci.* **1973**, *43*, 409–420.
77. Yates, D. E.; Levine, S.; Healy, T. W. Site-binding model of electrical double-layer at oxide-water interface. *J. Chem. Soc., Faraday Trans. 1* **1974**, *70*, 1807–1818.
78. Hohl, H.; Stumm, W. Interaction of Pb^{2+} with hydrous γ-Al_2O_3. *J. Colloid Interface Sci.* **1976**, *55*, 281–288.
79. Sposito, G. On the surface complexation model of the oxide-aqueous interface. *J. Colloid Interface Sci.* **1983**, *91*, 329–340.
80. Westall, J. C. Reactions at the oxide-solution interface: Chemical and electrostatic models. In *Geochemical Processes at Mineral Surfaces*; Davis, J. A.; Hayes, K. F., Eds.; American Chemical Society: Washington, DC, 1986; Vol. 323, pp 54−78.
81. Langmuir, I. The adsorption of gases on plane surfaces of glass, mica and platinum. *J. Am. Chem. Soc.* **1918**, *40*, 1361–1403.

82. Piasecki, W. Theoretical description of the kinetics of proton adsorption at the oxide/electrolyte interface based on the statistical rate theory of interfacial transport and the 1pK model of surface charging. *Langmuir* **2003**, *19*, 9526–9533.
83. Liger, E.; Charlet, L.; Van Cappellen, P. Surface catalysis of uranium(VI) reduction by iron(II). *Geochim. Cosmochim. Acta* **1999**, *63*, 2939–2955.
84. Dixit, S.; Hering, J. G. Comparison of arsenic(V) and arsenic(III) sorption onto iron oxide minerals: implications for arsenic mobility. *Environ. Sci. Technol.* **2003**, *37*, 4182–4189.
85. Cowan, C. E.; Zachara, J. M.; Resch, C. T. Cadmium adsorption on iron-oxides in the presence of alkaline-earth elements. *Environ. Sci. Technol.* **1991**, *25*, 437–446.
86. Venema, P.; Hiemstra, T.; van Riemsdijk, W. H. Multisite adsorption of cadmium on goethite. *J. Colloid Interface Sci.* **1996**, *183*, 515–527.
87. Arai, Y.; Sparks, D. L. ATR-FTIR spectroscopic investigation on phosphate adsorption mechanisms at the ferrihydrite-water interface. *J. Colloid Interface Sci.* **2001**, *241*, 317–326.
88. Dzombak, D. A.; Morel, F. M. M. Sorption of cadmium on hydrous ferric-oxide at high sorbate/sorbent ratios - Equilibrium, kinetics, and modeling. *J. Colloid Interface Sci.* **1986**, *112*, 588–598.
89. Evans, L. J. Chemistry of metal retention by soils - Several processes are explained. *Environ. Sci. Technol.* **1989**, *23*, 1046–1056.
90. Zhang, Y.; Charlet, L.; Schindler, P. W. Adsorption of protons, Fe(II) and Al(III) on lepidocrocite (γ-FeOOH). *Colloids Surf.* **1992**, *63*, 259–268.
91. Rose, A. W.; Bianchi-Mosquera, G. C. Adsorption of Cu, Pb, Zn, Co, Ni, and Ag on goethite and hematite; a control on metal mobilization from red beds into stratiform copper deposits. *Econ. Geol.* **1993**, *88*, 1226–1236.
92. Misawa, T.; Asami, K.; Hashimoto, K.; Shimodiara, S. The mechanism of atmospheric rusting and the protective amorphous rust on low alloy steel. *Corros. Sci.* **1974**, *14*, 279–289.
93. Tamaura, Y.; Ito, K.; Katsura, T. Transformation of γ-FeO(OH) to Fe_3O_4 by adsorption of iron(II) ion on γ-FeO(OH). *J. Chem. Soc., Dalton Trans.* **1983**, *2*, 189–194.
94. Lovley, D. R.; Stolz, J. F.; Gordon, L. N.; Phillips, E. J. P. Anaerobic production of magnetite by dissimilatory iron-reducing microarganism. *Nature* **1987**, *330*, 252–254.
95. Lovley, D. R.; Baedecker, M. J.; Lonergan, D. J; Cozzarelli, I. M.; Phillips, E. J. P.; Siegel, D. I. Oxidation of aromatic contaminants coupled to microbial iron reduction. *Nature* **1989**, *339*, 297–300.
96. Nealson, K. H.; Myers, C. R. Microbial reduction of manganese and iron: New approaches to carbon cycling. *Appl. Environ. Microbiol.* **1992**, *58*, 439–443.
97. Pennisi, E. Geobiologists: As diverse as the bugs they study. *Science* **2002**, *296*, 1058–1060.
98. Cui, D. Q.; Eriksen, T. E. Reduction of pertechnetate by ferrous iron in solution: Influence of sorbed and precipitated Fe(II). *Environ. Sci. Technol.* **1996**, *30*, 2259–2262.

99. Amonette, J. E.; Workman, D. J.; Kennedy, D. W.; Fruchter, J. S.; Gorby, Y. A. Dechlorination of carbon tetrachloride by Fe(II) associated with goethite. *Environ. Sci. Technol.* **2000**, *34*, 4606–4613.
100. Gregory, K. B.; Larese-Casanova, P.; Parkin, G. F.; Scherer, M. M. Abiotic transformation of hexahydro-1,3,5-trinitro-1,3,5-triazine by Fe^{II} bound to magnetite. *Environ. Sci. Technol.* **2004**, *38*, 1408–1414.
101. Stumm, W.; Sulzberger, B. The cycling of iron in natural environments: Considerations based on laboratory studies of heterogeneous redox processes. *Geochim. Cosmochim. Acta* **1992**, *56*, 3233–3257.
102. Charlet, L.; Silvester, E.; Liger, E. N-compound reduction and actinide immobilization in surficial fluids by Fe(II): the surface Fe(III)OFe(II)OH degrees species, as major reductant. *Chem. Geol.* **1998**, *151*, 85–93.
103. Jeon, B. H.; Dempsey, B. A.; Burgos, W. D. Kinetics and mechanisms for reactions of Fe(II) with iron(III) oxides. *Environ. Sci. Technol.* **2003**, *37*, 3309–3315.
104. Jeon, B. H.; Dempsey, B. A.; Burgos, W. D.; Royer, R. A. Sorption kinetics of Fe(II), Zn(II), Co(II), Ni(II), Cd(II), and Fe(II)/Me(II) onto hematite. *Water Res.* **2003**, *37*, 4135–4142.
105. Jeon, B. H.; Dempsey, B. A.; Burgos, W. D.; Royer, R. A. Reactions of ferrous iron with hematite. *Colloids Surf., A* **2001**, *191*, 41–55.
106. Khader, M. M.; Vurens, G. H.; Kim, I. K.; Salmeron, M.; Somorjai, G. A. Photoassisted catalytic dissociation of H_2O to produce hydrogen on partially reduced α-Fe_2O_3. *J. Am. Chem. Soc.* **1987**, *109*, 3581–3585.
107. Quinn, R. K.; Nasby, R. D.; Baughman, R. J. Photoassisted electrolysis of water using single crystal α-Fe_2O_3 anodes. *Mater. Res. Bull.* **1976**, *11*, 1011–1018.
108. Dare-Edwards, M. P.; Goodenough, J. B.; Hamnett, A.; Trevellick, P. R. Electrochemistry and photoelectrochemistry of iron(III) oxide. *J. Chem. Soc., Faraday Trans. 1* **1983**, *79*, 2027–41.
109. Sulzberger, B.; Laubscher, H.-U.; Karametaxas, G. Photoredox reactions at the surface of iron(III) (hydro)oxides. In *Aquatic and Surface Photochemistry*; Helz, G. R., Zepp, R. G., Crosby, D. G., Eds.; Lewis: Boca Raton, FL, 1994; pp 53−73.
110. Sulzberger, B.; Laubscher, H. Photochemical reductive dissolution of lepidocrocite - Effect of pH. In *Aquatic Chemistry - Interfacial and Interspecies Processes*; Huang, C. P., Omelia, C. R., Morgan, J. J., Eds.; American Chemical Society: Washington, DC, 1995; Vol. 244, pp 279−290.
111. Faust, B. C.; Hoffmann, M. R. Photoinduced reductive dissolution of α-Fe_2O_3 by bisulfite. *Environ. Sci. Technol.* **1986**, *20*, 943–948.
112. Jaegermann, W.; Tributsch, H. Interfacial properties of semiconducting transition-metal chalcogenides. *Prog. Surf. Sci.* **1988**, *29*, 1–167.
113. Hoffmann, M. R.; Martin, S. T.; Choi, W.; Bahnemann, D. W. Environmental applications of semiconductor photocatalysis. *Chem. Rev.* **1995**, *95*, 69–96.
114. Amonette, J. E. Iron redox chemistry of clays and oxides: Environmental applications. In *Electrochemical Properties of Clays*; Clay Mineral Society: Aurora, CO, 2002; Vol. 10, pp 89−146.

115. Gorski, C. A.; Scherer, M. M. Influence of magnetite stoichiometry on Fe^{II} uptake and nitrobenzene reduction. *Environ. Sci. Technol.* **2009**, *43*, 3675–3680.
116. Larese-Casanova, P.; Scherer, M. M. Fe(II) sorption on hematite: New insights based on spectroscopic measurements. *Environ. Sci. Technol.* **2007**, *41*, 471–477.
117. Cwiertny, D. M.; Handler, R. M.; Schaefer, M. V.; Grassian, V. H.; Scherer, M. M. Interpreting nanoscale size-effects in aggregated Fe-oxide suspensions: Reaction of Fe(II) with goethite. *Geochim. Cosmochim. Acta* **2008**, *72*, 1365–1380.
118. Schaefer, M. V.; Gorski, C. A.; Scherer, M. M. Spectroscopic evidence for interfacial Fe(II)-Fe(III) electron transfer in a clay mineral. *Environ. Sci. Technol.* **2010**, *45*, 540–545.
119. Amthauer, G.; Groczicki, M.; Lottermoser, W.; Redhammer, G. Mössbauer spectroscopy: Basic principles. In *Spectroscopic Methods in Mineralogy*; Beran, A., Libowitzky, E., Eds.; 2004; Vol. 6, pp 345−367.
120. Dyar, M. D.; Agresti, D. G.; Schaefer, M. W.; Grant, C. A.; Sklute, E. C. Mossbauer spectroscopy of earth and planetary elements. *Annu. Rev. Earth Planet. Sci.* **2006**, *34*, 83–125.
121. Rancourt, D. G. Mossbauer spectroscopy in clay science. *Hyperfine Interact.* **1998**, *117*, 3–38.
122. Murad, E.; Cashion, J. *Mossbauer Spectroscopy of Environmental Materials and their Industrial Utilization*; Kluwer Academic Publishers: 2004.
123. Murad, E.; Cashion, J. *Mossbauer Spectroscopy of Environmental Materials and their Industrial Utilization*; Kluwer: Dordrecht, 2006.
124. Gorski, C. A.; Scherer, M. M. Determination of nanoparticulate magnetite stoichiometry by Mossbauer spectroscopy, acidic dissolution, and powder X-ray diffraction: A critical review. *Am. Mineral.* **2010**, *95*, 1017–1026.
125. Silvester, E.; Charlet, L.; Tournassat, C.; Gehin, A.; Greneche, J. M.; Liger, E. Redox potential measurements and Mössbauer spectrometry of Fe(II) adsorbed onto Fe(III) (oxyhydr)oxides. *Geochim. Cosmochim. Acta* **2005**, *69*, 4801–4815.
126. Condon, N. G.; Murray, P. W.; Leibsle, F. M.; Thornton, G.; Lennie, A. R.; Vaughan, D. J. Fe_3O_4(111) termination of α-Fe_2O_3(0001). *Surf. Sci.* **1994**, *310*, L609–L613.
127. Eggleston, C. M.; Hochella, M. F. The structure of hematite (001) surfaces by scanning tunneling microscopy - Image interpretation, surface relaxation, and step structure. *Am. Mineral.* **1992**, *77*, 911–922.
128. Eggleston, C. M.; Stack, A. G.; Rosso, K. M.; Higgins, S. R.; Bice, A. M.; Boese, S. W.; Pribyl, R. D.; Nichols, J. J. The structure of hematite (α-Fe_2O_3) (001) surfaces in aqueous media: Scanning tunneling microscopy and resonant tunneling calculations of coexisting O and Fe terminations. *Geochim. Cosmochim. Acta* **2003**, *67*, 985–1000.
129. Becker, U.; Hochella, M. F.; Apra, E. The electronic structure of hematite{001} surfaces: Applications to the interpretation of STM images and heterogeneous surface reactions. *Am. Mineral.* **1996**, *81*, 1301–1314.

130. Kennedy, J. H.; Frese, K. W. Flatband potentials and donor densities of polycrystalline α-Fe_2O_3 determined from Mott-Schottky plots. *J. Electrochem. Soc.* **1978**, *125*, 723–726.
131. Mulvaney, P.; Grieser, F.; Meisel, D. Redox reactions on colloidal metals and metal oxides. In *Kinetics and catalysis in microheterogeneous systems*; Taylor & Francis, Inc.: 1991; Vol. 38, pp 303−373.
132. Mulvaney, P.; Swayambunathan, V.; Grieser, F.; Meisel, D. Effect of the zeta-potential on electron-transfer to colloidal iron-oxides. *Langmuir* **1990**, *6*, 555–559.
133. Jolivet, J. P.; Belleville, P.; Tronc, E.; Livage, J. Influence of Fe(II) on the formation of the spinel iron oxide in alkaline medium. *Clays Clay Miner.* **1992**, *40*, 531–539.
134. Larese-Casanova, P.; Scherer, M. M. Morin transition suppression in polycrystalline ^{57}hematite (α-Fe_2O_3) exposed to ^{56}Fe(II). *Hyperfine Interact.* **2007**, *174*, 111–119.
135. Siffert, C.; Sulzberger, B. Light-induced dissolution of hematite in the presence of oxalate - A case-study. *Langmuir* **1991**, *7*, 1627–1634.
136. Postma, D. The reactivity of iron-oxides in sediments - A kinetic approach. *Geochim. Cosmochim. Acta* **1993**, *57*, 5027–5034.
137. Beard, B. L.; Johnson, C. M. High precision iron isotope measurements of terrestrial and lunar materials. *Geochim. Cosmochim. Acta* **1999**, *63*, 1653–1660.
138. Van Ammel, R.; Pomme, S.; Sibbens, G. Half-life measurement of ^{55}Fe. *Appl. Radiat. Isot.* **2006**, *64*, 1412–1416.
139. Beard, B. L.; Johnson, C. M.; Skulan, J. L.; Nealson, K. H.; Cox, L.; Sun, H. Application of Fe isotopes to tracing the geochemical and biological cycling of Fe. *Chem. Geol.* **2003**, *195*, 87–117.
140. Beard, B. L.; Johnson, C. M. Fe isotope variations in the modern and ancient earth and other planetary bodies. In *Reviews in Mineralogy and Geochemistry: Geochemistry of Non-traditional Stable Isotopes*; Johnson, C. M., Beard, B. L., Albarede, F., Eds.; The Mineralogical Society of America: Washington, DC, 2004; Vol. 55, pp 319−357.
141. Johnson, C. M.; Beard, B. L.; Albarede, F. Overview and general concepts. In *Geochemistry of Non-Traditional Stable Isotopes*; 2004; Vol. 55, pp 1−24.
142. Rea, B. A.; Davis, J. A.; Waychunas, G. A. Studies of the reactivity of the ferrihydrite surface by iron isotopic exchange and Mossbauer-spectroscopy. *Clays Clay Miner.* **1994**, *42*, 23–34.
143. Icopini, G. A.; Anbar, A. D.; Ruebush, S. S.; Tien, M.; Brantley, S. L. Iron isotope fractionation during microbial reduction of iron: The importance of adsorption. *Geology* **2004**, *32*, 205–208.
144. Teutsch, N.; von Gunten, U.; Porcelli, D.; Cirpka, O. A.; Halliday, A. N. Adsorption as a cause for iron isotope fractionation in reduced groundwater. *Geochim. Cosmochim. Acta* **2005**, *69*, 4175–4185.
145. Johnson, C. M.; Beard, B. L.; Klein, C.; Beukes, N. J.; Roden, E. E. Iron isotopes constrain biologic and abiologic processes in banded iron formation genesis. *Geochim. Cosmochim. Acta* **2008**, *72*, 151–169.

146. Wu, L. L.; Beard, B. L.; Roden, E. E.; Kennedy, C. B.; Johnson, C. M. Stable Fe isotope fractionations produced by aqueous Fe(II)-hematite surface interactions. *Geochim. Cosmochim. Acta* **2010**, *74*, 4249–4265.
147. Crosby, H. A.; Johnson, C. M.; Roden, E. E.; Beard, B. L. Coupled Fe(II)-Fe(III) electron and atom exchange as a mechanism for Fe isotope fractionation during dissimilatory iron oxide reduction. *Environ. Sci. Technol.* **2005**, *39*, 6698–6704.
148. Wiederhold, J. G.; Teutsch, N.; Kraemer, S. M.; Halliday, A. N.; Kretzschmar, R. Iron isotope fractionation in oxic soils by mineral weathering and podzolization. *Geochim. Cosmochim. Acta* **2007**, *71*, 5821–5833.
149. Schwertmann, U.; Thalmann, H. Influence of Fe(II) , Si , and pH on formation of lepidocrocite and ferrihydrite during oxidation of aqueous $FeCl_2$ solutions. *Clay Miner.* **1976**, *11*, 189–200.
150. Cornell, R. M.; Schwertmann, U. *The iron oxides: Structure, properties, reactions, occurrence, and uses*; VCH: New York, 2003.
151. Wiesli, R. A.; Beard, B. L.; Johnson, C. M. Experimental determination of Fe isotope fractionation between aqueous Fe(II), siderite and "green rust" in abiotic systems. *Chem. Geol.* **2004**, *211*, 343–362.
152. Johnson, C. M.; Roden, E. E.; Welch, S. A.; Beard, B. L. Experimental constraints on Fe isotope fractionation during magnetite and Fe carbonate formation coupled to dissimilatory hydrous ferric oxide reduction. *Geochim. Cosmochim. Acta* **2005**, *69*, 963–993.
153. Beard, B. L.; Handler, R. M.; Scherer, M. M.; Wu, L. L.; Czaja, A. D.; Heimann, A.; Johnson, C. M. Iron isotope fractionation between aqueous ferrous iron and goethite. *Earth Planet. Sci. Lett.* **2010**, *295*, 241–250.
154. Rosso, K. M.; Yanina, S. V.; Gorski, C. A.; Larese-Casanova, P.; Scherer, M. M. Connecting observations of hematite (a-Fe_2O_3) growth catalyzed by Fe(II). *Environ. Sci. Technol.* **2010**, *44*, 61–67.
155. Gerischer, H. Electrochemical photo and solar cells - Principles and some experiments. *J. Electroanal. Chem.* **1975**, *58*, 263–274.
156. Duonghong, D.; Ramsden, J.; Gratzel, M. Dynamics of interfacial electron-transfer processes in colloidal semiconductor systems. *J. Am. Chem. Soc.* **1982**, *104*, 2977–2985.
157. Ward, M. D.; White, J. R.; Bard, A. J. Electrochemical investigation of the energetics of particulate titanium-dioxide photocatalysts - The methyl viologen acetate system. *J. Am. Chem. Soc.* **1983**, *105*, 27–31.
158. Gratzel, M. Photoelectrochemical cells. *Nature* **2001**, *414*, 338–344.
159. Wood, A.; Giersig, M.; Mulvaney, P. Fermi level equilibration in quantum dot-metal nanojunctions. *J. Phys. Chem. B* **2001**, *105*, 8810–8815.
160. Dimitrijevíc, N. M.; Savíc, D.; Mícíc, O. I.; Nozik, A. J. Interfacial electron-transfer equilibria and flat-band potentials of α-Fe_2O_3 and TiO_2 colloids studied by pulse radiolysis. *J. Phys. Chem.* **1984**, *88*, 4278–4283.
161. Nozik, A. J.; Memming, R. Physical chemistry of semiconductor-liquid interfaces. *J. Phys. Chem.* **1996**, *100*, 13061–13078.
162. Nozik, A. J. Photochemical diodes. *Appl. Phys. Lett.* **1977**, *30*, 567–569.

163. Buxton, G. V.; Rhodes, T.; Sellers, R. M. Radiation-induced dissolution of colloidal haematite. *Nature* **1982**, *295*, 583–585.
164. Gratzel, M.; Frank, A. J. Interfacial electron-transfer reactions in colloidal semiconductor dispersions - kinetic-analysis. *J. Phys. Chem.* **1982**, *86*, 2964–2967.
165. Kiwi, J.; Gratzel, M. Protection, size factors, and reaction dynamics of colloidal redox catalysts mediating light-induced hydrogen evolution from water. *J. Am. Chem. Soc.* **1979**, *101*, 7214–7217.
166. Koval, C. A.; Howard, J. N. Electron-transfer at semiconductor electrode liquid electrolyte interfaces. *Chem. Rev.* **1992**, *92*, 411–433.
167. Rakeshwar, K. Fundamentals of semiconductor electrochemistry and photoelectrochemistry. In *Encyclopedia of Electrochemistry, Volume 6, Semiconductor Electrodes and Photoelectrochemistry*; Bard, A. J., Stratmann, M., Licht, S., Eds.; Wiley-VCH: 2002; Vol. 6, pp 1−51.
168. Cesar, I.; Kay, A.; Martinez, J. A. G.; Gratzel, M. Translucent thin film Fe_2O_3 photoanodes for efficient water splitting by sunlight: Nanostructure-directing effect of Si-doping. *J. Am. Chem. Soc.* **2006**, *128*, 4582–4583.
169. Kay, A.; Cesar, I.; Gratzel, M. New benchmark for water photooxidation by nanostructured α-Fe_2O_3 films. *J. Am. Chem. Soc.* **2006**, *128*, 15714–15721.
170. Sartoretti, C. J.; Alexander, B. D.; Solarska, R.; Rutkowska, W. A.; Augustynski, J.; Cerny, R. Photoelectrochemical oxidation of water at transparent ferric oxide film electrodes. *J. Phys. Chem. B* **2005**, *109*, 13685–13692.
171. Hu, Y. S.; Kleiman-Shwarsctein, A.; Forman, A. J.; Hazen, D.; Park, J. N.; McFarland, E. W. Pt-doped α-Fe_2O_3 thin films active for photoelectrochemical water splitting. *Chem. Mater.* **2008**, *20*, 3803–3805.
172. Balko, B. A.; Clarkson, K. M. The effect of doping with Ti(IV) and Sn(IV) on oxygen reduction at hematite electrodes. *J. Electrochem. Soc.* **2001**, *148*, E85–E91.
173. Bard, A. J.; Bocarsly, A. B.; Fan, F. R. F.; Walton, E. G.; Wrighton, M. S. The concept of fermi level pinning at semiconductor-liquid junctions - Consequences for energy-conversion efficiency and selection of useful solution redox couples in solar devices. *J. Am. Chem. Soc.* **1980**, *102*, 3671–3677.
174. Hobbs, C. C.; Fonseca, L. R. C.; Knizhnik, A.; Dhandapani, V.; Samavedam, S. B.; Taylor, W. J.; Grant, J. M.; Dip, L. G.; Triyoso, D. H.; Hegde, R. I.; Gilmer, D. C.; Garcia, R.; Roan, D.; Lovejoy, M. L.; Rai, R. S.; Hebert, E. A.; Tseng, H. H.; Anderson, S. G. H.; White, B. E.; Tobin, P. J. Fermi-level pinning at the polysilicon/metal oxide interface - Part I. *IEEE Trans. Electron Devices* **2004**, *51*, 971–977.
175. Tai, Y. L.; Dempsey, B. A. Nitrite reduction with hydrous ferric oxide and Fe(II): Stoichiometry, rate, and mechanism. *Water Res.* **2009**, *43*, 546–552.
176. Park, B.; Dempsey, B. A. Heterogeneous oxidation of Fe(II) on ferric oxide at neutral pH and a low partial pressure of O_2. *Environ. Sci. Technol.* **2005**, *39*, 6494–6500.

177. Jang, J. H.; Brantley, S. L. Investigation of wustite (FeO) dissolution: Implications for reductive dissolution of ferric oxides. *Environ. Sci. Technol.* **2009**, *43*, 1086–1090.
178. Ford, R. G.; Bertsch, P. M.; Farley, K. J. Changes in transition and heavy metal partitioning during hydrous iron oxide aging. *Environ. Sci. Technol.* **1997**, *31*, 2028–2033.
179. Nico, P. S.; Stewart, B. D.; Fendorf, S. Incorporation of oxidized uranium into Fe (hydr)oxides during Fe(II) catalyzed remineralization. *Environ. Sci. Technol.* **2009**, *43*, 7391–7396.
180. Boland, D. D.; Collins, R. N.; Payne, T. E.; Waite, T. D. Effect of amorphous Fe(III) oxide transformation on the Fe(II)-mediated reduction of U(VI). *Environ. Sci. Technol.* **2011**, *45*, 1327–1333.
181. Ilton, E. S.; Boily, J. F.; Buck, E. C.; Skomurski, F. N.; Rosso, K. M.; Cahill, C. L.; Bargar, J. R.; Felmy, A. R. Influence of dynamical conditions on the reduction of U-VI at the magnetite-solution interface. *Environ. Sci. Technol.* **2010**, *44*, 170–176.
182. Cwiertny, D. M.; Hunter, G. J.; Pettibone, J. M.; Scherer, M. M.; Grassian, V. H. Surface chemistry and dissolution of α-FeOOH nanorods and microrods: Environmental implications of size-dependent interactions with oxalate. *J. Phys. Chem. C* **2009**, *113*, 2175–2186.
183. Anschutz, A. J.; Penn, R. L. Reduction of crystalline iron(III) oxyhydroxides using hydroquinone: Influence of phase and particle size. *Geochem. Trans.* **2005**, *6*, 60–66.
184. Vikesland, P. J.; Heathcock, A. M.; Rebodos, R. L.; Makus, K. E. Particle size and aggregation effects on magnetite reactivity toward carbon tetrachloride. *Environ. Sci. Technol.* **2007**, *41*, 5277–5283.
185. Mayo, J. T.; Yavuz, C.; Yean, S.; Cong, L.; Shipley, H.; Yu, W.; Falkner, J.; Kan, A.; Tomson, M.; Colvin, V. L. The effect of nanocrystalline magnetite size on arsenic removal. *Sci. Technol. Adv. Mater.* **2007**, *8*, 71–75.
186. Madden, A. S.; Hochella, M. F. A test of geochemical reactivity as a function of mineral size: Manganese oxidation promoted by hematite nanoparticles. *Geochim. Cosmochim. Acta* **2005**, *69*, 389–398.
187. Leygraf, C.; Hendewerk, M.; Somorjal, G. A. Photocatalytic production of hydrogen from water by a p-type and n-type polycrystalline iron-oxide assembly. *J. Phys. Chem.* **1982**, *86*, 4484–4485.
188. Turner, J. E.; Hendewerk, M.; Somorjai, G. A. The photodissociation of water by doped iron-oxides - The unbiased p/n assembly. *Chem. Phys. Lett.* **1984**, *105*, 581–585.
189. Sivula, K.; Zboril, R.; Le Formal, F.; Robert, R.; Weidenkaff, A.; Tucek, J.; Frydrych, J.; Gratzel, M. Photoelectrochemical water splitting with mesoporous hematite prepared by a solution-based colloidal approach. *J. Am. Chem. Soc.* **2010**, *132*, 7436–7444.
190. Eggleston, C. M. Toward new uses for hematite. *Science* **2008**, *320*, 184–185.

Chapter 16

Redox Driven Stable Isotope Fractionation

Jay R. Black,[1,*] Jeffrey A. Crawford,[1] Seth John,[3] and Abby Kavner[1,2]

[1]Institute for Geophysics and Planetary Physics, [2]Earth and Space Sciences, University of California, Los Angeles, 595 Charles E. Young Drive East, Los Angeles, CA 90095
[3]Department of Earth and Ocean Sciences, University of South Carolina, Columbia, SC 29208
[*]jayblack@ucla.edu

Electrochemical reduction/oxidation (redox) reactions have been observed to drive stable isotope fractionation in many metal systems, where the lighter isotopes of a metal are typically partitioned into the reduced chemical species. While physical processes such as diffusive mass transport lead to small isotope fractionations, charge transfer processes can lead to isotope fractionations up to ten times larger and twice that predicted by equilibrium stable isotope theory. Control over physical and chemical variables during electrochemical experiments, such as overpotential and temperature, allow for the isotopic composition of deposited metals to be fine tuned to a specific value.

Introduction

Redox reactions drive many chemical transformations in the environment, and are vital in biological cycles for energy production. Many elements are redox sensitive, with multiple accessible oxidation states stable in a variety of environments. The transition metal elements are of particular interest, as some of the largest stable isotope fractionations seen in these metal systems are associated with redox transformations. Therefore, stable isotope signatures may be used for a wide range of applications, from studying the transport and deposition of metals in the environment for the purposes of monitoring and remediation of contaminants, to reconstructing the chemistry of our planet in the past. In order to do this, an understanding of what drives stable isotope fractionation during a redox reaction is needed.

The history behind the discovery of deuterium is a good starting point, as it was after its discovery by Harold Urey (*1*) that the first electrolytically produced sample of heavy water was reported by Gilbert N. Lewis and Ronald T. MacDonald (*2*) who concentrated 20 liters of normal abundance water down to 0.5 ml of 66% deuterium through electrolysis. The electrochemical separation of stable isotopes subsequently became one of the earliest industrialized forms of stable isotope purification, with pure deuterium being produced using electrolytic cells in cascade reactors (*3–5*). The fractionation factor of industrial electrolytic cells varied as a function of the cell geometry and physical parameters, such as temperature and overpotential used, from $\alpha_{liquid\text{-}vapor} = 3$ to 13 (*4, 5*) (where $\alpha = D/H_{liquid}/D/H_{vapor}$). The electrochemical separation of ^{6}Li and ^{7}Li was later developed showing smaller fractionation factors of approximately $\alpha_{aqueous\text{-}metal} = 1.020\text{-}1.079$, where $\alpha = {}^{7}Li/{}^{6}Li_{aqueous}/{}^{7}Li/{}^{6}Li_{metal}$ (*6, 7*). The accepted theory to explain these fractionations was that they were due to a difference in the equilibrium potential for the reduction of one isotopologue species over another, leading to a 'kinetic isotope effect' far from equilibrium (*5*). Quantum-mechanical tunneling may also play a role in the case of hydrogen fractionations (*8*), altering the reaction coordinate for isotopologue species. Interest in separating lithium isotopes for the nuclear fuel industry has sparked recent interest in using electrochemical methods for the task (*9–13*). Recent experiments by Kavner and co-workers (*14–19*) have developed our understanding of the physical and chemical controls upon isotope fractionation during electrochemical experiments and form the basis for discussion in this chapter.

Figure 1 illustrates the electric double layer at an electrode surface and the physical and chemical processes that may contribute to the fractionation of stable isotopes during an electrochemical reaction. These include equilibria between different chemical species in solution or at the surface of the electrode, diffusion of reactant to the electrode surface, and finally the charge transfer kinetics at the electrode surface. Stable isotope ratios are reported here in a delta notation (Figure 1), with units of permil (‰) expressing the difference in isotopic ratio of a sample to a standard:

$$\delta^{X/Y}M = \left(\frac{{}^{X/Y}M_{sample}}{{}^{X/Y}M_{standard}} - 1\right) \cdot 1000 \tag{1}$$

$$\Delta^{X/Y}M_{A\text{-}B} = \delta^{X/Y}M_{A} - \delta^{X/Y}M_{B}$$

where $^{X/Y}M$ is often reported as the ratio of a heavy isotope over a light isotope, (e.g., $^{56/54}Fe$, $^{66/64}Zn$ and $^{7/6}Li$) therefore, positive and negative $\delta^{X/Y}M$ indicate a sample which contains more of the heavy and light isotope of an element relative to a standard, respectively. The capital delta notation reflects the isotopic difference (in permil) between two substances, A and B.

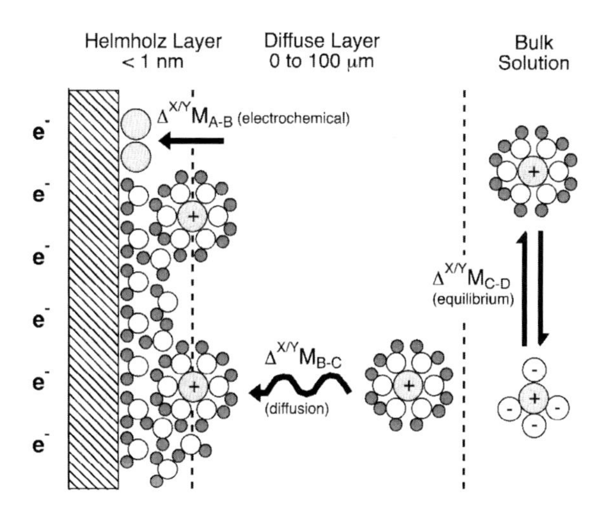

Figure 1. Schematic of the electric double layer adjacent to the surface of an electrode. The processes that may contribute to the overall isotopic signature in the plated (substance A) metal are labeled next to the arrows (19). Substance B = surface adsorbed reactant species; C = diffusing reactant species; D = secondary bulk solution species.

Combining Marcus Theory with Equilibrium Stable Isotope Theory

In a series of papers investigating iron and zinc stable isotope fractionation during electrodeposition Kavner et al. (*14*, *15*) derived an equation which combined equilibrium stable isotope theory (*20–22*) with Marcus's statistical mechanic description of charge transfer processes at an electrode (*23*):

$$\ln\alpha_{metal-aqueous} = \ln\left(\frac{k'}{k}\right) = \ln\left(\frac{\nu'}{\nu}\right) + \left(\frac{\Delta G^{*} - \Delta G^{*\prime}}{k_B T}\right)$$

$$= \frac{1}{2}\ln\left(\frac{m}{m'}\right) + \left(\frac{\ln\alpha_{eq}}{2} - \frac{k_B T(\ln\alpha_{eq})^2 + 2k_B T\ln(Q_P/Q_R)\ln\alpha_{eq} + 2\eta ze\ln\alpha_{eq}}{4\lambda}\right) \quad (2)$$

where isotopic fractionation ($\ln\alpha_{metal-aqueous}$) depends on the charge transfer rate (k) for metal deposition ($^{light}M^{z+}_{(aq)} + ze^- = {}^{light}M_{(s)}$), collision frequency ($\nu$), activation free energy (ΔG^*), Boltzmann's constant (k_B), temperature (T), mass in motion (m), equilibrium fractionation factor (α_{eq}), partition function ratio of abundant isotopologues of product and reactant (Q_P/Q_R), number of electrons (z), charge of an electron (e), and Marcus reorganization energy (λ) (*9*). The primed symbols are for the corresponding heavy isotopologue charge transfer reaction ($^{heavy}M^{z+}_{(aq)} + ze^- = {}^{heavy}M_{(s)}$).

Equation 2 makes a number of testable predictions about how stable isotopes are fractionated during a charge transfer reaction:

- Fractionation is linearly dependent upon applied voltage (overpotential = $\eta = E-E^0$) or, in other words, the driving force or extent of disequilibrium.
- The magnitude and slope of the fractionation scales with the equilibrium fractionation factor between product and reactant divided by the Marcus reorganization energy.
- Fractionation decreases with increasing temperature, primarily controlled by the change in magnitude of the equilibrium fractionation factor, which scales proportionally with $1/T^2$ under most circumstances (*24*), but not in all cases (*25*).

To test these predictions, electrochemical experiments were conducted where a number of the variables of interest could be controlled including the solution chemistry, temperature, and applied overpotential. Other properties such as the effect of mass transport of reactant to the electrode surface were tested by the use of planar electrodes in unstirred solutions compared with controlled hydrodynamic conditions using a rotating disc electrode. Here we present and summarize the results of previous electrochemical experiments (see following section) and the results of new experiments where copper is electroplated on rotating disc electrodes as a function of temperature, overpotential and electrode rotation rate.

Prior Experimental Results

Results from previously published electrodeposition experiments where metal was deposited on planar electrodes are shown in Figure 2, which plots results for $\Delta^{66/64}Zn_{metal\text{-}aqueous}$ and $\Delta^{56/54}Fe_{metal\text{-}aqueous}$ on the left hand axis and $\Delta^{7/6}Li_{metal\text{-}solution}$ on the right hand axis versus the applied overpotential. The following observations and interpretations can be made:

- In all cases the lighter isotope is preferentially partitioned into the deposited metal, with much larger fractionation factors seen for lithium compared to iron and zinc.
- The observed electrochemical fractionations for iron and zinc are much larger than calculated equilibrium fractionation factors ($\Delta^{56/54}Fe_{metal\text{-}aqueous}$ = -0.9 ‰ (*26*) and $\Delta^{66/64}Zn_{metal\text{-}aqueous}$ = -2.5 ‰ (*27*) at 25°C), especially at lower deposition rates near zero overpotential (Fig. 2). For lithium, fractionation approaches its equilibrium value ($\Delta^{7/6}Li_{metal\text{-}solvated}$ = -29.6 ‰ at room temperature (*28*)) near zero overpotential where fractionation should be controlled by equilibrium processes (Fig. 2). Equilibrium stable isotope theory predicts that fractionation should scale roughly with the difference in mass over the product of the mass of the two isotopes ($\Delta m/m^2$, (*24*)). Therefore it makes sense that the fractionation in lithium deposits is larger on an absolute scale than the Fe and Zn fractionations. However, when the results shown in Figure 2 are normalized to this ratio ($\Delta m/m^2$), the lithium fractionations, which do not exceed the predicted equilibrium

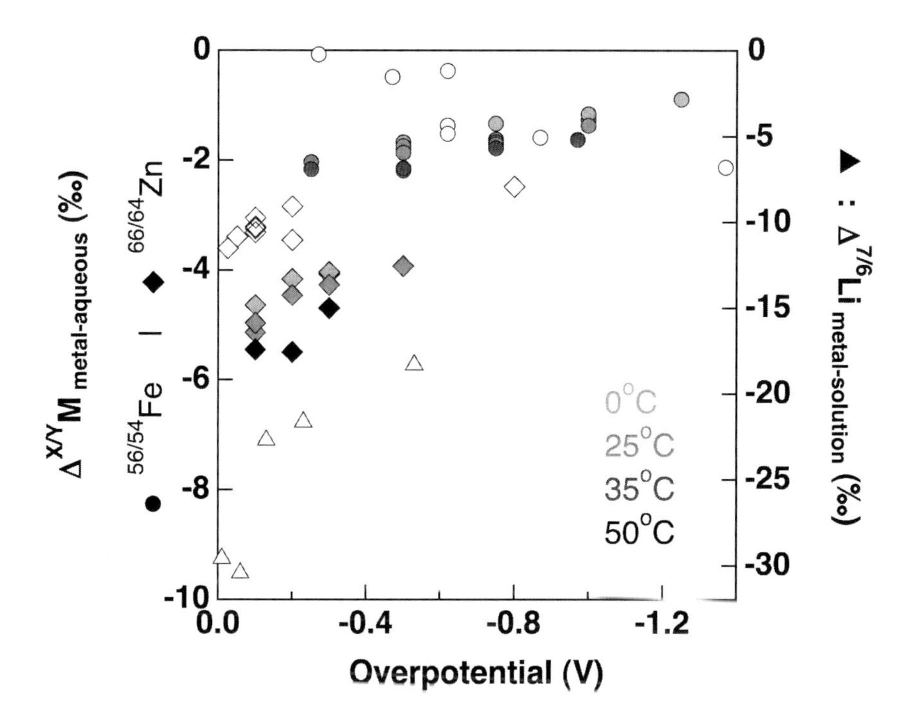

Figure 2. Stable isotope composition of various metals electrodeposited on planar electrodes including Zn (diamonds) (15, 19), Fe (circles) (14, 19) and Li (triangles) (17). Hollow symbols indicate experiments that were not temperature controlled. Filled symbols are temperature controlled experiments.

value (*28*), show smaller mass-relative electrochemical effects than do zinc and iron, which do exceed their equilibrium values.

- When examining the iron and zinc results side by side (Fig. 2) larger fractionations are seen in zinc metal deposits. This result is partly explained by the solvation structure of the metals in solution, as calculated equilibrium fractionation factors for the hexaquo-iron and -zinc complexes (predominant forms of aqueous iron and zinc) relative to metal are $\Delta^{56/54}Fe_{metal\text{-}aqueous}$ = -0.9 ‰ (*26*) and $\Delta^{66/64}Zn_{metal\text{-}aqueous}$ = -2.5 ‰ (*27*) at 25°C. Other discrepancies between experiments arise from different experimental designs.
- Fractionations were different in stagnant solutions (hollow markers, Fig. 2) compared to stirred solutions (solid markers, Fig. 2). Comparing trends in the zinc data (diamonds, Fig. 2) between the hollow (*15*) and solid marker samples (*19*), where the same solution chemistry was used, shows an almost 1.5 ‰ difference in the isotopic composition of samples. Temperature cannot account for this large discrepancy between the experiments and so this indicates that the stirring of solutions may have a large effect on the observed fractionation.
- Temperature trends in the solid marker data sets (Fig. 2) are unusual, with fractionation increasing with increasing temperature in both the iron and

zinc experiments, while equilibrium stable isotope theory predicts that fractionation should decrease at higher temperatures as $1/T^2$ (*24*).

- The trend in fractionation versus overpotential for iron samples (circles, Fig. 2) changes depending on experimental conditions, with fractionation increasing at increasing applied overpotential in stagnant experiments (*14*) and decreasing with increasing applied overpotential in stirred experiments (*19*). The solution chemistry may partly explain the differences in the case of iron, as the pH of solutions in stagnant experiments was lower (pH < 1) compared to stirred experiments (pH ~ 2.5) and this affected the iron deposition efficiency relative to $H_{2(g)}$ production at the cathode.

Experimental Methods

Sample Solution Preparation

A 1 L stock solution of 0.7 M copper sulfate was prepared from a $CuSO_4.5H_2O$ salt (pH = 3.05). Fresh 50 mL aliquots were taken from this stock for every second to fourth experiment (stock solution compositions before and after samples runs were monitored to ensure there was no Rayleigh-type evolution of the reservoirs).

Electrodeposition Experiments

Electrochemical parameters were controlled by an Autolab potentiostat (PGSTAT30) using a platinum rotating disc working electrode (3 mm diameter), a platinum counter electrode, and a Ag/AgCl double junction reference electrode (filled with 3 m KCl). The electrodes were immersed in ~50 mL of degassed sample solution and the temperature of this solution was controlled by an external water bath. The equilibrium potential (E^0) of sample solutions in the electrochemical cell were experimentally determined using cyclic voltammetry with variable sweep rates. A series of electrodeposition experiments were performed as a function of: applied overpotential ($E-E^0$); time (one Coulomb of charge being delivered in all cases, regardless of the rate of reduction which varied with overpotential); temperature; and electrode rotation rate, summarized in Table 1.

Samples of electroplated copper were dissolved in warm 2% (w/v) HNO_3 in Teflon cups. The samples were then evaporated to dryness and diluted with 4 mL of 2% HNO_3. Before isotopic analysis (see below) the samples were diluted to a final concentration of 20 ppm copper. Dilutions were measured for Cu concentrations to provide an estimate of the total amount of metal electroplated. A comparison between this value and the total amount of charge passed provides a lower bound estimate of the efficiency of the plating reaction and in all cases was ~108 % ± 7% (2σ); therefore efficiencies are taken as 100% in all cases.

Table 1. Summary of electrochemical experiments and MC-ICP-MS results

Sample	*Temp. (°C)*	*η (V)*[a]	*Stir Rate (rpm)*	*Average Current (mA)*	*$Δ^{65/63}Cu$ (‰) ± 2σ*[b]
Cu01	0	-0.2	5000	-0.98	-2.17 ± 0.02
Cu02	0	-0.2	10000	-0.91	-2.31 ± 0.02
Cu03	0	-0.3	5000	-1.92	-2.27 ± 0.02
Cu04	0	-0.3	10000	-1.90	-2.22 ± 0.02
Cu05	0	-0.5	2500	-4.25	-2.02 ± 0.02
Cu06	0	-0.5	5000	-4.24	-2.11 ± 0.02
Cu07	0	-0.5	7500	-4.31	-2.10 ± 0.02
Cu08	0	-0.5	10000	-4.36	-2.13 ± 0.02
Cu09	25	-0.1	5000	-1.57	-2.83 ± 0.02
Cu10	25	-0.1	10000	-1.64	-2.74 ± 0.02
Cu11	25	-0.15	5000	-2.75	-2.62 ± 0.02
Cu12	25	-0.15	10000	-2.39	-2.33 ± 0.02
Cu13	25	0.2	5000	-4.03	-2.28 ± 0.02
Cu14	25	-0.2	10000	-3.05	-2.09 ± 0.02
Cu15	25	-0.3	5000	-5.32	-1.84 ± 0.02
Cu16	25	-0.3	10000	-1.69	-2.48 ± 0.02
Cu17	25	-0.5	2500	-9.25	-1.70 ± 0.02
Cu18	25	-0.5	5000	-9.34	-1.76 ± 0.02
Cu19	25	-0.5	7500	-9.61	-1.84 ± 0.02
Cu20	25	-0.5	10000	-9.51	-1.84 ± 0.02
Cu21	25	-1.0	2500	-21.23	-1.34 ± 0.02
Cu22	25	-1.0	5000	-20.79	-1.54 ± 0.02
Cu23	25	-1.0	7500	-20.79	-1.61 ± 0.02
Cu24	25	-1.0	10000	-20.79	-1.68 ± 0.02
Cu25	50	-0.2	5000	-6.49	-2.40 ± 0.02
Cu26	50	-0.2	10000	-6.57	-2.41 ± 0.02
Cu27	50	-0.3	5000	-10.30	-2.45 ± 0.02
Cu28	50	-0.3	10000	-9.99	-2.29 ± 0.02
Cu29	50	-0.5	2500	-16.92	-1.90 ± 0.02
Cu30	50	-0.5	5000	-18.15	-2.22 ± 0.02

Continued on next page.

Table 1. (Continued). Summary of electrochemical experiments and MC-ICP-MS results

Sample	*Temp. (°C)*	*η (V)*[a]	*Stir Rate (rpm)*	*Average Current (mA)*	*Δ^65/63^Cu (‰) ± 2σ*[b]
Cu31	50	-0.5	7500	-17.21	-2.10 ± 0.02
Cu32	50	-0.5	10000	-17.83	-2.33 ± 0.02

[a] Overpotential, η, (E – E⁰) relative to measured E⁰ from CV spectra; [b] The isotopic composition of all samples were measured in triplicate with the exception of Cu27 which was only measured in duplicate.

Isotope Analysis

Isotopic analyses were performed using a Thermo-Finnigan Neptune Multi-collector Inductively Coupled Plasma Mass Spectrometer. Samples were introduced via a cyclonic spray chamber. Samples and standards were diluted to a Cu concentration of 20 ppm and spiked with 20 ppm of an in-house Zn standard. Samples were run alternately with a NIST SRM 682 Zn standard, also spiked with Cu, which has previously been compared to JMC 3–0749L Zn (*29*). Signal intensity on ^{60}Ni, ^{63}Cu, ^{64}Zn, ^{65}Cu, ^{66}Zn and ^{68}Zn were monitored on cups L3, L2, L1, C, H1, and H2, respectively. Samples and standards were run alternately for three minutes each, with 3 minutes rinsing in between. Monitoring of the measured Zn isotope ratio in samples showed that there was no significant mass bias due to matrix effects in copper samples compared to standards. Samples were therefore corrected for instrumental mass bias using only sample-standard bracketing. Table 1 summarizes the isotopic composition of samples relative to stock solutions. In all experiments the isotopic composition of the stock solution was unchanged within experimental error before and after an electroplating experiment. This establishes a constant isotope reservoir, with no effects from Rayleigh-type compositional evolution.

Results and Discussion

To better understand the trends described in the previous experiments (Figure 2), the experiments presented here were designed with control over several different parameters including the rate at which solution was advected over the electrode (controlled by using a rotating disc electrode), solution chemistry, the applied overpotential and temperature. Figure 3 presents the isotopic composition of copper metal samples as a function of the temperature (Fig. 3A) and electrode rotation rate (Fig. 3B). The calculated equilibrium fractionation between a copper hexaaquo complex and metallic copper (RPFR estimated from frequency calculations for the optimized aqueous complex and a literature report of the vibrational density of states for metallic copper (*30*)) is also plotted (dashed line, Fig. 3). The following sections discuss the trends observed in terms of the physical and chemical variables examined in the experiments.

Effect of Temperature

Figure 3A plots the stable isotope composition of copper metal samples from this study as a function of temperature. There is a decrease in the sample fractionation with increasing temperature from 0 to 25°C of ~0.5 ‰, as predicted by the plotted equilibrium fractionation (dashed line, Fig. 3), and the predictions of Equation 2. However, an increase in fractionation is observed at 50°C and may indicate a change in the electrodepostion mechanism or surface chemistry at the electrode.

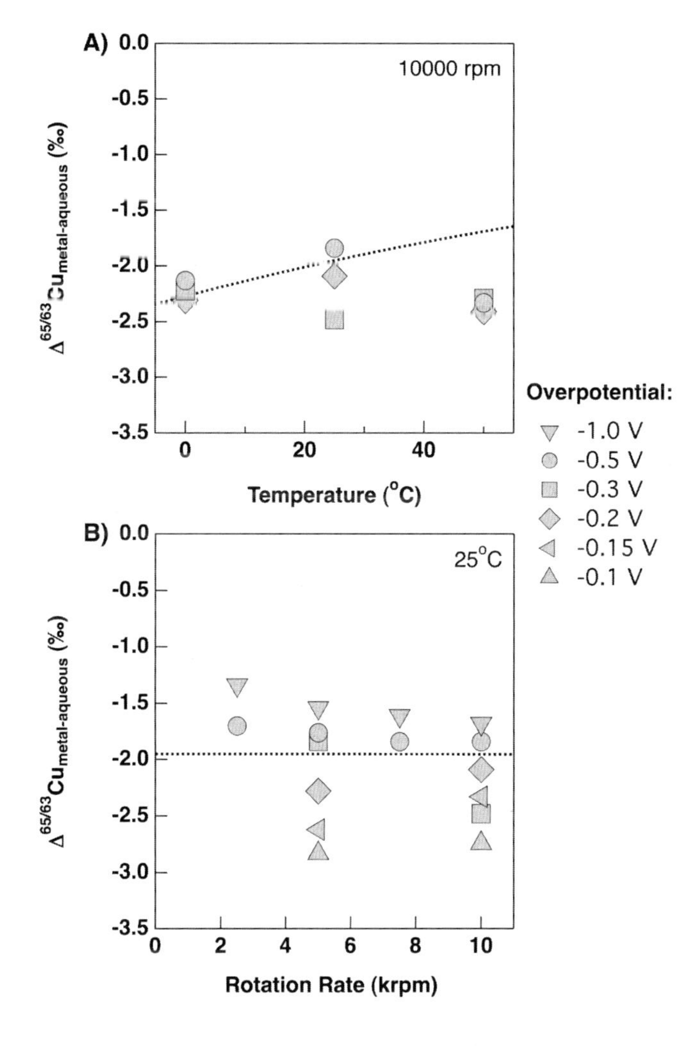

*Figure 3. Stable isotope composition of electrodeposited copper as a function of **A)** temperature; **B)** RDE rotation rate. Dashed line = calculated equilibrium.*

Effect of Mass Transport

Figure 3B plots the stable isotope composition of copper metal samples from this study as a function of the electrode rotation rate at 25°C. At higher overpotentials the same trend is observed of increasing fractionation with increasing rotation rate, and the same trends were observed at higher and lower temperatures and in another study of iron (*18*). Calculation of the magnitude of fractionation due to diffusion across the mass-transport boundary layer (diffuse layer, Fig. 1) at the electrode surface indicate that diffusive limitations would lead to much smaller fractionations for iron and zinc (e.g. $\Delta^{66/64}Zn_{diffusion}$ = -0.27 ‰ (*19*, *31*, *32*)), and copper with a similar solution chemistry and diffusion coefficient is likely to exhibit a similar small diffusive fractionation. This explains some of the trends observed with temperature and rotation rate. Under diffusive control the fractionation in metal samples will be attenuated by the smaller fractionation associated with diffusion to the reactive interface. As temperature increases and rotation rate increases fractionation will also increase because diffusion occurs more quickly at higher temperatures and the width of the diffusive sublayer decreases at higher rotation rates, so that diffusion no longer attenuates the large electrochemical isotope effects as much. This effect can be quantified by plotting isotope fractionation against the ratio of observed current at the cathode versus a calculated diffusion limiting current (Figure 4) given by the Levich equation (*32*). Figure 4 shows that the data falls along a trend of increasing fractionation with decreasing current ratio (i.e. as reaction kinetics dominate over diffusion limitations fractionation increases) and resolves the temperature and rotation rate effects observed.

Effect of Electrochemical Variables

Amount Plated

In the copper deposition experiments presented here one Coulomb of metal was plated in all cases. However, it was shown in previous studies (*9*, *10*, *14*) that varying the amount of charge delivered from 5 (0.5 minute) to 50 (5 minutes) Coulombs with all other variables fixed, led to no change in the observed fractionation in metal within the experimental error.

Overpotential

Figure 3 plots data collected at different applied overpotentials as different symbols, showing a general trend towards smaller fractionations in the copper metal with increasing overpotential. The same trends were seen in previous experiments (Fig. 2 (*15*, *17*, *19*)) with the exception of one study (hollow circular symbols, Fig. 2 (*14*)). Figure 3 clearly illustrates that at low overpotentials the fractionation in samples is larger than predicted by equilibrium stable isotope theory (dashed line, Fig. 3), and is therefore suggestive of a kinetic isotope effect. However, with increasing overpotential (i.e., increasing disequilibrium) the fractionation in samples decreases below this equilibrium fractionation line. As discussed in the introduction, Equation 2 predicts that fractionation scales linearly with the applied overpotential, however, closer inspection of the trends in the data do not show a perfectly linear dependence. This may be partly explained by concurrent isotope effects, such as diffusive limitations of reactant to the electrode interface where reaction occurs. Diffusive limitations may also explain the smaller fractionations observed further from equilibrium at higher applied overpotentials.

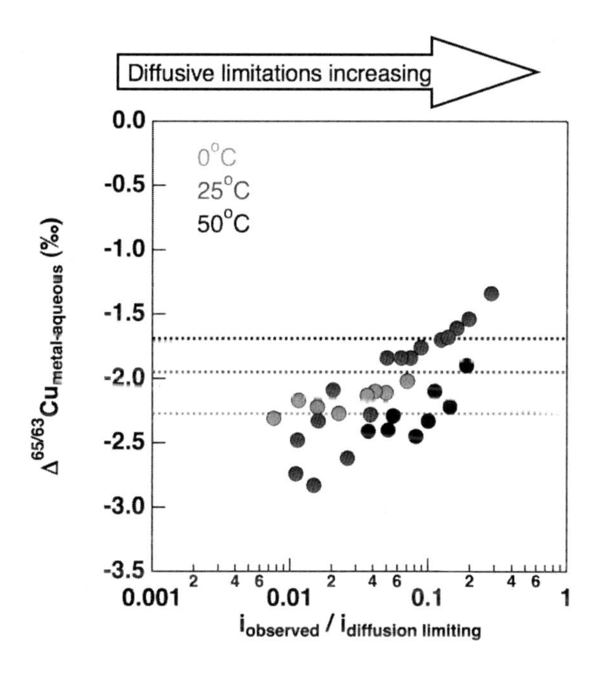

Figure 4. Stable isotope composition of electrodeposited copper as a function of the current ratio of observed to calculated diffusion limiting current. Dashed lines represent calculated equilibrium fractionation at different temperatures.

Conclusions

Large fractionation of metal stable isotopes are observed during redox reactions. These fractionations are not controlled by a single process, but instead result from the interaction between several different processes, each of which can fractionate isotopes on their own and/or moderate the fractionations caused by other processes. Under experimental conditions, three chemical and physical processes dominate these stable isotope fractionations:

1) Equilibrium isotope effects between different chemical species, leading to smaller fractionations at higher temperatures, scaling as $1/T^2$ under most circumstances.

2) Electrochemical kinetic isotope effects, where differences in the activation energy for charge transfer lead to large isotope effects.
3) Diffusive isotope effects, which are small in comparison to these other effects and which increase at higher temperatures.

The relative difference in magnitude of these effects can be used to predict and produce isotopic products of metals of specifically tuned composition by controlling the mass transport, temperature and reaction kinetics of a given system. Understanding the interaction between these processes also provides a framework for interpreting isotopic signals in nature, where the magnitude of fractionation may be used to distinguish between kinetic and equilibrium effects, or to determine under what natural conditions reactions become diffusion limited. Many of the chapters in this volume discuss various pathways involving redox transformations of iron. Chapters 8 (*33*) and 9 (*34*) discuss reactions of iron with reactive oxygen species in aquatic systems, chapter 14 (*35*) discusses how complexation of iron to organic ligands mediates redox reactions between the Fe(II)-Fe(III) couple, chapters 13 (*36*) and 15 (*37*) discuss sorption of iron to FeS adlayers and ferric oxide minerals followed by redox activity, chapter 17 (*38*) looks at how the structural properties of clays change with the oxidation and reduction of iron within the structure, and chapter 18 (*39*) looks at the reactivity of zero valent iron in the environment and its importance in reactive-transport models. All of these processes may induce their own unique isotopic signature in the separate redox phases of iron which potentially could be used to study these reaction pathways. Many other redox active metals have their own suite of stable isotopes whose ratios will be affected by their redox cycle.

Acknowledgments

We thank Professor Jess Adkins for access to facilities at Caltech. This work was funded by NASA Exobiology NNG05GQ92G (A.K.) and the U.S. Department of Energy: Division of Chemical Sciences, Geosciences, & Biosciences DE-FG02-10ER16136 (A.K.).

References

1. Urey, H. C.; Brickwedde, F. G.; Murphy, G. M. The hydrogen isotope of mass 2 and its concentration. *Phys. Rev.* **1932**, *40*, 1–15.
2. Lewis, G. N.; Macdonald, R. T. Concentration of H2 isotope. *J. Chem. Phys.* **1933**, *1* (6), 341–344.
3. Rae, H. K. Selecting Heavy Water Processes. In *Separation of Hydrogen Isotopes*; Rae, H. K., Ed.; Americal Chemical Society: Washington, DC, 1978; Vol. 68, pp 1−26.
4. Villani, S. *Separation of Isotopes*; American Nuclear Society: Hinsdale, IL, 1974; p 499.

5. Ishida, T.; Fujii, Y. Enrichment of isotopes. In *Isotope Effects in Chemistry and Biology*; Kohen, A., Limbach, H.-H., Eds.; CRC Press, Taylor & Francis Group: Boca Raton, FL, 2006; pp 41−87.
6. Taylor, T. I.; Urey, H. C. The electrolytic and chemical-exchange methods for the separation of the lithium isotopes. *J. Chem. Phys.* **1937**, *5*, 597–8.
7. Johnston, H. L.; Hutchison, C. A. Efficiency of the electrolytic separation of lithium isotopes. *J. Chem. Phys.* **1940**, *8*, 869–77.
8. Weston, R. E., Jr. Isotope Effects and Quantum-Mechanical Tunneling. In *Isotopes and Chemical Principles*; Rock, P. A., Ed.; American Chemical Society: Washington, DC, 1975; pp 44−63.
9. Fujie, M.; Fujii, Y.; Nomura, M.; Okamoto, M. Isotope effects in electrolytic formation of lithium amalgam. *J. Nucl. Sci. Technol.* **1986**, *23* (4), 330–7.
10. Mouri, M.; Yanase, S.; Oi, T. Observation of Lithium Isotope Effect Accompanying Electrochemical Insertion of Lithium into Zinc. *J. Nucl. Sci. Technol.* **2008**, *45* (5), 384–389.
11. Yanase, S.; Hayama, W.; Oi, T. Lithium Isotope Effect Accompanying Electrochemical Intercalation of Lithium into Graphite. *Z. Naturforsch.* **2003**, *58a*, 306–312.
12. Yanase, S.; Oi, T.; Hashikawa, S. Observation of Lithium Isotope Effect Accompanying Electrochemical Insertion of Lithium into Tin. *J. Nucl. Sci. Technol.* **2000**, *37* (10), 919–923.
13. Zenzai, K.; Yanase, S.; Zhang, Y.-H.; Oi, T. Lithium isotope effect accompanying electrochemical insertion of lithium into gallium. *Prog. Nucl. Energy* **2008**, *50* (2−6), 494–498.
14. Kavner, A.; Bonet, F.; Shahar, A.; Simon, J.; Young, E. The isotopic effects of electron transfer: An explanation for Fe isotope fractionation in nature. *Geochim. Cosmochim. Acta* **2005**, *69* (12), 2971–2979.
15. Kavner, A.; John, S. G.; Sass, S.; Boyle, E. A. Redox-driven stable isotope fractionation in transition metals: Application to Zn electroplating. *Geochim. Cosmochim. Acta* **2008**, *72* (7), 1731–1741.
16. Kavner, A.; Shahar, A.; Black, J. R.; Young, E. Iron Isotopes at an Electrode: Diffusion-Limited Fractionation. *Chem. Geol.* **2009**, *267* (3−4), 131–138.
17. Black, J. R.; Umeda, G.; Dunn, B.; McDonough, W. F.; Kavner, A. The Electrochemical Isotope Effect and Lithium Isotope Separation. *J. Am. Chem. Soc.* **2009**, *131*, 9904–9905.
18. Black, J. R.; Young, E.; Kavner, A. Electrochemically controlled iron isotope fractionation. *Geochim. Cosmochim. Acta* **2010**, *74*, 809–817.
19. Black, J. R.; John, S.; Young, E. D.; Kavner, A. Effect of Temperature and Mass-Transport on Transition Metal Isotope Fractionation During Electroplating. *Geochim. Cosmochim. Acta* **2010**, *74*, 5187–5201.
20. Bigeleisen, J.; Mayer, M. G. Calculation of equilibrium constants for isotopic exchange reactions. *J. Chem. Phys.* **1947**, *15*, 261–7.
21. Urey, H. C. Thermodynamic properties of isotopic substances. *J. Chem. Soc.* **1947**, 562–81.
22. Bigeleisen, J. Nuclear Size and Shape Effects in Chemical Reactions. Isotope Chemistry of the Heavy Elements. *J. Am. Chem. Soc.* **1996**, *118* (15), 3676–80.

23. Marcus, R. A. Electron-transfer reactions in chemistry: theory and experiment (Nobel lecture). *Angew. Chem.* **1993**, *105* (8), 1161–72.
24. Schauble, E. A. Applying stable isotope fractionation theory to new systems. In *Geochemistry of Non-traditional Stable Isotopes*; Johnson, C. M., Beard, B. L., Albarede, F., Eds.; Mineralogical Society of America: Washington, DC, 2004; Vol. 55, pp 65−111.
25. Bigeleisen, J. Temperature dependence of the isotope chemistry of the heavy elements. *Proc. Natl. Acad. Sci. U.S.A.* **1996**, *93* (18), 9393–9396.
26. Schauble, E. A.; Rossman, G. R.; Taylor, H. P. Theoretical estimates of equilibrium Fe-isotope fractionations from vibrational spectroscopy. *Geochim. Cosmochim. Acta* **2001**, *65* (15), 2487–2497.
27. Black, J. R.; Kavner, A.; Schauble, E. A. Calculation of Equilibrium Stable Isotope Partition Function Ratios for Aqueous Zinc Complexes and Metallic Zinc. *Geochim. Cosmochim. Acta* **2011**, *75*, 769–783.
28. Singh, G.; Hall, J. C.; Rock, P. A. Thermodynamics of lithium isotope exchange reactions. II. Electrochemical investigations in diglyme and propylene carbonate. *J. Chem. Phys.* **1972**, *56* (5), 1855–62.
29. John, S. G.; Park, J. G.; Zhang, Z.; Boyle, E. A. The isotopic composition of some common forms of anthropogenic zinc. *Chem. Geol.* **2007**, *245* (1−2), 61–69.
30. Durukanoglu, S.; Kara, A.; Rahman, T. S. Vibrational modes and relative stability of stepped surfaces of copper. *NATO ASI Ser., Ser. B* **1997**, *360*, 599–605.
31. Rodushkin, I.; Stenberg, A.; Andren, H.; Malinovsky, D.; Baxter, D. C. Isotopic Fractionation during Diffusion of Transition Metal Ions in Solution. *Anal. Chem.* **2004**, *76* (7), 2148–2151.
32. Bard, A. J.; Faulkner, L. R. *Electrochemical Methods: fundamentals and applications*, 2nd ed.; John Wiley & Sons, Inc.: USA, 2001.
33. Garg, S.; Rose, A. L.; Waite, T. D. Pathways Contributing to the Formation and Decay of Ferrous Iron in Sunlit Natural Waters. In *Aquatic Redox Chemistry*; Tratnyek, P. G., Grundl, T. J., Haderlein, S. B., Eds.; ACS Symposium Series; American Chemical Society: Washington, DC, 2011; Vol. 1071, Chapter 8, pp 153−176.
34. Remucal, C. K.; Sedlak, D. L. The Role of Iron Coordination in the Production of Reactive Oxidants from Ferrous Iron Oxidation by Oxygen and Hydrogen Peroxide. In *Aquatic Redox Chemistry*; Tratnyek, P. G., Grundl, T. J., Haderlein, S. B., Eds.; ACS Symposium Series; American Chemical Society: Washington, DC, 2011; Vol. 1071, Chapter 9, pp 177−197.
35. Strathman, T. J. Redox Reactivity of Organically Complexed Iron(II) Species with Aquatic Contaminant. In *Aquatic Redox Chemistry*; Tratnyek, P. G., Grundl, T. J., Haderlein, S. B., Eds.; ACS Symposium Series; American Chemical Society: Washington, DC, 2011; Vol. 1071, Chapter 14, pp 283−313.
36. Helz, G. R.; Ciglenecki, I.; Krznaric, D.; Bura-Nakic, E. Voltammetry of Sulfide Nanoparticles and the FeS(aq) Problem. In *Aquatic Redox Chemistry*; Tratnyek, P. G., Grundl, T. J., Haderlein, S. B., Eds.; ACS Symposium Series;

American Chemical Society: Washington, DC, 2011; Vol. 1071, Chapter 13, pp 265−282.

37. Gorski, C. A.; Scherer, M. M. Fe^{2+} Sorption at the Fe Oxide-Water Interface: A Revised Conceptual Framework. In *Aquatic Redox Chemistry*; Tratnyek, P. G., Grundl, T. J., Haderlein, S. B., Eds.; ACS Symposium Series; American Chemical Society: Washington, DC, 2011; Vol. 1071, Chapter 15, pp 315−343.
38. Neumann, A.; Sander, M.; Hofstetter, T. B. Redox properties of structural Fe in smectite clay minerals. In *Aquatic Redox Chemistry*; Tratnyek, P. G., Grundl, T. J., Haderlein, S. B., Eds.; ACS Symposium Series; American Chemical Society: Washington, DC, 2011; Vol. 1071, Chapter 17, pp 361−379.
39. Tratnyek, P. G.; Salter-Blanc, A. J.; Nurmi, J. T.; Baer, D. R.; Amonette, J. E.; Liu, J.; Dohnalkova, A. Reactivity of zerovalent metals in aquatic media: Effects of organic surface coatings. In *Aquatic Redox Chemistry*; Tratnyek, P. G., Grundl, T. J., Haderlein, S. B., Eds.; ACS Symposium Series; American Chemical Society: Washington, DC, 2011; Vol. 1071, Chapter 18, pp 381−406.

Chapter 17

Redox Properties of Structural Fe in Smectite Clay Minerals

Anke Neumann,[1] Michael Sander,[1] and Thomas B. Hofstetter[1,2,*]

[1]Institute of Biogeochemistry and Pollutant Dynamics (IBP), Swiss Federal Institute of Technology (ETH) Zürich, Universitätsstr. 16, 8092 Zürich, Switzerland
[2]Eawag, Swiss Federal Institute of Aquatic Science and Technology, Überlandstr. 133, 8600 Dübendorf, Switzerland
[*]corresponding author: thomas.hofstetter@eawag.ch

Redox reactions of structural Fe in clay minerals play important roles in biogeochemical processes and for the fate of contaminants in the environment. Many of the redox properties of Fe in clay minerals are, however, poorly understood, thus limiting the knowledge of the factors that make structural Fe participate in electron transfer reactions. This chapter summarizes the current state of knowledge on the redox properties of structural Fe in clay minerals. In the first part, we review the various spectroscopic observations associated with structural Fe reduction and oxidation and how changes in Fe oxidation state affect the clay mineral structure and the binding environment of Fe in the octahedral sheet of planar 2:1 clay minerals. In the second part, we show how information on the structural alterations and arrangement of Fe can be interpreted to assess the apparent reactivity and the thermodynamic redox properties of structural Fe in clay minerals.

Introduction

The Fe^{2+}/Fe^{3+} redox couple plays an important role in the biogeochemical cycling of elements, and is of direct relevance for the remediation of environmental systems contaminated with organic and inorganic pollutants (*1–4*). Clay-mineral-bound Fe is of particular importance in environmental electron transfer reactions because Fe-containing clay minerals are ubiquitous in subsurface environments and structural Fe in clay minerals is not subject to the same dissolution and re-precipitation processes as Fe in (oxyhydr)oxides (*5–7*). Structural Fe in clay minerals therefore can act as renewable source of redox equivalents in soils and sediments for the natural or enhanced attenuation of pollutants at contaminated sites (Figure 1, (*2*, *8*)). Structural Fe^{3+} in clay minerals can be reduced to Fe^{2+} by microorganisms and surface-bound Fe^{2+} (*9–11*), which is a viable reductant for many organic pollutants (e.g., chlorinated solvents, nitroaromatic explosives; (*12–14*)) and metals (e.g., U, Tc, Cr; (*15–17*)). The reduction of organic compounds leads to transient products, which often are more susceptible to complete microbial degradation via oxidative pathways. Reduction of metals from radioactive waste repositories or re-processing sites by Fe^{2+} often results in the formation of sparingly soluble, and hence, less mobile metal species.

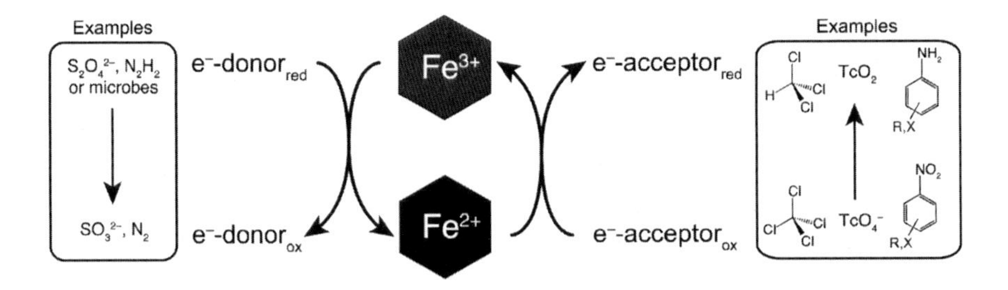

Figure 1. Schematic representation of redox cycling of structural Fe in clay minerals. The colored hexagons symbolize the octahedral binding environment of Fe^{3+} (red) and Fe^{2+} (blue) in the clay mineral structure. Structural Fe^{3+} can be reduced by electron donors such as dithionite, hydrazine, or microbes. The resulting structural Fe^{2+}, in turn, reduces oxidized organic contaminants (e.g., carbon tetrachloride or nitroaromatic compounds) or metals (e.g., Tc^{7+}).

Reduction of structural Fe^{3+} to Fe^{2+} increases the net negative charge in the clay mineral structure (*6*), which alters important mineral properties such as its cation exchange capacity (*18*, *19*) and its swelling pressure (*18*, *20*). These oxidation-state induced changes are highly relevant for the bioavailability of soil nutrient such as K^+, Ca^{2+}, Cu^{2+}, Zn^{2+}, and NH_4^+ (*21*) as well as for maintaining the integrity and stability of radioactive waste repositories that rely on clay mineral-based backfill material for radionuclide retention (*22*). A comprehensive understanding of Fe reduction and oxidation processes and their effects on clay mineral properties is therefore essential for assessing clay mineral mediated natural and engineered processes.

The characterization of the redox properties of structural Fe in clay minerals has proven to be challenging. Some of the most fundamental properties are only poorly understood, including the fraction of total structural Fe available for reduction/oxidation and estimates of structural Fe^{3+}/Fe^{2+}-reduction potentials. The redox properties of structural Fe are affected by its bonding environment in the clay mineral's lattice, which itself depends on the layer composition (*23*), the ordering of structural cations (*24*, *25*), the total Fe content (*26*), and the Fe oxidation state (*27*, *28*). While these structural parameters are well studied for natural clay samples, the redox reactions of structural Fe remain poorly understood on the molecular level. As a consequence, numerous macroscopic observations lack mechanistic explanations. For example, it is unclear why Fe-containing clay minerals of similar composition cannot be reduced to comparable extents (*21*, *29*, *30*) or why the extent of structural Fe reduction differs between chemical and microbial reduction (*10*, *31–33*). Advances towards a more holistic understanding of electron transfer processes involving structural Fe in clay minerals requires linking Fe^{2+} and Fe^{3+} structural arrangements to Fe^{3+}/Fe^{2+}-reduction potentials and apparent reactivity of clay mineral particles.

The objective of this chapter is to review the current state of knowledge on the effects of structural Fe reduction and oxidation on the clay mineral redox properties and structure, and on the reactivity of structural Fe in electron transfer reactions. We first summarize the most commonly used analytical approaches for elucidating the binding environment of structural Fe in clay minerals. The second part focuses on the effects of Fe redox changes on the Fe-binding environment. In the third section, spectroscopic observations are linked to the apparent reactivities and the thermodynamic redox properties of structural Fe in clay minerals.

The discussion primarily focuses on smectites because these planar 2:1 clay minerals have been most intensely investigated. In these clay minerals, structural Fe is located in the octahedral and/or the tetrahedral sheets. Most experimental data is available on electron transfer to and from octahedrally coordinated Fe. We will therefore elaborate on the effect of octahedral sheet properties including the di- versus trioctahedral site occupancy, *cis/trans* vacancies in dioctahedral clay minerals, as well as the distribution of Fe and its neighboring cations Al, Mg, and Fe on electron transfer reactions to and from structural Fe (*34–36*). A schematic representation of the clay mineral properties addressed in this chapter is depicted in Figure 2.

Spectroscopic Approaches Used To Elucidate the Structural Environment of Fe in Clay Minerals

Several spectroscopic techniques facilitate the study of clay mineral structures and, specifically, the binding environments of structural Fe in clay minerals. The most widely used techniques are ^{57}Fe Mössbauer, X-ray absorption, infrared, and visible spectroscopy. Spectroscopic analyses are generally complemented with the quantification of the total structural Fe content as well as Fe^{2+}/Fe^{3+} or Fe^{2+}/total Fe ratios after acid digestion of clay minerals. Notice that erroneous digestion procedures may have led to inaccurate estimates for redox active Fe^{2+}/Fe^{3+}-species and total Fe (*39*).

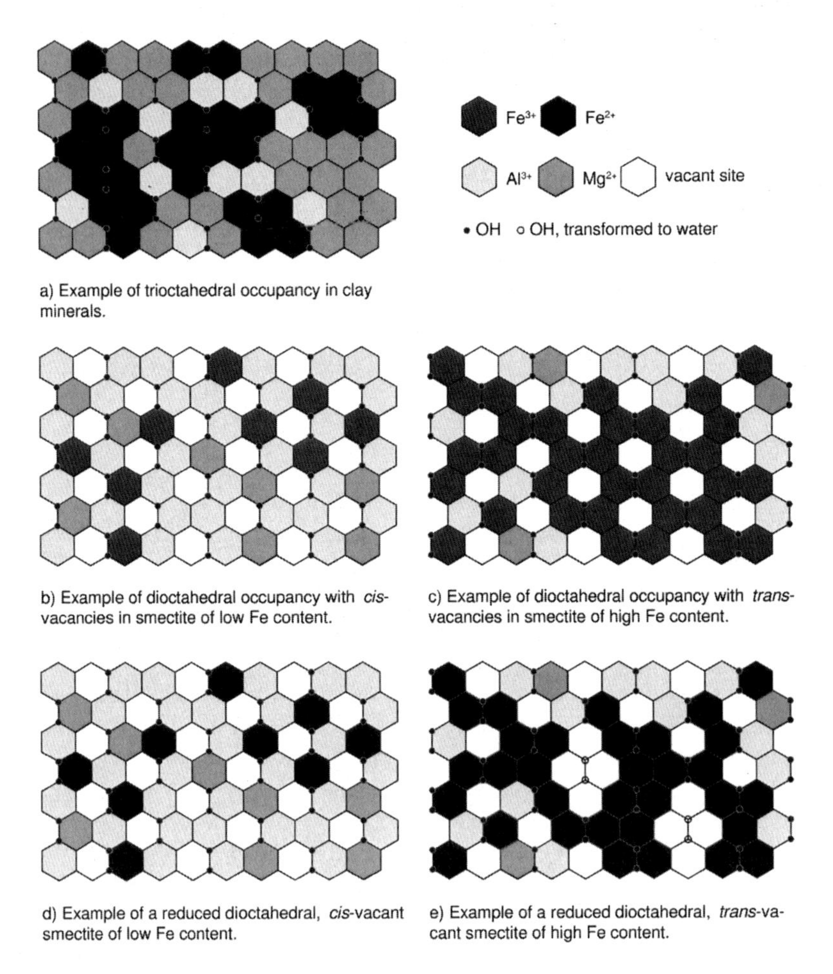

Figure 2. Schematic representations of theoretical octahedral sheet compositions of, smectites showing a) trioctahedral and b, c) dioctahedral occupancies of the octahedral sheet with cations. Dioctahedral clay minerals (b, c) can be categorized as cis- and trans-vacant, depending on whether the octahedral OH groups are on the same or on opposite sides of the octahedral vacancies, respectively. Smectites with low Fe content are cis-vacant (b), whereas Fe-rich smectites are trans-vacant (c). In panel a) local clustering of Fe^{2+} is indicated, whereas panel b) depicts a cation distribution following an exclusion rule with Fe^{3+} cations neighboring Al^{3+} but not Mg^{2+} or Fe^{3+}. For the Fe-rich clay mineral, c), a largely random cation distribution was chosen (25). Panels d) and e) depict structural changes in the octahedral sheet after Fe^{3+}reduction for the model clay minerals given in panels b) and c), respectively. In cis-vacant dioctahedral smectite of low Fe content only small structural changes are observed (d; (37)). Fe reduction of Fe-rich, trans-vacant dioctahedral smectite leads to structural rearrangements through the formation of trioctahedral Fe^{2+} groups enclosing domains of vacancies and to the dehydroxylation of the octahedral sheet (indicated as open circles in panel d, (26, 38)).

^{57}Fe Mössbauer spectroscopy allows for the determination of the bulk Fe^{2+}/Fe^{3+} ratio and can distinguish between octahedrally and tetrahedrally coordinated Fe atoms (*40*, *41*). Additionally, Mössbauer spectroscopy can be used on mixed-phase samples, as clay minerals can be distinguished from iron oxides and other impurities. Spectra collected at low temperatures (77 K to 4 K) of clay minerals with high structural Fe contents provide information on electronic interactions between neighboring Fe atoms, such as magnetic ordering (*33*, *41*, *42*). For some clay minerals, the population of Fe atoms in *cis* and *trans* configuration can be quantified, although this remains a controversial topic (*43–46*). One notable shortcoming is that fitting of clay mineral spectra can often result in non-unique fits, requiring secondary techniques (*47*). An additional drawback of Mössbauer spectroscopy is that the exact identity of the Fe-neighboring cations cannot be determined at an atomic level (*41*).

An alternative Fe-sensitive spectroscopic technique is Fe K-edge X-ray absorption spectroscopy (XAS). These spectra are collected by measuring the absorbance of 7112 eV X-rays available in synchrotron facilities. Near edge spectra (XANES, energy range 7090-7210 eV) allow determining the Fe oxidation state and can distinguish between tetrahedral and octahedral Fe coordination (*26*, *28*). EXAFS spectra are typically interpreted relative to computed spectra for model structures of clay minerals. The extended X-ray absorption fine structure (EXAFS, energy range 7000-8300 eV) contains information on the nearest octahedral and tetrahedral neighbors (*25*, *26*, *28*), but the spectra do not typically yield reliable information on Fe-neighboring cations exceeding the second coordination sphere due to multiple scattering and the structural heterogeneity of natural clay minerals (*25*, *48*).

Various forms of IR spectroscopy monitor absorbance of IR light by hydroxyl (OH) groups bound to octahedral cations in clay minerals. The positions of OH absorption bands in IR spectra are indicative of octahedral cations attached to the OH groups as well as to the oxidation state and the structural environment of octahedral Fe (*37*, *49–54*). Absorption bands of OH bending vibrations (spectral range 600-950 cm^{-1}) are usually better resolved than absorption bands observed in the stretching region (3500-3700 cm^{-1}). The near IR region contains bands resulting from the combination of bending and stretching modes (spectral range 4100-4600 cm^{-1}) as well as the first overtones of the stretching vibrations (spectral range 6900-7400 cm^{-1}). The combined analysis of the absorption bands in these four IR regions facilitates a rigorous and accurate characterization of Fe structural environments (*55–57*). In contrast to Mössbauer spectroscopy and XAS, IR spectroscopy cannot detect tetrahedrally bound Fe because it is not directly bound to hydroxyl groups. Clay minerals with high amounts of tetrahedral Fe do, however, exhibit a characteristic absorption band for tetrahedral Fe-O entities (*58*).

Visible light absorption spectroscopy is limited to studying the intervalence electron transition between Fe^{2+} and Fe^{3+} in adjacent octahedral sites at 720 nm (*59*). This method provides information on the oxidation state of octahedral Fe (*29*) but yields no insight into the Fe binding environment. Furthermore, this technique is limited to clay minerals with high structural Fe contents (see further discussion below).

Reduction and Oxidation of Structural Fe and Their Effects on Clay Mineral Properties

Numerous studies investigated the effect of structural Fe oxidation state on clay mineral properties in the context of Fe biogeochemistry in soils and sediments. For manipulating the structural Fe oxidation state, a variety of reductants and oxidants have been used that result in different extents of reduction and oxidation, and differences in redox cycling. Different mechanisms of structural Fe redox processes have been proposed based on spectroscopic analysis of clay minerals with manipulated Fe oxidation states.

Reduction of Fe^{3+}

Various reducing agents, including sulfide species (*60*), tetraphenyl boron (*61*, *62*), and hydrazine (*63*, *64*) have been used to reduce structural Fe^{3+} in clay minerals. The most frequently used reduction methods are chemical reduction by dithionite (*65*, *66*), and microbial reduction using pure cultures and enrichment cultures (*9*, *10*, *31*).

Chemical Reduction Using Dithionite

The sulfoxylate radical species, $SO_2^{\bullet-}$, resulting from the disproportionation of dithionite in aqueous solution, has a reduction potential, E_h ($SO_3^{2-}/SO_2^{\bullet-}$) of -0.66 V (vs. SHE; pH 7) (*66*). This potential is so low that >90% of the structural Fe in smectites can be reduced (*29*, *30*, *50*). Dithionite reduction in buffered citrate-bicarbonate solutions leads to minor, incongruent (i.e., not uniform) smectite dissolution with <5% of the structural Fe being released into solution (*6*). Dithionite reduces Fe^{3+} in the octahedral sheet whereas the extent of reduction of tetrahedrally coordinated Fe^{3+} is still debated (*37*, *39*, *64*, *67–69*).

Reduction of octahedral Fe^{3+} in clay minerals results in an initially equimolar increase in the negative excess charge. This increase is compensated by the uptake of solution cations and protons, with the latter generally leading to structural dehydroxylation (*28*, *70*). Spectroscopic analysis of dithionite reduced clay minerals identified dioctahedral Fe^{2+} neighboring Al, Mg and Fe^{2+} (*28*, *32*, *37*, *50*, *51*). The reduction of Fe-rich dioctahedral smectites resulted in the formation of trioctahedral Fe^{2+} entities (*28*, *37*, *50*, *51*), whereas the reduction of dioctahedral smectites with lower Fe contents did not form trioctahedral Fe^{2+} groups (*37*). Trioctahedral Fe^{2+} domains were found after the reduction of *trans*- but not *cis*-vacant smectites (see Figure 2), presumably due to higher activation energy for the involved structural rearrangements of *cis*-vacant smectites (*71*, *72*). The location of the negative excess charge in the octahedral sheet can additionally limit the formation of trioctahedral Fe^{2+} domains during reduction of *trans*-vacant smectites such as in Mg- and Fe-rich Ölberg montmorillonite (*37*). Furthermore, when trioctahedral Fe^{3+} domains are present in the unaltered state of Fe-rich smectites, the extent of octahedral Fe reduction is smaller compared to entirely dioctahedral smectites (*27*) because charge compensation via protonation and dehydroxylation are energetically unfavorable (*73*).

Monitoring of the Fe^{2+}-O-Fe^{3+} intervalence electron transfer band in Fe-rich dioctahedral smectites during reduction by dithionite indicated that Fe^{3+}-O-Fe^{2+} arrangements are formed below Fe^{2+}/total Fe ratios of 45%. Fe^{3+}-O-Fe^{2+} groups are further reduced to Fe^{2+}-O-Fe^{2+} above this threshold (*29*, *59*). Fe^{2+}-O-Fe^{3+} groups could not be observed with IR spectroscopy in the structure of partially reduced Fe-rich smectites (*37*), whereas their existence was confirmed by Mössbauer spectroscopy (*33*). The differences between the two techniques may reflect their sensitivity to the temperature-dependent extents of electron delocalization. IR spectra were obtained at room temperature, possibly leading to delocalized electrons in distinct partly reduced domains, which then appear to be only Fe^{2+}-containing in the IR spectra. In contrast, Mössbauer spectra were obtained at low temperatures of 77K and 4K (*33*), limiting the delocalization of electrons and thus facilitating the detection of localized Fe^{2+} adjacent to Fe^{3+} cations (*74*).

Microbial Reduction

Several studies showed that certain microorganisms reduce structural Fe in clay minerals (e.g., (*9*, *10*, *31*, *75*)). In contrast to reduction by dithionite, microbial reduction typically does not exceed extents of 45% of total Fe (*31*), resulting also in less extensive alterations of the clay mineral structure, such as dehydroxylation or structural rearrangements to trioctahedral Fe domains (*32*, *33*). It is debated whether electron transfer from microbes and dithionite to Fe^{3+} proceeds via the same mechanism and whether the resulting structural changes are comparable. An IR spectroscopic study revealed similar structural changes for smectite samples reduced to the same extents by microorganisms and by dithionite (*32*). In contrast, Mössbauer spectroscopy indicated that the structure of microbially reduced smectite contained distinctly separate domains of Fe^{2+} and Fe^{3+}, whereas dithionite-reduced smectites showed mixed-valent Fe entities (*33*). This finding led to the hypothesis that microbial Fe reduction proceeds via the edge surfaces of the smectite, resulting in a moving front of Fe^{2+} within the structure with a clear transition to the non-reduced part of the smectite. In contrast, dithionite is thought to reduce structural Fe in a random manner, which can only be achieved from the basal surfaces (*33*, *38*).

Oxidation of Fe^{2+}

Oxidation of structural Fe^{2+} in clay minerals is much less investigated compared to reduction of Fe^{3+}. Re-oxidation is commonly studied to assess the reversibility of structural changes induced during a previous reduction step. Common oxidants are O_2 gas or air (*6*, *29*, *32*, *33*, *37*, *50*, *51*, *76*, *77*). More than 90% of the structural Fe^{2+} can be reoxidized to Fe^{3+} by this treatment (*32*, *33*, *37*, *50*, *51*, *77*). The availability and reactivity of structural Fe^{2+} has also been determined by the reduction of organic and metal contaminants. The probe contaminants include nitroaromatic compounds (*12*, *30*, *78*), chlorinated solvents (*14*), heavy metals, such as Cr^{6+} (*15*, *79*, *80*), and radionuclides, such as U^{6+} (*16*, *81*) or Tc^{7+} (*17*). To quantify contaminant transformation kinetics as proxies for the Fe^{2+} reactivity, experiments are conducted so that only small amounts of Fe^{2+} are consumed.

The reversibility of structural alterations of clay minerals and the extent of Fe^{2+} re-oxidation have been found to depend primarily on: (i) the total structural Fe content, (ii) the degree of Fe^{3+} reduction, and (iii) the octahedral cation composition and mineralogy of the clay mineral. These observations applied regardless of whether structural Fe^{3+} reduction was achieved chemically or microbially (*31*). Qualitatively, re-hydroxylation of the mineral structure is considered complete once IR spectra of re-oxidized smectites resembled those of the original, unaltered smectite (*29*, *32*, *33*, *37*, *50*, *51*, *76*). This observation was made for clay minerals, in which Fe^{3+} reduction did not exceed approximately 1 mmol Fe^{2+} per gram clay mineral. In contrast, reduction above this Fe^{2+} concentration correlated with irreversible structural changes. For some clay minerals, octahedral cation composition and distribution was shown to impede reversibility to the initial structure despite smaller extent of Fe reduction than the above Fe^{2+} concentration (*37*, *77*). Mössbauer analysis and wet-chemical quantification of the Fe^{2+} content, however, indicate that most re-oxidized samples, even after prolonged exposure to oxygen at elevated temperatures, contained 5-10% residual Fe^{2+} in the mineral structure (*33*, *37*, *77*, *82*, *83*). This observation was made regardless of the total Fe content (*33*, *37*, *77*). Similar findings regarding the reversibility were obtained after microbial Fe^{3+} reduction (*31*).

In analogy to the formation of various Fe^{2+}-entities during reduction, an oxidation sequence for distinct groups of Fe^{2+} is observed in Fe-rich smectites. Fe^{3+}-O-Fe^{3+} entities are first reduced to Fe^{2+}-O-Fe^{3+} groups, which are then further reduced to Fe^{2+}-O-Fe^{2+} groups. The oxidation occurs in the reverse order, i.e., Fe^{2+}-O-Fe^{2+} groups are oxidized first to Fe^{2+}-O-Fe^{3+} entities, which are further oxidized to Fe^{3+}-O-Fe^{3+} groups, as derived from intensity changes of the intervalence electron transfer bands (*29*, *59*). IR spectra of a variety of reactive structural Fe^{2+} entities corroborated these findings (*37*, *50*, *51*). In smectites with low Fe contents, only one type of structural Fe^{2+} was observed, whereas Fe-rich smectites exhibited a variety of reactive structural Fe^{2+} entities. Depending on the type of Fe-rich smectite, trioctahedral Fe^{2+} was found to be more readily oxidized than $Fe^{2+}Fe^{2+}$ and $AlFe^{2+}$ groups. In other clay minerals, no trioctahedral Fe^{2+} entities were observed at all and Fe^{2+} in $Fe^{2+}Fe^{2+}$ and $MgFe^{2+}$ entities was oxidized prior to Fe^{2+} in $AlFe^{2+}$ groups (*37*).

Mechanism of Fe Reduction in Clay Minerals

Mechanistic studies have exclusively focused on the reduction of structural Fe^{3+} because of the lack of 'standardized' reference materials for investigating Fe^{2+} oxidation. In the reduction formalism ((*84*) and references therein), proton and cation uptake from solution as well as dehydroxylation reactions of structural hydroxyl groups are explicitly taken into account. These processes counterbalance the increase in negative charge of the clay mineral particle upon Fe^{3+} reduction. In empirical models, eq. 1 have been used to correlate the observed increase in cation .exchange capacity with the increase in structural Fe(II) content (*84*) while maintaining the octahedral Fe coordination environment (*26*, *28*).

$$n \cdot Fe^{3+}_{cm} + n \cdot e^{-} + w_0 \cdot C^{+}_{cm} + p \cdot C^{+}_{s} + m \cdot H^{+}_{s} + m \cdot OH^{-}_{cm} \rightarrow n \cdot Fe^{2+}_{cm} + w \cdot C^{+}_{cm} + m \cdot H_2O_{cm} \quad (1)$$

In eq. 1, n, m, and p are stoichiometric coefficients for the number of electrons, protons, and cations, respectively, consumed during structural Fe^{3+} reduction and the subscripts *cm* and *s* indicate that the protons and cations are in the structure of the clay mineral or in solution. The uptake of cations, C^+, from solution results in an increase of interlayer cation content from $w_0 \cdot C^{+}_{cm}$ in the unaltered clay mineral to $w \cdot C^{+}_{cm}$ in the reduced clay mineral. The additional negative charge induced by the electron transfer to structural Fe^{3+} (n) has to correspond to the uptake of cations (p) and protons (m) , as in eq 2.

$$n = p + m \quad (2)$$

It is assumed that the ratio of proton to cation uptake is proportional to the degree of Fe^{3+} reduction, (n/n_{tot} in eq. 3, (*84*)):

$$\frac{m}{p} = K_0 \frac{n}{n_{tot}} \quad (3)$$

where n_{tot} is the total amount of structural Fe, n is the amount of Fe^{2+}, and K_0 is an empirical, mineral-dependent fitting parameter of unclear physical meaning (*84*) obtained from experimental data. K_0 contains information on the extent of cation migration and dehydroxylation *in dioctahedral smectites*, which depend on the total Fe content and the configuration of octahedral OH groups. In Fe-rich, *trans*-vacant smectites, the uptake of protons exceeds that of cations at high degree of reduction, leading to extensive dehydroxylation and octahedral cation migration. In contrast, reduction of *cis*-vacant, Fe-poor smectites results in less extensive dehydroxylation and octahedral cation mobility due to the higher activation energy of these processes, leading to a predominant uptake of cations over protons to compensate the induced charge (*84*). Notice that such general chemical equations cannot reflect rearrangement of Fe^{2+} and migration of other cations in the octahedral sheet which occur during the reduction of Fe-rich dioctahedral smectites (*84*).

The proposed chemical equations for Fe^{3+} reduction can be used to rationalize changes in the clay mineral X-ray absorption spectra related to alterations of the structure and the bonding environment of Fe as well as changes in macroscopic clay mineral properties, such as the changes in cation exchange capacity. To date there is, unfortunately, no comprehensive conceptual model that allows inferring macroscopically observable reactivity of Fe and important redox properties of clay minerals such as reduction potential, electron-donating and accepting capacity from data on Fe binding and mineralogical parameters.

Linking Mineralogical Observations to the Apparent Reactivity and Redox Properties of Fe in Clay Minerals

Visible and IR spectroscopy of smectites exhibiting variable degrees of Fe^{3+} reduction reveal preferential reduction and re-oxidation of different structural Fe entities. These observations illustrate the challenge of assessing Fe reactivity and redox properties of clay minerals based on clay mineralogy and Fe binding. Many studies imply that not all Fe in the clay mineral structure is chemically equivalent. Different approaches are currently undertaken to elucidate the kinetics and thermodynamics of electron transfer to and from structural Fe as a function of clay mineral and solution properties.

Kinetics of Electron Transfer to and from Structural Fe Revealed by Reactive Probe Compounds

Investigations of the *apparent* reactivity of structural Fe in clay minerals involved the use of reactive probe compounds. With such approaches, the disappearance kinetics of selected oxidants (usually environmental contaminants) are evaluated quantitatively to obtain operational measures of the reactivity of structural Fe^{2+} entities, which are inferred from spectroscopy or, less rigorously, from structural Fe^{2+}/total Fe measurements. A generalized kinetic scheme (eqs. 4-6) relies on the quantification of the concentration of dissolved probe compound, $[P]$, and its reduction rate constant, $k^{J(n)}{}_{obs}$, by various operationally defined reactive Fe^{2+} entities J_n. Thus, the apparent pseudo-first order rate constant, $k^{J(n)}{}_{obs}$ (eq. 4), represents a weighted average of n second-order rate constants $k_{j(n)}$ for the reduction of probe compounds P by each Fe^{2+} entity (eq. 5 (*37*)).

$$\frac{d[P]}{dt} = k_{obs}^{J_n} \cdot [P] \tag{4}$$

$$k_{obs}^{J_n} = \sum_{i=1}^{n} k_{j_n} \cdot [J_n] \tag{5}$$

where $[P]$ and $[J_n]$ are the concentrations of the probe compound and the Fe^{2+} entities, respectively. The dynamics of Fe^{2+} entity formation and rearrangement is included in this conceptual model by interconversion reactions between different structural Fe^{2+} entities during the overall Fe^{2+} oxidation by the probe compound. The second and third terms in eq. 6 take into account the empirical decay and formation of structural Fe^{2+} entities with pseudo-first order rates $k_{j(n)\rightarrow j(m)}$ and $k_{j(m)\rightarrow j(n)}$, respectively.

$$\frac{d[J_n]}{dt} = k_{j_n} \cdot [J_n] \cdot [P] - \sum_{i=1}^{n} k_{j_n \rightarrow j_m} \cdot [J_n] + \sum_{i=1}^{m} k_{j_m \rightarrow j_n} \cdot [J_m] \tag{6}$$

The Fe^{2+} entity-specific rate constants, $k_{j(n)}$, were estimated, for example, for the reduction of a series of nitroaromatic compounds by solving equations 4-6 numerically on the basis of measured probe compound concentrations and structural Fe^{2+} contents (*30*). This procedure revealed that in Fe-rich smectites (>12 wt% Fe) the observable biphasic probe compound reduction kinetics could be explained by the presence of two distinct and interconvertable Fe^{2+} entities, exhibiting rate constants, which differed by three orders of magnitude. The same biphasic reduction kinetics of probe compounds was found for chlorinated solvents (*14*) and Cr^{6+} (*79*). In the case of smectites exhibiting low structural Fe content (3 wt% Fe), the above kinetic scheme simplifies to a pseudo-first order rate law of probe compound reduction, indicating the presence of only one type of reactive Fe^{2+}, which is of low reactivity (*12*, *30*).

The above observations confirm spectroscopic findings that the structural Fe content is the predominant factor responsible for the formation of various structural Fe^{2+} entities with different reactivity. The reactive probe compound approach has mostly included only a small portion of the available structural Fe (i.e., <10% (*12*, *15*, *30*, *79*, *85*)) and the number of reactive Fe entities invoked was smaller than the number of Fe arrangements observed spectroscopically (*28*, *50*, *51*, *70*, *76*, *84*). It is unclear to date, to what extent proton transfer and dehydroxylation reactions contributed to the apparent reactivity of structural Fe. Moreover, the use of inorganic compounds as reactive probes such as Cr^{6+} (*15*, *79*, *80*), U^{6+} (*16*, *81*) and Tc^{7+} (*17*) gave rise to the formation of secondary phases of the oxidized metal, confounding the analysis of the apparent reaction rates.

Thermodynamics of Electron Transfer to and from Structural Fe

The free energy change of an electron transfer reaction, ΔG_{ET}, involving structural Fe and an environmental reductant or oxidant X (e.g., an organic contaminant, metal, or electron transfer mediator), is given as

$$\Delta G_{ET} = -nF\left(E_h^{Fe} - E_h^{X}\right) \qquad (7)$$

where n is the number of electrons transferred, F is the Faraday constant, and E_h^{Fe} and E_h^{X} are the reduction potentials of a structural Fe^{3+}/Fe^{2+} couple and X, respectively. The sign of ΔG_{ET} defines the tendency of the electron transfer reaction to proceed in a certain direction (i.e., reduction or oxidation of structural Fe). Assuming that E_h^{X} is known, ΔG_{ET} calculation requires knowledge of E_h^{Fe}. The half reaction given in equation 1 can be expressed in terms of E_h^{Fe} by using the Nernst equation (*86*) if one implies that electron transfer to and from structural Fe is reversible.

$$E_h^{Fe} = E_h^{0\ Fe} - \frac{RT}{nF}\ln\frac{\left[Fe_{cm}^{2+}\right]^n \cdot \left[H_2O_{cm}\right]^m \cdot \left[C_{cm}^{+}\right]^w}{\left[Fe_{cm}^{3+}\right]^n \cdot \left[OH_{cm}^{-}\right]^m \cdot \left[H_s^{+}\right]^m \cdot \left[C_{cm}^{+}\right]^{w_0} \cdot \left[C_s^{+}\right]^p} \qquad (8)$$

where $E_h^{0\ Fe}$ is the standard reduction potential of structural Fe. For *trans*-vacant dioctahedral smectites, the increase in negative excess charge due to Fe^{3+} reduction to Fe^{2+} is compensated by both the uptake of m H^+, part of which cause structural dehydroxylation, and p cations C^+ (*84*, *86*). The ratio of m/p increases with increasing extent of Fe reduction (*84*). Assuming extensive reduction (i.e., $m >> p$) and unit activity of structural water, OH^-, and cations (*84*), equation 8 simplifies to

$$E_h^{Fe} = E_h^{0\ Fe} - 0.059 \cdot m \cdot pH - 0.059 \cdot \log \frac{\left[Fe_{cm}^{2+}\right]}{\left[Fe_{cm}^{3+}\right]} \qquad (9)$$

Equations (8) and (9) have been proposed to relate E_h^{Fe} to the oxidation state of structural Fe. These equations, however, were derived assuming a constant $E_h^{0\ Fe}$ for all structural Fe in a given clay mineral, reversible electron transfer, and independent electron transfer to and from each structural Fe atom. These assumptions are unlikely fulfilled for most clay minerals for numerous reasons. (i) Structural Fe atoms in clay minerals are expected to be bound in various, chemically different microenvironments, resulting in a distribution of E_h^{Fe} rather than a constant E_h^{Fe} value. (ii) Fe^{3+} reduction in *trans*-vacant smectites involves dehydroxylation reactions and cation migration in the octahedral sheets (Figure 2; (*26*, *28*, *84*)). It is likely that some of these structural changes are at least partially irreversible, resulting in irreversible alterations of the binding environment of structural Fe in a reduction-oxidation cycle. Electron transfer is then no longer fully reversible and E_h^{Fe} at a given Fe oxidation state will depend on the "redox history" and the direction of electron transfer (i.e., reduction or oxidation). (iii) Changes in the oxidation state of a given structural Fe atom in clay minerals with high structural iron content may affect E_h^{Fe} of adjacent Fe^{3+}/Fe^{2+} couples. This interdependence was suggested from monitoring the intervalence electron transfer band in Fe-rich dioctahedral smectites which showed sequential reduction of Fe^{3+}-O-Fe^{3+} to Fe^{3+}-O-Fe^{2+} up to Fe^{2+}/total Fe ratios of 45%, followed by Fe^{2+}-O-Fe^{2+} formation upon further reduction (*29*, *38*). Similar to Fe (oxyhydr-)oxides, there may also be electron transfer between adjacent Fe sites in Fe-rich smectites (*11*, *87*), which would violate the above assumption of independent electron transfer to and from individual structural Fe atoms. These considerations suggest that trends in E_h^{Fe} predicted by equations (8) and (9) in dependence of the extent of reduction and solution chemistry may deviate from experimental trends of E_h^{Fe}, which are yet to be measured.

Advances towards improving our understanding of electron transfer to and from structural Fe rely on experimental approaches to directly quantify E_h^{Fe} as a function of the oxidation state, the redox history, and solution chemistry. Based on the coordination chemistry of structural Fe in layer silicates, the standard reduction potential for structural Fe ($E_h^{0\ Fe}$) was estimated to range between $E_h^{0\ Fe}$ of 0.71 to 0.74 V (pH 0, SHE, (*88*, *89*)). Experimental validation of these $E_h^{0\ Fe}$ is, however, still scarce. Attempts to characterize E_h^{Fe} by using redox-active surfactants adsorbed to the clay mineral interlayers were unsuccessful (*90*). In principle, traditional batch reactivity assays may be used to estimate E_h^{Fe}. However, the electron transfer in these systems is only indirectly monitored via the kinetics and extent of probe compound transformation. Instead, homogeneous electrocatalysis, successfully used to characterize the redox properties of natural organic matter (*91*), has great promise. In this approach, electron transfer to and from particulate environmental phases at desired reduction potentials and solution pH is directly quantified by chronocoulometry (i.e., integration of reductive and oxidative currents). Electron transfer between the solid phase and the working electrode is mediated by mobile, redox-active organic radicals. Current work is directed towards exploring the possibilities of homogeneous electrocatalysis to elucidate the electron transfer mechanism to and from structural Fe in clay minerals.

Conclusion and Outlook

Assessing the role of Fe-containing clay minerals in biogeochemical processes and pollution dynamics requires a fundamental understanding of the mineral properties that make structural Fe participate in environmental electron transfer reactions. While the available spectroscopic approaches for the description of Fe binding environments are elaborate, methods for the characterization of the clay particle's essential redox properties are currently being developed. Quantifying the Fe^{3+}/Fe^{2+} reduction potential as well as the electron donating and accepting capacities, for example by electrochemical methods, will be key to improve the understanding of the redox chemistry of structural Fe in clay minerals.

Future work should also include a wider range of clay mineral types and structures. Most studies to date have addressed Fe-rich, planar 2:1 clay minerals to illustrate the relevance of Fe-mediated redox processes of clay minerals. In soils, sediments, and waste repositories, however, a variety of Fe-containing clay minerals are present or form as a result of weathering processes and their contribution to the redox cycling of Fe is essentially unknown. Finally, the proposed characterization of clay mineral redox properties should provide the basis for delineating the relevance of redox processes catalyzed by Fe in clay minerals versus that involving Fe (oxyhydr)oxides and other redox active species in the aquatic environments (see other chapters in this volume).

References

1. Christensen, T. H.; Kjeldsen, P.; Bjerg, P. L.; Jensen, D. L.; Christensen, J. B.; Baun, A.; Albrechtsen, H. J.; Heron, G. Biogeochemistry of landfill leachate plumes. *Appl. Geochem.* **2001**, *16*, 659–718.
2. Ernstsen, V.; Gates, W. P.; Stucki, J. W. Microbial reduction of structural iron in clays - A renewable source of reduction capacity. *J. Environ. Qual.* **1998**, *27*, 761–766.
3. Kenneke, J. F.; Weber, E. J. Reductive dehalogenation of halomethanes in iron- and sulfate-reducing sediments. 1. Reactivity pattern analysis. *Environ. Sci. Technol.* **2003**, *37*, 713–720.
4. Rugge, K.; Hofstetter, T. B.; Haderlein, S. B.; Bjerg, P. L.; Knudsen, S.; Zraunig, C.; Mosbaek, H.; Christensen, T. H. Characterization of predominant reductants in an anaerobic leachate-contaminated aquifer by nitroaromatic probe compounds. *Environ. Sci. Technol.* **1998**, *32*, 23–31.

5. Kostka, J. E.; Haefele, E.; Viehweger, R.; Stucki, J. Respiration and dissolution of iron(III)-containing clay minerals by bacteria. *Environ. Sci. Technol.* **1999**, *33*, 3127–3133.
6. Stucki, J. W.; Golden, D. C.; Roth, C. B. Effects of reduction and reoxidation of structural iron on the surface charge and dissolution of dioctahedral smectites. *Clays Clay Miner.* **1984**, *32*, 350–356.
7. Weber, K. A.; Achenbach, L. A.; Coates, J. D. Microorganisms pumping iron: anaerobic microbial iron oxidation and reduction. *Nat. Rev. Microbiol.* **2006**, *4*, 752–764.
8. Schwarzenbach, R. P.; Egli, T.; Hofstetter, T. B.; von Gunten, U.; Wehrli, B. Global water pollution and human health. *Annu. Rev. Environ. Res.* **2010**, *35*, 109–136.
9. Kostka, J. E.; Dalton, D. D.; Skelton, H.; Dollhopf, S.; Stucki, J. W. Growth of iron(III)-reducing bacteria on clay minerals as the sole electron acceptor and comparison of growth yields on a variety of oxidized iron forms. *Appl. Environ. Microbiol.* **2002**, *68*, 6256–6262.
10. Stucki, J. W.; Komadel, P.; Wilkinson, H. T. Microbial reduction of structural iron(III) in smectites. *Soil Sci. Soc. Am. J.* **1987**, *51*, 1663–1665.
11. Schaefer, M. V.; Gorski, C. A.; Scherer, M. M. Spectroscopic evidence for interfacial Fe(II)–Fe(III) electron transfer in a clay mineral. *Environ. Sci. Technol.* **2011**, *45*, 540–545.
12. Hofstetter, T. B.; Neumann, A.; Schwarzenbach, R. P. Reduction of nitroaromatic compounds by Fe(II) species associated with iron-rich smectites. *Environ. Sci. Technol.* **2006**, *40*, 235–242.
13. Lee, W. J.; Batchelor, B. Reductive capacity of natural reductants. *Environ. Sci. Technol.* **2003**, *37*, 535–541.
14. Neumann, A.; Hofstetter, T. B.; Skarpeli-Liati, M.; Schwarzenbach, R. P. Reduction of polychlorinated ethanes and carbon tetrachloride by structural Fe(II) in smectites. *Environ. Sci. Technol.* **2009**, *43*, 4082–4089.
15. Brigatti, M. F.; Franchini, G.; Lugli, C.; Medici, L.; Poppi, L.; Turci, E. Interaction between aqueous chromium solutions and layer silicates. *Appl. Geochem.* **2000**, *15*, 1307–1316.
16. Ilton, E. S.; Haiduc, A.; Moses, C. O.; Heald, S. M.; Elbert, D. C.; Veblen, D. R. Heterogeneous reduction of uranyl by micas: Crystal chemical and solution controls. *Geochim. Cosmochim. Acta* **2004**, *68*, 2417–2435.
17. Peretyazhko, T.; Zachara, J. M.; Heald, S. M.; Jeon, B. H.; Kukkadapu, R. K.; Liu, C.; Moore, D.; Resch, C. T. Heterogeneous reduction of Tc(VII) by Fe(II) at the solid-water interface. *Geochim. Cosmochim. Acta* **2008**, *72*, 1521–1539.
18. Lear, P. R.; Stucki, J. W. Effects of iron oxidation state on the specific surface area of nontronite. *Clays Clay Miner.* **1989**, *37*, 547–552.
19. Stucki, J. W.; Lee, K.; Zhang, L. Z.; Larson, R. A. Effects of iron oxidation state on the surface and structural properties of smectites. *Pure Appl. Chem.* **2002**, *74*, 2145–2158.
20. Yan, L. B.; Stucki, J. W. Structural perturbations in the solid-water interface of redox transformed nontronite. *J. Colloid Interface Sci.* **2000**, *225*, 429–439.

21. Khaled, E. M.; Stucki, J. W. Iron oxidation-state effects on cation fixation in smectites. *Soil Sci. Soc. Am. J.* **1991**, *55*, 550–554.
22. Anastacio, A. S.; Aouad, A.; Sellin, P.; Fabris, J. D.; Bergaya, F.; Stucki, J. W. Characterization of a redox-modified clay mineral with respect to its suitability as a barrier in radioactive waste confinement. *Appl. Clay Sci.* **2008**, *39*, 172–179.
23. Brigatti, M. F.; Galan, E.; Theng, B. K. G. Structures and Mineralogy of Clay Minerals. In *Developments in Clay Science*; Bergaya, F., Theng, B. K. G., Lagaly, G., Eds.; Elsevier: 2006; Vol. 1; pp 19−86.
24. Manceau, A.; Bonnin, D.; Stone, W. E. E.; Sanz, J. Distribution of Fe in the octahedral sheet of trioctahedral micas by polarized EXAFS - Comparison with NMR results. *Phys. Chem. Miner.* **1990**, *17*, 363–370.
25. Vantelon, D.; Montarges-Pelletier, E.; Michot, L. J.; Briois, V.; Pelletier, M.; Thomas, F. Iron distribution in the octahedral sheet of dioctahedral smectites. An FeK-edge X-ray absorption spectroscopy study. *Phys. Chem. Miner.* **2003**, *30*, 44–53.
26. Manceau, A.; Lanson, B.; Drits, V. A.; Chateigner, D.; Gates, W. P.; Wu, J.; Huo, D.; Stucki, J. W. Oxidation-reduction mechanism of iron in dioctahedral smectites: I. Crystal chemistry of oxidized reference nontronites. *Am. Mineral.* **2000**, *85*, 133–152.
27. Komadel, P.; Madejova, J.; Laird, D. A.; Xia, Y.; Stucki, J. W. Reduction of Fe(III) in griffithite. *Clay Miner.* **2000**, *35*, 625–634.
28. Manceau, A.; Drits, V. A.; Lanson, B.; Chateigner, D.; Wu, J.; Huo, D.; Gates, W. P.; Stucki, J. W. Oxidation-reduction mechanism of iron in dioctahedral smectites: II. Crystal chemistry of reduced Garfield nontronite. *Am. Mineral.* **2000**, *85*, 153–172.
29. Komadel, P.; Lear, P. R.; Stucki, J. W. Reduction and reoxidation of nontronite - extent of reduction and reaction rates. *Clays Clay Miner.* **1990**, *38*, 203–208.
30. Neumann, A.; Hofstetter, T. B.; Lussi, M.; Cirpka, O. A.; Petit, S.; Schwarzenbach, R. P. Assessing the redox reactivity of structural iron in smectites using nitroaromatic compounds as kinetic probes. *Environ. Sci. Technol.* **2008**, *42*, 8381–8387.
31. Kostka, J. E.; Wu, J.; Nealson, K. H.; Stucki, J. W. The impact of structural Fe(III) reduction by bacteria on the surface chemistry of smectite clay minerals. *Geochim. Cosmochim. Acta* **1999**, *63*, 3705–3713.
32. Lee, K.; Kostka, J. E.; Stucki, J. W. Comparisons of structural Fe reduction in smectites by bacteria and dithionite: An infrared spectroscopic study. *Clays Clay Miner.* **2006**, *54*, 195–208.
33. Ribeiro, F. R.; Fabris, J. D.; Kostka, J. E.; Komadel, P.; Stucki, J. W. Comparisons of structural iron reduction in smectites by bacteria and dithionite: II. A variable-temperature Mossbauer spectroscopic study of Garfield nontronite. *Pure Appl. Chem.* **2009**, *81*, 1499–1509.
34. Drits, V. A.; McCarty, D. K.; Zviagina, B. B. Crystal-chemical factors responsible for the distribution of octahedral cations over trans- and cis-sites in dioctahedral 2 : 1 layer silicates. *Clays Clay Miner.* **2006**, *54*, 131–152.

35. Guggenheim, S.; Adams, J. M.; Bain, D. C.; Bergaya, F.; Brigatti, M. F.; Drits, V. A.; Formoso, M. L. L.; Galan, E.; Kogure, T.; Stanjek, H. Summary of recommendations of nomenclature committees relevant to clay mineralogy: Report of the Association Internationale pour l'Etude des Argiles (AIPEA) Nomenclature. *Clays Clay Miner.* **2006**, *54*, 761–772.
36. Wolters, F.; Lagaly, G.; Kahr, G.; Nüesch, R.; Emmerich, K. A comprehensive Characterization of dioctahedral Smectites. *Clays Clay Miner.* **2009**, *57*, 115–133.
37. Neumann, A.; Petit, S.; Hofstetter, T. B. Evaluation of redox-active iron sites in smectites using middle and near infrared spectroscopy. *Geochim. Cosmochim. Acta* **2011**, *75*, 2336–2355.
38. Komadel, P.; Madejova, J.; Stucki, J. W. Structural Fe(III) reduction in smectites. *Appl. Clay Sci.* **2006**, *34*, 88–94.
39. Anastacio, A. S.; Harris, B.; Yoo, H. I.; Fabris, J. D.; Stucki, J. W. Limitations of the ferrozine method for quantitative assay of mineral systems for ferrous and total iron. *Geochim. Cosmochim. Acta* **2008**, *72*, 5001–5008.
40. Murad, E. Mössbauer spectroscopy of clays and clay minerals. In *Developments in Clay Science*; Bergaya, F., Theng, B. K. G., Lagaly, G., Eds.; Elsevier: 2006; Vol. 1; pp 755–764.
41. Rancourt, D. G. Mössbauer spectroscopy in clay science. *Hyperfine Interact.* **1998**, *117*, 3–38.
42. Rancourt, D. G.; Christie, I. A. D.; Lamarche, G.; Swainson, I.; Flandrois, S. Magnetisms of synthetic and natural annite mica - Ground state and nature of excitation in an exchange-wise 2-dimensional easy-plane ferromagnet with disorder. *J. Magn. Magn. Mater.* **1994**, *138*, 31–44.
43. Rancourt, D. G.; Ping, J. Y.; Boukili, B.; Robert, J. L. Octahedral-site Fe^{2+}-quadrupole splitting distributions from Mössbauer spectroscopy along the (OH, F)-annite join. *Phys. Chem. Miner.* **1996**, *23*, 63–71.
44. Cardile, C. M.; Johnston, J. H. ^{57}Fe Mossbauer-spectroscopy of montmorillonites - A new Interpretation. *Clays Clay Miner.* **1986**, *34*, 307–313.
45. Johnston, J. H.; Cardile, C. M. Iron sites in nontronite and the effect of interlayer cations from Mossbauer-spectra. *Clays Clay Miner.* **1985**, *33*, 21–30.
46. Besson, G.; Bookin, A. S.; Dainyak, L. G.; Rautureau, M.; Tsipursky, S. I.; Tchoubar, C.; Drits, V. A. Use of diffraction and Mossbauer methods for the structural and crystallochemical characterization of nontronites. *J. Appl. Crystallogr.* **1983**, *16*, 374–383.
47. Heller-Kallai, L.; Rozenson, I. The use of Mossbauer-spectroscopy of iron in clay mineralogy. *Phys. Chem. Miner.* **1981**, *7*, 223–238.
48. Manceau, A. Distribution of cations among the octahedra of phyllosilicates - Insight from EXAFS. *Can. Mineral.* **1990**, *28*, 321–328.
49. Farmer, V. C. The Infrared Spectra of Minerals. In *The Layer Silicates*; Farmer, V. C., Ed.; Mineralogical Society: London, 1974; pp 331–364.
50. Fialips, C. I.; Huo, D.; Yan, L. B.; Wu, J.; Stucki, J. W. Infrared study of reduced and reduced-reoxidized ferruginous smectite. *Clays Clay Miner.* **2002**, *50*, 455–469.

51. Fialips, C. I.; Huo, D. F.; Yan, L. B.; Wu, J.; Stucki, J. W. Effect of Fe oxidation state on the IR spectra of Garfield nontronite. *Am. Mineral.* **2002**, *87*, 630–641.
52. Gates, W. P. Infrared spectroscopy and the chemistry of dioctahedral smectites. In *Application of Vibrational Spectroscopy to Clay Minerals and Layered Double Hydroxides, Workshop of the Clay Minerals Society*; Kloprogge, J. T., Ed.; 2005; Vol. 13; pp 125−168.
53. Madejova, J. FTIR techniques in clay mineral studies. *Vib. Spectrosc.* **2003**, *31*, 1–10.
54. Petit, S.; Caillaud, J.; Righi, D.; Madejova, J.; Elsass, F.; Koster, H. M. Characterization and crystal chemistry of an Fe-rich montmorillonite from Olberg, Germany. *Clay Miner.* **2002**, *37*, 283–297.
55. Petit, S.; Decarreau, A.; Martin, F.; Buchet, R. Refined relationship between the position of the fundamental OH stretching and the first overtones for clays. *Phys. Chem. Miner.* **2004**, *31*, 585–592.
56. Petit, S.; Madejova, J.; Decarreau, A.; Martin, F. Characterization of octahedral substitutions in kaolinites using near infrared spectroscopy. *Clays Clay Miner.* **1999**, *47*, 103–108.
57. Post, J. L.; Noble, P. N. The near-infrared combination band frequencies of dioctahedral smectites, micas, and illites. *Clays Clay Miner.* **1993**, *41*, 639–644.
58. Decarreau, A.; Petit, S.; Martin, F.; Farges, F.; Vieillard, P.; Joussein, E. Hydrothermal synthesis, between 75 and 150°C, of high-charge, ferric nontronites. *Clays Clay Miner.* **2008**, *56*, 322–337.
59. Lear, P. R.; Stucki, J. W. Intervalence electron transfer and magnetic exchange in reduced nontronite. *Clays Clay Miner.* **1987**, *35*, 373–378.
60. Rozenson, I.; Heller-Kallai, L. Reduction and oxidation of Fe^{3+} in dioctahedral smectites. 2. Reduction with sodium sulfide solutions. *Clays Clay Miner.* **1976**, *24*, 283–288.
61. Hunter, D. B.; Bertsch, P. M. In-situ measurements of tetraphenylboron degradation kinetics on clay mineral surfaces by IR. *Environ. Sci. Technol.* **1994**, *28*, 686–691.
62. Hunter, D. B.; Gates, W. P.; Bertsch, P. M.; Kemner, K. M. Degradation of tetrapheynlboron at hydrated smectite surfaces studied by time-resolved IR and X-ray absorption spectroscopies. In *Kinetics and Mechanisms of Reactions at the Mineral/Water Interface*; Sparks, D. L., Grundl, T. J., Eds.; ACS: Washington, DC, 1999; Vol. 715; pp 282−300.
63. Rozenson, I.; Heller-Kallai, L. Reduction and oxidation of Fe^{3+} in dioctahedral smectites. 1. Reduction with hydrazine and dithionite. *Clays Clay Miner.* **1976**, *24*, 271–282.
64. Russell, J. D.; Goodman, B. A.; Fraser, A. R. Infrared and Mössbauer studies of reduced nontronites. *Clays Clay Miner.* **1979**, *27*, 63–71.
65. Gan, H.; Stucki, J. W.; Bailey, G. W. Reduction of structural iron in ferruginous smectite by free-radicals. *Clays Clay Miner.* **1992**, *40*, 659–665.
66. Mayhew, S. G. Redox potential of dithionite and SO_2^- from equilibrium reactions with flavodoxins, methyl viologen and hydrogen plus hydrogenase. *Eur. J. Biochem.* **1978**, *85*, 535–547.

67. Dong, H. L.; Kostka, J. E.; Kim, J. Microscopic evidence for microbial dissolution of smectite. *Clays Clay Miner.* **2003**, *51*, 502–512.
68. Jaisi, D. P.; Dong, H. L.; Morton, J. P. Partitioning of Fe(II) in reduced nontronite (NAu-2) to reactive sites: Reactivity in terms of Tc(VII) reduction. *Clays Clay Miner.* **2008**, *56*, 175–189.
69. Jaisi, D. P.; Kukkadapu, R. K.; Eberl, D. D.; Dong, H. L. Control of Fe(III) site occupancy on the rate and extent of microbial reduction of Fe(III) in nontronite. *Geochim. Cosmochim. Acta* **2005**, *69*, 5429–5440.
70. Stucki, J. W.; Roth, C. B. Oxidation-reduction mechanism for structural iron in nontronite. *Soil Sci. Soc. Am. J.* **1977**, *41*, 808–814.
71. Drits, V. A.; Besson, G.; Muller, F. An improved model for structural transformations of heat-treated aluminous dioctahedral 2:1 layer silicates. *Clays Clay Miner.* **1995**, *43*, 718–731.
72. Muller, F.; Drits, V.; Plancon, A.; Robert, J. L. Structural transformation of 2:1 dioctahedral layer silicates during dehydroxylation-rehydroxylation reactions. *Clays Clay Miner.* **2000**, *48*, 572–585.
73. Stucki, J. W. Properties and Behavior of Iron in Clay Minerals. In *Handbook of Clay Science*; Bergaya, F., Theng, B. K. G., Lagaly, G., Eds.; Elsevier: 2006; Vol. 1; pp 423–477.
74. Coey, J. M. D.; Moukarika, A.; Mcdonagh, C. M. Electron hopping in Cronstedtite. *Solid State Commun.* **1982**, *41*, 797–800.
75. Dong, H. L.; Kukkadapu, R. K.; Fredrickson, J. K.; Zachara, J. M.; Kennedy, D. W.; Kostandarithes, H. M. Microbial reduction of structural Fe(III) in illite and goethite. *Environ. Sci. Technol.* **2003**, *37*, 1268–1276.
76. Komadel, P.; Madejova, J.; Stucki, J. W. Reduction and reoxidation of nontronite - Questions of reversibility. *Clays Clay Miner.* **1995**, *43*, 105–110.
77. Yan, L. B.; Stucki, J. W. Effects of structural Fe oxidation state on the coupling of interlayer water and structural Si-O stretching vibrations in montmorillonite. *Langmuir* **1999**, *15*, 4648–4657.
78. Yan, L. B.; Bailey, G. W. Sorption and abiotic redox transformation of nitrobenzene at the smectite-water interface. *J. Colloid Interface Sci.* **2001**, *241*, 142–153.
79. Brigatti, M. F.; Lugli, C.; Cibin, G.; Marcelli, A.; Giuli, G.; Paris, E.; Mottana, A.; Wu, Z. Y. Reduction and sorption of chromium by Fe(II)-bearing phyllosilicates: Chemical treatments and X-ray absorption spectroscopy (XAS) studies. *Clays Clay Miner.* **2000**, *48*, 272–281.
80. Taylor, R. W.; Shen, S. Y.; Bleam, W. F.; Tu, S. I. Chromate removal by dithionite-reduced clays: Evidence from direct X-ray adsorption near edge spectroscopy (XANES) of chromate reduction at clay surfaces. *Clays Clay Miner.* **2000**, *48*, 648–654.
81. Ilton, E. S.; Heald, S. M.; Smith, S. C.; Elbert, D.; Liu, C. X. Reduction of uranyl in the interlayer region of low iron micas under anoxic and aerobic conditions. *Environ. Sci. Technol.* **2006**, *40*, 5003–5009.
82. Gates, W. P.; Stucki, J. W.; Kirkpatrick, R. J. Structural properties of reduced Upton montmorillonite. *Phys. Chem. Miner.* **1996**, *23*, 535–541.

83. Mermut, A. R.; Cano, A. F. Baseline studies of The Clay Minerals Society Source Clays: Chemical analyses of major elements. *Clays Clay Miner.* **2001**, *49*, 381–386.
84. Drits, V. A.; Manceau, A. A model for the mechanism of Fe^{3+} to Fe^{2+} reduction in dioctahedral smectites. *Clays Clay Miner.* **2000**, *48*, 185–195.
85. Hofstetter, T. B.; Schwarzenbach, R. P.; Haderlein, S. B. Reactivity of Fe(II) species associated with clay minerals. *Environ. Sci. Technol.* **2003**, *37*, 519–528.
86. Favre, F.; Stucki, J. W.; Boivin, P. Redox properties of structural Fe in ferruginous smectite. A discussion of the standard potential and its environmental implications. *Clays Clay Miner.* **2006**, *54*, 466–472.
87. Rosso, K. M.; Ilton, E. S. Charge transport in micas: The kinetics of Fe-II/III electron transfer in the octahedral sheet. *J. Chem. Phys.* **2003**, *119*, 9207–9218.
88. Amonette, J. E. Iron redox chemistry of clays and oxides: environmental applications. In *Electrochemical Properties of Clays*; Fitch, A., Ed.; The Clay Minerals Society: Aurora, CO, 2002; Vol. 10; pp 89−146.
89. White, A. F.; Yee, A. Aqueous oxidation-reduction kinetics associated with coupled electron cation transfer from iron-containing silicates at 25°C. *Geochim. Cosmochim. Acta* **1985**, *49*, 1263–1275.
90. Swearingen, C.; Wu, J.; Stucki, J.; Fitch, A. Use of ferrocenyl surfactants of varying chain lengths to study electron transfer reactions in native montmorillonite clay. *Environ. Sci. Technol.* **2004**, *38*, 5598–5603.
91. Aeschbacher, M.; Sander, M.; Schwarzenbach, R. P. Novel electrochemical approach to assess the redox properties of humic substances. *Environ. Sci. Technol.* **2010**, *44*, 87–93.

Chapter 18

Reactivity of Zerovalent Metals in Aquatic Media: Effects of Organic Surface Coatings

Paul G. Tratnyek,[1,*] Alexandra J. Salter-Blanc,[1] James T. Nurmi,[1] James E. Amonette,[2] Juan Liu,[2] Chongmin Wang,[3] Alice Dohnalkova,[3] and Donald R. Baer[3]

[1]Division of Environmental and Biomolecular Systems, Oregon Health & Science University, 20000 NW Walker Road, Beaverton, OR 97006
[2]Fundamental and Computational Sciences Directorate,
[3]Environmental Molecular Sciences Laboratory, Pacific Northwest National Laboratory, P.O. Box 999, Richland, WA 99352
[*]tratnyek@ebs.ogi.edu

Granular, reactive zerovalent metals (ZVMs)—especially iron (ZVI)—form the basis for model systems that have been used in fundamental and applied studies of a wide variety of environmental processes. This has resulted in notable advances in many areas, including the kinetics and mechanisms of contaminant reduction reactions, theory of filtration and transport of colloids in porous media, and modeling of complex reactive-transport scenarios. Recent emphasis on nano-sized ZVI has created a new opportunity: to advance the understanding of how coatings of organic polyelectrolytes—like natural organic matter (NOM)—influence the reactivity of environmental surfaces. Depending on many factors, organic coatings can be activating or passivating with respect to redox reactions at particle-solution interfaces. In this study, we show the effects of organic coatings on nZVI vary with a number of factors including: (*i*) time (i.e., "aging" is evident not only in the structure and composition of the nZVI but also in the interactions between nZVI and NOM) and (*ii*) the type of organic matter (i.e., suspensions of nZVI are stabilized by NOM and the model polyelectrolyte carboxymethylcellulose (CMC), but NOM stimulates redox reactions involving nZVI while CMC inhibits them).

Introduction

The redox reactivity of zerovalent metals (ZVMs) in aquatic media is relevant in a diverse array of contexts, ranging from corrosion of ferruginous metals in all sorts of structures (bridges, ships, nuclear reactors, etc.) to activity of catalysts for organic synthesis (*1–4*), bioavailability of nutritional supplements (*5*, *6*), weathering of chondritic meteorites (*7*, *8*), metal/oxide cycling for solar hydrogen production (*9*, *10*), lowering carbon emissions from fuel combustion (*11*, *12*), and the recently proposed role of blood-borne metal nanoclusters as nucleating sites for a variety of physiological phenomena (*13*, *14*). However, the context with the most direct relevance to environmental science and engineering is the use of ZVMs (mainly zerovalent iron, ZVI) for water treatment. This technology takes several forms, the most important involving filters for removing metals (where the process is sometimes referred to as "cementation" (*15*, *16*) or "electrocoagulation" (*17*)) and reactive treatment zones for remediation of contaminated groundwater (typically called permeable reactive barriers (*18–21*)). There is also a wide range of variations on these technologies where ZVI is used, including reactive impermeable barriers (*22*), mechanically-mixed soils and sediments (*23*, *24*), constructed wetlands (*25*, *26*), reactive caps for sediments (*27*, *28*), and many combined or sequential remedies (e.g., (*29*, *30*)).

Starting in the early 1990s, interest in ground water remediation applications of ZVMs grew rapidly, and the technology soon became well established (*31*). Simultaneously, a large body of scientific literature developed on many aspects of remediation with ZVMs (*32*), and some of these papers have earned exceptionally high numbers of citations. The quantity—and especially the high citation impact—of this research suggests significance that extends beyond the practical applications of ZVMs to remediation of contamination. The main reason for this is that granular ZVI in aquatic media has become a preferred model system for investigating aspects of many processes related to contaminant fate and redox processes in the aquatic environment. Some of these studies have produced significant advances, and these have fueled further interest in what can be learned using ZVI model systems.

A prominent example of a fundamental advance from early work done using ZVI model systems is the determination of how branching between hydrogenolysis and reductive-elimination pathways of dechlorination determine the final distribution of products from this important contaminant degradation process (*33–35*). The data originally used to demonstrate the dynamics of this complex set of reactions were obtained with batch model systems containing ZVI or zerovalent zinc (ZVZ), but the conceptual model is now used widely in interpreting the outcome of dehalogenation in other systems (*36–40*). Another fundamental advance that derived from early work on dechlorination with ZVMs concerns the relationship between dechlorination rates and the structure of chlorinated aliphatic parent compounds. Using rate constants from batch experiments performed with ZVI, the first quantitative structure-activity relationships (QSARs) were derived for this contaminant degradation pathway (*41*), which has lead to numerous efforts to obtain additional or alternative QSARs for dechlorination (*42–47*). Other areas where studies using ZVM model systems have produced notable advances of fundamental significance include diagnosing mechanisms of dechlorination from carbon isotope fractionation (*36*) and molecular modeling (*48*), kinetic modeling of complex surface reactions (*49*), and reactive transport in heterogeneous media (*50*).

Since the late 1990s, much of the research on ZVMs has focused on nano-sized materials (e.g., nZVI). Model systems based on nZVI have lead to new emphasis on particle-size dependent effects, and significant advances have been made as a result of this work. The most prominent example of this is advancement of classical filtration theory to accommodate a wider range of particle interactions (*51–54*). Among the particle interaction effects that have proven to be most important to the colloidal behavior of nZVI are those mediated by organic surface coatings. In addition to the effects of organic surface coatings on nZVI aggregation, attachment, and transport (*55–60*), they influence reactivity. This chapter provides a perspective on results regarding the effects of organic matter on reactivity of nZVI and other particulate ZVMs.

Background

Organic coatings can influence the reactivity of surfaces in many ways that could be significant under environmental, aquatic conditions. With respect to reactivity of the granular ZVI that traditionally has been used for groundwater remediation, four types of effects on contaminant degradation have been proposed (*61–63*). The first is *enhanced solubilization* of hydrophobic organic contaminants by organic polyelectrolytes, making the contaminants more mobile and possibly more available for reaction with the particle surface. Second, *enhanced sorption* might result if the organic polyelectrolyte coats the particle surface first, making sorption of the contaminant to the modified surface more favorable. Third, *competitive sorption* may arise if the organic coating that forms on the particle surface inhibits reaction with organic contaminants. And, fourth, *mediated electron transfer* could be important if the organic polyelectrolyte (either as a surface coating or as a solute) serves as a catalyst by shuttling electrons between the ZVM and the organic contaminant. Recently, additional studies have added further evidence that both inhibition and acceleration of contaminant removal are possible, depending on various operational factors (*64–66*).

The four-part conceptual model described above was formulated for systems where reaction occurs on the surface of ZVM particles that are large relative to other features of the system. Allowing for nano-sized ZVMs—and micelles or other relatively large features formed from amphiphilic organic polyelectrolytes—introduces additional considerations (*67*, *68*). As the ZVM particles become small relative to the organic polyelectrolyte, their relationship evolves from molecules of adsorbed organic polyelectrolyte distributed on a ZVM particle surface, to a thin film of organic polyelectrolyte coating ZVM particles, an aggregate of nZVI bound together by an organic polyelectrolyte phase, and, finally, a particle composed primarily of organic polyelectrolyte that is embedded or encrusted with nZVI. The organic coating conceptual model is prototypical and illustrated in Fig. 1., although the interpretation of real experimental data (such as presented below) usually requires accommodation for non-uniform or incomplete surface coatings, complex and heterogeneous aggregate compositions, etc.

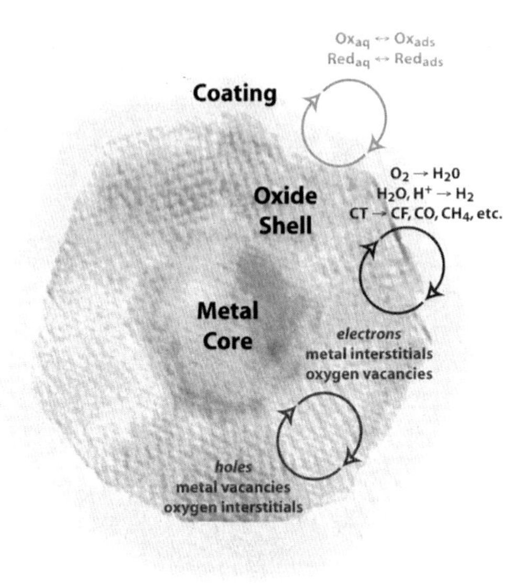

Figure 1. Organic coating model for the effects of organic polyelectrolytes on the structure and properties of nano ZVMs that have core-shell structure consisting of metal and metal oxide.

With respect to nZVI, the organic coatings of primary significance are those that are deliberately added to impart properties to nZVI that make the material more suitable for groundwater remediation applications. The main concern here has been with making nZVI more mobile in porous geological materials by decreasing its tendency to aggregate with itself and other colloids or to stick to the mineral grains of the aquifer matrix. This has been achieved by coating the particles with a variety of organics, including anionic polyelectrolytes like polyacrylate, polyaspartate, and carboxymethyl cellulose (*56*, *58*, *69–78*); other polysaccharides like guar, xanthan gum, and cyclodextrins (*74–77*, *79–82*); and synthetic surfactants or water-soluble polymers like the triblock copolymers and polyvinylpyrrolidone (*55*, *75*, *83–85*). Coatings of this type are now part of almost all formulations of nZVI that are used in field scale remediation applications (*69*).

Of the alternative materials for coating nZVI, carboxymethyl cellulose (CMC) probably has been used in the widest range of laboratory and field studies of nZVI behavior. In addition to being inexpensive, nontoxic, and biodegradable, CMC is highly effective at controlling the aggregation and sedimentation of nZVI (*58*, *76–78*, *86–88*). Synthesizing nZVI in the presence of CMC can produce primary particles with Fe^0/Fe-oxide core-shell structure that are uniform, spherical, and <20 nm in size; unaggregated and homogeneously dispersed, and non-settling over time periods of hours to days. This is similar to the benefits that have been reported for synthesis of other types of nanoparticles in CMC, including various noble metals (Ag, Au, Pd, Pt, etc.) and iron oxides (*89*, *90*). However, the effects of CMC on reactivity of nZVI are variable and less well understood. For example, rates of TCE reduction increase when palladized nZVI is synthesized in CMC (*68*, *91*) but decrease when nZVI is prepared separately and then dispersed in CMC (*67*). The former could be due to smaller particle size (and therefore greater reactive surface area) and the latter is, at least in part, due to blockage of reactive surface sites by adsorbed polymer. However, there are many potentially complicating factors, such as differences in the arrangement of the polymer chains on the particle surface (*67*).

Another type of organic coating that could be very significant involves natural organic matter (NOM), which is ubiquitous in environmental waters at concentrations ranging from tenths of a mg/L in seawater and groundwater to tens of mg/L in some surface waters (*92*). The surfactancy of NOM is well known (*93*, *94*) and the resulting organic coatings can profoundly alter the properties of mineral surfaces (*95*, *96*). NOM and closely related materials such as green tea extract are known to affect the colloidal properties (aggregation, etc.) of nZVI in ways that are similar to the surfactants that are used in formulation of nZVI for groundwater remediation (*59*, *97*), but the effect of such materials on reactivity of nZVI is less well characterized. The characteristics of these systems present some interesting possibilities, such as alteration of product distributions from contaminant degradation due to the local concentration of H-donors at the surface. This hydrogen could be present as sorbed atomic or molecular H, as part of surface hydroxyl groups, or as part of organic structures (alkyl, amide, alcohol, carboxylate, etc.) sorbed at the surface. Atomic or molecular H, if available near the surface, could participate directly in contaminant reduction reactions, potentially resulting in significant changes in the distribution of degradation products (*98*).

Another possible effect on nZVI reactivity by adsorbed surfactants involves the availability of specific surface sites to solvent and solute molecules. Organic molecules containing amino (*99*), phenol (*100*), and carboxyl (*101*, *102*) functional groups can form surface complexes, thereby blocking other adsorbates from accessing these sites. If the adsorbed moieties are redox active (as are the hydroquinone moieties associated with NOM), they may function as electron shuttles that acquire electrons from strong reductants (e.g., iron-reducing bacteria), and then release electrons to Fe^{III} oxides upon sorption (*103*). A complex organic molecule, such as natural organic matter, that contains quinone as well as other functional groups, thus, could remain sorbed to an oxide surface and potentially serve as an electron-transfer mediator with the bulk solution (Fig. 1). These interactions of organic molecules with oxide surfaces have long been recognized, but their influence on the reactivity of environmental nanoparticles has become a focus of research only recently (*64*, *104–111*).

Looking beyond the recent but limited progress on understanding the primary interactions between nZVI and its organic coatings, essentially no work has been done on the higher order questions of how the coatings age, how they are affected by nZVI aging, and how the fate and effects of the coating vary with environmental conditions such as the concentrations of potential competitive adsorbants. Even for nZVI without deliberate organic coatings, the fundamental processes controlling the effects of organic coatings on nZVI reactivity are relevant because organic matter is ubiquitous in the environment and commonly found adsorbed to surfaces of all types. Finally, we note that this issue has special relevance for possible application of nZVI to remediation at field sites—such as those in the U.S. Department of Energy complex—that contain high concentrations of surfactants, complexing agents, and other organic co-contaminants.

Methods

Reagents

The nZVI used in this study was RNIP-10DP (Toda America Inc., Schaumburg, IL, Lot 160804), which was shipped and stored as a powder under dry, anoxic conditions. Properties of this material—which we have designated $Fe^{H2(D)}$, for nZVI prepared by H_2 reduction and stored dry—are described in previous work (*112–114*). Note that the manufacturer's name for this material is RNIP-10DP, but it was erroneously designated as RNIP-10DS (a similar product shipped in the form of an aqueous slurry) in two early papers (*112*, *113*).

NOM was provided by Baohua Gu (Oak Ridge National Laboratory, Oak Ridge, TN), who obtained the raw material (NOM-GT) by reverse osmosis of brown water from a wetland pond in Georgetown, SC (*101*). The carbon content of NOM-GT is 48.3% (*115*), and it has been further characterized by a variety of other methods (*116–119*). Carboxymethyl cellulose (CMC) with molecular weight ~90,000 g/mol was obtained from Aldrich. All solutions were prepared in deoxygenated deionized water (DO/DI) and/or were deoxygenated following preparation (dexoxygenation was performed by sparging with nitrogen or argon gas). Carbonate buffer was prepared by dissolving reagent-grade sodium bicarbonate in DO/DI water. Saturated stock solutions of carbon tetrachloride (CT) were prepared in DI water.

Reactivity of nZVI in Solution

Changes in the properties of nZVI while in solution were determined using $Fe^{H2(D)}$ at a solid/solution ratio of 0.83 mg mL^{-1} in serum vials (30 or 60 mL) with minimal headspace. Experiments were run for up to 63 d, under a N_2 atmosphere at 22°C, in DO/DI water containing 0, 20, or 200 μg ml^{-1} NOM-GT.

Zeta-potential measurements were performed by taking 1-mL aliquots from 60-mL suspensions, diluting the sample with 2 mL of 5 mM KCl, sonicating for 5 minutes to disperse the particles, and analyzing immediately with a ZetaPALS zeta potential analyzer (Brookhaven Instruments Corp., Holtsville, NY). Three successive measurements were made within a total time of about 90-150 s. High resolution transmission electron microscopy (TEM) was carried out using a JOEL JEM 2010 microscope with a specified point-to-point resolution of 0.194 nm. The operating voltage was 200 kV. Sample handling for TEM was performed as described previously (*113*, *120*, *121*).

Micro-X-ray diffraction (μXRD) analysis was performed on samples from the 30-mL serum vials, which were preserved by immediately flash drying using the method (FDv2) described by Nurmi et al. (*114*). Specimens for micro-X-ray diffraction (μXRD) were prepared under N_2 by mixing a few mg of the dried solid with a couple drops of tricaprylylmethylammonium chloride (to prevent oxidation during the analysis). A small drop of this mixture was transferred to an Al stub and then analyzed immediately using a Rigaku MicroMax-007 HF microfocus generator (Rigaku Americas, The Woodlands, TX), operating with CrKα radiation and a curved-image plate detector.

Electrochemical characterization included linear sweep voltammograms (LSVs) obtained in two-electrode cells, containing a powder disk electrode (PDE) made with $Fe^{H2(D)}$ and a Ag/AgCl reference electrode. Details of the design and electrochemical properties of the PDE used in this study have been published previously (*122–124*). All potentials are reported relative to the Ag/AgCl reference.

Reactivity of Contaminants with nZVI

Reactivity of nZVI with contaminants was determined using carbon tetrachloride (CT) as a model chlorinated aliphatic compound. Batch experiments were performed in 160-mL serum vials sealed with Hycar® septa (Thermo Scientific) and aluminum crimp caps. Reactors were prepared in an anoxic chamber, and filled with 100 mg $Fe^{H2(D)}$ and sufficient solution to allow for only minimal headspace. Following preparation, reactors were sonicated until the nZVI visually dispersed (solution appeared black), which took approximately 2-5 minutes. Reactors were then allowed to equilibrate for ~48 hr while rotating end-over-end at ~9 RPM. Following the "preexposure" period, reactors where spiked with 100 μL of a saturated CT stock solution for a nominal concentration of ~4 μM. Aliquots (1 mL) were removed from the reactor vial—while replacing the solution volume with a second needle through the septum to minimize the formation of headspace—for analysis by gas chromatography. Gas chromatography was performed with a DB-624 column (J&W/Agilent) and electron capture detection.

Results and Discussion

Structure and Composition of nZVI

The structure and composition of nZVI—and its evolution with time upon exposure to water—is similar to that of any surface of a metal passivated with a metal oxide, but with potentially important differences that may be regarded as nano-size effects. As a dry powder, nZVI typically consists of an α-Fe^0 core that is passivated by a 2-3 nm thick iron oxide shell (*113*, *125*, *126*). In this state, oxidation of the Fe^0 can proceed only by adsorption, solid-state diffusion, and reduction of (atmospheric) oxygen to form iron oxide (a process whose rate decreases exponentially with the thickness of the oxide film). In an inert atmosphere (absent oxygen and moisture), the core-shell structure of dry nZVI is stable for months (*114*, *125*).

Upon immersion of nZVI into water, breakdown of the passive film occurs, presumably by approximately the same mechanism as iron depassivation in general (*127*). Protons adsorb at the surface and diffuse into the oxide layer, thereby weakening Fe-O bonds and progressively changing the structure and composition of the passive layer from an Fe(III)-oxide to a mixed Fe(II)-Fe(III) hydroxide. Enhanced diffusion of both positive (H^+, Fe^{2+} cations) and negative (electrons) charges through the hydroxide layer—together with diffusion and removal of neutral reduction products such as H^0 by the aqueous solution—presumably account for the faster rate of core Fe^0 oxidation than occurs with nZVI in the dry state. As the thickness of the new hydroxide layer increases, however, the rate of core Fe^0 oxidation decreases relative to the peak oxidation rate reached shortly after immersion. For nZVI, it appears that a new steady-state condition—where corrosion is limited primarily by the rate of proton diffusion through the layer—is reached within a few hours ((*114*, *123*) and electrochemical results presented below).

With time, the Fe(II)-Fe(III) hydroxide shell—which has variable and usually indeterminant composition sometimes characterized by the metastable phase green rust (*128–130*)—transforms to magnetite (Fe_3O_4) or a similar spinel type structure (*131*, *132*). Because these phases are highly conductive, and dissolution of aqueous Fe(II) species is favorable, a stable and strongly passivating oxide layer analogous to the dry state may never form. Instead, corrosion is sustained by electrons conducted through the oxide layer to reduce H_2O and O_2 in cathodic regions of the surface, while Fe(II) forms and may dissolve at anodic regions that typically are located at pits or other defects in the oxide layer. It is unclear at this time if there are true nano-size effects that alter the mechanism of aqueous corrosion of nZVI relative to flatter iron surfaces.

The processes summarized above cause the transformations of nZVI in DO/DI water that we have described as "aging" in previous work (*120*, *123*). Some of the data from (*120*) are also used here in Figs. 2-3 to provide a basis for interpreting the effect of NOM on nZVI aging. The TEMs in Fig. 2 show that the nZVI particles have relatively smooth surfaces and a thin shell of oxide before contact with water (Fig. 2, 0 d). Over the first week of exposure to DO/DI water, Fe(II) dissolves, and subsequently precipitates during preparation for TEM as the euhedral sheets of green rust that are seen in the 3 day sample in Fig. 2. During the same period, some surface roughening is evident (Fig. 2, 3 d), and the Fe^0 content of the solid ($\chi_{Fe(0)}$, determined from XRD as described previously (*123*)) decreases by about one fourth (Fig. 3).

With further aging, the quantity of green-rust and roughness of the surface seem to stabilize (Fig. 2, 11 d and 44 d). The Fe^0 content continues to decrease after the first week, but at a rate of less than a mole % per week (Fig. 3). Analysis of dissolved Fe(II) levels, even during the first few days of aging, indicates only very small amounts ($\ll$ 1% of the total) in solution at any time (data not shown).

In the presence of 20 mg/L NOM-GT, some changes in the aging of nZVI are evident by TEM (Fig 2). The granularity of the oxide shell seems to increase more than it does in DO/DI water, and there is no indication of green rust precipitation

(Fig. 2, right column). Rather, a secondary polymer or thin film is seen coating the particles, especially after one week of contact (Fig. 2, 11 d and 44 d). The Fe^0 content in crystalline phases decreases more slowly than in the absence of NOM, although after 15 d it approaches that observed in the absence of NOM (Fig. 3). When the concentration of NOM is increased to 200 mg/L, the Fe^0 content in the crystalline phases decreases even more slowly, only about 7 mole % over one month (Fig. 3).

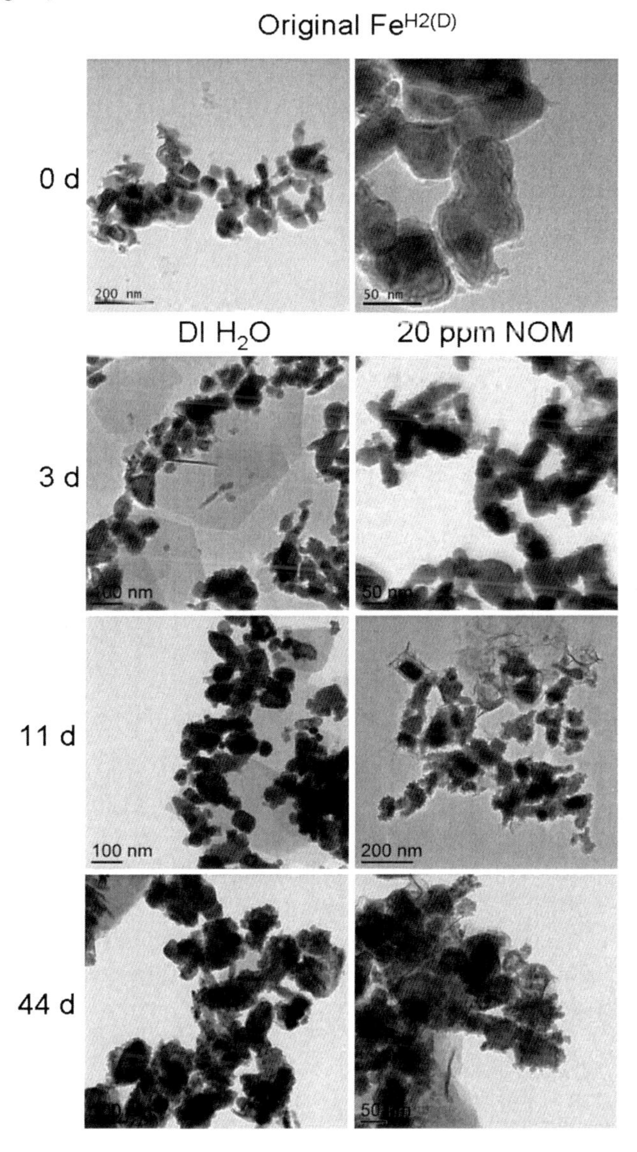

Figure 2. TEM images of the original dry $Fe^{H2(D)}$ and of this material after suspension in either DO/DI water (left) or 20 mg/L NOM-GT in DO/DI water (right) for periods of 3, 11, and 44 days.

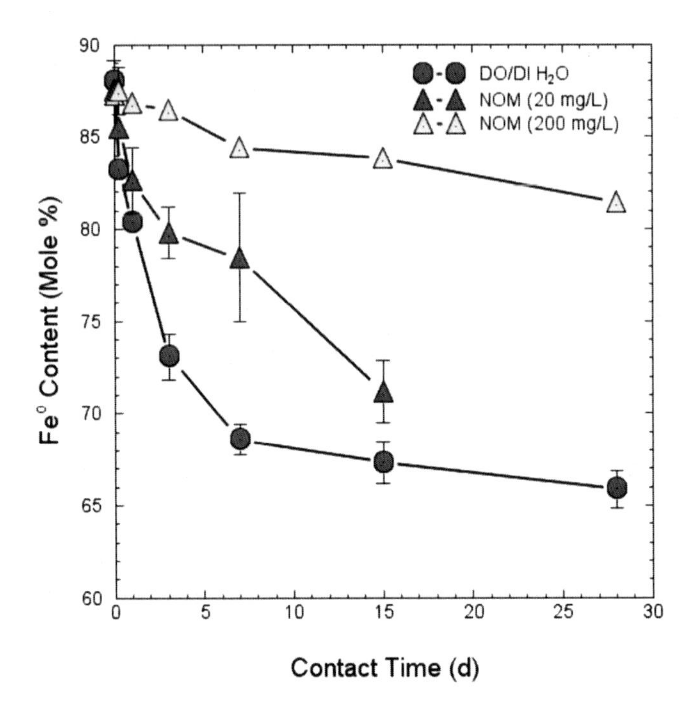

Figure 3. Changes in the Fe^0 content ($\chi^{Fe(0)}$ in mole percent) of $Fe^{H2(D)}$ after immersion in DO/DI water, or solutions containing 20 or 200 mg/L NOM in DO/DI water.

The long-term effect of NOM suggested by the XRD data in Fig. 3 seems to be inhibition of reaction between the nZVI and H_2O by coating the surface with relatively unreactive organic matter. This interpretation, however, rests on an assumption that all the Fe in the NOM system is reflected in the XRD data. Although noncrystalline phases were not evident in the XRD data, complexation of Fe(II) with NOM is expected and is suggested by the films associated with the nZVI particles that were seen in the TEM data in Fig. 2. Additional evidence for Fe-NOM complexation was presented by Baer et al. (*133*) in the form of X-ray photoelectron (XPS) spectra showing that 20 mg/L NOM resulted in large increases in surface carbon and corresponding decreases in surface iron (atomic %). Allowing for this evidence that XRD data likely do not represent all of the Fe in this system, it is possible that the main effect of NOM in these experiments was simply to inhibit the formation of green rust and magnetite from the Fe(II) released by reaction with H_2O.

Surface Properties of nZVI

Surface-specific properties, such as zeta potential, should be highly sensitive to the formation of coatings of organic matter on nanoparticles and, therefore, may be essential to evaluating their environmental transport and fate (*53*, *134*). Zeta potential measurements of nZVI are challenging to make and interpret, however, because of the material's strong tendency to aggregate due to magnetic and other attractive forces (*51*, *56*, *107*, *113*). Nevertheless, several recent studies have used changes in zeta potential measurements to interpret effects of organic coatings on properties of nZVI (*60*, *107*).

We measured zeta potential of $Fe^{H2(D)}$ in DO/DI water, with 0 or 20 mg/L NOM and up to ~1 month of contact time, to test for coupled effects of NOM and aging on nZVI (Fig. 4). In the absence of NOM, the initial zeta potential of $Fe^{H2(D)}$ is positive, as expected for a synthetic ferric oxide surface equilibrated at neutral pH (*135*, *136*). However, the zeta potential decreased exponentially with time, with reversal of the surface charge occurring after ~35 d of aging. We attribute this change to gradual formation of a nearly complete magnetite surface layer, which lowers the isoelectric point below pH 7 (*135*, *136*), and to an increase in solution pH to well above 7, resulting in the deprotonation of most of the surface sites. Other neutral-pH measurements of zeta potential on nZVI have typically yielded negative values, but these have either been on samples of nZVI that have been in contact with water for several months before measurement (*51*, *58*, *107*) or on specimens prepared by reduction with borohydride (*60*, *106*).

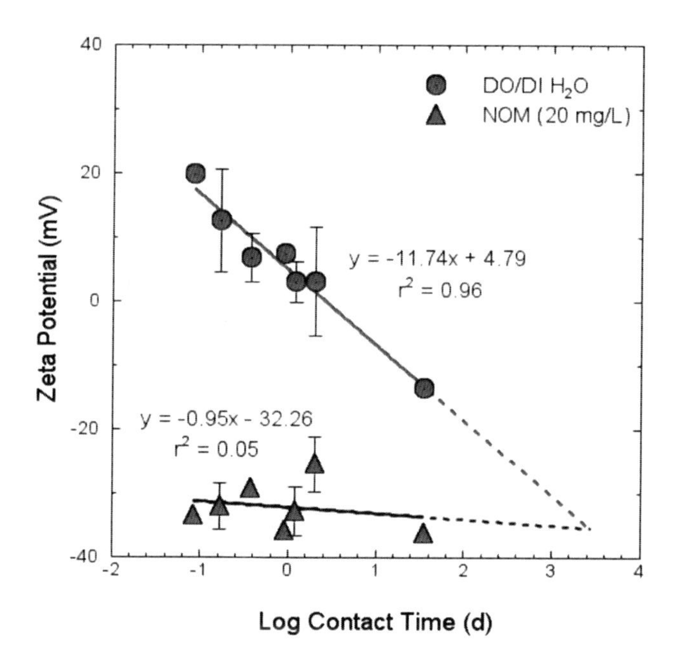

Figure 4. Changes in the zeta potential of $Fe^{H2(D)}$ nanoparticles after immersion in DO/DI water, or solutions containing 20 mg/L NOM in DO/DI water. Error bars represent standard deviation (n = 3).

In the presence of 20 mg/L NOM, initial measurements of zeta potential were about 50 mV more negative than we obtained for pure DO/DI water, and there was no significant change in the zeta potential with aging time (Fig. 4). One possible explanation for these results is that the zeta potential measurements in the presence of NOM are dominated by the polyelectrolytic properties of the NOM (rather than $Fe^{H2(D)}$ surfaces). Although complexation between dissolved Fe(II) and NOM eventually produces a polymeric phase that appears distinct—as shown in Fig. 2—the external surfaces of this material likely retained similar charge characteristics to the original NOM, which could account for the lack of an aging effect on zeta potential measurements in the presence of NOM. Similar decreases in zeta potential of nanoparticles after contact with NOM have been reported for nZVI (*60*, *106*, *107*) and other metal oxide (*137*, *138*) nanoparticles after contact with NOM solutions. In at least one instance, the zeta potential of the nZVI-NOM particles was even more negative than that of the original NOM (*106*).

Fitting the trends in zeta potential shown in Fig. 4 facilitates extrapolation to longer aging times, which suggests that the zeta potential of both systems might converge after a number of years. Complete convergence, however, is unlikely due to the different acidity properties of the dominant functional groups in the two systems. In every experiment conducted to date with nZVI particles where zeta potential has been measured, contact with NOM has resulted in a decrease in zeta potential, regardless of the age of the nZVI. The main result then, is that the zeta potential values for $Fe^{H2(D)}$ particles in DO/DI water clearly showed an evolution consistent with that expected for iron surfaces, whereas the values in the presence of NOM were consistently negative and did not evolve with time.

Electrochemical Reactivity of nZVI

Since we are particularly interested in changes in the redox activity of nZVI in solutions, a useful way to characterize this is with time resolved electrochemical measurements. This can be done by packing nZVI into powder disk electrodes (PDEs), as we have described in two recent publications (*122*, *124*). Using the same materials and solutions as in other parts of this study, we obtained the open-circuit chronopotentiograms (CPs) shown in Fig. 5.

The measured parameter in these experiments (E_{corr}) is the corrosion potential of the nZVI that comprises the PDE. E_{corr} measured under such conditions is a mixed potential reflecting the equilibrium potential of the dominant redox couples, the kinetics of charge transfer between these species and the electrode, and mass transport of species into the pore space and across the passive film (*124*). Despite this complexity, CPs provide a sensitive indicator of changes in redox reactions at the surface of the material that comprises the PDE (*122*). For $Fe^{H2(D)}$ in DO/DI water, we have shown previously that E_{corr} starts around -0.6 V vs. Ag/AgCl, decreases steeply to a relative plateau of about -0.75 V at 8 hr, then decreases steeply again (at the Flade potential, corresponding to passive film breakdown) to a final plateau around -0.85 V at 16 hr (*123*). The final value of Ecorr is typical of nZVI in the active state (*113*), although it is lower than we have observed with PDEs made of micron-sized particles of ZVI. The new data presented in Fig. 5 are for $Fe^{H2(D)}$ in carbonate buffer, and they show a minimum E_{corr} of about -0.7

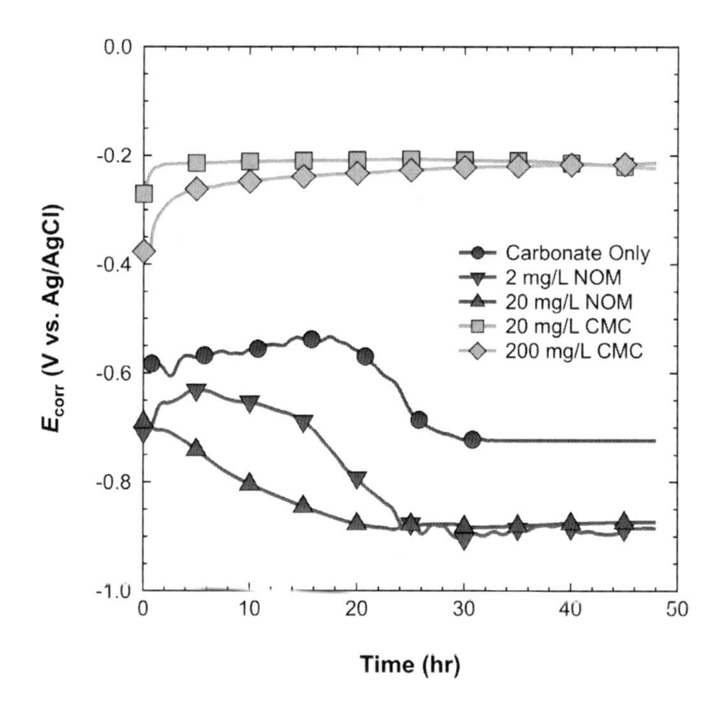

Figure 5. Effect of organic polyelectrolytes on the open circuit chronopotentiograms (CPs) obtained with stationary powder disk electrodes (PDEs) made with Fe $^{H2(D)}$. All experiments in 200 mg/L carbonate at pH ~7.4. Symbols mark every 300th data point.

V vs. Ag/AgCl (at around 20 hr). This anodic shift in the minimum E_{corr} suggests a degree of passivation of the nZVI, possibly due to precipitation of carbonate phases, which we have observed previously with micro ZVI (*122*).

Adding NOM to carbonate buffer shifts the CP to more negative E_{corr}'s and also makes the Flade potential appear earlier. This suggests that NOM favors depassivation of the nZVI, consistent with previous results obtained with micro ZVI (*122*). If this is because complexation of Fe(II) with NOM inhibits formation of passivating surface layers, it would be consistent with the discussion of the TEM data presented above. Other mechanisms are possible, however, such as localized acidity from functional groups associated with the adsorbed NOM. Under circumstances where NOM adsorbs before the nZVI makes contact with an inert electrode (e.g., such as in oxidation-reduction potential, ORP, measurements made on Pt), this has an insulating effect that results in less negative electrode potentials (*104*).

Using CMC as polyelectrolyte has an effect that is similar to that of NOM on ORP (*104*), but opposite the effect reported here for E_{corr}. The CPs in Fig 5 for two concentrations of CMC in carbonate buffer start at relatively positive E_{corr}'s and then increase quickly to a plateau at around -0.2 V vs. Ag/AgCl. These results suggest passivation of $Fe^{H2(D)}$, most likely due to a combination of factors: (*i*) formation of organic film on the nZVI surface that is less permeable and/or electrically conductive than with NOM and (*ii*) lack of redox active or strongly acid functional groups in CMC to facilitate interfacial reactions. These considerations might suggest greater overall stability of nZVI in solutions of polyelectrolytes such as CMC. However, the effect of CMC on the overall activity of nZVI due to corrosion reactions, as reflected by changing $\chi_{Fe(0)}$ (Fig. 3) or E_{corr}, (Fig. 5) is not necessarily the same as CMC's effect on reduction of specific solutes (esp. contaminants). The latter is addressed by the results presented next.

Reactivity of nZVI with Contaminants

The effect of polyelectrolytes on the kinetics of contaminant degradation by nZVI was investigated using carbon tetrachloride (CT) so that the results could be compared using the large body of kinetic data for CT versus ZVI that we and others have reported in past work (*113*, *123*, *139–141*). Figure 6 summarizes rate constants for reaction of CT with several types nZVI, with emphasis on $Fe^{H2(D)}$ aged for various time periods in DO/DI water and solutions of NOM, bicarbonate, and NOM with bicarbonate. The data are displayed in a plot of log k_{SA} (surface area normalized rate constant) vs. log k_M (mass normalized rate constant), where k_{SA} is calculated from k_M assuming one value of specific surface area (a_s) for each material (represented by the diagonal lines). This format provides a flexible framework for comparing the effects of multiple experimental variables (*142*) and has proven particularly useful for diagnosing interaction effects between variables, such as ZVM type vs. contaminant type (*112*) or background solution chemistry (*139*). In this case (Fig. 6), we use the graph to put the effects of organic coatings in the context of other factors (ZVI type, but not other ZVMs or bimetallics; various aging times, but all prepared in DO/DI water; and CT as the probe compound, not other contaminants).

A compilation of literature data on CT disappearance in batch experiments with nZVI is shown in Fig. 6 with unfilled circles enclosed with a gray oval. This dataset does not include the rate constants for construction grade or high-purity micron-sized ZVI that we showed in previous versions of this figure (*139*, *143*), but it includes additional kinetic data on the effect of aging Fe^{H2} in water for up to ~7 months (*123*). Therefore, the shaded region in Fig. 6 defines the range of kinetic data that is typical for CT degradation by simple model systems involving unmodified nZVI in DO/DI water without added organic polyelectrolytes. Superimposed onto this are new data (represented with filled symbols) for CT degradation by $Fe^{H2(D)}$ aged for two days in DO/DI water or solutions of DO/DI water containing carbonate, NOM or CMC, and combinations of carbonate and NOM or CMC.

The new data in Fig. 6 show that the rate of CT degradation by $Fe^{H2(D)}$ is decreased by the addition of carbonate, NOM, and CMC. The inhibitory effect of carbonate (alone) has been seen with other contaminants (*144*, *145*) when enough contact time is allowed for precipitation of carbonates to coat the iron surfaces. The inhibitory effect of CMC (without carbonate) and of NOM (with or without carbonate) on CT degradation is up to two orders of magnitude in the rate constants, which is considerably larger than the inhibitory effects of organic coating reported previously for CT and trichloroethylene (TCE) with NOM (*61*, *107*, *146*, *147*), TCE with organic polymers (*109*), or arsenic and chromate with

NOM (*106, 110*). A likely explanation of this relatively large inhibitory effect is the two-day preexposure period used in this study, which may have allowed time for greater adsorption of organic matter to the ZVI surfaces and perturbation of dissolution/precipitation processes such as the formation of secondary iron oxide phases (as suggested by the TEM data in Fig. 2). The data for 20 mg/L NOM suggest a large synergistic effect of carbonate, which would not be surprising given the complexities we have noted previously for both effects (*61, 122, 144*), but carbonate does not affect the results at 200 mg/L NOM, so it is unclear if this interaction effect is significant.

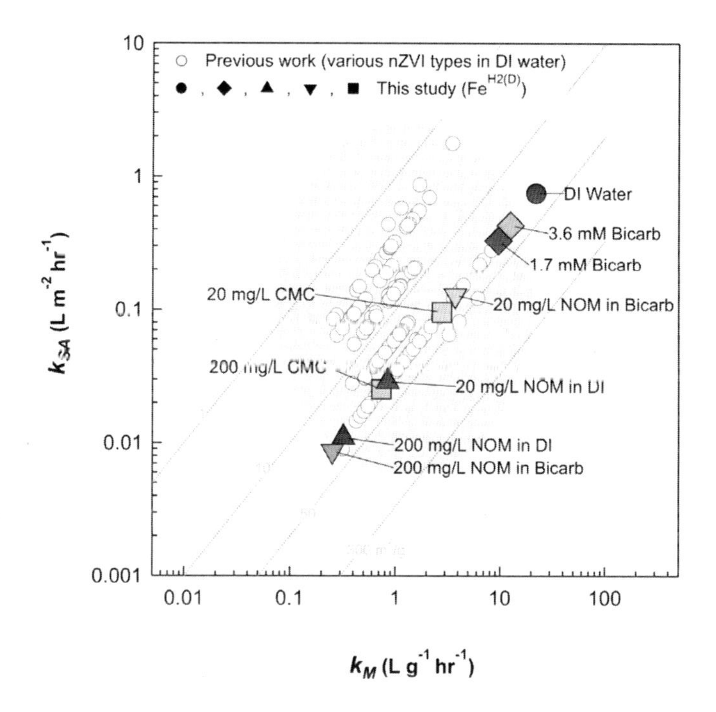

Figure 6. Disappearance of carbon tetrachloride (CT) in the presence of various types of nZVI aged in DO/DI water solutions of NOM, sodium bicarbonate, or NOM with sodium bicarbonate. Data for $Fe^{H2(W)}$ and $Fe^{H2(D)}$ (123) and other nZVI materials (141) in DO/DI water are enclosed in the gray oval, which represents the rage of rate constants typical for nZVI aged in DO/DI water up to ~7 months. New data for $Fe^{H2(D)}$ aged in solution for ~48 hr are shown with labels specifying each condition.

Conclusions

Just as zerovalent iron has been the model system used in recent studies that have made fundamental advances in a number of areas of aquatic redox chemistry, studies with nano-sized zerovalent iron are playing a large role in advancing our understanding of the environmental fate and effects of nanoparticles. An emerging example of these trends concerns the role of organic surface coatings in influencing the reactivity of environmental particles, which has increasingly become the focus of recent studies on zerovalent metals in aquatic media. The problem has many facets, reflecting the variety of organic amphiphiles and their properties, the dynamic nature of the oxide film on reactive metals in solution, and the numerous potential interaction effects between the constituents of the system. As a consequence, experimental results have shown effects of organic coatings on reactivity of zerovalent metals that range from negligible to large, depending on a host of factors that can usually be rationalized but often were not anticipated. Despite this complexity, it appears that the main effect of most organic coatings is partial passivation of particle surfaces by forming a semi-permeable barrier to reactants from solution. This is the only effect of surfactants that lack acidic or reactive functional groups, but NOM contains both types of moieties and therefore can stimulate surface reaction through specific interactions at the metal-oxide interface. The potential remains to engineer organic coatings to modify the surface properties of zerovalent metals for a wide range of applications, which will continue to motivate additional work in this area.

Acknowledgments

We acknowledge P. Nachimuthu and C. K. Russell for assistance with the XRD measurements and sample preparation. Samples of $Fe^{H2(D)}$ were donated by Toda America, Inc. This work was supported by the U.S. Department of Energy (DOE), Offices of Biological and Environmental Research (Environmental Management Sciences Program DE-FG07-02ER63485) and Basic Energy Sciences (Chemical Sciences, Geosciences, and Biosciences Division). A portion of the work was conducted in the William R. Wiley Environmental Molecular Sciences Laboratory (EMSL), a DOE Scientific User Facility operated by Battelle for DOE's Office of Biological and Environmental Research and located at the Pacific Northwest National Laboratory (PNNL). PNNL is operated for the DOE by Battelle under Contract DE-AC06-76RLO 1830. This report has not been reviewed by any of the sponsors and therefore does not necessarily reflect their views and no official endorsement should be inferred.

References

1. Enthaler, S.; Junge, K.; Beller, M. Sustainable metal catalysis with iron: From rust to a rising star? *Angew. Chem., Int. Ed.* **2008**, *47*, 3317–3321.
2. Huber, D. L. Synthesis, properties, and applications of iron nanoparticles. *Small* **2005**, *1*, 482–501.
3. Schloegl, R.; Schoonmaker, R. C.; Muhler, M.; Ertl, G. Bridging the "material gap" between single crystal studies and real catalysts. *Catal. Lett.* **1988**, *1*, 237–241.
4. Fürstner, A. From oblivion into the limelight: Iron (domino) catalysis. *Angew. Chem., Int. Ed.* **2009**, *48*, 1364–1367.

5. Hoppe, M.; Hulthen, L.; Hallberg, L. The relative bioavailability in humans of elemental iron powders for use in food fortification. *Eur. J. Nutr.* **2006**, *45*, 37–44.
6. Swain, J. H.; Newman, S. M.; Hunt, J. R. Bioavailability of elemental iron powders to rats is less than bakery-grade ferrous sulfate and predicted by iron solubility and particle surface area. *J. Nutr.* **2003**, *133*, 3546–3552.
7. Berry, F. J.; Bland, P. A.; Oats, G.; Pillinger, C. T. Iron-57 Moessbauer spectroscopic studies of the weathering of L-chondrite meteorites. *Hyperfine Interact.* **1994**, *91*, 577–581.
8. Tackett, S. L.; Goudy, A. J. Potentiostatic study of iron meteorite corrosion. *Meteoritics* **1972**, *7*, 487–494.
9. Azad, A.-M.; Kesavan, S.; Al-Batty, S. A closed-loop proposal for hydrogen generation using steel waste and a prototype solar concentrator. *Int. J. Energy Res.* **2009**, *33*, 481–498.
10. Ernst, F. O.; Steinfeld, A.; Pratsinis, S. E. Hydrolysis rate of submicron Zn particles for solar H_2 synthesis. *Int. J. Hydrogen Energy* **2009**, *34*, 1166–1175.
11. Beach, D. B.; Rondinone, A. J.; Sumpter, B. G.; Labinov, S. D.; Richards, R. K. Solid-state combustion of metallic nanoparticles: New possibilities for an alternative energy carrier. *J. Energy Resour. Technol.* **2007**, *129*, 29–32.
12. Wen, D., Song, P.; Zhang, K.; Qian, J. Thermal oxidation of iron nanoparticles and its implication for chemical-looping combustion. *J. Chem. Technol. Biotechnol.* **2010**, 375–380.
13. Vodyanoy, V. Zinc nanoparticles interact with olfactory receptor neurons. *BioMetals* **2010**, *23*, 1097–1103.
14. Samoylov, A. M.; Samoylova, T. I.; Pustovyy, O. M.; Samoylov, A. A.; Toivio-Kinnucan, M. A.; Morrison, N. E.; Globa, L. P.; Gale, W. F.; Vodyanoy, V. Novel metal clusters isolated from blood are lethal to cancer cells. *Cells Tissues Organs* **2005**, *179*, 115–124.
15. Rangsivek, R.; Jekel, M. R. Removal of dissolved metals by zero-valent iron (ZVI): Kinetics, equilibria, processes and implications for stormwater runoff treatment. *Water Res.* **2005**, *39*, 4153–4163.
16. Hussam, A.; Munir, A. K. M. A simple and effective arsenic filter based on composite iron matrix: Development and deployment studies for groundwater of Bangladesh. *J. Environ. Sci. Health, Part A* **2007**, *42*, 1869–1878.
17. Moreno, H. A.; Cocke, D. L.; Gomes, J. A. G.; Morkovsky, P.; Parga, J. R.; Peterson, E.; Garcia, C. Electrochemical reactions for electrocoagulation using iron electrodes. *Ind. Eng. Chem. Res.* **2009**, *48*, 2275–2282.
18. Scherer, M. M.; Richter, S.; Valentine, R. L.; Alvarez, P. J. J. Chemistry and microbiology of permeable reactive barriers for in situ groundwater clean up. *Crit. Rev. Environ. Sci. Technol.* **2000**, *30*, 363–411.
19. Naftz, D. L.; Morrison, S. J.; Davis, J. A.; Fuller, C. C. *Handbook of Groundwater Remediation using Permeable Reactive Barriers: Applications to Radionuclides, Trace Metals, and Nutrients*; Academic Press: San Diego, CA, 2002.

20. Tratnyek, P. G.; Scherer, M. M.; Johnson, T. J.; Matheson, L. J. Permeable reactive barriers of iron and other zero-valent metals. In *Chemical Degradation Methods for Wastes and Pollutants: Environmental and Industrial Applications*; Tarr, M. A., Ed.; Marcel Dekker: New York, 2003; pp 371−421.
21. Gillham, R. W.; Vogan, J.; Gui, L.; Duchene, M.; Son, J. Iron barrier walls for chlorinated solvent remediation. In *In Situ Remediation of Chlorinated Solvent Plumes*; Stroo, H. F., Ward, C. H., Eds.; Springer: New York, 2010; pp 537−571.
22. Rabideau, A. J.; Shen, P.; Khandelwal, A. Feasibility of amending slurry walls with zero-valent iron. *J. Geotech. Geoenviron. Eng.* **1999**, *125*, 330–333.
23. Comfort, S.; Shea, P. J.; Machacek, T. A.; Statapanajaru, T. Pilot-scale treatment of RDX-contaminated soil with zerovalent iron. *J. Environ. Qual.* **2003**, *32*, 1717–1725.
24. Wadley, S. L. S.; Gillham, R. W.; Gui, L. Remediation of DNAPL source zones with granular iron: Laboratory and field tests. *Ground Water* **2005**, *43*, 9–18.
25. Kaplan, D. I.; Knox, A. S. Wet/dry cycling effects on soil contaminant stabilization with apatite and Fe(0). *J. Mater. Civ. Eng.* **2007**, *19*, 49–57.
26. Shokes, T. E.; Möller, G. Removal of dissolved heavy metals from acid rock drainage using iron metal. *Environ. Sci. Technol.* **1999**, *33*, 282–287.
27. Gu, C.; Jia, H.; Li, H.; Teppen, B. J.; Boyd, S. A. Synthesis of highly reactive subnano-sized zero-valent iron using smectite clay templates. *Environ. Sci. Technol.* **2010**, *44*, 4258–4263.
28. Choi, H.; Agarwal, S.; Al-Abed, S. R. Adsorption and simultaneous dechlorination of PCBs on GAC/Fe/Pd: Mechanistic aspects and reactive capping barrier concept. *Environ. Sci. Technol.* **2008**, *43*, 488–493.
29. Morkin, M.; Devlin, J. F.; Barker, J. F.; Butler, B. J. In situ sequential treatment of a mixed contaminant plume. *J. Contam. Hydrol.* **2000**, *45*, 2833–2302.
30. Bayer, P.; Finkel, M. Modelling of sequential groundwater treatment with zero valent iron and granular activated carbon. *J. Contam. Hydrol.* **2005**, *78*, 129–146.
31. *Interstate Technology and Regulatory Council (ITRC) Permeable Reactive Barriers: Lessons Learned/New Directions*; Interstate Technology and Regulatory Council (ITRC): Washington, DC, 2005.
32. Tratnyek, P. G. Keeping up with all that literature: The IronRefs database turns 500. *Ground Water Monit. Rem.* **2002**, *22*, 92–94.
33. Roberts, A. L.; Totten, L. A.; Arnold, W. A.; Burris, D. R.; Campbell, T. J. Reductive elimination of chlorinated ethylenes by zero-valent metals. *Environ. Sci. Technol.* **1996**, *30*, 2654–2659.
34. Arnold, W. A.; Roberts, A. L. Pathways of chlorinated ethylene and chlorinated acetylene reaction with Zn(0). *Environ. Sci. Technol.* **1998**, *32*, 3017–3025.

35. Arnold, W. A.; Roberts, A. L. Pathways and kinetics of chlorinated ethylene and chlorinated acetylene reaction with Fe(0) particles. *Environ. Sci. Technol.* **2000**, *34*, 1794–1805.
36. VanStone, N.; Elsner, M.; Lacrampe-Couloume, G.; Mabury, S.; Sherwood Lollar, B. Potential for identifying abiotic chloroalkane degradation mechanisms using carbon isotopic fractionation. *Environ. Sci. Technol.* **2008**, *42*, 126–132.
37. Lim, T.-T.; Feng, J.; Zhu, B.-W. Kinetic and mechanistic examinations of reductive transformation pathways of brominated methanes with nano-scale Fe and Ni/Fe particles. *Water Res.* **2007**, *41*, 875–883.
38. Liang, X.; Dong, Y.; Kuder, T.; Krumholz, L. R.; Philp, R. P.; Butler, E. C. Distinguishing abiotic and biotic transformation of tetrachloroethylene and trichloroethylene by stable carbon isotope fractionation. *Environ. Sci. Technol.* **2007**, *41*, 7094–7100.
39. Butler, E. C.; Dong, Y.; Krumholz Lee, R.; Liang, X.; Shao, H.; Tan, Y. Rate controlling processes in the transformation of tetrachloroethylene and carbon tetrachloride. In *Aquatic Redox Chemistry*; Tratnyek, P. G., Grundl, T. J., Haderlein, S. B., Eds.; ACS Symposium Series; American Chemical Society: Washington, DC, 2011; Vol. 1071, Chapter 23, pp 519−538.
40. Elsner, M.; Hofstetter, T. B. Current perspectives on the mechanisms of chlorohydrocarbon degradation in subsurface environments: Insight from kinetics, product formation, probe molecules and isotope fractionation. In *Aquatic Redox Chemistry*; Tratnyek, P. G., Grundl, T. J., Haderlein, S. B., Eds.; ACS Symposium Series; American Chemical Society: Washington, DC, 2011; Vol. 1071, Chapter 19, pp 407−439.
41. Scherer, M. M.; Balko, B. A.; Gallagher, D. A.; Tratnyek, P. G. Correlation analysis of rate constants for dechlorination by zero-valent iron. *Environ. Sci. Technol.* **1998**, *32*, 3026–3033.
42. Burrow, P. D.; Aflatooni, K.; Gallup, G. A. Dechlorination rate constants on iron and the correlation with electron attachment energies. *Environ. Sci. Technol.* **2000**, *34*, 3368–3371.
43. Tratnyek, P. G.; Weber, E. J.; Schwarzenbach, R. P. Quantitative structure-activity relationships for chemical reductions of organic contaminants. *Environ. Toxicol. Chem.* **2003**, *22*, 1733–1742.
44. Onanong, S.; Comfort, S. D.; Burrow, P. D.; Shea, P. J. Using gas-phase molecular descriptors to predict dechlorination rates of chloroalkanes by zerovalent iron. *Environ. Sci. Technol.* **2007**, *41*, 1200–1205.
45. Cwiertny, D. M.; Arnold, W. A.; Kohn, T.; Rodenburg, L. A.; Roberts, A. L. Reactivity of alkyl polyhalides toward granular iron: Development of QSARs and reactivity cross correlations for reductive dehalogenation. *Environ. Sci. Technol.* **2010**, *44*, 7928–7936.
46. Chen, J.; Pei, J.; Quan, X.; Zhao, Y.; Chen, S.; Schramm, K. W.; Kettrup, A. Linear free energy relationships on rate constants for dechlorination by zero-valent iron. *SAR QSAR Environ. Res.* **2002**, *13*, 597–606.
47. Song, H.; Carraway, E. R. Reduction of chlorinated ethanes by nanosized zero-valent iron: Kinetics, pathways, and effects of reaction conditions. *Environ. Sci. Technol.* **2005**, *39*, 6237–6245.

48. Arnold, W. A.; Winget, P.; Cramer, C. J. Reductive dechlorination of 1,1,2,2-tetrachloroethane. *Environ. Sci. Technol.* **2002**, *36*, 3536–3541.
49. Bandstra, J. Z.; Tratnyek, P. G. Applicability of single-site rate equations for reactions on inhomogenous surfaces. *Ind. Eng. Chem. Res.* **2004**, *43*, 1615–1622.
50. Jeen, S.-W.; Mayer, K. U.; Gillham, R. W.; Blowes, D. W. Reactive transport modeling of trichloroethene treatment with declining reactivity of iron. *Environ. Sci. Technol.* **2007**, *41*, 1432–1438.
51. Phenrat, T.; Saleh, N.; Sirk, K.; Tilton, R. D.; Lowry, G. V. Aggregation and sedimentation of aqueous nanoscale zerovalent iron dispersions. *Environ. Sci. Technol.* **2007**, *41*, 284–290.
52. Hong, Y.; Honda, R. J.; Myung, N. V.; Walker, S. L. Transport of iron-based nanoparticles: Role of magnetic properties. *Environ. Sci. Technol.* **2009**, *43*, 8834–8839.
53. Petosa, A. R.; Jaisi, D. P.; Quevedo, I. R.; Elimelech, M.; Tufenkji, N. Aggregation and deposition of engineered nanomaterials in aquatic environments: Role of physicochemical interactions. *Environ. Sci. Technol.* **2010**, *44*, 6532–6549.
54. Phenrat, T.; Kim, H.-J.; Fagerlund, F.; Illangasekare, T.; Lowry, G. V. Empirical correlations to estimate agglomerate size and deposition during injection of a polyelectrolyte-modified Fe^0 nanoparticle at high particle concentration in saturated sand. *J. Contam. Hydrol.* **2010**, *118*, 152–164.
55. Saleh, N.; Sirk, K.; Liu, Y.; Phenrat, T.; Dufour, B.; Matyjaszewski, K.; Tilton, R. D.; Lowry, G. V. Surface modifications enhance nanoiron transport and NAPL targeting in saturated porous media. *Environ. Eng. Sci.* **2007**, *24*, 45–57.
56. Hydutsky, B. W.; Mack, E. J.; Beckerman, B. B.; Skluzacek, J. M.; Mallouk, T. E. Optimization of nano- and microiron transport through sand columns using polyelectrolyte mixtures. *Environ. Sci. Technol.* **2007**, *41*, 6418–6424.
57. Tiraferri, A.; Chen, K. L.; Sethi, R.; Elimelech, M. Reduced aggregation and sedimentation of zero-valent iron nanoparticles in the presence of guar gum. *J. Colloid Interface Sci.* **2008**, *324*, 71–79.
58. Phenrat, T.; Saleh, N.; Sirk, K.; Kim, H.-J.; Tilton, R., D.; Lowry, G., V. Stabilization of aqueous nanoscale zerovalent iron dispersions by anionic polyelectrolytes: Adsorbed anionic polyelectrolyte layer properties and their effect on aggregation and sedimentation. *J. Nanoparticle Res.* **2008**, *10*, 795–814.
59. Johnson, R. L.; O'Brien Johnson, R.; Nurmi, J. T.; Tratnyek, P. G. Natural organic matter enhanced mobility of nano zero-valent iron. *Environ. Sci. Technol.* **2009**, *43*, 5455–5460.
60. Fatisson, J.; Ghoshal, S.; Tufenkji, N. Deposition of carboxymethylcellulose-coated zero-valent iron nanoparticles onto silica: Roles of solution chemistry and organic molecules. *Langmuir* **2010**, *26*, 12832–12840.
61. Tratnyek, P. G.; Scherer, M. M.; Deng, B.; Hu, S. Effects of natural organic matter, anthropogenic surfactants, and model quinones on the reduction of contaminants by zero-valent iron. *Water Res.* **2001**, *35*, 4435–4443.

62. Loraine, G. A. Effects of alcohols, anionic and nonionic surfactants on the reduction of PCE and TCE by zero-valent iron. *Water Res.* **2001**, *35*, 1453–1460.
63. Alessi, D. S.; Li, Z. Synergistic effect of cationic surfactants on perchloroethylene degradation by zero-valent iron. *Environ. Sci. Technol.* **2001**, *35*, 3713–3717.
64. Liu, T.; Lo, I. M. C. Influences of humic acid on Cr(VI) removal by zero-valent iron from groundwater with various constituents: Implication for long-term PRB performance. *Water, Air, Soil Pollut.* **2011**, *216*, 473–483.
65. Rangsivek, R.; Jekel, M. R. Natural organic matter (NOM) in roof runoff and its impact on the Fe^0 treatment system of dissolved metals. *Chemosphere* **2008**, *71*, 18–29.
66. Shin, M.-C.; Choi, H.-D.; Kim, D.-H.; Baek, K. Effect of surfactant on reductive dechlorination of trichloroethylene by zero-valent iron. *Desalination* **2008**, *223*, 299–307.
67. Phenrat, T.; Liu, Y.; Tilton, R. D.; Lowry, G. V. Adsorbed polyelectrolyte coatings decrease fe nanoparticle reactivity with TCE in water: Conceptual model and mechanisms. *Environ. Sci. Technol.* **2009**, *43*, 1507–1514.
68. He, F.; Zhao, D. Hydrodechlorination of trichloroethene using stabilized Fe-Pd nanoparticles: Reaction mechanism and effects of stabilizers, catalysts and reaction conditions. *Appl. Catal. B: Environ.* **2008**, *84*, 533–540.
69. Schrick, B.; Hydutsky, B. W.; Blough, J. L.; Mallouk, T. E. Delivery vehicles for zerovalent metal nanoparticles in soil and groundwater. *Chem. Mater.* **2004**, *16*, 2187–2193.
70. Arias, J. L.; Gallardo, V.; Linares-Molinero, F.; Delgado, A. V. Preparation and characterization of carbonyl iron/poly(butyl cyanoacrylate) core/shell nanoparticles. *J. Colloid Interface Sci.* **2006**, *299*, 599–607.
71. Kanel, S. R.; Choi, H. Transport characteristics of surface-modified nanoscale zero-valent iron in porous media. *Water Sci. Technol.* **2007**, *55*, 157–162.
72. Wu, L.; Shamsuzzoha, M.; Ritchie, S. M. C. Preparation of cellulose acetate supported zero-valent iron nanoparticles for the dechlorination of trichloroethylene in water. *J. Nanoparticle Res.* **2005**, *7*, 469–476.
73. Raychoudhury, T.; Naja, G.; Ghoshal, S. Assessment of transport of two polyelectrolyte-stabilized zero-valent iron nanoparticles in porous media. *J. Contam. Hydrol.* **2010**, *118*, 143–151.
74. Tiraferri, A.; Sethi, R. Enhanced transport of zerovalent iron nanoparticles in saturated porous media by guar gum. *J. Nanoparticle Res.* **2009**, *11*, 635–645.
75. Sakulchaicharoen, N.; O'Carroll, D. M.; Herrera, J. E. Enhanced stability and dechlorination activity of pre-synthesis stabilized nanoscale FePd particles. *J. Contam. Hydrol.* **2010**, *118*, 117–127.
76. Lin, Y.-H.; Tseng, H.-H.; Wey, M.-Y.; Lin, M.-D. Characteristics of two types of stabilized nano zero-valent iron and transport in porous media. *Sci. Total Environ.* **2010**, *408*, 2260–2267.
77. Bishop, E. J.; Fowler, D. E.; Skluzacek, J. M.; Seibel, E.; Mallouk, T. E. Anionic homopolymers efficiently target zerovalent iron particles to

hydrophobic contaminants in sand columns. *Environ. Sci. Technol.* **2010**, *44*, 9069–9074.
78. Wang, Q.; Qian, H.; Yang, Y.; Zhang, Z.; Naman, C.; Xu, X. Reduction of hexavalent chromium by carboxymethyl cellulose-stabilized zero-valent iron nanoparticles. *J. Contam. Hydrol.* **2010**, *114*, 35–42.
79. Comba, S.; Sethi, R. Stabilization of highly concentrated suspensions of iron nanoparticles using shear-thinning gels of xanthan gum. *Water Res.* **2009**, *43*, 3717–3726.
80. Shirin, S.; Buncel, E.; VanLoon, G. W. Effect of cyclodextrins on iron-mediated dechlorination of trichloroethylene - A proposed new mechanism. *Can. J. Chem.* **2004**, *82*, 1674–1685.
81. Comba, S.; Dalmazzo, D.; Santagata, E.; Sethi, R. Rheological characterization of xanthan suspensions of nanoscale iron for injection in porous media. *J. Hazard. Mater.* **2011**, *185*, 598–605.
82. Zhan, J.; Sunkara, B.; Le, L.; John, V. T.; He, J.; McPherson, G. L.; Piringer, G.; Lu, Y. Multifunctional colloidal particles for in situ remediation of chlorinated hydrocarbons. *Environ. Sci. Technol.* **2009**, *43*, 8616–8621.
83. Saleh, N.; Kim, H.-J.; Phenrat, T.; Matyjaszewski, K.; Tilton, R. D.; Lowry, G. V. Ionic strength and composition affect the mobility of surface-modified Fe nanoparticles in water-saturated sand columns. *Environ. Sci. Technol.* **2008**, *42*, 3349–3355.
84. Saleh, N.; Phenrat, T.; Sirk, K.; Dufour, B.; Ok, J.; Sarbu, T.; Matyjaszewski, K.; Tilton, R. D.; Lowry, G. V. Adsorbed triblock copolymers deliver reactive iron nanoparticles to the oil/water interface. *Nano Lett.* **2005**, *5*, 2489–2494.
85. Sun, Y.-P.; Li, X.-Q.; Zhang, W.-X.; Wang, H. P. A method for the preparation of stable dispersion of zero-valent iron nanoparticles. *Colloids Surf., A* **2007**, *308*, 60–66.
86. He, F.; Zhao, D. Manipulating the size and dispersibility of zerovalent iron nanoparticles by use of carboxymethyl cellulose stabilizers. *Environ. Sci. Technol.* **2007**.
87. He, F.; Zhao, D.; Liu, J.; Roberts, C. B. Stabilization of Fe-Pd nanoparticles with sodium carboxymethyl cellulose for enhanced transport and dechlorination of trichloroethylene in soil and groundwater. *Ind. Eng. Chem. Res.* **2007**, *46*, 29–34.
88. He, F.; Zhang, M.; Qian, T.; Zhao, D. Transport of carboxymethyl cellulose stabilized iron nanoparticles in porous media: Column experiments and modeling. *J. Colloid Interface Sci.* **2009**, *334*, 96–102.
89. Si, S.; Kotal, A.; Mandal, T. K.; Giri, S.; Nakamura, H.; Kohara, T. Size-controlled synthesis of magnetite nanoparticles in the presence of polyelectrolytes. *Chem. Mater.* **2004**, *16*, 3489–3496.
90. Chang, P. R.; Yu, J.; Ma, X.; Anderson, D. P. Polysaccharides as stabilizers for the synthesis of magnetic nanoparticles. *Carbohydr. Polym.* **2011**, *83*, 640–644.
91. Cho, Y.; Choi, S.-I. Degradation of PCE, TCE and 1,1,1-TCA by nanosized FePd bimetallic particles under various experimental conditions. *Chemosphere* **2010**, *81*, 940–945.

92. Thurman, E. M. Humic substances in groundwater. *Humic Substances in Soil, Sediment, and Water: Geochemistry, Isolation, and Characterization*; John Wiley & Sons: 1985; pp 87−103.
93. Klavins, M.; Purmalis, O. Humic substances as surfactants. *Environ. Chem. Lett.* **2010**, *8*, 349–354.
94. Gibson, C. T.; Turner, I. J.; Roberts, C. J.; Lead, J. R. Quantifying the dimensions of nanoscale organic surface layers in natural waters. *Environ. Sci. Technol.* **2007**, *41*, 1339–1344.
95. Baalousha, M. Aggregation and disaggregation of iron oxide nanoparticles: Influence of particle concentration, pH and natural organic matter. *Sci. Total Environ.* **2009**, *407*, 2093–2101.
96. Baalousha, M.; Manciulea, A.; Cumberland, S.; Kendall, K.; Lead, J. R. Aggregation and surface properties of iron oxide nanoparticles: influence of pH and natural organic matter. *Environ. Toxicol. Chem.* **2008**, *27*, 1875–1882.
97. Nadagouda, M. N.; Castle, A. B.; Murdock, R. C.; Hussain, S. M.; Varma, R. S. In vitro biocompatibility of nanoscale zerovalent iron particles (NZVI) synthesized using tea polyphenols. *Green Chem.* **2010**, *12*, 114–122.
98. Balko, B. A.; Tratnyek, P. G. Photoeffects on the reduction of carbon tetrachloride by zero-valent iron. *J. Phys. Chem. B* **1998**, *102*, 1459–1465.
99. Wielant, J.; Hauffman, T.; Blajiev, O.; Hausbrand, R.; Terryn, H. Influence of the iron oxide acid-base properties on the chemisorption of model epoxy compounds studied by XPS. *J. Phys. Chem. C* **2007**, *111*, 13177–13184.
100. McBride, M. B.; Kung, K. H. Adsorption of phenol and substituted phenols by iron oxides. *Environ. Toxicol. Chem.* **1991**, *10*, 441–448.
101. Gu, B.; Schmitt, J.; Chen, Z.; Liang, L.; McCarthy, J. F. Adsorption and desorption of different organic matter fractions on iron oxide. *Geochim. Cosmochim. Acta* **1995**, *59*, 219–229.
102. Gu, B.; Mehlhorn, T. L.; Liang, L.; McCarthy, J. F. Competitive adsorption, displacement, and transport of organic matter on iron oxide: I. Competitive adsorption. *Geochim. Cosmochim. Acta* **1996**, *60*, 1943–1950.
103. Liu, C.; Zachara, J. M.; Foster, N. S.; Strickland, J. Kinetics of reductive dissolution of hematite by bioreduced anthraquinone-2,6-disulfonate. *Environ. Sci. Technol.* **2007**, *41*, 7730–7735.
104. Shi, Z.; Nurmi, J. T.; Tratnyek, P. G. Effects of nano zero-valent Iron (nZVI) on oxidation-reduction potential (ORP). *Environ. Sci. Technol.* **2011**, *45*, 1586–1592.
105. Mylon, S. E.; Sun, Q.; Waite, T. D. Process optimization in use of zero valent iron nanoparticles for oxidative transformations. *Chemosphere* **2010**, *81*, 127–131.
106. Giasuddin, A. B. M.; Kanel, S. R.; Choi, H. Adsorption of humic acid onto nanoscale zerovalent iron and its effect on arsenic removal. *Environ. Sci. Technol.* **2007**, *41*, 2022–2027.
107. Chen, J.; Xiu, Z.; Lowry, G. V.; Alvarez, P. J. J. Effect of natural organic matter on toxicity and reactivity of nano-scale zero-valent iron. *Water Res.* **2011**, *45*, 1995–2001.

108. Wang, W.; Zhou, M.; Jin, Z.; Li, T. Reactivity characteristics of poly(methyl methacrylate) coated nanoscale iron particles for trichloroethylene remediation. *J. Hazard. Mater.* **2010**, *173*, 724–730.
109. Wang, W.; Zhou, M. Degradation of trichloroethylene using solvent-responsive polymer coated Fe nanoparticles. *Colloids Surf., A* **2010**, *369*, 232–239.
110. Liu, T.; Tsang, D. C. W.; Lo, I. M. C. Chromium(VI) reduction kinetics by zero-valent iron in moderately hard water with humic acid: Iron dissolution and humic acid adsorption. *Environ. Sci. Technol.* **2008**, *42*, 2092–2098.
111. Zhu, B.-W.; Lim, T.-T.; Feng, J. Influences of amphiphiles on dechlorination of a trichlorobenzene by nanoscale Pd/Fe: Adsorption, reaction kinetics, and interfacial interactions. *Environ. Sci. Technol.* **2008**, *42*, 4513–4519.
112. Sarathy, V.; Tratnyek, P. G.; Salter, A. J.; Nurmi, J. T.; Johnson, R. L.; O'Brien Johnson, G. Degradation of 1,2,3-trichloropropane (TCP): Hydrolysis, elimination, and reduction by iron and zinc. *Environ. Sci. Technol.* **2010**, *44*, 787–793.
113. Nurmi, J. T.; Tratnyek, P. G.; Sarathy, V.; Baer, D. R.; Amonette, J. E.; Pecher, K.; Wang, C.; Linehan, J. C.; Matson, D. W.; Penn, R. L.; Driessen, M. D. Characterization and properties of metallic iron nanoparticles: Spectroscopy, electrochemistry, and kinetics. *Environ. Sci. Technol.* **2005**, *39*, 1221–1230.
114. Nurmi, J. T.; Sarathy, V.; Tratnyek, P. G.; Baer, D. R.; Amonette, J. E.; Linehan, J. C.; Karkamkar, A. Recovery of iron/iron oxide nanoparticles from aqueous media: A comparison of methods and their effects. *J. Nanoparticle Res.* **2011**, *13*, 1937–1952.
115. Royer, R. A.; Burgos, W. D.; Fisher, A. S.; Jeon, B.-H.; Unz, R. F.; Dempsey, B. A. Enhancement of hematite bioreduction by natural organic matter. *Environ. Sci. Technol.* **2002**, *36*, 2897–2904.
116. Nurmi, J. T.; Tratnyek, P. G. Electrochemical properties of natural organic matter (NOM), fractions of NOM, and model biogeochemical electron shuttles. *Environ. Sci. Technol.* **2002**, *36*, 617–624.
117. Chen, J.; Gu, B.; LeBoeuf, E. J.; Pan, H.; Dai, S. Spectroscopic characterization of structural and functional properties of natural organic matter fractions. *Chemosphere* **2002**, *48*, 59–68.
118. Chen, J.; Gu, B.; Royer, R. A.; Burgos, W. D. The roles of natural organic matter in chemical and microbial reduction of ferric iron. *Sci. Total Environ.* **2003**, *307*, 167–178.
119. Chen, J.; Gu, B.; LeBoeuf, E. J. Fluorescence spectroscopic studies of natural organic matter fractions. *Chemosphere* **2003**, *50*, 639–647.
120. Baer, D. R.; Amonette, J. E.; Engelhard, M. H.; Gaspar, D. J.; Karakoti, A. S.; Kuchibhatla, S.; Nachimuthu, P.; Nurmi, J. T.; Qiang, Y.; Sarathy, V.; Seal, S.; Sharma, A.; Tratnyek, P. G.; Wang, C. M. Characterization challenges for nanomaterials. *Surf. Interface Anal.* **2008**, *40*, 529–537.
121. Baer, D. R.; Tratnyek, P. G.; Qiang, Y.; Amonette, J. E.; Linehan, J.; Sarathy, V.; Nurmi, J. T.; Wang, C.; Anthony, J. Synthesis, characterization, and properties of zero-valent iron nanoparticles. In *Environmental Applications*

of Nanomaterials: Synthesis, Sorbents, and Sensors; Fryxell, G. E., Ed.; Imperial College Press: London, 2007; pp 49−86.

122. Nurmi, J. T.; Tratnyek, P. G. Electrochemical studies of packed iron powder electrodes: Effects of common constituents of natural waters on corrosion potential. *Corros. Sci.* **2008**, *50*, 144–154.
123. Sarathy, V.; Tratnyek, P. G.; Nurmi, J. T.; Baer, D. R.; Amonette, J. E.; Chun, C.; Penn, R. L.; Reardon, E. J. Aging of iron nanoparticles in aqueous solution: effects on structure and reactivity. *J. Phys. Chem. C* **2008**, *112*, 2286–2293.
124. Nurmi, J. T.; Bandstra, J. Z.; Tratnyek, P. G. Packed powder electrodes for characterizing the reactivity of granular iron in borate solutions. *J. Electrochem. Soc.* **2004**, *151*, B347–B353.
125. Kim, H.-S.; Ahn, J.-Y.; Hwang, K.-Y.; Kim, I.-K.; Hwang, I. Atmospherically stable nanoscale zero-valent iron particles formed under controlled air contact: Characteristics and reactivity. *Environ. Sci. Technol.* **2010**, *44*, 1760–1766.
126. Martin, J. E.; Herzing, A. A.; Yan, W.; Li, X.-Q.; Koel, B. E.; Kiely, C. J.; Zhang, W.-X. Determination of the oxide layer thickness in core-shell zerovalent iron nanoparticles. *Langmuir* **2008**, *24*, 4329–4334.
127. Cohen, M. The passivity and breakdown of passivity on iron. In *Passivity of Metals*; Frankenthal, R. P., Kruger, J., Eds.; The Electrochemical Society: Princeton, NJ, 1978; pp 521−545.
128. Suzuki, S.; Waseda, Y., Structural characterization of iron corrosion products formed in aqueous solution. In *Progress in Corrosion Research*; Nova Science Publishers, Inc.: 2007, pp 99−132.
129. Genin, J. M. R. Fe(II-III) hydroxysalt green rusts; from corrosion to mineralogy and abiotic to biotic reactions by Moessbauer spectroscopy. *Hyperfine Interact.* **2004**, *156/157*, 471–485.
130. Bonin, P. M. L.; Odziemkowski, M. S.; Reardon, E. J.; Gillham, R. W. In situ identification of carbonate-containing green rust on iron electrodes in solutions simulating groundwater. *J. Solution Chem.* **2000**, *29*, 1061–1074.
131. Davenport, A. J.; Oblonsky, L. J.; Ryan, M. P.; Toney, M. F. The structure of the passive film that forms on iron in aqueous environments. *J. Electrochem. Soc.* **2000**, *147*, 2162–2173.
132. Odziemkowski, M. S.; Schuhmacher, T. T.; Gillham, R. W.; Reardon, E. J. Mechanism of oxide film formation on iron in simulating groundwater solutions: Raman spectroscopic studies. *Corros. Sci.* **1998**, *40*, 371–389.
133. Baer, D. R.; Gaspar, D. J.; Nachimuthu, P.; Techane, S. D.; Castner, D. G. Application of surface chemical analysis tools for characterization of nanoparticles. *Anal. Bioanal. Chem.* **2010**, *396*, 983–1002.
134. Loux, N. T.; Savage, N. An assessment of the fate of metal oxide nanomaterials in porous media. *Water, Air, Soil Pollut.* **2008**, *194*, 227–241.
135. Sverjensky, D. A. Prediction of surface charge on oxides in salt solutions: Revisions for 1:1 (M^+L^-) electrolytes. *Geochim. Cosmochim. Acta* **2005**, *69*, 225–257.
136. Parks, G. A. The isolectric points of solid oxides, solid hydroxides, and aqueous hydroxo complex systems. *Chem. Rev.* **1965**, *65*, 177–198.

137. Zhang, Y.; Chen, Y.; Westerhoff, P.; Crittenden, J. Impact of natural organic matter and divalent cations on the stability of aqueous nanoparticles. *Water Res.* **2009**, *43*, 4249–4257.
138. Xu, X. Q.; Shen, H.; Xu, J. R.; Xie, M. Q.; Li, X. J. The colloidal stability and core-shell structure of magnetite nanoparticles coated with alginate. *Appl. Surf. Sci.* **2006**, *253*, 2158–2164.
139. Tratnyek, P. G.; Salter, A. J.; Nurmi, J. T.; Sarathy, V. Environmental applications of zerovalent metals: Iron vs. zinc. In *Nanoscale Materials in Chemistry: Environmental Applications*; Erickson, L. E., Koodali, R. T., Richards, R. M., Eds.; American Chemical Society: Washington, DC, 2010; Vol. 1045; pp 165−178.
140. Támara, M.; Butler, E. C. Effects of iron purity and groundwater characteristics on rates and products in the degradation of carbon tetrachloride by iron metal. *Environ. Sci. Technol.* **2004**, *38*, 1866–1876.
141. Lien, H.-L.; Zhang, W.-X. Transformation of chlorinated methanes by nanoscale iron particles. *J. Environ. Eng.* **1999**, *125*, 1042–1047.
142. Tratnyek, P., G.; Sarathy, V.; Kim, J.-H.; Chang, Y.-S.; Bae, B. Effects of particle size on the kinetics of degradation of contaminants. *International Environmental Nanotechnology Conference: Applications and Implications (7-9 October 2008)*; EPA 905-R09-032; U.S. Environmental Protection Agency: Chicago, IL, 2009; pp 67−72.
143. Tratnyek, P. G.; Johnson, R. L. Nanotechnologies for environmental cleanup. *NanoToday* **2006**, *1*, 44–48.
144. Agrawal, A.; Ferguson, W. J.; Gardner, B. O.; Christ, J. A.; Bandstra, J. Z.; Tratnyek, P. G. Effects of carbonate species on the kinetics of dechlorination of 1,1,1-trichloroethane by zero-valent iron. *Environ. Sci. Technol.* **2002**, *36*, 4326–4333.
145. Klausen, J.; Vikesland, P. J.; Kohn, T.; Burris, D. R.; Ball, W. P.; Roberts, A. L. Longevity of granular iron in groundwater treatment processes: solution composition effects on reduction of organohalides and nitroaromatic compounds. *Environ. Sci. Technol.* **2003**, *37*, 1208–1218.
146. Tsang, D. C. W.; Graham, N. J. D.; Lo, I. M. C. Humic acid aggregation in zero-valent iron systems and its effects on trichloroethylene removal. *Chemosphere* **2009**, *75*, 1338–1343.
147. Marconetto, S.; Gui, L.; Gillham, R. Adsorption of natural organic matter and its effects on TCE degradation by iron PRBs. In *Bringing Groundwater Quality Research to the Watershed Scale (Proceedings of GQ2004, the 4th International Groundwater Quality Conference, held at Waterloo, Canada, July 2004)*; IAHS Publ. 297; Thomson, N. R., Ed.; International Association of Hydrological Sciences: Wallingford, U.K., 2005; pp 389−397.

Chapter 19

Current Perspectives on the Mechanisms of Chlorohydrocarbon Degradation in Subsurface Environments: Insight from Kinetics, Product Formation, Probe Molecules, and Isotope Fractionation

Martin Elsner[1,*] and Thomas B. Hofstetter[2]

[1]Helmholtz Zentrum München, German Research Center for Environmental Health (GmbH), Ingolstaedter Landstr. 1 85764 Neuherberg, Germany
[2]Swiss Federal Institute of Aquatic Science and Technology (Eawag), Überlandstr. 133, 8600 Dübendorf, Switzerland
*martin.elsner@helmholtz-muenchen.de

Degradation of chlorinated organic contaminants by natural and engineered reductive dechlorination reactions can occur via numerous biotic and abiotic transformation pathways giving rise to either benign or more toxic products. To assess whether dechlorination processes may lead to significant detoxification (a) the thermodynamic feasibility of a reaction, (b) rates of transformation, and (c) product formation routes need to be understood. To this end, fundamental knowledge of chlorohydrocarbon (CHC) reaction mechanisms is essential. We review insight from reaction thermodynamics, structure-reactivity relationships, and applications of radical and carbene traps, as well as of synthetic probe molecules. We summarize the state-of-knowledge about intermediates and reductive dechlorination pathways of vicinal and geminal haloalkanes, as well as of chlorinated ethenes. Transformation conditions are identified under which problematic products may be avoided. In an outlook, we discuss the potential of stable carbon and chlorine isotope fractionation to identify initial transformation mechanisms, competing transformation pathways, and common branching points.

Chlorohydrocarbon Contamination of Soils and Groundwater

Chlorinated hydrocarbons (CHC) are the most relevant point-source groundwater contaminants besides mineral oils, hydrocarbon fuels, and propellants (*1*, *2*). Millions of tons of chlorinated ethenes and alkanes are produced worldwide per year as degreasing and dry cleaning agents or as intermediates of chemical synthesis (*3*). These compounds enter the environment through leakage in storage tanks, improper disposal, and accidents. Despite increasingly efficient control of industrial CHC use (*4*), we are faced with a legacy of more than ten thousand contaminated sites in Europe alone (*1*, *2*).

Among the many cases of soil and groundwater contamination with CHCs, a large share is caused by a legacy of compounds from industrial applications including chlorinated methanes and ethanes such as tetrachloromethane (CCl_4), 1,1,1-trichloroethane (1,1,1-TCA), 1,2-dichloroethane (1,2-DCA), chlorinated ethenes such as tetrachloroethene (perchloroethylene, PCE) and trichloroethene (TCE), as well as their degradation products *cis*-dichloroethene (*cis*-DCE) and vinyl chloride (VC). These compounds exhibit a series of properties that have made them some of the most widespread anthropogenic contaminants and thus threats to human health. The above mentioned CHC are quite persistent in oxic environments, soluble in water in millimolar concentrations, and show only a moderate affinity for interactions with organic material and mineral surfaces (*5*). As a consequence, CHC are quite mobile in the aquatic subsurface and therefore pose a serious risk of drinking water contamination.

The large amounts of spilled CHC usually form non-aqueous phase liquids in the subsurface, from which the contaminants are released over years to decades. Accessing and removing these contaminant sources is often difficult, always expensive, and thus not a feasible solution to avoid water contamination for the vast majority of sites. Neither has the off-site treatment of pumped water proven to be an efficient alternative. Transformation of CHCs to nontoxic, less or non-chlorinated products therefore represents the only meaningful mitigation option.

Assessing the degradation of CHCs in the environment as well as in engineered systems proposed for water treatment is a major challenge. Over the last two decades, researchers have shown that microbial and abiotic transformation of CHC is, in principle, possible. Direct degradation of contaminants in the subsurface can be facilitated by stimulation or inoculation of microbial activity (biostimulation / bioaugmentation), or by *in situ* treatment with abiotic reagents such as zero-valent iron. Nevertheless, many aspects of the underlying reaction mechanisms remain imperfectly understood giving rise to significant uncertainties within the CHC degradation assessment. It is unclear to date how molecular-level interactions between CHC and reactive moieties of enzymes or catalytic metals determine the route of transformation. Consequently, predictions regarding the probability of toxic product formation can hardly be made and processes cannot be manipulated systematically to generate benign compounds. Moreover, as CHC transformation can occur along different, sometimes competing pathways, reliable concepts, and tools are required that allow one to infer the predominant degradation reaction.

In this review, we provide an account of the current state of research undertaken to elucidate the mechanisms of CHC degradation with a focus on reduction of polychlorinated methanes, ethanes, and ethenes. In this context, we do not cover aspects of microbiology (*6*) or field studies, but emphasize studies of the underlying chemical mechanisms in well-defined laboratory model systems. We discuss how advances in the fundamental understanding of these reactions can contribute to an improved assessment of dechlorination processes in the contaminated subsurface. To this end, we briefly introduce the general pathways by which bonds in CHC are broken. We then focus on the kinetic and product studies that have lead to mechanistic insights of CHC transformation, as well as on evidence from studies with model reactants. Finally, we elucidate the potential of using stable isotope fractionation for future mechanistic studies and identification of transformation pathways.

A Brief Survey of Reactions Applicable to Polychlorinated Methanes, Ethanes, and Ethenes and of the Thermodynamics of Reductive Dechlorination

Types of Transformation Reactions

Owing to their highly oxidized carbon skeleton, initial reaction steps during the degradation of polychlorinated methanes, ethanes, and ethenes in the subsurface are mostly reductive. Reductive dechlorination mechanisms, that is, hydrogenolysis, reductive α- or β-elimination (*geminal* and *vicinal* dihaloelimination, respectively), are distinguished by the number and position of Cl atoms removed from the reactant (Figure 1a). In principle, reductive dechlorination can proceed until a contaminant is fully dechlorinated, which would be equivalent to detoxification. Compounds of the same substance class react faster if they contain a higher number of Cl atoms at the reactive C atom and thus rates of reduction often decrease with decreasing number of Cl atoms. Dechlorination leads to less chlorinated, more electron-rich compounds, which are often of greater toxicity than their parent compound (e.g., vinyl chloride being the most toxic chloroethene generated from PCE and TCE, (*7*)). Instead of reduction, further transformation via oxidation of one or two carbon atoms (α-hydroxylation or epoxidation, Figure 1b) becomes a favored route of degradation for less chlorinated hydrocarbons. For higher chlorinated compounds, in contrast, oxidative degradation in subsurface environments is disfavored and observed only during engineered treatment (e.g., with permanganate or Fenton's reagent, Figure 1b). Substitution and elimination processes (Figure 1c/d) are only observed at saturated carbon atoms under conditions typical for environmental transformations. These reactions typically do not change the oxidation state of the compounds, except for substitutions by hydride, which lead to a reduction via hydrogenolysis (Figure 2a). Because these mechanisms are well-known from the (bio)chemical literature, they are not discussed in this chapter. In the following, we focus on mechanisms of reductive dechlorination.

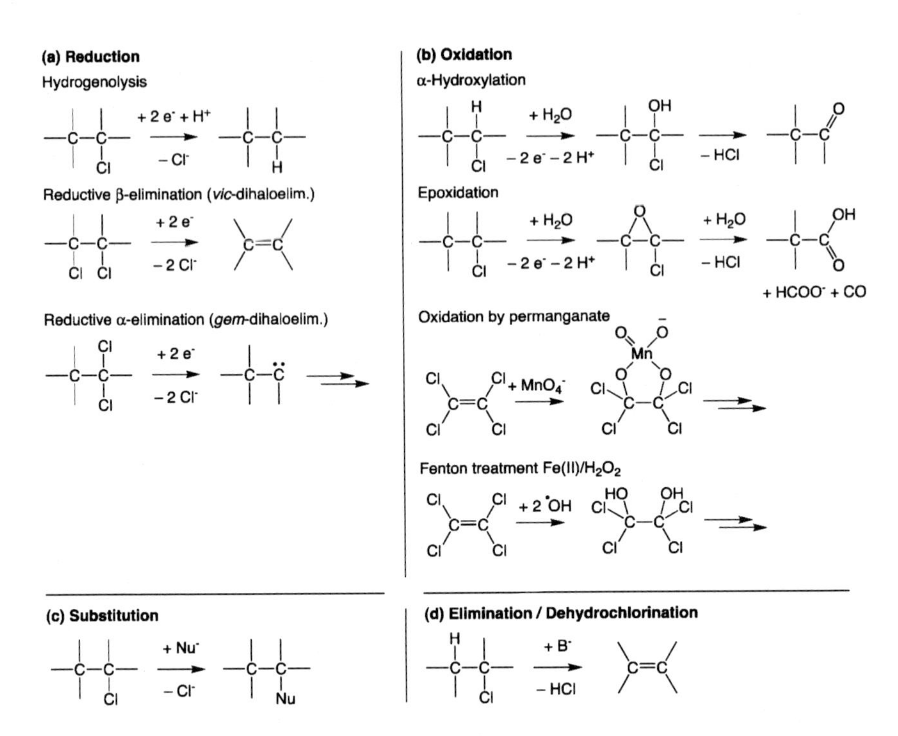

Figure 1. Principal initial steps of reductive, oxidative, and non-reductive dechlorination mechanisms of chlorohydrocarbons (5, 8). Note that for the sake of simplicity, most reactions are shown for alkanes. Reductive and oxidative dechlorinations also apply for chlorinated ethenes.

Reductive Dechlorination of CHCs in Natural and Engineered Systems

Reductive degradation can be facilitated by either biotic or abiotic transformations (*9, 10*). Figure 2 groups typical biological and biogeochemical reductants by their approximate (ranges of) reduction potentials and puts them in perspective to the thermodynamics of the initial one-electron transfer steps of CHC reductions. In the biotic case, microorganisms can dechlorinate chlorinated hydrocarbons either co-metabolically, or by using them as terminal electron acceptor (dehalorespiration) (*6*). The latter organisms require H_2 as electron donor, which arises from fermentative transformation of organic material. Although almost all dehalogenases, the enzymes responsible for reductive dechlorination, known to date contain cobalamin (Vitamin B_{12}, Figure 2) as common cofactor (*6*), microorganisms may perform sequential dechlorination to a different extent. For example, in most organisms reductive dechlorination of PCE stalls at the stage of more problematic *cis*-DCE or VC, whereas complete dechlorination has been observed with some strains of the genus *Dehalococcoides* (*7*). The mechanistic reasons on the enzymatic level are as yet unknown.

Abiotic transformations occur on mineral surfaces, by metal-organic complexes in solution, or, in engineered approaches, by zero-valent metals (usually iron, Figure 2). While metals are deliberately introduced into the subsurface for *in situ* remediation, natural reductants arise from the microbial activity in biogeochemical cycles. These reductants include Fe^{2+} species at the surface or within the structure of minerals (surface complexes at iron oxides, within the structure of sulfide or clay minerals), as well as dissolved Fe^{2+} complexes and (mercapto-) quinone moieties in natural organic matter.

Assessing chlorohydrocarbon reduction in the environment involves three general research questions, which all require a fundamental understanding of the underlying reaction mechanism. As will be outlined in greater detail in the following sections, these questions include (i) the thermodynamic feasibility of a reaction, (ii) whether rates of transformation are significant, and (iii) which products are formed in these processes.

Thermodynamic Considerations

Indeed, overall reductive dechlorination, which involves the transfer of two electrons to the CHC is often *thermodynamically favorable* (*5*). Reductive dechlorination is, however, kinetically limited by the initial, dissociative electron transfer to a C–Cl bond, while the reactions of the resulting intermediates are usually significantly faster (*11*, *12*). Whether reduction of a CHC will take place or not is therefore frequently evaluated on the basis of one-electron reduction potentials generating a chloroorganic radical and chloride. Because these numbers are not accessible experimentally, computational approaches have been necessary (*13*). Such computations have continuously improved over years (*13–16*), especially for chloroethanes and -methanes, where most recent calculations include specific conformations of parent compounds and radical intermediates (e.g., *syn* versus *anti*) (*17*).

As illustrated in Figure 2, dissociative electron transfer is often thermodynamically feasible and it is frequently the initial step of chlorinated alkanes reduction by a variety of biotic and abiotic reductants (*16*). For chlorinated ethenes, in contrast, such analyses do not seem to favor an initial outer-sphere electron transfer, but rather imply that one needs to identify the specific interaction between the CHC and the reductants that make this reaction possible. However, computations of reduction potentials are very challenging (e.g., (*23*, *24*)) and might not yet be fully applicable to assess reactions of highly chlorinated ethenes. Uncertainties pertinent to the measurement and calculation of accurate reduction potentials apply in a similar manner for the reductants involved in dechlorination reactions, especially for the various Fe species bound at surfaces or within the structure of minerals and metals (*19*, *22*, *25*, *26*). The picture given in Figure 2 is therefore meant to provide a semi-quantitative, rather than an exact, overview based on current-day knowledge.

As will be discussed in the following section, one-electron reduction potentials can nevertheless be very useful to understand or even predict the (relative) *rates of reductive dechlorination* of various classes of CHCs. Knowledge of the molecular-level interactions of reductant and CHC can help one to assess activation energies, and, hence, reaction rates of compounds in a given transformation.

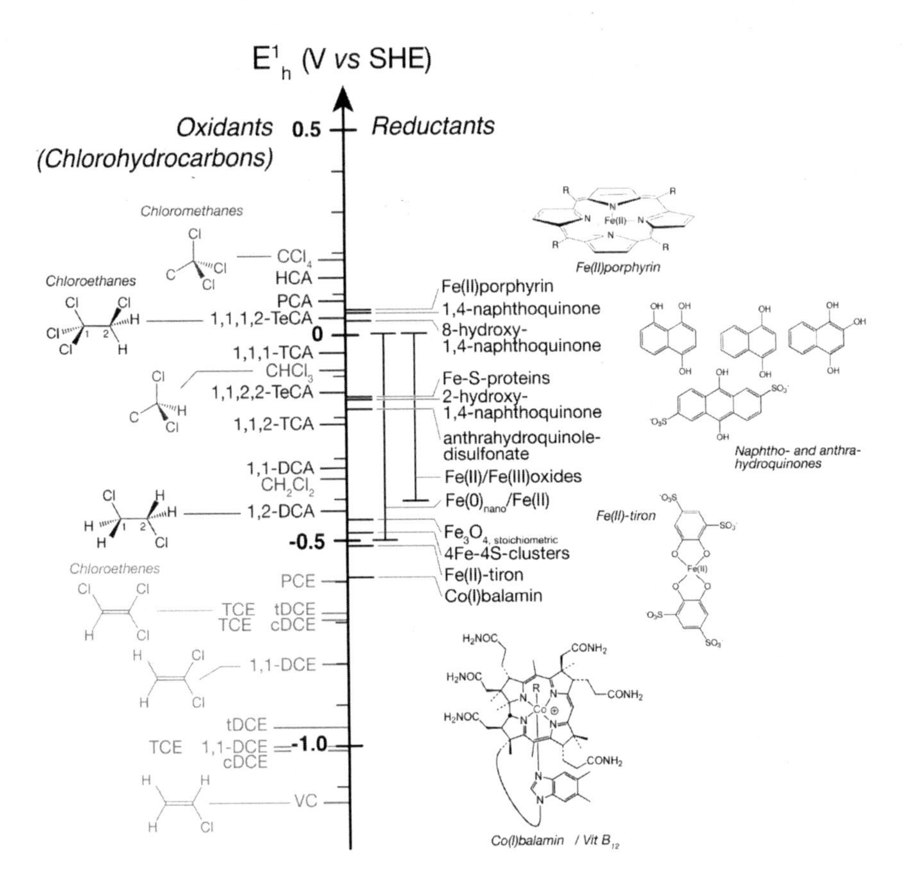

Figure 2. Comparison of one-electron reduction potentials for dissociative electron transfer to chlorinated methanes (red), ethanes (blue, HCA hexachloroethane; PCA pentachloroethane, TeCA tetrachloroethane, TCA trichloroethane), and ethenes (green, PCE tetrachloroethene; TCE trichloroethene; 1,1-/cis/trans-DCE dichloroethenes, VC vinyl chloride) with measured (ranges of) reduction potentials of a selection of typical reductants from biotic/abiotic systems (13, 17–22).

Insight from Kinetic Studies and Linear Free Energy Relationships (LFER): The Rate-Determining Step

Several studies have established quantitative structure-activity relationships (QSARs) in order to predict CHC transformation rates and to infer reaction mechanisms from information about the rate-determining step of a reaction (*17*, *27–30*)). As illustrated in Figure 3, the QSAR approach is based on a mechanistic assumption, for example, a dissociative electron transfer. In linear free energy relationships (LFERs) thermodynamic (e.g., reduction potentials) or kinetic descriptors (e.g., rates of CHC reduction in a well-defined reference reaction) are correlated with measured dechlorination rate constants to learn how the structural features of a CHC (e.g., the number and position of Cl atoms) determine its reactivity and thus whether an assumed mechanism is valid.

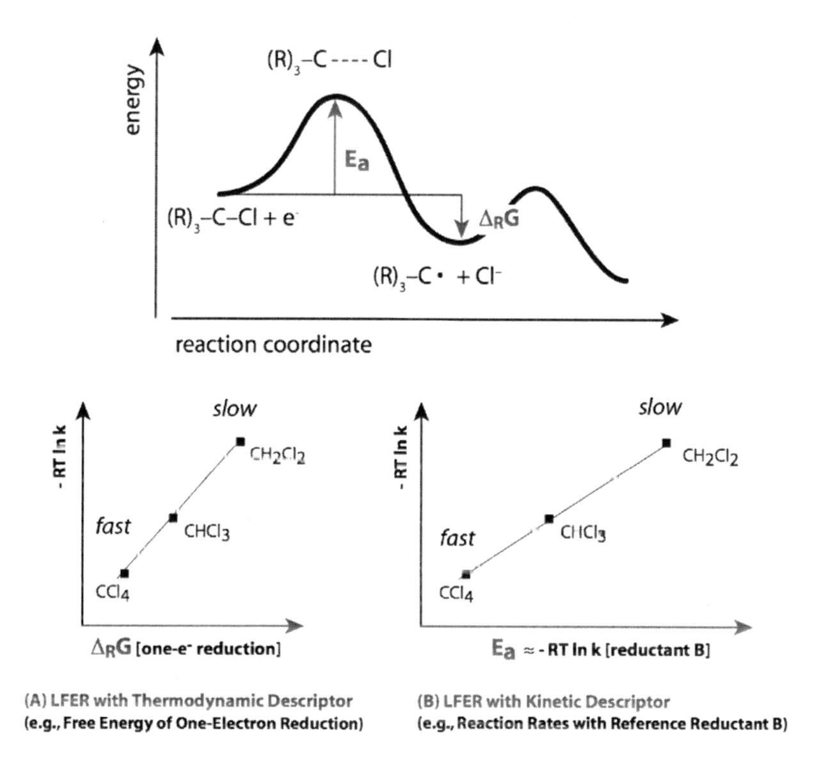

Figure 3. Illustration of Linear Free Energy Relationships (LFERs) for the example of chlorinated methanes. The y-axis represents dechlorination rate constants k for a transformation of interest, given in logarithmic form as -RT ln k (R: universal gas constant, T: Temperature in Kelvin). Correlations are performed with either thermodynamic descriptors (e.g., free energies of one-electron reduction, lower left panel), or with kinetic descriptors (e.g., reaction rates in a reference system, lower right panel). The upper diagram illustrates the nature of these descriptors, that is the free energy of one-electron reduction, $\Delta_R G[1\text{-}e^- \text{ reduction}]$ and the activation energy E_a of a given (reference) reaction.

Transformation rates of chlorinated alkanes have been investigated in a variety of systems of environmental relevance, most prominently (a) zero-valent iron (*17*, *31–33*); (b) iron(II) porphyrin (*34*), iron(II)-phenol complexes (*35*), or mercaptojuglone (*36*) as surrogates for natural organic matter; and (c) mineral-bound Fe(II) (*29*, *37*) and iron sulfides (*38*, *39*) as representatives of reducing mineral phases. Consistent trends that are observed in most systems allow deriving some *general qualitative rules* (*17*, *34*) about CHC structure and reactivity.

- *Identity of the carbon-chlorine bond.* Due to their lower bond dissociation energy, C-Br bonds are broken more easily than C-Cl bonds, while C-F bonds are essentially non-reactive in subsurface environments.
- *Degree of chlorination.* Higher chlorinated substances are generally degraded more quickly, as indicated in Figure 3. In dechlorination of chlorinated ethenes, for example, the typical trend is k(PCE) > k(TCE) > k(*cis*-DCE) > k(VC). In the case of chlorinated alkanes in addition the position of substitution is important. This leads to the following rules.
- *Number of α-halogen substituents.* Reaction rates are faster if more electron-withdrawing substituents are present in α-position. For example, k(1,1,1-trichloroethane) >> k(1,1,2-trichloroethane) (*17*). Fluorine substituents are an exception, due to their ability to stabilize adjacent C-Cl bonds (α-fluorine effect).
- *Presence of β-halogen substituents.* Reaction rates are further favored by the presence of β-halogen substituents, due to their virtue of stabilizing intermediates (*40*). For example, k(1,1,2-tribromoethane) > k(1,1-dibromoethane), and k(1,1,1,2-tetrachoroethane) > k(1,1,1-trichloroethane) (*17*, *34*).

The principal aims of LFER studies have been (i) to develop a ***quantitative tool*** (*27*) to predict the reaction rate of compound X (e.g., CH_2Cl_2) in a given experimental system if the rate of compound Y (e.g., CCl_4) is known and (ii) to gain ***mechanistic insight*** from the information on the rate-determining step.

LFER as Quantitative Tool

Of the descriptors tested, the best quantitative correlations have been obtained with one-electron reduction potentials (*17*, *30*), bond dissociation energies (*28*, *35*), lowest unoccupied molecular orbitals (LUMO) (*32*, *35*) or transformation rates with single electron transfer reagents such as iron porphyrin and Cr(II) (*17*, *28*, *30*). Correlations were generally best when compound subsets were considered separately (e.g., chlorinated alkanes and ethenes, chloro- and bromoalkanes) (*17*, *28*, *35*). This agrees with the finding that overarching relationships for different compound classes are intrinsically difficult to establish (*41*, *42*).

Mechanistic Insight from LFERs

As indicated in Figure 3, reactivity LFERs reflect activation energies and so may give insight into the nature of the rate-determining step. In this context the good correlation of dechlorination rates with one-electron potentials indicates that in most experimental systems the reductive C-Cl bond cleavage is rate-limiting (*17*, *32*). The absence of such a relationship, on the other hand, can give strong evidence for the importance of other processes, as shown by the following examples.

i. Chlorinated ethene reduction has been observed to give a *reverse* reactivity trend with some types of zero-valent iron (k(VC) > k(*cis*-DCE)

> k(TCE) > k(PCE)) (*43–45*). This has provided strong evidence for rate-limitation by a different chemical process, possibly surface association via π-complexes (*43*).

ii. Transformation rates of 1,1,1-TCA with iron-based bimetallic reductants were found to correlate with the enthalpy of hydrogen *absorption* into the respective metals. This was interpreted as evidence that a process involving absorbed atomic hydrogen rather than electron transfer may be rate-determining (*46*).

iii. No discernible correlation between transformation rates and one-electron reduction potentials is commonly observed if diffusive mass transfer to reactive surfaces (*17*, *47*), or the availability of reactive surface-associated Fe(II) (*48*) is rate-limiting.

Based on expectations from Marcus theory (*28*), a more subtle interpretation of LFER has been attempted with the aim to distinguish even different types of electron transfer such as one- versus two-electron transfer mechanisms. (*28*, *29*, *39*). To this end, correlations between experimental systems were established such as shown in Figure 3, lower right panel. If the slope was unity, this was taken as evidence that the same mechanism prevailed; otherwise different mechanisms were assumed. However, Kohn et al. (*30*) have shown that the co correlation of relative dechlorination rate constants from different reductants merely reflects the sensitivity of activation energies to structural changes of the CHCs, which may or may not be attributable to the occurrence of a common mechanism. A slope of unity provides, therefore, neither necessary nor sufficient evidence for a common mechanism so that great care must be taken in such interpretations.

Mechanistic Insight from Product Studies and Synthesis of Model Reactants

Survey of Reactive Intermediates in Reductive Dechlorination Reactions

Whereas LFERs give evidence about the rate determining step, the key to product formation lies in understanding the formation and reaction of short-lived intermediates. The following schemes introduce the important intermediates and their reactions generated by electron transfer (dashed boxes in the schemes below). Initial reductive C-Cl bond cleavage has been conceptualized to produce organohalide radicals and carbanions (e.g., (*49*, *50*)).

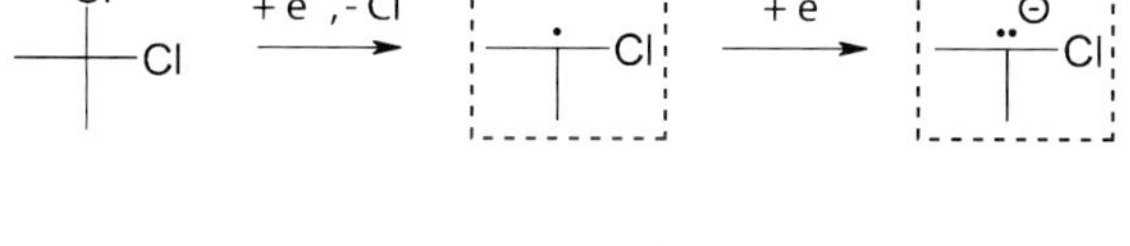

radical carbanion

$+ e^-, - Cl^-$ $+ e^-$

Scheme 1

Alternatively, carbanions may be directly formed by a nucleophilic attack at the halogen atom ("X-philic reaction") according to (*51*, *52*).

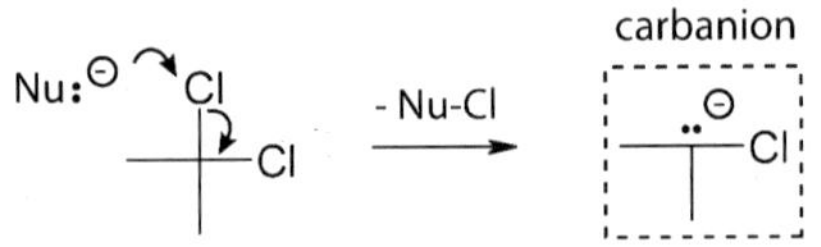

Scheme 2

Radicals can be trapped to give, among others, (a) hydrogenolysis products with H radical donors, (b) sulfur adducts with R-S- species (*53–55*), and (c) oxidized products with molecular oxygen (*56*, *57*).

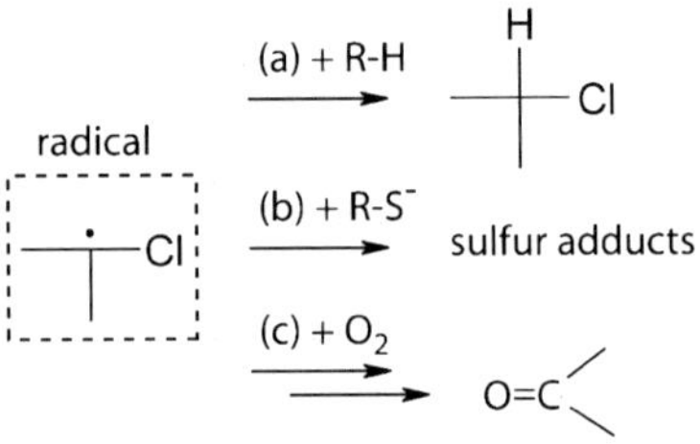

Scheme 3

Carbanions, in turn, may (d) be protonated to hydrogenolysis products or (e) undergo α-(*gem*)elimination to carbenes and (f) β-(*vic*)elimination to unsaturated products (*58*).

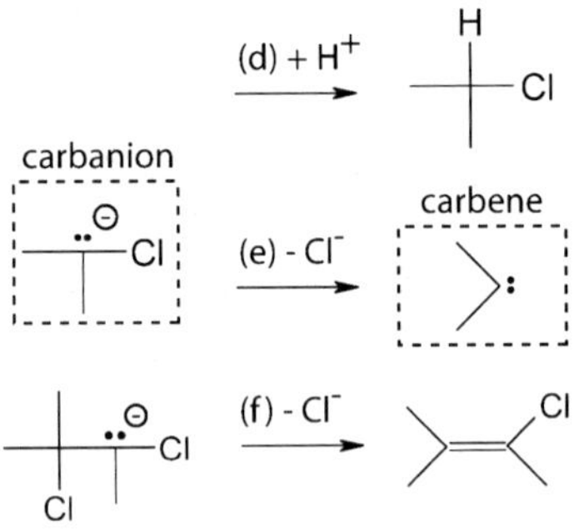

Scheme 4

Carbenes, finally, may be either hydrolyzed (carbene hydrolysis) or reduced (carbene reduction) under reducing conditions in aqueous solution (*49*).

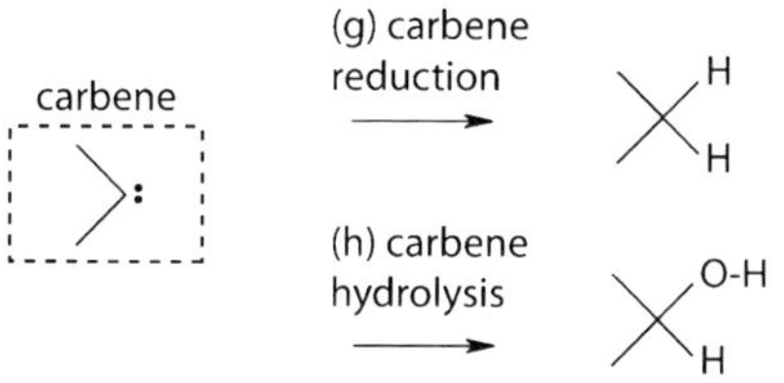

Scheme 5

It is noteworthy that in the case of 1,1,1-trihaloalkanes or tetrahalomethane the carbene hydrolysis further leads to ketones and carbon monoxide, respectively (*59*, *60*).

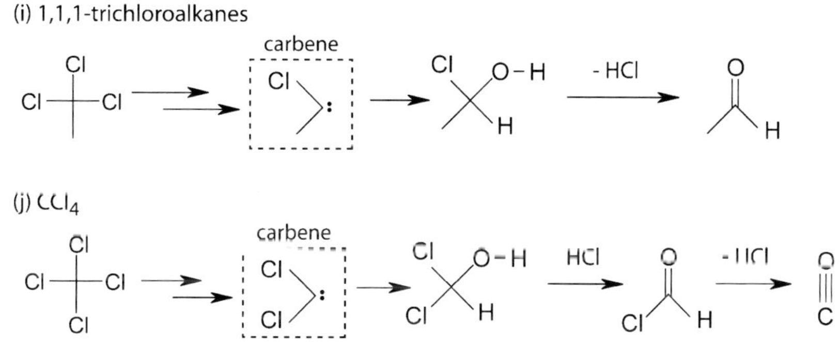

Scheme 6

Strategies for the Analysis of Reaction Intermediates

The schemes demonstrate that identification of reaction mechanisms in environmental transformations can be complicated, because *the same products may form in different ways*. Specifically, the typically more problematic hydrogenolysis products can be generated from radicals (a), carbanions (d) or carbenes (g) alike. Likewise, typically benign oxo products may arise either from radicals (c), or a sequence of carbanions and carbenes (e, h-j). To elucidate the pathways through which transformations are channelled, targeted mechanistic investigations are necessary. Circumstantial evidence may be obtained from systematic changes in reaction conditions (e.g., pH, Fe(II) concentration, temperature) and their effect on reaction rates and product formation (*61–64*). However, it is desirable to observe the short-lived intermediates (dashed boxes in the schemes) also by more direct means. To this end, organic model compounds have been synthesized as summarized in Table 1: (A) traps for intermediates, (B) probe compounds to mimic specific reaction pathways, and (C) putative intermediates of environmental transformations. Table 1 gives an overview of the organic chemists' toolbox, together with the expected mechanistic information. We will refer to this evidence when discussing hypothesized reaction mechanisms for chlorinated alkanes and alkenes below.

Reaction Mechanisms of Chlorinated Alkanes and Ethenes

The following section summarizes available mechanistic insight on reductive dechlorination of (i) *vicinal* haloalkanes, (ii) *geminal* haloalkanes, and (iii) chlorinated ethenes, with a particular focus on the circumstances that can give rise to the formation of (eco)toxic vs. benign dechlorination products.

Table 1. Survey of model reactants to elucidate intermediates of dechlorination reactions

(A) Radical, Carbanion and Carbene Traps		
Model Reactant	**Reaction Scheme**	**Mechanistic Insight (Selection)**
(1) Deuterated isopropanol (radical trap)	$^{\bullet}CCl_3$ + $(CD_3)_2CDOD$ → $DCCl_3$ + $(CD_3)_2C^{\bullet}OD$	Radical intermediates in reactions of CCl_4 + Fe(II)/iron oxides (*65, 66*), CCl_4 + Fe(II) porphyrin (*53*), Vitamin B_{12} + PCE, TCE, but not *cis*-DCE, VC (*67, 68*).
(2) Nitrones (radical spin traps, detection with EPR spectroscopy)	N-*tert*-butyl phenylnitrone (PBN) + $^{\bullet}CCl_3$ → spin adduct	Radical intermediates in reaction of CCl_4 + pdtc (see Table 1) (*55*) Problematic: Applications are limited by *in situ* decomposition of trap (*67*).
(3) Substituted alkenes (radical and carbene trap) (*53, 69*)	$^{\bullet}CCl_3$ + alkene → CCl_3 adduct; $:CCl_2$ + alkene → dichlorocyclopropane (Cl, Cl)	Radical and carbene intermediates in reactions of CCl_4 + Fe(II)/iron oxides (*70*), CCl_4 + Fe(II) porphyrin (*53*).
(4) Deuterated water (carbanion trap)	$^{\ominus}:CCl_3$ $\xrightarrow{D_2O}$ $DCCl_3$	No protonation in reaction of CCl_4 + Fe(II)/goethite (*29*). Protonation, or hydrogen abstraction from Fe-OH surface groups, in reaction of CCl_4 + Fe(II)/iron oxides (*66*).

Continued on next page.

Table 1. (Continued). Survey of model reactants to elucidate intermediates of dechlorination reactions

B) Reactive Probe Compounds		
Model Reactant	**Reaction Scheme**	**Mechanistic Insight (Selection)**
(5) Chloro-acetophenone (probe for electron transfer versus hydride transfer)	hydride transfer; electron transfer	Hypotheses: Dechlorination indicates electron transfer: Keto group reduction indicates hydride transfer; stereospecific keto group reduction indicates biotransformation (*71*).
(6) Probe for outer sphere versus inner sphere ET in reaction with Vitamin B_{12} (*72*)	Vitamin B_{12}; Ti(III)citrate	Hypotheses: Outer sphere electron transfer (ET) reduces the C_2Cl_3 group (more electron deficient, sterically hindered); inner sphere ET reduces the arylalkene group (better accessible, but less electron-deficient), observed dimerization therefore suggests inner sphere electron transfer with Vitamin B_{12}.
(7) 2,3-dibromo-pentanes (probes for single vs. two electron transfer) (*73*)	mixed stereochemistry; Single electron transfer; X-philic reaction (two electron transfer); stereospecific	Approach was validated with (a) I^- (X-philic reaction): stereospecific product formation. (b) Cr(II) (single electron transfer, SET): mixed stereochemistry. (c) Reaction with goethite/Fe(II) gave stereospecific products, despite an SET mechanism. This gives evidence for hindered rotation in surface-associated (rather than free) radical intermediates (*73*).
(8) Arylalkenes (intramolecular radical traps) (*74*)	Vitamin B_{12}; Ti(III)citrate	Cyclisation gives evidence for radical intermediate in reaction with Vitamin B12 (*74*).

Continued on next page.

Table 1. (Continued). Survey of model reactants to elucidate intermediates of dechlorination reactions

(C) Synthesis of Putative Intermediates		
Model Reactant	**Reaction Scheme**	**Mechanistic Insight (Selection)**
(9) Mass spectrometric evidence for chlorovinylcobalamin complexes in reaction of Vitamin B_{12} with chlorinated ethenes (*75*)		
(10) Synthesis of chlorovinylcobaloximes to mimic the reactivity of putative intermediates of chloroethene reaction with Vitamin B_{12}	(a) TCE → (2 [Co(I)], H+; -[Co(II)], HCl) → dichlorovinyl-[Co] (b) dichlorovinyl-[Co] → (e-) → dichlorovinyl carbanion → protic solvent: dichloroethene; aprotic solvent: Cl–≡ (c) Cl–≡ → (2 [Co(I)], H+; -[Co(II)]) → Cl(vinyl)-[Co] + Cl-vinyl-[Co] [Co] = Cobaloxim	(a) Formation of dichlorovinylcobaloxime from TCE and reduced cobaloxime. Stability towards acid or H_2 (*76, 77*). (b) Carbanionic intermediate in further reduction to hydrogenolysis or (*vic*)dichloroelimination products (*78*). (c) Chlorinated acetylene as intermediate in transformation to less reduced chlorovinylcobaloximes
(11) Synthesis of chlorovinylcobalamin to mimic the reactivity of putative intermediates of chloroethene reaction with Vitamin B_{12}	[Co(I)] ≡–Cl → (e-, H+) → Cl-vinyl-[Co] Cl-vinyl-[Co] → (D_2O, Ti(III)) → trideuterovinyl-[Co] [Co(I)] ≡ → (e-, D+; D_2O) → D-vinyl-[Co] [Co] = Cobalamin (Vitamin B_{12})	Evidence that transformation of chlorovinylcobalamin involves acetylene as intermediate (*77*).

Vicinal Haloalkanes

Product formation from *vicinal* haloalkanes is to a great extent predetermined by the CHC's molecular structure. These compounds carry a chlorine substituent in β-position, which is a favourable leaving group. Consequently, *vicinal* chloroalkanes show a strong tendency to *vic*-dichloroelimination producing chlorinated alkenes, which are again problematic compounds (Figure 1a). As illustrated in Figure 4, possible pathways are (i) concerted X-philic reactions (Table 1 (7)), (ii) carbanion formation (Scheme 1 and 2) followed by rapid elimination of a halide ion in β-position (Scheme 4, f), or (iii) hydrogenolysis

(Figure 1a) followed by elimination of HX (Figure 1d) (*51*, *79*). Circumventing the above products (I and II in Figure 4) requires alternative reactions of the hydrogenolysis product (e.g., substitution, oxidation, etc.) or an initial SET (single electron transfer, Scheme 1) after which the second SET is sufficiently slow so that radicals can be trapped in alternative reactions (Scheme 3, b and c, see discussion below).

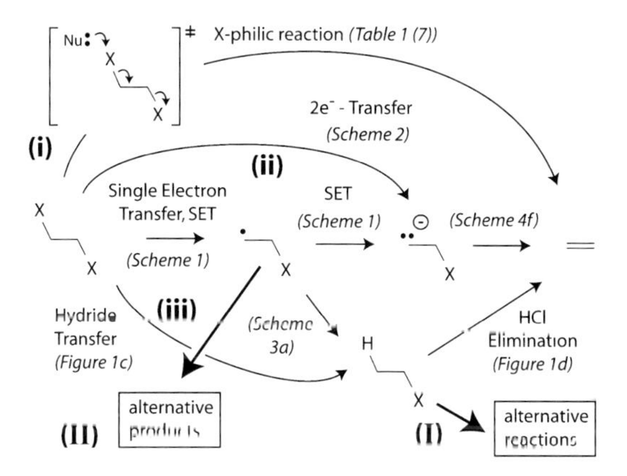

Figure 4. Concurring mechanisms for reductive dechlorination of vicinal haloalkanes. While pathways (i, ii, iii) produce the same vicinal dechlorination product, reaction paths (I) and (II) delineate potential "escape routes" to non-toxic substances (e.g., reaction with radical traps in the case of (II), and substitution (Figure 1c) or hydrogenolysis (Figure 1a) in the case of (I)).

Geminal Haloalkanes

The situation is different for geminal haloalkanes because these compounds lack a good leaving group in β-position, while they possess chlorine in α-position which can stabilize intermediates (e.g., radicals, carbanions). These intermediates have longer lifetimes and may be scavenged in alternative reactions. A recurring theme in transformation of these compounds is the parallel formation of problematic hydrogenolysis products (1,1-dichloroethane from 1,1,1-TCA, $CHCl_3$ from CCl_4) versus benign α-elimination products (ethane, ethene (*46*, *80*, *81*)), or acetaldehyde (*35*) from 1,1,1-TCA; CO (*64*, *65*, *70*, *82*), CO_2 and CS_2 (*54*, *55*, *83*), formate (*29*, *65*), or CH_4 (*70*) from CCl_4. Based on the detection of radicals ((*55*, *65*, *70*), Table 1 (1, 2, 3)) and carbenes ((*70*), Table 1 (3)) in the respective systems, the conceptual reaction scheme given in Figure 5 has been suggested (*70*, *80*, *81*).

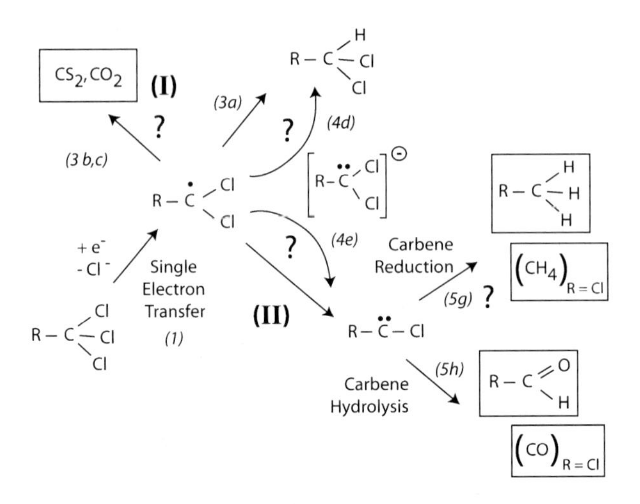

Figure 5. Concurring mechanisms for reductive dechlorination of geminal haloalkanes. Italic numbers in brackets correspond to Schemes 1-6. Boxes are drawn around completely dechlorinated products. (I) and (II) indicate potential "escape routes" to less problematic substances. Question marks indicate steps for which the mechanism / pathway is imperfectly understood.

Figure 5 shows that it is again the fate of short-lived radicals or carbanions, respectively, that determines the formation of less chlorinated versus completely dechlorinated products. From the available literature two possible "strategies" of natural transformations can be discerned to circumvent formation of problematic hydrogenolysis products:

a. The Presence of Good S- or O-Based Radical Scavengers. Several studies report that transformations are directed towards complete dechlorination in the presence of reactive sulfur or oxygen species. Buschmann et al. (*53*) observed negligible formation of $CHCl_3$, but large quantities of N-formyl cysteine when CCl_4 was reduced in the presence of cystine. Along the same lines, appreciable amounts of CO_2 and CS_2 were observed in CCl_4 transformation by sulfide-containing minerals (*54, 84*).

A most intriguing case is the biotic transformation of CCl_4 by *Pseudomonas stutzeri* Strain KC (*85*). Although CCl_4 is toxic to organisms, and although to date no microbial system has been described which can use CCl_4 as sole carbon source (*86*), this bacterium was found to completely transform CCl_4 without appreciable formation of $CHCl_3$ (*85*). The reaction was shown to be accomplished by an extracellular mediator, Cu(II) pyridine-2,6-bis(thiocarboxylate) (Cu(II):PDTC). Based on the detection of radical intermediates (Table 1, Entry 2) the following mechanism has been postulated (*55*):

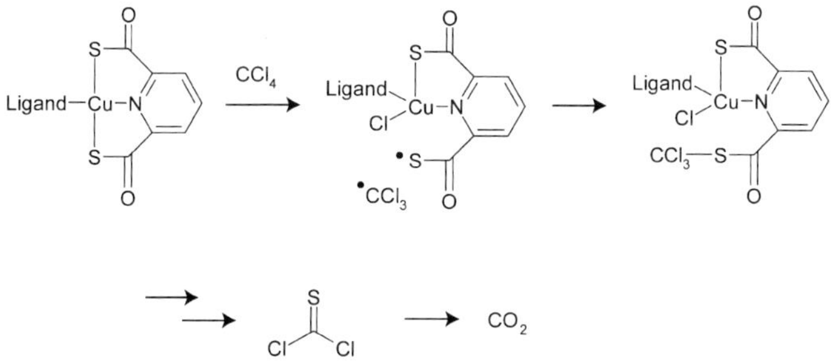

Scheme 7

According to this mechanism, the reactant Cu(II):PDTC does not only provide the reducing agent, but also carries a built-in radical trap. Concomitant with initial single electron transfer, a sulfur radical is generated, in time to catch the freshly produced $^{\bullet}CCl_3$ radical and channel the reaction towards formation of thiophosgene, which is further hydrolyzed to non-toxic products! A similar selectivity has been observed with iron(II) oxygen complexes: only acetaldehyde, but no 1,1-dichloroethane was observed in reaction of 1,1,1-trichloroethane with Fe(II)-tiron complexes (*35*), see Figure 2. Insight from future studies may show whether a common mechanism applies with such "intelligent" reducing agents.

b. Stabilization of Intermediates in (Surface) Complexes. Several studies have proposed that some short-lived intermediates are not present in free form, but stabilized as complexes, e.g., carbenes as carbenoids (*49*, *50*, *80*). In recent years, three different groups (*64*, *65*, *70*) have independently suggested such a complex formation at Fe(II) bearing mineral surfaces, driven by the motivation to rationalize the following observations: (i) the absence of deuterated haloform when tetrahalomethane is reacted in the presence of D_2O at pH 7 (Table 1, Entry 4, (*29*)) as well as (ii) the strong influence of mineral surface properties (surface charge, type of mineral) on product distribution, which contradicts the hypothesis of free CCl_3^- species in solution (*64*, *70*). Specifically, Elsner et al. (*65*) suggested that radicals may be stabilized according to

$$\left[\mathrm{S-Fe(II)\cdots CCl_3^{\bullet} \leftrightarrow S-Fe(III)-CCl_3}\right] \xrightarrow{+e^- \, -Cl^-} \left[\mathrm{S-Fe(II)=CCl_2}\right]$$

so that formation of free CCl_3^- in solution would essentially be circumvented. Future studies will be needed to substantiate such hypotheses and to provide more information about surface properties that are instrumental in stabilizing intermediates and thus circumventing problematic products.

a. Putative Initial Reaction Steps. Contrasting with chlorinated alkanes, the π-electron system of chlorinated ethenes, together with their less cramped coordination environment, lends these compounds additional features for the initial transformation step. Figure 6 illustrates different routes that have been postulated for chloroethene transformation by reduced cobalamin (Vitamin B_{12}). Besides (1) dissociative electron transfer, (2) a nucleophilic (S_N2) substitution mechanism and (3) nucleophilic addition of Co(I) at the alkene double bond have been suggested, as discussed in the following (Figure 6).

Evidence for the first scenario (**dissociative outer sphere electron transfer**) is delivered from experiments with radical traps that clearly demonstrate the existence of dichlorovinyl radicals (*67*, *68*, *74*); (Entry 1, 6, 8 in Table 1). The second scenario (**nucleophilic substitution**) is motivated by kinetic models (*87*), by mass balance considerations (*88*) as well as by the fact that dichlorovinyl cobalamin complexes have been detected in mass spectrometric analysis of TCE dechlorination reaction mixtures (*75*), (Entry 9 of Table 1). The third scenario (**nucleophilic addition**), finally, has been postulated for reactions of dichloroethylenes and of vinyl chloride, based on the pH dependence observed in reactions of these substrates (*61*).

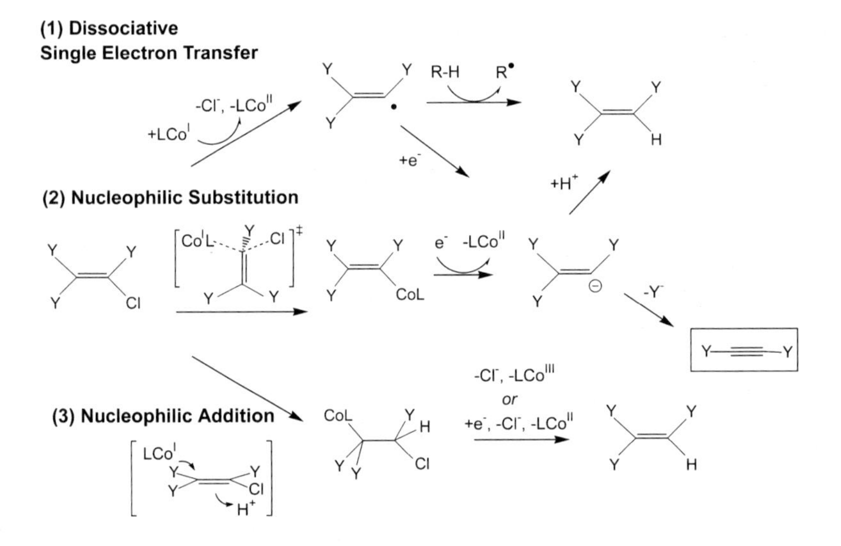

Figure 6. Possible transformation pathways proposed for reaction of chlorinated ethenes with cobalamin (Vitamin B_{12}) where Y = H or Cl (20). A box is drawn around the vic dichloroelimination intermediate, which rapidly reacts on to non-toxic products (Figure 7).

b. Product Formation. Hydrogenolysis of chlorinated ethenes typically generates more problematic products (the exception is ethene from VC), while vicinal dihaloelimination gives chlorinated acetylenes, which are quickly further transformed to benign ethene and ethane (see Figure 7). Interestingly, microbial transformation gives only the problematic hydrogenolysis pathway (*7*), whereas both pathways are reported with zero-valent metals (*43–45*).

Vitamin B_{12}. Also cob(I)alamin, the model system for microbial dechlorination, generates chlorinated acetylenes (*61*, *67*, *89*)) indicating that the *vicinal* dichloroelimination pathway is in principle operative with Vitamin B_{12} (lower part of Figure 7). The same insight is obtained from experiments with model complexes for putative intermediates, i.e., chlorovinylcobalamins / chlorovinylcobaloximes (Entries 10 and 11 of Table 1) (*76*, *77*, *90–94*). The selective hydrogenolysis pathway during biotransformation is, therefore, still not well understood. In principle, hydrogenolysis products may again be produced in more than one way. Nucleophilic substitution (Figure 6, pathway 2) results in hydrogenolysis products in the presence of protic solvents; otherwise they form via vicinal dichloroelimination. Outer sphere electron transfer (Figure 6, pathway 1) may lead to hydrogenolysis either through protonation or owing to the presence of H radical donors. Only pathway 3 in Figure 6 (nucleophilic addition) results almost completely in the hydrogenolysis product. It remains to be elucidated what factors exactly are responsible for the selective product formation observed during biotransformation.

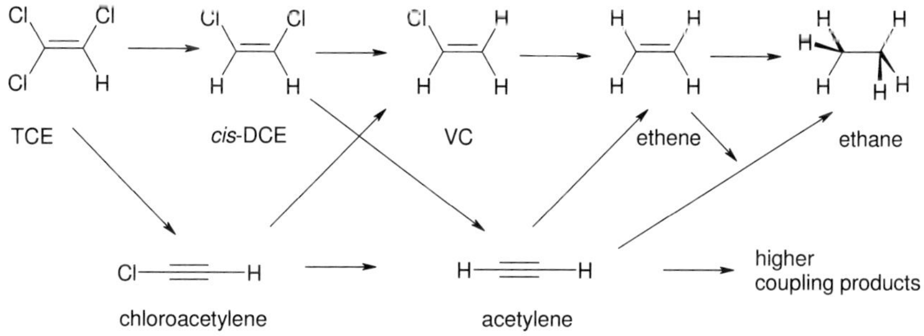

Figure 7. Concurring pathways in chlorinated ethene dechlorination. Biodegradation generally involves sequential hydrogenolysis to ethene (upper pathway), whereas abiotic dechlorination includes also vicinal dichloroelimination (lower pathway). Adapted from (45).

Zero-Valent Iron (ZVI). Compared to the insight accomplished with vitamin B_{12}, even less is known about the reductive dechlorination mechanism at iron metal surfaces. Available evidence even disagrees about reactivity trends: some studies report slower reaction rates for less chlorinated ethenes (*31*, *32*, *95*), and others report exactly the opposite trend (*43–45*, *96*). As discussed in the section "Mechanistic Insight from LFER" above, the first observation agrees with expectations from one-electron reduction potentials (*32*), whereas the latter may be explained by π-complexes of chlorinated ethenes at the metal surface, which would favour less chlorinated compounds (*43*). Also the relative contribution of hydrogenolysis and vicinal dichloroelimination has been found to be variable between studies (*43–45*).

In a similar way, knowledge gaps exist with respect to the nature of the reactive site at the iron surface. Reduction by Fe^0, Fe^{2+}, or by Fe-H surface species have been brought forward as alternative hypotheses (*97*, *98*). Iron metal materials may be manufactured in different ways and may have different properties. Nanoparticulate iron, for example, may be manufactured either through reductive precipitation of $Fe^{2+}_{(aq)}$ by $NaBH_4$ leading to boron inclusions ("Fe^{BH}"), or through high temperature reduction of iron oxides by hydrogen gas leading to magnetite impurities ("Fe^{H2}") (*25*). Current evidence indicates that reactivity trends of chlorinated ethylenes differ between both types of minerals, where less chlorinated ethylenes react more rapidly with Fe^{BH} and the inverse trend is observed with Fe^{H2}. Despite such insights it remains to be understood how properties of the metal (synthesis, presence of impurities) affect reactivity and pathways of reductive chlorinated ethene transformation: how they may change the mechanism of the initial transformation step, and how they affect branching points of product formation. An assessment is made difficult by the fact that the same product (ethene) may again originate from either the sequential hydrogenolyis or the dichloroelimination pathway.

Outlook: Insight from Stable Isotope Fractionation

As discussed so far, studies in model systems have made it possible to hypothesize well-defined, alternative mechanisms for reductive dechlorination. To improve their assessment in the contaminated subsurface, however, additional efforts are needed that (1) enable distinction of different initial transformation mechanisms leading to different short-lived intermediates and (2) allow for the deconvolution of simultaneously occurring product formation routes. In the following section, we highlight how these questions can be approached by stable isotope fractionation measurements.

Bridging the Gap between Model System and Reality: Identifying Dechlorination Mechanisms from Measuring (Multi Element) Reactant Isotope Fractionation

The measurement of kinetic isotope effects (KIE) is a well-established tool to elucidate reaction mechanisms of chemical reactions (*99*, *100*). Isotope effects express how a rate of transformation changes when a light isotope (e.g., ^{12}C, ^{35}Cl) is replaced by a heavy isotope of the same element (e.g., ^{13}C, ^{37}Cl) in a particular position of an organic compound

$$KIE = {}^{l}k/{}^{h}k$$

where ^{l}k and ^{h}k are the rate constants for molecules carrying the light and heavy isotope, respectively. The magnitude of a KIE depends on the element involved as well as the position within the molecule. For example, chlorine kinetic isotope effects ($^{35}k/^{37}k$) are largest if a C-Cl bond is broken and are expected to be smaller otherwise. KIE, therefore, make it possible to use isotopes as intramolecular probes to test which elements are involved in a reaction. Even more important is the fact that KIEs also differ for distinct reaction mechanisms. During a reductive dechlorination reaction, the isotopic composition (e.g., $^{13}C/^{12}C$, $^{37}Cl/^{35}Cl$, $^{2}H/^{1}H$) of the reactant is therefore altered (i.e., fractionated) in a characteristic way that allows one to infer the underlying transformation pathway. The same principles also apply to trends of isotope compositions of reaction products.

Isotope ratios of CHCs are typically analyzed by gas chromatography-isotope ratio mass spectrometry (GC-IRMS). Single chloroorganic compounds can be separated from relatively complex environmental samples, and the chemical substances may be analyzed directly for their stable isotope composition even at low concentrations. At natural isotopic abundance, this method measures compound-specific rather than position-specific enrichment of isotopes. The ease with which relevant reactions can be investigated has triggered a renaissance of isotope investigations on reactions of organic substances in recent years (*101–103*).

In fact, a considerable number of investigations have revealed significant C isotope fractionation associated with biotic and abiotic reductive dechlorination transformations, including hydrogenolyses as well as vic and *gem* dichoroeliminations (*45, 104–108*). However, as reported ranges of ^{13}C-KIEs overlap and/or are difficult to interpret due to the kinetic complexity of dechlorination processes, there is considerable interest in the simultaneous isotope analysis of multiple elements (*103, 109*) The advantage of this approach is that different mechanisms can be discerned simply by correlating the changes in isotope ratios for the two elements (typically an enrichment of heavy isotopes such as ^{13}C, ^{37}Cl, or ^{2}H in the reactant) (*109–112*). For example, as illustrated in Figure 8, when changes in isotope ratios of carbon are plotted relative to those of chlorine, different dual isotope slopes may be expected, due to different underlying kinetic isotope effects of the two elements in different initial reactions (e.g., dissociative electron transfer versus nucleophilic substitution versus nucleophilic addition, see above).

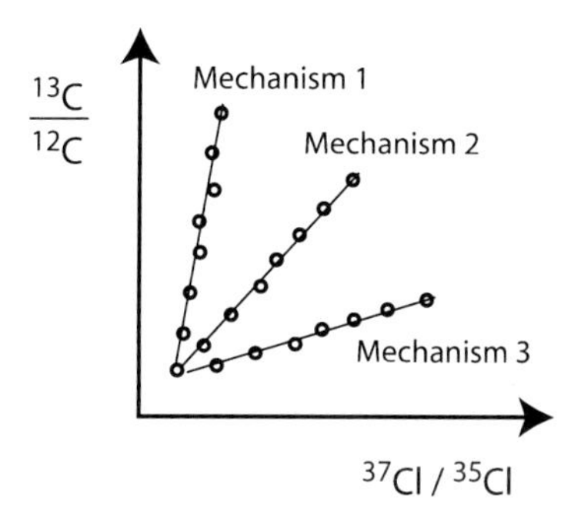

Figure 8. Potential use of dual (C and Cl) isotope plots to probe for different mechanisms of the initial dechlorination step (e.g., dissociative electron transfer versus nucleophilic substitution versus nucleophilic addition, see Figure 6).

This approach holds great promise to bridge the gap between model systems and the environment because the isotope ratios in a contaminant are altered in a characteristic way by bond cleavage and formation processes and do not necessarily require additional characterization of the environmental conditions and/or even an evaluation of transformation products. Therefore, if the dual isotope plot observed with a well-characterized model reactant can be reproduced during natural transformation, direct evidence may be obtained that both reactions share the same mechanism in the initial reaction step.

As summarized recently (*103*), this approach has been successfully applied to numerous organic compounds. Applications to CHCs, in contrast, have been hampered by the difficulty of analyzing isotope ratios for multiple elements in these compounds. Very recent developments have, for the first time enabled, compound-specific chlorine isotope analysis of chlorinated ethenes and ethanes (*113–115*)). The power of the dual isotope approach has been indicated by degradation studies (*107*, *116*) as well as theoretical considerations (*117*). Future studies may be expected to provide unique new insight into the initial mechanisms of dechlorination reactions.

Deconvolution of Reaction Pathways: Putting Product Isotope Ratios to Use

In contrast to the evaluation of reactant isotope fractionation, product isotope ratios can be indicative of further reaction steps following up on any irreversible bond cleavage in the reactant. This information is particularly useful not only to track precursor compounds and elucidate competing product formation pathways, but also to characterize branching points that lead to benign vs. toxic dechlorination products.

How is Ethene Formed? – Distinguishing Sequential Hydrogenolysis from Vicinal Dichloroelimination

As discussed above, conventional approaches are not able to deconvolute competing pathways when they give rise to the same product. Figure 9 gives an illustration of the added insight that may be obtained from isotope effect studies. In dechlorination reactions, the product ethene may either originate from hydrogenolysis or from vicinal dichloroelimination (Figure 7). Sequential hydrogenolysis forms ethene at the end of a cascade of consecutive reactions. In such a reaction cascade, products are initially depleted in ^{13}C compared to their precursor, but subsequently become enriched in ^{13}C when the isotope effect of their own degradation starts to affect their isotope ratio (Figure 9, left panel). In contrast, dichloroelimination can also form ethene by a much faster, alternative route via chlorinated acetylenes (Figure 7). In this case, *cis*-DCE and ethene are formed as parallel products so that both show a parallel isotope fractionation trend (Figure 9, right panel). Figure 9 compares the very different isotope ratio trends that are typical of either case and illustrates that isotope fractionation analysis allows the distinction of the two pathways. Since vicinal dichloroelimination typically occurs in abiotic dechlorination reactions (*43*, *45*,

63), whereas hydrogenolysis is the established biodegradation pathway, this approach has made it possible to distinguish abiotic from biotic chloroethylene transformation (*45*, *103*).

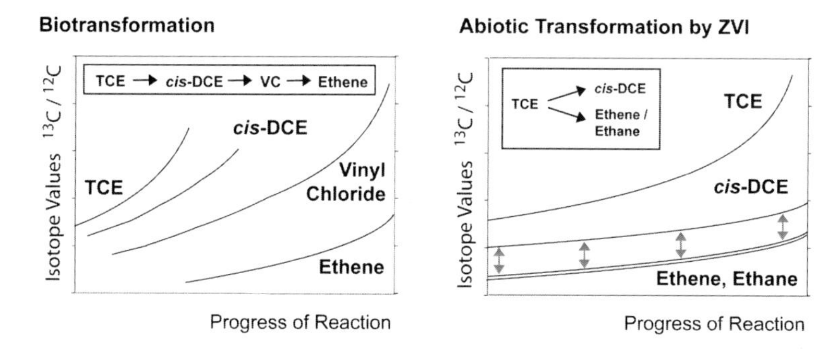

Figure 9. Different trends in product isotope ratios reveal whether cis-DCE and ethane are formed as parallel (right) or consecutive (left) products. The sketch is based on trends typically observed in these reactions (45, 104, 105, 118).

Where Is the Branching Point? – Probing for Common Intermediates

In addition, to understand the formation of problematic products, the identification of branching points and common intermediates is important. Specifically, to assess CHC reduction, it is of interest whether hydrogenolysis and dichloroelimination represent two entirely different transformation mechanisms – potentially occurring at different reactive sites – or whether they share a common intermediate. A well-established approach to address this issue is the variation of reaction conditions (e.g., pH) as a means to induce changes in the product distribution (*29*, *63*, *64*). Unfortunately, even with this approach it is typically impossible to obtain conclusive evidence about branching points. As demonstrated recently (*45*), such insight can be obtained if isotope values in reactant *and* products are measured in the same experiments and if the corresponding kinetic isotope effects are determined (Figure 10). The reason is that these observable KIEs change when the product yield shifts (e.g., from 10% dichloroelimination to 90% dichloroelimination), and that the manner of this change is highly indicative of either of two scenarios. As illustrated in Figure 10, if two independent pathways prevail, each one exhibits its specific isotope effect. Consequently, the KIE that is determined from measuring reactant isotope values only is a weighted average of both pathways and changes depending on product yield. In the case of a common intermediate, in contrast, the KIE in the reactant is invariable, because it reflects the initial irreversible step. Here, it is the KIE that is determined from the difference in isotope values of reactant and either product which changes depending on product yield. In a recent study on chloroethene dechlorination by nanoscale zero-valent iron, such lines of evidence could provide a first indication for the second case, that is, a common irreversible step was shared by both pathways (*45*).

Two Independent Rections

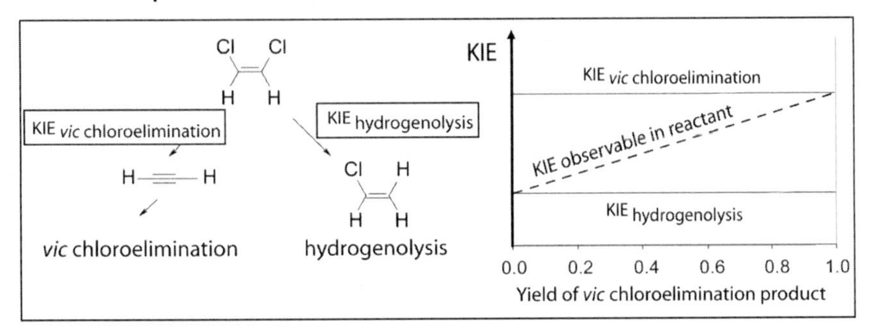

Common Intermediate

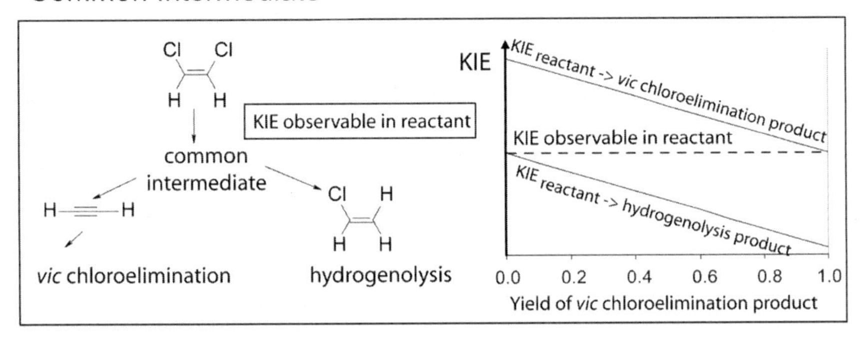

Figure 10. The occurrence of a common intermediate can be revealed by investigating how KIE values change when the product distribution shifts. For explanations see text. Adapted from (45).

Conclusion

The complexity of the various biotic and abiotic reductive dechlorinations of chloroethenes, -ethanes, and -methanes renders their assessment in the environment a formidable task. Over the past decades, the understanding of underlying mechanisms has been significantly advanced by approaches with structure-activity relationships, product studies, intermediate traps, and probe molecules. The field is at a point where well-established mechanistic hypotheses have been brought forward for different initial transformation steps (e.g., dissociative single electron transfer, nucleophilic substitution, etc., Figure 6), where evidence for different reaction intermediates has been established (e.g., radicals, carbanions, carbenes, Vitamin B_{12} complexes, etc.) and where different, simultaneously occurring product formation pathways have been hypothesized (Figures 4-6). Currently, however, interpretations of environmental transformations face the challenge that the mechanism of initial transformation steps is not easily distinguished, that branching points of product formation are difficult to identify, and that the same products can be formed in different ways. Our outlook shows how stable isotope fractionation studies may close some of these prevalent research gaps by distinguishing different initial transformation mechanisms, by deconvolving simultaneously occurring product formation routes, and by detecting branching points of product formation. Future developments of versatile analytical methods for carbon, chlorine, and hydrogen isotope analysis together with complementary computational approaches may substantially improve our ability to assess dechlorination processes in subsurface environments.

References

1. Schwarzenbach, R. P.; Egli, T.; Hofstetter, T. B.; von Gunten, U.; Wehrli, B. Global water pollution and human health. *Annu. Rev. Environ. Resour.* **2010**, *35*, 109–136.
2. European Environment Agency (EEA) Progress in management of contaminated sites. http://www.eea.europa.eu/data-and-maps/indicators/progress-in-management-of-contaminated-sites/progress-in-management-of-contaminated-1, Document CSI 015, Copenhagen, 2007.
3. EuroChlor Chlorine Industry Review CEFIC European Chemical Industry Council, Brussels. www.eurochlor.org, 2009−2010.
4. Von Grote, J.; Hurlimann, C.; Scheringer, M.; Hungerbühler, K. Reduction of occupational exposure to perchloroethylene and trichloroethylene in metal degreasing over the last 30 years: influences of technology innovation and legislation. *J. Exposure Anal. Environ. Epidemiol.* **2003**, *13*, 325–340.
5. Schwarzenbach, R. P.; Gschwend, P. M.; Imboden, D. M. *Environmental Organic Chemistry*; 2nd ed.; John Wiley & Sons: New York, 2003.
6. Smidt, H.; de Vos, W. M. Anaerobic microbial dehalogenation. *Annu. Rev. Microbiol.* **2004**, *58*, 43–73.
7. Löffler, F. E.; Edwards, E. A. Harnessing microbial activities for environmental cleanup. *Curr. Opin. Biotechnol.* **2006**, *17*, 274–284.
8. Vogel, T. M.; Criddle, C. S.; McCarty, P. L. Transformations of halogenated aliphatic compounds. *Environ. Sci. Technol.* **1987**, *21*, 722–735.
9. McCormick, M. L.; Bouver, E. J.; Adriaens, P. Carbon tetrachloride transformation in an model iron-reducing culture: relative kinetics of biotic and abiotic reactions. *Environ. Sci. Technol.* **2002**, *36*, 403–410.
10. Butler, E. C.; Dong, Y.; Krumholz, L. R.; Liang, X.; Shao, H.; Tan, Y. Rate controlling processes in the transformation of tetrachloroethylene and carbon tetrachloride. In *Aquatic Redox Chemistry*; Tratnyek, P. G., Grundl, T. J., Haderlein, S. B., Eds.; ACS Symposium Series; American Chemical Society: Washington, DC, 2011; Vol. 1071, Chapter 23, pp 519−538.
11. Costentin, C.; Robert, M.; Saveant, J. M. Successive removal of chloride ions from organic polychloride pollutants. Mechanisms of reductive electrochemical elimination in aliphatic gem-polychlorides, α,β-polychloroalkenes, and α,β-polychloroalkanes in mildly protic medium. *J. Am. Chem. Soc.* **2003**, *125*, 10729–10739.

12. Costentin, C.; Robert, M.; Saveant, J. M. Fragmentation of aryl halide pi anion radicals. Bending of the cleaving bond and activation vs driving force relationships. *J. Am. Chem. Soc.* **2004**, *126*, 16051–16057.
13. Totten, L. A.; Roberts, A. L. Calculated one- and two-electron reduction potentials and related molecular descriptors for reduction of alkyl and vinyl halides in water. *Crit. Rev. Environ. Sci. Technol.* **2001**, *31*, 175–221.
14. Curtis, G. P.; Reinhard, M. Reductive dehalogenation of hexachloroethane, carbon tetrachloride, and bromoform by anthrahydroquinone disulfonate and humic acid. *Environ. Sci. Technol.* **1994**, *28*, 2393–2401.
15. Valiev, M.; Bylaska, E. J.; Dupuis, M.; Tratnyek, P. G. Combined quantum mechanical and molecular mechanics studies of the electron-transfer reactions involving carbon tetrachloride in solution. *J. Phys. Chem. A* **2008**, *112*, 2713–2720.
16. Bylaska, E. J.; Salter-Blanc, A. J.; Tratnyek, P. G. One-electron reduction potentials from chemical structure theory calculations. In *Aquatic Redox Chemistry*; Tratnyek, P. G., Grundl, T. J., Haderlein, S. B., Eds.; ACS Symposium Series; American Chemical Society: Washington, DC, 2011; Vol. 1071, Chapter 3, pp 37−64.
17. Cwiertny, D. M.; Arnold, W. A.; Kohn, T.; Rodenburg, L. A.; Roberts, A. L. Reactivity of alkyl polyhalides toward granular iron: Development of QSARs and reactivity cross correlations for reductive dehalogenation. *Environ. Sci. Technol.* **2010**, *44*, 7928–7936.
18. Aeschbacher, M.; Sander, M.; Schwarzenbach, R. P. Novel electrochemical approach to assess the redox properties of humic substances. *Environ. Sci. Technol.* **2010**, *44*, 87–93.
19. Silvester, E.; Charlet, L.; Tournassat, C.; Gehin, A.; Greneche, J.-M.; Liger, E. Redox potential measurements and Mössbauer spectrometry of Fe^{II} adsorbed onto Fe^{III} (oxyhydr)oxides. *Geochim. Cosmochim. Acta* **2005**, *69*, 4801–4815.
20. Kliegman, S.; McNeill, K. Dechlorination of chloroethylenes by cob(I) alamin and cobalamin model complexes. *Dalton Trans.* **2008**, 4191–4201.
21. Schwarzenbach, R. P.; Stierli, R.; Lanz, K.; Zeyer, J. Quinone and iron porphyrin mediated reduction of nitroaromatic compounds in homogeneous aqueous solution. *Environ. Sci. Technol.* **1990**, *24*, 1566–1574.
22. Shi, Z.; Nurmi, J. T.; Tratnyek, P. G. Effects of nano zero-valent iron on oxidation-reduction potential. *Environ. Sci. Technol.* **2011**, *45*, 1586–1592.
23. Bylaska, E. J.; Dixon, D. A.; Felmy, A. R.; Tratnyek, P. G. One-electron reduction of substituted chlorinated methanes as determined from ab initio electronic structure theory. *J. Phys. Chem. A* **2002**, *106*, 11581–11593.
24. Bylaska, E. J.; Dupuis, M.; Tratnyek, P. G. Ab initio electronic structure study of one-electron reduction of polychlorinated ethylenes. *J. Phys. Chem. A* **2005**, *109*, 5905–5916.
25. Nurmi, J. T.; Tratnyek, P. G.; Sarathy, V.; Baer, D. R.; Amonette, J. E.; Pecher, K.; Wang, C. M.; Linehan, J. C.; Matson, D. W.; Penn, R. L.; Driessen, M. D. Characterization and properties of metallic iron nanoparticles: Spectroscopy, electrochemistry, and kinetics. *Environ. Sci. Technol.* **2005**, *39*, 1221–1230.

26. Nurmi, J. T.; Tratnyek, P. G. Electrochemical studies of packed iron powder electrodes: Effects of common constituents of natural waters on corrosion potential. *Corros. Sci.* **2008**, *50*, 144–154.
27. Tratnyek, P. G.; Weber, E. J.; Schwarzenbach, R. P. Quantitative structure-activity relationships for chemical reductions of organic contaminants. *Environ. Toxicol. Chem.* **2003**, *22*, 1733–1742.
28. Perlinger, J. A.; Venkatapathy, R.; Harrison, J. F. Linear free energy relationships for polyhalogenated alkane transformation by electron-transfer mediators in model aqueous systems. *J. Phys. Chem. A* **2000**, *104*, 2752–2763.
29. Pecher, K.; Haderlein, S. B.; Schwarzenbach, R. P. Reductive of polyhalogenated methanes by surface-bound Fe(II) in aqueous suspensions of iron oxides. *Environ. Sci. Technol.* **2002**, *36*, 1734–1741.
30. Kohn, T.; Arnold, W. A.; Roberts, A. L. Reactivity of substituted benzotrichlorides toward granular iron, Cr(II), and an iron(II) porphyrin: A correlation analysis. *Environ. Sci. Technol.* **2006**, *40*, 4253–4260.
31. Johnson, T. L.; Scherer, M. M.; Tratnyek, P. G. Kinetics of halogenated organic compound degradation by iron metal. *Environ. Sci. Technol.* **1996**, *30*, 2634–2640.
32. Scherer, M. M.; Balko, B. A.; Gallagher, D. A.; Tratnyek, P. G. Correlation analysis of rate constants for dechlorination by zero-valent iron. *Environ. Sci. Technol.* **1998**, *32*, 3026–3033.
33. Song, H.; Carraway, E. R. Reduction of chlorinated ethanes by nanosized zero-valent iron: Kinetics, pathways, and effects of reaction conditions. *Environ. Sci. Technol.* **2005**, *39*, 6237–6245.
34. Perlinger, J. A.; Buschmann, J.; Angst, W.; Schwarzenbach, R. P. Iron porphyrin and mercaptojuglone mediated reduction of polyhalogenated methanes and ethanes in homogeneous aqueous solution. *Environ. Sci. Technol.* **1998**, *32*, 2431–2437.
35. Bussan, A. L.; Strathmann, T. J. Influence of organic ligands on the reduction of polyhalogenated alkanes by Iron(II). *Environ. Sci. Technol.* **2007**, *41*, 6740–6747.
36. Perlinger, J. A.; Angst, W.; Schwarzenbach, R. P. Kinetics of the reduction of hexachloroethane by juglone in solutions containing hydrogen sulfide. *Environ. Sci. Technol.* **1996**, *30*, 3408–3417.
37. Neumann, A.; Hofstetter, T. B.; Skarpeli-Liati, M.; Schwarzenbach, R. P. Reduction of polychlorinated ethanes and carbon tetrachloride by structural Fe(II) in smectites. *Environ. Sci. Technol.* **2009**, *43*, 4082–4089.
38. Butler, E. C.; Hayes, K. F. Kinetics of the transformation of halogenated aliphatic compounds by iron sulfide. *Environ. Sci. Technol.* **2000**, *34*, 422–429.
39. Kenneke, J. F.; Weber, E. J. Reductive dehalogenation of halomethanes in iron- and sulfate- reducing sediments. 1. Reactivity pattern analysis. *Environ. Sci. Technol.* **2003**, *37*, 713–720.
40. Ulstrup, J. Relationship between energy of activation and overall free-energy of bridge-assisted electron-transfer reactions in polar media. *Acta Chem. Scand.* **1973**, *27*, 1067–1072.

41. Elsner, M.; Schwarzenbach, R. P.; Haderlein, S. B. Reactivity of Fe(II)-bearing minerals toward reductive transformation of organic contaminants. *Environ. Sci. Technol.* **2004**, *38*, 799–807.
42. Miehr, R.; Tratnyek, P. G.; Bandstra, J. Z.; Scherer, M. M.; Alowitz, M. J.; Bylaska, E. J. Diversity of contaminant reduction reactions by zerovalent iron: Role of the reductate. *Environ. Sci. Technol.* **2004**, *38*, 139–147.
43. Arnold, W. A.; Roberts, A. L. Pathways and kinetics of chlorinated ethylene and chlorinated acetylene reaction with Fe(0) particles. *Environ. Sci. Technol.* **2000**, *34*, 1794–1805.
44. Liu, Y. Q.; Majetich, S. A.; Tilton, R. D.; Sholl, D. S.; Lowry, G. V. TCE dechlorination rates, pathways, and efficiency of nanoscale iron particles with different properties. *Environ. Sci. Technol.* **2005**, *39*, 1338–1345.
45. Elsner, M.; Chartrand, M.; VanStone, N.; Lacrampe Couloume, G.; Sherwood Lollar, B. Identifying abiotic chlorinated ethene degradation: Characteristic isotope patterns in reaction products with nanoscale zero-valent iron. *Environ. Sci. Technol.* **2008**, *42*, 5963–5970.
46. Cwiertny, D. M.; Bransfield, S. J.; Livi, K. J. T.; Fairbrother, D. H.; Roberts, A. L. Exploring the influence of granular iron additives on 1,1,1-trichloroethane reduction. *Environ. Sci. Technol.* **2006**, *40*, 6837–6843.
47. Arnold, W. A.; Ball, W. P.; Roberts, A. L. Polychlorinated ethane reaction with zero-valent zinc: pathways and rate control. *J. Contam. Hydrol.* **1999**, *40*, 183–200.
48. Heijman, C. G.; Grieder, E.; Holliger, C.; Schwarzenbach, R. P. Reduction of nitroaromatic compounds coupled to microbial iron reduction in laboratory aquifer columns. *Environ. Sci. Technol.* **1995**, *29*, 775–783.
49. Castro, C. E.; Kray, W. C. Carbenoid intermediates from polyhalomethanes and chromium(II). Homogeneous reduction of geminal halides by chromous sulfate. *J. Am. Chem. Soc.* **1966**, *88*, 4447–4455.
50. Ahr, H. J.; King, L. J.; Nastainczyk, W.; Ullrich, V. The mechanism of chloroform and carbon-monoxide formation from carbon-tetrachloride by microsomal cytochrome P450. *Biochem. Pharmacol.* **1980**, *29*, 2855–2861.
51. Roberts, L. A.; Gschwend, P. M. Interaction of abiotic and microbial processes in hexachloroethane reduction in groundwater. *J. Contam. Hydrol.* **1994**, *16*, 157–174.
52. Zefirov, N. S.; Makhonkov, D. I. X-Philic Reactions. *Chem. Rev.* **1982**, *82*, 615–624.
53. Buschmann, J.; Angst, W.; Schwarzenbach, R. P. Iron porphyrin and cysteine mediated reduction of ten polyhalogenated methanes in homogeneous aqueous solution: Product analyses and mechanistic considerations. *Environ. Sci. Technol.* **1999**, *33*, 1015–1020.
54. Kriegman-King, M. R.; Reinhard, M. Transformation of carbon tetrachloride by pyrite in aqueous solution. *Environ. Sci. Technol.* **1994**, *28*, 692–700.
55. Lewis, T. A.; Paszczynski, A.; Gordon-Wylie, S. W.; Jeedigunta, S.; Lee, C. H.; Crawford, R. L. Carbon tetrachloride dechlorination by the bacterial transition metal chelator pyridine-2,6-bis(thiocarboxylic acid). *Environ. Sci. Technol.* **2001**, *35*, 552–559.

56. Monig, J.; Krischer, K.; Asmus, K. D. One-electron reduction of halothane and formation of halide-ions in aqueous solutions. *Chem.-Biol. Interact.* **1983**, *45*, 43–52.
57. Wagner, A. J.; Vecitis, C.; Fairbrother, D. H. Electron-stimulated chemical reactions in carbon tetrachloride/water (ice) films. *J. Phys. Chem. B* **2002**, *106*, 4432–4440.
58. Hine, J.; Ehrenson, S. J. The effect of structure on the relative stability of dihalomethylenes. *J. Am. Chem. Soc.* **1958**, *80*, 824–830.
59. Pliego, J. R.; DeAlmeida, W. B. Reaction paths for aqueous decomposition of CCl_2. *J. Phys. Chem.* **1996**, *100*, 12410–12413.
60. Phillips, D. L.; Zhao, C. Y.; Wang, D. Q. A theoretical study of the mechanism of the water-catalyzed HCl elimination reactions of CHXCl(OH) (X = H, Cl) and HClCO in the gas phase and in aqueous solution. *J. Phys. Chem. A* **2005**, *109*, 9653–9673.
61. Glod, G.; Brodman, U.; Angst, W.; Holliger, C.; Schwarzenbach, R. P. Cobalamin-mediated reduction of cis-and trans-dichloroethene, 1,1-dichloroethene, and vinyl chloride in homogeneous aqueous solution: reaction kinetics and mechanistic considerations. *Environ. Sci. Technol.* **1997**, *31*, 3154–3160.
62. Amonette, J. E.; Workman, D. J.; Kennedy, D. W.; Fruchter, J. S.; Gorby, Y. A. Dechlorination of carbon tetrachloride by Fe(II) associated with goethite. *Environ. Sci. Technol.* **2000**, *34*, 4606–4613.
63. Butler, E. C.; Hayes, K. F. Factors influencing rates and products in the transformation of trichloroethylene by iron sulfide and iron metal. *Environ. Sci. Technol.* **2001**, *35*, 3884–3891.
64. Danielsen, K. M.; Hayes, K. F. pH-dependence of carbon tetrachloride reductive dechlorination by magnetite. *Environ. Sci. Technol.* **2004**, *38*, 4745–4752.
65. Elsner, M.; Haderlein, S. B.; Kellerhals, T.; Luzi, S.; Zwank, L.; Angst, W.; Schwarzenbach, R. P. Mechanisms and products of surface-mediated reductive dehalogenation of carbon tetrachloride by Fe(II) on goethite. *Environ. Sci. Technol.* **2004**, *38*, 2058–2066.
66. Danielsen, K. M.; Gland, J. L.; Hayes, K. F. Influence of amine buffers on carbon tetrachloride reductive dechlorination by the iron oxide magnetite. *Environ. Sci. Technol.* **2005**, *39*, 756–763.
67. Glod, G.; Angst, W.; Holliger, C.; Schwarzenbach, R. P. Corrinoid mediated reduction of tetrachloroethene, trichloroethene, and trichlorofluoroethene in homogeneous aqueous solution: Reaction kinetics and reaction mechanisms. *Environ. Sci. Technol.* **1997**, *31*, 253–260.
68. Kliegman, S.; McNeill, K. Reconciling disparate models of the involvement of vinyl radicals in cobalamin-mediated dechlorination reactions. *Environ. Sci. Technol.* **2009**, *43*, 8961–8967.
69. Choi, W. Y.; Hoffmann, M. R. Kinetics and mechanism of CCl_4 photoreductive degradation on TiO_2: The role of trichloromethyl radical and dichlorocarbene. *J. Phys. Chem.* **1996**, *100*, 2161–2169.

70. McCormick, M. L.; Adriaens, P. Carbon tetrachloride transformation on the surface of nanoscale biogenic magnetite particles. *Environ. Sci. Technol.* **2004**, *38*, 1045–1053.
71. Smolen, J. M.; Weber, E. J.; Tratnyek, P. G. Molecular probe techniques for the identification of reductants in sediments: Evidence for reduction of 2-chloroacetophenone by hydride transfer. *Environ. Sci. Technol.* **1999**, *33*, 440–445.
72. Shey, J.; McGinley, C. M.; McCauley, K. M.; Dearth, A. S.; Young, B. T.; van der Donk, W. A. Mechanistic investigation of a novel vitamin B-12-catalyzed carbon-carbon bond forming reaction, the reductive dimerization of arylalkenes. *J. Org. Chem.* **2002**, *67*, 837–846.
73. Totten, L. A.; Jans, U.; Roberts, A. L. Alkyl bromides as mechanistic probes of reductive dehalogenation: Reactions of vicinal dibromide stereoisomers with zerovalent metals. *Environ. Sci. Technol.* **2001**, *35*, 2268–2274.
74. Shey, J.; van der Donk, W. A. Mechanistic studies on the vitamin B_{12}-catalyzed dechlorination of chlorinated alkenes. *J. Am. Chem. Soc.* **2000**, *122*, 12403–12404.
75. Lesage, S.; Brown, S.; Millar, K. A different mechanism for the reductive dechlorination of chlorinated ethenes: Kinetic and spectroscopic evidence. *Environ. Sci. Technol.* **1998**, *32*, 2264–2272.
76. Rich, A. E.; DeGreeff, A. D.; McNeill, K. Synthesis of (chlorovinyl) cobaloxime complexes, model complexes of proposed intermediates in the B-12-catalyzed dehalogenation of chlorinated ethylenes. *Chem. Commun.* **2002**, 234–235.
77. McCauley, K. M.; Pratt, D. A.; Wilson, S. R.; Shey, J.; Burkey, T. J.; van der Donk, W. A. Properties and reactivity of chlorovinylcobalamin and vinylcobalamin and their implications for vitamin B_{12}-catalyzed reductive dechlorination of chlorinated alkenes. *J. Am. Chem. Soc.* **2005**, *127*, 1126–1136.
78. Follett, A. D.; McNeill, K. Evidence for the formation of a cis-dichlorovinyl anion upon reduction of cis-1,2-dichlorovinyi(pyridine)cobaloxime. *Inorg. Chem.* **2006**, *45*, 2727–2732.
79. Roberts, A. L.; Gschwend, P. M. Mechanisms of pentachloroethane dehydrochlorination to tetrachloroethylene. *Environ. Sci. Technol.* **1991**, *25*, 76–86.
80. Fennelly, J. P.; Roberts, A. L. Reaction of 1,1,1-trichloroethane with zero-valent metals and bimetallic reductants. *Environ. Sci. Technol.* **1998**, *32*, 1980–1988.
81. O'Loughlin, E. J.; Burris, D. R. Reduction of halogenated ethanes by green rust. *Environ. Toxicol. Chem.* **2004**, *23*, 41–48.
82. Krone, U. E.; Thauer, R. K.; Hogenkamp, H. P. C.; Steinbach, K. Reductive formation of carbon monoxide from CCl_4 and Freon11, Freon 12, and Freon13 catalyzed by corrinoids. *Biochemistry* **1991**, *30*, 2713–2719.
83. Devlin, J. F.; Müller, D. Field and laboratory studies of carbon tetrachloride transformation in a sandy aquifer under sulfate reducing conditions. *Environ. Sci. Technol.* **1999**, *33*, 1021–1027.

84. Kriegman-King, M. R.; Reinhard, M. Transformation of carbon tetrachloride in the presence of sulfide, biotite, and vermiculite. *Environ. Sci. Technol.* **1992**, *26*, 2198–2206.
85. Criddle, C. S.; Dewitt, J. T.; Grbicgalic, D.; Mccarty, P. L. Transformation of carbon tetrachloride by *Pseudomonas* Sp. Strain Kc under denitrification conditions. *Appl. Environ. Microbiol.* **1990**, *56*, 3240–3246.
86. Penny, C.; Vuilleumier, S.; Bringel, F. Microbial degradation of tetrachloromethane: mechanisms and perspectives for bioremediation. *FEMS Microbiol. Ecol.* **2010**, *74*, 257–275.
87. Burris, D. R.; Delcomyn, C. A.; Deng, B. L.; Buck, L. E.; Hatfield, K. Kinetics of tetrachloroethylene reductive dechlorination catalyzed by vitamin B-12. *Environ. Toxicol. Chem.* **1998**, *17*, 1681–1688.
88. Semadeni, M.; Chiu, P. C.; Reinhard, M. Reductive transformation of trichloroethene by cobalamin: Reactivities of the intermediates acetylene, chloroacetylene, and the DCE isomers. *Environ. Sci. Technol.* **1998**, *32*, 1207–1213.
89. Burris, D. R.; Delcomyn, C. A.; Smith, M. H.; Roberts, A. L. Reductive dechlorination of tetrachloroethylene and trichloroethylene catalyzed by vitamin b 12 in homogeneous and heterogeneous systems. *Environ. Sci. Technol.* **1996**, *30*, 3047–3052.
90. Fritsch, J. M.; McNeill, K. Aqueous reductive dechlorination of chlorinated ethylenes with tetrakis(4-carboxyphenyl)porphyrin cobalt. *Inorg. Chem.* **2005**, *44*, 4852–4861.
91. Fritsch, J. M.; Retka, N. D.; McNeill, K. Synthesis, structure, and unusual reactivity of β-halovinyl cobalt porphyrin complexes. *Inorg. Chem.* **2006**, *45*, 2288–2295.
92. McCauley, K. M.; Wilson, S. R.; van der Donk, W. A. Synthesis and characterization of chlorinated alkenylcobaloximes to probe the mechanism of vitamin B-12-catalyzed dechlorination of priority pollutants. *Inorg. Chem.* **2002**, *41*, 393–404.
93. McCauley, K. M.; Wilson, S. R.; van der Donk, W. A. Characterization of chlorovinylcobalamin, a putative intermediate in reductive degradation of chlorinated ethylenes. *J. Am. Chem. Soc.* **2003**, *125*, 4410–4411.
94. Arguello, J. E.; Costentin, C.; Griveau, S.; Saveant, J. M. Role of protonation and of axial ligands in the reductive dechlorination of alkyl chlorides by vitamin B_{12} complexes. Reductive cleavage of chloroacetonitrile by Co(I) cobalamins and cobinamides. *J. Am. Chem. Soc.* **2005**, *127*, 5049–5055.
95. Gillham, R. W.; O'Hannesin, S. F. Enhanced degradation of halogenated aliphatics by zero-valent iron. *Ground Water* **1994**, *32*, 958–967.
96. Farrell, J.; Melitas, N.; Kason, M.; Li, T. Electrochemical and column investigation of iron-mediated reductive dechlorination of trichloroethylene and perchloroethylene. *Environ. Sci. Technol.* **2000**, *34*, 2549–2556.
97. Matheson, L. J.; Tratnyek, P. G. Reductive dehalogenation of chlorinated methanes by iron metal. *Environ. Sci. Technol.* **1994**, *28*, 2045–2053.
98. Scherer, M. M.; Richter, S.; Valentine, R. L.; Alvarez, P. J. J. Chemistry and microbiology of permeable reactive barriers for in situ groundwater clean up. *Crit. Rev. Environ. Sci. Technol.* **2000**, *30*, 363–411.

99. Kohen, A., Limbach, H., Eds. *Isotope Effects in Chemistry and Biology*; CRC Press / Taylor & Francis: New York; 2006.
100. Wolfsberg, M.; van Hook, A.; Paneth, P.; Rebelo, L. P. N. *Isotope Effects in Chemical, Geological, and Bio Sciences*; Springer: Heidelberg, 2010.
101. Elsner, M.; Zwank, L.; Hunkeler, D.; Schwarzenbach, R. P. A new concept linking observable stable isotope fractionation to transformation pathways of organic pollutants. *Environ. Sci. Technol.* **2005**, *39*, 6896–6916.
102. Hofstetter, T. B.; Schwarzenbach, R. P.; Bernasconi, S. M. Assessing transformation processes of organic compounds using stable isotope fractionation. *Environ. Sci. Technol.* **2008**, *42*, 7737–7743.
103. Elsner, M. Stable isotope fractionation to investigate natural transformation mechanisms of organic contaminants: principles, prospects and limitations. *J. Environ. Monit.* **2010**, *12*, 2005–2031.
104. Bloom, Y.; Aravena, R.; Hunkeler, D.; Edwards, E.; Frape, S. K. Carbon isotope fractionation during microbial dechlorination of trichloroethene, cis-1,2-dichloroethene, and vinyl chloride: Implications for assessment of natural attenuation. *Environ. Sci. Technol.* **2000**, *34*, 2768–2772.
105. Slater, G. F.; Sherwood Lollar, B.; Sleep, B. E.; Edwards, E. A. Variability in carbon isotopic fractionation during biodegradation of chlorinated ethenes: Implications for field applications. *Environ. Sci. Technol.* **2001**, *35*, 901–907.
106. Elsner, M.; Cwiertny, D. M.; Roberts, A. L.; Sherwood Lollar, B. 1,1,2,2-Tetrachloroethane reactions with OH-, Cr(II), granular Iron, and a copper-iron bimetal: Insights from product formation and associated carbon isotope fractionation. *Environ. Sci. Technol.* **2007**, *41*, 4111–4117.
107. Hofstetter, T. B.; Reddy, C. M.; Heraty, L. J.; Berg, M.; Sturchio, N. C. Carbon and chlorine isotope effects during abiotic reductive dechlorination of polychlorinated ethanes. *Environ. Sci. Technol.* **2007**, *41*, 4662–4668.
108. VanStone, N.; Elsner, M.; Lacrampe-Couloume, G.; Mabury, S.; Sherwood Lollar, B. Potential for identifying abiotic chloroalkane degradation mechanisms using carbon isotopic fractionation. *Environ. Sci. Technol.* **2008**, *42*, 126–132.
109. Zwank, L.; Berg, M.; Elsner, M.; Schmidt, T. C.; Schwarzenbach, R. P.; Haderlein, S. B. New evaluation scheme for two-dimensional isotope analysis to decipher biodegradation processes: Application to groundwater contamination by MTBE. *Environ. Sci. Technol.* **2005**, *39*, 1018–1029.
110. Meyer, A. H.; Penning, H.; Elsner, M. C and N isotope fractionation suggests similar mechanisms of microbial atrazine transformation despite involvement of different enzymes (AtzA and TrzN). *Environ. Sci. Technol.* **2009**, *43*, 8079–8085.
111. Hofstetter, T. B.; Spain, J. C.; Nishino, S. F.; Bolotin, J.; Schwarzenbach, R. P. Identifying competing aerobic nitrobenzene biodegradation pathways using compound-specific isotope analysis. *Environ. Sci. Technol.* **2008**, *42*, 4764–4770.
112. Penning, H.; Sørensen, S. R.; Meyer, A. H.; Aamand, J.; Elsner, M. C, N, and H isotope fractionation of the herbicide isoproturon reflects different

microbial transformation pathways. *Environ. Sci. Technol.* **2010**, *44*, 2372–2378.

113. Shouakar-Stash, O.; Drimmie, R. J.; Zhang, M.; Frape, S. K. Compound-specific chlorine isotope ratios of TCE, PCE and DCE isomers by direct injection using CF-IRMS. *Appl. Geochem.* **2006**, *21*, 766–781.
114. Sakaguchi-Söder, K.; Jager, J.; Grund, H.; Matthäus, F.; Schüth, C. Monitoring and evaluation of dechlorination processes using compound-specific chlorine isotope analysis. *Rapid Commun. Mass Spectrom.* **2007**, *21*, 3077–3084.
115. Elsner, M.; Hunkeler, D. Evaluating chlorine isotope effects from isotope ratios and mass spectra of polychlorinated molecules. *Anal. Chem.* **2008**, *80*, 4731–4740.
116. Abe, Y.; Aravena, R.; Zopfi, J.; Shouakar-Stash, O.; Cox, E.; Roberts, J. D.; Hunkeler, D. Carbon and chlorine isotope fractionation during aerobic oxidation and reductive dechlorination of vinyl chloride and cis-1,2-dichloroethene. *Environ. Sci. Technol.* **2009**, *43*, 101–107.
117. Hunkeler, D.; Van Breukelen, B. M.; Elsner, M. Modeling chlorine isotope trends during sequential transformation of chlorinated ethenes. *Environ. Sci. Technol.* **2009**, *43*, 6750–6756.
118. Elsner, M.; Couloume, G. L.; Mancini, S.; Burns, L.; Lollar, B. S. Carbon Isotope Analysis to Evaluate Nanoscale Fe(0) Treatment at a Chlorohydrocarbon Contaminated Site. *Ground Water Monit. Rem.* **2010**, *30*, 79–95.

Chapter 20

Degradation Routes of RDX in Various Redox Systems

Annamaria Halasz and Jalal Hawari*

Biotechnology Research Institute, National Research Council Canada, 6100 Royalmount Ave., Montreal (QC), Canada, H4P 2R2
***Jalal.Hawari@cnrc-nrc.gc.ca**

RDX, hexahydro-1,3,5-trinitro-1,3,5-triazine, is one of the most widely used nitro-organic explosives that presently contaminate various terrestrial and aquatic systems. It is a highly oxidized molecule whose degradation is governed by the electronic structure and redox chemistry of its $–CH_2–N(NO_2)–$ multifunctional group. In the present chapter, we discuss how the chemical reacts in various redox systems with particular emphasis on the initial steps involved in RDX decomposition. Three important redox reactions are analyzed and commented: 1) 1e–transfer to $–N–\underline{NO_2}$ leading to denitration, 2) 2e–transfer to $–\underline{NO_2}$ leading to the formation of the corresponding nitroso derivatives (MNX, DNX and TNX), and 3) H•-abstraction from one of the $–CH_2–$ groups by OH• and $O_2^{\bullet -}$. We will analyze and identify knowledge gaps in the transformation pathways of RDX to highlight future research needs.

Introduction

Polynitroorganic compounds are highly energetic chemicals that rapidly release large amounts of gaseous products and energy upon detonation. Because of their explosive properties these chemicals are extensively used by the military and in the construction and mining industry (*1*). It has been estimated that in the US alone close to 313 million kg of N-containing explosives had been released to the environment by 1992 (*2*). Both TNT (2,4,6-trinitrotoluene) and RDX (hexahydro-1,3,5-trinitro-1,3,5-triazine) are the most widely used explosives. TNT reportedly undergoes biotransformation in most redox conditions, under

Published 2011 by the American Chemical Society

both oxidative (aerobic) and reductive (anaerobic) conditions, and gives monoamino-dinitrotoluene (ADNTs) and diamino-nitrotoluene (DANTs) without breaking the aromatic ring (*3*). Without denitration it will be difficult to mineralize TNT because the resulting amino products through their nucleophilic $-NH_2$ groups can irreversibly bind with acidic-containing sites, e.g. humic acids abundant in aquifer systems. Indeed it has been suggested that stabilizing TNT through amide, –NH–CO–, linkages with soil humics can be considered as a passive remediation technology (*4*). More recently it has been reported that hydride anion attack on TNT can lead to denitration through the intermediary formation of Meisenheimer dihydride complexes. However since our last review on the biotransformation pathways of TNT (*3*) there has been remarkable progress made towards enhancing mineralization of the aromatic explosive based on the development of microbial and other biological tools capable of denitrating TNT (*5–8*).

Recently RDX has replaced TNT as a primary explosive in various munitions formulations. We thus selected RDX as a model nitrogenous explosive and sought to understand its environmental behavior in various redox systems. RDX, a cyclic nitroamine $(CH_2{-}N(NO_2))_3$, is of particular environmental concern because of its negligible sorption affinity to soil minerals and organic matter (*9*). The chemical thus can migrate through subsurface soil causing the contamination of aquifer environments including groundwater and sediment. The nitroamine is toxic (*10*, *11*) and is on the US Environmental Protection Agency's Drinking Water Contaminant Candidate List. Several reports describing degradation and fate of RDX in the environment have already been published (*3*, *12–14*) but the actual degradation mechanism(s) (especially the redox chemistry and the initial e-transfer processes involved in degrading RDX) are not fully understood. Due to the presence of three $-NO_2$ groups, RDX is highly oxidized and consequently its reactions are governed by the ability of the nitroamine to receive electrons from an adjacent e-donor (reductant). Natural redox systems, e.g. biogeochemical iron redox couples, are abundant in the environment and play crucial roles in determining the fate of many organic and inorganic contaminants in terrestrial and aquatic systems (*15–17*). Biogeochemical redox cycles often involve microorganisms using certain elements for respiration, e.g. Fe(III), to gain energy. Fe(III)-respiring bacteria e.g., *Shewanella* are ubiquitous in nature and can be found in various terrestrial and aquatic systems (*18*). Considering iron as the most abundant heavy metal on Earth, the presence of Fe(III)-respiring bacteria adjacent to other chemicals, e.g. RDX, can thus constitute an effective means of degrading these chemicals (Figure 1). Due to the importance of biogeochemical processes in determining the fate (mobility and transformation) of contaminants, the Journal Environmental Science & Technology has dedicated its 1st January 2010 issue to explain why and how biogeochemical redox processes play important roles in determining the fate of contaminants in the environment.

In this review we will summarize the most reported abiotic and biotic redox reactions of RDX that are relevant to natural biogeochemical redox processes. Understanding the response of RDX to the flow of electrons associated with a redox system and the chemical and biochemical changes that follow will help elucidate the fate of the nitroamine and also aid in the design of strategies to

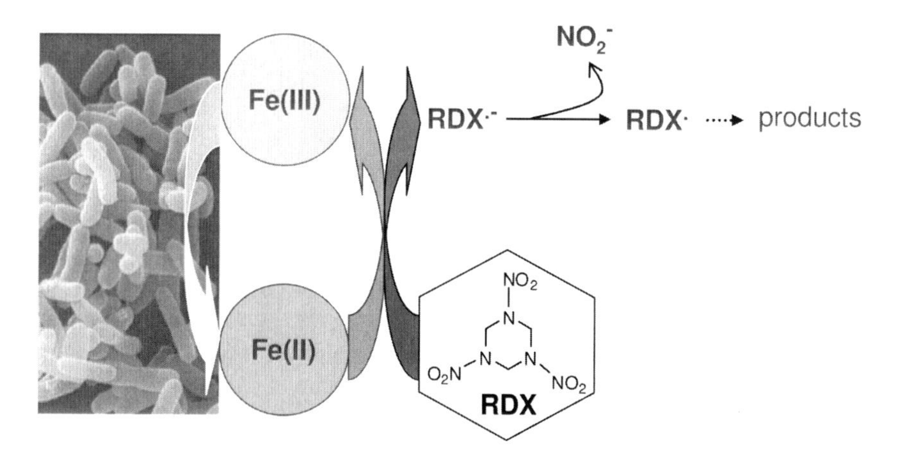

Figure 1. A simplified schematic representation of Fe(III)/Fe(II) as a model biogeochemical redox couple showing: microbial bioreduction of Fe(III) to Fe(II) by Shewanella and abiotic oxidation of Fe(II) to Fe(III) driven by RDX reduction.

enhance *in situ* remediation. In the present review we will specifically analyze how the highly oxidized nitroamine reacts in various redox systems with particular emphasis on the initial steps involved in RDX decomposition. We will discuss the current state of knowledge and identify knowledge gaps to highlight future research needs.

Degradation of RDX in Various Redox Systems

Abiotically RDX is reported to degrade using alkaline hydrolysis (*19*), photolysis (*20*, *21*), Fenton reagent (Fe^{+2}/H_2O_2) (*22*), or permanganate (*23*). In addition the cyclic nitroamine can degrade in the presence of metals including zero valent iron, ZVI (*24–28*), iron nanoparticles, nZVI (*29*), nickel (*30*), and mechanically alloyed bimetals such as Na(Hg) (*31*), or Fe(Ni) (*32*). Some of the problems associated with the use of metals such as ZVI are the corrosion of the metal and the formation of a layer of metal oxide at the surface of the metal that may stop the flow of electrons to RDX. In military shooting ranges, the major source of Fe would be warhead casting and fragments. The very reactive superoxide radical anions, $O2^{\bullet -}$, and hydroxyl radicals, $OH^{\bullet}$, produced by Fenton and permanganate reagents or formed in water exposed to sunlight can lead to the degradation of many chemicals including RDX. These highly oxidative systems can thus be used to complete RDX degradation following its reduction with ZVI (*28*).

Several other studies have described degradation of RDX using special redox systems that are more relevant to biogeochemical environments and thus are suitable for in situ remediation. For example, biogenic Fe(II) (*33*, *34*), black C/H_2S (*35*), and Fe(II) bound to magnetite (*36*) have been successfully employed to degrade RDX.

Biotically the nitroamine degrades under both aerobic and anaerobic conditions. For example, RDX degrades with anaerobic sludge (*37–39*), sulfate-

(*40*), nitrate- (*41*), and manganese-reducing bacteria (*42*), *Acetobacterium malicum* (*43*), *Shewanella halifaxensis* (*44*) and *Geobacter metallireducens* (*45, 46*). Aerobic bacteria such as *Rhodococcus* sp (*47–49*) and *Gordonia* and *Williamsia* (*50*) are also described as RDX degraders. Furthermore enzymes including diaphorase (*51*), nitroreductase (*52*), cytochrome P450, and XplA enzymatic system are also too reported to degrade RDX under aerobic, anaerobic, or anoxic conditions (*53, 54*).

Among reported RDX reactions, the ones involving iron-containing species bear special environmental significance. Iron is the most abundant transition metal on Earth and through its reversible F(III)-Fe(II) redox cycles participates in several other important redox cycles including NO_3^-/NO_2^- and CO_2/CH_4 cycles (*15, 55*). Electron flow generated through redox cycles plays an important role in environmental biogeochemistry and in determining the fate of many contaminants in the environment (*56*). Iron deposits in an anoxic marine and freshwater environment (via Fe(III)-Fe(II) anaerobic redox cycling) can accelerate transformation of several organic pollutants (*15, 55, 57–60*) including RDX (*34, 61*). For example, Fe(III)-containing deposits in anaerobic aquifers are constantly subjected to potential microbial reduction to Fe(II) which in turn acts as an electron-donor to regenerate Fe(III) causing the reduction of other contaminants nearby. We found that sediments collected from an unexploded ordnance (UXO) site near Halifax Harbour, Canada, contained high amounts of insoluble iron (40 g kg^{-1} sediment) in the form of mainly Fe(III) and Fe(II) (*62*). Indeed several anaerobic bacteria including *Clostridium* sp. EDB2 (*34*) that have been successfully isolated from Halifax sediment are found capable of reducing Fe(III) to Fe(II), which in turn acts as electron-transfer agent to RDX leading to its degradation (Figure 1).

Degradation Products of RDX and Transformation Pathways

Table I summarizes RDX degradation in various aquatic redox systems under both abiotic and biotic conditions. Product distributions clearly show that when RDX degrades it gives two distinct sets of products depending on conditions used: one set marked by the transformation of RDX to nitroso derivatives MNX (hexahydro-1-nitroso-3,5-dinitro-1,3,5-triazine), DNX (hexahydro-1,3-dinitroso-5-nitro-1,3,5-triazine) and TNX (hexahydro-1,3,5-trinitroso-1,3,5-triazine); and another by the formation of ring cleavage products including NO_2^-, NO_3^-, NH_3, N_2O, HCHO, and HCOOH, and the two characteristic intermediates methylenedinitramine (MEDINA, $NO_2NHCH_2NHNO_2$) and 4-nitro-2,4-diazabutanal (NDAB, NO_2NHCH_2NHCHO) (Table I). As we will discuss in the following section the extent of the formation of MEDINA and NDAB depends on the actual redox conditions used to initiate RDX degradation and on whether RDX degrades via the loss of only one or two nitrite ions before ring cleavage. Analyses of product distributions from RDX degradation under various redox conditions indicate that RDX can degrade *via* two distinctive pathways. One pathway involves sequential reduction ($2e/H^+$) of the $-N-NO_2$

to the corresponding –N–NO functional groups giving MNX, DNX, and TNX (*24*, *39*, *76*, *77*). A second pathway involves the cleavage of the –N–NO_2 bond(s) leading to decomposition of RDX and the formation of benign products such as NO_2^-, NO_3^-, NH_3, HCHO and HCOOH (*34*, *35*, *47*, *65*).

Table I. Degradation of RDX in aqueous systems under aerobic/ anaerobic and abiotic/biotic conditions: Intermediates and final products

Reactions	*Intermediates / Products*	*References*
Anaerobic / abiotic		
ZVI – Zerovalent iron	MNX, DNX, TNX, NO_2^-, NO_3^-, N_2O	(*24*)
	MNX, DNX, TNX, MEDINA	(*25*)
	Not determined	(*26*)
	MNX, DNX, TNX, MEDINA, N_2O	(*29*)
	MNX, DNX, TNX	(*27*)
nZVI – Zerovalent iron nanoparticles	MNX, DNX, TNX, MEDINA, NH_4^+, N_2, N_2O, H_2NNH_2, HCHO	(*29*)
Fe^{II} bound to magnetite	MNX, DNX, TNX, NH_4^+, N_2O, HCHO	(*36*)
Fe^{II}-organic complex	HCHO, NH_4^+, N_2O	(*65*)
Green rust	MNX, DNX, TNX, MEDINA, NH_4^+, N_2O, HCHO	(*66*)
Biogenic Fe^{II}	MNX, DNX, TNX, MEDINA, NDAB, NH_4^+,	(*33*, *61*)
	NO_2^-, MEDINA, NH_4^+, N_2O, HCHO, CO_2	(*34*)
Sulfides/black carbon	NDAB, HCHO, NO_2^-	(*35*)
Na(Hg)/THF/H_2O	1,3-dinitro-5-hydro-1,3,5-triazine, MEDINA, HCHO	(*31*)
Aerobic / abiotic		
$KMnO_4$	NDAB, CO_2, N_2O	(*23*)
Photolysis	NDAB, MEDINA, HCHO, HCOOH, N_2O, NO_2^-, NO_3^-	(*21*)
	1,3-dinitro-1,3,5-triazacyclohex-5-ene, 1-nitro-1,3,5-triaza-cyclohex-3,5-ene, HCO-NH_2, HCO-NHOH, HCO-NO_2	(*20*)

Continued on next page.

Table I. (Continued). Degradation of RDX in aqueous systems under aerobic/anaerobic and abiotic/biotic conditions: Intermediates and final products

Reactions	*Intermediates / Products*	*References*
Electrochemical reduction	MNX, MEDINA, HCHO, N_2O, CH_3OH, NO_2^-	(*67, 68*)
Alkaline hydrolysis	NO_2^-, N_2O, NH_4^+, HCHO, HCOOH, NDAB	(*19, 67*)
Aerobic / enzymatic		
XplA/XplB	NDAB, NO_2^-, HCHO	(*53*)
Anaerobic / enzymatic		
XplA/XplB	MEDINA, NO_2^-, HCHO	(*53*)
Cytochrome P450 2B4	NDAB, NO_2^-, HCHO	(*54*)
Cytochrome P450 reductase	MNX, MEDINA, N_2O, HCHO, NH_4^+	(*54*)
Diaphorase EC 1.8.1.4	MEDINA, NO_2^-, HCHO, NH_4^+, N_2O	(*51*)
Nitrate reductase EC 1.6.6.2	MNX, MEDINA, N_2O, HCHO, NH_4^+	(*52*)
Aerobic / microbial		
Stenotrophomonas maltophilia PB1	Methylene-*N*-(hydroxymethyl)-hydroxylamine-*N'*-(hydroxymethyl)-nitroamine	(*69*)
Rhodococcus sp. DN22	NDAB, MEDINA, NO_2^-, NH_4^+, HCHO, CO_2	(*47, 70*)
Gordonia KTR4/*Williamsia* KTR9	NDAB, NO_2^-, HCHO, CO_2	(*50*)
Acremonium	MNX,DNX,TNX, MEDINA, N_2O, HCHO, CO_2	(*71*)
Phanerochaete chrysosporium	MNX, N_2O, CO_2	(*72*)
Anaerobic / microbial		
Shewanella halifaxensis	MNX, DNX, TNX, MEDINA, NDAB, HCHO, N_2O	(*44, 63, 64*)
γ-Proteobacteria	MNX, MEDINA, N_2O, CO_2	(*73*)
Clostridium sp. EDB2	NO_2^-, N_2O, HCHO, HCOOH, CO_2	(*62*)
Acetobacterium malicum	MNX, MEDINA, N_2O, HCHO	(*43*)
Clostridium bifermentants	MNX, DNX, N_2O, HCHO, CH_3OH	(*74*)
Klebsiella pneumoniae	MEDINA, N_2O, HCHO, CH_3OH, CO_2	(*75*)

Continued on next page.

Table I. (Continued). Degradation of RDX in aqueous systems under aerobic/anaerobic and abiotic/biotic conditions: Intermediates and final products

Reactions	*Intermediates / Products*	*References*
Aquifer bacteria	MNX, DNX, TNX; CO_2	(*42*, *76*)
Municipal sludge	MNX, DNX, TNX, MEDINA, N_2O, HCHO, CO_2	(*38*)
Enterobacteriaceae	MNX, DNX, TNX	(*77*)
Sewage sludge	MNX, DNX, TNX, HCHO, hydrazines	(*39*)

- In the nitroso route all N and all C atoms that are originally present in RDX stay intact in MNX, DNX and TNX which reportedly are known to be more toxic than RDX (*78*).

The formation of RDX nitroso products is reductive and each step involves a 2e–transfer process (Figure 2). The nitroso pathway is dominant under highly reducing conditions as observed during RDX degradation using anaerobic sludge (*79*) and with marine isolates such as *Shewanella halifaxensis* (*44*, *64*). RDX reduction to nitroso products has also been observed during RDX reduction with ZVI (*24–29*) and biogenic Fe(II) (*33*, *34*).

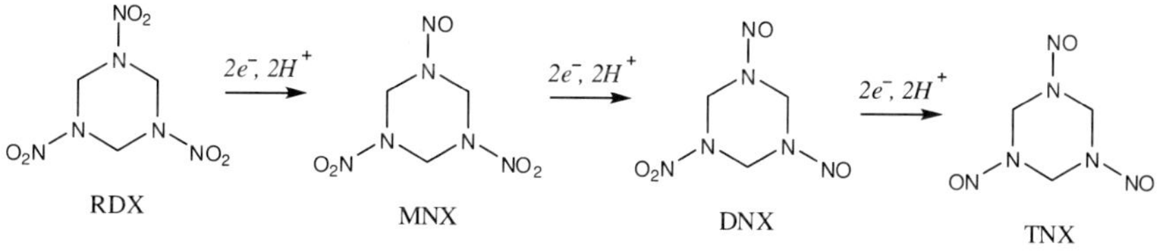

Figure 2. Reductive transformation of RDX to the corresponding nitroso derivative via sequential 2e-trasfer process.

- In the denitration route two key RDX ring cleavage products, NDAB and MEDINA, are formed (Figures 3 and 4).

Denitration of RDX proceeds *via* several mechanisms which will be described later. Following denitration the intermediate can cleave to produce NDAB and/or MEDINA with ratios depending on several factors: 1) the type of C–N bond(s) that cleaves following denitration; and 2) whether the ring cleavage occurs following mono-denitration or di-denitration.

The formation of NDAB and MEDINA depends on the stoichiometry of the denitration step: the loss of two nitrite anions prior to RDX ring cleavage produces only NDAB (Figure 3) whereas the loss of only one nitrite anion prior to ring cleavage produces MEDINA and/or NDAB (Figure 4). RDX denitration with *Rhodococcus* sp. DN22 under aerobic conditions showed a nitrite stoichiometry reaching approx. 2 with NDAB and MEDINA being formed in approx. 9:1 ratio

(*47, 70*). When RDX was degraded using the purified XplA cytochrome P450 system isolated from *Rhodococcus rhodochrous* strain 11Y we obtained two nitrite anions and only NDAB under aerobic conditions and one nitrite anion and MEDINA under anaerobic conditions (*53*).

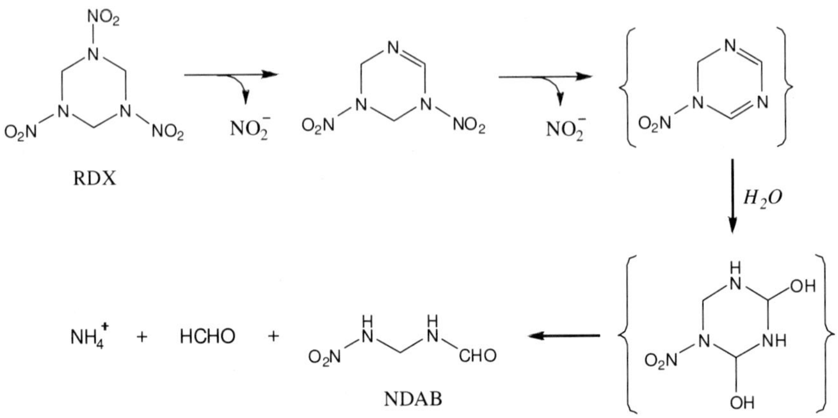

Figure 3. Schematic representation of RDX denitration: formation of NDAB

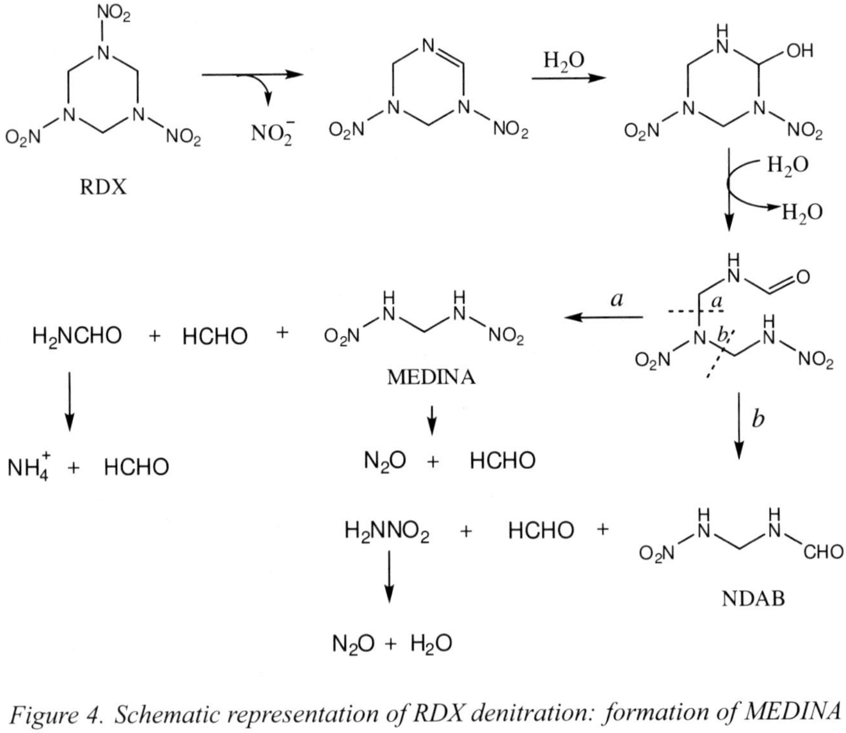

Figure 4. Schematic representation of RDX denitration: formation of MEDINA and/or NDAB

Generally speaking, mono-denitration leads to the predominant formation of MEDINA as observed following RDX reduction under anaerobic conditions with zero valent iron, ZVI (*28*), nano iron particles (*29*), diaphorase (*51*), nitrate reductase (*52*), *Shewanella sediminis* and *Shewanella halifaxensis* (*44*, *63*), or *Geobacter metallireducens* (*45*) (Figure 4). The absence of MEDINA in some of the cited RDX reactions (Table I) should not exclude its formation as an intermediate of RDX degradation. MEDINA is unstable in water and decomposes to N_2O and HCHO through the formation of NH_2NO_2 (*80*, *81*).

The indicator for the potential presence of MEDINA during RDX degradation is thus the detection of N_2O in almost stoichiometric amounts, i.e. $2N_2O$ for each degrading molecule of MEDINA (Figure 4). In contrast, NDAB is more stable in water under ambient conditions and if it completely decomposes the yield of N_2O would be half of that coming from MEDINA. NDAB was reported to degrade with *Methylobacterium* (*82*), *Phanerochaete chrysosporium* (*83*), and by alkaline hydrolysis at pH 12 (*19*). In the latter case, NDAB hydrolyzes to produce N_2O, HCHO, NH_3, and HCOOH (*70*). Several other studies have reported the formation of both NDAB and MEDINA together under both aerobic and anaerobic conditions including photodenitration (*21*), denitration by *Shewanella* (*44*), biogenic Fe(II) (*33*), and *Rhodococcus* sp. DN22 under microaerophilic conditions (*84*).

MNX formation constitutes an important point in the reduction of RDX; it can either continue reduction to produce DNX and TNX (Figure 2), denitrate, or denitrosate leading to ring cleavage and decomposition (Figure 5). In general little information is available on the degradation of MNX (Table II), but recently we studied its potential degradation with *Shewanella* isolates from Halifax Harbour (*74*, *75*) and with the soil isolate *Rhodococcus* sp. DN22 (*70*). Although DN22 degrades MNX at a slower rate than that of RDX its disappearance was accompanied with the formation of NO_2^-, NO_3^-, and ring cleavage products including NH_3, N_2O, HCHO, and HCOOH similar to those observed following RDX denitration. We also detected NDAB and a trace amount of an intermediate with a $[M-H]^-$ at *m/z* 102 Da, matching an empirical formula of $C_2H_5N_3O_2$, that we tentatively identified as 4-nitroso-2,4-diazabutanal (4-NO-NDAB, $ONNHCH_2NHCHO$) based on LC/MS(ES-) analysis coupled with the use of ring labeled ^{15}N-MNX (*70*). The formation of NDAB and NO-NDAB clearly demonstrates that MNX degrades *via* initial cleavage of the N–NO and $N–NO_2$ bonds, respectively.

As we mentioned above RDX biodegradation can be accompanied by the formation of NO_2^- and NO_3^-. To understand the mechanism of their formation we biodegraded RDX with DN22 in the presence of $^{18}O_2$ and searched for ^{18}O involvement in their composition using LC/MS (*70*).

Using abiotic and biotic controls we concluded that: 1) nitrite formed following $–N–NO_2$ bond cleavage without oxygen (gaseous) involvement, and 2) NO_3^- resulted from bio-oxidation of the initially formed NO_2^-, consistent with the previously reported biological oxidation of NO_2^- to NO_3^- (*85*).

Table II. Degradation of MNX in aqueous systems under aerobic/ anaerobic and abiotic/biotic conditions: Intermediates and final products

Reactions	*Intermediates / Products*	*References*
Anaerobic / abiotic		
nZVI – Zerovalent iron nanoparticles	NH_4^+, N_2O, H_2NNH_2, HCHO	(*29*)
Fe^{II}-organic complex	HCHO, NH_4^+, N_2O	(*65*)
Biogenic Fe^{II}	NDAB, HCHO	(*33*)
AH_2QDS	HCHO	(*33*)
Sulfides/black carbon	Not determined	(*35*)
Aerobic / abiotic		
Electrochemical reduction	MEDINA, HCHO, N_2O	(*68*)
Alkaline hydrolysis	NO_2^-, NDAB NH_4^+, HCHO, HCOOH	(*19*)
Anaerobic / enzymatic		
Nitrate reductase EC 1.6.6.2	MEDINA, N_2O, HCHO	(*52*)
Aerobic / microbial		
Acremonium	N_2O, HCHO	(*71*)
Rhodococcus sp. DN22	NDAB, NO_2^-, NO_3^-, N_2O, HCHO	(*70*)
Anaerobic / microbial		
Clostridium bifermentans	DNX, N_2O, HCHO, CH_3OH	(*74*)
Klebsiella pneumoniae	NO_2^-, N_2O, HCHO, CH_3OH, CO_2	(*75*)

Insights into Initial Reaction Steps Involved in RDX Denitration

Reductive transformation of RDX to MNX, DNX, and TNX involves sequential $2e/H^+$ reduction steps to convert the $N–NO_2$ to the corresponding N–NO functional groups (Figure 2). As for denitration, knowing the mechanism of $N–NO_2$ bond cleavage is more challenging. Part of the answer may be found by understanding the behavior of the key characteristic $–CH_2–N(NO_2)–$ functional group in RDX which is marked by the presence of several positions for possible attack in various chemical or biological redox systems. In the following section we will discuss three potential mechanisms for denitration that can lead to RDX decomposition.

Hydrogen-Atom Abstraction

One possible denitration mechanism may involve a H• atom abstraction from the methylene group to first produce the carbenyl centered free radical the denitration of which would produce intermediate **I** (pentahydro-3,5-dinitro-

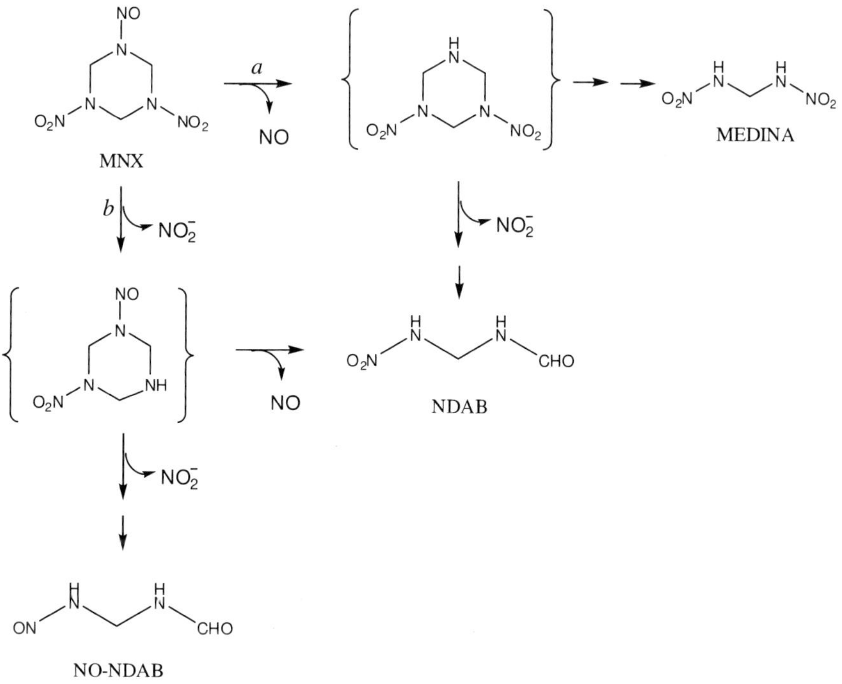

Figure 5. Constructed degradation pathway of MNX; a: denitrosohydrogenation; b: denitrohydrogenation

1,3,5-triazacyclohex-1-ene) (Figure 6, a). A second denitration of intermediate **I** by the same mechanism would give **II** (1-nitro-1,3,5-triaza-cyclohex-3,5-ene) (Figure 6, a). Both intermediates **I** and **II** have been observed by Bose et al. (*20*) during photolysis of RDX under UV conditions. Subsequent reactions of **I** or **II** with water produce the unstable α-hydroxylamine derivatives **III** and **IV**. The generated unstable α-hydroxylamines will hydrolyze rapidly across the –NH–CH– bond to eventually produce NO_2^-, NH_3, N_2O, HCHO, and HCOOH. Recently Guengerich reported an enzymatic H• abstraction/oxygen rebound mechanism for the hydroxylation of alkyl amines which subsequently decompose to produce aldehydes and amines (*86*). The reported H• abstraction/oxygen rebound mechanism resembles the present hydroxylation of the $–CH_2–$ group in RDX. In our case, however, we found that the oxygen participating in the hydroxylation of RDX originated from water and not from air as confirmed by $^{18}O_2$-labeling experiments (*70*). The denitration/α-hydroxylation mechanism is applicable to RDX degradation with the enzymatic system XplA (*53*), *Rhodococcus* sp. strain DN22 (*47*), and cytochrome P450 (*54*) under aerobic conditions (Figure 6, a). To better understand the stoichiometry of MEDINA and NDAB formation we suggest further research, e.g. degrading RDX in more defined redox systems such as using purified enzymes.

Electron-Transfer to $-NO_2$

A second denitration mechanism may involve an e–transfer to the nitro group in $-N-NO_2$ leading to the formation of a radical anion which upon denitration would give the aminyl radical **V** (Figure 6, b). H• atom abstraction by the aminyl radical **V** from a nearby substrate would give 1,3-dinitro-perhydro-1,3,5-triazine **VI** by a process henceforth named denitrohydrogenation (Figure 6, b). McHugh et al. (*31*) reported the synthesis of **VI** by reducing RDX with amalgamated Na in THF. **VI** is unstable and decomposes to MEDINA, HCHO and NH_3. In this context we propose that the e–transfer process responsible for RDX denitration and decomposition is best applicable to the following reactions: RDX degradation with ZVI, nZVI, biogenic Fe(II), nitrate reductase, and the XplA enzymatic system under anaerobic conditions (Figure 6, b). Photolysis causing $N-NO_2$ homolytic bond cleavage also results in the formation of the aminyl radical (**V**) which after H• atom abstraction would lead to **VI**, a precursor to MEDINA (*21*).

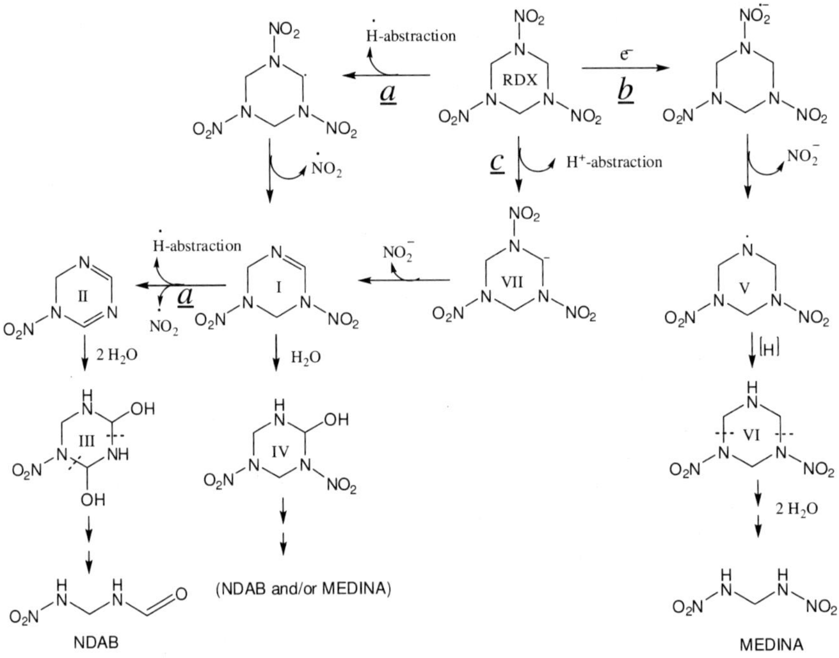

Figure 6. Proposed denitration mechanisms of RDX: a H•-abstraction followed by N-NO₂ bond cleavage; b electron-transfer to NO₂ followed by loss of NO₂⁻; c H⁺-abstraction followed by denitration

Proton-Abstraction from $-CH_2-$

A nucleophile, e.g. OH^- or H^-, can abstract one of the two acidic hydrogens from the methylene group in RDX giving the corresponding carbenyl anion (**VII**). The resulting carbenyl anion **VII** would force the elimination of nitrite ($-N-NO_2$ bond cleavage) to generate intermediate **I** (Figure 6, *c*). Once again **I** (or its second denitrated intermediate **II**) would decompose in water to give products distribution similar to that obtained in the hydrogen-atom abstraction pathway (Figure 6, *a*). This mechanism of denitration is best represented by the alkaline hydrolysis of RDX (*19*).

Conclusions and Research Needs

RDX can perform many abiotic and biotic reactions because of its characteristic multifunctional group $-CH_2-N(NO_2)-$. Depending on how $-CH_2-N(NO_2)-$ is initially attacked, i.e. 1e– *vs.* 2e–transfer to $-N-\underline{NO_2}$, H•-atom abstraction from $-\underline{CH_2}-N(NO_2)$, or nucleophilic attack by H^- or OH^-, we realized that RDX can follow two major degradation routes: 1) sequential reduction, carried out by 2e–transfer processes, to produce MNX, DNX, and TNX keeping the nitroamine ring intact and the original N and C content of RDX preserved, and 2) denitration, initiated by either 1e transfer to the $-NO_2$ group or by H•-atom abstraction, followed by ring cleavage and decomposition to benign products including NO_2^-, NH_2CHO, NH_3, N_2O, HCHO, HCOOH, and NDAB. Therefore engineering a degradation pathway for RDX that involves denitration would be the preferred choice for the development of remediation technologies. Despite our laboratory's success in understanding how RDX reacts in defined redox systems, we are still facing some challenges on how actually these reactions take place in the field. New and emerging chemical and molecular tools and technologies suitable for on site monitoring and assessment are thus needed to provide new insights into how to monitor and optimize RDX in-situ degradation in the field.

Detection in the Field

Advances in sample preparation (Solid Phase Micro-Extraction, SPME), and preservation combined with the availability of sensitive instrumentation (MALDI-TOF MS/MS) allow the detection and identification of trace amounts of RDX products and thus help gain new insights into the degradation pathways of RDX. As described in the previous sections, MEDINA and NDAB are now established RDX intermediates that provide important information on the degradation pathway of the nitroamine. Both intermediates can be used as markers to monitor the fate of RDX in the environment under reductive or hydrolytic conditions. We detected NDAB in soil samples collected from an ammunition plant in Valleyfield, Quebec, Canada (*83*) and in groundwater samples taken from Iowa Army Ammunition Plant (IAAP) (unpublished results). In contrast, MEDINA is unstable in neutral aqueous media and thus hard to detect in field samples. Despite progress towards understanding degradation pathways of RDX in controlled laboratory experiments analytical tools suitable for *on site* detection of RDX and its products are needed. Recently Borch et al. (*15*) highlighted the actual research needs of monitoring biogeochemical redox systems in the environment and highlighted emerging developments for on site monitoring of redox processes, e.g. application of synchrotron-based spectroscopy and in situ electrode and gel probe techniques.

Electron–Transfer Processes Involved in RDX Denitration

Two of the suggested mechanisms of denitration for RDX imply the formation of free radicals, e.g. either carbenyl (Figure 6, *a* or *b*) or aminyl (Figure 6, b) centered radical. To confirm the presence of free radicals we need to do more research and to introduce new instrumentation, e.g. electron spin resonance spectroscopy (ESR) to directly detect the presence of free radicals that may arise from RDX reactions, e.g. the 1e–transfer processes.

Molecular Tools for in-Situ Monitoring

Transformation pathways that we described earlier for RDX clearly show the chemical bonds, i.e., N–N, N–C, N–O that can break during chemical or microbial attack. Compound specific isotope analysis (CSIA) (*87–89*) and stable isotope probing (SIP) have been successfully used to elucidate the degradation or transformation pathways of pollutants in the environment (*90–92*). Presently both CSIA and SIP are applied for in-situ monitoring of organic pollutants and for the identification and isolation of bacteria capable of their degradion (*87–92*). In the pervious Chapter, Elsner and Hofstetter described the use of Stable Isotope Fractionation in the elucidation of the mechanisms of chloro-hydrocarbon degradation in subsurface environments (*93*). CSIA has been recently used to quantify RDX biodegradation in groundwater (*88*), to quantify aerobic biodegradation of 2,4-dinitrotoluene (2,4-DNT) and 2,4,6-trinitrotoluene (TNT) (*94*), and to detect 2,4,6-trinitrotoluene-utilizing anaerobic bacteria in a harbor sediment (*95*). The second technique, SIP, is applied to detect microorganisms able to degrade explosives enriched with ^{13}C– or ^{15}N–isotopes providing that the microorganism can use the chemical, e.g. RDX, as C or as N source. The ^{13}C–DNA or ^{15}N–DNA produced during the growth of the microorganism on an isotopically-labeled ^{13}C– or ^{15}N– is then resolved from ^{12}C–DNA and ^{14}N–DNA, respectively, by density-gradient centrifugation (*90*). The isolated isotopically-labeled DNA can be used as a biomarker to identify and isolate microorganisms responsible for RDX degradation *in situ*. Recently Roh et al. (*91*) identified microorganisms responsible for RDX biodegradation using the SIP technique. Although SIP and CSIA are showing promise for identifying and isolating RDX degraders more research and optimization is needed to develop robust molecular probes capable of fast monitoring of the chemical and its transformation products to elucidate its transformation mechanisms. Such information would be extremely important to help understand the eventual fate of RDX in the environment and its ecological impact.

Acknowledgments

We would like to thank DRDC–DND Canada, ONR–US Navy and SERDP–US for financial support over the last few years.

References

1. Jenkins, T. F.; Hewitt, A. D.; Grant, C. L.; Thiboutot, S.; Ampleman, G.; Walsh, M. E.; Ranney, T. A.; Ramsey, C. A.; Palazzo, A. J.; Pennington, J. P. Identity and distribution of residues of energetic compounds at army live-fire training ranges. *Chemosphere* **2006**, *63*, 1280–90.
2. Xu, W.; Dana, K. E.; Mitch, W. A. Black carbon-mediated destruction of nitroglycerin and RDX by hydrogen sulfide. *Environ. Sci. Technol.* **2010**, *44*, 6409–15.
3. Hawari, J.; Beaudet, S.; Halasz, A.; Thiboutot, S.; Ampleman, G. Microbial degradation of explosives: Biotransformation versus mineralization. *Appl. Microbiol. Biotechnol.* **2000**, *54*, 605 618.
4. Thorn, K. A.; Pennington, J. C.; Kennedy, K. R.; Cox, L. G.; Hayes, C. A.; Porter, B. E. N-15 NMR study of the immobilization of 2,4- and 2,6-dinitrotoluene in aerobic compost. *Environ. Sci Technol.* **2008**, *42*, 2542–50.
5. Stenuit, B. A.; Agathos, S. N. Microbial 2,4,6-trinitrotoluene degradation: Could we learn from (bio)chemistry for bioremediation and vice versa? *Appl. Microbiol. Biotechnol.* **2010**, *88*, 1043–64.
6. Wittich, R.-M.; Ramos, J. L.; van Dillewijn, P. Microorganisms and explosives: Mechanisms of nitrogen release from TNT for use as an N-source for growth. *Environ. Sci. Technol.* **2009**, *43*, 2773–76.
7. van Dillewijn, P.; Wittich, R.-M.; Caballero, A.; Ramos, J.-L. Type II hydride transferases from different microorganisms yield nitrite and diarylamines from polynitroaromatic compounds. *Appl. Environ. Microbiol.* **2008**, *74*, 6820–23.
8. Stenuit, B.; Eyers, L.; El Fantroussi, S.; Agathos, N. Promising strategies for the mineralization of 2,4,6-trinitrotoluene. *Rev. Environ. Sci. Bio/Technol.* **2005**, *4*, 39–60.
9. Sheremata, T.; Halasz, A.; Paquet, L.; Thiboutot, S.; Ampleman, G.; Hawari, J. The fate of cyclic nitramine explosive RDX in natural soil. *Environ. Sci. Technol.* **2001**, *35*, 1037–40.
10. Gong, P.; Hawari, J.; Thiboutot, S.; Ampleman, G.; Sunahara, G. I. Ecotoxicological effects of hexahydro-1,3,5-trinitro-1,3,5-triazine on soil microbial activities. *Environ. Toxicol. Chem.* **2001**, *20*, 947–51.
11. Lachance, B.; Robidoux, P. Y.; Hawari, J.; Ampleman, G.; Thiboutot, S.; Sunahara, G. I. Cytotoxic and genotoxic effects of energetic compounds on bacterial and mammalian cells in vitro. *Mutat. Res.* **1999**, *444*, 25–39.
12. Monteil-Rivera, F.; Halasz, A.; Groom, C.; Zhao, J.-S.; Thiboutot, S.; Ampleman, G.; Hawari J. Fate and transport of explosives in the environment A Chemist's view. In *Ecotoxicology of Explosives and Unexploded*

Ordnance; Sunahara, G. I., Lutofo, G., Kuperman, R. G., Hawari, J., Eds.; CRC Press: Boca Raton, FL, 2009; pp 5–33.
13. Monteil-Rivera, F.; Paquet, L.; Giroux, R.; Hawari, J. Contribution of hydrolysis in the abiotic attenuation of RDX and HMX in coastal waters. *J. Environ. Qual.* **2008**, *37*, 858–64.
14. Hawari, J.; Halasz, A. Microbial degradation of explosives. In *The Encyclopedia of Environmental Microbiology*; Bitton, G., Ed.; John Wiley & Sons: Amsterdam, The Netherlands, 2002; pp 1979–93.
15. Borch, T.; Kretzschmar, R.; Kappler, A.; van Cappellen, P.; Ginder-Vogel, M.; Voegelin, A.; Campbell, K. Biogeochemical redox processes and their impact on contaminant dynamics. *Environ. Sci. Technol.* **2010**, *44*, 15–23.
16. Heimann, A.; Jakobsen, R.; Blodau, C. Energetic constraints on H_2-dependent terminal electron accepting processes in anoxic environments: A review of observations and model approaches. *Environ. Sci. Technol.* **2010**, *44*, 24–33.
17. Zhang, H.; Weber, E. J. Elucidating the role of electron shuttles in reductive transformations in anaerobic sediments. *Environ. Sci. Technol.* **2009**, *43*, 1042–48.
18. Hau, H. H.; Gralnick, J. A. Ecology and biotechnology of the genus *Shewanella*. *Annu. Rev. Microbiol.* **2007**, *61*, 237–58.
19. Balakrishnan, V.K.; Halasz, A.; Hawari, J. Alkaline hydrolysis of the cyclic nitramine explosives RDX, HMX, and CL-20: New insights into degradation pathways obtained by the observation of novel intermediates. *Environ. Sci. Technol.* **2003**, *37*, 1838–43.
20. Bose, P.; Glaze, W. H.; Maddox, D. S. Degradation of RDX by various advanced oxidation processes: II. Organic by-products. *Water Res.* **1998**, *32*, 1005–18.
21. Hawari, J.; Halasz, A.; Groom, C.; Deschamps, S.; Paquet, L.; Beaulieu, C.; Corriveau, A. Photodegradation of RDX in aqueous solution: A mechanistic probe for biodegradation with *Rhodococcus* sp. *Environ. Sci. Technol.* **2002**, *36*, 5117–23.
22. Oh, S.-Y.; Chiu, P. C.; Kim, B. J.; Cha, D. K. Enhancing Fenton oxidation of TNT and RDX through pretreatment with zero-valent iron. *Water Res.* **2003**, *37*, 4275–83.
23. Adam, M. L.; Comfort, S. D.; Morley, M. C.; Snow, D. D. Remediating RDX-contaminated ground water with permanganate: Laboratory investigations for the Pantex perched aquifer. *J. Environ. Qual.* **2004**, *33*, 2165–73.
24. Wanaratna, P.; Christodoulatos, C.; Sidhoum, M. Kinetics of RDX degradation by zero-valent iron (ZVI). *J. Hazard. Mater.* **2006**, *136*, 68–74.
25. Shrout, J. D.; Larese-Casanova, P.; Scherer, M. M.; Alvarez, P. J. Sustained and complete hexahydro-1,3,5-trinitro-1,3,5-triazine (RDX) degradation in zero-valent iron simulated barriers under different microbial conditions. *Environ. Technol.* **2005**, *26* (10), 1115–26.

26. Comfort, S. D.; Shea, P. J.; Machacek, T. A.; Satapanajaru, T. Pilot-scale treatment of RDX-contaminated soil with zerovalent iron. *J. Environ. Qual.* **2003**, *32*, 1717–25.
27. Wildman, M. J; Alvarez, P. J. J. RDX degradation using an integrated Fe(0)-microbial treatment approach. *Water Sci. Technol.* **2001**, *43* (2), 25–33.
28. Oh, B.-T.; Just, C. L.; Alvarez, P. J. J. Hexahydro-1,3,5-trinitro-1,3,5-triazine mineralization by zerovalent iron and mixed anaerobic cultures. *Environ. Sci. Technol.* **2001**, *35*, 4341–46.
29. Naja, G.; Halasz, A.; Thiboutot, S.; Ampleman, G.; Hawari, J. Degradation of hexahydro-1,3,5-trinitro-1,3,5-triazine (RDX) using zerovalent iron nanoparticles. *Environ. Sci. Technol.* **2008**, *42*, 4364–70.
30. Fuller, M. E.; Schaefer, C. E.; Lowey, J. M. Degradation of explosives-related compounds using nickel catalysts. *Chemosphere* **2007**, *67*, 419–27.
31. McHugh, C. J.; Smith, W. E.; Graham, D. The first controlled reduction of the high explosive RDX. *Chem. Commun.* **2002**, *21*, 2514–15.
32. Fidler, R.; Legron, T.; Carvalho-Knighton, K.; Geiger, C. L.; Sigman, M. E.; Clausen, C. A. *Environmental Applications of Nanoscale and Microscale Reactive Metal Particles*; ACS Symposium Series; American Chemical Society: Washington, DC, 2009; Vol. 1027, Chapter 7, pp 117–34.
33. Kwon, M. J.; Finneran, K. T. Electron shuttle-stimulated RDX mineralization and biological production of 4-nitro-2,4-diazabutanal (NDAB) in RDX-contaminated aquifer material. *Biodegradation* **2010**, *21*, 923–37.
34. Bhushan, B.; Halasz, A.; Hawari, J. Effect of iron(III), humic acids and anthraquinone-2,6-disulfonate on biodegradation of cyclic nitramines by *Clostridium* sp. EDB2. *J. Appl. Microbiol.* **2006**, *100*, 555–63.
35. Kemper, J. M.; Ammar, E.; Mitch, W. A. Abiotic degradation of hexahydro-1,3,5-trinitro-1,3,5-triazine in the presence of hydrogen sulfide and black carbon. *Environ. Sci. Technol.* **2008**, *42*, 2118–23.
36. Gregory, K. B.; Larese-Casanova, P.; Parkin, G. F.; Scherer, M. M. Abiotic transformation of hexahydro-1,3,5-trinitro-1,3,5-triazine by FeII bound to magnetite. *Environ. Sci. Technol.* **2004**, *38*, 1408–14.
37. An, C.; He, Y.; Huang, G. H; Yang, S. Degradation of hexahydro-1,3,5-trinitro-1,3,5-triazine (RDX) by anaerobic mesophilic granular sludge from a UASB reactor. *J. Chem. Technol. Biotechnol.* **2010**, *85*, 831–38.
38. Hawari, J.; Halasz, A.; Sheremata, T.; Beaudet, S.; Groom, C.; Paquet, L.; Rhofir, C.; Ampleman, G.; Thiboutot, S. Characterization of metabolites during biodegradation of hexahydro-1,3,5-trinitro-1,3,5-triazine (RDX) with municipal anaerobic sludge. *Appl. Environ. Microbiol.* **2000**, *66*, 2652–57.
39. McCormick, N. G.; Cornell, J. H.; Kaplan, A. M. Biodegradation of hexahydro-1,3,5-trinitro-1,3,5-triazine. *Appl. Environ. Microbiol.* **1981**, *42*, 817–23.
40. Boopathy, R.; Gurgas, M.; Ullian, J.; Manning, J. F. Metabolism of explosive compounds by sulfate-reducing bacteria. *Curr. Microbiol.* **1998**, *37*, 127–31.
41. Freedman, D. L.; Sutherland, K. W. Biodegradation of hexahydro-1,3,5-trinitro-1,3,5-triazine (RDX) under nitrate-reducing conditions. *Water Sci. Technol.* **1998**, *38*, 33–40.

42. Bradley, P. M.; Dinicola, R. S. RDX (hexahydro-1,3,5-trinitro-1,3,5-triazine) biodegradation in aquifer sediments under manganese-reducing conditions. *Bioremediation J.* **2005**, *9*, 1–8.
43. Adrian, N. R.; Arnett, C. M. Anaerobic biodegradation of hexahydro-1,3,5-trinitro-1,3,5-triazine (RDX) by *Acetobacterium malicum* strain HAAP-1 isolated from a methanogenic mixed culture. *Curr. Microbiol.* **2004**, *48*, 332–40.
44. Zhao, J.-S.; Manno, D.; Hawari, J. Regulation of hexahydro-1,3,5-trinitro-1,3,5-triazine (RDX) metabolism in *Shewanella halifaxensis* HAW-EB4 by terminal electron acceptor and involvement of *c*-type cytochrome. *Microbiology (Reading, U. K.)* **2008**, *154*, 1026–37.
45. Kwon, M. J.; Finneran, K. T. Biotransformation products and mineralization potential for hexahydro-1,3,5-trinitro-1,3,5-triazine (RDX) in abiotic versus biological degradation pathways with anthraquinone-2,6-disulfonate (AQDS) and *Geobacter metallireducens*. *Biodegradation* **2008**, *19*, 705–15.
46. Kwon, M. J.; Finneran, K. T. Microbially mediated biodegradation of hexahydro-1,3,5-trinitro-1,3,5-triazine by extracellular electron shuttling compounds. *Appl. Environ. Microbiol.* **2006**, *72*, 5933–41.
47. Fournier, D.; Halasz, A.; Spain, J.; Fiurasek, P.; Hawari, J. Determination of key metabolites during biodegradation of hexahydro-1,3,5-trinitro-1,3,5-triazine with *Rhodococcus* sp. strain DN22. *Appl. Environ. Microbiol.* **2002**, *68*, 166–72.
48. Seth-Smith, H. M. B.; Rosser, S. J.; Basran, A.; Travis, E. R.; Dabbs, E. R.; Nicklin, S.; Bruce, N. C. Cloning, sequencing, and characterization of the hexahydro1,3,5-trinitro-1,3,5-triazine degradation gene cluster from *Rhodococcus rhodochrous*. *Appl. Environ. Microbiol.* **2002**, *68*, 4764–71.
49. Coleman, N. V.; Nelson, D. R.; Duxbury, T. Aerobic biodegradation of hexahydro-1,3,5-trinitro-1,3,5-triazine (RDX) as a nitrogen source by a *Rhodococcus* sp., strain DN22. *Soil Biol. Biochem.* **1998**, *30*, 1159–67.
50. Thompson, K. T.; Crocker, F. H.; Fredrickson, H. L. Mineralization of the cyclic nitramine explosive hexahydro1,3,5-trinitro-1,3,5-triazine by *Gordonia* and *Williamsia* spp. *Appl. Environ. Microbiol.* **2005**, *71*, 8265–72.
51. Bhushan, B.; Halasz, A.; Spain, J. C.; Hawari, A. Diaphorase catalyzed biotransformation of RDX via N-denitration mechanism. *Biochem. Biophys. Res. Commun.* **2002**, *296*, 779–84.
52. Bhushan, B.; Halasz, A.; Spain, J.; Thiboutot, S.; Ampleman, G.; Hawari, J. Biotransformation of hexahydro-1,3,5-trinitro-1,3,5-triazine catalyzed by a NAD(P)H: Nitrate oxidoreductase from *Aspergillus niger*. *Environ. Sci. Technol.* **2002**, *36*, 3104–08.
53. Jackson, R. G.; Rylott, E. L.; Fournier, D.; Hawari, J.; Bruce, N. C. Exploring the biochemical properties and remediation applications of the unusual explosive-degrading P450 system XplA/B. *Proc. Natl. Acad. Sci. U.S.A.* **2007**, *104*, 16822–27.
54. Bhushan, B.; Trott, S.; Spain, J. C.; Halasz, A.; Paquet, L.; Hawari, J. Biotransformation of hexahydro1,3,5-trinitro-1,3,5-triazine (RDX) by rabbit liver cytochrome P450: Insight into the mechanism of RDX biodegradation

by *Rhodococcus* sp. strain DN22. *Appl. Environ. Microbiol.* **2003**, *69*, 1347–51.
55. Weber, K. A.; Urrutia, M. M.; Churchill, P. F.; Kukkadapu, R. K.; Roden, E. E. Anaerobic redox cycling of iron by freshwater sediment microorganisms. *Environ. Microbiol.* **2006**, *8*, 100–13.
56. Rügge, K.; Hofstetter, T. B.; Haderlein, S. B.; Bjerg, P. L.; Knudsen, S.; Zraunig, C.; Mosbæk, H; Christensen, T. H. Characterization of predominant reductants in an anaerobic leachate-contaminated aquifer by nitroaromatic probe compounds. *Environ. Sci. Technol.* **1998**, *32*, 23–31.
57. Lovley, D. R. Microbial Fe(III) reduction in subsurface environments. *FEMS Microbiol. Rev.* **1997**, *20*, 305–13.
58. Curtis, G.P.; Reinhard, M. Reductive dehalogenation of hexachloroethane, carbon tetrachloride, and bromoform by anthrahydroquinone disulfonate and humic acid. *Environ. Sci. Technol.* **1994**, *28*, 2393–401.
59. Lovley, D. R.; Giovannoni, S. J.; White, D. C.; Champine, J. E.; Phillips, E. J. P.; Gorby, Y. A.; Goodwin, S. *Geobacter metallireducens* gen. nov. sp. nov., a microorganism capable of coupling the complete oxidation of organic compounds to the reduction of iron and other metals. *Arch. Microbiol.* **1993**, *159*, 336–44.
60. Schwarzenbach, R. P.; Stierli, R.; Lanz, K.; Zeyer, J. Quinone and iron mediated reduction of nitroaromatic compounds in homogeneous aqueous solution. *Environ. Sci. Technol.* **1990**, *24*, 1566–74.
61. Kwon, M. J.; Finneran, K. T. Hexahydro-1,3,5-trinitro-1,3,5-triazine (RDX) reduction is concurrently mediated by direct electron transfer from hydroquinones and resulting biogenic Fe(II) formed during electron shuttle-amended biodegradation. *Environ. Engineer. Sci.* **2009**, *26*, 961–71.
62. Bhushan, B.; Halasz, A.; Thiboutot, S.; Ampleman, G.; Hawari, J. Chemotaxis-mediated biodegradation of cyclic nitramine explosives RDX, HMX, and CL-20 by *Clostridium* sp. EDB2. *Biochem. Biophys. Res. Commun.* **2004**, *316*, 816–21.
63. Zhao, J.-S.; Deng, Y.; Manno, D.; Hawari, J. *Shewanella* spp. genomic evolution for a cold marine lifestyle and *in-situ* explosive biodegradation. *PLoS ONE* **2010**, *5* (2), e9109; doi:10.1371/journal.pone.0009109.
64. Zhao, J.-S.; Spain, J.; Thiboutot, S.; Ampleman, G.; Greer, C.; Hawari, J. Phylogeny of cyclic nitramine-degrading psychrophilic bacteria in marine sediment and their potential role in natural attenuation of explosives. *FEMS Microbiol. Ecol.* **2004**, *49*, 349–57.
65. Kim, D.; Strathmann, T. J. Role of organically complexed iron(II) species in the reductive transformation of RDX in anoxic environments. *Environ. Sci. Technol.* **2007**, *41*, 1257–64.
66. Larese-Casanova, P.; Scherer, M. M. Abiotic transformation of hexahydro-1,3,5-trinitro-1,3,5-triazine (RDX) by green rusts. *Environ. Sci. Technol.* **2008**, *42*, 3975–81.
67. Gent, D. B.; Wani, A. H.; Davis, J. L.; Alshawabkeh, A. Electrolytic redox and electrochemical generated alkaline hydrolysis of hexahydro-1,3,5-trinitro-1,3,5-triazine (RDX) in sand column. *Environ. Sci. Technol.* **2009**, *43*, 6301–07.

68. Bonin, P. M. L.; Bejan, D.; Schutt, L.; Hawari, J.; Bunce, N. J. Electrochemical reduction of hexahydro-1,3,5-trinitro-1,3,5-triazine in aqueous solutions. *Environ. Sci. Technol.* **2004**, *38*, 1595–99.
69. Binks, P. R.; Nicklin, S.; Bruce, N. C. Degradation of hexahydro-1,3,5-trinitro-1,3,5-triazine (RDX) by *Stenotrophomonas maltophilia* PB1. *Appl. Environ. Microbiol.* **1995**, *61*, 1318–22.
70. Halasz, A.; Manno, D.; Strand, S. E.; Bruce, N. C.; Hawari, J. Biodegradation of RDX and MNX with *Rhodococcus* sp. strain DN22: New insights into the degradation pathway. *Environ. Sci. Technol.* **2010**, *44*, 9330–36.
71. Bhatt, M.; Zhao, J.-S.; Halasz, A.; Hawari, J. Biodegradation of hexahydro-1,3,5-trinitro-1,3,5-triazine by novel fungi isolated from unexploded ordnance contaminated marine sediment. *J. Ind. Microbiol. Biotechnol.* **2006**, *33*, 850–58.
72. Sheremata, T. W.; Hawari, J. Mineralization of RDX by the white rot fungus *Phanerochaete chrysosporium* to carbon dioxide and nitrous oxide. *Environ. Sci. Technol.* **2000**, *34*, 3384–88.
73. Bhatt, M.; Zhao, J.-S.; Monteil-Rivera, F.; Hawari, J. Biodegradation of cyclic nitramines by tropical marine sediment bacteria. *J. Ind. Microbiol. Biotechnol.* **2005**, *32*, 261–67.
74. Zhao, J.-S.; Paquet, L.; Halasz, A.; Hawari, J. Metabolism of hexahydro-1,3,5-trinitro-1,3,5-triazine through initial reduction to hexahydro-1-nitroso-3,5-dinitro-1,3,5-triazine followed by denitration in *Clostridium bifermentants* HAW-1. *Appl. Microbiol. Biotechnol.* **2003**, *63*, 187–93.
75. Zhao, J.-S.; Halasz, A.; Paquet, L.; Beaulieu, C.; Hawari, J. Biodegradation of hexahydro-1,3,5-trinitro-1,3,5-triazine and its mononitroso derivative hexahydro-1-nitroso-3,5-dinitro-1,3,5-triazine by *Klebsiella pneumoniae* strain SCZ-1 isolated from an anaerobic sludge. *Appl. Environ. Microbiol.* **2002**, *68*, 5336–41.
76. Beller, H. R. Anaerobic biotransformation of RDX (hexahydro-1,3,5-trinitro-1,3,5-triazine) by aquifer bacteria using hydrogen as the sole electron donor. *Water Res.* **2002**, *36*, 2533–40.
77. Kitts, C. L.; Cunningham, D. P.; Unkefer, P. J. Isolation of three hexahydro-1,3,5-trinitro-1,3,5-triazine-degrading species of the family *Enterobacteriaceae* from nitramine explosive-contaminated soil. *Appl. Environ. Microbiol.* **1994**, *60*, 4608–711.
78. Zhang, B.; Kendall, R. J.; Anderson, T. A. Toxicity of the explosive metabolites hexahydro-1,3,5-trinitroso-1,3,5-triazine (TNX) and hexahydro-1-nitroso-3,5-dinitro-1,3,5-triazine (MNX) to the earthworm *Eisenia fetida*. *Chemosphere* **2006**, *64*, 86–95.
79. Shen, C. F.; Hawari, J.; Ampleman, G.; Thiboutot, S.; Guiot, S. R. Enhanced biodegradation and fate of hexahydro-1,3,5-trinitro-1,3,5-triazine (RDX) and octahydro-1,3,5,7-tetranitro-1,3,5,7-tetrazocine (HMX) in anaerobic soil slurry bioprocess. *Bioremed. J.* **2000**, *4*, 27–39.
80. Halasz, A.; Spain, J.; Paquet, L.; Beaulieu, C.; Hawari, J. Insights into the formation and degradation mechanisms of methylenedinitramine during the incubation of RDX with anaerobic sludge. *Environ. Sci. Technol.* **2002**, *36*, 633–38.

81. Lamberton, A. H. Some aspects of the chemistry of nitramines. *Q. Rev. Chem. Soc.* **1951**, *5*, 75–98.
82. Fournier, D.; Trott, S.; Spain, J.; Hawari, J. Metabolism of the aliphatic nitramine 4-nitro-2,4-diazabutanal by *Methylobacterium* sp. strain JS178. *Appl. Environ. Microbiol.* **2005**, *71*, 4199–202.
83. Fournier, D.; Halasz, A.; Spain, J. C.; Spanggord, R. J.; Bottaro, J. C.; Hawari, J. Biodegradation of the hexahydro-1,3,5-trinitro-1,3,5-triazine ring cleavage product 4-nitro-2,4-diazabutanal by *Phanerochaete chrysosporium*. *Appl. Environ. Microbiol.* **2004**, *70*, 1123–28.
84. Fuller, M. E.; Perreault, N.; Hawari, J. Microaerophilic degradation of hexahydro-1,3,5-trinitro-1,3,5-triazine (RDX) by three *Rhodococcus* strains. *Lett. Appl. Microbiol.* **2010**, *51*, 313–18.
85. Ignarro, L. J.; Fukuto, J. M.; Griscavage, J. M.; Rogers, N. E.; Byrns, R. E. Oxidation of nitric oxide in aqueous solution to nitrite but not nitrate : Comparison with enzymatically formed nitric oxide from L-arginine. *Proc. Natl. Acad. Sci. U.S.A.* **1993**, *90*, 8103–07.
86. Guengerich, F. P. Common and uncommon cytochrome P450 reactions related to metabolism and chemical toxicity. *Chem. Res. Toxicol.* **2001**, *14*, 611–50.
87. Sagi-Ben Moshe, S.; Ronen, Z.; Dahan, O.; Bernstein, A.; Weisbrod, N.; Gelman, F.; Adar, E. Isotopic evidence and quantification assessment of in situ RDX biodegradation in the deep unsaturated zone. *Soil Biol. Biochem.* **2010**, *42*, 1253–62.
88. Bernstein, A.; Adar, E.; Ronen, Z.; Lowag, H.; Stichler, W.; Meckenstock, R. U. Quantifying RDX biodegradation in groundwater using $\delta^{15}N$ isotope analysis. *J. Contam. Hydrol.* **2010**, *111*, 25–35.
89. Bernstein, A.; Ronen, Z.; Adar, E.; Nativ, R.; Lowag, H.; Stichler, W.; Meckenstock, R. U. Compound-specific isotope analysis of RDX and stable isotope fractionation during aerobic and anaerobic biodegradation. *Environ. Sci. Technol.* **2008**, *42*, 7772–77.
90. Radajewski, S.; Ineson, P.; Parekh, N. R.; Murrell, J. C. Stable-isotope probing as a tool in microbial ecology. *Nature* **2000**, *403*, 646–49.
91. Roh, H.; Yu, C.-P.; Fuller, M. E.; Chu, K. H. Identification of hexahydro-1,3,5-trinitro-1,3,5-triazine-degrading microorganisms via ^{15}N-stable isotope probing. *Environ. Sci. Technol.* **2009**, *43*, 2502–11.
92. Schmidt, T. C.; Zwank, L.; Elsner, M.; Berg, M.; Meckenstock, R. U.; Haderlein, S. B. Compound-specific stable isotope analysis of organic contaminants in natural environments: a critical review of the state of the art, prospects, and future challenges. *Anal. Bioanal. Chem.* **2004**, *378*, 283–300.
93. Elsner, M.; Hofstetter, T. B. Current perspectives on the mechanisms of chlorohydrocarbon degradation in subsurface.In *Aquatic Redox Chemistry*; Tratnyek, P. G., Grundl, T. J., Haderlein, S. B., Eds.; ACS Symposium Series; American Chemical Society: Washington, DC, 2011; Vol. 1071, Chapter 19, pp 407−439.
94. Amaral, H. I. F.; Fernandes, J.; Berg, M.; Schwarzenbach, R. P.; Kipfer, R. Assessing TNT and DNT groundwater contamination by compound-specific

isotope analysis and ^{3}H-^{3}He groundwater dating: A case study in Portugal. *Chemosphere* **2009**, *77*, 805–12.

95. Gallagher, E. M.; Young, L. Y.; McGuinness, L. M.; Kerkhof, L. J. Detection of 2,4,6-trinitrotoluene-utilizing anaerobic bacteria by ^{15}N and ^{13}C incorporation. *Appl. Environ. Microbiol.* **2010**, *76*, 1695–98.

Chapter 21

Role of Coupled Redox Transformations in the Mobilization and Sequestration of Arsenic

Janet G. Hering,[*,1] Stephan J. Hug,[1] Claire Farnsworth,[2] and Peggy A. O'Day[3]

[1]Eawag, Swiss Federal Institute of Aquatic Science and Technology, CH-8600 Dübendorf, Switzerland
[2]Division of Engineering and Applied Sciences, California Institute of Technology, Pasadena CA 91125
[3]School of Natural Sciences, University of California, Merced, 95343
[*]janet.hering@eawag.ch

Arsenic occurrence in groundwater, particularly in South and Southeast Asia, has had profoundly deleterious impacts on human health. To address this tragedy, extensive research has been conducted on the biogeochemical cycling of arsenic and its consequences for arsenic mobilization and sequestration. This research has elucidated a key role of microorganisms in redox transformations and the importance of iron and sulfur minerals as carrier phases for arsenic. Research gaps remain, particularly with regard to determining *in situ* rates of redox transformations and the coupled influence of hydrologic and biogeochemical processes on arsenic occurrence and mobility. Despite these gaps, the insights of this research can be applied to mitigate human exposure through improved water resources management as well as through treatment and remediation.

Introduction

The identification of chronic arsenicosis in West Bengal, India in the early-to-mid 1990's (*1*) motivated extensive studies of human exposure to arsenic and its health effects as well as of arsenic biogeochemistry in recent decades (*2*, *3*). The consumption of groundwater containing arsenic at levels 10 to 100-fold above drinking water standards has resulted in severe human health impacts including cancer mortality, and estimates of the exposed population throughout Asia number in the hundreds of millions.

Chronic exposure to arsenic, primarily through drinking water, poses the greatest risk for the largest populations. About 90% of dissolved inorganic arsenic is passed from the gut into the bloodstream, but most is also rapidly excreted, on the order of days to weeks, with a relatively small fraction retained in organs and tissues (*4*). The inorganic forms of arsenic, As(III) and As(V), that occur in drinking water are substantially more toxic to humans than organically-bound forms, although specific toxicity associated with the large number of organo-arsenic compounds found in food sources is mostly unknown (*5*). The extent of exposure and health risk from different arsenic species in food sources, particularly from rice, is receiving more study, but quantitative models of food bioavailability based on sufficiently large populations are lacking (*6*).

Large-scale campaigns to delineate the occurrence of elevated arsenic in groundwater – conducted initially in West Bengal, India and Bangladesh and later in Vietnam, Cambodia, and elsewhere in South and Southeast Asia – revealed some common features, specifically low arsenic concentrations in deep (Pleistocene) aquifers and substantial horizontal variability in arsenic concentrations in shallow (Holocene) aquifers (*7*). The spatial heterogeneity observed on a local scale poses a particular challenge to identify the geologic, biogeochemical and hydrologic controls on arsenic occurrence.

The well-known geochemistry of arsenic was suggestive of several possible mechanisms for As mobilization, including the reductive dissolution of iron(III) carrier phases, oxidative dissolution of sulfide carrier phases, and competitive desorption due to phosphate (*3*). In these settings, the reductive release mechanism was found to be the most generally dominant. This has also been observed at landfills and biostimulation sites, where inputs of organic carbon drive the reductive release of iron and arsenic (*8*).

In other settings, where sulfides are the primary carrier phase for arsenic, oxidative release can be the dominant mechanism; for example, elevated arsenic concentrations are often observed in acid mine drainage (*9*). Subsequent oxidative precipitation (i.e., of iron(III) oxyhydroxides) can sequester arsenic in such systems (*10*). Sequestration of arsenic can also be associated with authigenic sulfide precipitation, particularly in the presence of iron (*11*).

The availability of iron and sulfur, as well as the prevailing redox and pH conditions, can shift the balance between arsenic mobilization and sequestration. Understanding these processes is crucial both to the predication of arsenic occurrence and mobility in groundwater and to the design of remediation strategies and treatment systems.

The literature relevant to this topic is vast and cannot be comprehensively reviewed here. The reader is refered to other reviews (*3*, *7*, *12*, *13*). Here, we seek to provide the background needed to discuss some unresolved questions and issues in this field and to comment on the implications for water resource management, remediation, and mitigation of human exposure to arsenic.

Direct and Indirect Dependence of Arsenic Mobility on Redox Conditions

Redox conditions in aquatic systems generally reflect the dominant microbial respiratory processes, which are, in turn, influenced by the supply of electron donors (e.g., organic carbon) and electron acceptors (*14*). At the circumneutral pH values typical of aquatic systems, the range of redox conditions overlaps with the stability domains of multiple oxidation states of arsenic, iron and sulfur (as well as those of other elements). The redox transformations of arsenic itself (i.e., conversion between the +III and +V oxidation states common to aquatic systems) influence arsenic mobility directly, while those of other elements, particularly iron and sulfur, can influence arsenic mobility indirectly. These interactions are illustrated schematically in Fig. 1, which highlights sorption and precipitation in association with iron and sulfur species as the dominant processes that sequester arsenic.

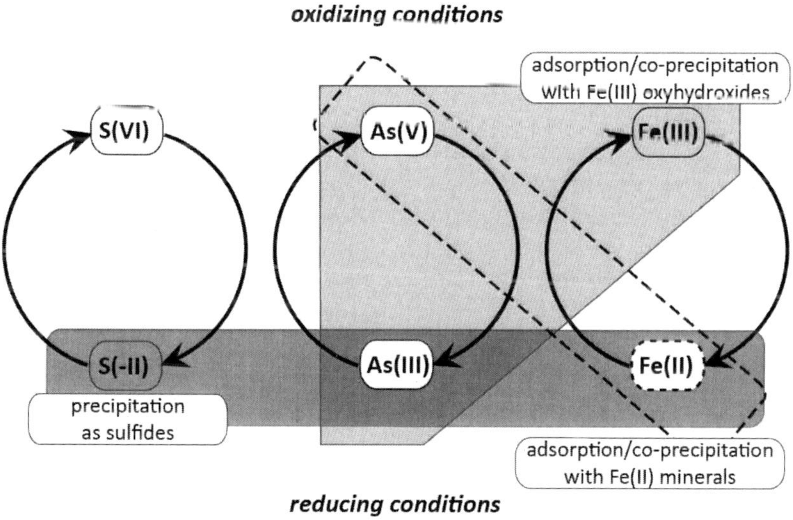

oxidizing conditions

adsorption/co-precipitation with Fe(III) oxyhydroxides

S(VI)

As(V)

Fe(III)

S(-II)

As(III)

Fe(II)

precipitation as sulfides

adsorption/co-precipitation with Fe(II) minerals

reducing conditions

Figure 1. Schematic illustration of interactions among oxidized and reduced forms of sulfer, arsenic, and iron. Shading indicates concomitant removal of chemical species through sorption and/or co-precipitation processes. Iron(II) minerals that may adsorb or co-precipitate with arsenic include phosphates (e.g. vivianite) or Fe(II,III)-oxides (e.g., green rust-type phases).

Minerals containing As(III) or As(V) as a constituent ion have widely varying solubilities and their stabilities also depend on the availability of other constituent ions, which is influenced by prevailing redox and pH conditions (*12*). In addition, As(III) and As(V) exhibit different, pH-dependent sorption behavior, particularly with regard to sorption on aluminum carrier phases (*15*) and to the effects of competing sorbates such as phosphate (*16*).

Slow kinetics, either of redox transformations or of dissolution-precipitation reactions, can, however, skew expectations based on thermodynamics. For example, precipitation of As(V), as arsenate, with Fe(II) in an analog of vivianite ($Fe^{II}_3(PO_4)_2 \cdot 8H_2O$) has been observed in microbial cultures that lack the capacity for As(V) reduction (*17*) even though concurrent reduction of As(V) and Fe(III) has been observed in other microbial systems and is consistent with the thermodynamics of these reactions (*18*).

Despite these complexities, and bearing in mind the variability due to local conditions, it is a reasonable generalization that, when Fe(III) oxyhydroxides are present, they are likely to be a dominant carrier phase for both As(V) and As(III), while under sulfate-reducing conditions, reduced arsenic is sequestered in the form of sulfides. Maximum mobility of arsenic is thus expected under reducing conditions (where Fe(III) oxyhydroxides are subject to dissolution) with limited sulfide availability (*19*, *20*).

Reductive Dissolution as a Mechanism for Arsenic Mobilization

Despite the extensive studies of arsenic mobilization, questions remain regarding the factors that control the rate and extent of the reductive dissolution of Fe(III) oxyhydroxides and the role that As(V) reduction, *per se*, plays in arsenic mobilization. Reviews on this topic have reported conflicting observations that may reflect the influence of factors such as the mineralogy of the Fe(III) oxyhydroxide substrates, the capacity of microorganisms for Fe(III) and/or As(V) reduction, and the specific conditions in experimental or field systems (*13*, *21*). There is no doubt, however, that microorganisms play a central role in arsenic mobilization, which will thus reflect the level of microbial activity and the inherent capabilities of the microbial community in a given system.

Because both Fe(III) and As(V) reduction are dissimilatory processes, it is difficult to identify the specific microorganisms within a microbial community that are responsible for these transformations. Incorporation of C-13 into DNA from labeled acetate concurrent with As(V) reduction has been used as the basis for the provisional identification of the microbes responsible for arsenic mobilization (*22*). A mass balance approach has been used to attribute organic carbon oxidation to various terminal electron acceptors (including manganese and iron oxides) in the incubation of marine sediments (*23*). This study highlighted the importance of metal recycling (i.e., consecutive oxidation and reduction transformations). For arsenic, it has been shown that the same biofilm community can both reduce As(V) and oxidize As(III) with the relative importance of these processes being dependent on light conditions (*24*).

Given the difficulty of determining *in situ* rates of Fe(III) and As(V) reduction directly, reactive transport modeling can be a useful approach to compare laboratory and field experiments and to assess transformation rates under field conditions. Although not yet widely applied, this approach has been used to elucidate the importance of arsenic recycling in lake sediments and its control by the supply of sulfate and organic matter (*25*). Application of reactive transport models requires better knowledge of site-specific rates associated with coupled oxidation-reduction reactions, in particular the relationship between rates and the composition and bioavailability of organic matter, and the importance of non-carbon electron acceptors and donors (*14*). Additional uncertainties include the thermodynamic stabilities and dissolution/precipitation rates of metastable Fe(III)-oxide phases such as ferrihydrite, or Fe(II,III) oxides such as green rust-type minerals, which typically have higher solubilities and surface areas than stable iron oxide minerals. Ideally, reactive transport modeling could be combined with methods to assses transformation rates *in situ* and/or in laboratory incubations to constrain processes occurring in field systems.

Role of Organic Substrates

It is to be expected that the availability and quality of organic substrates would significantly influence the rates of microbial Fe(III) and As(V) reduction. Many, if not most, incubation experiments documenting arsenic mobilization from sediments have used organic carbon amendments (usually acetate or lactate) to stimulate microbial processes.

At sites with localized, intense inputs of anthropogenic organic carbon, such as sanitary landfills (*26*) and biostimulation sites (*8*), it is clear that these organic inputs are the driver for release of naturally-occurring arsenic. In such systems, arsenic associated with aquifer sediments is immobile in the absence of anthropogenic perturbation.

The question of the source and quality of organic carbon supporting arsenic mobilization in Asian aquifer sediments has, however, evoked considerable controversy. Investigations linked arsenic mobilization initially with the mineralization of sedimentary organic carbon (*27*) or petroleum (*28*). However, isotopic studies have demonstrated the predominant influence of recent sources of organic carbon (*29*). Studies suggest that fresh organic carbon delivered to the subsurface by infiltration from surface ponds is important in stimulating arsenic mobilization (*30*). It has also been suggested that traditional agricultural waste management practices may supply organic carbon from fields (*31*).

Factors Affecting Arsenic Sequestration

Although arsenic sequestration in natural groundwater systems has been less of a focus of research than mobilization, it is clear that processes that lead to mobilization under reducing conditions are at least to some degree reversible under oxidizing conditions. Thus observed *net* mobilization and sequestration rates often reflect some degree of recycling. Processes in both directions are important in sediments close to the oxic-anoxic interface and particularly in soils that are subject to changing redox conditions (e.g., caused by seasonal or more frequent fluctuations of the water table).

An Example of the Spatial Variability in Arsenic Speciation at a Contaminated Site

The variability in arsenic speciation in solid phases reflects the spatial heterogeneity of the environment as a function of sediment composition and texture, water-table level and groundwater flow and, at contaminated sites, the source of arsenic contamination. In tidally-influenced sediments near a former pesticide manufacturing facility, arsenic oxidation state and sequestration in shallow sediments, as As-sulfide, As(III)-oxide, or As(V)-oxide, were related to lithologic horizons and depth to the water table on a cm-to-m scale (*20*). Arsenic sulfide and minor iron sulfide phases were identified in the upper unsaturated, organic-rich clay layer, while the lower sand-dominated, oxic zone contained only As(V) associated with iron oxide phases (Fig. 2). The zone of transition between the presence and absence of arsenic and iron sulfides corresponded to the approximate seasonal water table level associated with shallow groundwater, and also corresponded to a minimum concentration in sediment arsenic. This example demonstrates, on a relatively small scale, how vertical infiltration of surface salt water promotes sulfate reduction and sulfide precipitation in the upper zone, while the influx of oxygenated water in the aquifer sands leads to arsenic oxidation and sorption. In this setting, arsenic mobility is maximized in groundwater zones that fluctuate between reduced and oxidized conditions.

Arsenic Sequestration during Remediation and Treatment

Arsenic sequestration has received considerable attention in the context of treatment and remediation. A number of treatments in engineered or augmented systems take advantage of the same processes by which arsenic is sequestered under natural conditions. Reductive sequestration has been proposed for *in situ* remediation (*32*, *33*). This is likely to require amendment with sulfate to overcome the limitation by sulfate availability commonly observed in freshwater systems (*19*, *20*).

Formation of sulfide minerals has also been invoked as a mechanism for arsenic sequestration in subsurface permeable reactive barriers employing zerovalent iron, ZVI (*34*). While iron corrosion provides Fe(II)- and Fe(III)-phases for arsenic adsorption or co-precipitation, the availability of dissolved sulfate and organic carbon in field settings determines the extent to which sequestration in sulfide minerals can contribute to arsenic immobilization and hence to the effectiveness and longevity of the barrier.

In (*ex situ*) treatment systems, however, arsenic removal is more commonly based on sequestration under oxic conditions. For example, since groundwaters in South Asia often contain elevated concentrations of both Fe(II) and arsenic, the oxidative precipitation of Fe(III)-phases offers the possibility of concomitant removal of arsenic. The composition, mineralogy and reactivity of Fe(III)-phases formed by oxidation of dissolved Fe(II) by dissolved O_2 are, however, influenced by the solution composition. In waters with low phosphate and silicate concentrations, Fe(III)(hydr)oxides such as lepidocrocite, ferrihydrite, and goethtite are typically precipitated, while mixed Fe(III)phosphate-hydroxides and silicate-rich Fe(III)-phases are formed in water with higher dissolved phosphate and silicate concentrations (*35*, *36*).

Oxidative precipitation of Fe(III)-phases is also significant in that the oxidation of Fe(II) by O_2 can promote co-oxidation of As(III). The reactive oxygen intermediates, such as O_2^-, H_2O_2 and OH-radicals and possibly higher-valent (e.g., +IV) iron species that are formed during the reaction of Fe(II) with O_2, are also able to oxidize As(III) (*37*). The oxidation of As(III) to As(V) concomitant with Fe(II) oxidation tends to promote arsenic uptake into the Fe(III)-phases formed at circumneutral pH in the presence of phosphate and silicate (*38*).

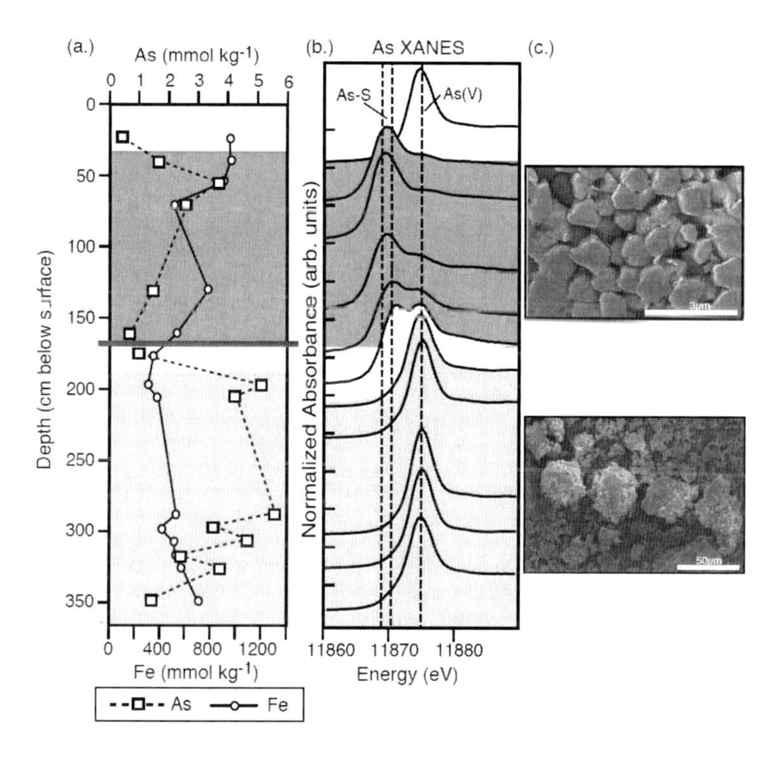

Figure 2. Results from a sediment core showing total As and Fe concentrations (a), As XANES spectra from corresponding depths (b), and example scanning electron micrographs (c) (at 61-76 cm from the same core and at 259-274 cm from a similar core ~15 m away). XANES results show the transition from sulfide-associated As in the upper unsaturated zone to a lower oxidized zone with only As(V), which corresponds to the approximate level of the seasonally-averaged groundwater table (horizonal line in (a)). The upper reduced sediments contain rare cubic Fe- and S-rich crystals, while the lower aquifer sediments contain abundant surface coatings on quartz grains. Modified from (20).

These processes are successfully exploited in the removal of arsenic using sand filters in Vietnam, where Fe(II) concentrations of up to 40 mg/L occur naturally in the groundwater (*39*). In addition to co-oxidation with Fe(II), As(III) oxidation in the top layer of sand is most likely accelerated by a population of microorganism that is typically established within days. Manganese (III,IV) oxides, formed by oxidative precipitation of Mn(II) within the filters, can also oxidize As(III).

In contrast, naturally-occurring Fe(II) in groundwater in Bangladesh is insufficient to achieve effective arsenic removal, particularly in the presence of elevated phosphate and silica concentrations (*40*). In the SONO-filter, a successful and widely-used household filter, corrosion of metallic iron serves as a continuous source of Fe(II) for several years (*41*). The continuous Fe(II) release supports co-oxidation of As(III) and arsenic removal during the oxidative precipitation of Fe(III)-phases. It is likely, however, that these filters are populated by (as yet uncharacterized) microorganisms that also promote microbial As(III) and Mn(II) oxidation. Furthermore, iron corrosion processes form magnetite, maghemite, hematite and other solid phases that can also sorb arsenic (*42*). Transport limitations at corroding iron surfaces resulting in anoxic regions with locally and/or temporally high Fe(II) concentrations can lead to transformations of As-containing Fe(III)-phases. Reduction of As(III)- or As(V)-containing ferrihydrite by microbially produced Fe(II), for example, forms magnetite with adsorbed As(III) or incorporated As(V), respectively (*43*). Transformation of Fe(III)-phases to more dense phases such as magnetite might prevent clogging of filters.

The strong affinity of arsenic both to sorb to and coprecipitate with iron oxides has also been exploited in the development of technologies for removing arsenic from soil or stabilizing it *in situ*. A variety of treatments with different forms of iron oxides, iron sulfates, and recycled iron metal have been used to immobilize arsenic in contaminated geomedia (*44*). The similarity in chemical behavior between arsenate and oxyanions such as sulfate and phosphate has also been exploited as a means to lower arsenic mobility under oxidizing conditions through addition of Portland cement and sulfate treatments to stabilize contaminated soils (*45*). Lime or organic amendments combined with stabilization treatments for arsenic have been used to promote plant growth for phytostabilization in addition to chemical stabilization.

Coupling of Hydrologic and Biogeochemical Processes

The influence of transport processes has often been neglected in biogeochemical studies of arsenic mobilization and sequestration. Investigations targeting the supply of organic carbon, sulfate, and dissolved oxygen have, however, drawn attention to the importance of transport processes and constraints. As already mentioned, arsenic mobilization is particularly sensitive to the supply of organic carbon (*46*, *47*); the localization of arsenic sequestration reflects the supply of sulfate (*19*) and dissolved oxygen (*20*, *48*, *49*). Groundwater seepage into (oxygenated) surface water creates a gradient in redox conditions conducive to arsenic sequestration (*48*). Water table fluctuations associated with irrigated agriculture (*29*, *49*) or intermittent pumping in riverbank filtration systems (*50*) can significantly perturb redox conditions and promote redox cycling.

An Example of Redox Cycling in Rice Fields

Successive mobilization and sequestration of arsenic under varying redox conditions has been observed in irrigated rice fields in Bangladesh. In yearly cycles, arsenic accumulates in top soils during irrigation and is partly remobilized and redistributed during monsoon flooding (*49*). On a shorter time scale, arsenic is also mobilized and sequestered during irrigation (*51*). After the rice is planted, the fields are kept flooded for several weeks, mainly to limit weed growth and to establish a robust population of rice plants. During this time, anoxic conditions develop in the topsoil (0-25 cm depth) and arsenic concentrations of up to 500 μg/L are measured in the soil porewater with >70% of the arsenic occurring as As(III); elevated concentrations of Fe(II) and Mn(II), up to 10 and 2 mg/L respectively, are also observed. Once the rice plants are established, the fields are irrigated less frequently and the soils become dry at the surface. Partly oxic conditions develop down to a depth of 20 cm; only low concentrations of dissolved Fe and Mn were observed with arsenic, predominantly (>70%) as As(V), occurring at concentrations below 50 μg/L. Although the limited time resolution of pore water samplers precludes the determination of rates of arsenic-mobilization and sequestration, the timescales for these processes may be constrained to days or less. Thus the practice of irrigation with alternating wet-dry cycles may significantly reduce the exposure of plant roots to arsenic and the subsequent uptake of arsenic by the rice plants.

Open Issues Related to Human Exposure, Treatment, Remediation and Water Resources Management

The extensive and multi-faceted studies of arsenic in recent decades were motivated, in large part, by the human tragedy in South and Southeast Asia. While much has been learned, the problem of human exposure has not been adequately addressed and open questions remain regarding the future consequences of management, remediation, and treatment strategies. Quantitative predictions of arsenic mobility, attenuation, or remediation effectiveness are limited by insufficient knowledge of the range of reaction rates associated with multiple redox processes, and of the coupling between biogeochemical and hydrologic factors at specific sites.

Minimizing the current human exposure to elevated concentrations of arsenic in drinking water is clearly the most pressing need. This can be accomplished either by treating the contaminated water to achieve an acceptable quality or substituting an alternate, uncontaminated water source. Well-switching and drilling of deep wells are the most common practices to avoid the arsenic contamination in shallow aquifers in South Asia (*52*), but the effectiveness of these strategies may be compromised by lack of convenience and public acceptance (*53*).

There is no shortage of effective technologies for the removal of arsenic from drinking water and many of these can also be used at the household scale (*3*). In developing countries, cost is a major barrier to implementation (*52*), and it must also be recognized that arsenic occurrence in drinking water is only one of many pressing public health concerns.

Even in developed countries, the option of drinking water treatment presents the issue of appropriately handling treatment residuals. This concern has been raised in regard to the use of Fe(III)-based coagulants or sorbents for arsenic removal; disposal of arsenic-rich treatment residuals in municipal landfills could result in arsenic mobilization under reducing conditions (*54*).

Preventing (further) degradation of both shallow and deep aquifers is an important concern for the future. In the long-term, use of deep wells may also lead to their contamination with arsenic (and potentially with other metals or organic contaminants) derived from the surface or shallow aquifers (*4, 55*). Intensive use of deep aquifers for irrigation is likely to lead to future contamination of deep wells and thus should be avoided (*7*). The question of whether the modification of current agricultural practices could influence the amount or composition of organic carbon that reaches shallow aquifers and thus arsenic mobilization deserves further investigation.

References

1. Mandal, B. K.; Chowdhury, T. R.; Samanta, G.; Basu, G. K.; Chowdhury, P. P.; Chanda, C. R.; Lodh, D.; Karan, N. K.; Dhar, R. K.; Tamili, D. K.; Das, D.; Saha, K. C.; Chakraborti, D. Arsenic in groundwater in seven districts of West Bengal, India - The biggest arsenic calamity in the world. *Curr. Sci.* **1996**, *70* (11), 976–986.
2. Mukherjee, A.; Fryar, A. E.; O'Shea, B. M.Major occurrences of elevated arsenic in groundwater and other natural waters. In *Arsenic: Environmental Chemistry, Health Threat and Waste Treatment*; Henke, K. R., Ed.; John Wily & Sons: Chichester, 2009; Vol. 303–349.
3. Ravenscroft, P.; Brammer, H.; Richards, K. *Arsenic Pollution: A Global Synthesis*; RBS-IBG Book Series; Ward, K., Bullard, J., Ed.; Wiley-Blackwell: Chichester, 2009; p 588.
4. Caussy, D.; Priest, N. D. Introduction to Arsenic Contamination and Health Risk Assessment with Special Reference to Bangladesh. In *Arsenic Pollution and Remediation: An International Perspective*; 2008; Vol. 197, pp 1–15.
5. Francesconi, K. A. Arsenic species in seafood: Origin and human health implications. *Pure Appl. Chem.* **2010**, *82* (2), 373–381.
6. Khan, N. I.; Owens, G.; Bruce, D.; Naidu, R. Human arsenic exposure and risk assessment at the landscape level: a review. *Environ. Geochem. Health* **2009**, *31*, 143–166.

7. Fendorf, S.; Michael, H. A.; van Geen, A. Spatial and Temporal Variations of Groundwater Arsenic in South and Southeast Asia. *Science* **2010**, *328* (5982), 1123–1127.
8. Hering, J. G.; O'Day, P. A.; Ford, R. G.; He, Y. T.; Bilgin, A.; Reisinger, H. J.; Burris, D. R. MNA as a Remedy for Arsenic Mobilized by Anthropogenic Inputs of Organic Carbon. *Ground Water Monit. Rem.* **2009**, *29* (3), 84–92.
9. Welch, A. H.; Westjohn, D. B.; Helsel, D. R.; Wanty, R. B. Arsenic in ground water of the United States: Occurrence and geochemistry. *Ground Water* **2000**, *38* (4), 589–604.
10. Romero, F. M.; Prol-Ledesma, R. M.; Canet, C.; Alvares, L. N.; Perez-Vazquez, R. Acid drainage at the inactive Santa Lucia mine, western Cuba: Natural attenuation of arsenic, barium and lead, and geochemical behavior of rare earth elements. *Appl. Geochem.* **2010**, *25* (5), 716–727.
11. Kirk, M. F.; Roden, E. E.; Crossey, L. J.; Brearley, A. J.; Spilde, M. N. Experimental analysis of arsenic precipitation during microbial sulfate and iron reduction in model aquifer sediment reactors. *Geochim. Cosmochim. Acta* **2010**, *74* (9), 2538–2555.
12. O'Day, P. A. Chemistry and mineralogy of arsenic. *Elements* **2006**, *2* (2), 77–83.
13. Reyes, C.; Lloyd, J. R.; Saltikov, C. W. Geomicrobiology of iron and arsenic in anoxic sediments. In *Arsenic Contamination of Groundwater: Mechanism, Analysis, and Remediation*; Ahuja, S., Ed.; John Wiley and Sons: Hoboken, 2008; Vol. 123−146.
14. McMahon, P. B.; Chapelle, F. H.; Bradley, P. M. Evolution of redox processes in groundwater. In *Aquatic Redox Chemistry*; Tratnyek, P. G., Grundl, T. J., Haderlein, S. B., Eds.; ACS Symposium Series; American Chemical Society: Washington, DC, 2011; Vol. 1071, Chapter 26, pp 581−597.
15. Hering, J. G.; Dixit, S. Contrasting sorption behavior of arsenic(III) and arsenic(V) in suspensions of iron and aluminum oxyhydroxides. In *Advances in Arsenic Research - Integration of Experimental and Observational Studies and Implications for Mitigation*; O'Day, P. A., Vlassopoulos, D., Meng, Z., Benning, L. G., Eds.; ACS Symposium Series; American Chemical Society: Washington, DC, 2005; Vol. 915, pp 8−24.
16. Dixit, S.; Hering, J. G. Comparison of arsenic(V) and arsenic(III) sorption onto iron oxide minerals: Implications for arsenic mobility. *Environ. Sci. Technol.* **2003**, *37* (18), 4182–4189.
17. Wang, X. J.; Chen, X. P.; Kappler, A.; Sun, G. X.; Zhu, Y. G. Arsenic binding to iron(II) miinerals produced by an iron(III)-reducing *Aeromonas* strain isolated from padddy soil. *Environ. Toxicol. Chem.* **2009**, *28* (11), 2255–2262.
18. Campbell, K. M.; Malasarn, D.; Saltikov, C. W.; Newman, D. K.; Hering, J. G. Simultaneous microbial reduction of iron(III) and arsenic(V) in suspensions of hydrous ferric oxide. *Environ. Sci. Technol.* **2006**, *40* (19), 5950–5955.
19. Buschmann, J.; Berg, M. Impact of sulfate reduction on the scale of arsenic contamination in groundwater of the Mekong, Bengal and Red River deltas. *Appl. Geochem.* **2009**, *24* (7), 1278–1286.

20. Root, R. A.; Vlassopoulos, D.; Rivera, N. A.; Rafferty, M. T.; Andrews, C.; O'Day, P. A. Speciation and natural attenuation of arsenic and iron in a tidally influenced shallow aquifer. *Geochim. Cosmochim. Acta* **2009**, *73* (19), 5528–5553.
21. Crowe, S. A.; Roberts, J. A.; Weisener, C. G.; Fowle, D. A. Alteration of iron-rich lacustrine sediments by dissimilatory iron-reducing bacteria. *Geobiology* **2007**, *5* (1), 63–73.
22. Hery, M.; van Dongen, B. E.; Gill, F.; Mondal, D.; Vaughan, D. J.; Pancost, R. D.; Polya, D. A.; Lloyd, J. R. Arsenic release and attenuation in low organic carbon aquifer sediments from West Bengal. *Geobiology* **2010**, *8* (2), 155–168.
23. Canfield, D. E.; Thamdrup, B.; Hansen, J. W. The anaerobic degradation of organic-matter in Danish coastal sediments - iron reduction, manganese reduction, and sulfate reduction. *Geochim. Cosmochim. Acta* **1993**, *57* (16), 3867–3883.
24. Hoeft, S. E.; Kulp, T. R.; Han, S.; Lanoil, B.; Oremland, R. S. Coupled Arsenotrophy in a Hot Spring Photosynthetic Biofilm at Mono Lake, California. *Appl. Environ. Microbiol.* **2010**, *76* (14), 4633–4639.
25. Couture, R. M.; Shafei, B.; Van Cappellen, P.; Tessier, A.; Gobeil, C. Non-Steady State Modeling of Arsenic Diagenesis in Lake Sediments. *Environ. Sci. Technol.* **2010**, *44* (1), 197–203.
26. Keimowitz, A. R.; Simpson, H. J.; Stute, M.; Datta, S.; Chillrud, S. N.; Ross, J.; Tsang, M. Naturally occurring arsenic: Mobilization at a landfill in Maine and implications for remediation. *Appl. Geochem.* **2005**, *20* (11), 1985–2002.
27. Nickson, R.; McArthur, J.; Burgess, W.; Ahmed, K. M.; Ravenscroft, P.; Rahman, M. Arsenic poisoning of Bangladesh groundwater. *Nature* **1998**, *395* (6700), 338–338.
28. Rowland, H. A. L.; Boothman, C.; Pancost, R.; Gault, A. G.; Polya, D. A.; Lloyd, J. R. The Role of Indigenous Microorganisms in the Biodegradation of Naturally Occurring Petroleum, the Reduction of Iron, and the Mobilization of Arsenite from West Bengal Aquifer Sediments. *J. Environ. Qual.* **2009**, *38* (4), 1598–1607.
29. Harvey, C. F.; Swartz, C. H.; Badruzzaman, A. B. M.; Keon-Blute, N.; Yu, W.; Ali, M. A.; Jay, J.; Beckie, R.; Niedan, V.; Brabander, D.; Oates, P. M.; Ashfaque, K. N.; Islam, S.; Hemond, H. F.; Ahmed, M. F. Arsenic mobility and groundwater extraction in Bangladesh. *Science* **2002**, *298* (5598), 1602–1606.
30. Polizzotto, M. L.; Kocar, B. D.; Benner, S. G.; Sampson, M.; Fendorf, S. Near-surface wetland sediments as a source of arsenic release to ground water in Asia. *Nature* **2008**, *454* (7203), 505–U5.
31. Farooq, S. H.; Chandrasekharam, D.; Berner, Z.; Norra, S.; Stüben, D. Influence of traditional agricultural practices on mobilization of arsenic from sediments to groundwater in Bengal delta. *Water Res.* **2010**, *44* (19), 5575–5588.
32. Keimowitz, A. R.; Mailloux, B. J.; Cole, P.; Stute, M.; Simpson, H. J.; Chillrud, S. N. Laboratory investigations of enhanced sulfate reduction as a

groundwater arsenic remediation strategy. *Environ. Sci. Technol.* **2007**, *41* (19), 6718–6724.

33. Lee, M. K.; Saunders, J. A.; Wilkin, R. T.; Mohammad, S. Geochemical modeling of arsenic speciation and mobilization: Implications for bioremediation. In *Advances in Arsenic Research - Integration of Experimental and Observational Studies and Implications for Mitigation*; Oday, P. A., Vlassopoulos, D., Meng, Z., Benning, L. G., Eds.; ACS Symposium Series; American Chemical Society: Washington, DC, 2005; Vol. 915, pp 398−413.
34. Beak, D. G.; Wilkin, R. T. Performance of a zerovalent iron reactive barrier for the treatment of arsenic in groundwater: Part 2. Geochemical modeling and solid phase studies. *J. Contam. Hydrol.* **2009**, *106* (1−2), 15–28.
35. Kaegi, R.; Voegelin, A.; Folini, D.; Hug, S. J. Effect of phosphate, silicate, and Ca on the morphology, structure and elemental composition of Fe(III)-precipitates formed in aerated Fe(II) and As(III) containing water. *Geochim. Cosmochim. Acta* **2010**, *74* (20), 5798–5816.
36. Voegelin, A.; Kaegi, R.; Frommer, J.; Vantelon, D.; Hug, S. J. Effect of phosphate, silicate, and Ca on Fe(III)-precipitates formed in aerated Fe(II)- and As(III)-containing water studied by X-ray absorption spectroscopy. *Geochim. Cosmochim. Acta* **2010**, *74* (1), 164–186.
37. Hug, S. J.; Leupin, O. Iron-catalyzed oxidation of arsenic(III) by oxygen and by hydrogen peroxide: pH-dependent formation of oxidants in the Fenton reaction. *Environ. Sci. Technol.* **2003**, *37* (12), 2734–2742.
38. Roberts, L. C.; Hug, S. J.; Ruettimann, T.; Billah, M.; Khan, A. W.; Rahman, M. T. Arsenic removal with iron(II) and iron(III) waters with high silicate and phosphate concentrations. *Environ. Sci. Technol.* **2004**, *38* (1), 307–315.
39. Berg, M.; Luzi, S.; Trang, P. T. K.; Viet, P. H.; Giger, W.; Stuben, D. Arsenic removal from groundwater by household sand filters: Comparative field study, model calculations, and health benefits. *Environ. Sci. Technol.* **2006**, *40* (17), 5567–5573.
40. Hug, S. J.; Leupin, O. X.; Berg, M. Bangladesh and Vietnam: Different groundwater compositions require different approaches to arsenic mitigation. *Environ. Sci. Technol.* **2008**, *42* (17), 6318–6323.
41. Hussam, A.; Munir, A. K. M. A simple and effective arsenic filter based on composite iron matrix: Development and deployment studies for groundwater of Bangladesh. *J. Environ. Sci. Health, Part A: Toxic/Hazard. Subst. Environ. Eng.* **2007**, *42* (12), 1869–1878.
42. Manning, B. A.; Hunt, M. L.; Amrhein, C.; Yarmoff, J. A. Arsenic(III) and arsenic(V) reactions with zerovalent iron corrosion products. *Environ. Sci. Technol.* **2002**, *36* (24), 5455–5461.
43. Coker, V. S.; Gault, A. G.; Pearce, C. I.; Van Der Laan, G.; Telling, N. D.; Charnock, J. M.; Polya, D. A.; Lloyd, J. R. XAS and XMCD evidence for species-dependent partitioning of arsenic during microbial reduction of ferrihydrite to magnetite. *Environ. Sci. Technol.* **2006**, *40* (24), 7745–7750.

44. Miretzky, P.; Cirelli, A. F. Remediation of Arsenic-Contaminated Soils by Iron Amendments: A Review. *Crit. Rev. Environ. Sci. Technol.* **2010**, *40* (2), 93–115.
45. O'Day, P. A.; Vlassopoulos, D. Mineral-based amendments for remediation. *Elements* **2010**, *6* (6), 375–381.
46. Kocar, B. D.; Polizzotto, M. L.; Benner, S. G.; Ying, S. C.; Ung, M.; Ouch, K.; Samreth, S.; Suy, B.; Phan, K.; Sampson, M.; Fendorf, S. Integrated biogeochemical and hydrologic processes driving arsenic release from shallow sediments to groundwaters of the Mekong delta. *Appl. Geochem.* **2008**, *23* (11), 3059–3071.
47. Polizzotto, M. L.; Harvey, C. F.; Sutton, S. R.; Fendorf, S. Processes conducive to the release and transport of arsenic into aquifers of Bangladesh. *Proc. Natl. Acad. Sci. U.S.A.* **2005**, *102* (52), 18819–18823.
48. Ford, R. G.; Wilkin, R. T.; Hernandez, G. Arsenic cycling within the water column of a small lake receiving contaminated ground-water discharge. *Chem. Geol.* **2006**, *228* (1−3), 137–155.
49. Roberts, L. C.; Hug, S. J.; Dittmar, J.; Voegelin, A.; Kretzschmar, R.; Wehrli, B.; Cirpka, O. A.; Saha, G. C.; Ali, M. A.; Badruzzaman, A. B. M. Arsenic release from paddy soils during monsoon flooding. *Nat. Geosci.* **2010**, *3* (1), 53–59.
50. Farnsworth, C.; Hering, J. G. Inorganic geochemistry and redox dynamics in bank filtration settings. *Environ. Sci. Technol.* **2011**, *45* (12), 5079–5087.
51. Roberts, L. C.; Hug, S. J.; Voegelin, A.; Dittmar, J.; Kretzschmar, R.; Wehrli, B.; Saha, G. C.; Badruzzaman, A. B. M.; Ali, M. A. Arsenic Dynamics in Porewater of an Intermittently Irrigated Paddy Field in Bangladesh. *Environ. Sci. Technol.* **2010**, *45* (3), 971–976.
52. Ahmed, M. F.; Ahuja, S.; Alauddin, M.; Hug, S. J.; Lloyd, J. R.; Pfaff, A.; Pichler, T.; Saltikov, C.; Stute, M.; van Geen, A. Epidemiology - Ensuring safe drinking water in Bangladesh. *Science* **2006**, *314* (5806), 1687–1688.
53. Mosler, H. J.; Blochliger, O. R.; Inauen, J. Personal, social, and situational factors influencing the consumption of drinking water from arsenic-safe deep tubewells in Bangladesh. *J. Environ. Manage.* **2010**, *91* (6), 1316–1323.
54. Ghosh, A.; Mukiibi, M.; Ela, W. TCLP underestimates leaching of arsenic from solid residuals under landfill conditions. *Environ. Sci. Technol.* **2004**, *38* (17), 4677–4682.
55. Winkel, L.; Trang, P. T. K.; Lan, V. M.; Stengel, C.; Amini, M.; Ha, N. T.; Viet, P. H.; Berg, M. Arsenic pollution of groundwater in Vietnam exacerbated by deep aquifer exploitation for more than a century. *Proc. Natl. Acad. Sci. U.S.A.* **2011**, *108* (4), 1246–1251.

Chapter 22

Redox Processes Affecting the Speciation of Technetium, Uranium, Neptunium, and Plutonium in Aquatic and Terrestrial Environments

Edward J. O'Loughlin,[1,*] Maxim I. Boyanov,[1] Dionysios A. Antonopoulos,[1,2] and Kenneth M. Kemner[1]

[1]Biosciences Division, Argonne National Laboratory, 9700 South Cass Avenue, Argonne, IL 60439
[2]The Institute for Genomics and Systems Biology, Argonne National Laboratory, 9700 South Cass Avenue, Argonne, IL 60439
*oloughlin@anl.gov

Understanding the processes controlling the chemical speciation of radionuclide contaminants is key for predicting their fate and transport in aquatic and terrestrial environments, and is a critical consideration in the design of nuclear waste storage facilities and the development of remediation strategies for management of nuclear legacy sites. The redox processes that influence the chemical speciation, and thus mobility, of Tc, U, Np, and Pu in surface and near-subsurface environments are reviewed, with a focus on coupled biotic-abiotic reactions driven by microbial activity. A case study of U^{VI} reduction under Fe^{III}- and sulfate-reducing conditions is presented, using a laboratory-based experimental system to simulate potential electron transfer pathways in natural systems. The results suggest that U^{VI} was reduced to nanoparticulate uraninite (UO_2) and complexed mononuclear U^{IV} via multiple pathways including direct microbial reduction and coupled biotic-abiotic processes. These results highlight the potential importance of coupled biotic-abiotic processes in determining the speciation and mobility of Tc, U, Np, and Pu in natural and engineered environments.

Introduction

Past defense-related activities at government facilities — in particular, activities associated with the mining, extraction, and processing of uranium for nuclear fuel and weapons — have resulted in the introduction of a broad range of chemical contaminants into the environment, including organic compounds, heavy metals, and radionuclides. As an example, environmental contamination due to past waste disposal practices has been identified at nearly 10,000 discrete locations within the vast network of United States Department of Energy (DOE) facilities and waste sites and has resulted in an estimated 75 million m^3 of contaminated soil and nearly 1,800 million m^3 of contaminated groundwater (*1*), with total life cycle cleanup costs estimated in excess of $200 billion over 70 y (*2*). As a group, radionuclides represent the largest fraction of contaminants in soil and groundwater at DOE sites (*1*), and nearly 70% of DOE facilities report groundwater contamination by radionuclides (*3*). At most of these sites the major risk drivers are the long-lived, mobile radionuclides that may pose risks to humans or the environment, such as ^{60}Co, ^{90}Sr, ^{129}I, ^{99}Tc, ^{137}Cs, ^{232}Th, ^{235}U, ^{238}U, ^{241}Np, ^{238}Pu, ^{239}Pu, ^{241}Am, and ^{243}Am (*3*); of these I, Tc, U, Np, and Pu are highly redox active within the range of conditions in typical surface and near-subsurface environments (Figure 1).

At most sites, the presence of U and other radionuclides with potentially high mobility in the subsurface (typically species with high aqueous solubility) can lead over time to highly dispersed contaminant plumes. Although excavation followed by reclamation is the preferred approach for remediation of radionuclide contamination of soils/sediments, the large volume of contaminated material at many sites makes this approach impractical both logistically and economically; thus the need to develop *in situ* remediation approaches. Techniques suitable for *in situ* remediation of radionuclides typically involve stabilization/immobilization by abiotic processes (via introduction of reactive chemical agents into the subsurface) or biotic (microbially mediated) processes. For radionuclides such as Tc, U, Np, and Pu, the chemical species that are stable in oxic environments (e.g., TcO_4^-, UO_2^{2+}, NpO_2^+, PuO_2^+, and PuO_2^{2+}) are generally thought to be more soluble/mobile than the more reduced species (e.g., TcO_2, UO_2, NpO_2, and PuO_2) (*4, 5*). Thus, conversion from an oxidized/potentially mobile form to a more reduced/potentially less mobile form is seen as an attractive approach for *in situ* stabilization of Tc, U, Np, and Pu — the radionuclides that are the focus of this paper. However, many key factors controlling the redox reactions affecting the fate and transport of Tc, U, Np, and Pu are not fully understood, particularly with respect to the interplay of relevant biological and geochemical processes.

Technetium

Technetium is the lightest of the elements that have no stable isotopes; all 25 Tc isotopes (and numerous nuclear isomers) are radioactive with half-lives ranging from < 1 s to 4.2×10^6 y (^{98}Tc). Since the age of Earth is ca. 4.5×10^9 y, any Tc present in the primordial Earth has long since decayed below detectable levels. Minute amounts of natural ^{99}Tc have been identified in high-grade U ores ($\sim 10^{-10}$

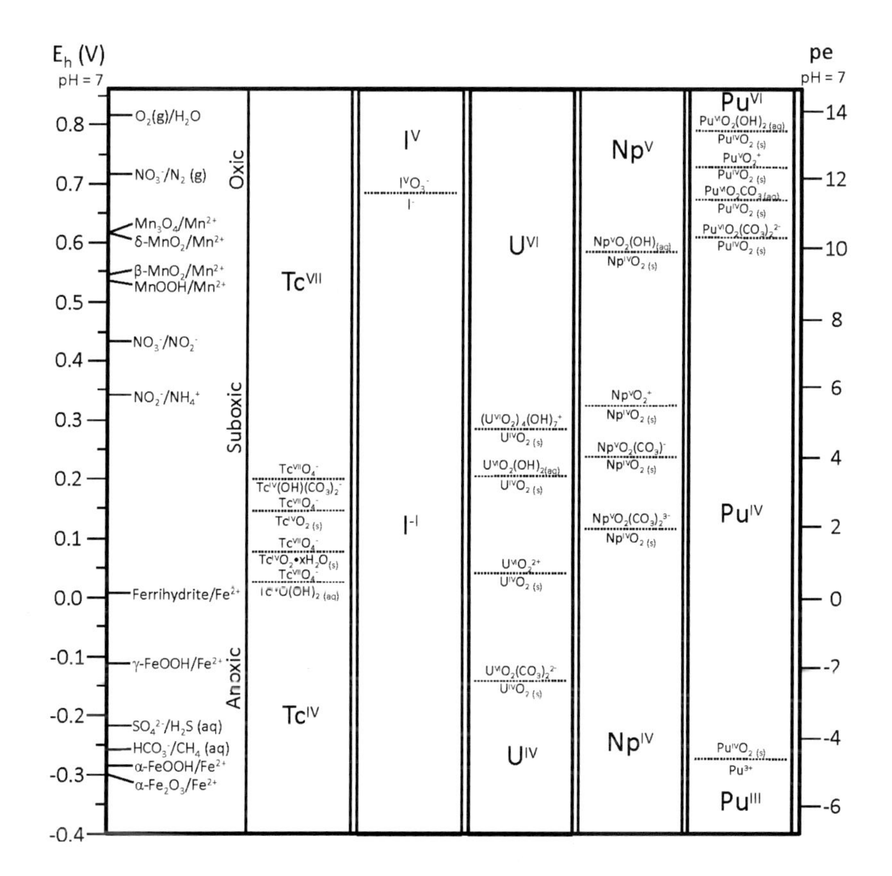

Figure 1. Comparison of redox potentials of dominant biogeochemical electron donor-acceptor couples with redox potentials of select environmentally relevant electron donor-acceptor couples of redox active elements having radionuclides that are major risk drivers at DOE sites. Reduction potentials of the dominant biogeochemical electron-acceptor couples were obtained from Langmuir (260) and Thamdrup (261) under the conditions identified therein. Reduction potentials for I, Tc, U, Np, and Pu electron-acceptor couples are from (262, 263) or calculated from thermodynamic data in (206, 264) under the following conditions: concentrations of soluble I, Tc, U, Np, and Pu species were set to 1 μM; total carbonate concentration was 1 mM; I = 0; and T = 298.15 K.

g Tc g^{-1} ore) as a result of spontaneous fission of ^{238}U, neutron-induced fission of ^{235}U, and possibly by neutron capture by ^{98}Mo in geologic environments with high concentrations of both U and Mo (*6–8*). Because of its extremely low natural abundance, the presence of Tc in the environment is essentially the result of the release of ^{99}Tc (a fission product of ^{235}U and ^{239}Pu with a half-life of 2.13×10^5 y) in radioactive wastes and in the fallout of fission bomb explosions.

Although Tc can exist in -1, 0, +3, +4, +5, +6, and +7 valence states, +4 and +7 are the most stable under conditions relevant to typical surface and near-subsurface environments, (Figure 1). In oxic to moderately suboxic environments Tc^{VII} is stable as pertechnetate (TcO_4^-), which is highly soluble, sorbs weakly to geological materials, and is not known to form significant aqueous complexes or discrete concentration-limiting phases under environmental conditions. As pertechnetate, Tc^{VII} is therefore highly mobile in aquatic and terrestrial systems. In anoxic environments, Tc^{IV} is stable and may be present in a variety of sparingly soluble forms including sorbed octahedral TcO_2 monomer-tetramer chains (*9–11*), $TcO_2 \cdot nH_2O/TcO_2$ (*12–15*), and TcS_2-like phases (*16*, *17*). There are also indications that Tc^{IV} may be incorporated into Fe^{III} oxides (*18–20*) and siderite (*21*). In these forms, Tc^{IV} is considered to be relatively immobile; however, Tc^{IV} can also form dissolved complexes with ligands such as carbonate (*22*, *23*) and humic substances/dissolved organic carbon (DOC) (*14*, *24–27*), potentially facilitating transport.

Uranium

Uranium is an element that has no stable isotopes, but (unlike Tc, Np, and Pu) U from natural sources is relatively abundant on Earth, with an average concentration of 2-4 ppm in Earth's crust and in excess of 10% in some high-grade U ores. All of the 23 isotopes of U are radioactive, having half-lives ($t_{½}$) ranging from 1 ns to 4.5×10^9 y (^{238}U) (*28*). The three most abundant isotopes of U include primordial ^{238}U (99.28% natural abundance), primordial ^{235}U (0.72%, $t_{½} = 7.0 \times 10^8$ y), and ^{234}U (0.005%, $t_{½} = 2.5 \times 10^5$ y), a decay product of ^{238}U. Uranium combines readily with other elements and is an essential structural constituent of over 200 distinct mineral phases (*29*, *30*). In addition to natural background levels, U is present in the environment as a consequence of activities associated with U ore mining and processing of U for use as nuclear fuel or in nuclear weapons, as well as the transport and storage of nuclear wastes generated as a result of these activities. In addition to the naturally abundant U isotopes, nuclear wastes can also contain significant amounts of ^{232}U, ^{233}U, and ^{236}U depending on the source. Additional inputs of U to the environment come from the application of phosphate fertilizers to agricultural lands (*31*) and the use of depleted U (in which the abundance of ^{235}U has been reduced to below 0.3%) for high-penetration projectiles.

Although U can exist in 0, +2, +3, +4, +5, and +6 valence states, +4 and +6 are the most relevant environmentally (Figure 1). Pentavalent U can exist under environmental conditions; however it is generally unstable and readily disproportionates to U^{VI} and U^{IV}, although it is occasionally observed as a transient or even stable species in certain systems (*32–34*). Hexavalent U is stable in oxic environments and as the uranyl cation (UO_2^{2+}), and its hydroxy, carbonato, and organic ligand complexes are relatively soluble (*35*) and thus potentially mobile. Under reducing conditions, U^{IV} is stable and is typically present as sparingly soluble minerals (e.g., UO_2 and U_3O_8), but it can also occur in solid-phase U^{IV} complexes, such as U^{IV} associated with Fe oxides or sorbed mononuclear U^{IV} (*36–42*). Although this low solubility tends to limit its mobility, U^{IV} may be mobilized via formation of soluble complexes with DOC (*43–47*).

Neptunium

Neptunium, the first of the transuranic elements, has 22 known isotopes/isomers, all of which are radioactive (*48*). The half-lives for Np range from < 100 ms to 2.1 × 10^6 y (^{237}Np), thus primordial Np no longer remains on Earth. However, trace amounts of ^{237}Np have been identified in U ore (*49*, *50*) where it is formed by capture of neutrons originating from U decay. Given its extremely low natural abundance, the presence of Np in the environment is due largely to atmospheric and underground nuclear weapons testing. However, Np is a byproduct of Pu production and localized Np contamination has occurred in conjunction with the handling and storage of high-level radioactive wastes resulting from the processing of spent nuclear fuel. In addition, trace amounts of Np may enter the environment from disposal of household smoke detectors with ionization chambers that contain ^{237}Np formed from the decay of ^{241}Am (*51*).

Neptunium exists in 0, +2. +3, +4, +5, +6, and +7 oxidation states, of which only the +4 and +5 valence states are relevant in typical surface and near-subsurface environments (Figure 1). Unlike U^V, pentavalent Np is highly stable under oxic to moderately suboxic conditions at environmentally relevant pH values (~ pH 4–9) and is typically present as the neptunyl cation NpO_2^+ or neptunyl carbonato complexes (*52*, *53*), which are relatively soluble and tend not to be strongly sorbed to common minerals (*54*, *55*). Therefore, Np^V is potentially mobile. Tetravalent Np is stable in anoxic environments; the fact that Np^{IV} solids such as $Np(OH)_4/NpO_2 \cdot nH_2O/NpO_2$ are less soluble than typical Np^V solids as well as the propensity for Np^{IV} to form strong surface complexes with a variety of mineral phases, tends to limit the mobility of Np^{IV} relative to Np^V (*56*). However, although thermodynamics predicts $NpO_2 \cdot nH_2O/NpO_2$ as the stable solid under anoxic conditions typical of natural surface and near subsurface environments, $NpO_2 \cdot nH_2O/NpO_2$ has not been unambiguously identified as a solid phase Np^{IV} species resulting from the reduction of Np^V under environmentally relevant conditions (*56*), and complexation of Np^{IV} by organic ligands may limit reductive precipitation (*57*, *58*). Moreover, the formation of colloidal metastable Np^{IV} hydroxy polymers alone, or as colloidal Np^{IV}-humic complexes, can enhance Np mobility (*4*, *59*).

Plutonium

Plutonium is the second transuranic element. Twenty isotopes and 8 meta states of Pu have been characterized. All of these are radioactive, with half-lives ranging from 1 s to 8.1 × 10^7 y (^{244}Pu) (*60*). Because of its relatively long half-life, traces of primordial ^{244}Pu have been identified on Earth (*61*). In addition, traces of natural ^{239}Pu resulting from the decay of ^{238}U have been identified in U ores (*6*, *7*, *49*, *62*), and traces of ^{238}Pu formed from the double beta decay of ^{238}U have been identified in a natural U sample (*63*). Plutonium is produced in larger quantities than any other synthetic element; currently > 2000 metric tons exist throughout the world (*64*), the majority being ^{239}Pu. Because of its extremely low natural abundance, the presence of Pu in the environment is effectively due to nuclear weapons testing and accidental releases during activities associated with nuclear weapons production and spent nuclear fuel processing. With their relative abundance and long half-lives, ^{238}Pu ($t_{½}$ = 87.7 y), ^{239}Pu ($t_{½}$ = 24,100 y), and ^{240}Pu ($t_{½}$ = 6,560 y) are considered to be of the greatest concern from the perspective of environmental contamination (*60*).

Pu can exist in 0, +2, +3, +4, +5, +6, and +7 valence states but +3, +4, +5, and +6 are most relevant for natural environments (Figure 1). The redox chemistry of Pu in natural systems is much more complex than that of Tc, U, or Np and and in-depth discussion of this topic is beyond the scope of this paper; however, many comprehensive reviews of the subject are available (*60, 65–69*). Given the number of Pu valence states that are stable within the Eh-pH range of typical surface and near-subsurface environments, the redox speciation of Pu in any particular system is controlled by multiple factors including Eh, pH, ionic strength, concentrations of organic and inorganic ligands, and the kinetics of disproportionation reactions of Pu^{V} and Pu^{IV} (*67, 70–72*). The speciation of Pu is further complicated because Pu often coexists in multiple valence states in natural systems. In brief, Pu^{III} and Pu^{IV} are generally less stable than Pu^{V} and Pu^{VI} under oxic conditions, within a pH range typical of most terrestrial and aquatic environments (i.e., ~ pH 4–9). Pu^{III} and Pu^{IV} species also tend to have lower solubility — $Pu(OH)_4/PuO_2 \cdot nH_2O/PuO_2$ in particular — and hence potentially lower mobility than Pu_V and Pu_{VI} species (*65, 71–73*). However, Pu^{IV} can form intrinsic colloids of polymeric Pu^{IV} hydroxide or $Pu(OH)_4/PuO_2 \cdot nH_2O/PuO_2$ and also readily forms complexes with organic and inorganic colloids; this may increase overall Pu mobility in some environments (*74–79*). The complex redox chemistry of Pu in aquatic and terrestrial systems makes predicting Pu fate and transport particularly challenging (*68, 73, 80, 81*).

Abiotic Redox Processes

Reduction Reactions

Typical suboxic and anoxic environments contain a suite of potential reductants — including Fe^{II} species, S^{-II} species, and reduced natural organic matter (NOM) — for the reduction of Tc^{VII}, U^{VI}, Np^{V}, and $Pu^{V,VI}$ to lower, and generally less soluble, oxidation states.

Ferrous iron (Fe^{II}), one of the most abundant reductants typically present in aquatic and terrestrial environments under suboxic and anoxic conditions (*82–84*), is an effective reductant for a wide range of organic and inorganic contaminants; however, Fe^{II} redox reactivity strongly depends on Fe^{II} speciation (see Chapter 14). The homogeneous reduction of Tc^{VII}, U^{VI}, Np^{V}, and Pu^{VI} by $Fe^{2+}_{(aq)}$ is highly variable, ranging from extensive reduction of Pu^{VI} to Pu^{IV} (*85, 86*) to limited or no reduction of Tc^{VII}, U^{VI}, and Np^{V} (*20, 37, 87–90*). Given the range of conditions in these experimental systems, it is unclear whether the extent of reduction is due to thermodynamic or kinetic constraints, although limited to no apparent reduction has been reported under conditions that were ostensibly thermodynamically favorable for reduction (*87–89*). In addition, assessment of homogeneous reduction by $Fe^{2+}_{(aq)}$ is often complicated by the formation of particulate phases as the reaction proceeds, thus providing heterogeneous reaction pathways (*20*). Indeed, Fe^{2+} species sorbed to a variety of surfaces including carboxyl-functionalized polystyrene beads; Fe^{III}, Al^{III}, and Si^{IV} oxides; phyllosilicate clays; and natural sediments can be effective reductants for U^{VI} and Tc^{VII} (*11*, *12*, *20*, *37*, *39*, *87–89*, *91–93*).

In surface and near-subsurface environments, Fe^{II} is typically present as both a major and minor structural component of various mineral phases. Green rust, a mixed $Fe^{II/III}$ layered hydroxide, is a particularly effective reductant for Tc^{VII}, U^{VI}, and Np^{V} (*94–97*). Magnetite (Fe_3O_4), another mixed $Fe^{II/III}$ phase, reduces Tc^{VII}, Np^{V}, and Pu^{V} (and presumably Pu^{VI}) (*14*, *90*, *98–100*). However, the effectiveness of magnetite as a reductant for U^{VI} is highly variable, ranging from no apparent reduction to complete reduction to U^{IV} (*32*, *42*, *91*, *92*, *96*, *101–104*). These differences in magnetite reactivity may be due, in part, to differences in the Fe^{II}:Fe^{III} ratio of the magnetites used in these studies, as the effectiveness of magnetite as a reductant decreases with decreasing Fe^{II} content (*105*, *106*). Compared to green rust and magnetite, siderite ($FeCO_3$) is typically not reported to be an effective reductant for common environmental contaminants (*107–109*); however, there is evidence for limited reduction of U^{VI} in the presence of siderite (*96*, *110*). Reports of U^{VI} reduction by vivianite [$Fe_3PO_4)_2$] are mixed; no reduction by abiotically formed vivianite (*42*, *111*) but complete reduction to U^{IV} by biogenic vivianite (*42*). Structural Fe^{II} in phyllosilicate minerals is another potential source of reducing equivalents. Chlorite, mica (celadonite), nontronite (NAu-2), and a mixture of clays (vermiculite, illite, and muscovite) have been shown to reduce Tc^{VII} to Tc^{IV} (*5*, *10*, *11*, *98*). Structural Fe^{II} in reduced NAu-2 nontronite does not reduce U^{VI} (*112*), but mica (biotite) reduces U^{VI} to U^{V} (*113*), with the pentavalent state stabilized on the mica surface (*33*). In addition, Fe^{II}-bearing minerals in granite and basalt reduce Tc^{VII} and Np^{V} (*114*).

In anoxic environments, sulfidogenic conditions can lead to the accumulation of significant levels of reduced sulfur species, sulfides in particular. Dissolved sulfides (H_2S, HS^-, and S^{2-}) are highly reactive and efficient at scavenging divalent transition metals by forming sparingly soluble metal sulfides (FeS, HgS, ZnS, etc.). Many sulfide species (both dissolved and solid phases) can be effective reductants for a range of environmental contaminants. Dissolved sulfide reduces Pu^{VI} to Pu^{V} under neutral to mildly alkaline conditions (*115*). Similarly, laboratory studies indicate that dissolved sulfide reduces U^{VI} to UO_2, with the kinetics of U^{VI} reduction largely controlled by the pH and carbonate concentration (*116*, *117*); however, U^{VI} has been shown to be highly stable in anoxic marine environments with high dissolved sulfide concentrations such as the Black Sea (*118*). In contrast, Tc^{VII} is reported to precipitate primarily as $Tc^{VII}{}_2S_7$ in the presence of dissolved sulfide, with only minor amounts (> 10%) of TcO_2 observed (*119*). However, extensive reduction of Tc^{VII} is reported during mackinawite precipitation and in the presence of preformed mackinawite (FeS), greigite (Fe_3S_4), or pyrite (FeS_2) (*13*, *16*, *17*, *26*, *120*), resulting in the formation of either a $Tc^{IV}S_2$-like phase or $TcO_2 \cdot nH_2O/TcO_2$. Reduction of U^{VI} to UO_2 and U_3O_8 is reported for many sulfide minerals including mackinawite, pyrite, and galena (PbS) (*16*, *116*, *121–126*). Similarly, Np^{V} is reduced by mackinawite, forming a Np^{IV} surface complex (*16*, *127*).

Living biomass aside, the pool of NOM in typical surface and near-subsurface environments consists primarily of low-molecular-mass components (< 1 kDa) that usually include fatty acids and free sugars and amino acids, along with high-molecular-mass (> 1 kDa) components that consist largely of humic substances with minor amounts of soluble proteins, carbohydrates, and other non-humic macromolecules (*128*). These components of NOM may be dissolved or surface-associated, or they may constitute discrete bulk solid phases. Many constituents of the NOM pool can provide reducing equivalents for geochemically important redox reactions. Low-molecular-mass aliphatic acids including acetate, oxalate, lactate, succinate, and citrate can reduce Pu^{VI} to Pu^{V}, but not U^{VI} (*86*). Humic acids readily reduce Pu^{VI} and Pu^{V} to Pu^{IV} (*129–133*), but reduction of Np^{V} is only observed under acidic conditions (pH 4.7, but not pH 7.4) (*133*, *134*) and reduction of U^{VI} requires above-ambient temperatures (120–400 °C) (*135*). The redox activity of humic substances is largely attributed to quinoid groups within their structures (see Chapters 6 and 7), consistent with the enhanced reduction of Pu^{V} and Np^{V} by quinoid-enriched humic acids (*133*, *134*). Anthrahydroquinone-2,6-disulfonate (AH_2QDS), the reduced form of 9,10-anthraquinone-2,6-disulfonate (AQDS, a synthetic quinone that has been used extensively as a model for the redox activities of quinone groups in humic substances) reduces U^{VI} to U^{IV} and PuO^2 to Pu^{III} (*88*, *136*, *137*). Similarly, the reduced form of flavinmononucleotide (FMN, a quinone-containing biomolecule) can reduce U^{VI} to U^{IV} (*138*).

Although it is not typically found naturally in surface and near-subsurface environments, Fe^0 has been investigated extensively in laboratory and field-scale studies as a potential material for construction of *in situ* permeable reactive barriers. This use reflects the ability of Fe^0 to reductively transform a wide range of organic and inorganic contaminants, including Tc^{VII}, U^{VI}, and Pu^{VI} (*85*, *139–145*).

Oxidation Reactions

The precipitation of reduced species of Tc, U, Np, and Pu from groundwater can limit the migration of these contaminants; however, many laboratory- and field-based studies suggest that these contaminants may be remobilized in the presence of suitable oxidants (*5*, *146–152*). In typical surface and near-subsurface environments, O_2, $Mn^{III,IV}$, and Fe^{III} species are generally the most abundant potential oxidants for Tc^{IV}, U^{IV}, Np^{IV}, and $Pu^{III,IV,V}$ species, but relatively little is known regarding the rates, mechanisms, and specific geochemical conditions under which reduced Tc, U, Np, and Pu species are oxidized.

By definition, O_2 is an abundant potential oxidant for reduced Tc, U, Np, and Pu species in oxic environments. Uraninite, a commonly observed product of abiotic and microbial U^{VI} reduction, is readily oxidized to U^{VI} by O_2 (*37*, *153–157*), although the rate of uraninite oxidation is decreased by mono- and divalent cation impurities (*157*, *158*). The stability of other relevant U^{IV} forms (e.g., mononuclear complexes) in the presence of O_2 is largely unknown. In addition, several processes have been identified that can limit the rate and perhaps the overall extent of U^{IV} oxidation by O_2, including redox buffering by stoichiometric excesses of alternate electron donors (e.g., Fe^{II} or sulfide phases)

(*159*, *160*) or constraints on O_2 diffusion by compartmentalization of U^{IV} (*161*, *162*).

The oxidation of Tc^{IV} phases by O_2 appears to be highly variable. Bulk $TcO_2 \bullet nH_2O/TcO_2$ is readily oxidized to Tc^{VII} (*10*, *20*), while Tc^{IV} in association with Fe^{III} is resistant to oxidation by O_2 (*5*, *20*, *97*), and the TcS_2-like phase that forms during the reduction of Tc^{VII} by mackinawite is converted to a stable TcO_2-like phase in the presence of O_2 (*16*, *17*). The oxidation of Np^{IV} by O_2 is also quite variable, ranging from partial to no oxidation in simple laboratory systems and natural sedimentary materials (*57*, *94*, *152*). In the form of PuO_2, Pu^{IV} is generally quite stable in the presence of O_2 under conditions typical of surface and near-subsurface environments but it may undergo limited oxidation to form a solid-solution PuO_{2+x} ($x \leq 0.06$) phase that accommodates Pu^V (*163*, *164*).

Oxic or even moderately suboxic environments often contain a variety of $Mn^{III,III/IV,IV}$ and Fe^{III} phases that can oxidize reduced Tc, U, Np, and Pu species. Native $Mn^{III/IV}$ oxides in natural sediments reportedly oxidize $TcO_2 \bullet nH_2O$ to Tc^{VII} (*12*). However, not surprisingly, the reactivity is dependent on $Mn^{IV,III}$ speciation; for example, K^+-birnessite [$K_x(Mn^{4+}, Mn^{3+})_2O_4$, where $x < 1$] oxidized Tc, but manganite (MnOOH) did not (*10*). Similarly, δ-MnO_2 oxidizes Pu^{IV} and Pu^V to Pu^{VI} or mixtures of Pu^V and Pu^{VI} (*165–168*), while no net oxidation of Pu^{IV} was observed with pyrolusite (β-MnO_2) (*81*). The oxidation of uraninite by $Mn^{III,III/IV,IV}$ and Fe^{III} phases has received considerable attention in recent years. Bixbyite (Mn_2O_3) and pyrolusite (*161*); Fe^{III} oxides such as ferrihydrite, goethite (α-FeOOH), and hematite (α-Fe_2O_3) (*155*, *169–172*); and structural Fe^{III} in nontronite (*112*) all oxidize UO_2 to U^{VI} with various degrees of effectiveness. Comparative studies examining U^{IV} oxidation by a broad range of $Mn^{III,III/IV,IV}$ and Fe^{III} phases under similar experimental conditions are lacking, therefore it is difficult to assess the trends in the reactivity of Fe and Mn oxides in oxidizing UO_2 or to identify the electron transfer mechanism between uraninite and these metal oxides. However, Ginder-Vogel et al. have observed that ferrihydrite is far more effective at oxidizing biogenic UO_2 than goethite or hematite (*169*) and that the reaction proceeds through a soluble U_{IV} intermediate (*170*).

Biological Redox Processes

Until rather recently, the general belief was that the redox transformations of Tc, U, Np, and Pu in natural systems were the result of abiotic processes. However, over the past 30 y, direct microbial reduction of Tc^{VII}, U^{VI}, Np^V, and $Pu^{IV,V,VI}$ has been unequivocally demonstrated [(*173–178*) and references therein], with microbial U^{VI} reduction being the most extensively investigated. Although the specific factors controlling the microbial reduction of Tc^{VII}, U^{VI}, Np^V, and $Pu^{IV,V,VI}$ are complex (e.g., kinetics of reduction; chemical speciation of Tc^{VII}, U^{VI}, Np^V, and $Pu^{IV,V,VI}$; type and availability of electron donor; species-specific microbial physiology, etc), some general observations can be made.

The ability to reduce U^{VI} to U^{IV} has been reported for a phylogenetically diverse range of microorganisms in the domains *Bacteria* and *Archaea*; however, the majority of identified U^{VI}-reducing microorganisms are dissimilatory Fe^{III}- or sulfate-reducing bacteria (DIRB and DSRB, respectively) within the *γ-Proteobacteria* (e.g., *Shewanella* spp.), *δ-Proteobacteria* (e.g., *Anaeromyxobacter* spp., *Geobacter* spp., and *Desulfovibrio* spp.) and *Firmicutes* (e.g., *Desulfitobacterium* spp., *Desulfosporosinus* spp., and *Desulfotomaculum* spp.) that can couple the reduction of U^{VI} to the oxidation of H_2 or reduced organic matter (*34*, *36*, *40*, *44*, *153*, *179–194*). The microbial reduction of Tc^{VII}, Np^{V}, and $Pu^{IV,V,VI}$ has been less extensively investigated. However, as with U^{VI}, the majority of identified Tc^{VII}, Np^{V}, and $Pu^{VI,V,IV}$ reducers have been DIRB and DSRB (*15*, *57*, *58*, *86*, *184*, *188*, *195–206*). The specific biomolecular mechanisms involved in microbial Tc^{VII}, U^{VI}, Np^{V}, and $Pu^{IV,V,VI}$ reduction have not been well characterized, but they appear to involve periplasmic and membrane-bound reductases (e.g., hydrogenases and *c*-type cytochromes) that are components of electron transport chains involved in anaerobic respiration using Fe^{III} or sulfate as terminal electron acceptors (*15*, *187*, *196*, *200*, *207*, *208*).

The successful application of *in situ* biostimulation for the reductive immobilization of Tc, U, Np, and Pu requires that the reduced Tc, U, Np, and Pu species have lower solubility/mobility than the more oxidized species. Sparingly soluble uraninite, often in the form of nanometer-sized crystals, is commonly observed as a product of the microbial reduction of U^{VI} (*34*, *44*, *153*, *155*, *157*, *179–181*, *184–193*, *209*); however, the presence of strong complexing agents such as citrate, NTA, EDTA, or humic acids can inhibit the precipitation of UO_2 by the formation of soluble U^{IV} complexes (*44*, *194*, *210*, *211*). In the presence of sufficient phosphate, ningyoite [$CaU(PO_4)_2 \cdot H_2O$] and other U^{IV} orthophosphates have been identified (*36*, *182*). Recently, solids-associated mononuclear or molecular U^{IV} species have been reported as products of microbial U^{VI} reduction (*36*, *38*, *40*, *212*). The long-term stability of mononuclear U^{IV} and nanoparticulate uraninite has not been determined.

Microbial reduction of Tc^{VII} typically results in the precipitation of $TcO_2 \cdot nH_2O$ (*15*, *188*, *202*, *206*); however soluble Tc^{IV} carbonate complexes can form if carbonate concentrations are high enough (*206*). Precipitation of uncharacterized Np^{IV} solids during microbial Np^{V} reduction has been observed (*58*), though Np^{IV} can remain in solution in the presence of suitable complexants (*57*, *58*, *203*). As with most aspects of Pu redox chemistry, the products of microbial reduction of $Pu^{IV,V,VI}$ are highly variable. Microbial reduction of Pu^{IV} or Pu^{V} can result in the precipitation of sparingly soluble Pu^{IV} oxide, which can be further reduced to sorbed and dissolved Pu^{III} complexes, depending on the organism and experimental conditions (*86*, *195*, *197*, *204*, *205*, *213*); this variability makes it difficult to generalize with respect to the final distribution of reduced Pu species.

Compared with microbial reduction, direct microbial oxidation of reduced Tc, U, Np, and Pu species has been largely unexplored, and what little is known is based on a limited number of studies focused on microbial oxidation of U^{IV}. *Acidithiobacillus ferrooxidans* (formerly *Thiobacillus ferrooxidans*) is an autotrophic bacterium that can couple the reduction of O_2 to the oxidation of U^{IV} to U^{VI} at relatively low pH (1.5) (*214*). *Thiobacillus denitrificans, Geobacter metallireducens, Geothrix fermentans* HS, *Pseudogulbenkiania* sp. 2002, *Acidovorax ebreus* TPSY, *Pseudomonas* sp. PK, and *Magnetospirillum* sp. VDY oxidize U^{IV} to U^{VI} at circumneutral pH with nitrate as the electron acceptor,

(*215–217*); the nitrate-dependent oxidation of U^{IV} by *G. metallireducens* is particularly interesting, given that it is a well-studied, U^{VI}-reducing bacterium (*180*, *184*, *185*). The oxidation of U^{IV} with either nitrate or O_2 as the terminal electron acceptor is thermodynamically favorable under environmentally relevant conditions and could yield enough energy to support microbial growth; however, a microorganism that can couple U^{IV} oxidation to growth has not yet been identified, although there is strong evidence for the involvement of electron transport chains with U^{IV} serving as the sole electron donor in *A. ferrooxidans* and *Pseudogulbenkiania* sp. 2002 (*214*, *217*).

Coupled Biotic-Abiotic Processes

Coupled Biotic-Abiotic Redox Transformations

By definition, redox transformations entail the transfer of electrons between chemical species. The inherent complexity of natural systems offers multiple pathways for electron transfer with a potential for coupling of biotic and abiotic processes. For example, the major "abiotic" reductants of Tc^{VII}, U^{VI}, Np^{V}, and $Pu^{V,VI}$ in natural systems (i.e., Fe^{II}, S^{-II}, and reduced organic matter) are typically present as products of microbial processes. Indeed, in the last 20-30 y the overarching role of microbial activity in controlling the redox chemistry (and thus the potential for redox transformations of contaminants) of surface and near-subsurface environments has been recognized, if not fully characterized.

Microbial activity can produce a suite of reductants that can reduce contaminants directly and also provide redox buffering to natural systems. The occurrence of Fe^{II} in suboxic and anoxic environments is commonly attributed to the action of DIRB and Fe^{III}-reducing archaea, phylogenetically diverse microorganisms that can obtain energy by coupling oxidation of organic compounds or H_2 to reduction of Fe^{III} to Fe^{II}. The reduction of Fe^{III} by DIRB can yield a suite of Fe^{II} species including soluble Fe^{II} complexes, Fe^{II} complexes with the surfaces of organic and inorganic solid phases, and a host of mineral phases containing structural Fe^{II} (e.g., magnetite, siderite, green rust) (*218–223*). Likewise, DSRB are anaerobes that can obtain energy by coupling the oxidation of organic compounds or molecular H_2 with the reduction of sulfate to sulfide. The production of sulfide by DSRB — and its subsequent reaction with Fe^{2+} resulting from dissimilatory Fe^{III} reduction — leads to the formation of insoluble ferrous sulfides such as mackinawite, greigite, pyrite, and pyrrhotite (*224–229*). This pathway is an important link coupling the Fe and S biogeochemical cycles (*230–232*). Moreover, NOM and humic substances in general can serve as electron acceptors for anaerobic respiration. Indeed, humic substances and low-molecular-mass quinones (e.g., AQDS) can be reduced by a phylogenetically diverse range of microorganisms in the domains *Bacteria* and *Archaea,* including most DIRB and DSRB [(*233*) and references therein].

Microbial activity can also produce many of the most abundant potential oxidants for Tc^{IV}, U^{IV}, Np^{IV}, and $Pu^{III,IV,V}$ species in natural environments. Although most of the O_2 in surface and near subsurface environments is due to bulk diffusion from the atmosphere, in photic zones the activity of photosynthetic microorganisms provides the potential for *in situ* O_2 production. In addition to abiotic oxidation of Fe^{II} and Mn^{II} by O_2 (which may be kinetically limited), Fe^{III} oxides (e.g., ferrihydrite, α-FeOOH, γ-FeOOH) and $Mn^{III,III/IV,IV}$ oxides (e.g., δ-MnO_2, Mn_3O_4, H^+-birnessite) can be produced by direct (enzymatic) microbial activity or non-metabolic biosurface-catalyzed processes with O_2, nitrate, or bicarbonate as electron acceptors (*172*, *234–240*). As discussed previously, many of these Fe^{III} and $Mn^{III,III/IV,IV}$ oxides can effectively oxidize reduced Tc, U, Np, and Pu species.

Recent studies of Tc^{VII}, U^{VI}, Np^{V}, and Pu^{V} reduction by biogenic Fe^{II} and the indirect microbial nitrate-dependent U^{IV} oxidation highlight the potential significance of coupled biotic-abiotic processes in redox transformations of Tc, U, Np, and Pu in natural and engineered environments. For example, the DIRB *Geobacter sulfurreducens* is unable to use acetate as an electron donor for enzymatic reduction of Tc^{VII}; however, acetate is a suitable electron donor for Fe^{III} reduction by this organism. Therefore, by observing the reduction of Tc^{VII} in systems containing acetate and ferrihydrite, Lloyd et al. (*202*) were able to unambiguously demonstrate that *G. sulfurreducens* reduces Tc^{VII} indirectly via the production of Fe^{II} species from the bioreduction of Fe^{III}. Several other studies further demonstrate the potential for Tc^{VII} reduction by biogenic Fe^{II} species under a range of experimental conditions (*12*, *27*, *188*, *241*). Similarly, biogenic Fe^{II} has been shown to reduce Np^{V} (*152*), and Pu^{V} (*86*). However, the potential significance of biogenic Fe^{II} as a reductant for U^{VI} is less clear. In a study of U^{VI} reduction in aqueous suspensions of goethite in the presence of *Shewanella putrefaciens* CN32, Fredrickson et al. (*88*) indicated that reduction of U^{VI} in Fe^{III}-reducing environments may result from coupled biotic-abiotic processes; in addition to direct microbial reduction of U^{VI}, U^{VI} was also reduced, to various extents, by dissolved Fe^{II} complexes and Fe^{II} sorbed on goethite. Similar results were reported by Behrends and Van Cappellen (*242*) for the reduction of U^{VI} in systems containing hematite and *Shewanella putrefaciens* 200. In contrast, a study by Finneran et al. (*243*) suggested that in systems containing reducible Fe^{III}, reduction of U^{VI} by Fe^{II} species is negligible compared to direct microbial reduction. Likewise, results of a study by Jeon et al. (*91*) suggested that accumulated biogenic Fe^{II} in subsurface environments is unlikely to foster long-term abiotic reduction of U^{VI}.

The effects of nitrate on U^{IV} oxidation provide another example of the importance of coupled biotic-abiotic redox processes. Nitrate is not generally an effective oxidant for the oxidation of U^{IV} to U^{VI} (*147*, *216*), however in the presence of nitrate-reducing microorganisms, U^{IV} can be effectively oxidized by nitrite, nitrous oxide, and nitric oxide (*147*, *155*, *172*), which are intermediates in the dissimilatory reduction of nitrate to N_2. Furthermore, Fe^{III} resulting from the oxidation of Fe^{II} by nitrite produced during dissimilatory nitrate reduction or Fe^{III} produced by nitrate-dependent Fe^{II}-oxidizing bacteria can oxidize U^{IV} to U^{VI} (*172*, *244–246*).

A Case Study of Coupled Processes Resulting in U^{VI} Reduction

Recent studies suggest that electron shuttles such as low-molecular-mass quinones and humic substances may play a role in many redox reactions involved in contaminant transformations and the biogeochemical cycling of redox-active elements in aquatic and terrestrial environments. In an effort to better define the role(s) of electron shuttles in biogeochemical processes in natural and engineered subsurface environments, we investigated the effects of AQDS (a synthetic electron shuttle often used as a surrogate for quinone moieties in humic substances) on Fe and S transformations under conditions favorable for dissimilatory Fe^{III} and sulfate reduction and subsequent effects on the reduction of U^{VI}.

The following study provides an illustration of the overall contributions of multiple coupled biotic and abiotic processes leading to the reduction of U^{VI} to U^{IV} under Fe^{III}-reducing and sulfidogenic conditions in the presence and absence of an exogenous electron shuttle.

Experimental System

The experimental systems were sterile 500-mL serum vials containing 400 mL of sterile defined mineral medium (pH 6.8) consisting of silica (72 g L^{-1}), natural sienna (7.6 g L^{-1}; 30 mM Fe^{III}), HEPES buffer (20 mM), PIPES buffer (20 mM), sodium acetate (10 mM), Na_2SO_4 (5 mM), $CaCl_2$ (5 mM), $MgCl_2$ (1 mM), KCl (0.5 mM), NH_4Cl (1 mM), Na_2HPO_4 (10 μM), $NaHCO_3$ (30 mM), and 10 mL L^{-1} of trace minerals solution (*247*), with or without 100 μM AQDS. Iron(III) was provided as natural sienna (Earth Pigments Co.), an iron-rich earth mined from the Ochre deposits in the Provence region of France, consisting primarily of quartz and goethite (α-FeOOH), as determined by powder x-ray diffraction (pXRD) and Fe K-edge extended x-ray absorption fine-structure (EXAFS) spectroscopy (data not shown).

The serum bottles were sparged with Ar and inoculated with 10 g of sediment from a study site in the DOE Old Rifle Integrated Field-Scale Subsurface Research Challenge located at a U mill tailings site in Rifle, Colorado. The suspensions were placed on a roller drum and incubated at 25 °C in the dark. Samples of the suspensions for measurement of Fe^{II} production and acetate and sulfate consumption were collected over time by using sterile syringes. The reduction of Fe^{III} was monitored by the ferrozine assay (*248*) with HEPES-buffered ferrozine reagent (*249*) to measure the total Fe^{II} content of acid extractions (0.75 M HCl) of the suspensions. Sulfate and acetate were measured by ion chromatography on a Dionex 3000 instrument with an IonPac AS11 analytical column (250 × 2 mm, Dionex) and a 1–20mM KOH gradient at a flow rate of 0.5 mL min^{-1}. Unless otherwise indicated, sample collection and processing were conducted in a glove box containing an anoxic atmosphere (95% N_2 with 5% H_2).

After Fe^{II} production reached steady state, aliquots of suspension were removed for Fe EXAFS analysis, microbial community analysis, and U^{VI} reduction experiments. Samples for Fe K-edge EXAFS analysis were prepared under anoxic conditions by filtering 10 mL of suspension through 0.22-μm PTFE filters and sealing the solids remaining on the filter membrane in Kapton film.

For microbial community analysis, 35 mL of mixed suspension was drawn from each sample under anaerobic conditions and allowed to settle overnight. The supernatant was then removed and DNA was extracted from the sediment by using the MOBIO UltraClean™ Mega Prep Soil DNA Kit. Multiple-displacement amplification was performed with phi29 (GenomiPhi V2 DNA Amplification Kit [GE Healthcare]). Amplicon libraries targeting the V4 region of the 16S rRNA encoding gene (563-802) were then constructed by using primers to target members of the *Eubacteria*. Permuted primers containing 8-bp sequences between the 454 Life Sciences A sequence adapter and the 16S rRNA primer sequence on the forward primer were used to sequence multiple libraries within the same run. The 454 Life Sciences Genome Sequencer FLX System with LR70 sequencing chemistry was used to generate sequence data in massively parallel fashion according to the manufacturer's protocols. All sequencing and data generation was performed by the High-throughput Genome Analysis Core of the Institute for Genomics and Systems Biology at Argonne National Laboratory. The 16S rRNA-based V4 amplicon libraries were analyzed with the Ribosomal Database Project's Pyrosequencing Pipeline (RDP-II) (*250*).

For the U^{VI} reduction experiment, one set of suspensions from each system (with and without AQDS) was pasteurized at 70 °C for 1 h; a second set was not pasteurized. The suspensions were spiked with a stock solution of uranyl chloride to achieve a U^{VI} concentration of 500 μM. After 48 h, the pH of each suspension was measured, the suspensions were centrifuged, and the supernatants were saved for determination of solution-phase U^{VI} concentrations by inductively coupled plasma-optical emission spectroscopy. Samples of the hydrated solids were mounted in holes machined in Plexiglas sample holders, covered with Kapton film, sealed with Kapton tape, and maintained under anoxic conditions during analysis by U L_{III}-edge x-ray absorption fine structure (XAFS) spectroscopy.

X-ray absorption near edge structure (XANES) and EXAFS data collection was carried out at sector 10-ID, Advanced Photon Source, Argonne National Laboratory, Illinois (*251*). The Fe K-edge (7,112 eV) and U L_{III}-edge (17,166 eV) XAFS scan procedures, and the data reduction procedures have been published previously (*37*, *252*). Briefly, the beamline undulator was tapered, and the incident photon energy was scanned by using the Si(111) reflection of a double-crystal monochromator in continuous-scanning mode (approximately 3 min per scan for the extended region and 40 s per scan for the near-edge region). Ambient-temperature transmission and fluorescence spectra were obtained by using gas-filled ionization detectors. Heterogeneity in the samples and beam-induced chemical changes were monitored by collecting spectra from 3-6 locations on the sample and by examining successive scans. Energy calibration was maintained at all times by simultaneously collecting a transmission spectrum from a uranyl phosphate mineral or an Fe^0 foil, with x-rays transmitted through the samples.

Over time, the light tan suspensions darkened and eventually turned dark gray. The darkening of the suspensions coincided with the consumption of acetate and sulfate and the production of Fe^{II} (Figure 2), consistent with the formation of FeS. Analysis of Fe K-edge XANES and EXAFS spectroscopy data confirmed the appearance of a minority FeS phase after incubation (Figure 3). Spectral analysis using linear combinations of standards indicated that about 20% of the total solid-phase Fe in the incubated systems was present as FeS and possibly 10% or less as other Fe^{II} species (sorbed Fe^{II}, siderite, etc.). No differences in the final Fe phase compositions were observed in the presence or absence of AQDS. As is often observed in experiments examining the bioreduction of Fe^{III} oxides, the presence of AQDS increased the initial rate but not the overall extent of Fe^{II} production (*223*, *247*). AQDS did not appear to affect the onset or rate of sulfate reduction.

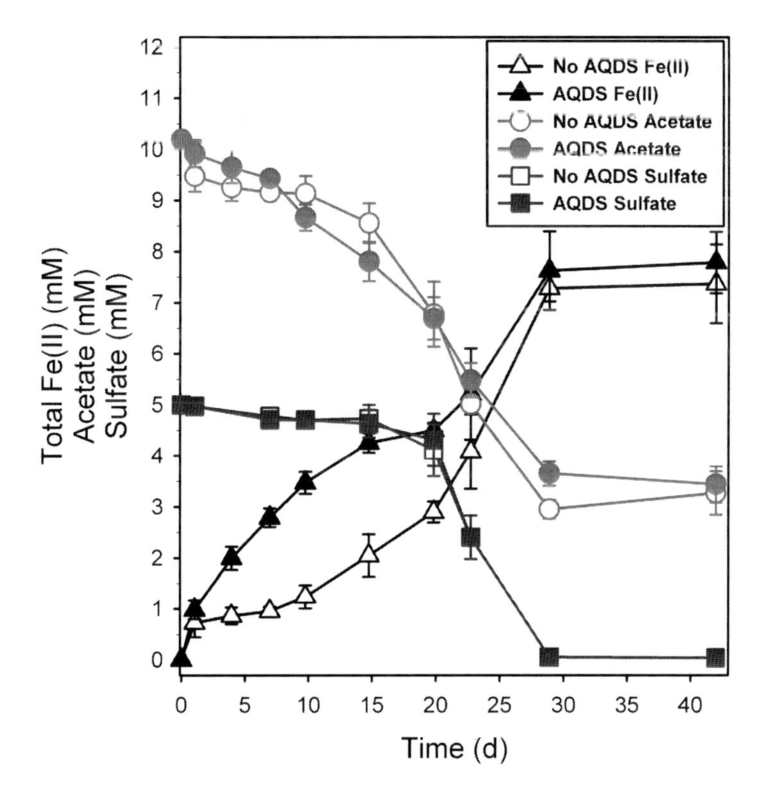

Figure 2. Production of Fe^{II} and consumption of acetate and sulfate in acetate amended microcosms with and without 100 μM AQDS.

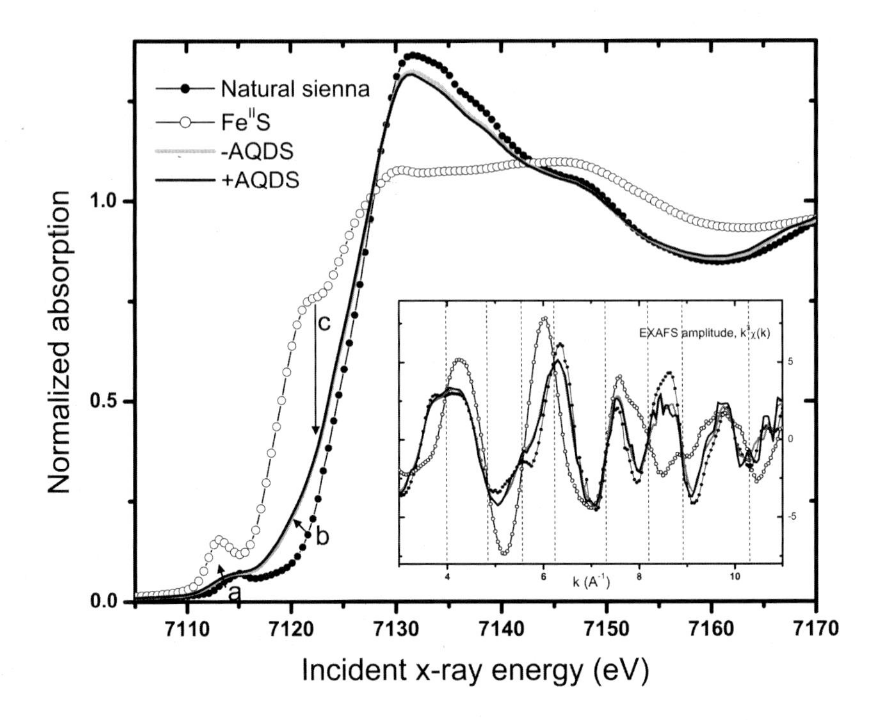

Figure 3. Fe K-edge XANES spectra from the incubations with and without AQDS incubations after they reached steady state, compared to the initial material and to an Fe^{II}S standard. Inset: EXAFS data. Arrow "a" points to the small spectral change from the initial material due to the specific pre-edge feature in Fe^{II}S; arrow "c" points to the edge region where feature "c" in Fe^{II}S contributes to the greater difference "b" between the original and the incubated material. The nearly-isosbestic spectral behavior in the EXAFS data (inset) is noted at the positions of the vertical dashed lines. Between the isosbestic points the spectra from the incubated samples (lines) are intermediate between the spectra from the initial material and the FeS standard (symbols), suggesting the predominance of these two end members in the incubated systems with and without AQDS. Linear combination analysis of the spectra corroborates this interpretation.

At the broadest taxonomic categorization (phylum level), members of the *Proteobacteria* dominated the subsurface sediment used to inoculate the batch cultures (Figure 4). *Sulfuricurvum*-like organisms (*ε-Proteobacteria*) were detected in the seed material at a high abundance that diminished over time in the batch systems with and without AQDS. *Sulfuricurvum* is a member of the *Thiovulum* subgroup of the *ε-Proteobacteria*; presently only one species of *Sulfuricurvum* has been described, *S. kujiense* (*253*). The diversity of this group of organisms became reduced in both biostimulated samples while the relative abundance of *Bacteroides* and *Firmicutes* increased. Specific *Firmicutes* detected included *Desulfitobacterium* spp., *Desulfosporosinus* spp., and *Desulfotomaculum* spp. Greater overall community diversity was observed in the presence of AQDS than in its absence (Figure 5). The greater microbial diversity in the presence of AQDS is perhaps not surprising, given the phylogenetic diversity of known AQDS-reducing microorganisms (*233*). The increase in overall diversity of the AQDS-amended system was also characterized by expansion of the *δ-Proteobacteria* (Figure 4), including *Geobacter* spp. Overall, the increased populations of known DIRB and DSRB observed in these systems are consistent with the types of communities reported to develop following *in situ* acetate or ethanol (which is oxidized to acetate) amendment of subsurface sediments particularly with respect to increases in *δ-Proteobacteria* and *Firmicutes* (*146, 254, 255*).

Within 48 h of the addition of the uranyl chloride stock solution, 100% of the added U was removed from solution in the non-pasteurized AQDS-amended system (Figure 6). However, only 58%, 25%, or 11% of added U was removed in the non-pasteurized no-AQDS, AQDS-amended pasteurized, or no-AQDS pasteurized system, respectively. The final pH of the suspensions ranged from 6.8 to 6.9.

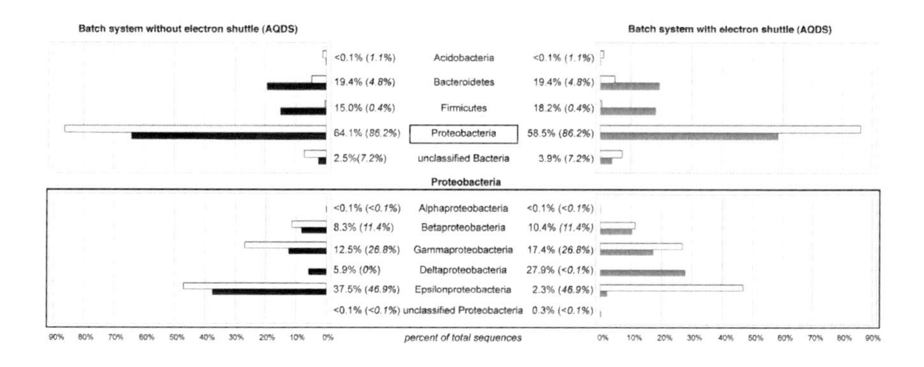

Figure 4. Community composition of biostimulated batch systems with and without the electron shuttle AQDS based on V4 tag sequence analysis. The upper set of bar graphs display the community composition of the batch systems at the phylum level based on the RDP Classifier. In both cases, the relative abundances of the Firmicutes increased from 0.4% in the inoculum. Also observed was a concomitant increase of the Bacteroidetes from 4.8% in the inoculum. In the lower bar graphs (Proteobacteria composition), the relative proportion of ε-Proteobacteria and δ-Proteobacteria are inverted, with the δ-Proteobacteria becoming more dominant in the AQDS-amended batch system.

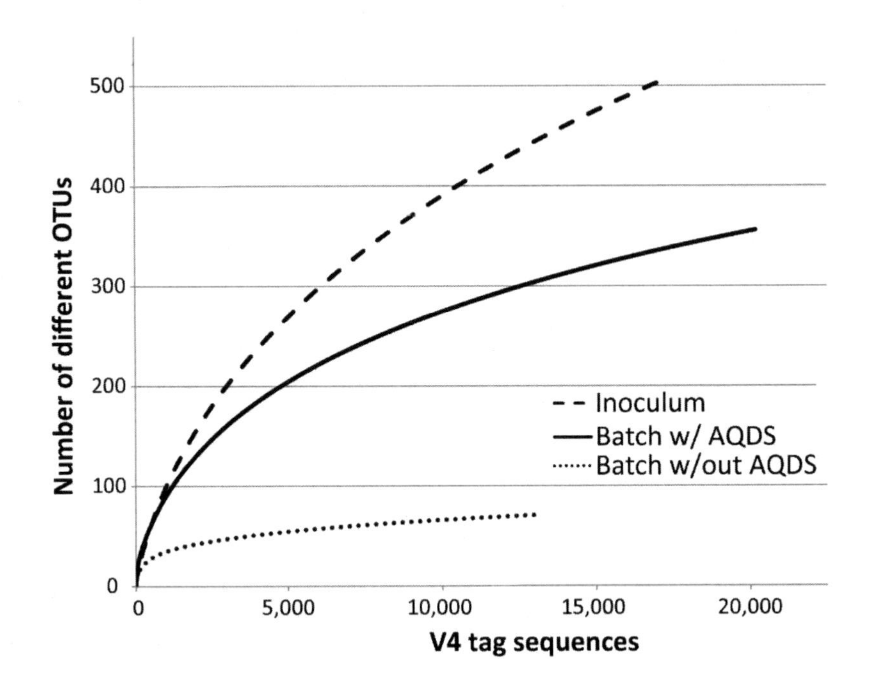

Figure 5. Rarefaction analysis of V4 tag sequences sampled from the inoculum and from biostimulated batch systems with and without the electron shuttle AQDS. Each rarefaction curve represents the number of operational taxonomic units (OTUs; clusters of sequences with >97% similarity in this case) detected based on sampling intensity of the libraries. The inoculum (dashed line) contains the greatest amount of 16S rRNA gene diversity (evidenced by the higher OTU count noted on the y-axis). This diversity is greatly diminished in the batch system without AQDS (dotted line), and increases with the inclusion of the electron shuttle (solid line).

Comparison and linear-combination analysis of the U L_{III}-edge XANES data for the experimental systems with data from U^{VI} and U^{IV} standards indicates that, except for the no-AQDS pasteurized system, most (85-95%) of the solid-phase-associated U was reduced to U^{IV} (Figure 7). The U L_{III}-edge EXAFS data from the solids in the non-pasteurized with and without AQDS systems indicate U^{IV}-U^{IV} coordination at ca. 3.8 Å in the Fourier transform, consistent with the presence of nanoparticulate uraninite (Figure 8). However, the amplitude of the U-U coordination peak is significantly smaller than that of biogenic 2–5-nm diameter uraninite (*153*), suggesting coexistence of the nanoparticulate uraninite with a substantial mononuclear U^{IV} phase or phases that lack U-U coordination, such as adsorbed/complexed or solid-incorporated U^{IV} species (*36*, *40*, *182*, *212*). The various possible U^{VI} reductants in the system appear to create two or more U^{IV} phases that produce the observed average EXAFS spectrum, suggesting that the U^{IV} phases were the result of multiple U^{VI} reduction pathways.

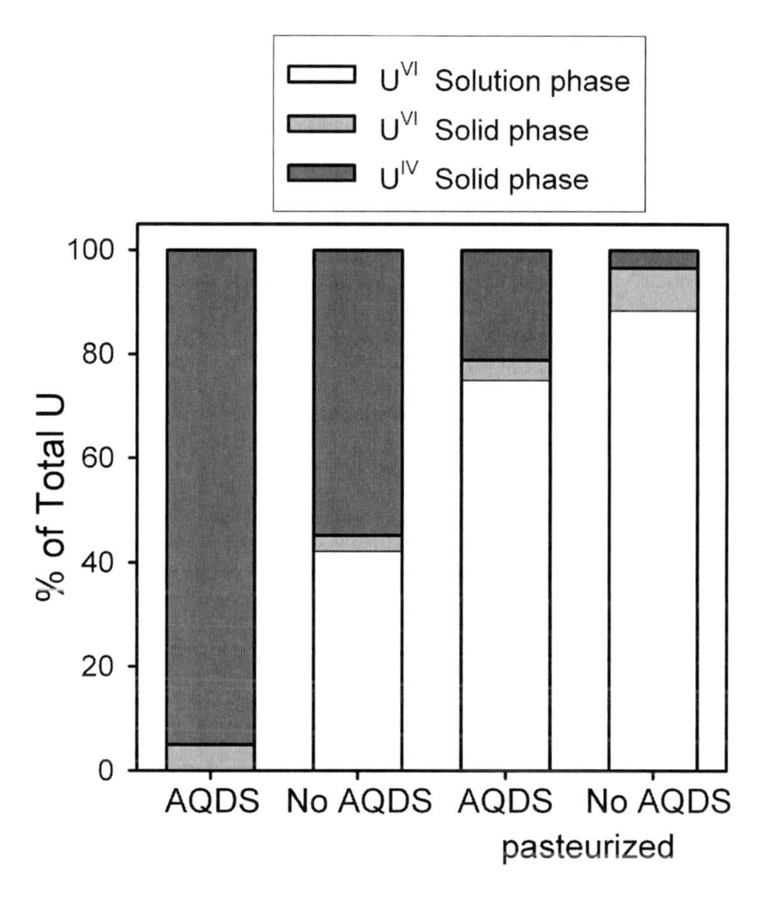

Figure 6. Distribution of U^{VI} and U^{IV} between solution and solid phases in live and pasteurized sediment suspensions with and without AQDS, 48 h after the addition of U^{VI}.

The complexity of our experimental systems provides a number of potential pathways for U^{VI} reduction. After Fe^{III} and sulfate reduction ceased, microbial community analysis of the sediments indicated substantial increases in populations of bacteria within taxa containing known U^{VI}-reducing microorganisms, particularly *δ-Proteobacteria* (*Geobacter*) and *Firmicutes* (*Desulfitobacterium*, and *Desulfosporosinus*), suggesting that direct enzymatic reduction of U^{VI} contributed to the greater accumulation of U^{IV} in non-pasteurized than in pasteurized sediments (Figure 6). In addition, higher levels of U^{IV} were observed in AQDS-amended suspensions than in the corresponding non-AQDS-amended suspensions, a result consistent with the ability of AH_2QDS to reduce U^{VI} (*88*, *137*). However, it is also possible that some of the U^{VI} reduction in the AQDS-amended non-pasteurized system was due to a greater proportion of *δ-Proteobacteria*. AQDS also provides a mechanism by which bacteria that cannot reduce U^{VI} directly, but can reduce AQDS, can contribute to overall U^{VI} reduction through production of AH_2QDS (*256*). As discussed previously, many Fe^{II} species that are products of dissimilatory Fe^{III} reduction can reduce U^{VI} to

U^{IV}; however, the significance of biogenic U^{VI} reduction by biogenic Fe^{II} species in natural systems is unresolved. Similarly, the effects on U^{VI} reduction by sulfide species resulting from dissimilatory sulfate reduction range from essentially no effect to being the dominant pathway (*116*, *257–259*). The observed reduction of < 5% of U^{VI} added to the no-AQDS, pasteurized sediment suggests that "abiotic" reduction of U^{VI} by FeS or other Fe^{II} or S^{-II} species played only a minor role in U^{VI} reduction in our experimental systems.

These results illustrate the importance of multiple pathways and coupled biotic-abiotic processes in the redox transformations of U and the inherent complexity of natural systems.

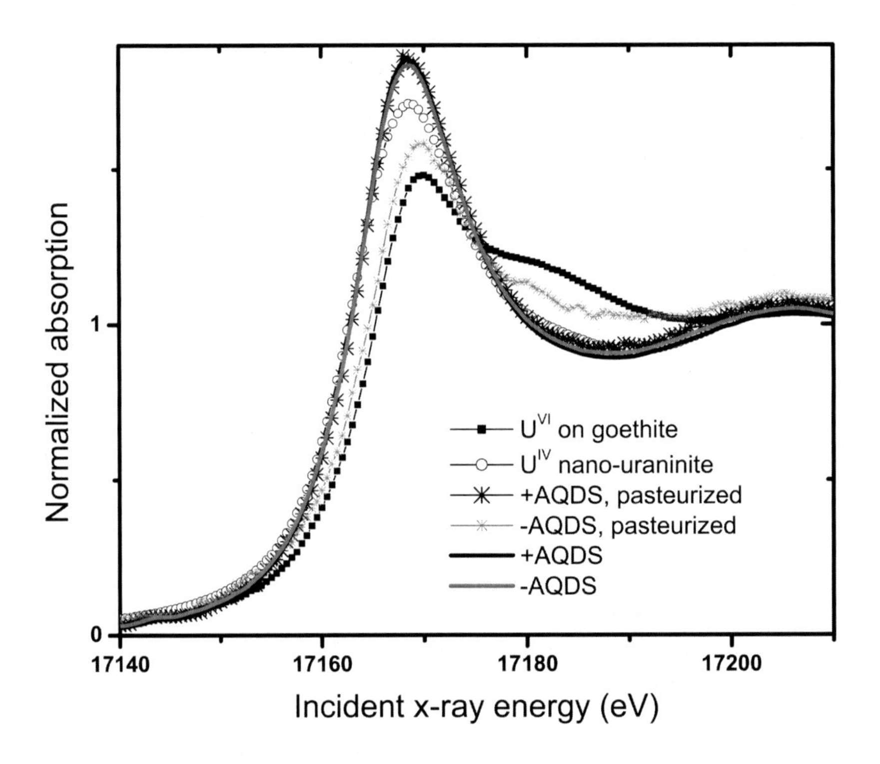

Figure 7. U L_{III}-edge XANES data from the pasteurized and non-pasteurized systems with and without AQDS. Data are compared to U^{IV} and U^{VI} standards, biogenic nanoparticulate uraninite (153) and U^{VI} sorbed to goethite, respectively.

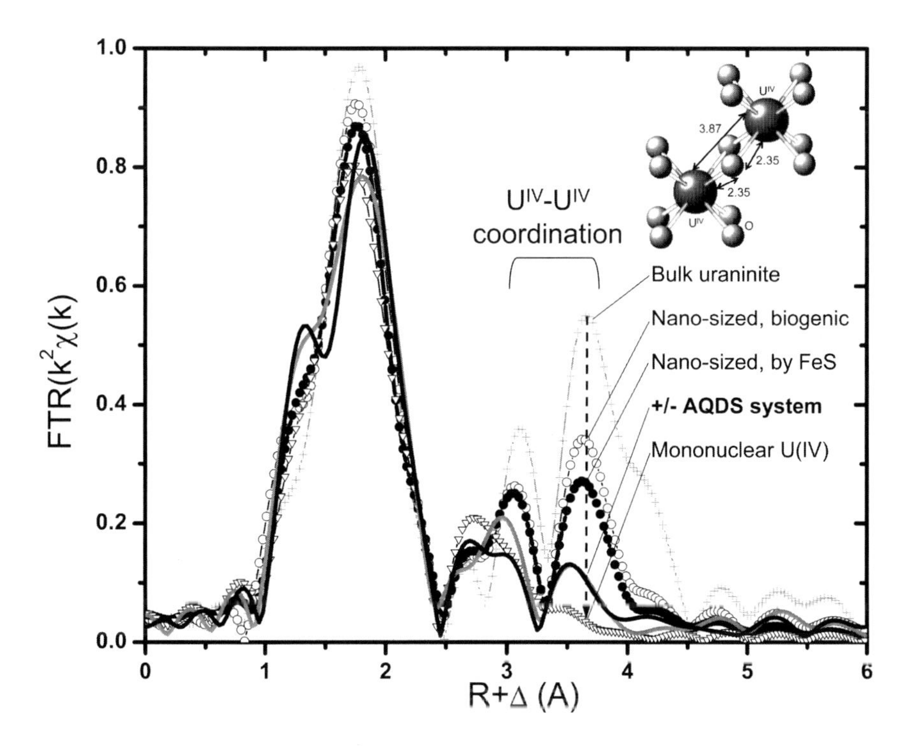

Figure 8. Fourier transforms of U L_{III}-edge EXAFS spectra after U^{VI} reaction with the non-pasteurized, with and without AQDS experimental systems (lines), compared to U^{IV} standards (symbols). The inset illustrates the U^{IV}-U^{IV} coordination in crystalline uraninite, where each U atom is surrounded by 12 other U atoms in this bidentate bonding configuration. The U-U coordination results in the peak noted by the vertical dashed line. The decrease in amplitude of this peak indicates the presence of nano-particulate uraninite (153) and the absence of this peak indicates lack of a bidentate U^{IV}-U^{IV} coordination, such as in biogenically-produced mononuclear U^{IV} (40).

Conclusions

Evidence to date suggests that, in general, the reduction of Tc, U, Np, and Pu to lower valence states decreases their solubility and may provide a mechanism by which to limit the migration of contaminant plumes. However, the ultimate fate of reduced Tc, U, Np, and Pu species depends on the long-term stability of these species, particularly with respect to the potential for oxidation to more mobile forms. Given the complexity of the biogeochemical processes occurring in typical surface and near subsurface environments, additional insight is clearly needed to definitively assess the contributions of coupled biotic-abiotic processes to the redox behavior of Tc, U, Np, and Pu.

Acknowledgments

We thank Philip Long, Dick Dayvault, and Ken Williams for providing subsurface sediments from the DOE Old Rifle Integrated Field-Scale Subsurface Research Challenge study site; Kelly Skinner for DNA extraction and preparation of samples for sequencing; Bhoopesh Mishra and Tomohiro Shibata for XAFS data collection/beamline support; and Karen Haugen and three anonymous reviewers for their thoughtful reviews of the manuscript. This research is part of the Subsurface Science Scientific Focus Area at Argonne National Laboratory supported by the DOE Subsurface Biogeochemical Research Program, Office of the Biological and Environmental Research, Office of Science, under contract DE-AC02-06CH11357. Use of the Advanced Photon Source, a User Facility operated for the DOE Office of Science by Argonne National Laboratory, was supported by the DOE under Contract No. DE-AC02-06CH11357. MRCAT/EnviroCAT operations are supported by DOE and the MRCAT/EnviroCAT member institutions.

References

1. DOE. *Linking Legacies: Connecting the Cold War Nuclear Weapons Production Processes to Their Environmental Consequences*; DOE/EM-0319; U.S. Department of Energy: Washington, DC, 1997.
2. DOE. *Status Report on Paths to Closure*; DOE/EM-0526; U.S. Department of Energy: Washington, DC, 2000.
3. Riley, R. G.; Zachara, J. M.; Wobber, F. J. *Chemical contaminants on DOE lands and selection of contaminant mixtures for subsurface science research*; DOE/ER 0547T; U.S. Department of Energy: Washington, DC, 1992; p 77.
4. Allard, B.; Olofsson, U.; Torstenfelt, B. Environmental actinide chemistry. *Inorg. Chim. Acta* **1984**, *94*, 205–221.
5. Fredrickson, J. K.; Zachara, J. M.; Plymale, A. E.; Heald, S. M.; McKinley, J. P.; Kennedy, D. W.; Liu, C.; Nachimuthu, P. Oxidative dissolution potential of biogenic and abiogenic TcO_2 in subsurface sediments. *Geochim. Cosmochim. Acta* **2009**, *73*, 2299–2313.
6. Curtis, D.; Fabryka-Martin, J.; Dixon, P.; Cramer, J. Nature's uncommon elements: Plutonium and technetium. *Geochim. Cosmochim. Acta* **1999**, *63*, 275–285.
7. Dixon, P.; Curtis, D. B.; Musgrave, J. A.; Roensch, F.; Roach, J.; Rokop, D. Analysis of naturally produced technetium and plutonium in geologic materials. *Anal. Chem.* **1997**, *69*, 1692–1699.
8. Kenna, B. T.; Kuroda, P. K. Technetium in nature. *J. Inorg. Nucl. Chem.* **1964**, *26*, 493–499.
9. Darab, J. G.; Amonette, A. B.; Burke, D. S. D.; Orr, R. D.; Ponder, S. M.; Schrick, B.; Mallouk, T. E.; Lukens, W. W.; Caulder, D. L.; Shuh, D. K. Removal of pertechnetate from simulated nuclear waste streams using supported zerovalent iron. *Chem. Mater.* **2007**, *19*, 5703–5713.
10. Jaisi, D. P.; Dong, H.; Plymale, A. E.; Frederickson, J. K.; Zachara, J. M.; Heald, S.; Liu, C. Reduction and longterm immobilization of technetium by

Fe(II) associated with clay mineral nontronite. *Chem. Geol.* **2009**, *264*, 127–138.

11. Peretyazhko, T.; Zachara, J. M.; Heald, S. M.; Jeon, B.-H.; Kukkadapu, R. K.; Liu, C.; Moore, D.; Resch, C. T. Heterogeneous reduction of Tc(VII) by Fe(II) at the solid-water interface. *Geochim. Cosmochim. Acta* **2009**, *72*, 1521–1539.
12. Fredrickson, J. K.; Zachara, J. M.; Kennedy, D. W.; Kukkadapu, R. K.; McKinley, J. P.; Heald, S. M.; Liu, C.; Plymale, A. E. Reduction of TcO_4^- by sediment-associated biogenic Fe(II). *Geochim. Cosmochim. Acta* **2004**, *68*, 3171–3187.
13. Liu, Y.; Terry, J.; Jurisson, S. Pertechnetate immobilization with amorphous iron sulfide. *Radiochim. Acta* **2008**, *96*, 823–833.
14. Maes, A.; Geraedts, K.; Bruggeman, C.; Vancluysen, J.; Rossberg, A.; Hennig, C. Evidence for the interaction of technetium colloids with humic substances by X-ray absorption spectroscopy. *Environ. Sci. Technol.* **2004**, *38*, 2044–2051.
15. Marshall, M. J.; Plymale, A. E.; Kennedy, D. W.; Shi, L.; Wang, Z.; Reed, S. B.; Dohnalkova, A. C.; Simonson, C. J.; Liu, C.; Saffarini, D. A.; Romine, M. F.; Zachara, J. M.; Beliaev, A. S.; Fredrickson, J. K. Hydrogenase- and outer membrane c-type cytochrome-facilitated reduction of technetium(VII) by *Shewanella oneidensis* MR-1. *Environ. Microbiol.* **2007**, 1–12.
16. Livens, F. R.; Jones, M. J.; Hynes, A. J.; Charnock, J. M.; Mosselmans, J. F. W.; Hennig, C.; Steele, H.; Collison, D.; Vaughan, D. J.; Pattrick, R. A. D.; Reed, W. A.; Moyes, L. N. X-ray adsorption spectroscopy studies of reactions of technetium, uranium and neptunium with mackinawite. *J. Environ. Radioact.* **2004**, *74*, 211–219.
17. Wharton, M. J.; Atkins, B.; Charnock, J. M.; Livens, F. R.; Pattrick, R. A. D.; Collison, D. An x-ray absorption spectroscopy study of the coprecipitation of Tc and Re with mackinawite (FeS). *Appl. Geochem.* **2000**, *15*, 347–354.
18. Skomurski, F. N.; Rosso, K. M.; Krupka, K. M.; McGrail, B. P. Technitium incorporation into hematite (α-Fe_2O_3). *Environ. Sci. Technol.* **2010**, *44*, 5855–5861.
19. Walton, F. B.; Paquette, J.; Ross, J. P. M.; Lawrence, W. E. Tc(IV) and Tc(VII) interactions with iron oxyhydroxides. *Nucl. Chem. Waste Manage.* **1986**, *6*, 121–126.
20. Zachara, J. M.; Heald, S. M.; Jeon, B.-H.; Kukkadapu, R. K.; Liu, C.; McKinley, J. P.; Dohnalkova, A. C.; Moore, D. A. Reduction of pertechnetate [Tc(VII)] by aqueous Fe(II) and the nature of the solid phase redox products. *Geochim. Cosmochim. Acta* **2007**, *71*, 2137–2157.
21. Llorens, I. A.; Deniard, P.; Gautron, E.; Olicard, A.; Fattahi, M.; Jobic, S.; Grambow, B. Structural inverstigation of coprecipitation of technitium-99 with iron phases. *Radiochim. Acta* **2008**, *96*, 569–574.
22. Alliot, I.; Alliot, C.; Vitorge, P.; Fattahi, M. Speciation of technetium(IV) in bicarbonate media. *Environ. Sci. Technol.* **2009**, *43*, 9174–9182.
23. Eriksen, T. E.; Ndalamba, P.; Bruno, J.; Caceci, M. The solubility of TcO_2 * nH_2O in neutral to alkaline solutions under constant pCO_2. *Radiochim. Acta* **1992**, *58/59*, 67–70.

24. Artinger, R.; Buckau, G.; Geyer, S.; Fritz, P.; Wolf, M.; Kim, J. I. Characterization of groundwater humic substances: Influence of sedimentary organic carbon. *Appl. Geochem.* **2000**, *15*, 97–116.
25. Boggs, M. A.; Minton, T.; Dong, W.; Lomasney, S.; Islam, M. R.; Gu, B.; Wall, N. A. Interactions of Tc(IV) with huimc substances. *Environ. Sci. Technol.* **2011**, *45*, 2718–2724.
26. Maes, A.; Bruggeman, C.; Geraedts, K.; Vancluysen, J. Quantification of the interaction of Tc with dissolved boom clay humic substances. *Environ. Sci. Technol.* **2003**, *37*, 747–753.
27. Wildung, R. E.; Li, S. W.; Murray, C. J.; Krupka, K. M.; Xie, Y.; Hess, N. J.; Roden, E. E. Technitium reduction in sediments of a shallow aquifer exhibiting dissimilatory iron reduction potential. *FEMS Microbiol. Ecol.* **2004**, *49*, 151–162.
28. Grenthe, I.; Drożdżyński, J.; Fujino, T.; Buck, E. C.; Albrecht-Schmitt, T. E.; Wolf, S. F.Uranium. In *The chemistry of the actinide and transactinide elements*, 3rd ed.; Morss, L. R., Edelstein, N. M., Fuger, J., Eds.; Springer: Dordrecht, Netherlands, 2006; Vol. 1, pp 253−698.
29. Back, M.; Mandarino, J. A. *Fleischers Glossary of Mineralogical Species*; Mineralogical Record Inc.: Tucson, AZ, 2008.
30. Burns, P. C. The crystal chemistry of uranium. In *Uranium: Mineralogy, geochemistry and the environment*; Burns, P. C., Finch, R., Eds.; Mineralogical Society of America: Washington, DC, 1999; Vol. 38, pp 23−90.
31. Kratz, S.; Schnug, E. Rock phosphates and P fertilizers as sources of U contamination in the environment. In *Uranium in the environment*; Merkel, B. J., Hasche-Berger, A., Eds.; Springer Berlin Heidelberg: New York, 2006; pp 57−67.
32. Ilton, E. S.; Boily, J.-F.; Buck, E. C.; Skomurski, F. N.; Rosso, K. M.; Cahill, C. L.; Bargar, J. R.; Felmy, A. R. Influence of dynamical conditions on the reduction of U^{VI} at the Magnetite-solution interface. *Environ. Sci. Technol.* **2010**, *44*, 170–176.
33. Ilton, E. S.; Haiduc, A.; Cahill, C. L.; Felmy, A. R. Mica surfaces stabilize pentavalent uranium. *Inorg. Chem.* **2005**, *44*, 2986–2988.
34. Renshaw, J. C.; Butchins, L. J. C.; Livens, F. R.; May, I.; Charnock, J. M.; Lloyd, J. R. Bioreduction of uranium: Environmental implications of a pentavalent intermediate. *Environ. Sci. Technol.* **2005**, *39*, 5657–5660.
35. Ragnarsdottir, K. V.; Charlet, L. Uranium behaviour in natural environments. In *Environmental Mineralogy: Microbial Interactions, Anthropogenic Influences, Contaminated Land and Waste Management*; Cotter-Howells, J. D., Campbell, L. S., Valsami-Jones, E., Batchelder, M., Eds.; The Mineralogical Society of Great Britain & Ireland: London, 2000; Vol. 9.
36. Bernier-Latmani, R.; Veeramani, H.; Vecchia, E. D.; Junier, P.; Lezama-Pacheco, J. S.; Suvorova, E. I.; Sharp, J. O.; Wigginton, N. S.; Bargar, J. R. Non-uraninite Products of Microbial U(VI) reduction. *Environ. Sci. Technol.* **2010**, *44*, 9456–9462.
37. Boyanov, M. I.; O'Loughlin, E. J.; Roden, E. E.; Fein, J. B.; Kemner, K. M. Adsorption of Fe(II) and U(VI) to carboxyl-functionalized microspheres:

The influence of speciation on uranyl reduction studied by titration and XAFS. *Geochim. Cosmochim. Acta* **2007**, *71*, 1898–1912.
38. Boyanov, M. I.; Fletcher, K. E.; Kwon, M. J.; Rui, X.; O'Loughlin, E. J.; Löffler, F. E.; Kemner, K. M. Solution and microbial controls on the formation of reduced U(IV) phases. *Environ. Sci. Technol.*, submitted for publication.
39. Chakraborty, S.; Favre, F.; Banerjee, D.; Scheinost, A. C.; Mullet, M.; Ehrhardt, J.-J.; Brendle, J.; Vidal, L.; Charlet, L. U(VI) sorption and reduction by Fe(II) sorbed on montmorillonite. *Environ. Sci. Technol.* **2010**, *44*, 3779–3785.
40. Fletcher, K. E.; Boyanov, M. I.; Thomas, S. H.; Wu, Q.; Kemner, K. M.; Löffler, F. E. U(VI) reduction to mononuclear U(IV) by *Desulfirobacterium* species. *Environ. Sci. Technol.* **2010**, *44*, 4705–4709.
41. Kelly, S. D.; Kemner, K. M.; Carley, J.; Criddle, C.; Jardine, P. M.; Marsh, T. L.; Phillips, D.; Watson, D.; Wu, W.-M. Speciation of uranium in sediments before and after in situ biostimulation. *Environ. Sci. Technol.* **2008**, *42*, 1558–1564.
42. Veeramani, H.; Alessi, D. S.; Suvorova, E. I.; Lezama-Pacheco, J. S.; Stubbs, J. E.; Sharp, J. O.; Dippon, U.; Kappler, A.; Bargar, J. R.; Bernier-Latmani, R. Products of abiotic U(VI) reduction by biogenic magnetite and vivianite. *Geochim. Cosmochim. Acta* **2011**, *75*, 2512–2528.
43. Frazier, S. W.; Kretzschmar, R.; Kraemer, S. M. Bacterial siderophores promote dissolution of UO_2 under reducing conditions. *Environ. Sci. Technol.* **2005**, *39*, 5709–5715.
44. Haas, J. R.; Abraham, N. Effects of aqueous complexation on reductive precipitation of uranium by *Shewanella putrefaciens*. *Geochem. Trans.* **2004**, *5*, 3.
45. Luo, W.; Gu, B. Dissolution and mobilization of uranium in a reduced sediment by natural humic substances under anaerobic conditions. *Environ. Sci. Technol.* **2009**, *43*, 152–156.
46. Luo, W.; Gu, B. Dissolution of uranium-bearing minerals and mobilization of uranium by organic ligands in a biologically reduced sediment. *Environ. Sci. Technol.* **2011**, *45*, 2994–2999.
47. Warwick, P.; Evans, N.; Hall, A.; Walker, G.; Steigleder, E. Stability constants of U(VI) and U(IV)-humic acid complexes. *J. Radioanal. Chem.* **2005**, *266*, 179–190.
48. Yoshida, Z.; Johnson, S. G.; Kimura, T.; Krsul, J. R. Neptunium. In *The chemistry of the actinide and transactinide elements*, 3rd ed.; Morss, L. R., Edelstein, N. M., Fuger, J., Eds.; Springer: Dordrecht, Netherlands, 2006; Vol. 2, pp 699−812.
49. Myers, W. A.; Lindner, M. Precise determination of the natural abundance of ^{237}Np and ^{239}Pu in Katanga pitchblende. *J. Inorg. Nucl. Chem.* **1971**, *33*, 3233–3238.
50. Peppard, D. F.; Mason, G. W.; Gray, P. R.; Mech, J. F. Occurrence of the (4*n* + 1) series in nature. *J. Am. Chem. Soc.* **1952**, *74*, 6081–6084.
51. Gray, T. *The elements: A visual exploration of every known atom in the universe*; Black Dog and Leventhal Publishers: New York, 2009.

52. Lieser, K. H.; Mühlenweg, U. Neptunium in the hydrosphere and in the geosphere. *Radiochim. Acta* **1988**, *43*, 27–35.
53. Clark, D. L.; Conradson, S. D.; Ekberg, S. A.; Hess, N. J.; Neu, M. P.; Palmer, P. D.; Runde, W.; Tait, C. D. EXAFS studies of pentavalent neptunium carbonato complexes. Structural elucidation of the principal constituents of neptunium in groundwater environments. *J. Am. Chem. Soc.* **1996**, *118*, 2089–2090.
54. Fröhlich, D. R.; Amayri, S.; Drebert, J.; Reich, T. Sorption of neptunium(V) on Opalinus Clay under aerobic/anaerobic conditions. *Radiochim. Acta* **2011**, *99*, 71–77.
55. Siegel, M. D.; Bryan, C. R. Environmental geochemistry of radioactive contamination. In *Environmental Geochemistry*; Lollar, B. S., Ed.; Elsevier: Amsterdam, 2003; Vol. 9, pp 205−262.
56. Kaszuba, J. P.; Runde, W. H. The aqueous geochemistry of neptunium: Dynamic control of soluble concentrations with applications to nuclear waste disposal. *Environ. Sci. Technol.* **1999**, *33*, 4427–4433.
57. Icopini, G. A.; Boukhalfa, H.; Neu, M. P. Biological reduction of Np(V) and Np(V) citrate by metal-reducing bacteria. *Environ. Sci. Technol.* **2007**, *41*, 2764–2769.
58. Rittmann, B. E.; Banaszak, J. E.; Reed, D. T. Reduction of Np(V) and precipitation of Np(IV) by an anaerobic microbial consortium. *Biodegradation* **2003**, *13*, 329–342.
59. Artinger, R.; Marquardt, C. M.; Kim, J. I.; Seibert, A.; Trautmann, N.; Kratz, J. V. Humic colloid-borne Np migration: influence of the oxidation state. *Radiochim. Acta* **2000**, *88*, 609–612.
60. Clark, D. L.; Hecker, S. S.; Jarvinen, G. D.; Neu, M. P. Plutonium. In *The chemistry of the actinide and transactinide elements*, 3rd ed.; Morss, L. R., Edelstein, N. M., Fuger, J., Eds.; Springer: Dordrecht, Netherlands, 2006; Vol. 2, pp 813−1264.
61. Hoffman, D. C.; Lawrence, F. O.; Mewherter, J. L.; Rourke, F. M. Detection of plutonium-244 in nature. *Nature* **1971**, *234*, 132–134.
62. Peppard, D. F.; Studier, M. H.; Gergel, M. V.; Mason, G. W.; Sullivan, J. C.; Mech, J. F. Isolation of microgram quantities of naturally-occurring plutonium and examination of its isotopic composition. *J. Am. Chem. Soc.* **1951**, *73*, 2529–2531.
63. Turkevich, A. L.; Economou, T. E.; Cowan, G. A. Double beta decay of ^{238}U. *Phys. Rev. Lett.* **1991**, *67*, 3211–3214.
64. Albright, D.; Kramer, K. Stockpiles still growing. *Bull. At. Sci.* **2004**, *60*, 14–16.
65. Choppin, G. R.; Bond, A. H.; Hromadka, P. M. Redox speciation of plutonium. *J. Radioanal. Nucl. Chem.* **1997**, *219*, 203–210.
66. Choppin, G. R.; Wong, P. J. The chemistry of actinide behavior in marine systems. *Aquat. Geochem.* **1998**, *4*, 77–101.
67. Clark, D. L. The chemical complexeties of plutonium. *Los Alamos Sci.* **2000**, *26*, 364–381.
68. Guillaumont, R.; Adloff, J. P. Behavior of environmental plutonium at very low concentration. *Radiochim. Acta* **1992**, *58/59*, 53–60.

69. Silva, R. J.; Nitsche, H. Actinide environmental chemistry. *Radiochim. Acta* **1995**, *70/71*, 377–396.
70. Newton, T. W. Redox reactions of plutonium ions in aqueous solutions. In *Advances in plutonium chemistry 1967-2000*; Hoffman, D. C., Ed.; American Nuclear Society: LaGrange Park, IL, 2002; pp 24−60.
71. Runde, W.; Conradson, S. D.; Efurd, D. W.; Lu, N.; VanPelt, C. E.; Tait, C. D. Solubility and sorption of redox sensitive radionuclides (Np, Pu) in J-13 water from the Yucca Mountain site: comparison between experiment and theory. *Appl. Geochem.* **2002**, *17*, 837–853.
72. Silva, R. J.; Nitsche, H. Environmental Chemistry. In *Advances in plutonium chemistry 1967-2000*; Hoffman, D. C., Ed.; American Nuclear Society: LaGrange Park, IL, 2002; pp 89−117.
73. Kaplan, D. I.; Powell, B. A.; Demirkanli, D. I.; Fjeld, R. A.; Molz, F. J.; Serkiz, S. M.; Coates, J. T. Influence of oxidation states on plutonium mobility during long-term transport through an unsaturated subsurface environment. *Environ. Sci. Technol.* **2004**, *38*, 5053–5058.
74. Artinger, R.; Buckau, G.; Zeh, P.; Geraedts, K.; Vancluysen, J.; Maes, A.; Kim, J. I. Humic colloid mediated transport of tetravalent actinides and technitium. *Radiochim. Acta* **2003**, *91*, 743–750.
75. Kersting, A. B.; Efurd, D. W.; Finnegan, D. L.; Rokop, D. J.; Smith, D. K.; THompson, J. L. Migration of plutonium in ground water at the Nevada Test Site. *Nature* **1999**, *397*, 56–59.
76. Marty, R. C.; Bennett, D.; Thullen, P. Mechanism of plutonium transport in a shallow aquifer in Mortandad Canyon, Los Alamos National Laboratory, New Mexico. *Environ. Sci. Technol.* **1997**, *31*, 2020–2027.
77. Penrose, W. R.; Polzer, W. L.; Essington, E. H.; Nelson, D. M.; Orlandini, K. A. Mobility of plutonium and americium through a shallow aquifer in a semiarid region. *Environ. Sci. Technol.* **1990**, *24*, 228–234.
78. Schwantes, J. M.; Santschi, P. H. Mechanisms of plutonium sorption to mineral oxide surfaces: new insights with implications for colloid-enhanced migration. *Radiochim. Acta* **2010**, *98*, 737–742.
79. Zhao, P.; Zavarin, M.; Leif, R. N.; Powell, B. A.; Singleton, M. J.; Lindvall, R. E.; Kersting, A. B. Mobilization of actinides by dissolved organic compounds at the Nevada Test Site. *Appl. Geochem.* **2011**, *26*, 308–318.
80. Duff, M. C.; Hunter, D. B.; Triay, I. R.; Bertsch, P. M.; Reed, D. T.; Sutton, S. R.; Shea-McCarthy, G.; Kitten, J.; Eng, P.; Chipera, S. J.; Vaniman, D. T. Mineral associations and average oxidation states of sorbed Pu on tuff. *Environ. Sci. Technol.* **1999**, *33*, 2163–2169.
81. Powell, B. A.; Duff, M. C.; Kaplan, D. I.; Fjeld, R. A.; Newville, M.; Hunter, D. B.; Bertsch, P. M.; Coates, J. T.; Eng, P.; Rivers, M. L.; Serkiz, S. M.; Sutton, S. R.; Triay, I. R.; Vaniman, D. T. Plutonium oxidation and subsequent reduction by Mn(IV) minerals in Yucca Mountain tuff. *Environ. Sci. Technol.* **2006**, *40*, 3508–3514.
82. Hering, J. G.; Stumm, W. Oxidative and reductive dissolution of minerals. In *Mineral-Water Interface Geochemistry*; Hochella, M. F. J., White, A. F.,

Eds.; American Mineralogical Society: Washington, DC, USA, 1990; Vol. 23, pp 427–465.

83. Lyngkilde, J.; Christensen, T. H. Redox zones of a landfill leachate pollution plume (Vejen, Denmark). *J. Contam. Hydrol.* **1992**, *10*, 273–289.
84. Rügge, K.; Hofstetter, T. B.; Haderlein, S. B.; Bjerg, P. L.; Knudsen, S.; Zraunig, C.; Mosbæk, H.; Christensen, T. H. Characterization of predominant reductants in an anaerobic leachate-contaminated aquifer by nitroaromatic probe compounds. *Environ. Sci. Technol.* **1998**, *32*, 23–31.
85. Reed, D. T.; Lucchini, J. F.; Aase, S. B.; Kropf, A. J. Reduction of plutonium(VI) in brine under subsurface conditions. *Radiochim. Acta* **2006**, *94*, 591–597.
86. Reed, D. T.; Pepper, S. E.; Richmann, M. K.; Smith, G.; Deo, R.; Rittmann, B. E. Subsurface bio-mediated reduction of higher-valent uranium and plutonium. *J. Alloys Compd.* **2007**, *444–445*, 376–382.
87. Cui, D.; Eriksen, T. E. Reduction of pertechnetate by ferrous iron in solution: Influence of sorbed and precipitated Fe(II). *Environ. Sci. Technol.* **1996**, *30*, 2259–2262.
88. Fredrickson, J. K.; Zachara, J. M.; Kennedy, D. W.; Duff, M. C.; Gorby, Y. A.; Li, S.-M. W.; Krupka, K. M. Reduction of U(VI) in goethite (α-FeOOH) suspensions by a dissimilatory metal-reducing bacterium. *Geochim. Cosmochim. Acta* **2000**, *64*, 3085–3098.
89. Liger, E.; Charlet, L.; Van Cappellen, P. Surface catalysis of uranium(VI) reduction by iron(II). *Geochim. Cosmochim. Acta* **1999**, *63*, 2939–2955.
90. Nakata, K.; Nagasaki, S.; Tanaka, S.; Sakamoto, Y.; Tanaka, T.; Ogawa, H. Reduction rate of neptunium(V) in heterogeneous solution with magnetite. *Radiochim. Acta* **2004**, *92*, 145–149.
91. Jeon, B.-H.; Dempsey, B. A.; Burgos, W. D.; Barnett, M. O.; Roden, E. E. Chemical reduction of U(VI) by Fe(II) at the solid-water interface using natural and synthetic Fe(III) oxides. *Environ. Sci. Technol.* **2005**, *39*, 5642–5649.
92. Regenspurg, S.; Schild, D.; Schäfer, T.; Huber, F.; Malmström, M. E. Removal of U(VI) from the aqueous phase by iron(II) minerals in presence of bicarbonate. *Appl. Geochem.* **2009**, *24*, 1617–1625.
93. Peretyazhko, T.; Zachara, J. M.; Heald, S. M.; Kukkadapu, R. K.; Liu, C.; Plymale, A. E.; Resch, C. T. Reduction of Tc(VII) by Fe(II) sorbed on Al (hydr)oxides. *Environ. Sci. Technol.* **2008**, *42*, 5499–5506.
94. Christiansen, B. C.; Geckeis, H.; Marquardt, C. M.; Bauer, A.; Römer, J.; Wiss, T.; Schild, D.; Stipp, S. L. S. Neptunyl (NpO_2^+) interaction with green rust, $GR_{Na,SO4}$. *Geochim. Cosmochim. Acta* **2011**, *75*, 1216–1226.
95. O'Loughlin, E. J.; Kelly, S. D.; Csencsits, R.; Cook, R. E.; Kemner, K. M. Reduction of uranium(VI) by mixed iron(II)/iron(III) hydroxide (green rust): Formation of UO_2 nanoparticles. *Environ. Sci. Technol.* **2003**, *37*, 721–727.
96. O'Loughlin, E. J.; Kelly, S. D.; Kemner, K. M. XAFS investigation of the interactions of U^{VI} with secondary mineralization products from the bioreduction of Fe^{III} oxides. *Environ. Sci. Technol.* **2010**, *44*, 1656–1661.
97. Pepper, S. E.; Bunker, D. J.; Bryan, N. D.; Livens, F. R.; Charnock, J. M.; Pattrick, R. A. D.; Collison, D. Treatment of radioactive wastes: An X-ray

adsorption spectroscopy study of the treatment of technetium with green rust. *J. Colloid Interface Sci.* **2003**, *268*, 408–412.
98. Cui, D.; Eriksen, T. E. Reduction of pertechnetate in solution by heterogeneous electron transfer from Fe(II)-containing geological material. *Environ. Sci. Technol.* **1996**, *30*, 2263–2269.
99. Haines, R. I.; Owen, D. G.; Vandergraff, T. T. Technetium-iron oxide reactions under anaerobic conditions: a fourier transform infrared, FTIR study. *Nucl. J. Canada* **1987**, *1*, 32.
100. Powell, B. A.; Fjeld, R. A.; Kaplan, D. I.; Coates, J. T.; Serkiz, S. M. $Pu(V)O_2^+$ adsorption and reduction by synthetic magnetite. *Environ. Sci. Technol.* **2004**, *38*, 6016–6024.
101. Duro, L.; Aamrani, S. E.; Rovira, M.; de Pablo, J.; Bruno, J. Study of the interaction between U(VI) and the anoxic corrosion products of carbon steel. *Appl. Geochem.* **2008**, *23*, 1094–1100.
102. Missana, T.; Maffiotte, C.; García-Gutiérrez, M. Surface reactions kinetics between nanocrystalline magnetite and uranyl. *J. Colloid Interface Sci.* **2003**, *261*, 154–160.
103. Rovira, M., El Aamrani, S.; Duro, L.; Giménez, J.; de Pablo, J.; Bruno, J. Interaction of uranium with *in situ* anoxically generated magnetite on steel. *J. Hazard. Mater.* **2007**, *147*, 726–731.
104. Scott, T. B., Allen, G. C.; Heard, P. J.; Randell, M. G. Reduction of U(VI) to U(IV) on the surface of magnetite. *Geochim. Cosmochim. Acta* **2005**, *69*, 5639–5646.
105. Gorski, C. A.; Scherer, M. M. Influence of magnetite stoichiometry on Fe^{II} uptake and nitrobenzene reduction. *Environ. Sci. Technol.* **2009**, *43*, 3675–3680.
106. Latta, D. E.; Gorski, C. A.; Boyanov, M. I.; O'Loughlin, E. J.; Kemner, K. M.; Scherer, M. M. Influence of magnetite stoichiometry on U^{VI} reduction. *Environ. Sci. Technol.*, submitted for publication.
107. Elsner, M.; Schwarzenbach, R. P.; Haderlein, S. B. Reactivity of Fe(II)-bearing minerals toward reductive transformation of organic contaminants. *Environ. Sci. Technol.* **2004**, *38*, 799–807.
108. Scheinost, A. C.; Charlet, L. Selenite reduction by mackinawite, magnetite and siderite: XAS characterization of nanosized redox products. *Environ. Sci. Technol.* **2008**, *42*, 1984–1989.
109. Scheinost, A. C.; Kirsch, R.; Banerjee, D.; Fernandez-Martinez, A.; Zaenker, H.; Funke, H.; Charlet, L. X-ray absorption and photoelectron spectroscopy investigation of selenite reduction by Fe^{II}-bearing minerals. *J. Contam. Hydrol.* **2008**, *102*, 228–245.
110. Ithurbide, A.; Peulon, S.; Miserque, F.; Beaucaire, C.; Chaussé, A. Interaction between uranium(VI) and siderite ($FeCO_3$) surfaces in carbonate solutions. *Radiochim. Acta* **2009**, *97*, 177–180.
111. Boyanov, M. I.; O'Loughlin, E. J.; Kwon, M. J.; Mishra, B.; Rui, X.; Shibata, T.; Kemner, K. M. Mineral nucleation and redox transformations of U(VI) and Fe(II) species at a carboxyl surface. *Geochim. Cosmochim. Acta* **2010**, *74*, A115.

112. Zhang, G.; Senko, J. M.; Kelly, S. D.; Tan, H.; Kemner, K. M.; Burgos, W. D. Microbial reduction of iron(III)-rich nontronite and uranium(VI). *Geochim. Cosmochim. Acta* **2009**, *73*, 3523–3538.
113. Ilton, E. S.; Haiduc, A.; Moses, C. O.; Heald, S. M.; Elbert, D. C.; Veblen, D. R. Heterogeneous reduction of uranyl by micas: Crystal chemical and solution controls. *Geochim. Cosmochim. Acta* **2004**, *68*, 2417–2435.
114. Bondietti, E. A.; Francis, C. W. Geologic migration potentials of technetium-99 and neptunium-237. *Science* **1979**, *203*, 1337–1340.
115. Nash, K. L.; Cleveland, J. M.; Sullivan, J. C.; Woods, M. Kinetics of reduction of plutonium(VI) and neptunium(VI) by sulfide in neutral and alkaline solutions. *Inorg. Chem.* **1986**, *25*, 1169–1173.
116. Beyenal, H.; Sani, R. K.; Peyton, B. M.; Dohnalkova, A. C.; Amonette, J. E.; Lewandowski, Z. Uranium immobilization by sulfate-reducing biofilms. *Environ. Sci. Technol.* **2004**, *38*, 2067–2074.
117. Hua, B.; Xu, H.; Terry, J.; Deng, B. Kinetics of uranium(VI) reduction by hydrogen sulfide in anoxic aqueous systems. *Environ. Sci. Technol.* **2006**, *40*, 4666–4671.
118. Anderson, R. F.; Fleisher, M. Q.; LeHuray, A. P. Concentration, oxidation state, and particulate flux of uranium in the Black Sea. *Geochim. Cosmochim. Acta* **1989**, *53*, 2215–2224.
119. Liu, Y.; Terry, J.; Jurisson, S. Pertechnetate immobilization in aqueous media with hydrogen sulfide under anaerobic and aerobic environments. *Radiochim. Acta* **2007**, *95*, 717–725.
120. Watson, J. H. P.; Croudace, I. W.; Warwick, P. E.; James, P. A. B.; Charnock, J. M.; Ellwood, D. C. Adsorption of radioactive metals by strongly magnetic iron sulfide nanoparticles produced by sulfate-reducing bacteria. *Sep. Sci. Technol.* **2001**, *36*, 2571–2607.
121. Aubriet, H.; Humbert, B.; Perdicakis, M. Interaction of U(VI) with pyrite, galena and their mixtures: A theoretical and multitechnique approach. *Radiochim. Acta* **2006**, *94*, 657–663.
122. Bruggeman, C.; Maes, N. Uptake of uranium(VI) by pyrite under boom clay conditions: Influence of dissolved organic carbon. *Environ. Sci. Technol.* **2010**, *44*, 4210–4216.
123. Descostes, M.; Schlegel, M. L.; Eglizaud, N.; Descamps, F.; Miserque, F.; Simoni, E. Uptake of uranium and trace elements in pyrite (FeS_2) suspensions. *Geochim. Cosmochim. Acta* **2010**, *74*, 1551–1562.
124. Hua, B.; Deng, B. Reductive Immobilization of uranium(VI) by amorphous iron sulfide. *Environ. Sci. Technol.* **2008**, *42*, 8703–8708.
125. Moyes, L. N.; Parkman, R. H.; Charnock, J. M.; Vaughan, D. J.; Livens, F. R.; Hughes, C. R.; Braithwaite, A. Uranium uptake from aqueous solution by interaction with goethite, lepidocrocite, muscovite, and mackinawite: An X-ray ansorption spectroscopy study. *Environ. Sci. Technol.* **2000**, *34*, 1062–1068.
126. Wersin, P.; Hochella, M. F. J.; Persson, P.; Redden, R.; Leckie, J. O.; Harris, D. Interaction between aqueous uranium(VI) and sulfide minerals: Spectroscopic evidence for sorption and reduction. *Geochim. Cosmochim. Acta* **1994**, *58*, 2829–2843.

127. Moyes, L. N.; Jones, M. J.; Reed, W. A.; Livens, F. R.; Charnock, J. M.; Mosselmans, J. F. W.; Hennig, C.; Vaughan, D. J.; Pattrick, R. A. D. An X-ray absorption spectroscopy study of neptunium(V) reactions with mackinawite. *Environ. Sci. Technol.* **2002**, *36*, 179–183.
128. Thurman, E. M. *Organic Geochemistry of Natural Waters*; Martinus Nijhoff/ Dr W. Junk Publishers: Boston, 1985; p 497.
129. Skovbjerg, L. L.; Christiansen, B. C.; Nedel, S.; Stipp, S. L. S. THe role of green rust in the migration of radionuclides: An overview of processes that can control mobility of radioactive elements in the environment using as examples Np, Se, and Cr. *Radiochim. Acta* **2010**, *98*, 607–612.
130. Blinova, O.; Novikov, A.; Perminova, I.; Goryachenkova, T.; Haire, R. Redox Interactions of Pu(V) in solutions containing different humic substances. *J. Alloys Compd.* **2007**, *444−445*, 486–490.
131. Nash, K. L.; Fried, S.; Friedman, A. M.; Sullivan, J. C. Redox behavior, complexing, and adsorption of hexavalent actinides by humic acid and selected clays. *Environ. Sci. Technol.* **1981**, *15*, 834–837.
132. Sanchez, A. L.; Murray, J. W.; Sibley, T. H. The adsorption of plutonium IV and V on goethite. *Geochim. Cosmochim. Acta* **1985**, *49*, 2297–2307.
133. Shcherbina, N. S.; Kalmykov, S. N.; Perminova, I. V.; Kovalenko, A. N. Reduction of actinides in higher oxidation states by hydroquinone-enriched humic derivatives. *J. Alloys Compd.* **2007**, *444-445*, 518–521.
134. Shcherbina, N. S.; Perminova, I. V.; Kalmykov, S. N.; Kovalenko, A. N.; Haire, R. G.; Novikov, A. P. Redox and complexation interactions of neptunium(V) with quinonoid-enriched humic derivatives. *Environ. Sci. Technol.* **2007**, *41*, 7010–7015.
135. Nakashima, S. Kinetics and thermodynamics of U reduction by natural and simple organic matter. *Adv. Org. Geochem.* **1991**, *19*, 421–430.
136. Rai, D.; Gorby, Y. A.; Fredrickson, J. K.; Moore, D. A.; Yui, M. Reductive dissolution of PuO_2(am): The effect of Fe(II) and hydroquinone. *J. Sol. Chem.* **2002**, *31*, 433–453.
137. Wang, Z.; Wagnon, K. B.; Ainsworth, C. C.; Liu, C.; Rosso, K. M.; Fredrickson, J. K. A spectroscopic study of the effect of ligand complexation on the reduction of uranium(VI) by anthraquinone-2,6-disulfonate (AH_2DS). *Radiochim. Acta* **2008**, *96*, 599–605.
138. Suzuki, Y.; Kitatsuji, Y.; Ohnuki, T.; Tsujimura, S. Flavin mononucleotide mediated electron pathway for microbial U(VI) reduction. *Phys. Chem. Chem. Phys.* **2010**, *12*, 10081–10087.
139. Blowes, D. W.; Ptacek, C. J.; Benner, S. G.; McRae, C. W. T.; Bennett, T. A.; Puls, R. W. Treatment of inorganic contaminants using permeable reactive barriers. *J. Contam. Hydrol.* **2000**, *45*, 123–137.
140. Cantrell, K. J.; Kaplan, D. I.; Wietsma, T. W. Zero-valent iron for the in situ remediation of selected metals in groundwater. *J. Hazard. Mater.* **1995**, *42*, 201–212.
141. Farrell, J.; Bostick, W. D.; Jarabeck, R. J.; Fiedor, J. N. Uranium removal from ground water using zero valent iron media. *Ground Water* **1999**, *37*, 618–624.

142. Gu, B.; Liang, L.; Dickey, M. J.; Yin, X.; Dai, S. Reductive precipatation of uranium(VI) by zero valent iron. *Environ. Sci. Technol.* **1998**, *32*, 3366–3373.
143. Liang, L.; Gu, B.; Yin, X. Removal of technetium-99 from contaminated groundwater with sorbents and reductive materials. *Sep. Technol.* **1996**, *6*, 111–122.
144. Riba, O.; Scott, T. B.; Ragnarsdottir, K. V.; Allen, G. C. Reaction mechanism of uranyl in the presence of zero-valent iron nanoparticles. *Geochim. Cosmochim. Acta* **2008**, *72*, 4047–4057.
145. Yan, S.; Hua, B.; Bao, Z.; Yang, J.; Liu, C.; Deng, B. Uranium(VI) removal by nanoscale zerovalent iron in anoxic batch systems. *Environ. Sci. Technol.* **2010**, *44*, 7783–7789.
146. Wu, W.-M.; Carley, J.; Luo, J.; Ginder-Vogel, M. A.; Cardenas, E.; Leigh, M. B.; Hwang, C.; Kelly, S. D.; Ruan, C.; Wu, L.; van Nostrand, J.; Gentry, T.; Lowe, K.; Melhorn, T.; Carroll, S.; Luo, W.; Fields, M. W.; Gu, B.; Watson, D.; Kemner, K. M.; Marsh, T.; Tiedje, J.; Zhou, J.; Fendorf, S.; Kitanidis, P. K.; Jardine, P. M.; Criddle, C. S. In situ bioreduction of uranium(VI) to submicromolar levels and reoxidation by dissolved oxygen. *Environ. Sci. Technol.* **2007**, *41*, 5716–5723.
147. Senko, J. M.; Istok, J. D.; Suflita, J. M.; Krumholz, L. R. In-situ evidence for uranium immobilization and remobilization. *Environ. Sci. Technol.* **2002**, *36*, 1491–1496.
148. Wan, J.; Tokunaga, T. K.; Brodie, E. L.; Wang, Z.; Zheng, Z.; Herman, D.; Hazen, T. C.; Firestone, M. K.; Sutton, S. R. Reoxidation of bioreduced uranium under reducing conditions. *Environ. Sci. Technol.* **2005**, *39*, 6162–6169.
149. Komlos, J.; Peacock, A.; Kukkadapu, R. K.; Jaffé, P. R. Long-term dymanics of uranium reduction/reoxidation under low sulfate conditions. *Geochim. Cosmochim. Acta* **2008**, *72*, 3603–3615.
150. Komlos, J.; Mishra, B.; Lanzirotti, A.; Myneni, S. C. B.; Jaffé, P. R. Real-time speciation of uranium during active bioremediation and U(IV) reoxidation. *J. Environ. Eng.* **2008**, *134*, 78–86.
151. Burke, I. T.; Boothman, C.; Lloyd, J. R.; Livens, F. R.; Charnock, J. M.; McBeth, J. M.; Mortimer, R. J. G.; Morris, K. Reoxidation behavior of technetium, iron, and sulfur in estuarine sediments. *Environ. Sci. Technol.* **2006**, *40*, 3529–3535.
152. Law, G. T. W.; Geissler, A.; Lloyd, J. R.; Livens, F. R.; Boothman, C.; Begg, J. D. C.; Denecke, M. A.; Rothe, J.; Dardenne, K.; Burke, I. T.; Charnock, J. M.; Morris, K. Geomicrobiological redox cycling of the transuranic element neptunium. *Environ. Sci. Technol.* **2010**, *44*, 8924–8929.
153. Burgos, W. D.; McDonough, J. T.; Senko, J. M.; Zhang, G.; Dohnalkova, A. C.; Kelly, S. D.; Gorby, Y.; Kemner, K. M. Characterization of uraninite nanoparticles produced by *Shewanella oneidensis* MR-1. *Geochim. Cosmochim. Acta* **2008**, *72*, 4901–4915.

154. Gu, B.; Yan, H.; Zhou, P.; Watson, D. B.; Park, M.; Istok, J. Natural humics impact uranium bioreduction and oxidation. *Environ. Sci. Technol.* **2005**, *39*, 5268–5275.
155. Senko, J. M.; Kelly, S. D.; Dohnalkova, A. C.; McDonough, J. T.; Kemner, K. M.; Burgos, W. D. The effect of U(VI) bioreduction kinetics on subsequent reoxidation of biogenic U(IV). *Geochim. Cosmochim. Acta* **2007**, *71*, 4644–4654.
156. Ulrich, K.-U.; Ilton, E. S.; Veeramani, H.; Sharp, J. O.; Bernier-Latmani, R.; Schofield, E. J.; Bargar, J. R.; Giammar, D. E. Comparative dissolution kinetics of biogenic and chemogenic uraninite under oxidizing conditions in the presence of carbonate. *Geochim. Cosmochim. Acta* **2009**.
157. Veeramani, H.; Schofield, E. J.; Sharp, J. O.; Suvorova, E. I.; Ulrich, K.-U.; Metha, A.; Giammar, D. E.; Bargar, J. R.; Bernier-Latmani, R. Effect of Mn(II) on the structure and reactivity of biogenic uraninite. *Environ. Sci. Technol.* **2009**, *43*, 6541–6547.
158. Finch, R. J.; Ewing, R. C. The corrosion of uraninite under oxidizing conditions. *J. Nuclear Mater.* **1992**, *190*, 133–156.
159. Abdclouas, A.; Lutze, W.; Nuttall, H. E. Oxidative dissolution of uraninite precipatated on Navajo sandstone. *J. Contam. Hydrol.* **1999**, *36*, 353–375.
160. Zhong, L.; Liu, C., Zachara, J. M.; Kennedy, D. W.; Szecsody, J. E.; Wood, B. Oxidative remobilization of biogenic uranium(IV) precipitates: Effects of iron(II) and pH. *J. Environ. Qual.* **2005**, *34*, 1763–1771.
161. Fredrickson, J. K.; Zachara, J. M.; Kennedy, D. W.; Liu, C.; Duff, M. C.; Hunter, D. B.; Dohnalkova, A. Influence of Mn oxides on the reduction of uranium(VI) by the metal-reducing bacterium *Shewanella putrefaciens*. *Geochim. Cosmochim. Acta* **2002**, *66*, 3247–3262.
162. Ilton, E. S.; Heald, S. M.; Smith, S. C.; Elbert, D.; Liu, C. Reduction of uranyl in the interlayer region of low iorn micas under anoxic and aerobic conditions. *Environ. Sci. Technol.* **2006**, *40*, 5003–5009.
163. Conradson, S. D.; Begg, B. D.; Clark, D. L.; den Auwer, C.; Ding, M.; Dorhout, P. K.; Espinosa-Faller, F. J.; Gordon, P. L.; Haire, R. G.; Hess, N. J.; Hess, R. F.; Keogh, W.; Morales, L. A.; Neu, M. P.; Paviet-Hartman, P.; Rundle, W.; Tait, C. D.; Veirs, D. K.; Villella, P. M. Local and nanoscale struture and speciation in the $PuO_{2+x-y}(OH)_{2y} \cdot zH_2O$ system. *J. Am. Chem. Soc.* **2004**, *126*, 13443–13458.
164. Neck, V.; Altmaier, M.; Seibert, A.; Yun, J. I.; Marquardt, C. M.; Fanghänel, T. Solubility and redox reactions of Pu(IV) hydrous oxide: Evidence for the formation of PuO_{2+x} (s, hyd). *Radiochim. Acta* **2007**, *95*, 193–207.
165. Keeney-Kennicutt, W. L.; Morse, J. W. The redox chemistry of $Pu(V)O_2^+$ interaction with common mineral surfaces in dilute solutions and seawater. *Geochim. Cosmochim. Acta* **1985**, *49*, 2577–2588.
166. Khasanova, A. B.; Shcherbina, N. S.; Kalmykov, S. N.; Teterin, Y. A.; Novikov, A. P. Sorption of Np(V), Pu(V), and Pu(IV) on colloids of Fe(III) oxides and hydrous oxides and MnO_2. *Radiochemistry* **2007**, *49*, 419–425.
167. Morgenstern, A.; Choppin, G. R. Kinetics of the oxidation of Pu(IV) by manganese dioxide. *Radiochim. Acta* **2002**, *90*, 69–74.

168. Tanaka, K.; Suzuki, Y.; Ohnuki, T. Sorption and oxidation of tetravalent plutonium on Mn oxide in the presence of citric acid. *Chem. Lett.* **2009**, *38*, 1032–1033.
169. Ginder-Vogel, M.; Criddle, C. S.; Fendorf, S. Thermodynamic constraints on the oxidation of biogenic UO_2 by Fe(III) (hydr)oxides. *Environ. Sci. Technol.* **2006**, *40*, 3544–3550.
170. Ginder-Vogel, M.; Stewart, B.; Fendorf, S. Kinetic and mechansitic constraints on the oxidation of biogenic uraninite by ferrihydrite. *Environ. Sci. Technol.* **2010**, *44*, 163–169.
171. Sani, R. K.; Peyton, B. M.; Dohnalkova, A.; Amonette, J. E. Reoxidation of reduced uranium with iron(III) (hydr)oxides under sulfate-reducing conditions. *Environ. Sci. Technol.* **2005**, *39*, 2059–2066.
172. Senko, J. M.; Mohamed, Y.; Dewers, T. A.; Krumholz, L. R. Role for Fe(III) minerals in nitrate-dependent microbial U(VI) oxidation. *Environ. Sci. Technol.* **2005**, *39*, 2529–2536.
173. Lloyd, J. R. Microbial reduction of metals and radionuclides. *FEMS Microbiol. Rev.* **2003**, *27*, 411–425.
174. Marshall, M. J.; Beliaev, A. S.; Fredrickson, J. K. Microbiological transformations of radionuclides in the subsurface. In *Environmental Microbiology*, 2nd ed.; Mitchell, R.; Gu, J.-D., Eds.; Wiley-Blackwell: Hoboken, NJ, 2010; pp 95−114.
175. Merroun, M. L.; Selenska-Pobell, S. Bacterial interactions with uranium: An environmental perspective. *J. Contam. Hydrol.* **2008**, *102*, 2008.
176. Suzuki, Y.; Suko, T. Geomicrobiological factors that control uranium mobility in the environment: Update on recent advances in the bioremediation of uranium-contaminated sites. *J. Mineral. Petrol. Sci.* **2006**, *101*, 299–307.
177. Wall, J. D.; Krumholz, L. R. Uranium reduction. *Ann. Rev. Microbiol.* **2006**, *60*, 149–166.
178. Wilkins, M. J.; Livens, F. R.; Vaughan, D. J.; Lloyd, J. R. The impact of Fe(III)-reducing bacteria on uranium mobility. *Biogeochemistry* **2006**, *78*, 125–150.
179. Francis, A. J.; Dodge, C. J.; Lu, F.; Halada, G. P.; Clayton, C. R. XPS and XANES studies of uranium reduction by *Clostridium* sp. *Environ. Sci. Technol.* **1994**, *28*, 636–639.
180. Gorby, Y. A.; Lovley, D. R. Enzymatic uranium precipitation. *Environ. Sci. Technol.* **1992**, *26*, 205–207.
181. Junier, P.; Junier, T.; Podell, S.; Sims, D. R.; Detter, J. C.; Lykidis, A.; Han, C. S.; Wigginton, N. S.; Terry, G.; Bernier-Latmani, R. The genome of the Gram-positive metal- and sulfate-reducing bacterium *Desulfotomaculum reducens* stain MI-1. *Environ. Microbiol.* **2010**.
182. Khijniak, T. V.; Slobodkin, A. I.; Coker, V.; Renshaw, J. C.; Livens, F. R.; Bonch-Osmolovskaya, E. A.; Birkeland, N.-K.; Medvedeva-Lyalikova, N. N.; Lloyd, J. R. Reduction of Uranium(VI) phosphate during growth of the thermophilic bacterium *Thermoterrabacterium ferrireducens*. *Appl. Environ. Microbiol.* **2005**, *71*, 6423–6426.

183. Lee, S. Y.; Baik, M. H.; Choi, J. W. Biogenic formation and growth of uraninite (UO_2). *Environ. Sci. Technol.* **2010**, *44*, 8409–8414.
184. Liu, C.; Gorby, Y. A.; Zachara, J. M.; Fredrickson, J. K.; Brown, C. F. Reduction kinetics of Fe(III), Co(III), U(VI), Cr(VI), and Tc(VII) in cultures of dissimilatory metal-reducing bacteria. *Biotechnol. Bioeng.* **2002**, *80*, 637–649.
185. Lovley, D. R.; Phillips, E. J. P.; Gorby, Y. A.; Landa, E. R. Microbial reduction of uramium. *Nature* **1991**, *350*, 413–416.
186. Lovley, D. R.; Phillips, E. J. P. Bioremediation of uranium contamination with enzymatic uranium reduction. *Environ. Sci. Technol.* **1992**, *26*, 2228–2234.
187. Marshall, M. J.; Beliaev, A. S.; Dohnalkova, A. C.; Kennedy, D. W.; Shi, L.; Wang, Z.; Boyanov, M. I.; Lai, B.; Kemner, K. M.; McLean, J. S.; Reed, S. B.; Culley, D. E.; Bailey, V. L.; Simonson, C. J.; Saffarini, D. A.; Romine, M. F.; Zachara, J. M.; Fredrickson, J. K. *c*-Type cytochrome-dependent formation of U(VI) nanoparticles by *Shewanella oneidensis*. *PLOS Biol.* **2006**, *4*, 1324–1333.
188. Marshall, M. J.; Dohnalkova, A. C.; Kennedy, D. W.; Plymale, A. E.; Thomas, S. H.; Löffler, F. E.; Sanford, R. A.; Zachara, J. M.; Fredrickson, J. K., Beliaev, A. S. Electron donor-dependent radionuclide reduction and nanoparticle formation by *Anaeromyxobacter dehalogenans* strain 2CP-C. *Environ. Microbiol.* **2009**, *11*, 534–543.
189. Sanford, R. A.; Wu, Q.; Sung, Y.; Thomas, S. H.; Amos, B. K.; Prince, E. K.; Löffler, F. E. Hexavalent uranium supports growth of *Anaeromyxobacter dehalogenans* and *Geobacter* spp. with lower than predicted biomass yields. *Environ. Microbiol.* **2007**, *9*, 2885–2893.
190. Schofield, E. J.; Veeramani, H.; Sharp, J. O.; Suvorova, E.; Bernier-Latmani, R.; Metha, A.; Stahlman, J.; Webb, S. M.; Clark, D. L.; Conradson, S. D.; Ilton, E. S.; Bargar, J. R. Structure of biogenic uraninite produced by *Shewanella oneidensis* strain MR-1. *Environ. Sci. Technol.* **2008**, *42*, 7898–7904.
191. Sharp, J. O.; Schofield, E. J.; Veeramani, H.; Suvorova, E. I.; Kennedy, D. W.; Marshall, M. J.; Metha, A.; Bargar, J. R.; Bernier-Latmani, R. Structural similarities between biogenic uraninites produced by phylogenetically and metabolically diverse bacteria. *Environ. Sci. Technol.* **2009**, *43*, 8295–8301.
192. Singer, D. A.; Farges, F.; Brown, G. E., Jr. Biogenic nanoparticulate UO_2: Synthesis, characterization, and factors affecting surface reactivity. *Geochim. Cosmochim. Acta* **2009**, *73*, 3593–3611.
193. Suzuki, Y.; Kelly, S. D.; Kemner, K. M.; Banfield, J. F. Enzymatic U(VI) reduction by *Desulfosporosinus* species. *Radiochim. Acta* **2004**, *92*, 11–16.
194. Suzuki, Y.; Tanaka, K.; Kozai, N.; Ohnuki, T. Effects of citrate, NTA, and EDTA on the reduction of U(VI) by *Shewanella putrefaciens*. *Geomicrobiol. J.* **2010**, *27*, 245–250.
195. Boukhalfa, H.; Icopini, G. A.; Reilly, S. D.; Neu, M. P. Plutonium(IV) reduction by the metal-reducing bacteria *Geobacter metallireducens* GS15 and *Shewanella oneidensis* MR1. *Appl. Environ. Microbiol.* **2007**, *73*, 5897–5903.

196. De Luca, G.; De Philip, P.; Dermoun, Z.; Rousset, M.; Verméglio, A. Reduction of technetium(VII) by *Desulfovibrio fructosovorans* is mediated by the nickel-iron hydrogenase. *Appl. Environ. Microbiol.* **2001**, *67*, 4583–4587.
197. Francis, A. J.; Dodge, C. J.; Gillow, J. B. Reductive dissolution of Pu(IV) by *Clostridium* sp. under anaerobic conditions. *Environ. Sci. Technol.* **2008**, *42*, 2355–2360.
198. Icopini, G. A.; Lack, J. G.; Hersman, L. E.; Neu, M. P.; Boukhalfa, H. Plutonium(V/VI) reduction by the metal-reducing bacteria *Geobacter metallireducens* GS-15 and *Shewanella oneidensis* MR-1. *Appl. Environ. Microbiol.* **2009**, *75*, 3641–3647.
199. Kashefi, K.; Lovley, D. R. Reduction of Fe(III), Mn(IV), and toxic metals at 100 °C by *Pyrobacterium islandicum. Appl. Environ. Microbiol.* **2000**, *66*, 1050–1056.
200. Lloyd, J. R.; Mabbett, A. N.; Williams, D. R.; MaCaskie, L. E. Metal reduction by sulphate-reducing bacteria: physiological diversity and metal specificity. *Hydrometallurgy* **2001**, *59*, 327–337.
201. Lloyd, J. R.; Ridley, J.; Khizniak, T.; Lyalikova, N. N.; MaCaskie, L. E. Reduction of technetium by *Desulfovibrio desulfuricans:* Biocatalyst characterization and use in a flowthrough bioreactor. *Appl. Environ. Microbiol.* **1999**, *65*, 2691–2696.
202. Lloyd, J. R.; Sole, V. A.; Van Praagh, C. V. G.; Lovley, D. R. Direct and Fe(II)-mediated reduction of technetium by Fe(III)-reducing bacteria. *Appl. Environ. Microbiol.* **2000**, *66*, 3743–3749.
203. Lloyd, J. R.; Yong, P.; Macaskie, L. E. Biological reduction and removal of Np(V) by two microorganisms. *Environ. Sci. Technol.* **2000**, *34*, 1297–1301.
204. Neu, M. P.; Icopini, G. A.; Boukhalfa, H. Plutonium speciation affected by environmental bacteria. *Radiochim. Acta* **2005**, *93*, 705–714.
205. Renshaw, J. C.; Law, N.; Geissler, A.; Livens, F. R.; Lloyd, J. R. Impact of the Fe(III)-reducing bacteria *Geobacter sulfurreducens* and *Shewanella oneidensis* on the speciation of plutonium. *Biogeochem.* **2009**, *94*, 191–196.
206. Wildung, R. E.; Gorby, Y. A.; Krupka, K. M.; Hess, N. J.; Li, S. W.; Plymale, A. E.; McKinley, J. P.; Fredrickson, J. K. Effect of electron donor and solution chemistry on products of dissimilatory reduction of technetium by *Shewanella putrefaciens. Appl. Environ. Microbiol.* **2000**, *66*, 2451–2460.
207. Bencheikh-Latmani, R.; Williams, S. M.; Haucke, L.; Criddle, C. S.; Wu, L.; Zhou, J.; Tebo, B. M. Global transcriptional profiling of *Shewanella oneidensis* MR-1 during Cr(VI) and U(VI) reduction. *Appl. Environ. Microbiol.* **2005**, *71*, 7453–7460.
208. Shelobolina, E. S.; Coppi, M. V.; Korenevsky, A. A.; DiDonato, L. N.; Sullivan, S. A.; Konishi, H.; Xu, H.; Leang, C.; Butler, J. E.; Kim, B. H.; Lovley, D. R. Importance of *c*-type cytochromes for U(VI) reduction by *Geobacter sulfurreducens. BMC Microbiol.* **2007**, *7*, 1–15.
209. Suzuki, Y.; Kelly, S. D.; Kemner, K. M.; Banfield, J. F. Nanometre-size products of uranium bioreduction. *Nature* **2002**, *419*, 134.

210. Burgos, W. D.; Senko, J. M.; Dempsey, B. A.; Roden, E. E.; Stone, J. J.; Kemner, K. M.; Kelly, S. D. Soil humimc acid decreases biological uranium(VI) reduction by *Shewanella putrefaciens* CN32. *Environ. Eng. Sci.* **2007**, *24*, 755–761.
211. Francis, A. J.; Dodge, C. J. Bioreduction of uranium(VI) complexes with citric acid by *Clostridia* affects its structure and solubility. *Environ. Sci. Technol.* **2008**, *42*, 8277–8282.
212. Sivaswamy, V.; Boyanov, M. I.; Peyton, B. M.; Viamajala, S.; Gerlach, R.; apel, W. A.; Sani, R. K.; Dohnalkova, A.; Kemner, K. M.; Borch, T. Multiple mechanisms of uranium immobilization by *Cellulomonas* sp. strain ES6. *Biotechnol. Bioeng.* **2011**, *108*, 264–276.
213. Rusin, P. A.; Quintana, L.; Brainard, J. R.; Strieteimeier, B. A.; Tait, C. D.; Ekberg, S. A.; Palmer, P. D.; Newton, T. W.; Clark, D. L. Solubilization of plutonium hydrous oxide by iron-reducing bacteria. *Environ. Sci. Technol.* **1994**, *28*, 1686–1690.
214. DiSpirito, A. A.; Tuovinen, O. H. Uranous ion oxidation and carbon dioxide fixation by *Thiobacillus ferrooxidans*. *Arch. Microbiol.* **1982**, *133*, 28–32.
215. Beller, H. R. Anaerobic, nitrate-dependent oxidation of U(VI) oxide minerals by the chemolithoautotrophic bacterium *Thiobacillus denitrificans*. *Appl. Environ. Microbiol.* **2005**, *71*, 2170–2174.
216. Finneran, K. T.; Housewright, M. E.; Lovley, D. R. Multiple influences of nitrate on uranium solubility during bioremediation of uranium-contaminated subsurface sediments. *Environ. Microbiol.* **2002**, *4*, 510–516.
217. Weber, K. A.; Thrash, J. C.; Van Trump, J. I.; Achenbach, L. A.; Coates, J. D. Environmental and taxanomic bacterial diversity of anaerobic uranium(IV) bio-oxidation. *Appl. Environ. Microbiol.* **2011**, *77*, 4693–4696.
218. Lovley, D. R.; Stolz, J. F.; Nord, G. L., Jr.; Phillips, E. J. P. Anaerobic production of magnetite by a dissimilatory iron-reducing microorganism. *Nature* **1987**, *330*, 252–254.
219. Kukkadapu, R. K.; Zachara, J. M.; Fredrickson, J. K.; Kennedy, D. W.; Dohnalkova, A. C.; Mccready, D. E. Ferrous hydroxy carbonate is a stable transformation product of biogenic magnetite. *Am. Mineral.* **2005**, *90*, 510–515.
220. O'Loughlin, E. J.; Larese-Casanova, P.; Scherer, M. M.; Cook, R. E. Green rust formation from the bioreduction of γ-FeOOH (lepidocrocite): Comparison of several *Shewanella* species. *Geomicrobiol. J.* **2007**, *24*, 211–230.
221. Ona-Nguema, G.; Abdelmoula, M.; Jorand, F.; Benali, O.; Géhin, A.; Block, J.-C.; Génin, J.-M. R. Microbial reduction of lepidocrocite γ-FeOOH by *Shewanella putrefaciens*; The formation of green rust. *Hyperfine Interact.* **2002**, *139/140*, 231–237.
222. Glasauer, S.; Weidler, P. G.; Langley, S.; Beveridge, T. J. Controls on Fe reduction and mineral formation by a subsurface bacterium. *Geochim. Cosmochim. Acta* **2003**, *67*, 1277–1288.
223. Fredrickson, J. K.; Zachara, J. M.; Kennedy, D. W.; Dong, H.; Onstott, T. C.; Hinman, N. W.; Li, S.-M. Biogenic iron mineralization accompanying the

dissimilatory reduction of hydrous ferric oxide by a groundwater bacterium. *Geochim. Cosmochim. Acta* **1998**, *62*, 3239–3257.

224. Benning, L. G.; Wilkin, R. T.; Konhauser, K. O. Iron monosulfide stability: Experiments with sulfate reducing bacteria. In *Geochemistry of the Earth's surface*, Ármannsson, H., Ed.; A. A. Balkema: Brookfield, Rotterdam, 1999; pp 429–432.
225. Donald, R.; Southam, G. Low temperature anaerobic bacterial diagenesis of ferrous monosulfide to pyrite. *Geochim. Cosmochim. Acta* **1999**, *63*, 2019–2023.
226. Herbert, R. B., Jr.; Benner, S. G.; Pratt, A. R.; Blowes, D. W. Surface chemistry and morphology of poorly crystalline iron sulfides precipitated in media containing sulfate-reducing bacteria. *Chem. Geol.* **1998**, *144*, 87–97.
227. Neal, A. L.; Techkarnjanaruk, S.; Dohnalkova, A.; Mccready, D. E.; Peyton, B. M.; Geesey, G. G. Iron sulfides and sulfur species produced at hematite surfaces in the presence of sulfate-reducing bacteria. *Geochim. Cosmochim. Acta* **2001**, *65*, 223–235.
228. Rickard, D. T. The microbiological formation of iron sulphides. *Stockholm Contrib. Geol.* **1969**, *20*, 49–66.
229. Vaughan, D. J.; Lennie, A. R. The iron sulfide minerals: their chemistry and role in nature. *Sci. Prog.* **1991**, *75*, 371–388.
230. Howarth, R. W.; Stewart, J. W. B. The interactions of sulphur with other element cycles in ecosystems. In *Sulphur cycling on the continents: Wetlands, terrestrial ecosystems, and associated water bodies*; Howarth, R. W., Stewart, J. W. B., Ivanov, M. V., Eds.; John Wiley & Sons: Chichester, 1992; Vol. SCOPE 48, pp 67–84.
231. Lovley, D. R. Dissimilatory metal reduction. *Annu. Rev. Microbiol.* **1993**, *47*, 263–290.
232. Nealson, K. H.; Saffarini, D. A. Iron and manganese in anaerobic respiration: Environmental significance, physiology, and regulation. *Annu. Rev. Microbiol.* **1994**, *48*, 311–343.
233. Field, J. A.; Cervantes, F. J. Microbial redox reactions mediated by humics and structurally related quinones. In *Use of humic substances to remediate polluted environments: From theory to practice*; Perminova, I. V., Hatfield, K., Hertkorn, N., Eds.; Springer: Dordrecht, 2005; Vol. 52, pp 343–352.
234. Bargar, J. R.; Tebo, B. M.; Bergmann, U.; Webb, S. M.; Glatzel, P.; Chiu, V. Q.; Villalobos, M. Biotic and abiotic products of Mn(II) oxidation by spores of the marine *Bacillus* sp. strain SG-1. *Am. Mineral.* **2005**, *90*, 143–154.
235. Kappler, A.; Newman, D. K. Formation of Fe(III)-minerals by Fe(II)-oxidizing photoautotrophic bacteria. *Geochim. Cosmochim. Acta* **2004**, *68*, 1217–1226.
236. Kappler, A.; Straub, K. L. Geomicrobiological cycling of iron. In *Molecular Geomicrobiology*; Banfield, J. F., Cervini-Silva, J., Nealson, K. H., Eds.; Mineralogical Society of America: Washington, DC, 2005; Vol. 59, pp 85–108.
237. Lack, J. G.; Chaudhuri, S. K.; Chakraborty, R.; Achenbach, L. A.; Coates, J. D. Anaerobic biooxidation of Fe(II) by *Dechlorosoma suillum*. *Microb. Ecol.* **2002**, *43*, 424–431.

238. Tebo, B. M.; Johnson, H. A.; McCarthy, J. K.; Templeton, A. S. Geomicrobiology of manganese(II) oxidation. *Trends Microbiol.* **2005**, *13*, 421–428.
239. Villalobos, M.; Toner, B.; Bargar, J.; Sposito, G. Characterization of the manganese oxide produced by *Pseudomonas putida* strain MnB1. *Geochim. Cosmochim. Acta* **2003**, *67*, 2649–2662.
240. Weber, K. A.; Achenbach, L. A.; Coates, J. D. Microorganisms pumping iron: anaerobic microbial iron oxidation and reduction. *Nat. Rev. Microbiol.* **2006**, *4*, 752–764.
241. Plymale, A. E.; Frederickson, J. K.; Zachara, J. M.; Dohnalkova, A. C.; Heald, S. M.; Moore, D. A.; Kennedy, D. W.; Marshall, M. J.; Wang, C.; Resch, C. T.; Nachimuthu, P. Competetive reduction of pertechnetate ($^{99}TcO_4^-$) by dissimilatory metal reducing bacteria and biogenic Fe(II). *Environ. Sci. Technol.* **2011**, *45*, 951–957.
242. Behrends, T.; Van Cappellen, P. Competition between enzymatic and abiotic reduction of uranium(VI) under iron reducing conditions. *Chem. Geol.* **2005**, *220*, 315–327.
243. Finneran, K. T.; Anderson, R. T.; Nevin, K. P.; Lovley, D. R. Potential for bioremediation of uranium-contaminated aquifers with microbial U(VI) reduction. *Soil Sediment Contam.* **2002**, *11*, 339–357.
244. Lack, J. G.; Chaudhuri, S. K.; Kelly, S. D.; Kemner, K. M.; O'Connor, S. M.; Coates, J. D. Immobilization of radionuclides and heavy metals through anaerobic bio-oxidation of Fe(II). *Appl. Environ. Microbiol.* **2002**, *68*, 2704–2710.
245. Senko, J. M.; Suflita, J. M.; Krumholz, L. R. Geochemical controls on microbial nitrate-dependent U(IV) oxidation. *Geomicrobiol. J.* **2005**, *22*, 371–378.
246. Wu, W.-M.; Carley, J.; Green, S. J.; Luo, J.; Kelly, S. D.; Van Nostrand, J.; Lowe, K.; Mehlhorn, T.; Carrol, S.; Boonchayaanant, B.; Löffler, F. E.; Watson, D.; Kemner, K. M.; Zhou, J.; Kitanidis, P. K.; Kostka, J. E.; Jardine, P. M.; Criddle, C. S. Effects of nitrate on the stability of uranium in a bioreduced region of the subsurface. *Environ. Sci. Technol.* **2010**, *44*, 5104–5111.
247. O'Loughlin, E. J. Effects of electron transfer mediators on the biodegradation of lepidocrocite (γ-FeOOH) by *Shewanella putrefaciens* CN32. *Environ. Sci. Technol.* **2008**, *42*, 6876–6882.
248. Stookey, L. L. Ferrozine-A new spectrophotometric reagent for iron. *Anal. Chem.* **1970**, *42*, 779–781.
249. Sørensen, J. Reduction of ferric iron in anaerobic, marine sediment and interaction with reduction of nitrate and sulfate. *Appl. Environ. Microbiol.* **1982**, *43*, 319–324.
250. Cole, J. R.; Wang, Q.; Fish, J.; Chai, B.; Farris, R. J.; Kulam, S. A.; McGarrell, D. M.; Marsh, T.; Garrity, G. M.; Tiedje, J. M. The Ribosomal Database Project: Improved alignments and new tools for rRNA analysis. *Nucleic Acids Res.* **2009**, *37*, D141–D145.
251. Segre, C. U.; Leyarovska, N. E.; Chapman, L. D.; Lavender, W. M.; Plag, P. W.; King, A. S.; Kropf, A. J.; Bunker, B. A.; Kemner, K. M.; Dutta,

P.; Duran, R. S.; Kaduk, J. The MRCAT insertion device beamline at the Advanced Photon Source. In *Synchrotron Radiation Instrumentation: Eleventh U.S. National Conference*; Pianetta, P. A., Arthur, J. R., Brennan, S., Eds.; American Institute of Physics: New York, 2000; Vol. CP5321, pp 419–422.

252. Kemner, K. M.; Kelly, S. D. Synchrotron-based techniques for monitoring metal transformation. In *Manual of Environmental Microbiology*, 3rd ed.; Hurst, C. J., Ed.; ASM Press: Washington, DC, 2007; pp 1183–1194.
253. Kodama, Y.; Watanabe, K. *Sulfuricurvum kujiense* gen. nov., sp. nov., a facultatively anaerobic, chemolithoautotrophic, sulfur-oxidizing bacterium isolated from an underground crude-oil storage cavity. *Int. J. Syst. Evol. Microbiol.* **2004**, *54*, 2297–2300.
254. Anderson, R. T.; Vrionis, H. A.; Ortiz-Bernad, I.; Resch, C. T.; Long, P. E.; Dayvault, R.; Karp, K.; Marutzky, S.; Metzler, D. R.; Peacock, A.; White, D. C.; Lowe, M.; Lovley, D. R. Stimulating the in situ activity of *Geobacter* species to remove uranium from the groundwater of a uranium-contaminated aquifer. *Appl. Environ. Microbiol.* **2003**, *69*, 5884–5891.
255. Burkhardt, E.-M.; Akob, D. M.; Bischoff, S.; Sitte, J.; Kostka, J. E.; Banerjee, D.; Scheinost, A. C.; Küsel, K. Impact of biostimulated redox processes on metal dynamics in an iron-rich creek soil in a former uranium mining area. *Environ. Sci. Technol.* **2010**, *44*, 177–183.
256. Fredrickson, J. K.; Kostandarithes, H. M.; Li, S. W.; Plymale, A. E.; Daly, M. J. Reduction of Fe(III), Cr(VI), U(VI), and Tc(VII) by *Deinococcus radiodurans* R1. *Appl. Environ. Microbiol.* **2000**, *66*, 2006–2011.
257. Boonchayaanant, B.; Gu, B.; Wang, W.; Ortiz, M. E.; Criddle, C. S. Can microbially-generated hydrogen sulfide account for the rates of U(VI) reduction by a sulfate-reducing bacterium? *Biodegradation* **2009**.
258. Suzuki, Y.; Kelly, S. D.; Kemner, K. M.; Banfield, J. F. Direct microbial reduction and subsequent preservation of uranium in natural near-surface sediment. *Appl. Environ. Microbiol.* **2005**, *71*, 1790–1797.
259. Yi, Z.-J.; Tan, K.-X.; Tan, A.-L.; Yu, Z.-X.; Wang, S.-Q. Influence of environmental factors on reductive bioprecipitation of uranium by sulfate reducing bacteria. *Int. Biodeterior. Biodegrad.* **2007**, *60*, 258–266.
260. Langmuir, D. *Aqueous environmental geochemistry*; Prentice-Hall, Inc.: Upper Saddle River, NJ, 1997; p 600.
261. Thamdrup, B. Bacterial manganese and iron reduction in aquatic sediments. *Adv. Microb. Ecol.* **2000**, *16*, 41–84.
262. Banaszak, J. E.; Rittmann, B. E.; Reed, D. T. Subsurface interactions of actinide species and microorganisms: Implications for the bioremediation of actinide-organic mixtures. *J. Radioanal. Nucl. Chem.* **1999**, *241*, 385–435.
263. Icopini, G. A.; Boukhalfa, H.; Neu, M. P. A brief review of the enzymatic reduction of Tc, U, Np, and Pu by bacteria. In *Recent Advances in Actinide Science*; May, I., Alvares, R., Bryan, N. D., Eds.; Royal Society of Chemistry: London, 2006; pp 20–25.
264. Guillaumont, R.; Fanghänel, T.; Neck, V.; Fuger, J.; Palmer, D. A.; Grenthe, I.; Rand, M. H. *Update on the chemical thermodynamics of uranium,*

neptunium, plutonium, americium and technetium; Elsevier: Amsterdam, 2003; Vol. 5.

Chapter 23

Rate Controlling Processes in the Transformation of Tetrachloroethylene and Carbon Tetrachloride under Iron Reducing and Sulfate Reducing Conditions

Elizabeth C. Butler,[*,1] Yiran Dong,[1,3] Lee R. Krumholz,[2] Xiaoming Liang,[1,4] Hongbo Shao,[1,5] and Yao Tan[1,6]

[1]School of Civil Engineering and Environmental Science, University of Oklahoma, 202 W. Boyd, Room 334, Norman, OK 73019, USA
[2]Department of Botany and Microbiology and Institute for Energy and the Environment, University of Oklahoma, 770 Van Vleet Oval, Norman, OK 73019, USA
[3]Current address: Energy Biosciences Institute, University of Illinois-Urbana Champaign, 1206 W. Gregory Drive, Urbana, IL 61801, USA
[4]Current address: Department of Geology, University of Toronto, 22 Russell Street Toronto, ON, M5S 3B1, Canada
[5]Current address: Geosciences Group, Pacific Northwest National Laboratory, P.O. Box 999, Richland, WA 99352, USA
[6]Current address: Foreign Economic Cooperation Office, Ministry of Environmental Protection, Houyingfanghutong #5, Xicheng District, Beijing, 100035, People's Republic of China
[*]ecbutler@ou.edu

While *in situ* dechlorination of chlorinated aliphatic contaminants such as tetrachloroethylene (PCE) and carbon tetrachloride (CT) has been studied extensively, rate controlling processes in the transformation of these compounds remain uncertain. The objectives of this work were (1) to compare the relative rates of abiotic and microbial transformation of PCE and CT in microcosms designed to simulate natural conditions, and (2) for CT, to measure the relative rates of reactive mineral formation and CT transformation by these reactive minerals. While the rates of microbial dechlorination exceeded the rates of abiotic dechlorination of PCE in the microcosms, the opposite trend was observed for CT. The times required for microbial sulfate reduction, the first step in formation of many reactive minerals, were significantly longer than those required for transformation of CT by these reactive minerals, indicating possible rate control of abiotic CT transformation by microbial respiration under natural conditions.

Introduction

Chlorinated aliphatic hydrocarbons are widespread ground water contaminants in the United States and other countries that have a legacy of industrial activity. Despite advances in treatment and remediation technologies for aquifers contaminated with such contaminants, however, there remains a knowledge gap about rate controlling processes in their transformation under a range of natural or engineered conditions. Contaminants such as tetrachloroethylene (PCE) and carbon tetrachloride (CT) can undergo both abiotic (Figure 1, process 1) and microbial (Figure 1, process 2) dechlorination via a series of electron transfer steps. For abiotic dechlorination, these steps include microbial respiration of Fe(III) oxides and/or sulfate (Figure 1, process 3), leading to formation of reduced mineral species, such as adsorbed Fe(II), FeS, and green rusts, that have the potential to reductively transform chlorinated aliphatic contaminants such as CT (*1–3*) (Figure 1, process 4). Steps 3 and 4 (Figure 1) may include sub steps not shown, for example, generation of intermediates such as dissolved S(-II) that causes reductive dissolution of Fe(III) oxides (Step 3) and transformation of reduced minerals containing Fe(II) and S(-II) to products less oxidized than Fe(III) oxides and sulfate (Step 4).

Both abiotic and microbial dechlorination of CT can lead to a range of dechlorination products (*4*), none of which is considered a definitive marker of either abiotic or microbial degradation. Geochemical conditions have been shown to strongly influence the abiotic CT product distribution (*5*). For PCE, the pathways and products of abiotic versus microbial dechlorination are more distinct than those for CT (*6*). Microbial reductive dechlorination of PCE proceeds primarily by sequential hydrogenolysis that forms trichloroethylene (TCE), dichloroethylene isomers (particularly *cis* 1,2-dichloroethylene (*cis*-DCE)), vinyl chloride (VC), and ethene (*7*), while abiotic mineral-mediated reductive dechlorination of PCE typically proceeds by reductive β-elimination (*8*), yielding the reaction product acetylene via unstable intermediates (Figure 2). Acetylene, in turn, can be transformed rapidly by microbial (*9, 10*) and abiotic (*11*) processes to products including ethene, ethanol, acetaldehyde, and acetate. Abiotic reductive transformation of the PCE analog TCE also led to formation of glycolate, acetate, and formate, among other products (*12*), possibly via an acetylene intermediate. Detection of acetylene is not common at contaminated sites, and the presence of small organic acids, alcohols, and aldehydes in the subsurface cannot be attributed to abiotic PCE or TCE transformation, since such compounds are formed naturally as a result of microbial metabolism. While formation of TCE, *cis*-DCE, and/or VC are likely to arise primarily from microbial reductive dechlorination of PCE, mineral mediated transformation of PCE can also lead to small yields of these products (*6*). Nonetheless, disappearance of PCE from contaminated sites without the stoichiometric production of hydrogenolysis products is commonly attributed to abiotic dechlorination.

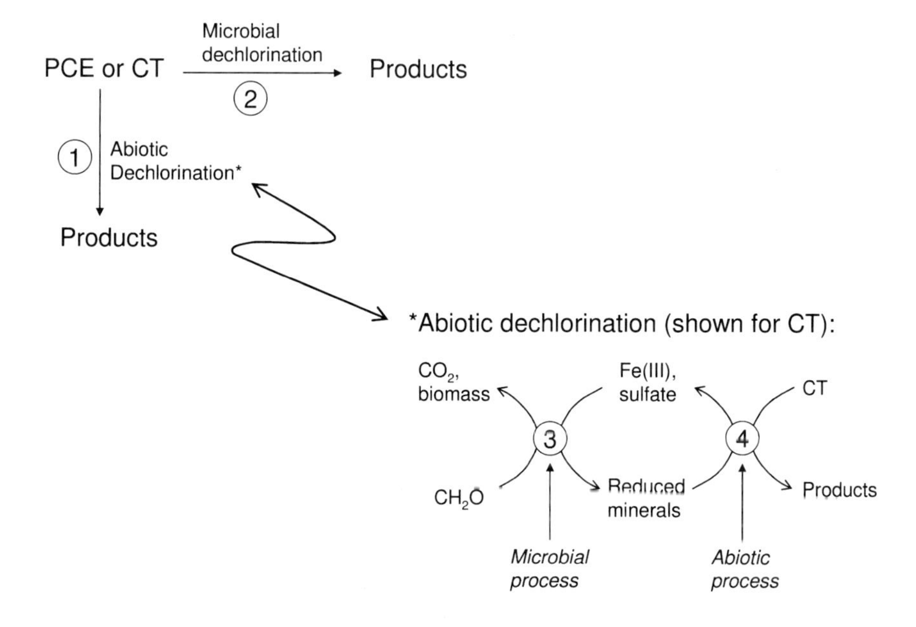

Figure 1. PCE and CT transformation processes.

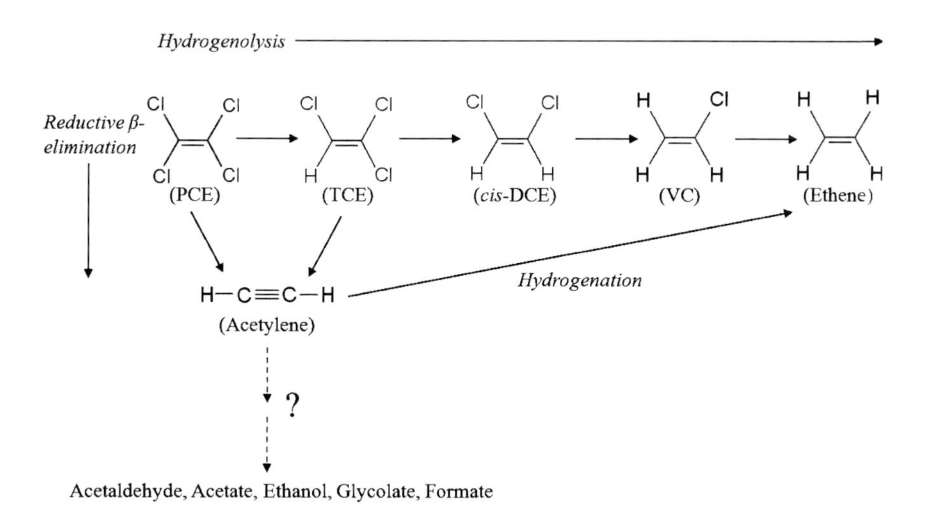

Figure 2. Pathways of PCE dechlorination. Pathways shown by dashed arrows are uncertain. Adapted with permission from ref. (65) (copyright 2001 American Chemical Society), and updated with information from refs. (11) and (12).

The specific abiotic minerals that are most reactive in abiotic reductive dechlorination vary with the chlorinated contaminant. Minerals observed to react most rapidly with PCE (in order of decreasing surface area normalized rate constants or percent PCE removed per time) are FeS (*13–15*) followed by green rusts and pyrite (*11*, *16*, *17*), for which overlapping rate constants have been reported. Magnetite (*11*, *16*) was less reactive with PCE than were FeS, chloride green rust, or pyrite. $MgCl_2$-extractable or "weakly bound" Fe(II) associated with the surface of the Fe(III) oxide goethite did not show significant reactivity with PCE over 7-8 months (*11*).

For CT, the largest surface area normalized rate constants have been observed in the presence of FeS (*2*, *18*) followed by pyrite (*5*), chloride and sulfate green rusts (*19*, *20*) and magnetite (*18*, *21*, *22*), with pyrite, green rusts, and magnetite having a range of overlapping rate constants. The Fe(II) mineral least reactive with CT appears to be siderite (*18*).

In addition to the well-defined minerals listed above, CT–unlike PCE–reacts rapidly with Fe(II) adsorbed to otherwise less reactive Fe minerals. For example, in a series of experiments performed under the same conditions with respect to pH and other solution amendments, goethite equilibrated with dissolved Fe(II) ("Fe(II)-goethite") showed both higher a concentration of weakly bound Fe(II) and a higher rate of CT transformation than did FeS (*23*). Surface area normalized rates or rate constants were not reported for these systems and so cannot be exactly compared. But while the total concentration of solid-phase Fe(II) in the Fe(II)-goethite system was approximately three times higher than in the FeS system, the concentration of weakly bound Fe(II) was almost 200 times larger and the initial rate of CT transformation was almost 600 times larger in the presence of Fe(II)-goethite versus FeS (*23*), pointing to the high reactivity of weakly bound Fe(II) with CT. This highly reactive weakly bound Fe(II) can be sorbed to the surfaces of Fe(III) oxides (*1*, *18*, *23–26*), magnetite (*9*, *18*), pyrite (*5*), FeS (*23*), and presumably other minerals. The role of the underlying Fe mineral in the reactivity of weakly bound Fe(II) may be to regenerate or re-reduce sorbed Fe(II) (*23*). Consistent with this, Williams and Scherer (*27*) used Mössbauer spectroscopy to demonstrate electron transfer between adsorbed Fe(II) and bulk Fe(III) oxides.

The suite of chlorinated aliphatic compounds detected at contaminated sites includes those that degrade relatively quickly, such as CT, and those that degrade more slowly, such as PCE. Rates for both abiotic and microbial transformation of chlorinated aliphatic contaminants vary widely with environmental properties such as redox conditions (*28–30*), which influences both microbial and geochemical conditions, and contaminant properties such as reduction potentials and carbon-chlorine bond strengths (*30–32*). But rates and rate constants for abiotic transformation of contaminants such as CT and PCE appear to vary more than those for microbial degradation of these contaminants. We compared reported first order rate constants for abiotic reductive dechlorination of CT and PCE by chloride green rust, pyrite, and FeS from studies conducted under similar conditions in the same laboratory. The results (Table I) indicate that abiotic CT transformation by a given mineral is two to almost five orders of magnitude faster than PCE transformation by the same mineral. (Note: the units and experimental conditions for the apparent rate constants reported Table I are consistent within rows, but not within the columns displaying log *k* values, meaning that only log *k* values in the same row can be fairly compared. The dimensionless values in the last column can be fairly compared.) The range of apparent rate constants for microbial transformation of PCE and CT appears to be smaller than those for abiotic transformation (Table I). For example, a review of 11 studies of microbial CT transformation and 36 studies of microbial PCE degradation found that the mean pseudo first order rate constants for the two contaminants differed by less than an order of magnitude (*29*). Another study (*28*) found a factor of 6-7 difference in normalized rate constants for PCE versus CT transformation in methanogenic biofilm reactors, which is also less than the typical difference between rate constants for PCE and CT transformation observed in abiotic systems (Table I). The fact that rate constants for mineral mediated reductive dechlorination vary over a greater range than those for microbial reductive dechlorination suggests the possibility that microbial transformations may be more important than abiotic transformations for slower reacting compounds (e.g., PCE), but not for faster reacting compounds (e.g., CT). Half-lives for CT transformation by abiotic mineral reductants range from approximately one-half to three days (*1*, *24*, *25*, *33*, *34*).

The overall objective of this chapter is to review rate limiting processes for the transformation of chlorinated aliphatic contaminants in natural systems. The first specific objective is to discuss the relative importance of abiotic versus microbial reductive dechlorination for two model contaminants; one that reacts relatively slowly (PCE) and one that reacts more quickly (CT). The second specific objective is to discuss the relative rates of mineral formation and subsequent reaction of these minerals with a relatively fast reacting contaminant, CT. Finally, the results are integrated to make recommendations for remediation applications. These results are relevant for systems under near natural conditions as well as engineered systems in which microbial activity or the mass loading of reactive minerals is enhanced above background levels through, for example, biostimulation.

Methods

General Procedures

The sources and purities of all chemical reagents are given in refs. (*10*) and (*35*). All experiments were conducted in an anaerobic chamber containing an atmosphere of 95% N_2 and 5% H_2 and a Pd catalyst to remove trace O_2 (Coy Laboratory Products, Grass Lake, MI). Water was from a Barnstead (Dubuque, IA) Ultrapure water system with a resistivity of 18 MΩ cm. Anoxic water was prepared by sparging boiled water with N_2 for at least 30 minutes. Materials used to prepare microcosms (e.g., serum bottles, aqueous solutions, and pipette tips) were autoclaved prior to microcosm preparation.

Table I. Log apparent first order rate constants for PCE and CT reductive transformation in mineral and microbial systems

System	Log k^a		Difference ($\log k_{CT} - \log k_{PCE}$)
	PCE	CT	
Mineral systems			
Chloride green rust (Study 1)[b]	-5.2	-0.82	4.4
Chloride green rust (Study 2)[c]	-1.9	-0.29	2.2
Pyrite[d]	-5.8	-1	4.8
FeS[e]	-1.6	-0.72	2.3
Microbial systems			
Field and laboratory studies[f]	-1.2	-0.91	0.29
Methanogenic biofilm reactors[g]	-1.0	-0.20	0.80

[a] Units of k are L m^{-2} d^{-1} for chloride green rust (study 1), pyrite, and FeS, d^{-1} for chloride green rust (study 2), and d^{-1} for "Field and laboratory studies". For "Methanogenic biofilm reactors", k is equal to k'/K_S, where k' = the Monod maximum specific substrate utilization rate and K_s = the Monod half velocity constant. [b] Data are from ref. (*11*) (PCE) and ref. (*20*) (CT); relevant conditions: pH 8 (HEPES) (for PCE), pH 7.6-8.0 (no buffer) (for CT), approx. 200 m^2/L chloride green rust; initial concentration of PCE or CT: approx. 3×10^{-5} M. [c] Data are from ref. (*3*); relevant conditions: pH 7.2 (HEPES), 1.5 g/L chloride green rust, initial concentration of PCE or CT: approx. 2×10^{-5} M; rate constant for PCE estimated from Fig. 3 in ref. (*3*). [d] Data are from ref. (*11*) (PCE) and ref. (*5*) (CT); relevant conditions: pH 8 (HEPES), 578 m^2/L pyrite (for PCE), 3.6 m^2/L pyrite (for CT); initial concentration of PCE: approx. 3×10^{-5} M, initial concentration of CT: approx. 2×10^{-5} M. [e] Data are from ref. (*13*) (PCE) and ref. (*2*) (CT); relevant conditions: pH 8.3 (TRIS), 0.5 m^2/L FeS, initial concentrations of PCE and CT: 2×10^{-5} M. [f] Data are from ref. (*29*), and include a range of field and laboratory conditions. [g] Data are from ref. (*28*).

Microcosm Setup

Microcosms were prepared inside the anaerobic chamber in duplicate or triplicate in sterilized 160 mL serum bottles that were closed with Teflon coated septa and aluminum crimp seals. Phase I microcosms contained 20 g sediment and 100 mL water from a pond in Norman, Oklahoma (the Duck Pond) and were buffered at pH 7.2 with 25 mM 4-(2-hydroxyethyl)-1-piperazineethanesulfonic acid (HEPES). The headspace volume was 44-50 mL and was flushed with N_2 prior to incubation. Further details about the sampling location and the sediment/water properties are given in refs. (*10*) and (*35*).

Microcosms were amended to promote either sulfate reducing or iron reducing conditions by addition of 30 mM $FeSO_4$ and 40 mM lactate (for sulfate reducing microcosms spiked with PCE), 5 mM Na_2SO_4 and 10 mM lactate (for sulfate reducing microcosms spiked with CT), and 50 mM hydrous ferric oxide (HFO) (*36*) and 20 mM acetate (for all iron reducing microcosms). To inhibit methanogenesis, microcosms were amended with 1 mM bromoethane sulfonic acid (BESA). Microcosms were preincubated until the concentrations of sulfate or dissolved Fe(II) leveled off, then spiked with PCE or CT. The initial concentrations of PCE or CT in Phase I microcosms were 24-103 μM (PCE) and 1.5-1.9 μM (CT). Prior to spiking microcosms with PCE, 5 mM of electron donor were added to support microbial reductive dechlorination. This was also done in selected microcosms spiked with CT, but did not significantly change initial rates of microbial reductive dechlorination (*35*). PCE and CT were quantified over time by gas chromatography (GC) as described in refs. (*10*) (PCE) and (*34*) (CT).

Phase II microcosms were also prepared in 160 mL serum bottles sealed with Teflon coated septa and aluminum crimp seals and contained 50 g of SIL-CO-SIL 63 ground silica (U.S. Silica, Berkeley Springs, WV) that was intended to model the inert (with respect to reductive dechlorination) fraction of soil. According to the manufacturer, SIL-CO-SIL 63 is 99.7% SiO_2, and 96.7% of the particle mass passes through a standard 325 mesh sieve. Each microcosm, prepared in duplicate, also contained 250 m^2 of either HFO, lepidocrocite (γ-FeOOH), goethite (α-FeOOH), or hematite (Fe_2O_3). The Fe(III) oxide sources or synthesis methods, modifications (if any), specific surface areas, and mass loadings in the microcosms (in mass Fe per total mass dry solids, as %) were as follows: HFO: synthesized by dissolving $FeCl_3$ in Nanopure water, raising the pH to 10 with concentrated NaOH, centrifuging at RCF = 873 × g, washing 6-7 times with 1 M $NaHCO_3$, and freeze drying, 305 m^2/g, 0.84 % Fe by mass; lepidocrocite: Laxness (Pittsburgh, PA), sieved (200-120 mesh), 16.0 m^2/g, 15% Fe by mass; goethite: Strem Chemicals, Newburyport, MA, sieved (200-140 mesh), 66.3 m^2/g, 4.4 % Fe by mass; and hematite: Noah Technologies Corporation, San Antonio, TX, sieved (360-200 mesh), 13.3 m^2/g, 19% Fe by mass. Although the mass loadings of Fe(III) oxides varied in the microcosms, the surface area loadings were equal at approximately 1850 m^2/L. BET specific surface area was measured using a Quantachrome Autosorb-1 instrument (Boynton Beach, FL).

Phase II microcosms contained 110 mL of synthetic aqueous medium (modified from the recipes reported in refs. (*37*) and (*38*)) that contained 1 g/L Bacto yeast extract (Becton, Dickinson, Sparks, MD), 2 mM Na_2SO_4, 30 mM sodium lactate, 8 mM $MgSO_4$, 5 mM NH_4Cl, 4.4 mM KCl, 0.6 mM $CaCl_2$, 2 mL/L vitamin solution (*39*), 12.5 mL/L trace metal solution (*40*), 0.62 mg/L resazurin, and 25 mM HEPES buffer (pH 7). Microcosms also contained approximately 80 μM Na_2S to scavenge dissolved oxygen remaining after boiling and sparging with N_2, as well as a small stir bar with an estimated volume of 3 mL. After autoclaving, sodium bicarbonate and cysteine were added from anaerobic stock solutions to final concentrations of 8 mM and 0.025% respectively. The total volume of the Phase II microcosm contents was approximately 135 mL, leaving approximately 25 mL of headspace. Microcosm medium was prepared under N_2 (*41*).

To initiate sulfate reduction, microcosms were inoculated with *Desulfovibrio desulfuricans* strain G20 (*42*) that was cultured in a separate stock solution until the sulfate was almost depleted and the optical density (600 nm) was approximately 0.3. The dilution ratio of stock solution in the microcosm medium was 1.5:110.

The stock solution used to culture *D. desulfuricans* strain G20 was identical to the microcosm aqueous medium, except that it contained K_2HPO_4 instead of KCl.

Six identical Phase II microcosms were prepared for each kind of Fe(III) oxide at the same time, then were divided into two groups of three. The first group was used to monitor in triplicate changes in potentially reactive mineral species over time while the second group was held in reserve under identical conditions. Potentially reactive mineral species were quantified operationally as weakly (1 M $MgCl_2$ extractable) and strongly (0.5 N HCl extractable) bound Fe(II), FeS, and Cr(II) reducible or extractable sulfur (CrES), which includes sulfur species in an oxidation state higher than (-II), such as FeS_2, S(0), and Fe_3S_4. Dissolved Fe(II) and SO_4^{2-} were also measured. The analytical procedures used to quantify these geochemical parameters are described in ref. (*23*).

After the measured geochemical parameters reached a constant level, indicating that sulfate reduction had stopped, the second group of three microcosms was heat treated in order to stop microbial activity and measure CT transformation by the abiotic components of the microcosms. (In some cases, microcosms were stored at 4 °C for several days to temporarily stop microbial activity and the corresponding geochemical changes before heat treatment.) Heat treated microcosms were then spiked with CT at an initial aqueous concentration of approximately 1 μM, and CT concentration was monitored over time by GC (*34*).

Heat Treatment

Selected microcosms spiked with CT were heat treated by placement in a boiling water bath for 10 minutes. This inhibited iron reduction for at least 16 days and sulfate reduction for at least 10 days in Phase I microcosms (not shown), and presumably inhibited microbial reductive dechlorination for a similar time period (*35*). Heat treatment resulted in only minor geochemical changes (no more than a 10% change in the concentrations of FeS, strongly bound Fe(II), and CrES, and a 37% decrease in the concentration of weakly bound Fe(II), all in Duck Pond microcosms prepared under iron reducing conditions at pH 8.2 (*10*)).

Results and Discussion

Treatment of Kinetic Data

Measured concentrations of PCE and CT for all experiments were corrected to account for partitioning between the aqueous and gas phases, and, for Phase I microcosms spiked with PCE, to account for partitioning to the solid phase (*10*, *35*). Partitioning calculations used the following dimensionless Henry's Law constants: 0.75 for PCE (*43*) and 1.13 for CT (*44*) and K_{oc} values: 231 mL/g for PCE (calculated as described in ref. (*10*)) and 71 mL/g for CT (*45*). Sorption of CT to the solid phase in Phase I microcosms (Duck Pond sediment) was estimated to account for no more than 3% of total CT (*35*) and was neglected, and sorption of CT to the solid phase in Phase II microcosms was assumed to be negligible based on its source (quarried, crushed silica). Unless otherwise noted, all concentrations reported here are "total concentrations" that equal the sum of the masses in all relevant phases (aqueous, gas, and, for Phase I PCE microcosms, sediment) divided by the aqueous volume. Example calculations are provided in refs. (*10*) and (*35*). Initial rates were calculated using data from the period when $[CT]/[CT]_0$ was greater than 60%, during which time plots of [CT] versus time were approximately linear. Reported initial rates and product yields were calculated from total concentrations.

Before spiking with CT, selected microcosms were heat treated to inhibit microbial activity and thereby differentiate between microbial and abiotic CT transformation. For PCE, abiotic and microbial transformations were distinguished by analysis of reaction kinetics and products as described below. In all cases, reductive dechlorination of CT in Phase II microcosms occurred in less than 10 days, so this method of heat treatment was most likely sufficient to inhibit microbial activity during the period when CT transformation took place, meaning that any CT degradation observed in heat treated microcosms can be attributed to abiotic processes.

Phase I. Comparison of Abiotic and Microbial Reductive Dechlorination of PCE and CT

Microcosms exhibited very different kinetic behavior depending on whether they were spiked with PCE or CT (Figures 3 and 4). First, as expected based on the overall relationship between the driving force for a reduction reaction (i.e., reduction potentials or free energies for the reduction reaction) and reaction rates or rate constants (*30–32*), CT was transformed much faster than PCE in unsterilized microcosms. Standard potentials for the one electron reduction reaction (i.e., $R\text{-}Cl + e^- \rightarrow R^\bullet + Cl^-$) are equal to -0.531 V for PCE (less favorable) and -0.160 V for CT (more favorable), and enthalpies for the related homolytic bond dissociation reaction $R\text{-}Cl \rightarrow R^\bullet + X^\bullet$ are equal to 334.6 kJ/mol for PCE (less favorable) and 304.4 kJ/mol for CT (more favorable) (*2*).

Second, the reaction kinetics indicated that the primary pathway of CT transformation was abiotic, while the primary pathway of PCE transformation was microbial. For CT, sterilized and unsterilized microcosms had nearly identical reaction profiles (Figures 3 and 4, insets) and statistically indistinguishable initial rates of CT transformation (*35*), indicating that CT transformation was primarily an abiotic process. Separate experiments with only the microcosm supernatant showed that, for the microcosms discussed here, the abiotic species responsible for CT dechlorination were associated with the solid and not the aqueous phase (*35*). In another study, autoclaved anaerobic water treatment sludges, consisting mainly of biomass, retained the ability to dechlorinate CT, but dechlorination rates were 2-3 times lower in the presence of autoclaved versus non-autoclaved sludges (*46*). A difference of this magnitude in the sterilized versus unsterilized microcosms spiked with CT was not observed here, pointing to mineral phases containing reduced iron or sulfur, and not lysed microbial cells or cellular biochemicals (which were likely present in the autoclaved sludge) as the primary reductants.

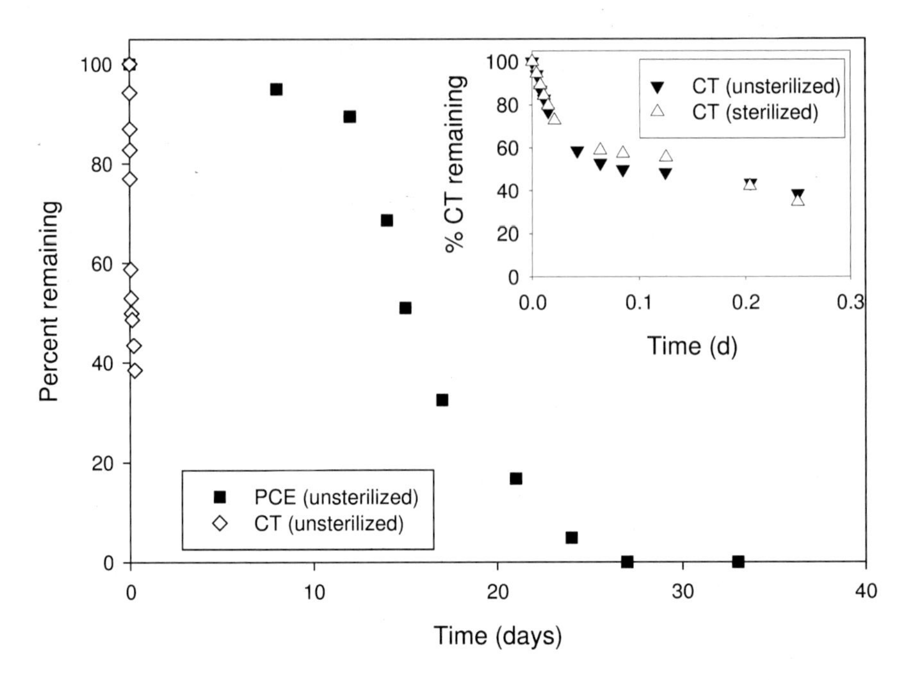

Figure 3. PCE and CT transformation in Duck Pond microcosms preincubated under sulfate reducing conditions. Although the CT data point partially obscures the PCE data point, the first y-axis value for both PCE and CT (100%) was measured at 0 days. The PCE data series is reprinted with permission from ref. (10). Copyright 2009 American Chemical Society.

For microcosms spiked with PCE, two lines of evidence indicate that microbial reductive dechlorination was the predominant transformation process. The first is the lag phase of approximately 10 days between the time when the preincubated microcosms were spiked with PCE and the onset of significant PCE disappearance, which was not observed in the microcosms spiked with CT (Figures 3 and 4). This lag phase most likely occurred as cells needed for PCE dechlorination grew to levels sufficient to transform PCE at detectable rates. The second line of evidence is that the reaction products for PCE transformation in the microcosms were primarily those of sequential hydrogenolysis (Figure 2). The acetylene yield never amounted to more than 1% of the transformed PCE in the microcosms shown in Figures 3 and 4 (*10*), supporting the conclusion that the main pathway for PCE transformation in the microcosms was microbial and not abiotic.

Shen and Wilson (*47*) conducted studies in columns filled with a mixture of sand, tree mulch, cotton gin trash, and in some cases hematite, through which ground water containing sulfate was flowed continuously for more than two years. They concluded that, on average, approximately half of the TCE transformation in the columns could be attributed to abiotic transformation by FeS after one year. The difference between their results and those reported here can be explained by the relative mass loadings of potentially reactive minerals in their column setup versus our batch microcosms. While our Phase I microcosms did not contain

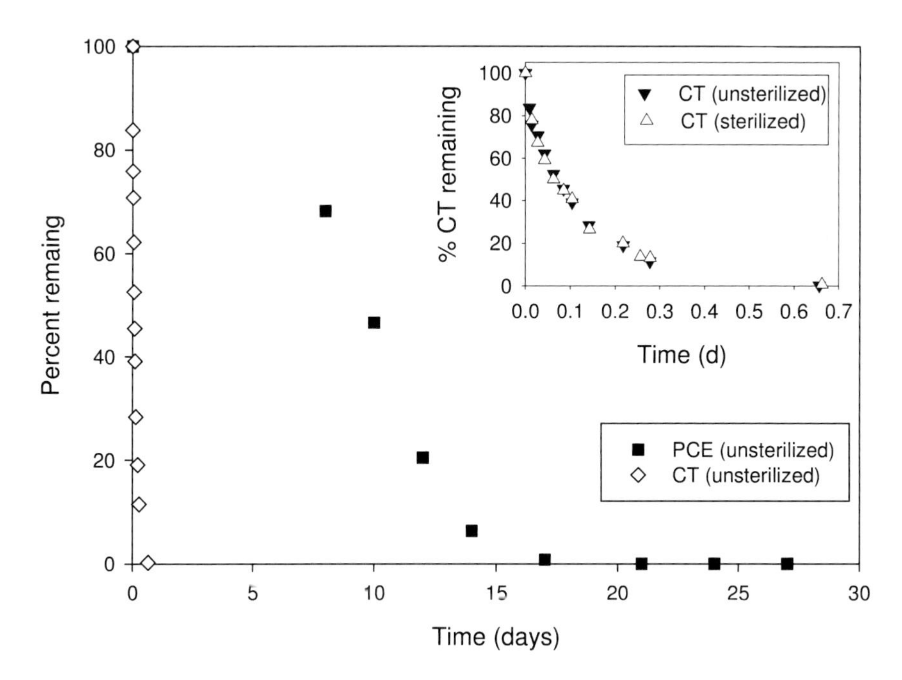

Figure 4. PCE and CT transformation in Duck Pond microcosms preincubated under iron reducing conditions. Although the CT data point partially obscures the PCE data point, the first y-axis value for both PCE and CT (100%) was measured at 0 days. The PCE data series is reprinted with permission from ref. (10). Copyright 2009 American Chemical Society.

FeS concentrations of more than 1 g/L, their estimated FeS concentration in the columns after approximately one year was 75-150 mM (6.6 – 13 g/L). Such a high mass loading of FeS was generated through continuous application of sulfate in flow through system containing native iron or hematite over an extended period. Thus, it appears that whether abiotic or microbial degradation of contaminants such as PCE or TCE predominates in a given system depends strongly on the mass loading of reactive minerals such as FeS. The mass loading, in turn, depends on the cumulative input of species needed for *in situ* generation of reactive minerals, specifically an electron donor and terminal electron acceptor such as Fe(III) or sulfate. Significant accumulation of FeS or other reactive minerals, as was observed in the column studies of Shen and Wilson (*47*), is most likely to occur in active remediation scenarios where there is a continual application or source of electron donors and acceptors needed to support iron and/or sulfate reduction.

Table I shows that rate constants for abiotic reductive dechlorination of PCE and CT by chloride green rust, pyrite, and FeS differ by more than two to almost five orders of magnitude. If the kinetics of microbial PCE and CT transformation in the microcosms varied over a similar range of values, we would expect that either abiotic or microbial reductive dechlorination would predominate in microcosms spiked with both PCE and CT, which was not observed. While there is a general relationship between the thermodynamics of pollutant reduction and microbial dechlorination rates (*28*, *30*, *32*), the results reported here suggest that thermodynamics alone cannot entirely explain the relative rates of microbial reductive dechlorination of PCE and CT. The availability of an enzyme-catalyzed respiratory pathway for microbial transformation of PCE (*48*), but not CT (*4*), probably serves to "equalize" the rates of microbial transformation of these compounds compared to rates expected from thermodynamic properties (for example, one-electron reduction potentials or C-Cl bond dissociation enthalpies) alone. While CT inhibits growth of some bacteria at concentrations as low as 0.1 μM (*49*), microorganisms capable of growing on PCE through the production of PCE dechlorinating enzymes appear to be common in natural environments, perhaps because naturally occurring chlorinated compounds similar in structure to PCE have existed for millennia, allowing these enzymes to evolve (*30*). CT, on the other hand, is transformed by cellular biochemicals such as cytochromes (*50*), hydroxycobalamin (*51*), the microbial metabolite pyridine-2,6-bis(thiocarboxylic acid) (*52*), or vitamin B_{12} (*53*) in a cometabolic process or processes not linked to microbial energy generation or growth. Rates of cometabolic reductive dechlorination can be orders of magnitude smaller than those for respiratory reductive dechlorination (*54*).

Phase II. Comparison of the Rates of Mineral Formation and Abiotic CT Reductive Dechlorination

The slowest and fastest rates of CT transformation were measured in microcosms containing HFO and goethite, respectively (Table II); results from both microcosms are illustrated in Figure 5. CT reduction rates were intermediate but closer to the low end of this range in microcosms containing hematite and lepidocrocite (Table II). Concentrations of dissolved Fe(II), weakly bound Fe(II), and FeS were below detection limits (2 μM for dissolved and weakly bound Fe(II) and 5 μM for FeS) at all times in all microcosms. The non-zero values for strongly bound Fe(II) and CrES at time zero in Figure 5 may be due to reactions between components of the microbial medium, which included initial dissolved concentrations of 19 μM $FeSO_4$ and 80 μM Na_2S.

The difference in rates of CT transformation in the two microcosms (Table II, Figure 5) cannot be explained by differences in the concentrations of CrES or strongly bound Fe(II), since the plots for CrES are very similar for microcosms containing HFO and goethite, and the microcosm with the higher concentration of strongly bound Fe(II) (HFO) had the slower rate of CT transformation (Figure 5). Weakly bound Fe(II) has been shown to strongly influence transformation rates when present at concentrations equal to or higher than initial CT concentrations (*5*, *23*). Because detection limits for weakly bound Fe(II) (2 μM) were slightly higher than the initial concentration of CT in these experiments (1 μM), it is possible that in some microcosms the concentration of weakly bound Fe(II) was below detection limits, but still high enough to influence the rates of CT transformation. If so, a higher concentration of weakly bound Fe(II) might explain the greater CT reactivity in the goethite versus HFO microcosm, but this speculation cannot be tested based on the data reported here.

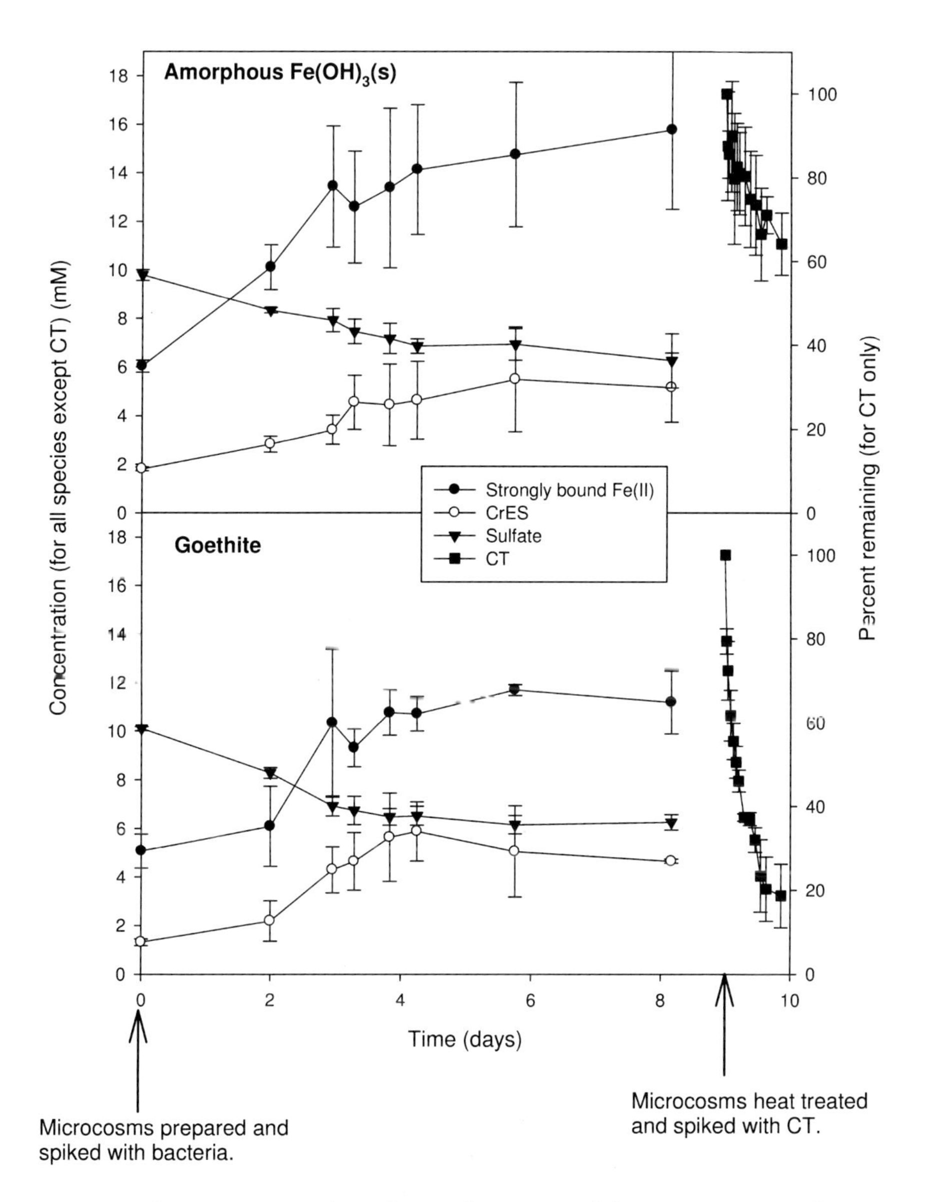

Figure 5. Concentrations of geochemical species and CT versus time in Phase II microcosms.

While it was not possible to distinguish among the rates of microbial sulfate reduction, abiotic Fe(III) reduction by microbially generated S(-II), and formation of mineral species such as strongly bound Fe(II), the time scales for depletion (for sulfate) or formation (for Fe(II) minerals) of these species are similar (Figure 5). The initial concentrations of CT and sulfate in these experiments differed significantly (micromolar for CT and millimolar for sulfate) and were chosen to reflect realistic conditions in the environment. Therefore, direct numerical comparison of rates for sulfate reduction and CT transformation is not justified since both these rates were functions of concentration (Figure 5). Nonetheless, Figure 5 illustrates that the time scale for sulfate reduction (days) was much faster than that for CT reduction (hours), even in the microcosms where CT reacted most slowly (HFO). This suggests that in natural systems having similar concentrations of reactive abiotic mineral species or their precursors (e.g., sulfate, Fe(III) minerals), the rate of formation of reactive abiotic mineral species may control the rate of abiotic CT transformation. Consistent with this, Heijman et al. (*55*) found that rate constants for abiotic reductive transformation of nitroaromatics (which, like CT, react relatively quickly with minerals (*56*)) were controlled by the rate of microbial iron reduction in soil columns, rather than thermodynamic properties associated with the particular nitroaromatic compound. In addition, analogous to the instantaneous reaction model (*57*) which assumes that oxidation of a substrate by oxygen is essentially instantaneous compared to the rate of oxygen diffusion into the system, and that the extent of substrate oxidation depends on the mass of oxygen that enters the system, the extent of abiotic CT degradation will be controlled by the cumulative mass loading of the geochemical precursors for reactive abiotic mineral species (namely an electron donor and Fe(III) and/or sulfate). Only upon continued stimulation of iron reducing and sulfate reducing bacteria is there likely to be significant accumulation of abiotic minerals in the subsurface.

Unlike CT, the rate of PCE degradation is not likely to be limited by the rate of abiotic mineral formation in natural systems, since it undergoes much slower abiotic transformation than does CT (Table I), and microbial PCE reductive dechlorination occurs on the same time scale as microbial sulfate reduction (compare Figures 3-5). Similar to CT, however, the extent of abiotic PCE degradation will also depend on the cumulative mass loading of electron donor and Fe(III) and/or sulfate needed for reactive mineral formation.

Table II. Kinetics of sulfate reduction and CT transformation in Phase II microcosms

Fe(III) oxide in microcosm	*Maximum sulfate disappearance rate (mM/day)*[a]	*Maximum CT disappearance rate (μM/day)*[b]
HFO	0.68±0.08	0.39±0.05
Lepidocrocite	6.2±2.3	0.79±0.15
Goethite	1.06±0.31	4.9±1.2
Hematite	1.44±0.96	0.65±0.15

[a] Calculated using data during the time period when zero order kinetics were observed and excluding any lag periods or periods when sulfate concentration leveled off. [b] Calculated using data from the period when $[CT]/[CT]_0 \geq 0.6$, during which time approximately zero order kinetics were observed. All uncertainties are 95% confidence intervals.

Conclusions

In microcosms set up to simulate natural conditions, the relative rates of abiotic versus microbial transformation of chlorinated aliphatic contaminants varied with the contaminant, with microbial dechlorination the predominant process for PCE transformation, and abiotic dechlorination the predominant process for CT transformation. Abiotic reductive dechlorination of PCE would likely be more significant in systems with higher mass loadings of reactive minerals. Practitioners seeking to stimulate abiotic transformation of PCE or the related contaminant TCE should aim to introduce or generate *in situ* significant mass loadings of reactive minerals. In practice, it may not be possible to generate *in situ* sufficient quantities of metastable phases such as green rusts that transform to magnetite over days to weeks (*58*, *59*). FeS may be the best choice for *in situ* abiotic remediation of PCE due to its high reactivity (*13–15*). The challenge associated with using FeS for *in situ* remediation is its natural tendency to age to pyrite in the presence of partially oxidized sulfur species (*60–64*), leading to a loss in reactivity (*5*, *64*, *65*). Thus, continuous *in situ* generation of FeS may be needed to maintain its reactivity with PCE.

The times required for sulfate reduction and the subsequent formation of abiotic mineral phases likely involved in CT abiotic reduction were significantly longer than those for abiotic CT transformation by these reactive mineral phases. For remediation scenarios in which abiotic minerals are generated *in situ* through microbial sulfate and iron reduction, the rate of abiotic CT transformation may be limited by the rates of these terminal electron accepting processes. For both PCE and CT, the extent of abiotic transformation (i.e., the fraction of total PCE or CT transformed via an abiotic pathway) will be proportional to the mass loading of reactive minerals, which in turn will depend on the cumulative inputs of electron donor and Fe(III) and/or sulfate to the system. While Fe(II) that is weakly adsorbed to Fe minerals is highly reactive with CT, it leads to the generation of the toxic and recalcitrant product chloroform (*5*). Application of dissolved Fe(II) salts to the subsurface at sites contaminated with CT should be minimized to limit generation of weakly bound Fe(II). The lowest chloroform yields were achieved in model and natural systems having the highest concentrations of chromium reducible sulfur (*5*), which consists of partially oxidized sulfur species, such as polysulfides, thiosulfate, and elemental sulfur (*66–69*), that arise naturally upon reduction of Fe(III) oxides by S(-II) generated by sulfate reducing bacteria. Thus, application of sulfate to Fe(III) oxide-rich or Fe(III) oxide-augmented systems may lead to acceptable CT transformation rates with lower accumulation of chloroform.

Acknowledgments

This work was supported by the National Science Foundation (CBET 0093332) and the U. S. Department of Defense through the Strategic Environmental Research and Development Program (SERDP). We thank Xiangzhen Li for guidance in the laboratory, and the reviewers and editor for insightful comments.

References

1. Amonette, J. E.; Workman, D. J.; Kennedy, D. W.; Fruchter, J. S.; Gorby, Y. A. Dechlorination of carbon tetrachloride by Fe(II) associated with goethite. *Environ. Sci. Technol.* **2000**, *34*, 4606–4613.
2. Butler, E. C.; Hayes, K. F. Kinetics of the transformation of halogenated aliphatic compounds by iron sulfide. *Environ. Sci. Technol.* **2000**, *34*, 422–429.
3. Maithreepala, R. A.; Doong, R. A. Enhanced Dechlorination of Chlorinated Methanes and Ethenes by Chloride Green Rust in the Presence of Copper(II). *Environ. Sci. Technol.* **2005**, *39*, 4082–4090.
4. Penny, C.; Vuilleumier, S.; Bringel, F. Microbial degradation of tetrachloromethane: mechanisms and perspectives for bioremediation. *FEMS Microbiol. Ecol.* **2010**, *74*, 257–275.
5. Shao, H.; Butler, E. C. The influence of soil minerals on the rates and products of abiotic transformation of carbon tetrachloride in anaerobic soils and sediments. *Environ. Sci. Technol.* **2009**, *43*, 1896–1901.
6. He, Y; Su, C.; Wilson, J. Wilkin, R.; Adair, C.; Lee, T.; Bradley, P.; Ferrey, M. *Identification and Characterization Methods for Reactive Minerals Responsible for Natural Attenuation of Chlorinated Organic Compounds in Ground Water*; EPA 600/R-09/115; U. S. EPA, Office of Research and Development, National Risk Management Research Laboratory: Ada, OK, 2009.
7. Freedman, D. L.; Gossett, J. M. Biological reductive dechlorination of tetrachloroethylene and trichloroethylene under methanogenic conditions. *Appl. Environ. Microbiol.* **1989**, *55*, 2144–2151.
8. Brown, R. A.; Mueller, J. G.; Seech, A. G.; Henderson, J. K.; Wilson, J. T. Interactions between biological and abiotic pathways in the reduction of chlorinated solvents. *Remediation* **2009**, *20* (1), 9–20.
9. Schink, B. Fermentation of acetylene by an obligate anaerobe, *Pelobacter acetylenicus* sp. nov. *Arch. Microbiol.* **1985**, *142*, 295–301.
10. Dong, Y.; Liang, X.; Krumholz, L. R.; Philp, R. P.; Butler, E. C. The relative contributions of abiotic and microbial processes to the transformation of tetrachloroethylene and trichloroethylene in anaerobic microcosms. *Environ. Sci. Technol.* **2009**, *43*, 690–697.
11. Liang, X.; Philp, R. P.; Butler, E. C. Kinetic and isotope analyses of tetrachloroethylene and trichloroethylene degradation by model Fe(II)-bearing minerals. *Chemosphere* **2009**, *75*, 63–69.
12. Darlington, R.; Lehmicke, L.; Andrachek, R. G.; Freedman, D. L. Biotic and abiotic anaerobic transformations of trichloroethene and *cis*-1,2-dichloroethene in fractured sandstone. *Environ. Sci. Technol.* **2008**, *42*, 4323–4330.
13. Butler, E. C.; Hayes, K. F. Kinetics of the transformation of trichloroethylene and tetrachloroethylene by iron sulfide. *Environ. Sci. Technol.* **1999**, *33*, 2021–2027.
14. Jeong, H. Y.; Kim, H.; Hayes, K. F. Reductive dechlorination pathways of tetrachloroethylene and trichloroethylene and subsequent transformation of

their dechlorination products by mackinawite (FeS) in the presence of metals. *Environ. Sci. Technol.* **2007**, *41*, 7736–7743.

15. Liang, X.; Dong, Y.; Kuder, T.; Krumholz, L. R.; Philp, R. P.; Butler, E. C. Distinguishing abiotic and biotic transformation of tetrachloroethylene and trichloroethylene by stable carbon isotope fractionation. *Environ. Sci. Technol.* **2007**, *41*, 7094–7100.
16. Lee, W.; Batchelor, B. Abiotic reductive dechlorination of chlorinated ethylenes by iron-bearing soil minerals. 1. pyrite and magnetite. *Environ. Sci. Technol.* **2002**, *36*, 5147–5154.
17. Lee, W.; Batchelor, B. Abiotic reductive dechlorination of chlorinated ethylenes by iron-bearing soil minerals. 2. green rust. *Environ. Sci. Technol.* **2002**, *36*, 5348–5354.
18. Zwank, L.; Elsner, M.; Aeberhard, A.; Schwarzenbach, R. P.; Haderlein, S. B. Carbon isotope fractionation in the reduction dehalogenation of carbon tetrachloride at iron (hydr)oxide and iron sulfide minerals. *Environ. Sci. Technol.* **2005**, *39*, 5634–5641.
19. O'Loughlin, E. J.; Kemner, K. M.; Burris, D. R. Effects of Ag^{I}, Au^{III}, and Cu^{II} on the reductive dechlorination of carbon tetrachloride by green rust. *Environ. Sci. Technol.* **2003**, *37*, 2905–2912.
20. Liang, X.; Butler, E. C. Effects of natural organic matter model compounds on the transformation of carbon tetrachloride by chloride green rust. *Water Res.* **2010**, *44*, 2125–2132.
21. McCormick, M. L.; Bouwer, E. J.; Adriaens, P. Carbon tetrachloride transformation in a model iron-reducing culture: relative kinetics of biotic and abiotic reactions. *Environ. Sci. Technol.* **2002**, *36*, 403–410.
22. Danielsen, K. M.; Hayes, K. F. pH dependence of carbon tetrachloride reductive dechlorination by magnetite. *Environ. Sci. Technol.* **2004**, *38*, 4745–4752.
23. Shao, H.; Butler, E. C. The influence of iron and sulfur mineral fractions on carbon tetrachloride transformation in model anaerobic soils and sediments. *Chemosphere* **2007**, *68*, 1807–1813.
24. Pecher, K.; Haderlein, S. B.; Schwarzenbach, R. P. Reduction of polyhalogenated methanes by surface-bound Fe(II) in aqueous suspensions of iron oxides. *Environ. Sci. Technol.* **2002**, *36*, 1734–1741.
25. Elsner, M.; Haderlein, S. B.; Kellerhals, T.; Luzi, S.; Zwank, L.; Angst, W.; Schwarzenbach, R. P. Mechanisms and products of surface-mediated reductive dehalogenation of carbon tetrachloride by Fe(II) on goethite. *Environ. Sci. Technol.* **2004**, *38*, 2058–2066.
26. Kenneke, J. F.; Weber, E. J. Reductive dehalogenation of halomethanes in iron- and sulfate-reducing sediments. 1. Reactivity pattern analysis. *Environ. Sci. Technol.* **2003**, *37*, 713–720.
27. Williams, A. G. B.; Scherer, M. M. Spectroscopic evidence for Fe(II)-Fe(III) electron transfer at the iron oxide-water interface. *Environ. Sci. Technol.* **2004**, *38*, 4782–4790.
28. Bouwer, E. J.; Wright, J. P. Transformations of trace halogenated aliphatics in anoxic biofilm columns. *J. Contam. Hydrol.* **1988**, *2*, 155–169.

29. Suarez, M. P.; Rifai, H. S. Biodegradation rates for fuel hydrocarbons and chlorinated solvents in groundwater. *Biorem. J.* **1999**, *3*, 337–362.
30. Bradley, P. M. History and ecology of chloroethenes biodegradation: A review. *Biorem. J.* **2003**, *7*, 81–109.
31. Vogel, T. M.; Criddle, C. S.; McCarty, P. L. Transformations of halogenated aliphatic compounds. *Environ. Sci. Technol.* **1987**, *21*, 722–736.
32. Bhatt, P.; Kumar, M. S.; Mudilar, S.; Chakrabarti, T. Biodegradation of chlorinated compounds—A review. *Crit. Rev. Environ. Sci. Technol.* **2007**, *37*, 165–198.
33. Maithreepala, R. A.; Doong, R. A. Synergistic effect of copper ion on the reductive dechlorination of carbon tetrachloride by surface-bound Fe(II) associated with goethite. *Environ. Sci. Technol.* **2004**, *38*, 260–268.
34. Hanoch, R. J.; Shao, H.; Butler, E. C. Transformation of carbon tetrachloride by bisulfide treated goethite, hematite, magnetite, and kaolinite. *Chemosphere* **2006**, *63*, 323–334.
35. Shao, H.; Butler, E. C. The relative importance of abiotic and biotic transformation of carbon tetrachloride in anaerobic soils and sediments. *Soil Sediment Contam.* **2009**, *18*, 455–469.
36. Schwertmann, U.; Cornell, R. M. *Iron Oxides in the Laboratory: Preparation and Characterization*, 2nd ed.; Wiley-VCH: Weinheim, Germany, 2000; pp 103–112.
37. Groh, J. L.; Luo, Q.; Ballard, J. D.; Krumholz, L. R. A method adapting microarray technology for signature tagged mutagenesis of *Desulfovibrio desulfuricans* G20 and *Shewanella oneidensis* MR-1 in anaerobic sediment survival experiments. *Appl. Environ. Microbiol.* **2005**, *71*, 7064–7074.
38. Rapp, B. J; Wall, J. D. Genetic transfer in *Desulfovibrio desulfuricans*. *Proc. Natl. Acad. Sci. U.S.A.* **1987**, *14*, 9128–9130.
39. Tanner, R. S. Cultivation of bacteria and fungi. In *Manual of Environmental Microbiology*, 3rd ed.; Garland, J. L., Ed.; ASM Press: Washington, DC, 2007.
40. Widdel, F.; Bak, F. Gram-negative mesophilic sulfate-reducing bacteria. In *The Prokaryotes - A Handbook on the Biology of Bacteria: Ecophysiology, Isolation, Identification, Applications*, 2nd ed.; Balows, A., Truper, H. G., Dworkin, M., Harder, W., Schleifer, K. H., Eds.; Springer-Verlag: New York, 1992; Vol. 4, pp 3353–3378.
41. Balch, W. E.; Wolf, R. S. New approach to the cultivation of methanogenic bacteria: 2-mercaptoethanesulfonic acid (HS-CoM)-dependent growth of *Methanobacterium ruminantium* in a pressurized atmosphere. *Appl. Environ. Microbiol.* **1976**, *32*, 781–791.
42. Weimer, P. J.; Van Kavelaar, M. J.; Michel, C. B.; Ng, T. K. Effect of phosphate on the corrosion of carbon steel and on the composition of the corrosion products in two-stage continuous cultures of *Desulfovibrio desulfuricans*. *Appl. Environ. Microbiol.* **1988**, *54*, 386–396.
43. Howard, P. H.; Meylan, W. M. *Physical Properties of Organic Chemicals*; CRC Lewis Publishers: Boca Raton, FL, 1997.
44. Warneck, P. A review of Henry's law coefficients for chlorine-containing C1 and C2 hydrocarbons. *Chemosphere* **2007**, *69*, 347–361.

45. US EPA, 1998. *National Primary Drinking Water Regulations*; EPA 811-F-95-004c-T; U.S. Environmental Protection Agency, Office of Water: Washington, DC, 1995.
46. Van Eekert, M. H. A.; Schröder, T. J.; Stams, A. J. M.; Schraa, G.; Field, J. A. Degradation and fate of carbon tetrachloride in unadapted methanogenic granular sludge. *Appl. Environ. Microbiol.* **1998**, *64*, 2350–2356.
47. Shen, H.; Wilson, J. T. Trichloroethylene removal from groundwater in flow-through columns simulating a permeable reactive barrier constructed with plant mulch. *Environ. Sci. Technol.* **2007**, *41*, 4077–4083.
48. Zhang, C.; Bennett, G. N. Biodegradation of xenobiotics by anaerobic bacteria. *Appl. Microbiol. Biotechnol.* **2005**, *67*, 600–618.
49. Futagami, T.; Yamaguchi, T.; Nakayama, S.; Goto, M.; Furukawa, K. Effects of chloromethanes on growth of and deletion of the *pce* gene cluster in dehalorespiring *Desulfitobacterium hafniense* strain Y51. *Appl. Environ. Microbiol.* **2006**, *72*, 5998–6003.
50. Picardal, F. W.; Arnold, R. G.; Couch, H.; Little, A. M.; Smith, M. E. Involvement of cytochromes in anaerobic biotransformation of tetrachloromethane by *Shewanella putrefaciens* 200. *Appl. Environ. Microbiol.* **1993**, *59*, 3763–3770.
51. Hashsham, S. A.; Freedman, D. L. Enhanced biotransformation of carbon tetrachloride by *Acetobacterium woodii* upon addition of hydroxocobalamin and fructose. *Appl. Environ. Microbiol.* **1999**, *65*, 4537–4542.
52. Lee, C.-H.; Lewis, T. A.; Paszczynski, A.; Crawford, R. L. Identification of an extracellular catalyst of carbon tetrachloride dehalogenation from *Pseudomonas stutzeri* Strain KC as pyridine-2,6-bis(thiocarboxylate). *Biochem. Biophys. Res. Commun.* **1999**, *261*, 562–566.
53. Zou, S.; Stensel, H. D.; Ferguson, J. F. Carbon tetrachloride degradation: Effect of microbial growth substrate and vitamin B_{12} content. *Environ. Sci. Technol.* **2000**, *34*, 1751–1757.
54. Fetzner, S. Bacterial dehalogenation. *Appl. Microbiol. Biotechnol.* **1998**, *50*, 633–657.
55. Heijman, C. G.; Grieder, E.; Holliger, C.; Schwarzenbach, R. P. Reduction of nitroaromatic compounds coupled to microbial iron reduction in laboratory aquifer columns. *Environ. Sci. Technol.* **1995**, *29*, 775–783.
56. Klausen, J.; Tröber, S. P.; Haderlein, S. B.; Schwarzenbach, R. P. Reduction of substituted nitrobenzenes by Fe(II) in aqueous mineral suspensions. *Environ. Sci. Technol.* **1995**, *29*, 2396–2404.
57. Rifai, H. S.; Bedient, P. B. Comparison of biodegradation kinetics with an instantaneous reaction model for groundwater. *Water Resour. Res.* **1990**, *26*, 637–645.
58. Génin, J.-M. R.; Bourrié, G.; Trolard, F.; Abdelmoula, M.; Jaffrezic, A.; Refait, P.; Maitre, V.; Humbert, B.; Herbillon, A. Thermodynamic equilibria in aqueous suspensions of synthetic and natural Fe(II)−Fe(III) green rusts: Occurrences of the mineral in hydromorphic soils. *Environ. Sci. Technol.* **1998**, *32*, 1058–1068.
59. Ona-Nguema, G.; Abdelmoula, M.; Jorand, F.; Benali, O.; Génin, A.; Block, J. C.; Génin, J.-M. R. Iron(II,III) hydroxycarbonate green rust

formation and stabilization from lepidocrocite bioreduction. *Environ. Sci. Technol.* **2002**, *36*, 16–20.

60. Sweeney, R. E.; Kaplan, I. R. Pyrite framboid formation: laboratory synthesis and marine sediments. *Econ. Geol.* **1973**, *68*, 618–634.
61. Schoonen, M. A. A.; Barnes, H. L. Reactions forming pyrite and marcasite from solution. I. Nucleation of FeS_2 below 100 °C. *Geochim. Cosmochim. Acta* **1991**, *55*, 1505–1514.
62. Wang, Q.; Morse, J. W. Pyrite formation under conditions approximating those in anoxic sediments I. Pathway and morphology. *Mar. Chem.* **1996**, *52*, 99–121.
63. Benning, L. G.; Wilkin, R. T.; Barnes, H. L. Reaction pathways in the Fe-S system below 100 °C. *Chem. Geol.* **2000**, *167*, 25–51.
64. He, Y.; Wilson, J. T.; Wilkin, R. T. Transformation of reactive iron minerals in a permeable reactive barrier (biowall) used to treat TCE in ground water. *Environ. Sci. Technol.* **2008**, *42*, 6690–6696.
65. Butler, E. C.; Hayes, K. F. Factors influencing rates and products in the transformation of trichloroethylene by iron sulfide and iron metal. *Environ. Sci. Technol.* **2001**, *35*, 3884–3891.
66. Berner, R. A. Iron sulfides formed from aqueous solution at low temperatures and atmospheric pressure. *J. Geol.* **1964**, *72*, 293–306.
67. Rickard, D. T. The chemistry of iron sulfide formation at low temperatures. *Stockholm Contrib. Geol.* **1969**, *26*, 67–95.
68. Pyzik, A. J.; Sommer, S. E. Sedimentary iron monosulfides: kinetics and mechanism of formation. *Geochim. Cosmochim. Acta* **1981**, *45*, 687–698.
69. Dos Santos Afonso, M.; Stumm, W. Reductive dissolution of iron(III) (hydr)oxides by hydrogen sulfide. *Langmuir* **1992**, *8*, 1671–1675.

Chapter 24

The Use of Chemical Probes for the Characterization of the Predominant Abiotic Reductants in Anaerobic Sediments

Huichun (Judy) Zhang,[1] Dalizza Colón,[2] John F. Kenneke,[2] and Eric J. Weber[2,*]

[1]Department of Civil and Environmental Engineering, Temple University, 1947 N. 12th Street, Philadelphia, PA 19122
[2]Ecosystems Research Division, National Exposure Research Laboratory, US Environmental Protection Agency, 960 College Station Rd, Athens, GA 30605
[*]weber.eric@epa.gov

Identifying the predominant chemical reductants and pathways for electron transfer in anaerobic systems is paramount to the development of environmental fate models that incorporate pathways for abiotic reductive transformations. Currently, such models do not exist. In this chapter we address the approaches based on the use of probe chemicals that have been successfully implemented for this purpose. The general approach has been to identify viable pathways for electron transfer based on the study of probe chemicals in well-defined model systems. The subsequent translation of these findings to natural systems has been based primarily on laboratory studies of probe chemicals in anaerobic sediments and aquifers. In summary, the results of these studies support a scenario in which pathways for reductive transformations in these systems are dominated by surface-mediated processes (i.e., reaction with Fe(II) associated with Fe(III) mineral oxides and clay minerals), and through the aqueous phase by reduced dissolved organic matter (DOM) (i.e., reduced quinone moieties) and Fe(II)/DOM complexes.

Introduction

The abiotic reduction of organic contaminants in anaerobic sediments continues to be a research area of much interest. The ability to predict the rates and pathways for these processes is critical to developing the necessary tools and models for estimating the environmental concentrations of the parent chemical and the reduction products of interest. This is of concern because unlike other transformation pathways such as hydrolysis and aerobic biodegradation, abiotic reduction often results in transformation products that are of more concern than the parent compound (e.g., aromatic amines resulting from the reduction of nitroaromatics and aromatic azo compounds (*1*, *2*) and halogenated ethenes from the reduction of halogenated ethanes) (*3*). Although significant progress has been made concerning the elucidation of these reaction pathways, little has been done to incorporate this knowledge into existing environmental fate and transport models.

The application of sophisticated analytical tools, such as Mössbauer spectroscopy and multicollector inductively coupled plasma mass spectrometry for Fe isotope analysis, have been used successfully to characterize the redox properties of Fe(II)-treated mineral oxides (*4*, *5*) and clay minerals (*6*). The application of these tools to the characterization of chemical reductants in natural sediments has not been reported yet, most likely because of the complex nature of these systems. Approaches for characterizing the predominant chemical reductants in sediments have been limited primarily to indirect methods. The reactivity of probe chemicals containing functional groups that are susceptible to abiotic reduction (i.e., aromatic nitro groups, aromatic azo groups, and halogenated aliphatics) have been measured in well-defined model systems and then compared to the probe chemical's reactivity measured in well-characterized anaerobic sediments. Compound-specific stable isotope analysis is a related approach to the study of probe chemicals in these reaction systems that is finding increasing applications in the elucidation of reductive transformation pathways. For a review of this topic, see (*7*).

The identification of the predominant reductants in anaerobic systems is the first step in determining the readily measurable environmental descriptors that can be used to parameterize fate models for predicting reductive transformation rates and pathways. An added challenge is the ability to provide these environmental descriptors for spatially and temporally explicit chemical exposure assessments. This need is required by the EPA Office of Pesticide Programs in support of the Endangered Species Act, which requires estimated environmental concentrations of pesticides for aquatic and soil ecosystems inhabited by species on the Endangered Species List (*8*). A similar need is required by the Department of Defense for prioritizing the multitude of training, testing and production sites contaminated with N-based munitions for cleanup (*9*).

This chapter provides an overview of the process science developed from studies conducted in well-defined model systems designed to mimic natural anaerobic systems and subsequent studies to determine the extent to which these results translate to natural systems. A generalized scheme for the dominant reductants and pathways for electron transfer consistent with the state of the current process science is presented at the end of the chapter. Of equal importance is the development of computational approaches for the molecular descriptors (e.g., one-electron reduction potentials and bond dissociation energies) necessary for predicting reduction rates of individual chemicals. Progress in this area is the focus of another chapter (*10*), and will not be addressed here.

Background

In this chapter, we define abiotic reduction as the process by which a chemical is reduced by reaction with an abiotic reductant versus direct reaction with a microbial species. We recognize, however, that the formation of abiotic reductants results from the microbial-mediated oxidation of biologically available organic carbon and the transfer of the resulting electrons to electron acceptors. Based on thermodynamic considerations, a sequence of redox zones (i.e. nitrate-reducing, iron-reducing, sulfate-reducing and methanogenic) in sediment and aquifers can develop that are characterized by the respective dominant terminal electron-accepting processes (TEAPs). Mapping of the redox zones in sediments and aquifers has been accomplished by measurement of the dissolved electron donors (i.e., H_2), electron acceptors (i.e., O_2, NO_3^-, and SO_4^{2-}), the reduced products of the electron acceptors (i.e., NH_4^+, HS^-, Fe(II) and CH_4), and redox species in the solid phase (extractable Fe(II), Fe(III) and S_2^-) (*11–14*). Knowing the identity and reactivity of chemical reductants as a function of the redox zones would greatly facilitate the development of models describing the reactive transport of redox-active contaminants through sediments and aquifers.

Surface-Mediated Pathways for Electron Transfer

Early studies of the reductive transformation of halogenated alkanes (*3*), azobenzenes (*15*), and methyl parathion (pesticide containing an aromatic nitro group) (*16*) in anaerobic sediments demonstrated that these were facile reactions (i.e., half lives typically on the order of minutes to hours), dominated by abiotic processes, and that the reactivity of the sediment slurries was associated primarily with the sediment phase. Speculation was provided that the Fe(II)/Fe(III) redox couple was the most likely source of electrons due to the ubiquitous occurrence of iron species in natural systems. Proposed pathways for electron transfer in these systems would have to account for the extremely fast reduction kinetics observed in the anaerobic sediments. Subsequent studies of probe chemicals in well-defined model systems of Fe(II) treated Fe(III) (hydr)oxides and iron-bearing clay minerals would provide viable pathways for such facile reactions.

The study of the reduction of a series of nitroaromatic compounds (NAC) in Fe(II) treated Fe(III) (hydr)oxide suspensions provided the initial evidence for the activation of Fe(II) through complexation to the Fe(III) (hydr)oxides (*17*). NAC reduction was not observed in the presence of magnetite or Fe(II) alone. The facile reduction of the NACs to their corresponding anilines was observed, however, in suspensions of Fe(III) mineral oxides (i.e., magnetite, goethite, and lepidocrocite) treated with Fe(II). The strong dependence of NAC reduction rates on pH (increasing rate with pH) in Fe(II) treated magnetite suspensions was consistent with the pH dependent formation of Fe(II) surface complexes (increasing adsorption of Fe(II) with pH). The generalized scheme for electron transfer that emerged from these early studies is initiated by the oxidation of bioavailable organic matter mediated by anaerobic bacteria and the transfer of the resulting electrons to Fe(III) (hydr)oxides resulting in the formation of surface associated Fe(II), which serves as the chemical reductant of the NAC (*17*).

$CO_2 + H_2O$ adsorbed Fe(II) NO_2
microbial Mineral Surface abiotic
$\{CH_2O\}$ Fe(III) lattice-bound NH_2

The application of ^{57}Fe(II) Mössbauer spectroscopy has recently provided further insight into the magnetite mediated reduction of nitrobenzene (*18*). The results of this study do not support the model representing surface complexed Fe(II), but rather an electron transfer process in which oxidation of Fe(II) results in the reduction of the octahedral Fe(III) atoms in the underlying magnetite to octahedral Fe(II) atoms. Rates of NAC reduction in magnetite suspensions were found to be strongly dependent on the magnetite particle Fe(II)/Fe(III) stoichiometry (*19*). NAC reduction increased nearly 5 orders of magnitude as Fe(II)/Fe(III) values increased from 0.31, representing a highly oxidized magnetite particle, to 0.50, representing a fully stoichiometric magnetite particle. These results indicate the need to consider particle stoichiometry when assessing reductive transformations mediated by magnetite.

Iron-Bearing Clay Minerals

The ubiquitous occurrence of iron-bearing clay minerals suggests that these sediment constituents must also be considered as potential abiotic reductants in natural systems (*20*). As with Fe(III) containing mineral oxides, the Fe(III) present in iron-bearing clays is also susceptible to microbial-mediated reduction (*21*). The situation with clay minerals is further complicated by the fact that iron reduction results in Fe(II) bound to surface hydroxyl groups, as well as structural Fe(II) in the octahedral layers of the minerals. Model studies have focused primarily on distinguishing the reactivity of the surface bound and structural Fe(II). Through an approach based on measuring the reduction kinetics for two probe chemicals, 2-acetylnitrobenzene, which had no selectivity for the potentially reactive Fe(II) sites, and 4-acetylnitrobenzene, a planar molecule that could be used to directly probe the reactivity of the structural Fe(II) in the octahedral layers, it was possible to determine that structural Fe(II) in the octahedral layers was the predominant form of reactive iron in reduced smectites and that electron transfer occurs via the basel siloxane planes (*22*).

Structural Fe(II) in smectites was also proposed as the dominant reactive site for the reduction of polychlorinated ethanes and carbon tetrachloride (*23*). The half lives for these chemicals spanned from 40 to 170 days in the Fe smectites suspensions. Because the half lives of these polychlorinated aliphatics in Fe(II) treated Fe(III) hydr(oxides) suspensions are on the order of minutes (*24*), it was concluded that Fe(II) in smectites would be a predominant abiotic reductant in anaerobic environments where Fe(III) hydr(oxides) have been reductively dissolved, and reduced DOM and solution phase complexed forms of Fe(II) are depleted.

As was observed for the sorption of ^{57}Fe(II) to Fe(III) mineral oxides, Mössbauer spectroscopy indicates that sorption of Fe(II) to an iron-bearing smectites clay surface results in oxidation of the Fe(II) and subsequent reduction of the underlying structural Fe(III) (*25*).

Aqueous Phase Pathways for Electron Transfer

Evidence for solution-phase reductants was provided initially in studies of the reduction of 4-chloronitrobenzene by quinones and iron porphyrin in the presence of redox buffers (*26*). In a subsequent study, the facile reduction of a series of substituted NACs to the aromatic amines occurred in aqueous solutions of DOM (natural organic matter) chemically reduced by treatment with sodium sulfide (*27*). Taken together, these studies provided the first evidence for the potential role of quinone-moieties present in DOM as electron carriers or shuttles in the reduction of NACs as illustrated in the scheme below:

(Bulk e^--Donor)$_{red}$ | -ne$^-$ | +ne$^-$ | (Bulk e^--Donor)$_{ox}$

O / O / OH / OH

(Contaminant)$_{red}$ | -ne$^-$ | +ne$^-$ | (Contaminant)$_{ox}$

This pathway for rapid turnover of electron equivalents is another plausible mechanism that could account for facile reduction observed for methyl parathion and halogenated ethanes in anaerobic sediments (*3*, *16*). Analysis of DOM isolates by ^{15}N-NMR (*28*), fluorescence spectroscopy (*29*) and electrochemical methods (*30–32*) has provided evidence for the presence of reducible quinone functional groups in DOM.

More recent studies in model systems have demonstrated that DOM can also facilitate electron transfer by lowering the redox potential of the Fe(II)/Fe(III) redox couple through formation of DOM/Fe(II) complexes. DOM contains a number of functional groups (i.e., carboxylic acids, catechols, amino, and thio groups) known to complex Fe(II) (*33*). The extent to which Fe(II) is activated through complexation is dependent on the nature of the complexing ligands (*34*). The reduction of oxamyl, an oxime carbamate pesticide, was found to vary over several orders of magnitude in Fe(II) solutions containing various carboxylate and aminocarboxylate ligands or DOM isolates. Reaction rates were dependent on the reactivity of the Fe(II) complex as described by the one-electron reduction potential of the corresponding Fe(II)/Fe(III) redox couples.

Model Systems for Assessing Biotic (Cell-Mediated) versus Abiotic (Mineral-Mediated) Pathways for Electron Transfer

The design of model systems has become increasingly complex in attempts to more closely simulate natural systems. The addition of a biological component (i.e., iron-reducing bacteria) to the abiotic model systems allows the ability to assess relative contributions of abiotic and biotic pathways for electron transfer. One such example was a study designed to assess the contributions of biotic (cell-mediated) versus abiotic (mineral-mediated) pathways for the reduction of carbon tetrachloride (CT) (*35*). This was accomplished by the addition of iron-reducing bacteria to an amorphous iron oxide suspension. Nano-scale magnetite particles were formed as the result of iron respiration. A comparison of the mineral-mediated and direct biological reduction rates indicated that the mineral-mediated reduction process occurred at rates significantly faster than direct biological reduction.

Luan et. al. (*36*) conducted a systematic investigation of the reduction of nitrobenzene by combinations of DOM, hematite, and iron-reducing bacteria (i.e., *Shewanella putrefaciens* strain CN32). Although CN32 was found to directly reduce nitrobenzene, the addition of either natural organic matter (NOM) or hematite to the reaction system enhanced the reduction of nitrobenzene. In reactions systems that contained CN32, NOM and hematite, it was demonstrated that NOM-mediated reduction of nitrobenzene was more important than reduction by surface associated Fe(II).

Studies in these relatively complex model systems serve to illustrate the challenges of translating the resulting process science to even more complex systems such as anaerobic sediments and aquifers.

Potential Candidates for Chemical Reductants in Anaerobic Sediments Based on Model Studies

The results of the studies in well-defined model systems suggest the suite of chemical reductants most likely to form in anaerobic sediments as a function of the TEAPs biogeochemical processes include surface-associated (Fe(II)), aqueous-phase complexed forms of Fe(II), and reduced quinone moieties associated with dissolved organic matter:

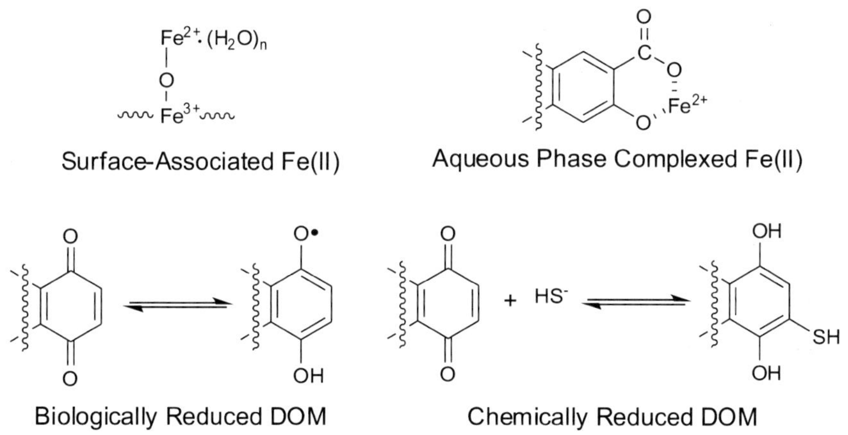

Surface-Associated Fe(II)   Aqueous Phase Complexed Fe(II)

Biologically Reduced DOM   Chemically Reduced DOM

The challenge remains to determine if these chemical reductants are actually representative of those occurring in anaerobic sediments and aquifers, and subsequently to determine the readily measurable environmental descriptors that would serve as a measure of reductant reactivity.

We recognize that the surface complexed model for the activation of Fe(II) by Fe(III) (hydr)oxides is an oversimplification of the interaction of Fe(II) with mineral oxide surfaces. Recent studies have proposed a 'redox-driven conveyor belt' to account for the significant exchange of aqueous Fe(II) and goethite (*5, 25*). This mechanism is initiated by the sorption of Fe(II) to Fe(III) at the oxide surface, subsequent oxidation of the sorbed Fe(II), followed by bulk conduction of the electrons through the oxide, and reductive dissolution of the oxide resulting in the reformation of aqueous Fe(II). As illustrated in the scheme above, we will use the term "surface-associated Fe(II)" in the subsequent discussion to refer to Fe(II) sorbed to the iron oxide surface or Fe(II) formed at the iron oxide surface due to electron transfer through the bulk crystal lattice.

Translation of the Process Science Generated from Model Studies to Anaerobic Sediments and Aquifers

Although a significant body of process science has been generated from the studies in well-defined model systems, the challenge is to determine to what extent this process science translates to natural anaerobic systems. Due to the complexity of these systems, such efforts have depended primarily on comparing the reactivity and transformation pathways of probe chemicals measured in the model systems vs. anaerobic sediments. For the following discussion, we have categorized these studies into 4 types: a) reactivity pattern analysis of a series of structurally related chemical probes to identify the predominant abiotic reductants in anaerobic systems, b) use of probe chemicals to identify readily measurable indicators of the predominant abiotic reductants in anaerobic sediments, c) use of customized probe chemicals designed to address specific questions concerning pathways for electron transfer, and d) an example of how studies in a well-defined model system were used to elucidate the predominant abiotic reductants for a chemical containing a functional group that had not been previously demonstrated to be susceptible to abiotic reduction in anaerobic sediments.

Reactivity Pattern Analysis

Reactivity pattern analysis has been demonstrated to be a very useful approach to the identification of the predominant abiotic reductants in anaerobic sediments and aquifers. This approach involves the measurement and comparison of the reactivity of a series of structurally related chemical probes (e.g., substituted nitrobenzenes or halogenated aliphatics) in different reaction systems. Similarities in the range and relative order of reactivity provide evidence for a common mechanism for reductive transformation in the reaction systems of interest.

One of the earliest demonstrations of the use of this approach was the study of the reduction of a series of nitroaromatic compounds (NACs) in a laboratory aquifer column (*37*). The relative reactivities of the NACs as a function of their one electron reduction potentials ($E_h1'(ArNO_2)$) are illustrated for a water soluble Fe(II) porphyrin (Figure 1A), microbially formed magnetite treated with Fe(II) in batch reactors (Figure 1B), and an aquifer column (Figure 1C). Competition quotients (Q_c), for the column studies were obtained from experiments with binary and ternary mixtures of NACs. The Qc values are a direct measure of the relative affinities of the different NACs for the reactive sites in the aquifer column. The similarities of the reactivity patterns for each of the reaction systems in Figure 1 provide strong evidence that iron species were the predominant reductants in the aquifer column. An interesting finding from the laboratory aquifer column studies was the observation that though the reduction rates (k_{obs}) for a series of substituted NACs were expected to vary as a function of their one-electron potentials, as had been observed in the Fe(II) treated magnetite suspensions (*17*), the NACs were reduced at the same rate, irrespective of their one-electron reduction values. The conclusion was that the regeneration of the reactive sites (i.e., surface associated Fe(II)) by iron-reducing bacteria became the rate controlling step in the electron transfer process. This finding illustrates one of the limitations in the direct translation of the process science generated from the studies in well-defined model systems to natural anaerobic systems.

In a subsequent study, the reductive transformation rates of a series of polynitroaromatic compounds ((P)NACs) were measured in sterile batch systems in either the presence of Fe(II)/Fe(III)(hydr)oxides or hydroquinones in the presence of H_2S, and in columns containing sand coated with FeOOH and a pure culture of an iron-reducing bacterium (*38*). The kinetic studies indicated that the Fe(II) treated Fe(III)(hydr)oxides were significantly more reactive towards the reduction of the (P)NACs than the chemically reduced quinones, and the process for the regeneration of Fe(II) at the surface of the Fe(III)(hydr)oxides (i.e., adsorption of Fe(II) from solution or the microbially-mediated reduction of the Fe(III)(hydr)oxides) had little effect on the reactivity patterns of the (P)NACs).

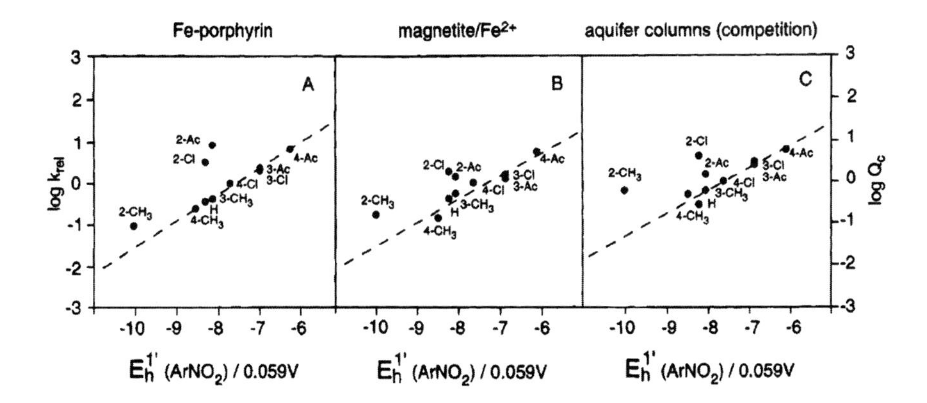

Figure 1. Plots of the logarithms of relative reduction rates of 10 model NACs in an aqueous suspension containing (A) water-soluble iron porphyrin, data from reference (26) or (B) microbially formed magnetite and Fe(II) (k_{obs} values represent initial pseudo-first-order reaction rate constants determine for the various compounds in batch reactors containing 10.6 mM $Fe(II)_{tot}$ versus $E_h1'(ArNO_2)$. (C) plot of logarithms of the competition quotients of the same compounds in the aquifer columns versus $E_h1'(ArNO_2)$. Reproduced from (37).

Reactivity pattern analysis for the characterization of chemical reductants in an anaerobic aquifer was reported for a series of substituted mono- and di-nitroaromatics measured in an iron-reducing model system (i.e., Fe(II)-treated goethite), sulfate-reducing model system (i.e., landfill-derived reduced DOM), landfill groundwater, and landfill groundwater/sediment (*39*). Based on comparison of measured reactivity patterns determined in the model and groundwater/sediment, it was concluded that surface-associated Fe(II) was the predominant reductant in the anaerobic region of the aquifer despite the presence of H_2S/HS^-, aqueous Fe(II) and reduced organic matter.

We have applied this approach to the study of the reductive transformation of a series of halogenated methanes in anaerobic sediments (*40*). The reduction kinetics for dibromochloromethane in iron- and sulfate-reducing sediments and the corresponding model systems (i.e., Fe(II) treated goethite and FeS suspensions) are illustrated in Figure 2 (left panel). The reactivity patterns for the series of halogenated methanes in each of the reactions systems are illustrated in Figure 2 (right panel).

Visual inspection of the data in Figure 2b indicate that the relative range and order of reactivity of the halogenated methanes is most comparable for the iron- and sulfate-reducing sediments and the Fe(II) treated goethite suspensions data sets. Significant differences are observed when these three data sets are compared to the FeS data set, an indication that Fe(II) associated with Fe(III) (hydr)oxides is the predominant reductant in both the iron- and sulfate-reducing sediments. Although significant formation of FeS occurred in the sulfate-reducing sediment, surface associated Fe(II) appears to still be the predominant reductant in these systems.

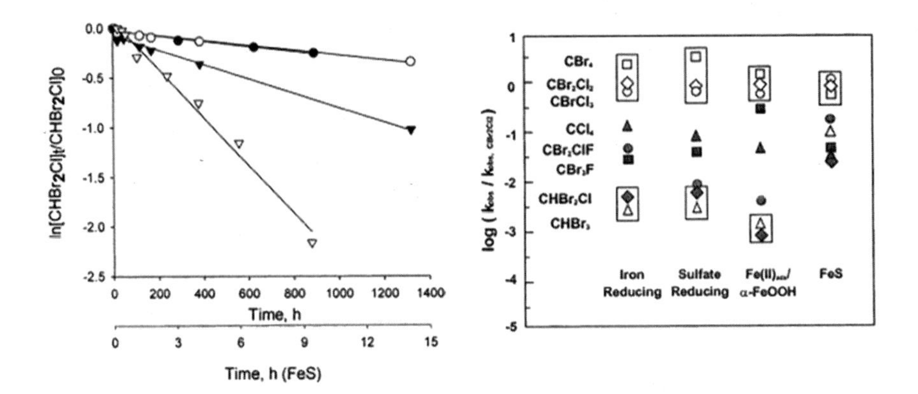

Figure 2. (a) Plot of ln [$CHBr_2Cl$]$_0$ time for the iron reducing (○) and sulfate reducing (●) sediments, and Fe(II) treated goethite (▾) and FeS (▵) model systems. (b) Reactivity patterns for halomethanes in iron- and sulfate-reducing sediments, Fe(II) treated goethite, and FeS model systems. Solid boxes provide a visual aid to group halomethanes that exhibited similar relative rates of transformation between systems. Source: Reproduced from (40).

Identification of Readily Measurable Indicators of Reductant Reactivity

Subsequent to the characterization of the predominant abiotic reductants is the need to identify readily measurable indicators of their reactivity. As stated previously, ability to estimate environmental concentrations of the parent chemical and potential transformation products requires knowledge of both the molecular descriptors and environmental descriptors necessary. Another example of this approach from our own work has been the study of 4-cyanonitrobenzene (4-CNB), in Fe(II)/goethite suspensions (*41*) and a pond sediment incubated with electron sources and acceptors to achieve dominant TEAPs (i.e., nitrate-reducing, iron-reducing, sulfate-reducing, and methanogenic) in both laboratory batch (*42*) and column studies (*43*). The choice of this particular probe chemical is 3 fold: a) The reduction pathway and intermediates, the *N*-Nitroso (pCNN), the *N*-hydroxylamine (pCNH), and the terminal product, aniline (pCNA), have been well characterized (*44*), b) pCNB does not adsorb to the sediment, thus, the observed disappearance kinetics can be attributed to reduction, and c) the cyano group, which is strongly electron withdrawing, activates the nitro towards reduction, and decreases the nucleophilicity of the amino group on pCNA (*45*). Nucleophilic addition of aromatic amines to quinone moieties in the natural organic carbon associated with the sediment is one of the primary pathways for covalent binding, which would result in low mass recoveries (*46*). High mass recoveries are critical if electron balances of potential readily measureable redox indicators are to be identified.

NO_2 | N=O | HN–OH | NH_2

2 e⁻ / 2 H⁺ → [ ] 2 e⁻ / 2 H⁺ → 2 e⁻ / 2 H⁺ →

CN CN CN CN

pCNB pCNN pCNH pCNA

Figure 3 illustrates the reaction kinetics for pCNB reduction in a) Fe(II) treated goethite system (Figure 3a), b) in a laboratory sediment column that was characterized with respect to redox zonation (Figure 3b), and in c) batch systems of an iron-reducing (Figure 3c) and d) sulfate-reducing (Figure 3d) pond sediment. A common feature of each of these data sets is reduction of pCNB with the concomitant loss of Fe(II). Reduction of pCNB in the Fe(II) goethite suspension is fast with complete conversion of pCNB to pCNA in 5 h (Figure 3a). Figure 3b illustrates the facile and complete reduction of pCNB to the hydroxylamine intermediate (pCNH) in the first 2 cm of the sediment column. The subsequent reduction of pCNH to the aniline (pCNA) coincided with the steep increase in the concentration of aqueous Fe(II). Whereas the electron equivalent balance for the reduction of pCNB in the Fe(II) treated goethite suspension can be accounted for with the loss of aqueous phase Fe(II), the measured losses of aqueous phase Fe(II) in iron-reducing and sulfate-reducing sediment slurries are significantly less than the theoretical loss of solution phase Fe(II) based on the reduction of pCNB. These data suggest that microbial-mediated regeneration of aqueous phase Fe(II) occurs at rates comparable to pCNB reduction. These results illustrate one of the challenges to identifying the predominant chemical reductants in microbially active systems.

Application of Customized Probe Chemicals for Elucidating Pathways for Electron Transfer

The of use customized probe chemicals to elucidating pathways for electron transfer in anaerobic systems has been quite limited. Studies in well-defined model systems have provided viable pathways for electron transfer through both surface-mediated and solution phase pathways. Although significant evidence from studies in both model and sediment systems had been developed for the surface- mediated pathway for electron transfer, it was still unclear if this pathway for electron transfer through aqueous phase electron shuttles was a contributing pathway for the abiotic reduction of chemical contaminants in anaerobic sediments. To this end, a chemical probe (i.e., 4-cyano-4′-aminoazobenzene (CNAAzB) covalently bound to an epoxide activated glass bead) was synthesized that allowed for differentiation between surface-associated and solution-phase electron-transfer processes (*47*) (Figure 4a). By measuring the formation of 4-cyanoaniline (4-CYA) it was then possible to determine the extent of the azo linkage reduction in the reaction systems of interest.

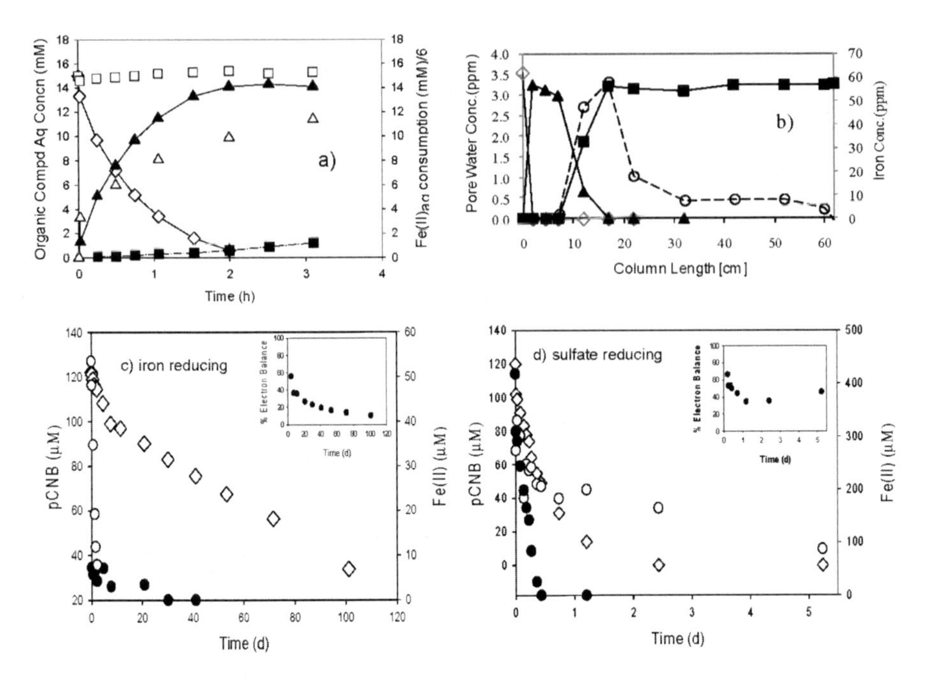

Figure 3. Reduction kinetics for pCNB reduction in a) Fe(II) treated goethite system, b) a laboratory sediment column as a function of [Fe(II)] and column length, and in batch systems of c) iron-reducing and d) sulfate-reducing pond sediment. Key: pCNB (◊), pCNH (▲), pCNA (■), mass balance (□), and Fe(II) consumption (△), Fe(II) measured (●) and Fe (II) theoretical (○) based on pCNB loss. The inserted graphs illustrate the % electron balance for the reduction of pCNB based on the loss of aqueous Fe(II). Reproduced from (41–43).

The utility of this chemical probe was demonstrated in well-defined model systems consisting of either Fe(II), Fe(II)/goethite, or chemically reduced juglone. Where no formation of 4-CYA was observed in the presence of Fe(II) or an Fe(II)/goethite suspension, significant and rapid formation of 4-CYA occurred in the presence of chemically reduced juglone (Figure 4b). It is important to note that juglone in the presence of Fe(II)/goethite did not promote reduction of the bound CNAAzB, an indication that the abiotic reduction of juglone was not occurring in this system (data not shown). The addition of a pure culture of *S. putrefaciens strain* CN32 to the abiotic model system containing Suwannee River NOM (SRNOM) and river sediment low in organic carbon content resulted in significant formation of 4-CYA (Figure 4c), consistent with a electron transfer process mediated by biologically reduced quinone moieties in the SRNOM. Studies of humic and fulvic acid isolates have demonstrated a significant increase in organic radical concentrations upon reduction with *Geobacter metallireducens* (*48*). ESR spectra were consistent with the formation of semiquinones, which are the one electron reduction products formed from the reduction of quinones. The subsequent addition of the bound azo chemical probe to several anaerobic sediment slurries resulted in the formation of 4-CYA (Figure 4d). The initial generation rate of CYA correlated best with the aqueous phase SUVA350nm, which is a measure of the fraction of quinone functional groups in DOM (*29*). These results provide direct evidence for a solution phase pathway for electron transfer. The significance of this pathway to the overall rate of abiotic reduction is the focus of on-going studies.

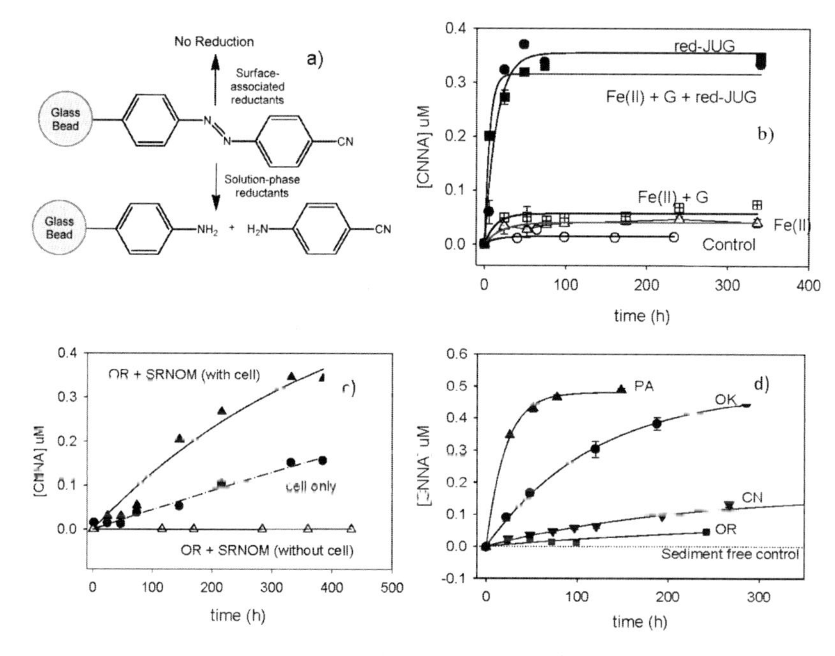

Figure 4. a) Scheme for the reduction of bound CNAAzB resulting in the formation of 4-cyanoaniline. Reduction kinetics of bound CNAAzB in b) abiotic model systems with goethite as the solid phase, c) biotic model system with Oconee River Sediment as the solid phase, and d) anaerobic sediment suspensions. Reaction conditions: 10 g/L anaerobic sediment, 25 mM pH 7 buffer, 25 mg bound CNAAzB. Points are experimental data and lines are model fits. Reproduced from (47).

Identifying New Functional Groups That Are Susceptible to Abiotic Reduction

Our understanding of the processes controlling abiotic reduction in anaerobic sediments and aquifer has resulted primarily from the systematic study of nitroaromatics, halogenated aliphatics, and aromatic azos to a lesser extent. It remains to be determined to what extent this body of process science generated from this work translates to the other reducible functional groups such as *N*-nitrosamines, sulphones, sulphoxides, and other reducible functional groups that have yet to be identified. The general approach of using the results of reduction kinetics measured in well-defined model systems to inform potential pathways for reduction of functional group in the latter category of "yet to be identified" was recently demonstrated in the study of the reduction of sulfamethoxazole (SMX) in anaerobic soil microcosms (*49*). SMX is a high volume antimicrobial used in livestock production. SMX reduction was found to occur most rapidly under iron-reducing condition, though facile reduction of SMX also was observed under sulfate-reducing and methanogenic conditions (Figure 5). Subsequent studies in Fe(II) treated goethite suspensions demonstrated that reduction of SMX is initiated by a one-electron reduction of the isoxazole N-O bond to form the unstable radical anion (Figure 5). As has been reported with NACs and halogenated aliphatics, SMX reduction rates were found to increase with surface-associated Fe(II) controlled by increasing goethite concentrations or solution pH. Although a limited example, these results indicate the potential for the general applicability of the process science generated based upon studies of the NACs and halogenated aliphatics.

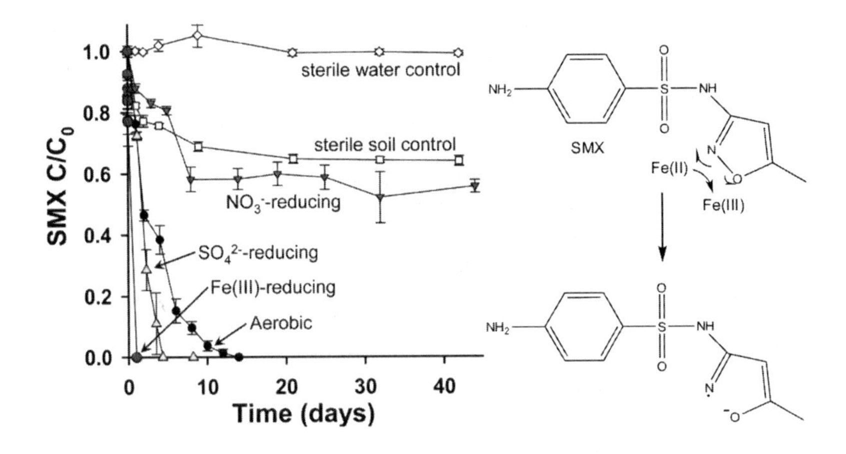

Figure 5. Observed time courses for dissipation of SMX in soil microcosms incubated under different TEAP conditions. Error bars represent 1σ uncertainty determined from triplicate microcosms. Reproduced from (49).

Environmental Implications

Based on the available process science, a generalized scheme emerges for the microbially-mediated formation of abiotic reductants and the subsequent reduction of contaminants with reducible functional groups in anaerobic sediments. This scheme is initiated by the oxidation of bioavailable organic carbon by anaerobic bacteria and the subsequent reduction of Fe(III) (hydr)oxides by electron transfer mediated by electron shuttles, most likely quinone moieties in dissolved organic matter. Likewise, contaminant reduction can occur through either direct contact with Fe(II) associated with iron-bearing mineral oxides and clay minerals, or indirectly by the quinone-based electron shuttles. The abiotic reduction of the quinone moiety by Fe(II)/Fe oxide appears to not be an energetically favorable process. Electron transfer through the aqueous phase by DOM complexed Fe(II) must also be considered as a viable pathway for reduction.

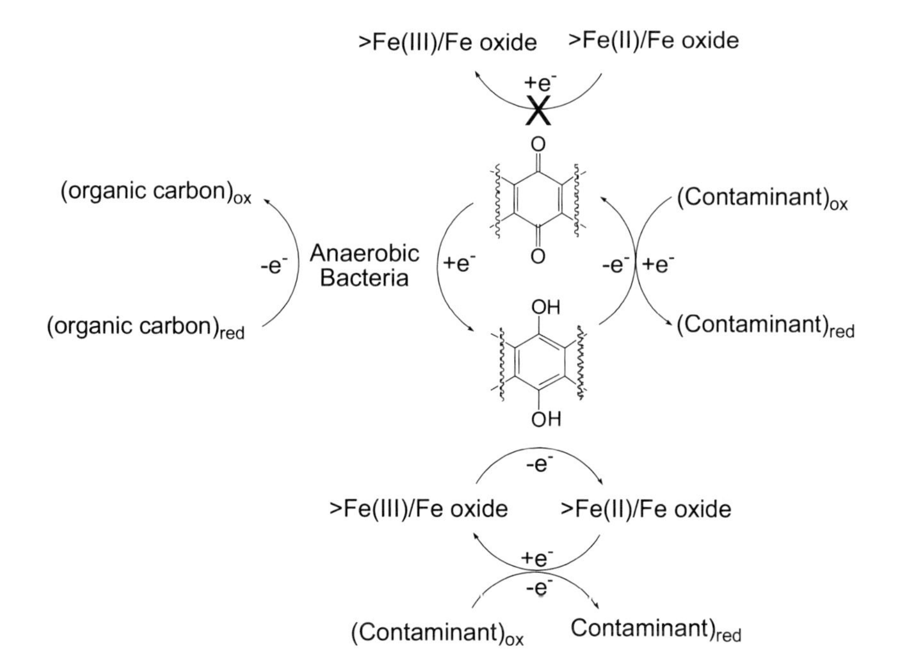

Disclaimer

This paper has been reviewed in accordance with the U.S. Environmental Protection Agency's peer and administrative review policies and approved for publication. Mention of trade names or commercial products does not constitute endorsement or recommendation for use.

References

1. Elovitz, M. S.; Weber, E. J. Sediment-mediated reduction of 2,4,6-trinitrotoluene and fate of the resulting aromatic (poly)amines. *Environ. Sci. Technol.* **1999**, *33*, 2617–2625.
2. Weber, E. J.; Adams, R.L Chemical- and sediment-mediated reduction of the azo dye disperse blue 79. *Environ. Sci. Technol.* **1995**, *29*, 1163–1170.
3. Jafvert, C. T.; Wolfe, N. L. Degradation of selected halogenated ethanes in anoxic sediment-water systems. *Environ. Toxicol. Chem.* **1987**, *6*, 827–837.
4. Williams, A. G. B.; Scherer, M. M. Spectroscopic evidence for Fe(II) – Fe(III) electron transfer at the iron oxide-water interface. *Environ. Sci. Technol.* **2004**, *38*, 4782–4790.
5. Handler, R. M.; Beard, B. L.; Johnson, C. M.; Scherer, M. M. Atom exchange between aqueous Fe(II) and goethite: An Fe isotope tracer study. *Environ. Sci. Technol.* **2009**, *43*, 1102–1107.
6. Schaefer, M. V.; Gorski, C. A.; Scherer, M. M. Spectroscopic evidence for interfacial Fe(II)−Fe(III) electron transfer in a clay mineral. *Environ. Sci. Technol.* **2011**, *45*, 540–545.

7. Elsner, M.; Hofstetter, T. B. Current perspectives on the mechanisms of chlorohydrocarbon degradation in subsurface environments: Insight from kinetics, product formation, probe molecules and isotope fractionation. In *Aquatic Redox Chemistry*; Tratnyek, P. G., Grundl, T. J., Haderlein, S. B., Eds.; ACS Symposium Series; American Chemical Society: Washington, DC, 2011; Vol. 1071, Chapter 19, pp 407−439.
8. U.S. Environmental Protection Agency. Pesticides: Endangered Species Protection Program. 2011. http://www.epa.gov/espp/.
9. Albright, R. D. *Cleanup of Chemical and Explosive munitions*; William Andrew Inc.: Norwich, NY, 2008.
10. Bylaska, E. J.; Tratnyek, P. G.; Salter-Blanc, A. J. One-electron reduction potentials from chemical structure theory calculations. In *Aquatic Redox Chemistry*; Tratnyek, P. G., Grundl, T. J., Haderlein, S. B., Eds.; ACS Symposium Series; American Chemical Society: Washington, DC, 2011; Vol. 1071, Chapter 3, pp 37−64.
11. Chapelle, F. H.; Haack, S. K.; Adriaens, P.; Henry, M. A.; Bradley, P. M. Comparison of Eh and H_2 measurements for delineating redox processes in a contaminated aquifer. *Environ. Sci. Technol.* **1996**, *30*, 3565–3569.
12. Bjerg, P. L.; Ruegge, K.; Pedersen, J. K.; Christensen, T. H. Distribution of redox-sensitive groundwater quality parameters downgradient of a landfill (Grindsted, Denmark). *Environ. Sci. Technol.* **1995**, *29*, 1387–1394.
13. Himmelheber, D. W.; Taillefert, M.; Pennell, K. D.; Hughes, J. B. Spatial and temporal evolution of biogeochemical processes following in situ capping of contaminated sediments. *Environ. Sci. Technol.* **2008**, *42*, 4113–4120.
14. Himmelheber, D. W.; Thomas, S. H.; Löffler, F. E.; Taillefert, M.; Hughes, J. B. Microbial colonization of an in situ sediment cap and correlation to stratified redox zones. *Environ. Sci. Technol.* **2008**, *43*, 66–74.
15. Weber, E. J.; Wolfe, N. L. Kinetic studies of the reduction of aromatic azo compounds in anaerobic sediment/water systems. *Environ. Toxicol. Chem.* **1987**, *6*, 911–919.
16. Wolfe, N. L.; Kitchens, B. E.; Macalady, D. L.; Grundl, T. J. Physical and chemical factors that influence the anaerobic degradation of methyl parathione in sediment systems. *Enrviron. Toxicol. Chem.* **1986**, *5*, 1019–1026.
17. Klausen, J.; Trober, S. P.; Haderlein, S. B.; Schwarzenbach, R. P. Reduction of substituted nitrobenzenes by Fe(II) in aqueous mineral suspensions. *Environ. Sci. Technol.* **1995**, *29*, 2396–2404.
18. Gorski, C. A.; Scherer, M. M. Influence of magnetite stoichiometry on Fe^{II} uptake and nitrobenzene reduction. *Environ. Sci. Technol.* **2009**, *43*, 3675–3680.
19. Gorski, C. A.; Nurmi, J. T.; Tratnyek, P. G.; Hofstetter, T. B.; Scherer, M. M. Redox behavior of magnetite: Implications for contaminant reduction. *Environ. Sci. Technol.* **2010**, *44*, 55–60.
20. Neumann, A.; Sander, M.; Hofstetter, T. B. Redox properties of structural Fe in smectite clay minerals. In *Aquatic Redox Chemistry*; Tratnyek, P. G., Grundl, T. J., Haderlein, S. B., Eds.; ACS Symposium Series; American

Chemical Society: Washington, DC, 2011; Vol. 1071, Chapter 17, pp 361−379.

21. Dong, H.; Kukkadapu, R. K.; Fredrickson, J. K.; Zachara, J. M.; Kennedy, D. W.; Kostandarithes, H. M. Microbial Reduction of Structural Fe(III) in Illite and Goethite. *Environ. Sci. Technol.* **2003**, *37*, 1268–1276.
22. Hofstetter, T. B.; Neumann, A.; Schwarzenbach, R. P. Reduction of nitroaromatic compounds by Fe(II) species associated with iron-rich smectites. *Environ. Sci. Technol.* **2006**, *40*, 235–242.
23. Neumann, A.; Hofstetter, T. B.; Skarpeli-Liati, M.; Schwarzenbach, R. P. Reduction of polychlorinated ethanes and carbon tetrachloride by structural Fe(II) in Smectites. *Environ. Sci. Technol.* **2009**, *43*, 4082–4089.
24. Elsner, M.; Schwarzenbach, R. P.; Haderlein, S. B. Reactivity of Fe(II)-bearing minerals toward reductive transformation of organic contaminants. *Environ. Sci. Technol.* **2004**, *38*, 799–807.
25. Gorski, C. A.; Scherer, M. M. Fe^{2+} Sorption at the Fe oxide-water interface: A revised conceptual framework. In *Aquatic Redox Chemistry*; Tratnyck, P. G., Grundl, T. J., Haderlein, S. B., Eds.; ACS Symposium Series; American Chemical Society: Washington, DC, 2011; Vol. 1071, Chapter 15, pp 315−343.
26. Schwarzenbach, R. P.; Stierli, R.; Lanz, K.; Zeyer, J. Quinone and iron porphyrin mediated reduction of nitroaromatic compounds in homogeneous aqueous solution. *Environ. Sci. Technol.* **1990**, *24*, 1566–1574.
27. Dunnivant, F. M.; Schwarzenbach, R. P.; Macalady, D. L. Reduction of substituted nitrobenzenes in aqueous solutions containing natural organic matter. *Environ. Sci. Technol.* **1992**, *26*, 2133–2141.
28. Thorn, K. A.; Arterburn, J. B.; Mikita, M. A. Nitrogen-15 and carbon-13 NMR investigation of hydroxylamine-derivatized humic substances. *Environ. Sci. Technol.* **1992**, *26*, 107–116.
29. Cory, R. M.; McKnight, D. M. Fluorescence spectroscopy reveals ubiquitous presence of oxidized and reduced quinones in dissolved organic matter. *Environ. Sci. Technol.* **2005**, *39*, 8142–8149.
30. Aeschbacher, M.; Sander, M.; Schwarzenbach, R. P. Novel electrochemical approach to assess the redox properties of humic substances. *Environ. Sci. Technol.* **2009**, *44*, 87–93.
31. Nurmi, J. T.; Tratnyek, P. G. Electrochemical properties of natural organic natter (NOM), fractions of NOM, and model biogeochemical electron shuttles. *Environ. Sci. Technol.* **2002**, *36*, 617–624.
32. Ratasuk, N.; Nanny, M. A. Characterization and quantification of reversible redox sites in humic substances. *Environ. Sci. Technol.* **2007**, *41*, 7844–7850.
33. Luther, G. W.; Shellenbarger, P. A.; Brendel, P. J. Dissolved organic Fe(III) and Fe(II) complexes in salt marsh porewaters. *Geochim. Cosmochim. Acta* **1996**, *60*, 951–960.
34. Strathmann, T. J.; Stone, A. T. Reduction of the pesticides oxamyl and methomyl by Fe^{II}: Effect of pH and inorganic ligands. *Environ. Sci. Technol.* **2002**, *36*, 653–661.

35. McCormick, M. L.; Bouwer, E. J.; Adriaens, P. Carbon tetrachloride transformation in a model iron-reducing culture: Relative kinetics of biotic and abiotic reactions. *Environ. Sci. Technol.* **2002**, *36*, 403–410.
36. Luan, F.; Burgos, W. D.; Xie, L.; Zhou, Q. Bioreduction of nitrobenzene, natural organic matter, and hematite by *Shewanella putrefaciens* CN32. *Environ. Sci. Technol.* **2010**, *44*, 184–190.
37. Heijman, C. G.; Grieder, E.; Holliger, C.; Schwarzenbach, R. P. Reduction of nitroaromatic compounds coupled to microbial iron reduction in laboratory aquifer columns. *Environ. Sci. Technol.* **1995**, *29*, 775–783.
38. Hofstetter, T. B.; Heijman, C. G.; Haderlein, S. B.; Holliger, C.; Schwarzenbach, R. P. Complete reduction of TNT and other (poly)nitroaromatic compounds under iron-reducing subsurface conditions. *Environ. Sci. Technol.* **1999**, *33*, 1479–1487.
39. Rügge, K.; Hofstetter, T. B.; Haderlein, S. B.; Bjerg, P. L.; Knudsen, S.; Zraunig, C.; Mosbek, H.; Christensen, T. H. Characterization of predominant reductants in an anaerobic leachate-contaminated aquifer by nitroaromatic probe compounds. *Environ. Sci. Technol.* **1998**, *32*, 23–31.
40. Kenneke, J. F.; Weber, E. J. Reductive dehalogenation of halomethanes in iron- and sulfate-reducing sediments. 1. Reactivity pattern analysis. *Environ. Sci. Technol.* **2003**, *37*, 713–720.
41. Colon, D.; Weber, E. J.; Anderson, J. L. QSAR study of the reduction of nitroaromatics by Fe(II) species. *Environ. Sci. Technol.* **2006**, *40*, 4976–4982.
42. Hoferkamp, L. A.; Weber, E. J. Nitroaromatic reduction kinetics as a function of dominant terminal electron acceptor processes in natural sediments. *Environ. Sci. Technol.* **2006**, *40*, 2206–2212.
43. Simon, R.; Colon, D.; Tebes-Stevens, C. L.; Weber, E. J. Effect of redox zonation on the reductive transformation of p-cyanonitrobenzene in a laboratory sediment column. *Environ. Sci. Technol.* **2000**, *34*, 3617–3622.
44. Colon, D.; Weber, E. J.; Anderson, J. L.; Winget, P.; Suarez, L. A. Reduction of nitrosobenzenes and N-hydroxylanilines by Fe(II) species: Elucidation of the reaction mechanism. *Environ. Sci. Technol.* **2006**, *40*, 4449–4454.
45. Colon, D.; Weber, E. J.; Baughman, G. L. Sediment-associated reactions of aromatic amines. 2. QSAR development. *Environ. Sci. Technol.* **2002**, *36*, 2443–2450.
46. Thorn, K. A.; Pettigrew, P. J.; Goldenberg, W. S.; Weber, E. J. Covalent binding of aniline to humic substances. 2. ^{15}N nmr studies of nucleophilic addition reactions. *Environ. Sci. Technol.* **1996**, *30*, 2764–2775.
47. Zhang, H.; Weber, E. J. Elucidating the role of electron shuttles in reductive transformations in anaerobic sediments. *Environ. Sci. Technol.* **2009**, *43*, 1042–1048.
48. Scott, D. T.; Mcknight, D. M.; Blunt-Harris, E. L.; Kolesar, S. E.; Lovley, D. R. Quinone moieties act as electron acceptors in the reduction of humic substances by humics-reducing microorganisms. *Environ. Sci. Technol.* **1998**, *32*, 2984–2989.

49. Mohatt, J. L.; Hu, L.; Finneran, K. T.; Strathmann, T. J. Microbially mediated abiotic transformation of the antimicrobial agent sulfamethoxazole under iron-reducing soil conditions. *Environ. Sci. Technol.* **2011**, 4793–4801.

Chapter 25

The Role of Transport in Aquatic Redox Chemistry

Wolfgang Kurtz[1] and Stefan Peiffer*

Department of Hydrology, University of Bayreuth, Germany

***s.peiffer@uni-bayreuth.de**
[1]present address: Forschungszentrum Jülich GmbH, Institute for Bio- and Geosciences, IBG-3: Agrosphere, Germany

Heterogeneous redox processes with mobile and immobile reactants can be significantly affected by transport processes in porous media. In this study, we have performed a sensitivity analysis on the effect of transport on the turnover of the reaction between H_2S and FeOOH that is based on a surface complexation model implemented into a reactive transport code. The analysis considered two reactive surface species ($\equiv FeS^-$ and $\equiv$FeSH), the amount of surface sites and the buffering capacity. Turnover depends on the ratio between residence time and characteristic reaction time, expressed as *Da* (Damköhler) numbers. Two completely different relationships between turnover and *Da* numbers were observed for the two reactive surface species which is due to the pH dependence of their speciation. Calibration of the kinetic model in a column experiment suggests the surface species $\equiv$FeSH to be responsible for the turnover. The characteristic reaction time depends on the concentration of reactive surface sites and the amount of ferric(hydr)oxides. In natural systems, spatial distributions of these parameters exist along with that of residence times. We, therefore, postulate that *Da* numbers are also spatially distributed reflecting zones of high and low turnover with implications for product accumulation and competition with other reactions involving iron oxides.

Introduction

The importance of transport phenomena for the turnover and pathways of geochemical reactions at the mineral-water interface has been stressed for many geochemical reactions such as dissolution of salts and subsequent sinkhole formation (*1*), establishing of redox fronts (*2*) or contaminant degradation (*3*, *4*). Predicting und quantifying such reactions in porous media in terms of simple upscaling of batch reactivities, however, very often fails (e.g., (*5*, *6*)) although on a molecular scale the reactions are the same. The reason for these observations has been related to the occurrence of dead-end pores and low-permeability zones in porous media that can only communicate with the primary flow paths through diffusion (*7*). As a result, the effective reactive surface area in a porous medium is much lower compared to batch systems. In addition, transport processes in heterogenous pore structures will also affect the supply of mobile reactants and the removal of mobile products from the immobile reactive sites and thus control the extent of the overall reaction kinetics.

Only a few attempts have been made to account for the effect of transport processes on turnover and pathways of redox processes, although these reactions very often reflect interactions at the mineral-water interface. For example, Hansel et al. (*8*) observed that for microbial iron reduction two different flow velocities resulted in a different mineral inventory downstream of the reduction zone, which they attributed to a higher removal rate of ferrous iron at higher flow velocities driving the transformation process. In another study on microbial iron reduction, variation of the flow rate in column experiments lead to an increase of ferrous iron release (*9*). Szecsody et al. (*10*) have found significant influence of flow and pore-scale heterogeneity on rates of a coupled redox network and postulated that development of transport-relevant reaction parameters are required to account for the non-linear behaviour of such redox systems in porous media. The little attention paid to transport phenomena in aquatic redox chemistry is even more surprising since experiments under flow conditions are widely used approaches to study redox processes (e.g., (*11–13*)).

In this study, we aim to derive a fundamental understanding to what extent the turnover of a surface controlled reaction and the product distribution are affected in an advective flow system. To these ends, a sensitivity analysis was performed based on numerical scenarios with a reactive transport code, which studies the single steps of the reaction between dissolved sulphide and ferric oxides in a column experiment. We have chosen this reaction because it is prominent in many anoxic systems (*14–16*), on the one hand, and its initial reaction steps and their kinetics are well established, on the other hand.

Since turnover and product formation are controlled by both, kinetic and transport parameters, we will discuss our data using dimensionless Damköhler numbers that compare hydraulic residence times with characteristic reaction times (*17*). The numerical results are used to evaluate and discuss the results of a laboratory column experiment.

Theoretical Background

Kinetics of Sulphide Oxidation by Ferric (Hydr)Oxides

H_2S reacts in an initial reaction with FeOOH to generate S° and Fe^{2+}

$$2\,FeOOH + H_2S + 4\,H^+ \rightarrow 2\,Fe^{2+} + S^0 + 4\,H_2O \qquad (1)$$

which may at higher pH form FeS (*18*) and ultimately lead to pyrite generation (*19*). The electron transfer reaction is preceded by the adsorption of sulphide to the oxide´s surface to form a reactive surface complex {≡FeX}, which is either {$\equiv FeS^-$}or {≡FeSH} (*20*). The reaction kinetics can be described by a first-order rate law that depends on the concentration of the surface species ≡FeX.

$$R = \frac{\partial c(H_2S)}{\partial t} = \frac{1}{2}\frac{\partial c(FeOOH)}{\partial t} = -k\{\equiv FeX\} \qquad (2)$$

The pH dependence of the surface speciation makes the reaction highly pH dependent with a maximum rate at circumneutral pH (*20, 21*). The rate constants of this reaction vary between 10^{-1} and 10^{-4} min^{-1} depending on the specific ferric (hydr)oxide (*19*).

Damköhler Numbers

For a certain range of rate constants, a decoupling between reaction and transport can occur, i.e. the residence time may not be long enough to ensure a complete turnover of the educts within a given flow distance.

The dependency of educt turnover on residence time can be parameterized with the dimensionless Damköhler number (*Da*) which relates the transport time scale to the reaction time scale in a given system (*22*). For an aquatic constituent that is removed from the system via a first order reaction, the corresponding *Da* number is readily derived from the Convection Dispersion Equation (*22*) CDE (eq 3) (c: concentration of the constituent, v: pore water velocity, x: length, t: time, D: dispersion coefficient).

$$\frac{\partial c}{\partial t} = \frac{\partial vc}{\partial x} + D\frac{\partial^2 c}{\partial x^2} - kc \qquad (3)$$

Normalization of the variables with the respective scales (L = characteristic length of the system, $\tau = L/v$: residence time, c_0: input concentration) yields the normalized CDE (eq 4) with dimensionless (relative) variables ($c' = c/c_0$, $x' = x/L$ and $t' = t/\tau = tv/L$).

$$\frac{\partial c'}{\partial t'} = \frac{\partial c'}{\partial x'} + \frac{D}{vL}\frac{\partial^2 c'}{\partial x'^2} - \frac{kL}{v}c' \qquad (4)$$

$$Da_I = k\tau = \frac{kL}{v} \qquad (5)$$

The *Da* number as defined in eq 5 is valid for advective systems and therefore termed Damköhler number I, contrary to the Damköhler number II that is applied to account for pore diffusion problems (cf. below). In the following discussion, we will use the term '*Da* number' synonymously for the Damköhler number I.

Eq 5 states that low *Da* numbers define a state where the residence time τ is low compared to the reaction time scale $1/k$, and, thus, turnover might not be complete within a given flow distance. Complementary to that, high *Da* numbers indicate a higher ratio of residence to reaction time, and thus, a higher removal rate of a specific constituent.

A theoretical relationship between turnover and *Da* numbers is depicted in Figure 1. It is derived from an analytical solution of eq 4 for simple boundary conditions (constant reactant supply, steady state, dispersivity α = 0.0125 m). It can be seen that turnover rises monotonically with *Da* numbers until turnover reaches completion for *Da* numbers >5. The *Da* number reflects different combinations of flow rate, reaction time and flow length.

Figure 1 implies that a certain geochemical reaction will show different turnover at different flow rates. The flow rate will also define the flow length at which a certain amount of reactant is depleted. The *Da* number at which 50% of the turnover is achieved approaches unity in the absence of a solid matrix, but is shifted to higher values in porous media with increasing dispersivity α.

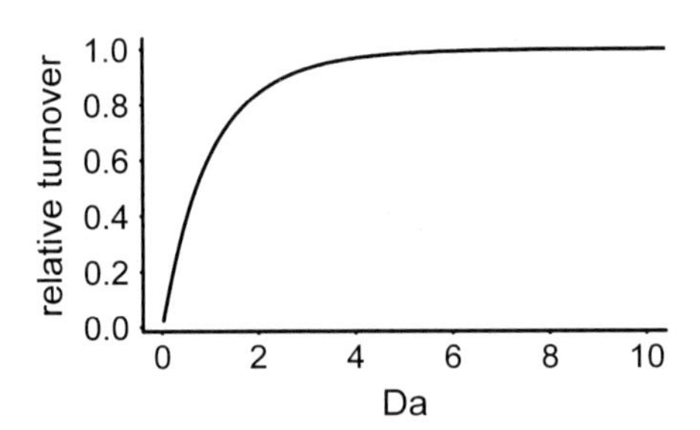

Figure 1. Theoretical relationship between Damköhler numbers and turnover for constant reactant supply and steady state with α *= 0.0125 m.*

The Damköhler concept has already been used in a variety of hydrochemical studies, which are mostly related to the fate of contaminants in aquatic systems. For example, it was demonstrated that double peaks due to kinetic sorption in simulation experiments occurred only at *Da* numbers <3 (*3*). Bold et al. (*4*) observed that kinetic sorption of organic contaminants onto mobile particles is mainly relevant for contaminant export at *Da* numbers between 0.01 and 100. Breakthrough of contaminants appeared to take place under non-equilibrium conditions at *Da* numbers <10 (*23*). Beyond the usage in sorption studies of organic contaminants, the *Da* concept has also been used in other fields of interest. Ocampo et al. (*24*) compiled the removal of nitrate in riparian zones of various catchments and demonstrated that nitrate removal was <50% for *Da* numbers <1 and increased to nearly 100% at *Da* numbers of 2-20. Similarly, *Da* numbers and residence time distributions were used to study removal extent of contaminants in treatment wetlands (*25*). Kern et al. (*26*) used *Da* numbers in a sensitivity analysis of the removal of Cd from rivers by adsorption onto settling particles and found a limitation of removal at *Da* numbers <10.

However, these applications were restricted to cases where the reaction itself did not show any feedback behaviour or dependency on additional reaction parameters. Both are true for the reaction of sulphide with iron (hydr)oxides, which is supposed to be a pH-dependent, surface controlled reaction. Here, turnover might be additionally affected by (reaction-related) pH-changes and surface properties of the iron (hydr)oxides (side density, adsorption constants, competition for surface sites).

Materials and Methods

The basic setup that was used for both approaches (numerical studies and laboratory experiments) consisted of a 25 cm long column (flow through area ~50 cm^2) that contained iron (hydr)oxide-coated sand. Columns were flushed with a sulphide solution for approximately 20 pore volumes (1 PV ~480 mL) at different flow rates (see Table I). Basic characteristics of the influent solution(s) and the bulk material are summarized in Table II. Variable values for the simulations refer to the sensitivity analysis (bold values mark the standard values used in the simulations).

Table I. Velocity in pore volumes v_{PV}, mean velocities $\bar{v}$, pore water velocities v and time step size Δt for simulations and column experiments

simulations											
v_{PV}	[PV d^{-1}]	0.25	0.5	1	2	4	6	8	12	16	24
$\bar{v}$	[cm d^{-1}]	2.5	5	10	20	40	60	80	120	160	240
v	[cm d^{-1}]	6.25	12.5	25	50	100	150	200	300	400	600
Δt	[s]	3456	1728	864	432	216	144	108	72	54	36
column experiments											
v_{PV}	[PV d^{-1}]		0.5	1.0	2.8		5.7		10.4	14.7	
$\bar{v}$	[cm d^{-1}]		5	10	27		55		100	142	
v	[cm d^{-1}]		12	25	69		139		254	360	

The sulphide concentration of 200 μmol L^{-1} is in the typical range of environmental concentrations. DIC was added to the system to provide a certain buffering capacity because the reaction between sulphide and ferric (hydr)oxide consumes protons. However, the concentration was not chosen too high in order to minimize the build-up of carbonate surface complexes. Sodium chloride was added to establish a constant ionic strength in the system.

Table II. Concentrations for inlet solutions

		simulations	*column experiments*
$c_{sulphide}$	[µmol L^{-1}]	200	188 ± 5 [a]
c_{DIC}	[µmol L^{-1}]	**18.5** / 200 / 400 [b]	200
c_{NaCl}	[mol L^{-1}]	~ 0.01	0.01
surface sites	[µmol]	27.4 / 83.3 / **163** / 247 / 547 [b]	40
pH	[-]	7	7.05 ± 0.06 [a]

[a] mean ± standard deviation [b] for sensitivity analysis

The large range of flow rates allowed to mimic a variety of subsurface conditions which are typical for heterogeneous aquifers. The different values for the total amount of surface sites (SSI) were used to model the variability of surface properties among iron (hydr)oxides and their content in environmental systems, which can show considerable variation (*27*).

Simulations

Simulations were performed with the transport-reaction-model TBC (*28*). The model column was discretized into 10 rectangular prisms with a length of 0.025 m and a width and height of 0.0709 m with 44 nodes. The volume of the column was separated into a bulk and a mobile phase with a porosity of 0.4 and a density of 2.65 g cm^{-3} for the bulk material. Transport was calculated one dimensional with a finite difference method. Boundary conditions (BCs) for flow were set as follows: first type BC for the last layer nodes (x = 0.25 m) with a fixed head value of 0; second type BC for the first layer nodes (x = 0 m) with the chosen flow rate. The inflow of solutes was also defined as a second type BC for the first layer nodes.

Dispersivity was set to a value of 0.0125 m for all simulations. A similar value was used by others for the simulation of column experiments (e.g., (*23*, *28*)), what allowed a reasonable computation time.

Temporal discretization was adapted for each flow velocity to 0.01 PVs in order to obtain the same amount of time steps for each simulation. All simulations met stability criteria for spatial and temporal discretization (Courant number, grid-Peclet number).

The equilibrium system was set up according to the component principle given in (*29*). Equilibrium reactions and the corresponding equilibrium constants are shown in Table III (equilibrium constants for the aquatic species were adapted to 20 °C).

With respect to the surface species it has to be mentioned that TBC does not account for surface charge as described in (*33*), and, thus, effects that are related to this parameter are not considered in the simulations.

Table III. Equilibrium reactions for TBC (with equilibrium constants K_{eq})

reaction			$log(K_{eq})$	*reference*
$H^+ + OH^-$	↔	H_2O	-13.998	(*30*)
H_2S	↔	$HS^- + H^+$	-6.994	(*30*)
HS^-	↔	$S^{2-} + H^+$	-12.918	(*30*)
H_2CO_3	↔	$HCO_3^- + H^+$	-6.151	(*30*)
HCO_3^-	↔	$CO_3^{2-} + H^+$	-10.13	(*30*)
$Ca^{2+} + CO_3^{2-}$	↔	$CaCO_3$	-8.475	(*30*)
$\equiv FeOH_2^+$	↔	$\equiv FeOH + H^+$	-6.7	(*31*)
$\equiv FeOH$	↔	$\equiv FeO^- + H^+$	-9.0	(*31*)
$\equiv FeOH + HS^-$	↔	$\equiv FeS^- + H_2O$	5.3	(*20*)
$\equiv FeOH + H_2S$	↔	$\equiv FeSH + H_2O$	3.8	(*20*)
$FeS + H^+$	↔	$Fe^{2+} + HS^-$	-3.195	(*32*)

For the 19 reactive species and the 11 associated reactions 8 components had to be chosen to describe the equilibrium system. The selected components were $TOTH^+$, $TOTH_2CO^3$, $TOTH_2S$, TOTFeOH, $TOTCaCO_3$ and $TOTFe^{2+}$. The two non-reactive species Na^+ and Cl^- were additionally balanced with the two components $TOTNa^+$ and $TOTCl^-$.

The reduction of FeOOH was defined as a first-order reaction according to the model proposed in (*20*) (eq 6) with either $\equiv FeS^-$ or $\equiv FeSH$ as the solely reducing species .The reduction reaction was implemented according to eq 6:

$$\frac{1}{2}\frac{\partial c(FeOOH)}{\partial t}=\frac{\partial\{\equiv FeX\}}{\partial t}=-\frac{\partial\{\equiv FeOH\}}{\partial t}=-\frac{1}{2}\frac{\partial c(Fe^{2+})}{\partial t}=Y\,\frac{\partial c(H^+)}{\partial t}=-k\{\equiv FeX\} \quad (6)$$

$\{\equiv FeX\}$ denotes the concentration of the reducing species $\equiv FFeS^-$ or $\equiv FeSH$ and Y was adapted to it respectively (5 for $\equiv FeS^-$ and 4 for $\equiv FeSH$). The adaption of proton consumption was necessary because the proton balance for the generation of both sulfur surface-species is different and had to be accounted for to meet the overall proton mass balance for the reduction. For every reduced quantity of FeOOH, one surface site of the reducing species was consumed and replaced by $\equiv FeOH$. This is in agreement with the production of new surface sites during the reduction, so the total amount of surface sites was kept constant over the simulation time. Produced Fe(II) was instantaneously released to the mobile phase, where it could equilibrate with sulphide. A second kinetic reaction was set up for the production of S^0 because, in TBC, only one species can be defined by a kinetic reaction. The rate constant for S^0 production was set to half of the rate constant for FeOOH reduction.

Siderite precipitation was also defined kinetically and was dependent on saturation index as described in (*34*). However, siderite did not form within all performed simulations and is therefore disregarded in the evaluation of results.

The initial speciation of influent solution, pore water and surface species was calculated with PhreeqC (*32*). The derived speciation of the surface groups was converted to volumetric concentrations by dividing them through the volume of solid phase present in the model column and the resulting species concentrations were then inserted into the model..

For the model runs of the sensitivity analysis two basic model setups were used which differed in the description of the reduction reaction (either $\equiv FeS^-$ or $\equiv FeSH$ as reducing species). Other kinds of sensitivity analysis just concerned model parameters.

For the different runs of the sensitivity analysis a set of ten flow rates per rate constant was simulated (see Table I). The rate constants were varied between $1 \cdot 10^{-5}$ and $1 \cdot 10^{-3}$ s^{-1} and thus allowed to cover a wide range of reactivities of ferric (hydr)oxides and, thus, also *Da* numbers. The range of rate constants was adapted from experimental data determined by (*19*).

Column Experiments

For the column experiments acrylic glass cylinders (∅ 8 cm, length 25 cm) were used that contained goethite-coated quartz sand. Influent solutions were pumped upwards through the columns via a peristaltic pump (Ismatec MCP Standard, Ismatec, Switzerland). The column was placed into a holding system in the bottom of which small channels were cut to distribute the inflowing water across the entire cross section of the column. Columns were equipped with glass fibre frits both at the bottom and the top to enable homogeneous flow conditions. Flow rate was controlled by weighting the collected outflow of the column. All components of the experimental setup were connected with PU-tubes (∅ 2.4 mm, Legris GmbH, Germany) and LuerLock®-connectors.

For the coating procedure of the quartz sand, Bayferrox® 920Z (Lanxess Deutschland GmbH, Germany), a synthetic commercially available goethite mineral, was used. Due to the rather low pH_{pzc} of this material (5.5), which was attributed to adherent sulfate on the surface, the utilized Bayferrox® 920Z was washed with a pH 10-solution (NaOH) several times before further use. This procedure raised the pH_{pzc} to a value of 7.4. The coating itself was done after the method of (*35*).

Water for the preparation of influent solutions was degassed with the help of two MiniModule®-degassing units (Membrana GmbH, Germany). All influent solutions were prepared and stored in gastight aluminized PP/PE-bags (Tesseraux Spezialverpackungen GmbH, Germany). Sulphide solution was prepared by injecting of appropriate amounts of 1M $NaHCO_3^-$, 1M NaCl and freshly produced 0.1M Na_2S with a syringe into a five liter aluminized PP/PE-bag to reach the final concentrations given in Table II.

The column experiments started by flushing the packed columns with ~2PV of degassed deionized water. Then, ~2PV of 0.01M NaCl-solution was pumped through the column and electric conductivity was measured at fixed intervals to register the breakthrough curve from which the column characteristics were derived with the code CXTfit (*36*). Then, sulphide solution was injected into the column for 20 PVs. Sulphide concentration was monitored every 0.5 PV by taking a probe with a syringe through inbuild stopcocks at the inflow and outflow (sampling volume: 0.5-2 mL). Manual sampling rate was adjusted to flow rate in order to not disturb flow conditions within the column. Sulphide concentrations were determined with the methylene-blue method (*37*).

For the determination of sulphide turnover of the column experiments the total amount of inflowing and outflowing sulphide was calculated from flow rate and measured sulphide concentrations. Relative turnover was calculated as the ratio of consumed to inflowing amount of sulphide.

Comparison between Experimental Data and Simulations

In order to compare the model with the experimental findings the surface properties of the ferric (hydr)oxide and the inlet concentration of sulphide had to be changed in the model. The inlet concentrations of sulphide were changed from 200 to 188 $\mu mol\ L^{-1}$ with the latter value being the mean inlet concentration of sulphide within the column experiments.

In order to adapt the surface properties of the model to the experimental conditions, data from the characterisation of the pure Bayferrox® goethite were used. The amount of surface sites within the experimental column was estimated to be about 40 µmol based on the amount of coated quartz sand, the specific surface area of the pure goethite (9.2 m^2/g) and a site density of $5 \times 10^{-6}\ mol/m^2$. The pure Bayferrox® goethite was additionally characterised with respect to the pH_{pzc} and the reactivity towards sulphide. The pH_{pzc} was estimated to be 7.4 (*38*) but no explicit information was available for the absolute values of the acidity constants. Since the pH_{pzc} is the mean value of the two pK_a constants (*31*), the two acidity constants of the model were shifted to meet the experimental pH_{pzc} and thus the original range between these two constants was preserved.

The intrinsic rate constant of the rate law in eq 2 was used either in terms of $\equiv FeS^-$ or $\equiv FeSH$ as the reactive species. Literature data suggest that $\equiv FeSH$ is the dominant species for the reduction of ferric (hyr)oxides (e.g., (*20*)). For comparison, we also performed simulations with a reduction in terms of $\equiv FeS^-$ since the both species completely differ in their pH-dependent speciation (Figure 2). $\equiv FeS^-$ concentration is nearly constant at pH>7, while $\equiv FeSH$ predominates in the circum-neutral pH range.

The rate constants were taken from experimentally derived rates (*19*). To these ends the experimental rates ($3.71 \cdot 10^{-8}\ mol\ min^{-1}\ m^{-2}$) were normalized to concentrations of either $\equiv FeS^-$ or $\equiv FeSH$ that were calculated with a PhreeqC simulation considering the experimental conditions. The intrinsic rate constants derived were $1 \cdot 10^{-2}\ s^{-1}$ for $\equiv FeSH$ and $8.5 \cdot 10^{-4}\ s^{-1}$ for $\equiv FeS^-$.

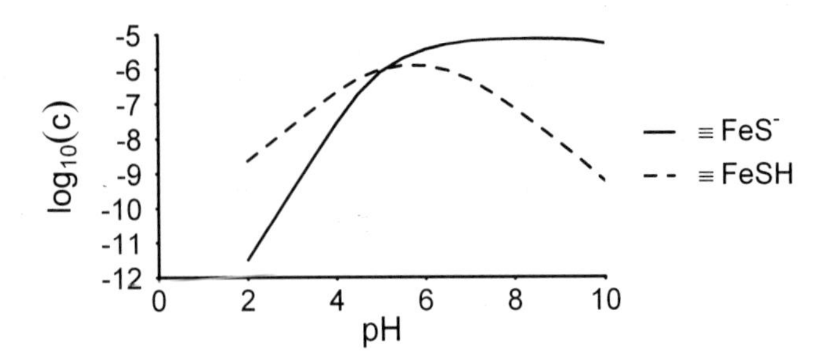

Figure 2. Calculation of surface speciation of ≡FeS⁻ and ≡FeSH for a batch system ($c(Na_2S) = 10^{-4}$ M, $c(FeOOH) = 2\ g\ L^{-1}$, $A_{spez} = 58\ m^2\ g^{-1}$, $I = 0.1$ M).

Results and Discussion

Effect of Surface Species

Da numbers have a significant effect on pH (Figure 3) which clearly depends on speciation. The pH increase is much higher with $\equiv FeS^-$ as the reducing species due its higher alkalinity. Nevertheless, the relationship between *Da* numbers and turnover behaves almost theoretical for this species (Figure 4) proposing equality between transport and reaction time scale for a *Da* number close to unity. In the simulations turnover is complete for *Da* numbers >1 and a fast decrease in turnover occurs for *Da* numbers <1. This pattern can be explained by the predominance of the species $\equiv FeS^-$ in the simulated pH range (Figure 2), so that the turnover is only marginally affected by changes of the pH during the reaction.

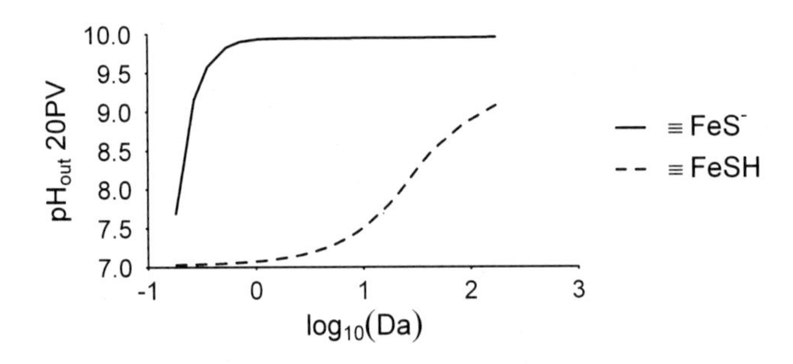

Figure 3. Comparison of simulated outflow pH after 20 PVs for the reducing species $\equiv FeS^-$ and $\equiv FeSH$.

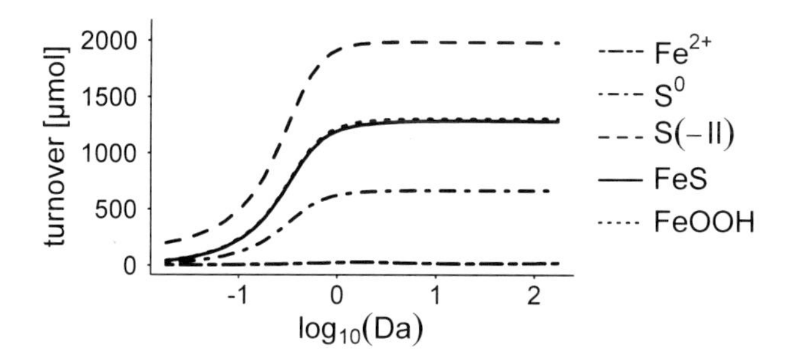

Figure 4. Simulated turnover of different model species for ≡FeS- as the reducing species.

In contrast, the concentration of the species ≡FeSH distinctly decreases with pH so that a significant effect on the relationship between *Da* numbers and turnover was observable (Figure 5), even though the pH range was much smaller (pH 7 to 9, Figure 3). Two features are observable. First, the total amount of consumed sulphide is much lower for ≡FeSH than for ≡FeS-. Second, the shape of the *Da*-turnover-curve for ≡FeSH is shifted and flattened towards higher *Da* numbers. The reason for this behaviour can be related to the lower capacity of the system to oxidize sulphide due to the pH dependency of ≡FeSH concentrations.

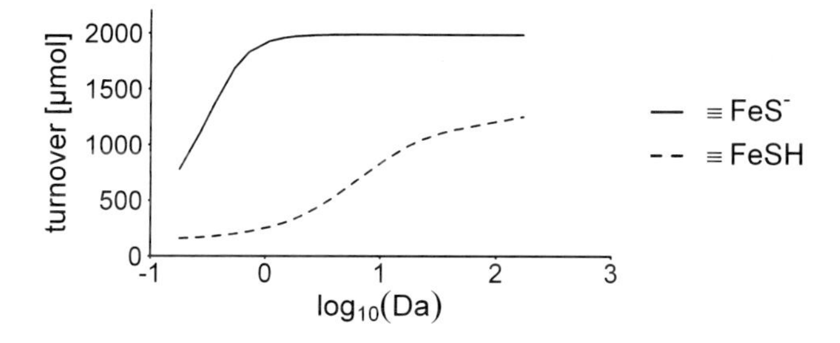

Figure 5. Comparison of simulated sulphide turnover for the reducing species ≡FeS- and ≡FeSH.

Also the spatial distribution of the reaction zone is affected by *Da* numbers. In Figure 6, the reaction rate is shown for three different runs that either shared the same flow rate or the same *Da* number (reaction rates were normalized to the highest occurring value respectively). For the two runs with the same flow rate but different values for *k*, a different spatial extent of the reaction zone can be observed, i.e. for the lower *Da* number the reaction zone is longer due to the higher reaction time (lower rate constant). In contrast, the two runs with different flow rates but the same *Da* number show a similar length of the reaction zone. This emphasizes that both the reaction and the residence time have to be accounted for when predicting the turnover within a sytem and that the *Da* number is more suitable for such a comparison than flow rate alone.

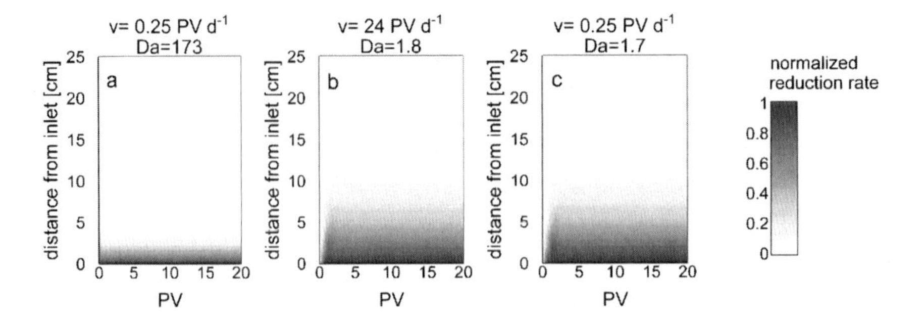

Figure 6. Longitudinal distribution of simulated relative reduction rates for three simulations with different flow rates and Damköhler numbers (≡FeS- as reducing species).

Correspondingly, the spatial distribution of sulphide and FeS were also dependent on *Da* numbers (Figure 7). The zone of FeS occurrence is longer for low *Da* numbers. It is also obvious that FeS formed at the same region where the highest sulphide consumption occurs, which means that Fe(II) was almost completely captured by FeS-precipitation right at the source region.

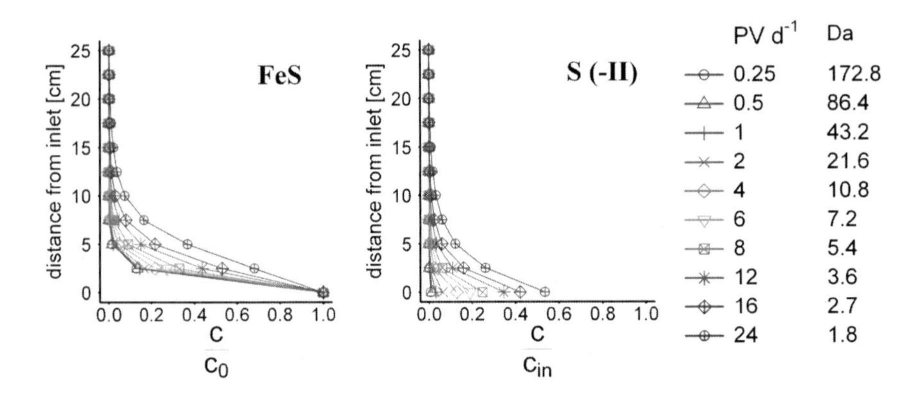

Figure 7. Longitudinal distribution of simulated relative concentrations of FeS and dissolved sulphide after 20PVs for ≡FeS- as reducing species.

Effect of the Amount of Surface Sites

In the next step, we investigated the effect of different amounts of reactive surface sites on the relationship between *Da* numbers and turnover with our simulations because this parameter may exhibit a considerable range in natural environments. It reflects the amount of iron (hydr)oxides in a system and the specific surface area of a certain Fe-mineral and both may be highly variable. The amount of surface sites (SSI) was varied over two orders of magnitude in order to mimic conditions ranging from low Fe-content, highly crystalline minerals (e.g., hematite, goethite) towards higher Fe-content, lower crystalline minerals (e.g., lepidocrocite, ferrihydrite). Results for ≡FeSH and ≡FeS^- are shown in Figure 8.

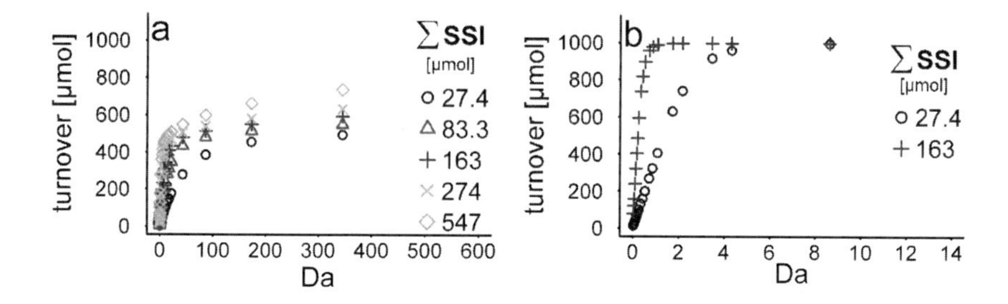

Figure 8. Simulated sulphide turnover between 10 and 20 PVs for a different amount of surface sites for ≡FeSH (left) and ≡FeS^- (right) as reducing species. In this simulation turnover was calculated only for the second half of the simulation period to avoid errors related to the initial sorption of sulphide so that the maximum turnover is 1000 µmol.

For ≡FeSH a relationship between *Da* numbers and turnover similar to that in Figure 4 can be observed, i.e. for simulations with a high amount of SSI the edge tends towards low *Da* numbers, the total turnover is higher than for simulations with low SSI and the curve approaches the theoretical *Da* relationship.

Figure 8a also shows that for low values of SSI for ≡FeSH, a complete turnover is never accomplished within the given flow rates and *k* values. Also, simulations with a lower value of SSI show a larger *Da* range between low and high turnover values and are generally shifted towards higher *Da* numbers. For ≡FeS^- (Figure 8b) this effect can also be observed but only for very low values of SSI. In addition, the magnitude of the shift in *Da* numbers to achieve constant turnover values is very low compared to ≡FeSH (note the scale on the x-axis in Figure 8).

A reason for the discrepancy between the two species is the difference in the amount of reactive surface sites, which determines the capacity of the system to oxidize sulphide, i.e. for low values of SSI the amount of surface sites are not sufficient to oxidize the whole amount of sulphide within a given flow distance even for long residence times.

Effect of Buffer Capacity

Since the relationship between *Da* numbers and turnover appears to be controlled by the pH-dependency of the surface speciation, we investigated the influence of the buffering capacity of the system. Therefore, three runs with different amounts of DIC were performed where the reaction was defined in terms of ≡FeSH due to its sensitive towards pH changes at neutral to alkaline conditions. Results for the three runs are given in Figure 9.

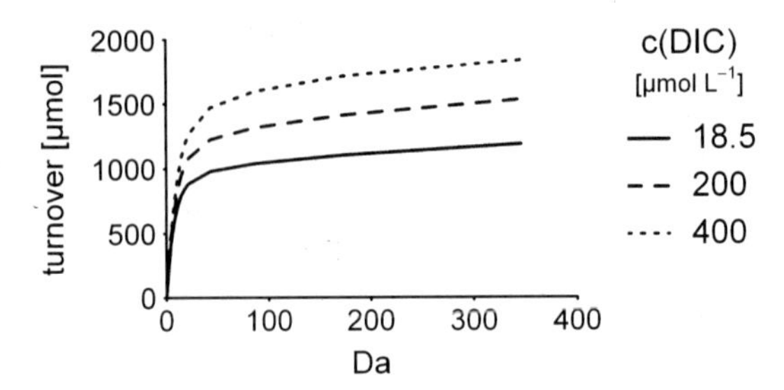

Figure 9. Simulated sulphide turnover for different DIC concentrations (≡FeSH as reducing species).

The maximum pH simulated dropped from 7.86 at 18.5 µmol L^{-1} DIC to 7.58 at 400 µmol L^{-1} DIC. For lower *Da* numbers, the curves are rather similar, whereas for higher *Da* numbers (and thus higher turnover) an almost constant difference between the three curves can be observed. Since the amount of ≡FeSH is dependent on pH and protons are consumed during the reaction, a higher buffering capacity stabilizes the ≡FeSH-concentration, and, thus, increases turnover at higher *Da* numbers. This feature might be of special interest in low buffered flow systems where a feedback of the reaction on the turnover can be expected.

Generalisation of the Dimensionless CDE Equation

The first order rate law for the consumption of sulphide (eq 2), which depends on the concentration of the surface species $\equiv FeS^-$ or ≡FeSH, can be reformulated with the equilibrium reaction of that surface complex, which yields for ≡FeSH (with equilibrium constant K_{FeSH}):

$$\frac{\partial c(FeOOH)}{\partial t} = 2\frac{\partial c(H_2S)}{\partial t} = -k\,K_{FeSH}\{\equiv FeOH\}c(H_2S) \qquad (7)$$

This equation provides a direct relationship between sulphide consumption and the sulphide concentration in the mobile phase, which can be incorporated in the dimensionless CDE (eq. 4):

$$\frac{\partial c'_{H_2S}}{\partial t'} = \frac{\partial c'_{H_2S}}{\partial x'} + \frac{D}{vL}\frac{\partial^2 c'_{H_2S}}{\partial x'^2} - \frac{k\,K_{FeSH}\{\equiv FeOH\}L}{v}c'_{H_2S} \qquad (8)$$

As a result, an apparent Damköhler number Da_{app} can be defined in terms of the capacity of the system to form reacting surface complexes with the help of the complexation constant (K_{FeSH}) and the amount of available surface sites {≡FeOH}, which depends on the reaction conditions.

$$Da_{app} = \frac{k\,K_{FeSH}\{\equiv FeOH\}L}{v} \qquad (9)$$

When introducing the equilibrium constant K_{FeSH} and the amount of ≡FeOH into the definition of the *Da* number, the individual curves for ≡FeSH of Figure 8 converge into one unique curve (Figure 10). The shape of this curve is closely related to the ideal Damköhler relationship (Figure 1) with the inflection point being close to 1. However, the curve tends to increase again at higher apparent *Da* numbers. This behaviour remains unresolved. A feed-back mechanism may be envisioned according to which the pH increases with increasing turnover on the one hand. On the other hand, a decrease in the oxidizing capacity occurs upon change of the surface speciation, decelerating the oxidation rate of sulphide.

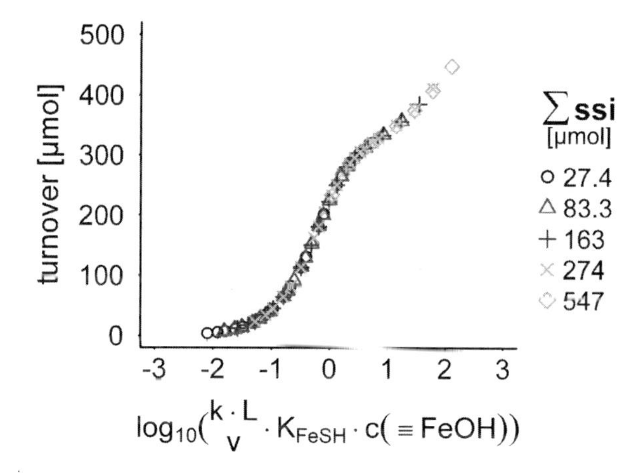

Figure 10. Normalization of simulated sulphide turnover using the apparent Da number from eq. 9 (≡FeSH as reducing species).

Comparison with Experimental Findings

The experimentally derived relative sulphide turnover is shown in Figure 11. Turnover is never complete for the chosen flow rates and decreases from approximately 90 to 70%, with increasing flow velocities in a non-linear way. Thus, a dependency of sulphide turnover on residence time is clearly visible.

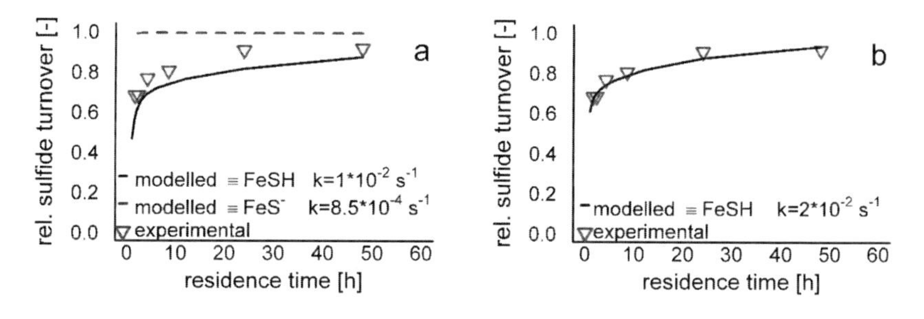

Figure 11. Experimental and simulated relative sulphide turnover for column experiments for the initial guess of rate constants for ≡FeS- and ≡FeSH (left) and the adapted rate constant for ≡FeSH (right).

Figure 11a also compares the measured sulphide turnovers with simulations in terms of ≡FeS^- and ≡FeSH using the rate constants as presented in the Materials and Methods section (i.e. $8.5 \cdot 10^{-4}$ s^{-1} and 10^{-2} s^{-1}, respectively). The correspondence between simulated and experimental turnovers was not satisfying for both reducing species. In case of ≡FeS^-, the turnover showed almost no dependency on residence times. Variation of the rate constant for the reduction with ≡FeS^- did not change the shape of the curve for ≡FeS^- (data not shown), which implies that the experimental curve can hardly be reproduced when using ≡FeS^- as the reducing species.

When using ≡FeSH as the reducing species, the simulated values for relative sulphide turnover reproduce the experimental values much better, but generally at a lower magnitude. It seems that the use of ≡FeSH as the reducing species could principally explain the experimental relationship between sulphide turnover and flow velocity, which would also be in accordance to suggestions made in literature (see above). A much closer fit between the simulated and the measured sulphide turnover could be obtained by increasing the rate constant by a factor of 2 to $2 \cdot 10^{-2}$ s^{-1} (Figure 11b). Duplication of the value derived from experimental data seems to be justified given the uncertainties inherent to the reactivities of ferric (hydr)oxides.

The simulated and experimental findings were also compared in terms of the sulphide export from the columns. Results of this comparison are shown in Figure 12. The simulated ouflow concentrations for the various pore volumes are rather similar. However, the simulated curves had a much sharper increase in turnover and achieved steady-state values already after 2-3 pore volumes. In contrast, the measured turnover increase was significantly delayed, which may be due to the formation of secondary products (e.g., pyrite) that were not considered in the simulations but that were observed in the solid phases of the columns as chromium reducible sulphur in excess to elemental S° (data not discussed) or to varying transport properties in the columns (e.g., dispersivity).

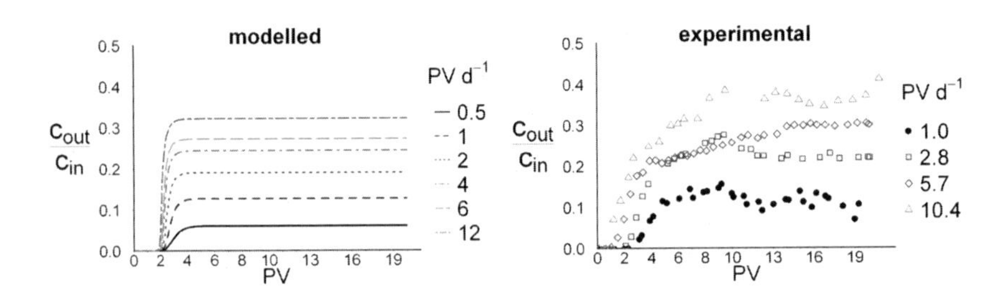

Figure 12. Simulated and experimental sulphide concentration at the outflow of the column.

Overall, the simulation underpins the need to account for the surface chemistry of the solid phase, and in particular for the specific interaction with reactants in the mobile phase, when discussing the effect of transport on the turnover of sulphide reacting at the immobile ferric oxide surface.

Implications

Geochemical Considerations

The experimental study and the results of the scenarios have clearly demonstrated that the turnover of a heterogeneous redox reaction, such as the oxidation of sulphide at the ferric (hydr)oxide surface in a flow-through system, depends on the ratio between residence time and characteristic reaction time expressed as *Da* numbers. From a geochemical perspective, the critical parameter in the *Da* number is the characteristic reaction time, which was shown not to be the inverse of the first order rate constant k. Rather, it is related to the apparent rate constant k_{app} being the product of k times the concentration of the reactive surface complex

$$k_{app} = k \cdot \{reactive\ surface\ complex\} \tag{10}$$

The concentration of the reactive surface complex determines the capacity of the system to oxidize sulphide. Depending on the reactive surface complex applied in the kinetic model, two completely different relationships between turnover and *Da* number could be observed. Simulations with $\equiv FeS^-$ as the reducing species showed a nearly theoretical behaviour of sulphide turnover and *Da* numbers, while $\equiv FeSH$ lead to a significant deviation from this behaviour (Figure 5). In this case the overall turnover decreased. Also the increase of turnover with rising *Da* numbers was affected by the choice of the reducing species. With $\equiv FeSH$ as the reactive species, turnovers comparable to those with $\equiv FeS^-$ were only achieved at higher *Da* numbers. The differences between these two kinetic models for the reduction is a result of the different supply of reducing surface species. Concentrations of $\equiv FeSH$ are generally lower at pH >7 compared to $\equiv FeS^-$ and also show a further decrease with increasing pH-values. The simulations further indicated that the spatial extent of the reaction zone is also defined by the *Da* number rather than by the flow velocity alone (Figure 6).

Calibration of the kinetic models in a column experiment suggests the surface species $\equiv FeSH$ to be responsible for the turnover in the experimental flow-system. Implementing the concentration of this species into the modified dimensionless CDE (eq. 8) to obtain an apparent rate constant allows to predict the turnover of sulphide in an ferric(hydr)oxide containing porous medium.

The dependency of turnover relationships on the amount of reactive surface sites is of paramount importance in natural systems. On the one hand, conditions are variable with respect to the iron content. Additionally, a large spectrum of ferric (hydr)oxides occurs in natural systems covering a wide range of specific surface areas, variable in time and in space as driven by fluctuating redox conditions. The amount of reactive surface sites can even be modulated by adsorption of substances inhibiting the reaction, such as phosphate or DOC (*39*). Further modification will arise from the buffering capacity of an aqueous system. An increase of the DIC concentration leads to a higher turnover especially for high *Da* numbers that is a cause of the pH dependency of the reaction, and, which is therefore also related to the change of the availability of surface sites. The impact of buffer capacity on the relationship between *Da* numbers and turnover is more pronounced in the lower range of DIC concentrations, although such values may be of minor importance to natural systems, which exhibit a higher buffering capacity under anaerobic conditions.

As a consequence a patchwork of reactive sites will exist in a natural system that controls the characteristic reaction time, and, thus, the turnover of sulphide at a certain location.

The Role of Transport

The implication of these considerations for natural systems is severe. On the one hand, a spatial distribution of characteristic reaction times can be expected, which causes a spatial variation of the product distribution such as FeS. Moreover, coupled chemical reaction networks will exert strong feedback on the breakthrough of individual species. In a sensitivity analysis of a reactive transport system in which a solute (Co(II)EDTA) could undergo oxidation (to Co(III)EDTA) or dissolution (to either Co^{2+} and Fe(III)EDTA) in contact with iron oxides, Szecsody et al. (*10*) were able to demonstrate that variation of the rate coefficients of individual reactions significantly affected the breakthrough of the various species due to change in speciation and their individual reactivities in the system. The highest iron dissolution rates (highest *Da* numbers (Da = 35) at the given residence time in their scenario, for example, lead to a higher breakthrough of Co^{2+} compared to Co(II)EDTA. In turn, this species had the highest breakthrough at the lowest *Da* number (Da = 1.4) simulated while only traces of Co^{2+} left the column under these conditions. Similar effects can be expected in regard to the formation of pyrite along with the reaction sequence between ferric oxides and dissolved sulphide as discussed in this article. Pyrite formation requires dissolution of FeS, formation of polysulphides and precipitation of FeS_2 (e.g., (*40*)), all of which operate parallel at different reaction rates.

On the other hand, transport in natural systems is typically not a piston-type flow as it is assumed to occur in columns, characterized by a unique residence time (even this assumption is only an approximation). Rather, a spatial distribution of residence times has been postulated to exist in many groundwater and wetland systems (*41*, *42*) and it is assumed that transport effects on geochemical reactions occur on the pore scale (e.g., (*7*)).

Combining the spatial distribution of characteristic reaction times with that of residence times, one obtains a distribution of *Da* numbers that will control the turnover and product distribution in a porous medium. Figure 13 shows a distribution of flowpaths in the saturated zone of a redox-chemically active wetland that has been simulated as a consequence of fluctuations of surface water-levels exposed to surface topography (*43*).

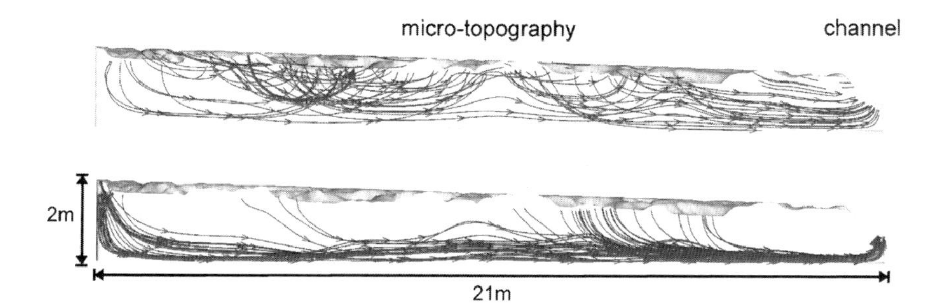

Figure 13. Simulated distribution of sub-surface flow paths (unpublished data based on (43)) for a redoxchemically very active environment (riparian wetland). Red arrows indicate different types of flow cells (deep + shallow flow cells) which are induced by micro-topography.

Each flowpath reflects also a different residence time of water that can be exposed to various redox processes. A water parcel moving along these flow-paths is being exposed to a multitude of immobile geochemical reactants. The extent of reaction depends on both, the exposure time at a certain location that can be expressed as the local residence time of the water parcel, and the apparent rate constant. Hot-Spots, i.e. zones of high turnover rates are then characterized by zones within flow paths that are subject to high *Da* numbers. Such zoning has been conceptualized by reactive transport modellers in terms of a mass transfer between mobile and immobile regions in which chemical and physical heterogeneities are reflected by distributions of kinetic rate constants and residence times in the immobile zone (*44*). It should be noted that at this scale the Damköhler number II (eq. 11), which accounts for pore diffusion (D_{diff} = diffusion coefficient), would be more appropriate to discuss transport effects.

$$Da_{II} = \frac{kL^2}{D_{diff}} \qquad (11)$$

One direct implication of these considerations is that *Da* numbers are specific for a certain redox process, which is due to specific apparent rate constants but also specific residence times of the reactants. Under conditions where competition exists for an electron acceptor, the preference for one pathway will be controlled by reaction specific *Da* numbers (cf. the discussion in (*45*)). For example, low *Da* numbers for the oxidation of sulphide might open the possibility for dissimilatory iron reducing bacteria. Minyard and Burgos (*9*) investigated the effect of different flow rates on dissimilatory iron reduction by bacteria in column experiments. They varied the flow rate from 0.6 to 12.4 pore volumes per day (PV d^{-1}) and found an increase of iron reduction with increasing flow rate, i.e. decreasing *Da* number As this study has demonstrated, flow-rate dependent sulphide turnover is then antagonistic to flow-rate dependent microbial reduction of ferric oxides, which opens biogeochemical niches for the predominance of one pathway over the other that is only dependent on transport properties. Competition needs, thus, to be expressed in terms of a relative difference between reaction specific *Da* numbers.

Acknowledgments

We are grateful to Katrin Hellige for her assistance in setting up the column experiments. The work was funded through DFG research unit 580 "Electron Transfer Processes in Anoxic Aquifers".

References

1. Shalev, E.; Lyakhovsky, V.; Yechieli, Y. Salt dissolution and sinkhole formation along the Dead Sea shore. *J. Geophys. Res.* **2006**, *111*, B03102.
2. Schlieker, M.; Schuring, J.; Hencke, J.; Schulz, H. D. The influence of advective transport on redox fronts in column experiments and their numeric modelling (part 1): the influence of variable flow velocities on turnover rates of primary redox processes. *Environ. Geol.* **2001**, *40* (11–12), 1353–1361.
3. Michalak, A. M.; Kitanidis, P. K. Macroscopic behavior and random-walk particle tracking of kinetically sorbing solutes. *Water Resour. Res.* **2000**, *36*, 2133–2146.
4. Bold, S.; Kraft, S.; Grathwohl, P.; Liedl, R. Sorption/desorption kinetics of contaminants on mobile particles: Modeling and experimental evidence. *Water Resour. Res.* **2003**, *39*.
5. Velbel, M. A. Geochemical mass balances and weathering rates in forested watersheds of the southern Blue Ridge. *Am. J. Sci.* **1985**, *285*, 904–930.
6. White, A. F.; Brantley, S. L. The effect of time on the weathering of silicate minerals: why do weathering rates differ in the laboratory and field? *Chem. Geol.* **2003**, *202* (3–4), 479–506.
7. Lichtner, P. C.; Kang, Q. Upscaling pore-scale reactive transport equations using a multiscale continuum formulation. *Water Resour. Res.* **2007**, *43*.
8. Hansel, C. M.; Benner, S. G.; Neiss, J.; Dohnalkova, A.; Kukkadapu, R. K.; Fendorf, S. Secondary mineralization pathways induced by dissimilatory iron reduction of ferrihydrite under advective flow. *Geochim. Cosmochim. Acta* **2003**, *67* (16), 2977–2992.
9. Minyard, M. L.; Burgos, W. D. Hydrologic flow controls on biologic iron(III) reduction in natural sediments. *Environ. Sci. Technol.* **2007**, *41* (4), 1218–1224.
10. Szecsody, J. E.; Zachara, J. M.; Chilakapati, A.; Jardine, P. M.; Ferrency, A. S. Importance of flow and particle-scale heterogeneity on CoII/IIIEDTA reactive transport. *J. Hydrol.* **1998**, *209* (1), 112–136.
11. Amirbahman, A.; Schonenberger, R.; Furrer, G.; Zobrist, J. Experimental study and steady-state simulation of biogeochemical processes in laboratory columns with aquifer material. *J. Contam. Hydrol.* **2003**, *64* (3–4), 169–190.
12. Heijman, C. G.; Grieder, E.; Hollinger, C.; Schwarzenbach, R. P. Reduction of nitroaromatic-compounds coupled to microbial iron reduction in laboratory aquifer columns. *Environ. Sci. Technol.* **1995**, *29* (3), 775–783.

13. Benner, S. G.; Hansel, C. M.; Wielinga, B. W.; Barber, T. M.; Fendorf, S. Reductive dissolution and biomineralization of iron hydroxide under dynamic flow conditions. *Environ. Sci. Technol.* **2002**, *36* (8), 1705–1711.
14. Canfield, D. E.; Raiswell, R.; Bottrell, S. The reactivity of sedimentary iron minerals toward sulphide. *Am. J. Sci.* **1992** (292), 659–683.
15. Rickard, D.; Schoonen, M. A. A.; Luther, G. W. Chemistry of iron sulfides in sedimentary environments. In *Geochemical Transformations of Sedimentary Sulfur*; Vairavamurthy, M. A., Schoonen, M. A. A., Eds.; ACS Symposium Series 612; American Chemical Society: Washington, DC, 1995; pp 168–193.
16. Peiffer, S. Reaction of H2S with ferric oxides In *Environmental chemistry of lakes and reservoirs*; Baker, L. A., Ed.; Advances in Chemistry Series 237; American Chemical Society: Washington, DC, 1994; pp 371–390.
17. Damköhler, G. Einfluss von Diffusion, Strömung und Wärmetransport auf die Ausbeute bei chemisch-technischen Reaktionen. In *Der Chemie-Ingenieur: Band III Chemische Operationen*; Arnold, E., Jakob, M., Eds.; Leipzig Akademische Verlagsgesellschaft M.B.H.: Leipzig, 1937; pp 359–485.
18. Rickard, D. The solubility of FeS. *Geochim. Cosmochim. Acta* **2006**, *70* (23), 5779–5789.
19. Hellige, K.; Peiffer, S. Reactivities of various ferric (hydr)oxides (goethite, lepidocrocite and ferrihydrite) against dissolved sulphide at pH 4 and 7. *Environ. Sci. Technol.* **2011**, submitted for publication.
20. Peiffer, S.; Dos Santos Afonso, M.; Wehrli, B.; Gächter, R. Kinetics and mechanism of the reaction of H_2S with lepidocrocite. *Environ. Sci. Technol.* **1992**, *26* (12), 2408–2413.
21. Yao, W. S.; Millero, F. J. Oxidation of hydrogen sulfide by hydrous Fe(III) oxides in seawater. *Mar. Chem.* **1996**, *52* (1), 1–16.
22. Domenico, P. A.; Schwartz, F. W. *Physical and Chemical Hydrogelogy*, 2nd ed.; Wiley & Sons: New York, 1997.
23. Wehrer, M.; Totsche, K. U. Detection of non-equilibrium contaminant release in soil columns: Delineation of experimental conditions by numerical simulations. *J. Plant Nutr. Soil Sci.* **2003**, *166* (4), 475–483.
24. Ocampo, C. J.; Oldham, C. E.; Sivapalan, M. Nitrate attenuation in agricultural catchments: Shifting balances between transport and reaction. *Water Resour. Res.* **2006**, *42*.
25. Carleton, J. N. Damkohler number distributions and constituent removal in treatment wetlands. *Ecol. Eng.* **2002**, *19* (4), 233–248.
26. Kern, U.; Li, C. C.; Westrich, B. Assessment of sediment contamination from pollutant discharge in surface waters. *Water Sci. Technol.* **1998**, *37* (6–7), 1–8.
27. Hanna, K. Adsorption of aromatic carboxylate compounds on the surface of synthesized iron oxide-coated sands. *Appl. Geochem.* **2007**, *22*, 2045–2053.
28. Schäfer, D.; Schäfer, W.; Kinzelbach, W. Simulation of reactive processes related to biodegradation in aquifers - 1. Structure of the three-dimensional reactive transport model. *J. Contam. Hydrol.* **1998**, *31* (1–2), 167–186.

29. Morel, F. M. M.; Hering, J. G. *Principles and applications of aquatic chemistry*; Wiley: New York, 1993.
30. Stumm, W.; Morgan, J. J. *Aquatic chemistry: Chemical equilibria and rates in natural waters*, 3rd ed.; Wiley: New York, 1996.
31. Cornell, R. M.; Schwertmann, U. *The iron oxides: Structure, properties, reactions, occurences and uses*, 2nd ed.; Wiley-VCH: Weinheim, 2003.
32. Parkhurst, D. L.; Appelo, C. A. J. *User's guide to PHREEQC--A computer program for speciation, batch-reaction, one-dimensional transport, and inverse geochemical calculations*; Water-Resources Investigations Report 99-4259; U.S. Geological Survey: 1999.
33. Dzombak, D. A.; Morel, F. M. M.: *Surface complexation modeling: Hydrous ferric oxide*; Wiley: New York, 1990.
34. Wang, Y. F.; Van Cappellen, P. A multicomponent reactive transport model of early diagenesis: Application to redox cycling in coastal marine sediments. *Geochim. Cosmochim. Acta* **1996**, *60* (16), 2993–3014.
35. Scheidegger, R. A.; Borkovec, M.; Sticher, H. Coating of silica sand with goethite - preparation and analytical identification. *Geoderma* **1993**, *58* (1–2), 43–65.
36. Toride, N.; Leji, F. J.; van Genuchten, M. Th. *The CXTFIT code for estimation transport parameters from laboratory of field tracer experiments*; Research Report 137; U.S. Salinity laboratory: Riverside, CA, 1995.
37. Cline, J. D. Spectrophotometric determination of hydrogen sulfide in natural waters. *Limnol. Oceanogr.* **1969**, *14* (3), 454–458.
38. Hellige, K. The Reactivity of Ferric (Hydr)oxides towards dissolved Sulphide; Dissertation, University of Bayreuth, 2011.
39. Biber, M. V.; Dos Santos Afonso, M.; Stumm, W. The coordination chemistry of weathering: IV. Inhibition of the dissolution of oxide minerals. *Geochim. Cosmochim. Acta* **1994**, *59*, 1999–2010.
40. Rickard, D.; Luther, G. W. Chemistry of iron sulfides. *Chem. Rev.* **2007**, *107* (2), 514–562.
41. Kirchner, J. W.; Feng, X. H.; Neal, C. Fractal stream chemistry and its implications for contaminant transport in catchments. *Nature* **2000**, *403* (6769), 524–527.
42. Cardenas, M. B.; Jiang, X.-W. Groundwater flow, transport, and residence times through topography-driven basins with exponentially decreasing permeability and porosity. *Water Resour. Res.* **2010**, *46*.
43. Frei, S.; Fleckenstein, J. H.; Lischeid, G. Effects of micro-topography on surface-subsurface exchange and runoff generation in a virtual riparian wetland - a modeling study. *Adv. Water Resour.* **2010**, *33* (11), 1388–1401.
44. Dentz, M.; Gouze, P.; Carrera, J. Effective non-local reaction kinetics for transport in physically and chemically heterogeneous media. *J. Contam. Hydrol.* **2011**, *120–121*, 222–236.
45. Blodau, C. Thermodynamic control on terminal electron transfer and methanogenesis. In *Aquatic Redox Chemistry*; Tratnyek, P. G., Grundl, T. J., Haderlein, S. B., Eds.; ACS Symposium Series; American Chemical Society: Washington, DC, 2011; Vol. 1071, Chapter 4, pp 65–83.

Chapter 26

Evolution of Redox Processes in Groundwater

Peter B. McMahon,*[,1] Francis H. Chapelle,[2] and Paul M. Bradley[2]

[1]U.S. Geological Survey, Denver Federal Center, Mail Stop 415, Lakewood, CO 80225
[2]U.S. Geological Survey, 720 Gracern Rd., Suite 129, Columbia, SC 29210
***pmcmahon@usgs.gov**

Reduction/oxidation (redox) processes affect the chemical quality of groundwater in all aquifer systems. The evolution of redox processes in groundwater is dependent on many factors such as the source and distribution of electron donors and acceptors in the aquifer, relative rates of redox reaction and groundwater flow, aquifer confinement, position in the flow system, and groundwater mixing. Redox gradients are largely vertical in the recharge areas of unconfined aquifers dominated by natural sources of electron donors, whereas substantial longitudinal redox gradients can develop in unconfined aquifers when anthropogenic sources of electron donors are dominant. Longitudinal redox gradients predominate in confined aquifers. Electron-donor limitations can result in the preservation of oxic groundwater over flow distances of many kilometers and groundwater residence times of several thousand years in some aquifers. Where electron donors are abundant, redox conditions can evolve from oxygen reducing to methanogenic over substantially shorter flow distances and residence times.

Introduction

The purpose of this chapter is to describe how reduction/oxidation (redox) processes evolve spatially and temporally in different groundwater systems. Redox processes, chemical reactions that transfer elections from donor compounds to acceptor compounds, are often catalyzed by microbial processes. Identifying the kinds of redox processes that occur in aquifers, documenting their spatial and temporal distribution, and understanding how they affect concentrations of natural or anthropogenic contaminants is central to assessing the chemical quality of groundwater (*1–8*). The chemical composition of mineral electron acceptors such as dissolved oxygen (O_2), nitrate (NO_3^-), manganese (Mn^{4+}), ferric iron (Fe^{3+}), sulfate (SO_4^{2-}), and carbon dioxide (CO_2) are relatively simple and analytical methods for quantifying them are well developed. Because microbial populations utilize these electron acceptors based on the free energy released by the electron donor-acceptor reaction, the sequence of electron acceptor utilization is often predictable and in the order $O_2 > NO_3^- > Mn^{4+} > Fe^{3+} > SO_4^{2-} > CO_2$. For these reasons, frameworks for describing redox processes in groundwater have traditionally been based on the sequential uptake of electron acceptors (*2*, *6*, *8*, *9*). This general framework forms the basis for the description of redox processes used in this chapter. Redox zonation in heterogeneous sediments, however, can be spatially complex due to variations in factors such as electron acceptor and donor reactivities, concentration of redox intermediate products, and chemical transport rates (*10–14*). Thus, redox zones may actually overlap in some settings or simply be indistinguisable at the scale sampled by well screens.

Sources of Electron Donors in Groundwater

Reduced chemical compounds, that is, compounds capable of donating electrons in redox reactions, have the potential for serving as electron donors in groundwater systems. By far the most common electron donor supporting microbial populations in groundwater systems is organic carbon. Aquifer materials deposited in sedimentary environments commonly contain particulate organic carbon that can support microbial metabolism. Particulate organic carbon can be present in aquifer recharge areas, discharge areas, or points in between, depending on the geologic framework of the aquifer. Dissolved organic carbon (DOC) can be delivered to aquifers by various processes such as water percolating through the unsaturated zone (*15–20*), or by diffusion into aquifers from adjacent confining layers (*21–23*). DOC in groundwater recharge can originate from natural or anthropogenic sources.

While natural DOC mobilized from surface sources is ultimately derived from decaying plant material (*17*, *24*), studies using both biochemical methods (*25*) and spectrofluorescence methods (*26*) indicate that DOC in groundwater systems is predominantly microbial in origin. This, in turn, reflects the cycling of dissolved and particulate organic carbon to microbial biomass, which is then recycled back to DOC. As a result of this continuous cycling of carbon, the natural DOC delivered to and/or produced in aquifers tends to have relatively low bioavailability (*25*). Confining layers adjacent to aquifers can sometimes be a source of DOC to aquifers through the process of diffusion. For example, confining layers of the South Carolina coastal-plain aquifer system contained groundwater with concentrations of organic acids greater than concentrations in the adjacent aquifers (*21*) (Figure 1). These organic acids, which are more bioavailable than most DOC in groundwater, appear to have been produced by acetogenic bacteria in the confining layers (*22*) and consumed by respiring bacteria in the aquifers. The concentration gradient that developed in response to these two processes drove a net diffusive flux of organic acids from the confining layers to the aquifers (*21*).

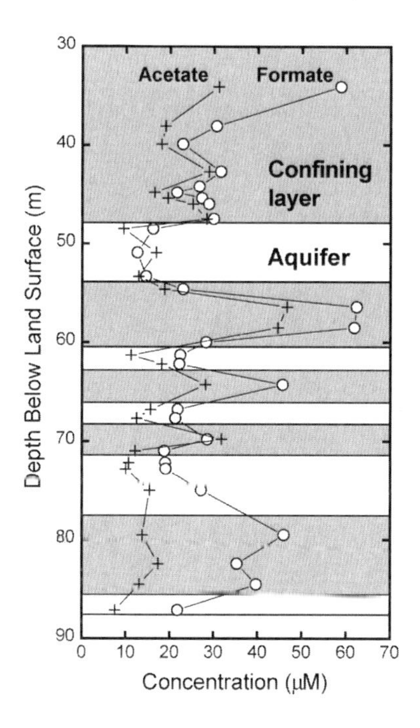

Figure 1. Concentrations of dissolved formate and acetate in groundwater of confining layers and aquifers in the coastal plain of South Carolina. (modified from reference (21)

Organic carbon may be the most common electron donor for microbial metabolism in groundwater systems, but mineral electron donors such as sulfide and ferrous iron minerals can be locally important. For example, it has been shown that oxidation of pyrite coupled to the reduction of nitrate is an important process limiting the migration of nitrate in a variety of hydrologic settings including a glacial-fluvial aquifer in southwestern Canada (*27*), an agriculturally-impacted glacial aquifer in Minnesota (*28*), and a coastal-plain aquifer in Maryland (*29*). These inorganic electron donors could be present in geologic materials at the time of deposition or formation, or they could be produced in aquifers as the result of previous reduction with organic electron donors.

Sources of Electron Acceptors in Groundwater

Of the commonly available electron acceptors, four of them (O_2, NO_3^-, SO_4^{2-}, and CO_2) are present in the atmosphere and in the unsaturated zone of many environments, soluble in near-neutral pH conditions of most groundwaters, and often enter the groundwater system through recharge processes. There also can be important anthropogenic sources for some electron acceptors at the land surface, such as nitrogen-fertilizer sources of nitrate in agricultural recharge (*30*) (Figure 2). Manganese and Fe^{3+}, the other commonly available electron acceptors, primarily exist in the solid phase of rocks and minerals, and are less likely to be present in groundwater recharge. Once recharge water reaches the water table it becomes isolated from the atmosphere, which could result in the depletion of electron acceptors along groundwater flow paths unless there are subsurface sources to replenish them.

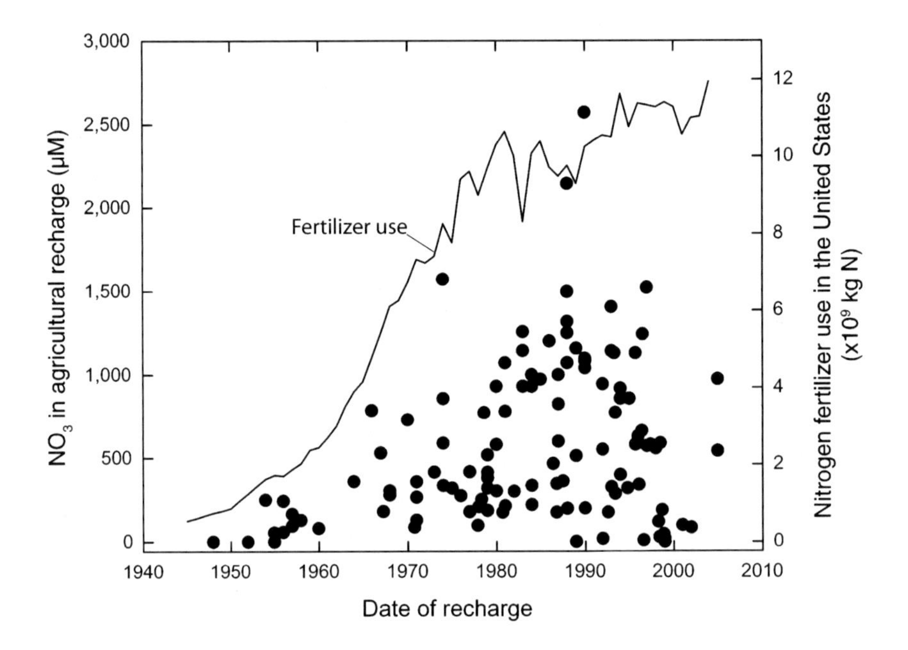

Figure 2. Nitrate concentrations in agricultural recharge and nitrogen fertilizer use in the United States, both in relation to estimated date of groundwater recharge. (data from references (28–35))

Manganese, Fe^{3+}, SO_4^{2-}, and CO_2 have important subsurface sources in many aquifers. Iron oxyhydroxide minerals such as goethite and amorphous Mn^{4+} and Fe^{3+} solid phases can occur as coatings on other minerals. Iron within the structure of minerals such as smectite also can serve as an electron acceptor (*36*, *37*). The minerals gypsum and anhydrite can serve as subsurface sources of SO_4^{2-}, particularly in carbonate-rock aquifers (*4*, *38*). Sulfate that has diffused from confining layers can also act as an electron acceptor for redox processes in marine sedimentary aquifers (*39*). Subsurface CO_2 can be produced by dissolution of carbonate minerals and by microbial respiration in many types of aquifers.

Evolution of Redox Processes along Groundwater Flow Paths

Flow Systems Dominated by Natural Sources of Electron

Unconfined Aquifers

Unconfined aquifers commonly receive spatially distributed recharge. This recharge water moves vertically downward across the water table and causes flow paths from upgradient areas to move deeper in the aquifer. Such a pattern of groundwater flow in the recharge area of unconfined aquifers results in vertical gradients in both groundwater age and redox processes. Flow systems characterized by an abundant supply of natural electron donors could evolve from O_2-reducing to methanogenic conditions in a relatively short distance below the water table. Flow systems characterized by a limited supply of electron donors may not evolve beyond O_2-reducing or mildly anoxic conditions. Figure 3 presents an example of the evolution from O_2-reducing to methanogenic conditions which occurred within a few meters below the water table in a glacial outwash aquifer in Minnesota (*40*). In contrast, figure 4 presents a fluvial aquifer in the High Plains of Kansas (*41*) where the system remained O_2 to NO_3^- reducing, even at depths of 200 m below the water table.

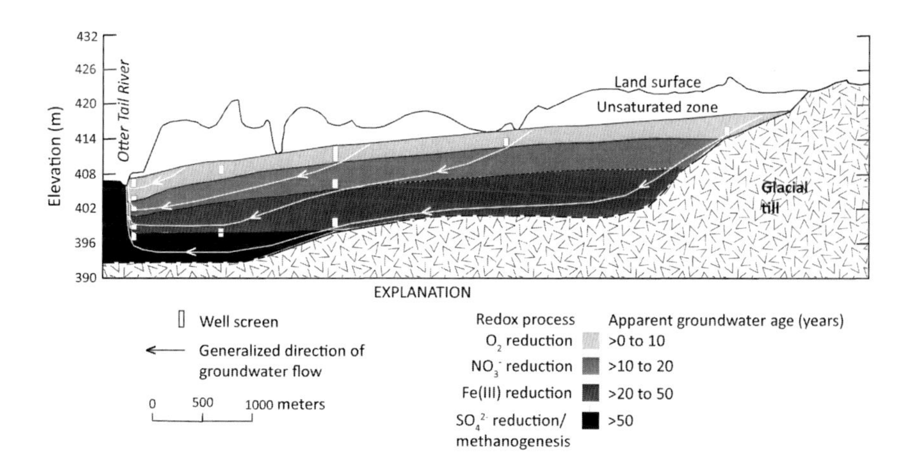

Figure 3. Distributions of redox processes and groundwater age along flow paths in a glacial outwash aquifer in Minnesota. (modified from reference (40))

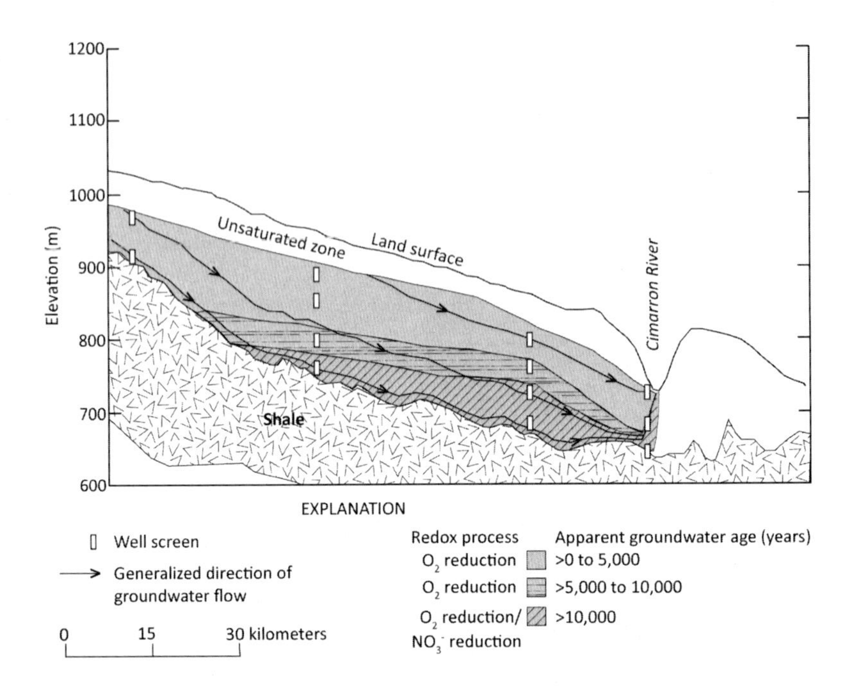

Figure 4. Distributions of redox processes and groundwater age along flow paths in a fluvial aquifer in Kansas. (data from reference (41))

The Minnesota aquifer was relatively thin and in an area of relatively high recharge rates so groundwater flow distances were less than a kilometer and residence times were on the order of decades or less. The Kansas aquifer was relatively thick and in an area of relatively low recharge rates so groundwater flow distances were tens of kilometers and residence times were on the order of millenia. Given these differences in hydrology and redox zonation, redox reaction rates in the Minnesota aquifer are assumed to have been much faster than those in the Kansas aquifer. Near discharge areas at streams or rivers in both systems, flow paths converged and turned upward with redox gradients becoming more horizontal (*42*). The ratio between groundwater residence time and chemical reaction time, sometimes expressed as the Damköhler number, has been used to quantitatively compare redox evolution in different flow systems (*13*, *14*, *43*). Whether an aquifer is transport or reaction dominated has important implications for any redox-sensitive chemical that has health and/or ecological concerns (*13*, *43*).

Confined Aquifers

Recharge in confined aquifers occurs in outcrop areas and flows laterally downgradient into confined parts of the aquifer. This pattern of groundwater flow generally results in longitudinal gradients in both groundwater age and redox processes in confined aquifers. Contributions of water and solutes from adjacent confining layers could alter these longitudinal patterns in the vicinity of aquifer/confining-layer interfaces (*44*). Thus, longitudinal gradients in groundwater age and redox processes represent a fundamental difference between confined and unconfined aquifers. A possible exception to this is contaminated unconfined aquifers, in which longitudinal redox gradients could develop, as noted below.

As in unconfined aquifers, the timescale for evolution of redox processes in confined aquifers is controlled in part by the availability of electron donors and acceptors and groundwater flow rates. For the example shown in Figures 5*A* and 5*B*, an electron-donor rich glacial aquifer in Ontario, Canada (*45*), the flow system evolved from Fe^{3+} reducing to methanogenic over a flow distance of about 12 km and a groundwater residence time of less than 20,000 years. In contrast, the example in Figures 5*C* and 5*D* shows an electron-donor poor sandstone aquifer in the Kalahari Desert, Namibia (*46*) where the flow system remained NO_3^- reducing over a distance of about 90 km and a residence time of more than 20,000 years. Figures 5*E* and 5*F* illustrates the effect of abundant supplies of a single electron acceptor on redox patterns in the Floridan carbonate aquifer in Florida (*38*), a system with relatively abundant supplies of electron donor. Oxygen and NO_3^- in recharge water were consumed relatively quickly in this flow system. Sulfate reduction became the next predominant redox process because of the limited supply of Fe^{3+} and abundant supply of SO_4^{2-} from the dissolution of gypsum and anhydrite. Thus, the flow system remained SO_4^{2-} reducing over a flow distance of about 100 km and a groundwater residence time of more than 10,000 years.

Redox Processes near Aquifer/Confining-Layer Interfaces

Aquifer/confining-layer interfaces represent mixing zones that sometimes are capable of supporting greater redox activity than either hydrogeologic unit alone. Groundwater in aquifers can be a source of electron acceptors for redox processes occurring in confining layers. Similarly, groundwater in confining layers can be a source of electron donors for redox processes occurring in aquifers (*44*). In both cases, redox processes have the potential to evolve over very short distances. In the example shown in Figure 6, a NO_3^- contaminated alluvial aquifer in Colorado overlying an organic-rich marine shale (*47*), O_2 and agricultural NO_3^- in groundwater from the alluvial aquifer diffused into the highly reducing shale and were consumed over a distance of less than 2 m.

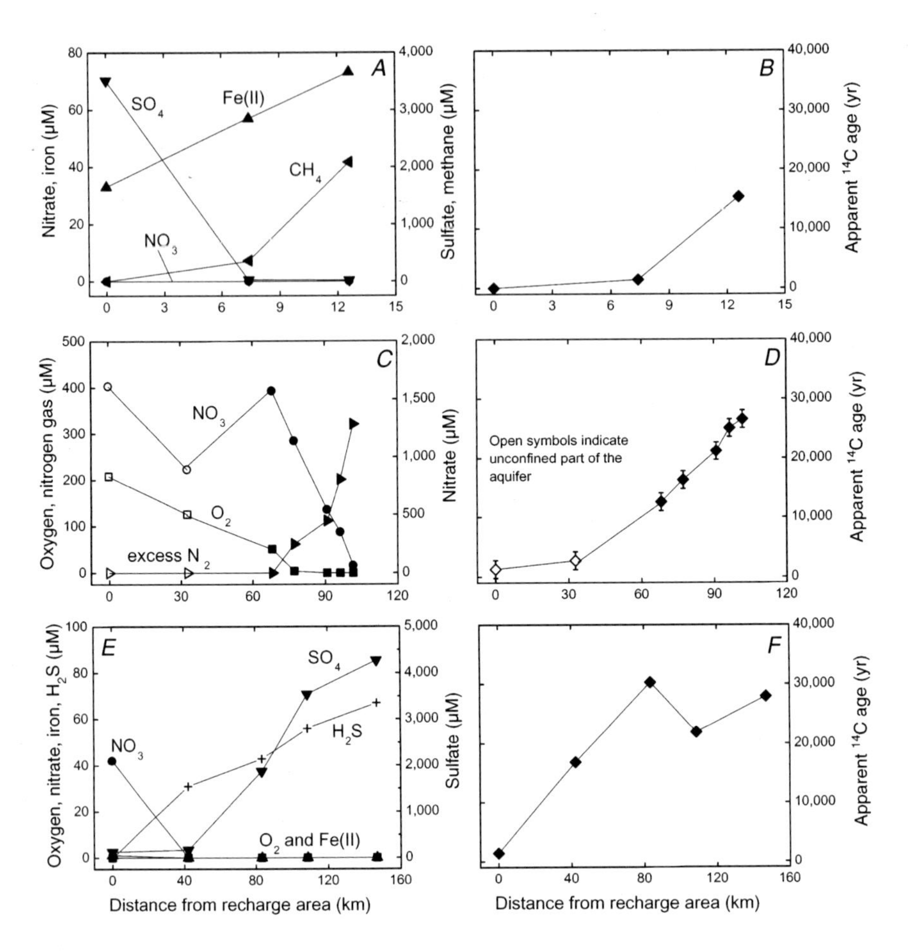

Figure 5. Distributions of redox reactants and products, and groundwater age, along groundwater flow paths in confined aquifers: (A and B) glacial aquifer, Ontario, Canada (data from reference 45), (C and D) sandstone aquifer, Kalahari Desert, Namibia (data from reference (46)), and (E and F) carbonate aquifer, Florida. (data from reference (38))

Flow Systems Dominated by Anthropogenic Sources of Electron Donors

Many pristine aquifers are primarily electron-donor limited. However, when large amounts of metabolizable organic carbon, such as gasoline, are introduced to an aquifer the carbon limitation is relieved and microbial metabolism then becomes limited by the availability of nutrients and electron acceptors. Numerous studies have described how redox processes in groundwater systems are affected by petroleum hydrocarbon spills (*1*, *5*, *48–51*). An example of how these redox processes change in space and time was described by (*52*) at a leaking gasoline underground storage tank site in South Carolina. The spill occurred in a relatively permeable, fully aerobic ($O_2 \sim 6.0$ mg/L), sandy, unconfined coastal-plain aquifer.

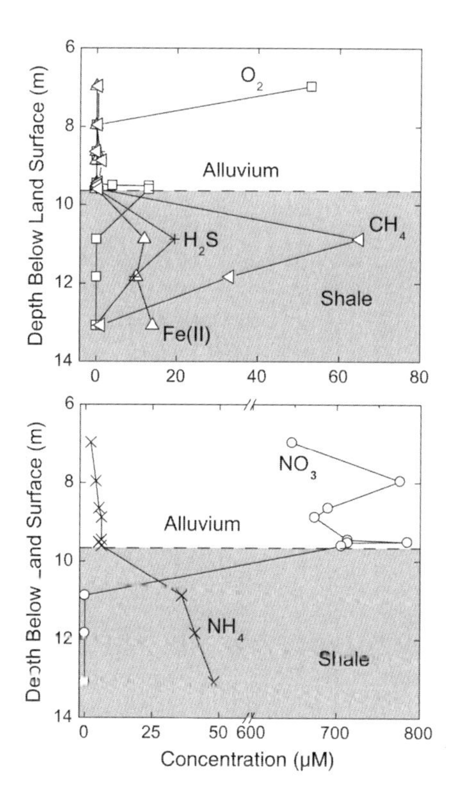

Figure 6. Distributions of redox reactants and products in groundwater near the interface of an alluvial aquifer and the Pierre Shale, northeastern Colorado. (modified from reference (47))

The spill is thought to have occurred sometime in 1992 or 1993. By 1994, the hydrocarbon plume immediately downgradient of the spill location had become anoxic and Fe^{3+} reduction had been initiated (Figure 7*A*). Four years later, however, the core of the plume was actively methanogenic, SO_4^{2-} reduction predominated downgradient of the methanogenic zone, and Fe^{3+} reduction predominated near the discharge area at a small ditch (Figure 7*B*). The rapid change of predominant redox processes from Fe^{3+} reduction to methanogenesis near the spill, a much more rapid change than observed at most sites (*1*), reflects the extremely low amount of Fe^{3+} present in this system. With so little available Fe^{3+}, the system rapidly shifted to SO_4^{2-} reduction and methanogenesis over just a few years. This process was accompanied by an increase in the density of methanogenic microoganisms in the core of the contaminant plume as well. This example illustrates how excess electron donors caused by anthropogenic contamination can radically alter ambient redox processes in groundwater systems and microbial ecology (*52*).

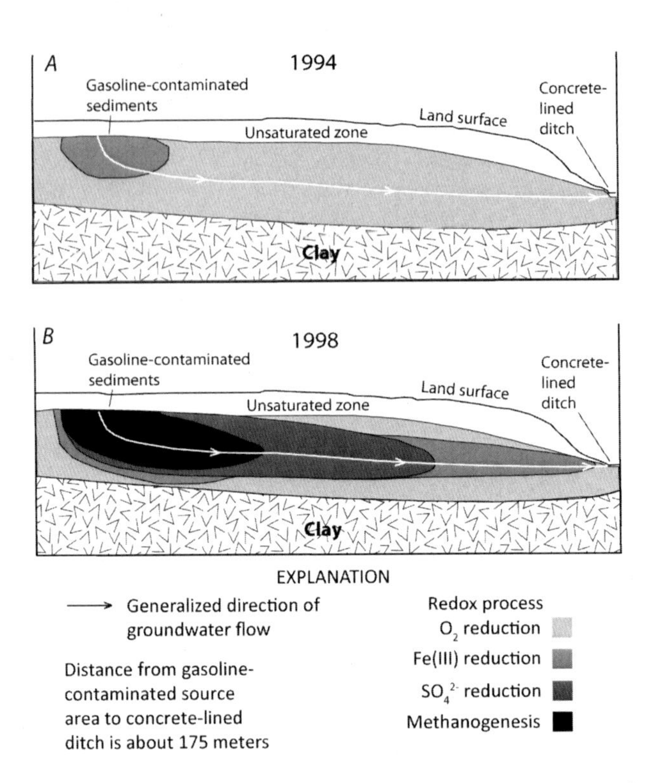

Figure 7. Distribution of redox processes in a gasoline-contaminated unconfined aquifer in South Carolina. (modified from reference (52))

The observed rapid shift from O_2-reducing to methanogenic conditions at this site is different from what commonly is observed in unconfined aquifers dominated by natural sources of electron donors (Figures 3 and 4), as is the spatial pattern of dominant redox processes. At the contaminated site in Figure 7, redox conditions were highly reducing at the upgradient end of flow paths and oxidizing at downgradient ends. Just the opposite pattern was observed in the uncontaminated flow system in Figure 3. Moveover, substantial longitudinal redox gradients developed in the contaminated system, whereas redox gradients in the uncontaminated flow system were largely vertical.

Effect of Groundwater Mixing on the Evolution of Redox Processes

In the above examples, evolution of redox processes occurred along groundwater flow paths that were generally unaffected by processes such as pumping, leakage through long well screens, or leakage through natural fractures that could mix groundwater of different ages, origins, and compositions. Such processes have the potential to alter the predicted spatial pattern of redox processes in aquifers by redistributing electron donors and acceptors in the flow system more quickly than would otherwise occur under flow conditions unaffected by mixing. Pumping and leakage through well screens, for example, introduced NO_3^- from shallow sources into deeper Fe^{3+} and SO_4^{2-}-reducing groundwater in Nebraska and Florida, thereby shifting redox conditions in those systems to NO_3^- reducing (*43*). Leakage through natural fractures introduced methane and possibly other electron donors from deep sources into a shallow, oxic aquifer in Colorado, thereby causing anoxic conditions to develop in the vicinity of fracture zones (*53*).

Recognizing the potential impacts of groundwater mixing also is critical for shallow, unconfined aquifer systems with anaerobic contaminant plumes. For example, natural attenuation of chloroethene-contaminated groundwater is complicated by the variation in redox character between different chloroethene compounds. Polychlorinated parent compounds, like tetrachloroethene and trichloroethene, are highly oxidized, tend to serve as electron acceptors during biodegradation, and attenuate most efficiently under reducing conditions. However, regulatory emphasis at chloroethene-contaminated sites is often on the production and accumulation of reduced daughter products like vinyl chloride (VC), which can serve as electron donors during biodegradation and attenuate most efficiently in the presence of O_2. Because VC is produced by microbial reductive dechlorination under anaerobic conditions, aerobic VC biodegradation is often deemed insignificant and the lack of accumulation of ethene, the product of VC reductive dechlorination, is interpreted as evidence of incomplete degradation, a so-called degradative "stall."

Trace concentrations of O_2 have been documented widely in shallow, anaerobically-active chloroethene plumes and generally dismissed as an artifact of atmospheric contamination during sampling or as the metabolically insignificant O_2 residual of aerobic consumption at the oxic/anoxic interface. However, a recent study of DOC and O_2 supply in shallow groundwater indicates that O_2 profiles can be explained by mixing of oxygenated recharge with shallow, nominally anoxic groundwater (*54*). This suggests that low O_2 concentrations observed in shallow, anaerobically-active aquifers may, in fact, reflect extensive precipitation-driven advection of O_2 into the "anoxic" plume. The potential impacts that a precipitation-driven flux of O_2 into nominally anoxic chloroethene plumes may have on contaminant attenuation are particularly relevant, because mineralization of VC to CO_2 can be substantial at O_2 concentrations well below the field standard for nominally "anoxic" conditions (*55–57*). Failure to recognize the impacts of mixing of oxygenated recharge in anaerobically-active chloroethene plumes can lead to the incorrect diagnosis of a degradative "stall" and adoption of expensive and ineffective remedial actions.

Redox Processes in Regional Aquifers of the United States

An analysis of water samples collected from 5,135 domestic wells was used to compare redox processes in regional aquifers of the United States (*9*, *58*). A regional analysis can be useful because of the unique perspective it provides regarding the effects of climate, geology, and hydrology on redox processes. The Snake River Plain basaltic-rock aquifer in Idaho exhibited the largest percentage

(100%) of oxic samples and the Silurian-Devonian carbonate-rock aquifer in the Midwest exhibited the smallest percentage (13%) (Figure 8). The high percentage of oxic samples in the basaltic-rock aquifer reflects the relatively fast rates of groundwater movement from fracture flow and limited supply of electron donors in that aquifer. The low percentage of oxic samples in the carbonate-rock aquifer reflects, in part, a relatively large supply of electron donors in that aquifer, but also electron-donor abundance in overlying glacial deposits that in some areas served as confining layers for the carbonate-rock aquifer. On average among the eight lithologic groups examined, sand and gravel aquifers in the western United States such as the High Plains, Basin and Range basin fill, and Central Valley aquifers contained the largest percentage (85%) of oxic samples. These western sand and gravel aquifers are largely oxic, even though they often contain old groundwater (Figure 4), because they are electron-donor limited. The glacial sand and gravel aquifers contained the smallest percentage (43%) of oxic samples because they contained relatively abundant supplies of electron donors such as organic carbon and pyrite. Some aquifers, such as the Coastal Lowlands aquifer system in the Gulf Coastal Plain, the Pennsylvanian sandstone aquifer in parts of Ohio, Pennsylvania, and West Virginia, and the glacial sand and gravel aquifers in the central United States, contained large percentages (25 to 39%) of samples with mixed redox conditions (Figure 8). In part, this redox heterogeneity reflects the heterogeneous nature of the sediment in these aquifers compared to aquifers like the Central Valley and Basin and Range aquifers (Figure 8). Even at the scale of this assessment, considering redox processes explained many of the water-quality trends observed in these regional aquifers (*9*, *58*).

Conclusions

The variability in the source and distribution of electron donors and acceptors in aquifers has important implications for the evolution of redox processes in groundwater systems. Highly reducing redox conditions generally indicate an abundance of electron donors compare to electron acceptors. The distance along groundwater flow paths over which redox processes evolve is largely a function of the relative rates of redox reaction and groundwater flow. If redox reaction rates are greater than groundwater flow rates, reducing conditions are expected to develop over relatively short flow distances. Vertical redox gradients predominate in the recharge areas of unconfined aquifers, whereas longitudinal redox gradients predominate in confined aquifers. An important exception is unconfined aquifers dominated by anthropogenic sources of electron donors, such as gasoline, where substantial longitudinal redox gradients can develop and the gradients are the reverse (more reducing to less reducing) of what is observed in uncontaminated flow systems. Groundwater mixing also has the potential to alter the predicted spatial pattern of redox process in aquifers by redistributing electron donors and acceptors in the flow system more quickly than would otherwise occur under flow conditions unaffected by mixing. Understanding mixing effects on redox processes can be particularly important for predicting spatial patterns of redox processes and contaminant degradation potential in groundwater systems. An

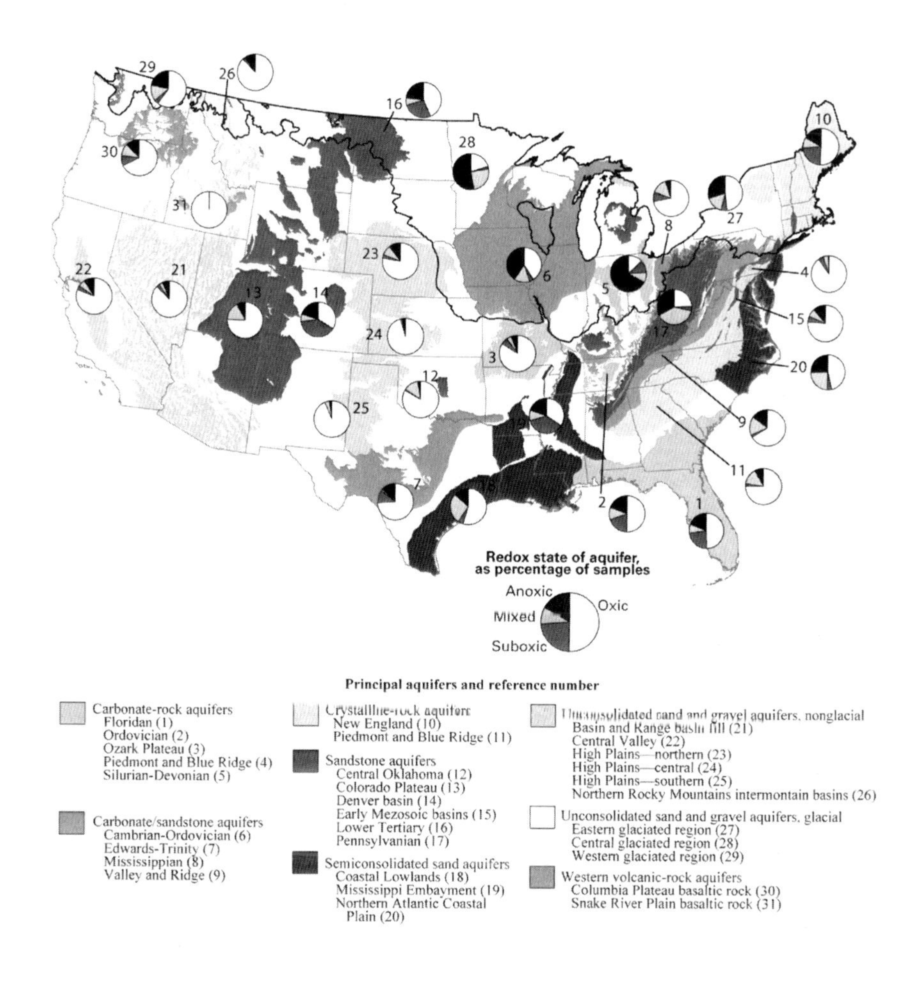

Figure 8. Pie diagrams indicating the percentages of domestic well samples that were oxic, suboxic, anoxic, or diagnostic of mixed redox processes in selected regional aquifers of the United States. (modified from reference (58)) (see color insert)

analysis of redox processes in regional aquifers of the United States indicated that, on average, sand and gravel aquifers of the western United States such as the High Plains, Basin and Range basin fill, and Central Valley aquifers where the most oxic and glacial sand and gravel aquifers of the northern United States were the least oxic. Some aquifers, such as the Coastal Lowlands aquifer system in the Gulf Coastal Plain, the Pennsylvanian sandstone aquifer in parts of Ohio, Pennsylvania, and West Virginia, and the glacial sand and gravel aquifers in the central United States, contained large percentages of samples with mixed redox conditions. In part, this redox heterogeneity reflects the heterogeneous nature of the sediment in these aquifers compared to aquifers like the Central Valley and Basin and Range aquifers.

References

1. Baedecker, M. J.; Cozzarelli, I. M.; Eganhouse, R. P.; Siegel, D. I.; Bennett, P. C. Crude oil in a shallow sand and gravel aquifer – III. Biogeochemical reactions and mass balance modeling in anoxic groundwater. *Appl. Geochem.* **1993**, *8*, 569–586.
2. Champ, D. R.; Gulens, J.; Jackson, R. E. Oxidation-reduction sequences in groundwater flow systems. *Can. J. Earth Sci.* **1979**, *16*, 12–23.
3. Thorstenson, D. C.; Fisher, D. W.; Croft, M. G. The geochemistry of the Fox Hills-Basal Hell Creek aquifer in southwestern North Dakota and northwestern South Dakota. *Water Resour. Res.* **1979**, *15*, 1479–1498.
4. Plummer, L. N.; Busby, J. F.; Lee, R. W.; Hanshaw, B. B. Geochemical modeling of the Madison aquifer in parts of Montana, Wyoming, and South Dakota. *Water Resour. Res.* **1990**, *26*, 1981–2014.
5. Vroblesky, D. A.; Chapelle, F. H. Temporal and spatial changes of terminal electron-accepting processes in a petroleum hydrocarbon-contaminated aquifer and the significance for contaminant biodegradation. *Water Resour. Res.* **1994**, *30*, 1561–1570.
6. Chapelle, F. H.; McMahon, P. B.; Dubrovsky, N. M.; Fujii, R. F.; Oaksford, E. T.; Vroblesky, D. A. Deducin the distribution of terminal electron-accepting processes in hydrologically diverse groundwater systems. *Water Resour. Res.* **1995**, *31*, 359–371.
7. Hunter, K. S.; Wang, Y.; Van Cappellen, P. Kinetic modeling of microbially-driven redox chemistry of subsurface environments: Coupling transport, microbial metabolism and geochemistry. *J. Hydrol.* **1998**, *209*, 53–80.
8. Christensen, T. H.; Bjerg, P. L.; Banwart, S. A.; Jakobsen, R.; Heron, G.; Albrechtsen, H.-H. Characterization of redox conditions in groundwater contaminant plumes. *J. Contam. Hydrol.* **2000**, *45*, 165–241.
9. McMahon, P. B.; Chapelle, F. H. Redox processes and water quality of selected principal aquifer systems. *Ground Water* **2008**, 259–285.
10. Jakobsen, R.; Postma, D. Redox zoning, rates of sulfate reduction and interactions with Fe-reduction and methanogenesis in a shallow sandy aquifer, Rømø, Denmark. *Geochim. Cosmochim. Acta.* **1999**, *63*, 137–151.
11. Jakobsen, R. Redox microniches in groundwater: A model study on the geometric and kinetic conditions required for concomitant Fe oxide reduction, sulfate reduction, and methanogenesis. *Water Resour. Res.* **2007**, *43*, doi:10.1029/2006WR005663.
12. Blodau, C. Thermodynamic control on terminal electron transfer and methanogenesis. In *Aquatic Redox Chemistry*; Tratnyek, P. G., Grundl, T. J., Haderlein, S. B., Eds.; ACS Symposium Series; American Chemical Society: Washington, DC, 2011; Vol. 1071, Chapter 4, pp 65−82.
13. Ocampo, C. J.; Oldham, C. E.; Sivapalan, M. Nitrate attenuation in agricultural catchments: Shifting balance between transport and reaction. *Water Resour. Res.* **2006**, *42*, doi:10.1029/2004WR003773.
14. Kurtz, W.; Peiffer, S. The role of transport in aquatic redox chemistry. *This volume*.

15. Thurman, E. M. *Organic Geochemistry of Natural Waters*; Martinus Nijhoff/DR W. Junk Publishers: Dordrecht/Boston/Lancaster, 1985; pp 71–80.
16. Cronan, C. S.; Aiken, G. R. Chemistry and transport of soluble humic substances in forested watersheds of the Adirondack Park, New York. *Geochim. Cosmochim. Acta.* **1985**, *49*, 1697–1705.
17. Hornberger, G. M.; Bencala, K. E.; McKnight, D. M. Hydrological controls on dissolved organic carbon during snowmelt in the Snake River near Montezuma, Colorado. *Biogeochemistry* **1994**, *25*, 147–165.
18. Boyer, E. W.; Hornberger, G. M.; Bencala, K. E.; McKnight, D. M. Response characteristics of DOC flushing in an alpine catchment. *Hydrol. Proc.* **1997**, *11*, 1635–1647.
19. Baker, M. A.; Valett, H. M.; Dahm, C. N. Organic carbon supply and metabolism in a shallow groundwater ecosystem. *Ecology* **2000**, *81*, 3133–3148.
20. Pabich, W. J.; Valiela, I.; Hemond, H. F. Relationship between DOC concentration and vadose zone thickness and depth below water table in groundwater of Cape Cod, USA. *Biogeochemistry* **2001**, *55*, 247–268.
21. McMahon, P. B.; Chapelle, F. H. Microbial production of organic acids in aquitard sediments and its role in aquifer geochemistry. *Nature* **1991**, *349*, 233–235.
22. Chapelle, F. H.; Bradley, P. M. Microbial acetogenesis as a source of organic acids in ancient Atlantic Coastal Plain sediments. *Geology* **1996**, *24*, 925–928.
23. Krumholz, L. R.; McKinley, J. P.; Ulrich, G. A.; Suflita, J. M. Confined subsurface microbial communities in Cretaceous rock. *Nature* **1997**, *386*, 64–66.
24. Lundquist, E. J.; Jackson, L. E.; Scow, K. M. Wet-dry cycles affect dissolved organic carbon in two California agricultural soils. *Soil Biol. Biochem.* **1999**, *31*, 1031–1038.
25. Chapelle, F. H.; Bradley, P. M.; Goode, D. J.; Tiedeman, C.; Lacombe, P. J.; Kaiser, K.; Benner, R. Biochemical indicators for the bioavailability of organic carbon in ground water. *Ground Water* **2009**, *47*, 108–121.
26. Birdwell, J. E.; Engle, A. S. Variability in terrestrial and microbial contributions to dissolved organic matter fluorescence in the Edwards aquifer, central Texas. *J. Cave Karst Stud.* **2009**, *71*, 144–156.
27. Tesoriero, A. J.; Liebscher, H.; Cox, S. E. Mechanism and rate of denitrification in an agricultural watershed: Electron and mass balance along groundwater flow paths. *Water Resour. Res.* **2000**, *36*, 1545–1559.
28. Böhlke, J. K.; Wanty, R.; Tuttle, M.; Delin, G.; Landon, M. Denitrification in the recharge area and discharge area of a transient agricultural nitrate plume in a glacial outwash sand aquifer, Minnesota. *Water Resour. Res.* **2002**, *38*, 1105–1131.
29. Böhlke, J. K.; O'Connell, M. E.; Prestegaard, K. L. Groundwater stratification and delivery of nitrate to an incised stream under varying flow conditions. *J. Environ. Qual.* **2007**, *36*, 664–680.
30. Böhlke, J. K.; Denver, J. M. Combined use of groundwater dating, chemical, and isotopic analyses to resolve the history and fate of nitrate contamination

in two agricultural watersheds, Atlantic Coastal Plain, Maryland. *Water Resour. Res.* **1995**, *31*, 2319–2339.

31. Böhlke, J. K.; Hatzinger, P. B.; Sturchio, N. C.; Gu, B.; Abbene, I.; Mroczkowski, S. J. Atacama perchlorate as an agricultural contaminant in groundwater: Isotopic and chronologic evidence from Long Island, New York. *Environ. Sci. Technol.* **2009**, *43*, 5619–5625.
32. Tesoriero, A. J.; Spruill, T. B.; Mew, H. E., Jr.; Farrell, K. M.; Harden, S. L. Nitrogen transport and transformations in a coastal plain watershed: Influence of geomorphology on flow paths and residence time. *Water Resour. Res.* **2005**, *41*, doi:10.1029/2003WR002953.
33. Puckett, L. J.; Hughes, W. B. Transport and fate of nitrate and pesticides: Hydrogeology and riparian zone processes. *J. Environ. Qual.* **2005**, *34*, 2278–2292.
34. Alexander, R. B.; Smith, R. A. County-level estimates of nitrogen and phosphorus fertilizer use in the United States, 1945 to 1985. *U.S. Geol. Surv. Open File Report 90-130*; 1990.
35. Ruddy, B. C.; Lorenz, D. L.; Mueller, D. K. County-level estimates of nutrient inputs to the land surface of the conterminous United States, 1982–2001. *U.S. Geol. Surv. Sci. Invest. Rep. 2006-5012*; 2006.
36. Kostka, J. E.; Stucki, J. W.; Nealson, K. H.; Wu, J. Reduction of structural Fe(III) in smectite by a pure culture of Shewanella putrefaciens strain MR-1. *Clays Clay Miner.* **1996**, *44*, 522–529.
37. Neumann, A.; Sander, M.; Hofstetter, T. B. Redox properties of structural Fe in smectite clay minerals. In *Aquatic Redox Chemistry*; Tratnyek, P. G., Grundl, T. J., Haderlein, S. B., Eds.; ACS Symposium Series; American Chemical Society: Washington, DC, 2011; Vol. 1071, Chapter 17, pp 359−377.
38. Plummer, L. N.; Sprinkle, C. L. Radiocarbon dating of dissolved inorganic carbon in groundwater from confined parts of the Upper Floridan aquifer, Florida, USA. *Hydrogeol. J.* **2001**, *9*, 127–150.
39. Chapelle, F. H.; McMahon, P. B. Geochemistry of dissolved inorganic carbon in a coastal plain aquifer: 1. Sulfate from confining beds as an oxidant in microbial CO_2 production. *J. Hydrol.* **1991**, *127*, 85–108.
40. Puckett, L. J.; Cowdery, T. K. Transport and fate of nitrate in a glacial outwash aquifer in relation to groundwater age, land use practices, and redox processes. *J. Environ. Qual.* **2002**, *31*, 782–796.
41. McMahon, P. B.; Böhlke, J. K.; Christensen, S. C. Geochemistry, radiocarbon ages, and paleorecharge conditions along a transect in the central High Plains aquifer, southwestern Kansas, USA. *Appl. Geochem.* **2004**, *19*, 1655–1686.
42. Puckett, L. J.; Cowdery, T. K.; McMahon, P. B.; Tornes, L. H.; Stoner, J. D. Using chemical, hydrologic, and age dating analysis to delineate redox processes and flow paths in the riparian zone of a glacial outwash aquifer-stream system. *Water Resour. Res.* **2002**, *38*, doi:10.1029/2001WR000396.
43. McMahon, P. B.; Böhlke, J. K.; Kauffman, L. J.; Kipp, K. L.; Landon, M. K.; Crandall, C. A.; Burow, K. R.; Brown, C. J. Source and transport controls on the movement of nitrate to public supply wells in selected

principal aquifers of the United States. *Water Resour. Res.* **2008**, *44*, doi:10.1029/2007WR006252.

44. McMahon, P. B. Aquifer/aquitard interfaces: Mixing zones that enhance biogeochemical reactions. *Hydrogeol. J.* **2001**, *9*, 34–43.
45. Aravena, R.; Wassenaar, L. I.; Plummer, L. N. Estimating ^{14}C groundwater ages in a methanogenic aquifer. *Water Resour. Res.* **1995**, *31*, 2307–2317.
46. Vogel, J. C.; Talma, A. S.; Heaton, T, H. E. Gaseous nitrogen as evidence for denitrification in groundwater. *J. Hydrol.* **1981**, *50*, 191–200.
47. McMahon, P. B.; Böhlke, J. K.; Bruce, B. W. Denitrification in marine shales in northeastern Colorado. *Water Resour. Res.* **1999**, *35*, 1629–1642.
48. Chapelle, F. H.; Bradley, P. M.; Lovley, D. R.; Vroblesky, D. A. Measuring rates of biodegradation in a contaminated aquifer using field and laboratory methods. *Ground Water* **1996**, *34*, 691–698.
49. Landmeyer, J. E.; Chapelle, F. H.; Bradley, P. M.; Pankow, J. F.; Church, C. D.; Tratnyek, P. C. Fate of MTBE relative to benzene in a gasoline-contaminated aquifer. *Ground Water Monit. Rem.* **1998**, *18*, 93–102.
50. Cozzarelli, I. M.; Herman, J. S.; Baedecker, M. J.; Fischer, S. M. Geochemical heterogeneity of a gasoline-contaminated aquifer. *J. Contam. Hydrol.* **1999**, *40*, 261–284.
51. Rooney-Varga, J. N.; Anderson, R. T.; Fraga, J. L.; Ringelburg, D. B.; Lovley, D. R. Microbial communities associated with anaerobic benzene degradation in a petroleum-contaminated aquifer. *Appl. Environ. Microbiol.* **1999**, *65*, 3056–3063.
52. Chapelle, F. H.; Bradley, P. M.; Lovley, D. R.; O'Neill, K.; Landmeyer, J. E. Rapid evolution of redox processes in a petroleum hydrocarbon-contaminated aquifer. *Ground Water* **2002**, *40*, 353–360.
53. McMahon, P. B.; Thomas, J. C.; Hunt, A. G. Use of diverse geochemical data sets to determine sources and sinks of nitrate and methane in groundwater, Garfield County, Colorado, 2009. *U.S. Geol. Surv. Sci. Invest. Rep. 2010-5215*; 2011.
54. Foulquier, A.; Malard, F.; Mermillod-Blondin, F.; Datry, T.; Simon, L.; Montuelle, B.; Gibert, J. Vertical change in dissolved organic carbon and oxygen at the water table region of an aquifer recharged with stormwater: Biological uptake or mixing? *Biogeochemistry* **2010**, *99*, 31–47.
55. Bradley, P. M. History and ecology of chloroethene biodegradation: A review. *Bioremed. J.* **2003**, *7*, 81–109.
56. Bradley, P. M.; Chapelle, F. H. In *In Situ Remediation of Chlorinated Solvent Plumes*; Stroo, H. F., Ward, C. H., Eds.; Springer: New York, NY, 2010; pp 39–67.
57. Gossett, J. M. Sustained aerobic oxidation of vinyl chloride at low oxygen concentrations. *Environ. Sci. Technol.* **2010**, *44*, 1405–1411.
58. McMahon, P. B.; Cowdery, T. K.; Chapelle, F. H.; Jurgens, B. C. Redox conditions in selected principal aquifers of the United States. *U.S. Geol. Surv. Fact Sheet 2009-3041*; 2009.

Editors' Biographies

Dr. Paul G. Tratnyek

Dr. Paul G. Tratnyek is currently Professor, and Associate Head, in the Division of Environmental and Biomolecular Systems (EBS), Institute of Environmental Health (formerly Oregon Graduate Institute, OGI), at the Oregon Health & Science University, Portland OR, USA.

He received his Ph.D. in Applied Chemistry from the Colorado School of Mines in 1987 (Major Professor: Donald L. Macalady); served as a National Research Council Postdoctoral Fellow at the U.S. Environmental Protection Agency Laboratory in Athens, GA, during 1988 (with N. Lee Wolfe); and as a Research Associate at the Swiss Federal Institute for Water Resources and Water Pollution Control (EAWAG) from 1989 to 1991 (with Jürg Hoigné).

His research concerns oxidation reduction reactions and other physico-chemical processes that control the fate and effects of environmental substances, including minerals, metals, organics, and nanoparticles.

Dr. Timothy J. Grundl

Dr. Timothy J. Grundl is currently Professor in the Geosciences Department and Chair in the School of Freshwater Sciences (formerly the Great Lakes WATER Institute) at the University of Wisconsin–Milwaukee, Milwaukee, WI, USA.

He received his Ph.D. in Geochemistry from the Colorado School of Mines in Golden CO in 1987 (Major Professor: Donald L. Macalady).

His research uses environmental tracers in the study of oxidation-reduction reactions, groundwater provenance and recharge dynamics and fate of pollutants in aquifer systems.

Dr. Stefan B. Haderlein

Dr. Stefan B. Haderlein is a full Professor at the Center for Applied Geosciences (ZAG) of the Eberhard-Karls University Tübingen, Germany.

He received his Ph.D. in Environmental Chemistry from the Swiss Federal Institute of Technology (ETH) Zurich, Switzerland in 1992 (Major Professor: René P. Schwarzenbach) and earned his habilitation and venia legendi from the same institution in 1998; he served as research scientist at the Swiss Federal Institute for Water Resources and Water Pollution Control (EAWAG) and ETH Zurich from 1993 to 1999 and was a Postdoctoral Fellow at the Massachusetts Institute of Technology (1996 to 1997; with Philip Gschwend).

His research concerns sorption processes, oxidation-reduction reactions and other physico-chemical processes that control the fate of contaminants in the subsurface as well as stable isotope techniques to trace their origin and fate.

Indexes

Author Index

Subject Index

B

C

D

E

F

G

H

I

J

K

L

M

N

O

P

Q

R

S

T